Trigonometry

[FOURTH EDITION]

Cynthia Y. Young

PROFESSOR OF MATHEMATICS

University of Central Florida

WILEY

VICE PRESIDENT & DIRECTOR Laurie Rosatone
ACQUISITIONS EDITOR Joanna Dingle
ASSISTANT DEVELOPMENT EDITOR Ryann Dannelly
EDITORIAL ASSISTANT Giana Milazzo
MARKETING MANAGER John LaVacca III
SENIOR PRODUCT DESIGNER David Dietz
PRODUCT DESIGN MANAGER Thomas Kulesa

FREELANCE PROJECT EDITOR Anne Scanlan-Rohrer
PRODUCTION SERVICES Cenveo® Publisher Services
SENIOR CONTENT MANAGER Valerie Zaborski
SENIOR PRODUCTION EDITOR Ken Santor
SENIOR DESIGNER Maureen Eide
SENIOR PHOTO EDITOR Billy Ray
COVER PHOTO Jupiter Images/Getty Images

This book was set in 10/12 Times by Cenveo® Publisher Services, and printed and bound by Quad/Graphics, Inc.

The cover was printed by Quad/Graphics, Inc.

This book is printed on acid free paper.

Founded in 1807, John Wiley & Sons, Inc. has been a valued source of knowledge and understanding for more than 200 years, helping people around the world meet their needs and fulfill their aspirations. Our company is built on a foundation of principles that include responsibility to the communities we serve and where we live and work. In 2008, we launched a Corporate Citizenship Initiative, a global effort to address the environmental, social, economic, and ethical challenges we face in our business. Among the issues we are addressing are carbon impact, paper specifications and procurement, ethical conduct within our business and among our vendors, and community and charitable support. For more information, please visit our website: www.wiley.com/go/citizenship.

Evaluation copies are provided to qualified academics and professionals for review purposes only, for use in their courses during the next academic year. These copies are licensed and may not be sold or transferred to a third party. Upon completion of the review period, please return the evaluation copy to Wiley. Return instructions and a free of charge return mailing label are available at www.wiley.com/go/returnlabel. If you have chosen to adopt this textbook for use in your course, please accept this book as your complimentary desk copy. Outside of the United States, please contact your local sales representative.

The inside back cover will contain printing identification and country of origin if omitted from this page. In addition, if the ISBN on the back cover differs from the ISBN on this page, the one on the back cover is correct.

ISBN: 978-1-119-32113-2 (ePub)
ISBN: 978-1-119-27331-8 (LLPC)

Printed in the United States of America

V10005689_103018

FOR
Christopher and Caroline

About the Author

University of Central Florida

University of Central Florida

Cynthia Y. Young is the Dean of the College of Science and Professor of Mathematics at Clemson University and the author of *College Algebra, Trigonometry, Algebra and Trigonometry*, and *Precalculus*. She holds a BA in Secondary Mathematics Education from the University of North Carolina (Chapel Hill), an MS in Mathematical Sciences from UCF, and both an MS in Electrical Engineering and a PhD in Applied Mathematics from the University of Washington. She has taught high school in North Carolina and Florida, developmental mathematics at Shoreline Community College in Washington, and undergraduate and graduate students at UCF.

Prior to her leadership role at Clemson, Dr. Young was a Pegasus Professor of Mathematics and the Vice Provost for Faculty Excellence and UCF Global at the University of Central Florida (UCF). Dr. Young joined the faculty at UCF in 1997 as an assistant professor of mathematics, and her primary research area was the mathematical modeling of the atmospheric effects on propagating laser beams. Her atmospheric propagation research was recognized by the Office of Naval Research Young Investigator Award, and in 2007 she was selected as a fellow of the International Society for Optical Engineering. Her secondary area of research centers on improvement of student learning in mathematics. She has authored or co-authored over 60 books and articles and has served as the principal investigator or co-principal investigator on projects with more than $2.5 million in federal funding. Dr. Young was on the team at UCF that developed the UCF EXCEL program, which was originally funded by the National Science Foundation to support the increase in the number of students graduating with a degree in science, technology, engineering, and mathematics (STEM). The EXCEL learning community approach centered around core mathematics courses has resulted in a significant increase in STEM graduation rates and has been institutionalized at UCF.

Preface

As a mathematics professor, I would hear my students say, "I understand you in class, but when I get home I am lost." When I would probe further, students would continue with "I can't read the book." As a mathematician, I always found mathematics textbooks quite easy to read—and then it dawned on me: Don't look at this book through a mathematician's eyes; look at it through the eyes of students who might not view mathematics the same way that I do. What I found was that the books were not at all like my class. Students understood me in class, but when they got home they couldn't understand the book.

It was then that the folks at Wiley lured me into writing. My goal was to write a book that is seamless with how we teach and is an ally (not an adversary) to student learning. I wanted to give students a book they could read without sacrificing the rigor needed for conceptual understanding. The following quote comes from a reviewer when asked about the rigor of the book:

> *I would say that this text comes across as a little less rigorous than other texts, but I think that stems from how easy it is to read and how clear the author is. When one actually looks closely at the material, the level of rigor is high.*

DISTINGUISHING FEATURES

Four key features distinguish this book from others, and they came directly from my classroom.

PARALLEL WORDS AND MATH

Have you ever looked at your students' notes? I found that my students were only scribbling down the mathematics that I would write—never the words that I would say in class. I started passing out handouts that had two columns: one column for math and one column for words. Each example would have one or the other; either the words were there and students had to fill in the math, or the math was there and students had to fill in the words. If you look at the examples in this book, you will see that the words (your voice) are on the left and the mathematics is on the right. In most math books, when the author illustrates an example, the mathematics is usually down the center of the page, and if the students don't know what mathematical operation was performed, they will look to the right for some brief statement of help. That's not how we teach; we don't write out an example on the board and then say, "Class, guess what I just did!" Instead we lead our students, telling them what step is coming and then performing that mathematical step *together*—and reading naturally from left to right. Student reviewers have said that the examples in this book are easy to read; that's because *your* voice is right there with them, working through problems *together*.

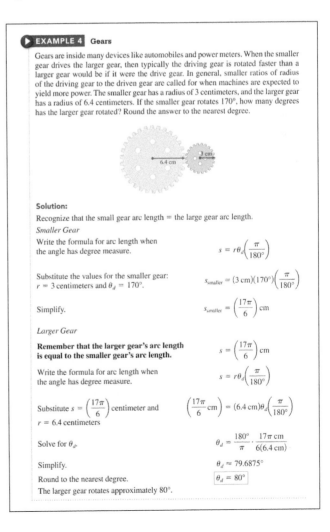

EXAMPLE 4 Gears

Gears are inside many devices like automobiles and power meters. When the smaller gear drives the larger gear, then typically the driving gear is rotated faster than a larger gear would be if it were the drive gear. In general, smaller ratios of radius of the driving gear to the driven gear are called for when machines are expected to yield more power. The smaller gear has a radius of 3 centimeters, and the larger gear has a radius of 6.4 centimeters. If the smaller gear rotates 170°, how many degrees has the larger gear rotated? Round the answer to the nearest degree.

Solution:

Recognize that the small gear arc length = the large gear arc length.

Smaller Gear

Write the formula for arc length when the angle has degree measure.

$$s = r\theta_d\left(\frac{\pi}{180°}\right)$$

Substitute the values for the smaller gear: $r = 3$ centimeters and $\theta_d = 170°$.

$$s_{\text{smaller}} = (3\text{ cm})(170°)\left(\frac{\pi}{180°}\right)$$

Simplify.

$$s_{\text{smaller}} = \left(\frac{17\pi}{6}\right)\text{cm}$$

Larger Gear

Remember that the larger gear's arc length is equal to the smaller gear's arc length.

$$s = \left(\frac{17\pi}{6}\right)\text{cm}$$

Write the formula for arc length when the angle has degree measure.

$$s = r\theta_d\left(\frac{\pi}{180°}\right)$$

Substitute $s = \left(\frac{17\pi}{6}\right)$ centimeter and $r = 6.4$ centimeters

$$\left(\frac{17\pi}{6}\text{ cm}\right) = (6.4\text{ cm})\theta_d\left(\frac{\pi}{180°}\right)$$

Solve for θ_d.

$$\theta_d = \frac{180°}{\pi} \cdot \frac{17\pi\text{ cm}}{6(6.4\text{ cm})}$$

Simplify.

$$\theta_d \approx 79.6875°$$

Round to the nearest degree.

$$\boxed{\theta_d = 80°}$$

The larger gear rotates approximately 80°.

SKILLS AND CONCEPTS
(LEARNING OBJECTIVES AND EXERCISES)

In my experience as a mathematics teacher/instructor/professor, I find skills to be on the micro level and concepts on the macro level of understanding mathematics. I believe that too often skills are emphasized at the expense of conceptual understanding. I have purposely separated *learning objectives* at the beginning of every section into two categories: *skills objectives*—what students should be able to do—and *conceptual objectives*—what students should understand. At the beginning of every class, I discuss the learning objectives for the day—both skills and concepts. These are reinforced with both skills exercises and conceptual exercises. Each subsection has a corresponding skills objective and conceptual objective.

2.3 EVALUATING TRIGONOMETRIC FUNCTIONS FOR NONACUTE ANGLES

SKILLS OBJECTIVES	CONCEPTUAL OBJECTIVES
▪ Determine the reference angle of a nonacute angle. ▪ Determine the ranges of the six trigonometric functions. ▪ Determine the reference angle for any angle whose terminal side lies in one of the four quadrants. ▪ Evaluate trigonometric functions for nonacute and quadrantal angles.	▪ Understand which trigonometric functions are positive or negative in each of the four quadrants. ▪ Understand why sine and cosine range between -1 and 1. ▪ Understand that reference angles are acute and have positive measure. ▪ Understand that the value of a trigonometric function at an angle is the same as the trigonometric value of its reference angle, except there may be a difference in its algebraic sign $(+/-)$.

CATCH THE MISTAKE

Have you ever made a mistake (or had a student bring you his or her homework with a mistake) and you've gone over it and over it and couldn't find the mistake? It's often easier to simply take out a new sheet of paper and solve it from scratch than it is to actually find the mistake. Finding the mistake demonstrates a higher level of understanding. I include a few *Catch the Mistake* exercises in each section that demonstrate a common mistake. Using these in class (with individuals or groups) leads to student discussion and offers an opportunity for formative assessment in real time.

• CATCH THE MISTAKE

In Exercises 53 and 54, explain the mistake that is made.

53. The terminal side of an angle θ, in standard position, passes through the point $(1, 2)$. Calculate $\sin\theta$.

Solution:

Label the coordinates.	$x = 1, y = 2$
Calculate r.	$r = 1^2 + 2^2 = 5$
Use the definition of sine.	$\sin\theta = \dfrac{y}{r}$
Substitute $y = 2, r = 5$.	$\sin\theta = \dfrac{2}{5}$

This is incorrect. What mistake was made?

54. Evaluate $\sec 810°$ exactly.

Solution:

Find a coterminal angle that lies between $0°$ and $360°$.	$810° - 360° = 450°$
$810°$ has the same terminal side as $90°$.	$450° - 360° = 90°$
Evaluate cosine for $90°$.	$\cos 90° = 0$
Evaluate secant for $90°$.	$\sec 90° = \dfrac{1}{\cos 90°} = \dfrac{1}{0} = 0$

This is incorrect. What mistake was made?

LECTURE VIDEOS BY THE AUTHOR

I authored the videos to ensure consistency in the students' learning experience. Throughout the book, wherever a student sees the video icon, that indicates a video. These videos provide mini lectures. The chapter openers and chapter summaries act as class discussions. The "Your Turn" problems throughout the book challenge the students to attempt a problem similar to a nearby example. The "worked-out example" videos are intended to come to the rescue for students if they get lost as they read the text and work problems outside the classroom.

NEW TO THE FOURTH EDITION

In the fourth edition, the main upgrades are updated applications throughout the text; new Skill and Conceptual objectives mapped to each subsection; new Concept Check questions in each subsection; and the substantially improved version of *WileyPLUS*, including ORION adaptive practice and interactive animations.

SKILL AND CONCEPTUAL OBJECTIVES

3.1 RADIAN MEASURE	
SKILLS OBJECTIVES	**CONCEPTUAL OBJECTIVES**
▪ Calculate the radian measure of an angle. ▪ Convert between degrees and radians.	▪ Realize that radian measure allows us to write trigonometric functions as functions of real numbers. ▪ Understand that degrees and radians are both units for measuring angles.

APPLICATIONS TO BUSINESS, ECONOMICS, HEALTH SCIENCES, AND MEDICINE

For Exercises 89 and 90, refer to the following:

An orthotic knee brace can be used to treat knee injuries by locking the knee at an angle θ chosen to facilitate healing. The angle θ is measured from the metal bar on the side of the brace on the thigh to the metal bar on the side of the brace on the calf (see the figure on the left below). To make working with the brace more convenient, rotate the image such that the thigh aligns with the positive *x*-axis (see the figure on the right below).

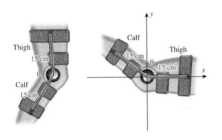

89. Health/Medicine. If $\theta = 165°$, find the measure of the reference angle. What is the physical meaning of the reference angle?

For Exercises 73–76, refer to the following:

When an airplane flies faster than the speed of sound, the sound waves that are formed take on a cone shape, and where the cone hits the ground, a sonic boom is heard. If θ is the angle of the vertex of the cone, then $\sin\left(\dfrac{\theta}{2}\right) = \dfrac{330 \text{ m/sec}}{V} = \dfrac{1}{M}$, where V is the speed of the plane and M is the mach number.

73. Sonic Booms. What is the speed of the plane if the plane is flying at mach 2?

74. Sonic Booms. What is the mach number if the plane is flying at 990 m/sec?

75. Sonic Booms. What is the speed of the plane if the cone angle is 60°?

76. Sonic Booms. What is the speed of the plane if the cone angle is 30°?

CONCEPT CHECK QUESTIONS

[CONCEPT CHECK]

Match the following: when the angle is formed using a clockwise/ counterclockwise rotation that corresponds to an angle with positive/negative measure.

▼

ANSWER Clockwise—Negative Counterclockwise—Positive

[CONCEPT CHECK]

TRUE OR FALSE If an angle α has measure 50°, then the angle β that has measure 310° is coterminal with angle α.

▼

ANSWER True

[CONCEPT CHECK]

In what quadrant are sine and cosine functions both negative?

▼

ANSWER III

FEATURE	BENEFIT TO STUDENT
Chapter-Opening Vignette	Piques the student's interest with a real-world application of material presented in the chapter. Later in the chapter, the concept from the vignette is reinforced.
Chapter Overview, Flowchart, and Learning Objectives	Allows students to see the big picture of how topics relate, and overarching learning objectives are presented.
Skills and Conceptual Objectives	Skills objectives represent what students should be able to do. Conceptual objectives emphasize a higher-level, global perspective of concepts.
Clear, Concise, and Inviting Writing Style, Tone, and Layout	Enables students to understand what they are reading, which reduces math anxiety and promotes student success.
Parallel Words and Math	Increases students' ability to read and understand examples with a seamless representation of their instructor's class (instructor's voice and what they would write on the board).
Common Mistakes	Addresses a different learning style: teaching by counter example. Demonstrates common mistakes so that students understand why a step is incorrect and reinforces the correct mathematics.
Color for Pedagogical Reasons	Particularly helpful for visual learners when they see a function written in red and then its corresponding graph in red or a function written in blue and then its corresponding graph in blue.
Study Tips	Reinforces specific notes that you would want to emphasize in class.
Author Videos	Gives students a mini class of several examples worked by the author.
Your Turn	Engages students during class, builds student confidence, and assists instructor in real-time assessment.
Concept Checks	Reinforce concept learning objectives, much as Your Turn features reinforce skill learning objectives.
Catch the Mistake Exercises	Encourages students to assume the role of teacher—demonstrating a higher mastery level.
Conceptual Exercises	Teaches students to think more globally about a topic.
Inquiry-Based Learning Project (online only)	Lets students *discover* a mathematical identify, formula, and the like that is derived in the book.
Modeling Our World (online only)	Engages students in a modeling project of a timely subject: global climate change.
Chapter Review	Presents key ideas and formulas section by section in a chart. Improves study skills.
Chapter Review Exercises	Improves study skills.
Chapter Practice Test	Offers self-assessment and improves study skills.
Cumulative Test	Improves retention.

INSTRUCTOR SUPPLEMENTS

INSTRUCTOR'S SOLUTIONS MANUAL (ISBN: 978-1-119-27333-2)

- Contains worked-out solutions to all exercises in the text.

INSTRUCTOR'S MANUAL

Authored by Cynthia Young, the manual provides practical advice on teaching with the text, including:

- sample lesson plans and homework assignments
- suggestions for the effective utilization of additional resources and supplements
- sample syllabi
- Cynthia Young's Top 10 Teaching Tips & Tricks
- online component featuring the author presenting these Tips & Tricks

ANNOTATED INSTRUCTOR'S EDITION (ISBN: 978-1-119-27336-3)

- Displays answers to the vast majority of exercise questions in the back of the book.
- Provides additional classroom examples within the standard difficulty range of the in-text exercises, as well as challenge problems to assess your students' mastery of the material.

POWERPOINT SLIDES

- For each chapter of the book, a corresponding set of lecture notes and worked-out examples are presented as PowerPoint slides, available on the Book Companion Site (www.wiley.com/college/young) and *WileyPLUS*.

TEST BANK

- Contains approximately 900 questions and answers from every section of the text.

COMPUTERIZED TEST BANK

Electonically enhanced version of the Test Bank that

- contains approximately 900 algorithmically generated questions.
- allows instructors to freely edit, randomize, and create questions.
- allows instructors to create and print different versions of a quiz or exam.
- recognizes symbolic notation.
- allows for partial credit if used within *WileyPLUS*.

BOOK COMPANION WEBSITE (WWW.WILEY.COM/COLLEGE/YOUNG)

- Contains all instructor supplements listed plus a selection of personal response system questions ("Clicker Questions").

WILEYPLUS

- *WileyPLUS* online homework features a full-service, digital learning environment, including additional resources for students, such as lecture videos by the author, self-practice exercises, tutorials, integrated links between the online text and supplements, and ORION adaptive practice.
- *WileyPLUS* has been substantially revised and improved since the third edition of *College Algebra*. It now includes ORION, an adaptive practice engine built directly into *WileyPLUS* that can connect directly into the *WileyPLUS* gradebook, or into your campus Learning Management System gradebook if you select that option. Wiley has been incorporating ORION into *WileyPLUS* courses for over five years, including the Young *Precalculus* program. ORION brings the power of adaptive learning, which will continue to help students and instructors "bridge the gap."

Other new *WileyPLUS* features include:

- Maple TA Math palette and question evaluator (compatible with tablets and Java-free)
- HTML 5 graphing questions
- Concept Check questions
- Secure testing enhancements, including IP restriction
- HTML 5 Show Work whiteboard

STUDENT SUPPLEMENTS

STUDENT SOLUTIONS MANUAL (ISBN: 978-1-119-27327-1)

- Includes worked-out solutions for all odd problems in the text.

BOOK COMPANION WEBSITE (WWW.WILEY.COM/COLLEGE/YOUNG)

- Provides additional resources for students to enhance the learning experience.

WILEYPLUS

- *WileyPLUS* online homework features a full-service, digital learning environment, including additional resources for students, such as lecture videos by the author, self-practice exercises, tutorials, integrated links between the online text and supplements, and ORION adaptive practice.
- *WileyPLUS* has been substantially revised and improved since the third edition of *College Algebra*. It now includes ORION, an adaptive practice engine built directly into *WileyPLUS* that can connect directly into the *WileyPLUS* gradebook, or into your campus Learning Management System gradebook if you select that option. Wiley has been incorporating ORION into *WileyPLUS* courses for over five years, including the Young *Precalculus* program. ORION brings the power of adaptive learning, which will continue to help students and instructors "bridge the gap."

ACKNOWLEDGMENTS

I want to express my sincerest gratitude to the entire Wiley team. I've said this before, and I will say it again: Wiley is the right partner for me. There is a reason that my dog is named Wiley—she's smart, competitive, a team player, and most of all, a joy to be around. There are several people within Wiley to whom I feel the need to express my appreciation: first and foremost to Laurie Rosatone, who convinced Wiley Higher Ed to invest in a young assistant professor's vision for a series and who has been unwavering in her commitment to student learning. To my editor Joanna Dingle, whose judgment I trust in both editorial and preschool decisions; thank you for surpassing my greatest expectations for an editor. To the rest of the math editorial team (Jennifer Lartz, Anne Scanlan-Rohrer, and Ryann Dannelly), you are all first class! This revision was planned and executed exceptionally well thanks to you. To the math marketing manager John LaVacca, thank you for helping reps tell my story: you are outstanding at your job. To product designer David Dietz, many thanks for your role in developing the online course and digital assets. To Mary Sanger, thank you for your attention to detail. To Maureen Eide, thank you for the new design! And finally, I'd like to thank all of the Wiley reps: thank you for your commitment to my series and your tremendous efforts to get professors to adopt this book for their students.

I would also like to thank all of the contributors who helped us make this *an even better book*. I'd first like to thank Mark McKibben. He is known as the author of the solutions manuals that accompany this series, but he is much more than that. Mark, thank you for making this series a priority, for being so responsive, and most of all for being my "go-to" person to think through ideas. Many thanks to Diane Cook for accuracy checking of the text, exercises, and solutions.

I'd also like to thank the following reviewers, whose input helped make this book even better.

Text Reviewers

Victor Akatsa, Chicago State University

Ari Aluthge, Marshall University

Marian Anton, Central Connecticut State University

Ioannis K. Argyros, Cameron University

Michelle Beard, Ventura College

Michael J. Carr, Mott Community College

Kaitlin Curry, Orange County Community College

Xin-Ran Duan, American River College

Mohammed Ganjizadeh, Tarrant County College

Sonja Godeken, University of Texas at Arlington

Keith Hubbard, Stephen F. Austin State University

Alan Jones, University of Central Oklahoma

Robin S. Kalder, Central Connecticut State University

Raja Khoury, Collin College

Frederic Latour, Central Connecticut State University

Rob Lewis, Linn-Benton Community College

Carrie L. Menard, SUNY Adirondack

Jim Stewart, Jefferson Community and Technical College

Sasha Townsend, Tulsa Community College

Phil Veer, Johnson County Community College

Contents

VitalyEdush/Getty Images, Inc.

Courtesy MGM Home Entertainment Distribution Corp.

SHANNON STAPLETON/ REUTERS / Newscom

Cn0ra/ Getty Images, Inc.

laurentiu iordache/Alamy StockPhotoE/JRC2

A Note from the Author
TO THE STUDENT

I wrote this text with careful attention to ways in which to make your learning experience more successful. If you take full advantage of the unique features and elements of this textbook, I believe your experience will be fulfilling and enjoyable. Let's walk through some of the special book features that will help you in your study of Trigonometry.

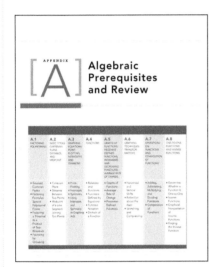

Prerequisites and Review (Appendix)

A comprehensive review of prerequisite knowledge (college algebra topics) provides a brush up on knowledge and skills necessary for success in the course.

Clear, Concise, and Inviting Writing

Special attention has been made to present an engaging, clear, precise narrative in a layout that is easy to use and designed to reduce any math anxiety you may have.

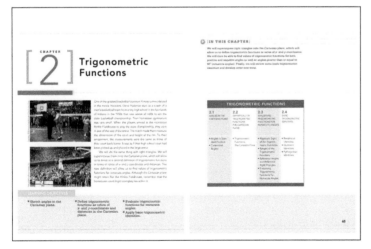

Chapter Introduction, Flowchart, Section Headings, and Objectives

An opening vignette, flowchart, list of chapter sections, and targeted chapter learning objectives give you an overview of the chapter to help you see the big picture, the relationships between topics, and clear objectives for learning in the chapter.

2.3 EVALUATING TRIGONOMETRIC FUNCTIONS FOR NONACUTE ANGLES

SKILLS OBJECTIVES	CONCEPTUAL OBJECTIVES
▪ Determine the reference angle of a nonacute angle.	▪ Understand which trigonometric functions are positive or negative in each of the four quadrants.
▪ Determine the ranges of the six trigonometric functions.	▪ Understand why sine and cosine range between -1 and 1.
▪ Determine the reference angle for any angle whose terminal side lies in one of the four quadrants.	▪ Understand that reference angles are acute and have positive measure.
▪ Evaluate trigonometric functions for nonacute and quadrantal angles.	▪ Understand that the value of a trigonometric function at an angle is the same as the trigonometric value of its reference angle, except there may be a difference in its algebraic sign $(+/-)$.

Skills and Conceptual Objectives

For every section, objectives are further divided by skills *and* concepts so you can see the difference between solving problems and truly understanding concepts.

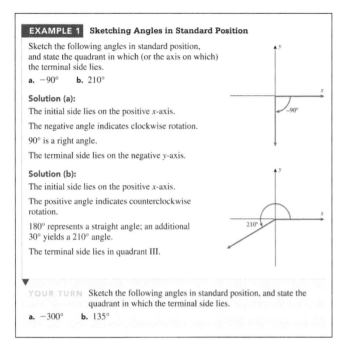

EXAMPLE 1 Sketching Angles in Standard Position

Sketch the following angles in standard position, and state the quadrant in which (or the axis on which) the terminal side lies.

a. $-90°$ **b.** $210°$

Solution (a):

The initial side lies on the positive x-axis.

The negative angle indicates clockwise rotation.

$90°$ is a right angle.

The terminal side lies on the negative y-axis.

Solution (b):

The initial side lies on the positive x-axis.

The positive angle indicates counterclockwise rotation.

$180°$ represents a straight angle; an additional $30°$ yields a $210°$ angle.

The terminal side lies in quadrant III.

▼

YOUR TURN   Sketch the following angles in standard position, and state the quadrant in which the terminal side lies.

a. $-300°$ **b.** $135°$

Examples

Examples pose a specific problem using concepts already presented and then work through the solution. These serve to enhance your understanding of the subject matter.

Your Turn

Immediately following many examples, you are given a similar problem to reinforce and check your understanding. This helps build confidence as you progress in the chapter. These are ideal for in-class activity or for preparing for homework later. Answers are provided in the margin for a quick check of your work.

Common Mistake/Correct versus Incorrect

In addition to standard examples, some problems are worked out both correctly and incorrectly to highlight common errors. Counter-examples like these are often an effective learning approach.

---

common mistake

A common mistake is forgetting to first put the radius and arc length in the same units.

✓CORRECT	✗INCORRECT

Write the formula relating radian measure to arc length and radius.

$$\theta \text{ (in radians)} = \frac{s}{r}$$

Substitute $s = 6$ centimeters and $r = 2$ meters into the radian expression.	Substitute $s = 6$ centimeters and $r = 2$ meters into the radian expression.
$\theta = \dfrac{6\,\text{cm}}{2\,\text{m}}$	$\theta = \dfrac{6\,\text{cm}}{2\,\text{m}}$
	$= 3$

Convert the radius (2) meters to centimeters: 2 meters = 200 centimeters

ERROR: (not converting both numerator and denominator to the same units)

$$\theta = \frac{6\,\text{cm}}{200\,\text{cm}}$$

The units, centimeters, cancel and the result is a unitless real number.

$$\boxed{\theta = 0.03 \text{ rad}}$$

Parallel Words and Math

This text reverses the common textbook presentation of examples by placing the explanation in words *on the left* and the mathematics in parallel *on the right*. This makes it easier to read through examples as the material flows more naturally from left to right and as commonly presented in class.

WORDS	MATH
Multiply 1 radian by $\dfrac{180°}{\pi}$.	$1\left(\dfrac{180°}{\pi}\right)$
Approximate π by 3.14.	$1\left(\dfrac{180°}{3.14}\right)$
Use a calculator to evaluate and round to the nearest degree.	$\approx 57°$
	$\boxed{1 \text{ rad} \approx 57°}$

Study Tips and Caution Notes

These marginal reminders call out important hints or warnings to be aware of related to the topic or problem.

STUDY TIP

To use the relationship

$$s = r\theta$$

the angle θ must be in radians.

▼

CAUTION

To correctly calculate radians from the formula $\theta = \dfrac{s}{r}$, the radius and arc length must be expressed in the same units.

> **[IN THIS CHAPTER]**
>
> You will learn a second way to measure angles using radians. You will
> convert between degrees and radians. You will calculate arc lengths,
> areas of circular sectors, and angular and linear speeds. Finally, the third
> definition of trigonometric functions using the unit circle approach will be

> ▶ **EXAMPLE 2** **Finding the Radian Measure of an Angle**
>
> What is the measure (in radians) of a central angle θ that intercepts an arc of length
> 6 centimeters on a circle with radius 2 meters?

common mistake

A common mistake is forgetting to first put the radius and arc length in the same
units.

> **[SECTION 2.1] SUMMARY**
>
> An angle in the Cartesian plane is in standard position if its initial side lies along the positive x-axis and its vertex is located at the origin. Angles in standard position have terminal sides that lie either in one of the four quadrants or along one of the two axes. The special triangles, 30°-60°-90° and 45°-45°-90° ... Symmetry was then used to locate similar pairs of coordinates in the other quadrants. All right triangles with one vertex (not the right angle vertex) located at the origin and hypotenuse equal to 1 have the other nonright angle vertex located along the unit circle ($x^2 + y^2 = 1$). Coterminal angles are angles in standard position ...

> **[CHAPTER 1 REVIEW]**
>
> | SECTION | CONCEPT | KEY IDEAS/FORMULAS |
> |---|---|---|
> | 1.1 | Angles, degrees, and triangles | |
> | | Angles and degree measure | 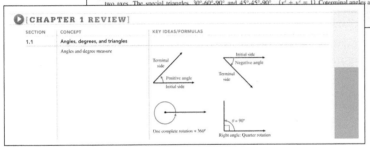 |

> **[SECTION 1.5] EXERCISES**
>
> • **SKILLS**
>
> In Exercises 1–4, determine the number of significant digits corresponding to each of the given angle measures and side lengths.
>
> 1. $\alpha = 37.5°$ 2. $\beta = 49.76°$ 3. $a = 0.37$ km   4. $b = 0.2$ mi
>
> In • **APPLICATIONS**
>
> 5.
>
> In an
>
> 49. **Golf.** If the flagpole that a golfer aims at on a green measures 5 feet from the ground to the top of the flag and a golfer
>
> **Exercises 51 and 52 illustrate a midair refueling scenario that military aircraft often enact. Assume the elevation angle that** ... $= 36°$.
>
> • **CATCH THE MISTAKE**
>
> For Exercises 79 and 80, refer to the right triangle diagram below and explain the mistake that is made.
>
> 80. If $\beta = 56°$ and $c = 15$ feet, find b and then find a.
>
> Solution:
>
> Write sine as the opposite side over the hypotenuse. $\sin 56° = \dfrac{b}{15}$
>
> Solve for b.
>
> 79.
>
> So • **CONCEPTUAL**
>
> In Exercises 81–88, determine whether each statement is true or false.
>
> 81. If you are given the lengths of two sides of a right triangle, you can solve the right triangle.
>
> 82. If you are given the length of one side and the measure of one acute angle of a right triangle, you can solve the right
>
> 83. If you are given the measures of the two acute angles of a right triangle, you can solve the right triangle.
>
> 84. If you are given the length of the hypotenuse of a right triangle and the measure of the angle opposite the hypotenuse, you
>
> • **CHALLENGE**
>
> 89. Use the information in the picture below to determine the height of the mountain.
>
> 91. From the top of a 12-foot ladder, the angle of depression to the far side of a sidewalk is 45°, while the angle of depression to the near side of the sidewalk is 65°. How wide is the sidewalk?
>
> 92. Find the perimeter of triangle A.
>
> • **TECHNOLOGY**
>
> 95. Use a calculator to find $\sin^{-1}(\sin 40°)$.
>
> 96. Use a calculator to find $\cos^{-1}(\cos 17°)$.
>
> 97. Use a calculator to find $\cos(\cos^{-1}0.8)$.
>
> 98. Use a calculator to find $\sin(\sin^{-1}0.3)$.
>
> 99. Based on the result from Exercise 95, what would $\sin^{-1}(\sin\theta)$ be for an acute angle θ?
>
> 100. Based on the result from Exercise 96, what would $\cos^{-1}(\cos\theta)$ be for an acute angle θ?

Video Icons

Video icons appear on all chapter
introductions, chapter and section
reviews, as well as selected
examples throughout the chapter
to indicate that the author has
created a video segment for that
element. These video clips help
you work through the selected
examples with the author as your
"private tutor."

Six Types of Exercises

Every text section ends with
**Skills, Applications, Catch
the Mistake, Conceptual,
Challenge, and Technology**
exercises. The exercises
gradually increase in difficulty
and vary in skill and conceptual
emphasis. Catch the Mistake
exercises increase the depth of
understanding and reinforce what
you have learned. Conceptual
and Challenge exercises
specifically focus on assessing
conceptual understanding.
Technology exercises enhance your
understanding and ability using
scientific and graphing calculators.

Concept Checks

Similar to how Your Turns reinforce skill learning objectives, Concept Checks reinforce concept learning objectives.

[CONCEPT CHECK]

Which of the following is true about the reference angle?
(A) It is the angle made by the terminal side and the x-axis (or negative x-axis);
(B) it is positive;
(C) it is an acute angle; or
(D) all of the above.

▼ ·

ANSWER (D) all of the above

Chapter Review, Review Exercises, Practice Test, Cumulative Test

At the end of every chapter, a summary review chart organizes the key learning concepts in an easy-to-use one- or two-page layout. This feature includes key ideas and formulas, as well as indicating relevant pages and review exercises so that you can quickly summarize a chapter and study smarter. Review Exercises, arranged by section heading, are provided for extra study and practice. A Practice Test, without section headings, offers even more self-practice before moving on. A new Cumulative Test feature offers study questions based on all previous chapters' content, thus helping you build upon previously learned concepts.

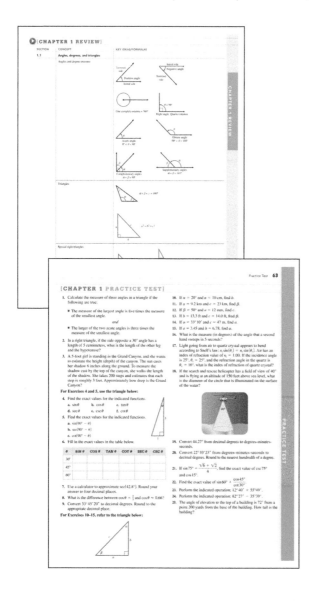

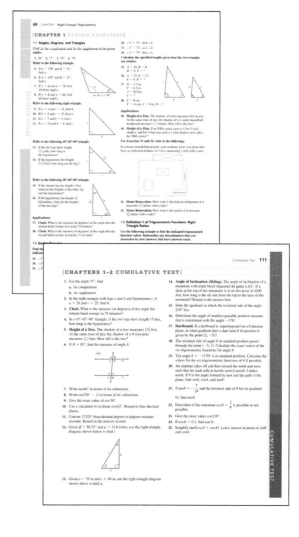

Trigonometry

[1]

Right Triangle Trigonometry

To the ancient Greeks, trigonometry was the study of right triangles. Trigonometric functions (sine, cosine, tangent, cotangent, secant, and cosecant) can be defined as right triangle ratios (ratios of the lengths of sides of a right triangle). Thousands of years later, we still find applications of right triangle trigonometry today in sports, surveying, navigation,* and engineering.

*Section 1.5, Example 7 and Exercises 66–68 and 73–74.

LEARNING OBJECTIVES

- Understand degree measure.
- Learn the conditions that make two triangles similar.

- Define the six trigonometric functions as ratios of lengths of the sides of right triangles.

- Evaluate trigonometric functions exactly and with calculators.
- Solve right triangles.

[IN THIS CHAPTER]

We will review angles, degree measure, and special right triangles. We will discuss the properties of similar triangles. We will use the concept of similar right triangles to define the six trigonometric functions as ratios of the lengths of the sides of right triangles (right triangle trigonometry).

RIGHT TRIANGLE TRIGONOMETRY

1.1 ANGLES, DEGREES, AND TRIANGLES	1.2 SIMILAR TRIANGLES	1.3 DEFINITION 1 OF TRIGONOMETRIC FUNCTIONS: RIGHT TRIANGLE RATIOS	1.4 EVALUATING TRIGONOMETRIC FUNCTIONS: EXACTLY AND WITH CALCULATORS	1.5 SOLVING RIGHT TRIANGLES
• Angles and Degree Measure • Triangles • Special Right Triangles	• Finding Angle Measures Using Geometry • Classification of Triangles	• Trigonometric Functions: Right Triangle Ratios • Cofunctions	• Evaluating Trigonometric Functions *Exactly* for Special Angle Measures: 30°, 45°, and 60° • Using Calculators to Evaluate (Approximate) Trigonometric Function Values • Representing Partial Degrees: DD or DMS	• Accuracy and Significant Digits • Solving a Right Triangle Given an Acute Angle Measure and a Side Length • Solving a Right Triangle Given the Lengths of Two Sides

1.1 ANGLES, DEGREES, AND TRIANGLES

SKILLS OBJECTIVES	CONCEPTUAL OBJECTIVES
■ Find the complement and supplement of an angle. ■ Use the Pythagorean theorem to solve a right triangle. ■ Solve 30°-60°-90° and 45°-45°-90° triangles.	■ Understand that degrees are a measure of an angle. ■ Understand that the Pythagorean theorem applies only to right triangles. ■ Understand that to solve a triangle means to find all of the angle measures and side lengths.

1.1.1 Angles and Degree Measure

1.1.1 SKILL

Find the complement and supplement of an angle.

1.1.1 CONCEPTUAL

Understand that degrees are a measure of an angle.

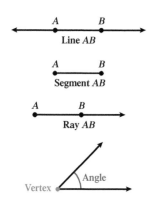

The study of trigonometry relies heavily on the concept of angles. Before we define angles, let us review some basic terminology. A **line** is the straight path connecting two points (*A* and *B*) and extending beyond the points in both directions. The portion of the line between the two points (including the points) is called a **line segment**. A **ray** is the portion of the line that starts at one point (*A*) and extends to infinity (beyond *B*). Point *A* is called the **endpoint of the ray**.

In geometry, an **angle** is formed when two rays share the same endpoint. The common endpoint is called the **vertex**. In trigonometry, we say that an **angle** is formed when a ray is rotated around its endpoint. The ray in its original position is called the **initial ray** or the **initial side** of an angle. In the Cartesian plane (the rectangular coordinate plane), we usually assume the initial side of an angle is the positive *x*-axis and the vertex is located at the origin. The ray after it is rotated is called the **terminal ray** or the **terminal side** of an angle. Rotation in a counterclockwise direction corresponds to a **positive angle**, whereas rotation in a clockwise direction corresponds to a **negative angle**.

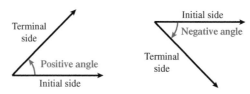

STUDY TIP

Positive angle: Counterclockwise
Negative angle: Clockwise

Lengths, or distances, can be measured in different units: feet, miles, and meters are three common units. To compare angles of different sizes, we need a standard unit of measure. One way to measure the size of an angle is with **degree measure**. We discuss *degrees* now, and in Chapter 3, we discuss another angle measure called *radians*.

[CONCEPT CHECK]

TRUE OR FALSE The measure of an acute angle is greater than the measure of an obtuse angle.

▼ ·

ANSWER False

DEFINITION	**Degree Measure of Angles**

An angle formed by one complete counterclockwise rotation has **measure 360 degrees,** denoted 360°.

One complete revolution = 360°

Therefore, a counterclockwise $\frac{1}{360}$ of a rotation has measure **1 degree**.

WORDS	MATH
360° represents 1 complete rotation.	1 complete rotation $= 1 \cdot 360° = 360°$
180° represents $\frac{1}{2}$ rotation.	$\frac{1}{2}$ complete rotation $= \frac{1}{2} \cdot 360° = 180°$
90° represents $\frac{1}{4}$ rotation.	$\frac{1}{4}$ complete rotation $= \frac{1}{4} \cdot 360° = 90°$

The Greek letter θ (theta) is the most common name for an angle in mathematics. Other common names for angles are α (alpha), β (beta), and γ (gamma).

WORDS   **MATH**

An angle measuring exactly 90° is called a **right angle**. A right angle is often represented by the adjacent sides of a rectangle, indicating that the two rays are **perpendicular**.

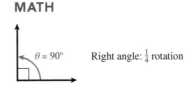

An angle measuring exactly 180° is called a **straight angle**.

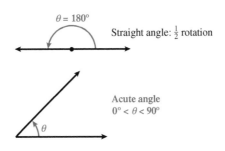

An angle measuring greater than 0°, but less than 90°, is called an **acute angle**.

An angle measuring greater than 90°, but less than 180°, is called an **obtuse angle**.

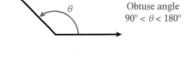

If the sum of the measures of two positive angles is 90°, the angles are called **complementary**. We say that α is the **complement** of β (and vice versa).

If the sum of the measures of two positive angles is 180°, the angles are called **supplementary**. We say that α is the **supplement** of β (and vice versa).

EXAMPLE 1   **Finding Measures of Complementary and Supplementary Angles**

Find the measure of each angle:

a. Find the complement of 50°.

b. Find the supplement of 110°.

c. Represent the complement of α in terms of α.

d. Find two supplementary angles such that the first angle is twice as large as the second angle.

Solution:

a. The sum of complementary angles is 90°. $\theta + 50° = 90°$

Solve for θ. $\theta = \boxed{40°}$

b. The sum of supplementary angles is 180°. $\theta + 110° = 180°$

Solve for θ. $\theta = \boxed{70°}$

c. Let β be the complement of α.

The sum of complementary angles is 90°. $\qquad$ $\alpha + \beta = 90°$

Solve for β. $\qquad$ $\beta = \boxed{90° - \alpha}$

d. The sum of supplementary angles is 180°. $\qquad$ $\alpha + \beta = 180°$

Let $\beta = 2\alpha$. $\qquad$ $\alpha + 2\alpha = 180°$

Solve for α. $\qquad$ $3\alpha = 180°$

$\qquad$ $\alpha = 60°$

Substitute $\alpha = 60°$ into $\beta = 2\alpha$. $\qquad$ $\beta = 120°$

The angles have measures $\boxed{60°}$ and $\boxed{120°}$.

ANSWER

The angles have measures 45° and 135°.

YOUR TURN Find two supplementary angles such that the first angle is three times as large as the second angle.

1.1.2 Triangles

1.1.2 SKILL

Use the Pythagorean theorem to solve a right triangle.

1.1.2 CONCEPTUAL

Understand that the Pythagorean theorem applies only to right triangles.

Trigonometry originated as the study of triangles, with emphasis on calculations involving the lengths of sides and the measures of angles. Triangles are three-sided closed-plane figures. An important property of triangles is that the sum of the measures of the three angles of any triangle is 180°.

ANGLE SUM OF A TRIANGLE

The sum of the measures of the angles of any triangle is 180°.

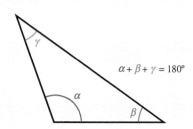

$\alpha + \beta + \gamma = 180°$

EXAMPLE 2 **Finding an Angle of a Triangle**

If two angles of a triangle have measures 32° and 68°, what is the measure of the third angle?

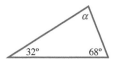

Solution:

The sum of the measures of all three angles is 180°. $\qquad$ $32° + 68° + \alpha = 180°$

Solve for α. $\qquad$ $\boxed{\alpha = 80°}$

ANSWER

68°

YOUR TURN If two angles of a triangle have measures 16° and 96°, what is the measure of the third angle?

In geometry, some triangles are classified as *equilateral*, *isosceles*, and *right*. An **equilateral triangle** has three equal sides and therefore has three equal angles (60°). An **isosceles triangle** has two equal sides (legs) and therefore has two equal angles opposite those legs. The most important triangle that we will discuss in this course is a *right triangle*. A **right triangle** is a triangle in which one of the angles is a right angle (90°). Since one angle is 90°, the other two angles must be complementary (sum to 90°) so that the sum of all three angles is 180°. The longest side of a right triangle, called the **hypotenuse**, is opposite the right angle. The other two sides are called the **legs** of the right triangle.

Right triangle: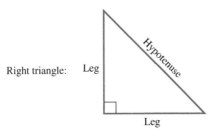

The *Pythagorean theorem* relates the sides of a right triangle. It is important to note that length (a synonym of distance) is always positive.

PYTHAGOREAN THEOREM

In any right triangle, the square of the length of the longest side (hypotenuse) is equal to the sum of the squares of the lengths of the other two sides (legs).

$$a^2 + b^2 = c^2$$

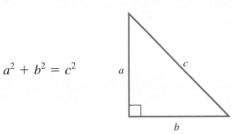

It is important to note that the **Pythagorean theorem applies** *only* **to right triangles**. In addition, it does not matter which leg is called a or b as long as the square of the longest side is equal to the sum of the squares of the smaller sides.

EXAMPLE 3 **Using the Pythagorean Theorem to Find the Side of a Right Triangle**

Suppose you have a 10-foot ladder and want to reach a height of 8 feet to clean out the gutters on your house. How far from the base of the house should the base of the ladder be?

Solution:

Label the unknown side as x.

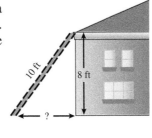

Apply the Pythagorean theorem. $x^2 + 8^2 = 10^2$

Simplify. $x^2 + 64 = 100$

Solve for x. $\qquad\qquad x^2 = 36$

$$x = \pm 6$$

Length must be positive. $\qquad\qquad \boxed{x = 6}$

The ladder should be $\boxed{6\text{ feet}}$ from the base of the house along the ground.

▼

YOUR TURN A steep ramp is being built for skateboarders. The height, horizontal ground distance, and ramp length form a right triangle. If the height is 9 feet and the horizontal ground distance is 12 feet, what is the length of the ramp?

When solving a right triangle exactly, simplification of radicals is often necessary. For example, if a side length of a triangle resulted in $\sqrt{17}$, the radical cannot be simplified any further. However, if a side length of a triangle resulted in $\sqrt{20}$, the radical would be simplified:

$$\sqrt{20} = \sqrt{4 \cdot 5} = \sqrt{4} \cdot \sqrt{5} = \sqrt{2^2} \cdot \sqrt{5} = 2\sqrt{5}$$

EXAMPLE 4 **Using the Pythagorean Theorem with Radicals**

Use the Pythagorean theorem to solve for the unknown side length in the given right triangle. Express your answer exactly in terms of simplified radicals.

Solution:

Apply the Pythagorean theorem. $\qquad 3^2 + x^2 = 7^2$

Simplify known squares. $\qquad\qquad 9 + x^2 = 49$

Solve for x. $\qquad\qquad\qquad\qquad x^2 = 40$

$$x = \pm\sqrt{40}$$

The side length x is a distance that is positive. $x = \sqrt{40}$

Simplify the radical. $\qquad x = \sqrt{4 \cdot 10} = \underset{2}{\underbrace{\sqrt{4}}} \cdot \sqrt{10} = \boxed{2\sqrt{10}}$

▼

YOUR TURN Use the Pythagorean theorem to solve for the unknown side length in the given right triangle. Express your answer exactly in terms of simplified radicals.

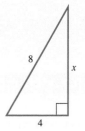

1.1.3 Special Right Triangles

Right triangles whose sides are in the ratios of 3-4-5, 5-12-13, and 8-15-17 are examples of right triangles that are special because their side lengths are equal to whole numbers that satisfy the Pythagorean theorem. A Pythagorean triple consists of three positive integers that satisfy the Pythagorean theorem.

$$3^2 + 4^2 = 5^2 \qquad 5^2 + 12^2 = 13^2 \qquad 8^2 + 15^2 = 17^2$$

There are two other special right triangles that warrant attention: a 30°-60°-90° triangle and a 45°-45°-90° triangle. Although in trigonometry we focus more on the angles than on the side lengths, we are interested in special relationships between the lengths of the sides of these right triangles. We will start with a 45°-45°-90° triangle.

WORDS	MATH
A 45°-45°-90° triangle is an isosceles (two legs are equal) right triangle.	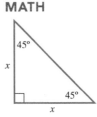

Apply the Pythagorean theorem.	$x^2 + x^2 = \text{hypotenuse}^2$
Simplify the left side of the equation.	$2x^2 = \text{hypotenuse}^2$
Solve for the hypotenuse.	$\text{hypotenuse} = \pm\sqrt{2x^2} = \pm\sqrt{2}\,|x|$
The x and the hypotenuse are both lengths and, therefore, must be positive.	$\text{hypotenuse} = \sqrt{2}x$

This shows that the hypotenuse of a 45°-45°-90° is $\sqrt{2}$ times the length of either leg.	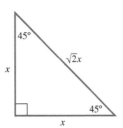

If we let $x = 1$, then the triangle will have legs with length equal to 1 and the hypotenuse will have length $\sqrt{2}$. Notice that these lengths satisfy the Pythagorean theorem: $1^2 + 1^2 = \left(\sqrt{2}\right)^2$, or $2 = 2$. Later when we discuss the unit circle approach, we will let the hypotenuse have length 1. The legs will then have lengths $\dfrac{\sqrt{2}}{2}$ and $\dfrac{\sqrt{2}}{2}$:

$$\left(\frac{\sqrt{2}}{2}\right)^2 + \left(\frac{\sqrt{2}}{2}\right)^2 = \frac{1}{2} + \frac{1}{2} = 1^2$$

▶ **EXAMPLE 5**   **Solving a 45°-45°-90° Triangle**

A house has a roof with a 45° pitch (the angle the roof makes with the house). If the house is 60 feet wide, what are the lengths of the sides of the roof that form the attic? Round to the nearest foot.

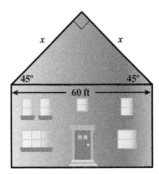

Solution:

Draw the 45°-45°-90° triangle.

Let x represent the length of the unknown legs.

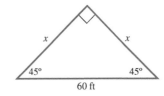

[CONCEPT CHECK]

TRUE OR FALSE In a right triangle, if you know one of the acute angles and one side length, then it is possible to solve the triangle.

▼ ···

ANSWER True

▼ ···

ANSWER
  $60\sqrt{2} \approx 85$ ft

Recall that the hypotenuse of a 45°-45°-90° triangle is $\sqrt{2}$ times the length of either leg.

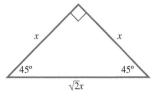

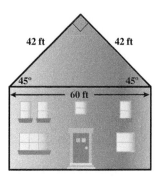

Let the hypotenuse equal 60 feet. $60 = \sqrt{2}x$

Solve for x. $x = \dfrac{60}{\sqrt{2}} \approx 42.42641$

Round to the nearest foot. $x \approx \boxed{42 \text{ ft}}$

▼

YOUR TURN   A house has a roof with a 45° pitch. If the sides of the roof are 60 feet, how wide is the house? Round to the nearest foot.

Let us now determine the relationship of the sides of a 30°-60°-90° triangle. We start with an equilateral triangle (equal sides and equal 60° angles).

WORDS	MATH
Draw an equilateral triangle with sides $2x$.	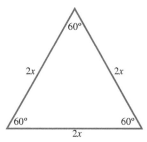

Draw a line segment from one vertex that is perpendicular to the opposite side; this line segment represents the height of the triangle h and bisects the base. There are now two identical 30°-60°-90° triangles.

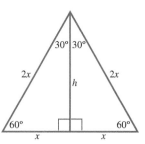

Notice that in each triangle the hypotenuse is twice the length of the shortest leg, which is opposite the 30° angle.

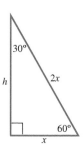

To find the length h, use the Pythagorean theorem.

$$h^2 + x^2 = (2x)^2$$

Solve for h.

$$h^2 + x^2 = 4x^2$$

$$h^2 = 3x^2$$

$$h = \pm\sqrt{3x^2} = \pm\sqrt{3}\,|x|$$

Both h and x are lengths and must be positive.

$$h = \sqrt{3}x$$

We see that the hypotenuse of a 30°-60°-90° triangle is two times the length of the leg opposite the 30° angle, the short side.

The leg opposite the 60° angle, the long leg, is $\sqrt{3}$ times the length of the leg opposite the 30° angle, the short leg.

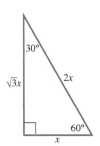

If we let $x = 1$, then the triangle will have legs with lengths 1 and $\sqrt{3}$ and hypotenuse of length 2. These lengths satisfy the Pythagorean theorem: $1^2 + (\sqrt{3})^2 = 2^2$, or $4 = 4$. Later when we discuss the unit circle approach, we will let the hypotenuse have length 1. The legs will then have lengths $\dfrac{1}{2}$ and $\dfrac{\sqrt{3}}{2}$.

$$\left(\frac{1}{2}\right)^2 + \left(\frac{\sqrt{3}}{2}\right)^2 = \frac{1}{4} + \frac{3}{4} = 1^2$$

EXAMPLE 6   **Solving a 30°-60°-90° Triangle**

Before a hurricane strikes, it is wise to stake down trees for additional support during the storm. If the branches allow for the rope to be tied 15 feet up the tree and a desired angle between the rope and the ground is 60°, how much total rope is needed? How far from the base of the tree should each of the two stakes be hammered?

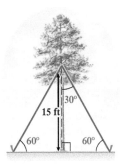

Solution:

Recall the relationship between the sides of a 30°-60°-90° triangle.

Label each side.

In this case, the leg opposite the 60° angle is 15 feet.

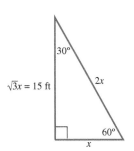

Solve for x.

$$\sqrt{3}x = 15 \text{ ft}$$

$$x = \frac{15}{\sqrt{3}} \approx 8.7 \text{ ft}$$

Find the length of the hypotenuse. $\text{hypotenuse} = 2x = 2\left(\dfrac{15}{\sqrt{3}}\right) \approx 17 \text{ ft}$

The ropes should be staked approximately $\boxed{8.7}$ feet from the base of the tree, and approximately $\boxed{2(17) = 34}$ total feet of rope will be needed.

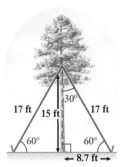

▼
ANSWER

Each rope should be staked approximately 12 feet from the base of the tree. Approximately 46 total feet of rope will be needed.

▼
YOUR TURN Rework Example 6 with a height (where the ropes are tied) of 20 feet. How far from the base of the tree should each of the ropes be staked, and how much total rope will be needed?

▶[SECTION 1.1] SUMMARY

In this section, you have practiced working with angles and triangles. One unit of measure of angles is degrees. An angle measuring exactly 90° is called a right angle. The sum of the three angles of any triangle is always 180°. Triangles that contain a right angle are called right triangles. With the Pythagorean theorem, you can solve for one side of a right triangle given the other two sides. The right triangles 30°-60°-90° and 45°-45°-90° are special because of the simple numerical relations of their sides.

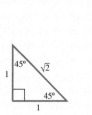

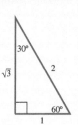

[SECTION 1.1] EXERCISES

• **SKILLS**

In Exercises 1–8, specify the measure of the angle in degrees using the correct algebraic sign (+ or −).

1. $\frac{1}{2}$ rotation counterclockwise
2. $\frac{1}{4}$ rotation counterclockwise
3. $\frac{1}{3}$ rotation clockwise
4. $\frac{2}{3}$ rotation clockwise
5. $\frac{5}{6}$ rotation counterclockwise
6. $\frac{7}{12}$ rotation counterclockwise
7. $\frac{4}{5}$ rotation clockwise
8. $\frac{5}{9}$ rotation clockwise

In Exercises 9–14, find (a) the complement and (b) the supplement of each of the given angles.

9. 18°
10. 39°
11. 42°
12. 57°
13. 89°
14. 75°

In Exercises 15–18, find the measure of each angle.

15.

$(6x)°$
$(4x)°$

16.

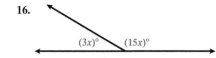

$(3x)°$ $(15x)°$

17. Supplementary angles with measures $8x$ degrees and $4x$ degrees

18. Complementary angles with measures $3x + 15$ degrees and $10x + 10$ degrees

In Exercises 19–24, refer to the triangle in the drawing.

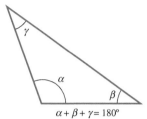

$\alpha + \beta + \gamma = 180°$

19. If $\alpha = 117°$ and $\beta = 33°$, find γ.

20. If $\alpha = 110°$ and $\beta = 45°$, find γ.

21. If $\gamma = \beta$ and $\alpha = 4\beta$, find all three angles.

22. If $\gamma = \beta$ and $\alpha = 3\beta$, find all three angles.

23. If $\beta = 53.3°$ and $\gamma = 23.6°$, find α.

24. If $\alpha = 105.6°$ and $\gamma = 13.2°$, find β.

In Exercises 25–34, refer to the right triangle in the drawing. Express lengths *exactly*.

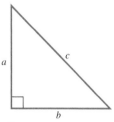

25. If $a = 4$ and $b = 3$, find c.
26. If $a = 3$ and $b = 3$, find c.

27. If $a = 6$ and $c = 10$, find b.
28. If $b = 7$ and $c = 12$, find a.

29. If $a = 8$ and $b = 5$, find c.
30. If $a = 6$ and $b = 5$, find c.

31. If $a = 7$ and $c = 11$, find b.
32. If $b = 5$ and $c = 9$, find a.

33. If $b = \sqrt{7}$ and $c = 5$, find a.
34. If $a = 5$ and $c = 10$, find b.

In Exercises 35–40, refer to the 45°-45°-90° triangle in the drawing. Express lengths *exactly*.

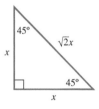

35. If each leg has length 10 inches, how long is the hypotenuse?

36. If each leg has length 8 meters, how long is the hypotenuse?

37. If the hypotenuse has length $2\sqrt{2}$ centimeters, how long is each leg?

38. If the hypotenuse has length $\sqrt{10}$ feet, how long is each leg?

39. If each leg has length $4\sqrt{2}$ inches, how long is the hypotenuse?

40. If the hypotenuse has length 6 meters, how long is each leg?

In Exercises 41–46, refer to the 30°-60°-90° triangle in the drawing. Express lengths *exactly*.

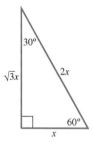

41. If the shortest leg has length 5 meters, what are the lengths of the other leg and the hypotenuse?

42. If the shortest leg has length 9 feet, what are the lengths of the other leg and the hypotenuse?

43. If the longer leg has length 12 yards, what are the lengths of the other leg and the hypotenuse?

44. If the longer leg has length n units, what are the lengths of the other leg and the hypotenuse?

45. If the hypotenuse has length 10 inches, what are the lengths of the two legs?

46. If the hypotenuse has length 8 centimeters, what are the lengths of the two legs?

• **APPLICATIONS**

47. **Clock.** What is the measure of the angle (in degrees) that the minute hand traces in 20 minutes?

48. **Clock.** What is the measure of the angle (in degrees) that the minute hand traces in 25 minutes?

49. **London Eye.** The London Eye (similar to a bicycle wheel) makes one rotation in approximately 30 minutes. What is the measure of the angle (in degrees) that a cart (spoke) will rotate in 12 minutes?

50. **London Eye.** The London Eye (similar to a bicycle wheel) makes one rotation in approximately 30 minutes. What is the measure of the angle (in degrees) that a cart (spoke) will rotate in 5 minutes?

51. **Revolving Restaurant.** If a revolving restaurant can rotate 270° in 45 minutes, how long does it take for the restaurant to make a complete revolution?

David Ball/Getty Images, Inc.

52. Revolving Restaurant. If a revolving restaurant can rotate 72° in 9 minutes, how long does it take for the restaurant to make a complete revolution?

53. Field Trial. In a Labrador retriever field trial, a dog is judged by the straightness of the line it takes to retrieve a fallen bird. The competitors are required to go *through* the water, not along the shore. If the judge wants to calculate how far a dog will travel along a straight path, she walks the two legs of a right triangle as shown in the drawing and uses the Pythagorean theorem. How far would this dog travel (run and swim) if it traveled along the hypotenuse? Round to the nearest foot.

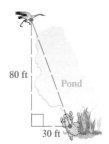

54. Field Trial. How far would the dog in Exercise 53 run and swim if it traveled along the hypotenuse if the judge walks 25 feet along the shore and then 100 feet out to the bird. Round to the nearest foot.

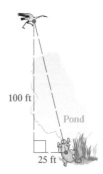

55. Christmas Lights. A couple want to put up Christmas lights along the roofline of their house. If the front of the house is 100 feet wide and the roof has a 45° pitch, how many linear feet of Christmas lights should the couple buy? Round to the nearest foot.

56. Christmas Lights. Repeat Exercise 55 for a house that is 60 feet wide. Round to the nearest foot.

57. Tree Stake. A tree needs to be staked down before a storm. If the ropes can be tied on the tree trunk 17 feet above the ground and the staked rope should make a 60° angle with the ground, how far from the base of the tree should each rope be staked?

58. Tree Stake. What is the total amount of rope (in feet) that extends to the two stakes supporting the tree in Exercise 57?

59. Tree Stake. A tree needs to be staked down before a storm. If the ropes can be tied on the tree trunk 10 feet above the ground and the staked rope should make a 30° angle with the ground, how far from the base of the tree should each rope be staked?

60. Tree Stake. What is the total amount of rope (in feet) that extends to the four stakes supporting the tree in Exercise 59?

61. Party Tent. Steve and Peggy want to rent a 40-foot by 20-foot tent for their backyard to host a barbecue. The base of the tent is supported 7 feet above the ground by poles, and then roped stakes are used for support. The ropes make a 60° angle with the ground. How large a footprint in their yard would they need for this tent (and staked ropes)? In other words, what are the dimensions of the rectangle formed by the stakes on the ground? Round to the nearest foot.

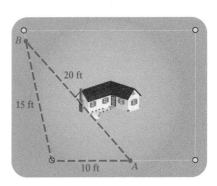

62. Party Tent. Ashley's parents are throwing a graduation party and are renting a 40-foot by 80-foot tent for their backyard. The base of the tent is supported 7 feet above the ground by poles, and then roped stakes are used for support. The ropes make a 45° angle with the ground. How large a footprint (see Exercise 61) in their yard will they need for this tent (and staked ropes)? Round to the nearest foot.

63. Fence Corner. Ben is checking to see if his fence measures 90° at the corner. To do so, he measures out 10 feet along the fence and places a stake marking it point *A*. He then goes back to the corner and measures out 15 feet along the fence in the other direction and places a stake marking it point *B* as shown in the figure below. Finally, he measures the distance between points *A* and *B* and finds it to be 20 feet. Does his corner measure 90°?

64. Fence Corner. For another corner of Ben's fence (see Exercise 63), he measures from the corner up one side of the fence 8 feet and marks it with a stake as point *A*. He then measures

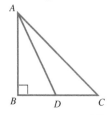

diagonally across the yard 17 feet from point A to the fence going in the other direction as shown and marks it as point B. If this corner is 90°, what is the length of the fence from the corner to point B?

65. Car Engine. Bill's car engine is said to run at 1700 RPMs (revolutions per minute) at idle. Through how many degrees does his engine turn each second?

66. Car Engine. Upon acceleration, Bill's car engine turns 300,000° in 15 seconds. At what RPM is the car engine turning?

• CATCH THE MISTAKE

In Exercises 67 and 68, explain the mistake that is made.

67. In a 30°-60°-90° triangle, find the length of the side opposite the 60° angle if the side opposite the 30° angle is 10 inches.

Solution:

The length opposite the 60° angle is twice the length opposite the 30° angle. $2(10) = 20$

The side opposite the 60° angle has length 20 inches.

This is incorrect. What mistake was made?

68. In a 45°-45°-90° triangle, find the length of the hypotenuse if each leg has length 5 centimeters.

Solution:

Use the Pythagorean theorem. $5^2 + 5^2 = \text{hypotenuse}^2$

Simplify. $50 = \text{hypotenuse}^2$

Solve for the hypotenuse. $\text{hypotenuse} = \pm 5\sqrt{2}$

The hypotenuse has length $\pm 5\sqrt{2}$ centimeters.

This is incorrect. What mistake was made?

• CONCEPTUAL

In Exercises 69–76, determine whether each statement is true or false.

69. The Pythagorean theorem can be applied to any equilateral triangle.

70. The Pythagorean theorem can be applied to all isosceles triangles.

71. The two angles opposite the legs of a right triangle are complementary.

72. In a 30°-60°-90° triangle, the length of the side opposite the 60° angle is twice the length of the side opposite the 30° angle.

73. The two acute angles in a right triangle must be complementary angles.

74. In a 45°-45°-90° triangle, the length of each side is the same.

75. Angle measure in degrees can be both positive and negative.

76. A right triangle cannot contain an obtuse angle.

• CHALLENGE

77. What is the measure (in degrees) of the smaller angle the hour and minute hands form when the time is 12:20?

78. What is the measure (in degrees) of the smaller angle the hour and minute hands form when the time is 9:10?

In Exercises 79–82, use the figure below:

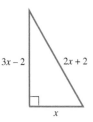

79. If $AB = 3$, $AD = 5$, and $AC = \sqrt{58}$, find DC.

80. If $AB = 4$, $AD = 5$, and $AC = \sqrt{41}$, find DC.

81. If $AC = 30$, $AB = 24$, and $DC = 11$, find AD.

82. If $AB = 60$, $AD = 61$, and $DC = 36$, find AC.

83. Given a square with side length x, draw the two diagonals. The result is four special triangles. Describe these triangles. What are the angle measures?

84. Solve for x in the triangle below.

• TECHNOLOGY

85. If the shortest leg of a 30°-60°-90° triangle has length 16.68 feet, what are the lengths of the other leg and the hypotenuse? Round answers to two decimal places.

86. If the longer leg of a 30°-60°-90° triangle has length 134.75 centimeters, what are the lengths of the other leg and the hypotenuse? Round answers to two decimal places.

1.2 SIMILAR TRIANGLES

SKILLS OBJECTIVES	CONCEPTUAL OBJECTIVES
■ Find angle measures using geometry. ■ Use similarity to determine the length of a side of a triangle.	■ Understand the properties of angles and how to apply those properties. ■ Understand the difference between similar and congruent triangles.

1.2.1 Finding Angle Measures Using Geometry

1.2.1 SKILL

Find angle measures using geometry.

1.2.1 CONCEPTUAL

Understand the properties of angles and how to apply those properties.

We stated in Section 1.1 that the sum of the measures of three angles of any triangle is 180°. We now review some angle relationships from geometry.

Vertical angles are angles of equal measure that are opposite one another and share the same vertex. In the figure below, angles **1** and **3** are vertical angles. Angles **2** and **4** are also vertical angles.

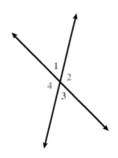

A **transversal** is a line that intersects two other coplanar lines. If a transversal intersects two parallel lines m and n, or $m \parallel n$, the corresponding angles have equal measure. Angles 3, 4, 5, and 6 are classified as **interior angles**, whereas angles 1, 2, 7, and 8 are classified as **exterior angles**.

In diagrams, there are two traditional ways to indicate angles having equal measure:

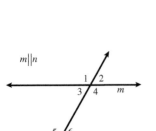

■ with the same number of arcs, *or*

■ with a single arc and the same number of hash marks.

In this text, we will use the same number of arcs.

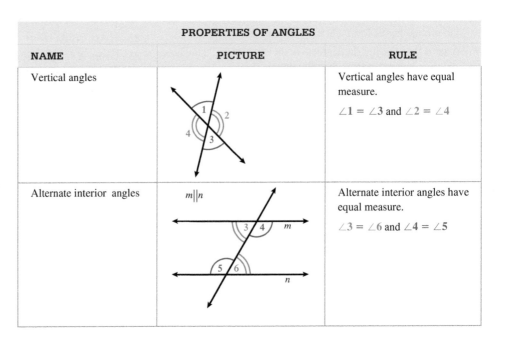

PROPERTIES OF ANGLES		
NAME	**PICTURE**	**RULE**
Vertical angles		Vertical angles have equal measure. $\angle 1 = \angle 3$ and $\angle 2 = \angle 4$
Alternate interior angles		Alternate interior angles have equal measure. $\angle 3 = \angle 6$ and $\angle 4 = \angle 5$

NAME	PICTURE	RULE
Alternate exterior angles	$m \| n$ 1 2 m 7 8 n	Alternate exterior angles have equal measure. $\angle 1 = \angle 8$ and $\angle 2 = \angle 7$
Corresponding angles	$m \| n$ 1 2 3 4 m 5 6 7 8 n	Corresponding angles have equal measure. $\angle 1 = \angle 5$ Note that the following are also corresponding angles: $\angle 2$ and $\angle 6$ $\angle 3$ and $\angle 7$ $\angle 4$ and $\angle 8$
Interior angles on the same side of transversal	$m \| n$ 3 4 m $\angle 3 + \angle 5 = 180°$ $\angle 4 + \angle 6 = 180°$ 5 6 n	Interior angles on the same side of the transversal have measures that sum to $180°$ (they are supplementary).

EXAMPLE 1   Finding Angle Measures

Given that $\angle 1 = 110°$ and $m \| n$, find the measure of angle 7.

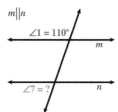

Solution:

Corresponding angles have equal measure, $\angle 1 = \angle 5$.

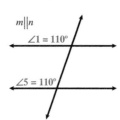

$\angle 5$ and $\angle 7$ are supplementary angles.

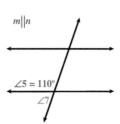

Measures of supplementary angles sum to $180°$.	$\angle 5 + \angle 7 = 180°$
Substitute $\angle 5 = 110°$.	$110° + \angle 7 = 180°$
Solve for $\angle 7$.	$\boxed{\angle 7 = 70°}$

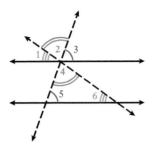

Draw two parallel lines (solid) and draw two transverse lines (dashed), which intersect at a point on one of the parallel lines.

The corresponding angles **1** and **6** have equal measure.

The vertical angles **2** and **4** have equal measure.

The corresponding angles **3** and **5** have equal measure.

The sum of the measures of angles **1**, **2**, and **3** is 180°. Therefore, the sum of the measures of angles **4**, **5**, and **6** is also 180°. So the sum of the measures of three angles of any triangle is 180°.

1.2.2 Classification of Triangles

1.2.2 SKILL

Use similarity to determine the length of a side of a triangle.

1.2.2 CONCEPTUAL

Understand the difference between similar and congruent triangles.

[CONCEPT CHECK]

TRUE OR FALSE Two similar triangles are congruent triangles, but two congruent triangles need not be similar triangles.

▼ ·

ANSWER False

Angles with equal measure are often labeled with the same number of arcs. Similarly, sides of equal length are often labeled with the same number of hash marks.

NAME	PICTURE	RULE
Equilateral triangle		All sides are equal.
Isosceles triangle		Two sides are equal.
Scalene triangle		No sides are equal.

The word *similar* in geometry means identical in shape, although not necessarily the same size. It is important to note that two triangles can have the exact same shape (same angles) but be of different sizes.

DEFINITION Similar Triangles

Similar triangles are triangles with equal corresponding angle measures (equal angles).

The word *congruent* means equal in all corresponding measures; therefore, congruent triangles have all corresponding angles of equal measure and all corresponding sides of equal measure. Two triangles are *congruent* if they have exactly the same shape *and* size. In other words, if one triangle can be picked up and situated on top of another triangle so that the two triangles coincide, they are said to be *congruent*.

DEFINITION Congruent Triangles

Congruent triangles are triangles with equal corresponding angle measures (equal angles) *and* corresponding equal side lengths.

All congruent triangles are also similar triangles, but not all similar triangles are congruent triangles.

Trigonometry (as you will see in Section 1.3) relies on the properties of similar triangles. Since similar triangles have the same shape (equal corresponding angles), the sides opposite the corresponding angles must be proportional.

Given any 30°-60°-90° triangle (see figure in the margin), if we let $x = 1$ correspond to triangle A and $x = 5$ correspond to triangle B, then we would have the following two triangles.

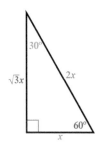

$$\text{Sides opposite } 30°: \quad \frac{\text{Triangle B}}{\text{Triangle A}} = \frac{5}{1} = 5$$

$$\text{Sides opposite } 60°: \quad \frac{\text{Triangle B}}{\text{Triangle A}} = \frac{5\sqrt{3}}{\sqrt{3}} = 5$$

$$\text{Sides opposite } 90°: \quad \frac{\text{Triangle B}}{\text{Triangle A}} = \frac{10}{2} = 5$$

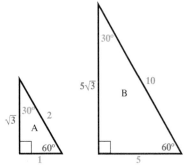

Notice that the sides opposite the corresponding angles are proportional. This means that we will find that all three ratios of each side of triangle B to its corresponding side of triangle A are equal. This proportionality property holds for all similar triangles.

CONDITIONS FOR SIMILAR TRIANGLES

One of the following must be verified in order for two triangles to be similar:

- Corresponding angles must have the same measure.

    or

- Corresponding sides must be proportional (ratios must be equal)

$$\frac{a}{a'} = \frac{b}{b'} = \frac{c}{c'}$$

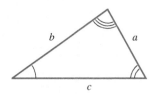

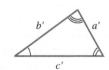

Let us now use the fact that corresponding sides of similar triangles are proportional to determine lengths of sides of similar triangles.

▶ **EXAMPLE 2** **Finding Lengths of Sides in Similar Triangles**

Given that the two triangles are similar, find the length of each of the unknown sides (b and c).

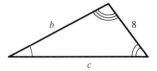

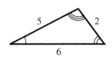

Solution:

Solve for b.

The corresponding sides are proportional.	$\dfrac{8}{2} = \dfrac{b}{5}$
Multiply by 5.	$b = 4(5)$
Solve for b.	$\boxed{b = 20}$

Solve for c.

The corresponding sides are proportional.	$\dfrac{8}{2} = \dfrac{c}{6}$
Multiply by 6.	$c = 4(6)$
Solve for c.	$\boxed{c = 24}$
Check that the ratios are equal:	$\dfrac{8}{2} = \dfrac{20}{5} = \dfrac{24}{6} = 4$

▼

ANSWER

$a = 5, b = 8$

▼

YOUR TURN Given that the two triangles are similar, find the length of each of the unknown sides (a and b).

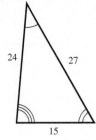

Applications Involving Similar Triangles

As you have seen, the common ratios associated with similar triangles are very useful. For example, you can quickly estimate the heights of flagpoles, trees, and any other tall objects by measuring their shadows along the ground because similar triangles are formed. Surveyors rely on the properties of similar triangles to determine distances that are difficult to measure.

EXAMPLE 3   **Calculating the Height of a Tree**

Billy wants to rent a lift to trim his tall trees. However, he must decide which lift he needs: one that will lift him 25 feet or a more expensive lift that will lift him 50 feet. His wife Jeanine hammered a stake into the ground and by measuring found its shadow to be 1.75 feet long and the tree's shadow to be 19 feet. (Assume both the stake and tree are perpendicular to the ground.) If the stake was standing 3 feet above the ground, how tall is the tree? Which lift should Billy rent?

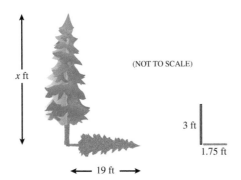

(NOT TO SCALE)

x ft

3 ft

1.75 ft

← 19 ft →

Solution:

Rays of sunlight are straight and parallel to each other. Therefore, the rays make the same angle with the tree that they do with the stake.

Draw and label the two similar triangles.

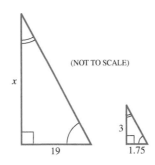

(NOT TO SCALE)

x

19

3

1.75

The ratios of corresponding sides of similar triangles are equal.

$$\frac{x}{3} = \frac{19}{1.75}$$

Solve for x.

$$x = \frac{3(19)}{1.75}$$

Simplify.

$$x \approx \boxed{32.57}$$

The tree is approximately 33 feet tall. Billy should rent the more expensive lift to be safe.

▼

YOUR TURN   Billy's neighbor decides to do the same thing. He borrows Jeanine's stake and measures the shadows. If the shadow of his tree is 15 feet and the shadow of the stake (3 feet above the ground) is 1.2 feet, how tall is Billy's neighbor's tree? Round to the nearest foot.

▼
ANSWER

approximately 38 ft

▶[SECTION 1.2]   SUMMARY

Similar triangles (the same shape) have corresponding angles with equal measure. Congruent triangles (the same shape and size) are similar triangles that also have equal corresponding side lengths. Similar triangles have the property that the ratios of their corresponding side lengths are equal.

[SECTION 1.2] EXERCISES

• SKILLS

In Exercises 1–12, find the measure of the indicated angles, using the diagram on the right.

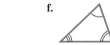

1. $C = 80°$, find B. **2.** $C = 80°$, find E. **3.** $C = 80°$, find F. **4.** $C = 80°$, find G.

5. $F = 75°$, find B. **6.** $F = 75°$, find A. **7.** $G = 65°$, find B. **8.** $G = 65°$, find A.

9. $A = (8x)°$ and $D = (9x - 15)°$, find the measures of A and D.

10. $B = (9x + 7)°$ and $F = (11x - 7)°$, find the measures of B and F.

11. $A = (12x + 14)°$ and $G = (9x - 2)°$, find the measures of A and G.

12. $C = (22x + 3)°$ and $E = (30x - 5)°$, find the measures of C and E.

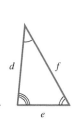

In Exercises 13–18, match the corresponding triangle with the appropriate name.

13. Equilateral triangle   **14.** Right triangle (nonisosceles)

15. Isosceles triangle (nonright)   **16.** Acute scalene triangle

17. Obtuse scalene triangle   **18.** Isosceles right triangle

a. **b.** **c.** **d.** **e.** **f.**

In Exercises 19–26, calculate the specified lengths, given that the two triangles are similar.

19. $a = 4, c = 6, d = 2, f = ?$ **20.** $a = 12, b = 9, e = 3, d = ?$

21. $d = 5, e = 2.5, b = 7.5, a = ?$ **22.** $e = 1.4, f = 2.6, c = 3.9, b = ?$

23. $d = 1.1$ m, $f = 2.5$ m, $c = 26.25$ km, $a = ?$ **24.** $e = 10$ cm, $f = 14$ cm, $c = 35$ m, $b = ?$

25. $b = \frac{4}{5}$ in., $c = \frac{5}{2}$ in., $f = \frac{3}{2}, e = ?$ **26.** $d = \frac{7}{16}$ m, $e = \frac{1}{4}$ m, $b = \frac{5}{4}$ mm, $a = ?$

• APPLICATIONS

27. Height of a Tree. The shadow of a tree measures $14\frac{1}{4}$ feet. At the same time of day, the shadow of a 4-foot pole measures $1\frac{1}{2}$ feet. How tall is the tree?

28. Height of a Flagpole. The shadow of a flagpole measures 15 feet. At the same time of day, the shadow of a stake 2 feet above ground measures $\frac{3}{4}$ foot. How tall is the flagpole?

29. Height of a Lighthouse. The Cape Hatteras Lighthouse at the Outer Banks of North Carolina is the tallest lighthouse in North America. If a 5-foot woman casts a $1\frac{1}{5}$-foot shadow and the lighthouse casts a 48-foot shadow, approximately how tall is the Cape Hatteras Lighthouse?

30. Height of a Man. If a 6-foot man casts a 1-foot shadow, how long a shadow will his 4-foot son cast?

For Exercises 31 and 32, refer to the following:

Although most people know that a list exists of the Seven Wonders of the Ancient World, only a few can name them: the Great Pyramid of Giza, the Hanging Gardens of Babylon, the Statue of Zeus at Olympia, the Temple of Artemis at Ephesus, the Mausoleum of Halicarnassus, the Colossus of Rhodes, and the Lighthouse of Alexandria.

31. Seven Wonders. One of the Seven Wonders of the Ancient World was a lighthouse on the Island of Pharos in Alexandria, Egypt. It is the first lighthouse in recorded history and was built about 280 BC. It survived for 1500 years until it was completely destroyed by an earthquake in the fourteenth

century. On a sunny day, if a 2-meter-tall man casts a shadow approximately 5 centimeters (0.05 meters) long, and the lighthouse approximately a 3-meter shadow, how tall was this fantastic structure?

Mary Evans Picture Library/Alamy

32. Seven Wonders. Only one of the great Seven Wonders of the Ancient World is still standing—the Great Pyramid of Giza. Each of the base sides along the ground measures 230 meters. If a 1-meter child casts a 90-centimeter shadow at the same time the shadow of the pyramid extends 16 meters along the ground (beyond the base), approximately how tall is the Great Pyramid of Giza?

In Exercises 33 and 34, use the drawing below:

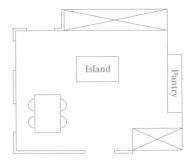

In a home remodeling project, your architect gives you plans that have an indicated distance of $2'4''$, which measures $\frac{1}{4}$ inch with a ruler on the blueprint.

33. **Measurement.** How long is the pantry in the kitchen if it measures $\frac{11}{16}$ inch with a ruler?

34. **Measurement.** How wide is the island if it measures $\frac{7}{16}$ inch with a ruler?

35. **Camping.** The front of Brian's tent is a triangle that measures 5 feet high and 4 feet wide at the base. The tent has one zipper down the middle that is perpendicular to the base of the tent and one zipper that is parallel to the base. With the tent unzipped, the opening and the outline of the tent are equilateral triangles as shown below. If the vertical zipper measures 3.5 feet, how wide is the opening at the base?

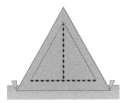

36. **Camping.** Zach sees a tree growing at an angle with the ground. If the angle between the tree and the ground is 60°, how far is it from the ground to the tree 15 feet from the base of the tree?

37. **Design.** Rita is designing a logo for her company's letterhead. As part of her design, she is trying to include two similar right triangles within a circle as shown. If the vertical leg of the large triangle measures 1.6 inches, while the hypotenuse measures 2.2 inches, what is the length of the vertical leg of the smaller triangle given that its hypotenuse measures 1.2 inches?

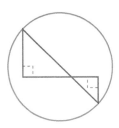

38. **Design.** Rita's supervisor asks that the triangles in Rita's design (shown in Exercise 37) be isosceles right triangles. If Rita decides to make the hypotenuse of the larger triangle 2 inches long, how long are its legs?

For Exercises 39–42, refer to the following:

A collimator is a device used in radiation treatment that narrows beams or waves, causing the waves to be more aligned in a specific direction. The use of a collimator facilitates the focusing of radiation to treat an affected region of tissue beneath the skin. In the figure, d_s is the distance from the radiation source to the skin, d_t is the distance from the outer layer of skin to the targeted tissue, $2f_s$ is the field size on the skin (diameter of the circular treated skin) and $2f_d$ is the targeted field size at depth d_t (the diameter of the targeted tissue at the specified depth beneath the skin surface).

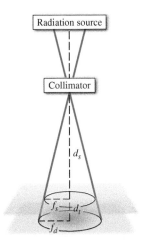

39. **Health/Medicine.** Radiation treatment is applied to a field size of 8 centimeters at a depth 2.5 centimeters below the skin surface. If the treatment head is positioned 80 centimeters from the skin, find the targeted field size to the nearest millimeter.

40. **Health/Medicine.** Radiation treatment is applied to a field size of 4 centimeters lying at a depth of 3.5 centimeters below the skin surface. If the field size on the skin is required to be 3.8 centimeters, find the distance from the skin that the radiation source must be located to the nearest millimeter.

41. **Health/Medicine.** Radiation treatment is applied to a field size on the skin of 3.75 centimeters to reach an affected region of tissue with field size of 4 centimeters at some depth below the skin. If the treatment head is positioned 60 centimeters from the skin surface, find the desired depth below the skin to the target area to the nearest millimeter.

42. **Health/Medicine.** Radiation treatment is applied to a field size on the skin of 4.15 centimeters to reach an affected area lying 4.5 centimeters below the skin surface. If the treatment head is positioned 60 centimeters from the skin surface, find the field size of the targeted area to the nearest millimeter.

● CATCH THE MISTAKE

In Exercises 43 and 44, explain the mistake that is made.

43. In the similar triangles shown, if $A = 8$, $B = 5$, and $D = 3$, find E.

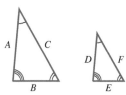

Solution:

Set up a ratio of corresponding sides. $\quad \dfrac{A}{E} = \dfrac{B}{D}$

Substitute $A = 8$, $B = 5$, and $D = 3$. $\quad \dfrac{8}{E} = \dfrac{5}{3}$

Cross multiply. $\quad\quad\quad\quad\quad 5E = 8(3)$

Solve for E. $\quad\quad\quad\quad\quad\quad E = \dfrac{24}{5}$

This is incorrect. What mistake was made?

44. In the similar triangles shown in Exercise 43, if $A = 8$, $B = 5$, and $D = 3$, find F.

Solution:

Set up a ratio of similar triangles. $\quad \dfrac{A}{B} = \dfrac{D}{F}$

Substitute $A = 8$, $B = 5$, and $D = 3$. $\quad \dfrac{8}{5} = \dfrac{3}{F}$

Cross multiply. $\quad\quad\quad\quad\quad 8F = 5(3)$

Solve for F. $\quad\quad\quad\quad\quad\quad F = \dfrac{15}{8}$

This is incorrect. What mistake was made?

● CONCEPTUAL

In Exercises 45–52, determine whether each statement is true or false.

45. Two similar triangles must have equal corresponding angles.

46. All congruent triangles are similar, but not all similar triangles are congruent.

47. Two angles in a triangle cannot have measures 82° and 67°.

48. Two equilateral triangles are similar but do not have to be congruent.

49. Alternate interior angles are supplementary.

50. Vertical angles are congruent.

51. All isosceles triangles are similar to each other.

52. Corresponding sides of similar triangles are congruent.

● CHALLENGE

53. Find x.

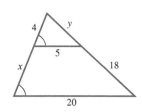

54. Find x and y.

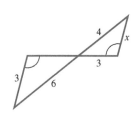

For Exercises 55–57, refer to the following:

The Lens law relates three quantities: the distance from the object to the lens, D_o; the distance from the lens to the image, D_i; and the focal length of the lens, f.

$$\frac{1}{D_o} + \frac{1}{D_i} = \frac{1}{f}$$

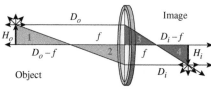

55. Explain why triangles 1 and 2 are similar triangles and why triangles 3 and 4 are similar triangles.

56. Set up the similar triangle ratios for

a. triangles 1 and 2   b. triangles 3 and 4

57. Use the ratios in Exercise 56 to derive the Lens law.

For Exercises 58–60, use the figure below:

$\overline{BF}$ is parallel to $\overline{CG}$, $\overline{CD}$ is congruent to $\overline{CG}$, and the measure of angle $A = 30°$.

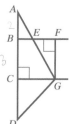

58. Name all similar triangles in the figure.

59. If $\overline{AB}$ measures 2 centimeters and $\overline{BC}$ measures 3 centimeters, find the measure of $\overline{EF}$ and $\overline{EG}$.

60. What is the measure of $\overline{DG}$?

1.3 DEFINITION 1 OF TRIGONOMETRIC FUNCTIONS: RIGHT TRIANGLE RATIOS

SKILLS OBJECTIVES	CONCEPTUAL OBJECTIVES
▪ Calculate trigonometric ratios of general acute angles. ▪ Express trigonometric functions in terms of their cofunctions.	▪ Understand that right triangle ratios are based on the properties of similar triangles. ▪ Understand that a trigonometric function of an angle is the cofunction of its complementary angle.

1.3.1 Trigonometric Functions: Right Triangle Ratios

The word **trigonometry** stems from the Greek words *trigonon*, which means "triangle," and *metrein*, which means "to measure." Trigonometry began as a branch of geometry and was utilized extensively by early Greek mathematicians to determine unknown distances. The *trigonometric functions* were first defined as ratios of side lengths in a right triangle. This is the way we will define them in this section. Since the two angles besides the right angle in a right triangle have to be acute, a second kind of definition was needed to extend the domain of trigonometric functions to nonacute angles in the Cartesian plane (Sections 2.2 and 2.3). Starting in the eighteenth century, broader definitions of the trigonometric functions came into use, under which the functions are associated with points along the unit circle (Section 3.4).

In Section 1.2, we reviewed the fact that triangles with equal corresponding angles also have proportional sides and are called similar triangles. The concept of similar triangles (one of the basic insights in trigonometry) allows us to determine the length of a side of one triangle if we know the length of one side of that triangle and the length of certain sides of a similar triangle.

Since the two right triangles to the right have equal angles, they are similar triangles, and the following ratios hold true:

$$\frac{a}{a'} = \frac{b}{b'} = \frac{c}{c'}$$

1.3.1 SKILL

Calculate trigonometric ratios of general acute angles

1.3.1 CONCEPTUAL

Understand that right triangle ratios are based on the properties of similar triangles.

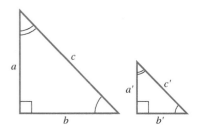

WORDS	MATH
Start with the first ratio.	$\dfrac{a}{a'} = \dfrac{b}{b'}$
Cross multiply.	$ab' = a'b$
Divide both sides by bb'.	$\dfrac{ab'}{bb'} = \dfrac{a'b}{bb'}$
Simplify.	$\dfrac{a}{b} = \dfrac{a'}{b'}$

Similarly, it can be shown that $\dfrac{b}{c} = \dfrac{b'}{c'}$ and $\dfrac{a}{c} = \dfrac{a'}{c'}$.

[CONCEPT CHECK]

How many possible ratios of the length of the sides of a right triangle are there for an acute angle in a right triangle?

ANSWER 6

Notice that even though the sizes of the triangles are different, since the corresponding angles are equal, the ratio of the lengths of the two legs of the large triangle is equal to the ratio of the lengths of the legs of the small triangle, or $\dfrac{a}{b} = \dfrac{a'}{b'}$. Similarly, the ratios of the lengths of a leg and the hypotenuse of the large triangle and the corresponding leg and hypotenuse of the small triangle are also equal; that is, $\dfrac{b}{c} = \dfrac{b'}{c'}$ and $\dfrac{a}{c} = \dfrac{a'}{c'}$.

For any right triangle, there are six possible ratios of the length of the sides that can be calculated for each acute angle θ:

$$\frac{b}{c} \quad \frac{a}{c} \quad \frac{b}{a} \quad \frac{c}{b} \quad \frac{c}{a} \quad \frac{a}{b}$$

These ratios are referred to as **trigonometric ratios** or **trigonometric functions**, since they depend on θ, and each is given a name:

FUNCTION NAME	ABBREVIATION
Sine	sin
Cosine	cos
Tangent	tan
Cosecant	csc
Secant	sec
Cotangent	cot

WORDS	MATH
The sine of θ	$\sin\theta$
The cosine of θ	$\cos\theta$
The tangent of θ	$\tan\theta$
The cosecant of θ	$\csc\theta$
The secant of θ	$\sec\theta$
The cotangent of θ	$\cot\theta$

Sine, cosine, tangent, cotangent, secant, and *cosecant* are names given to specific ratios of lengths of sides of a right triangle. These are the six trigonometric functions.

DEFINITION **Trigonometric Functions**

Let θ be an acute angle in a right triangle; then

$$\sin\theta = \frac{b}{c} \qquad \cos\theta = \frac{a}{c} \qquad \tan\theta = \frac{b}{a}$$

$$\csc\theta = \frac{c}{b} \qquad \sec\theta = \frac{c}{a} \qquad \cot\theta = \frac{a}{b}$$

In this right triangle, we say that:

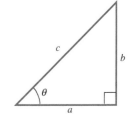

- Side c is the **hypotenuse**.
- Side b is the side (leg) **opposite** angle θ.
- Side a is the side (leg) **adjacent** to angle θ.

Also notice that since $\sin\theta = \dfrac{b}{c}$ and $\cos\theta = \dfrac{a}{c}$, then $\dfrac{\sin\theta}{\cos\theta} = \dfrac{\frac{b}{c}}{\frac{a}{c}} = \dfrac{b}{a} = \tan\theta$.

The three main trigonometric functions should be learned in terms of the ratios.

$$\sin\theta = \frac{\text{opposite}}{\text{hypotenuse}} \qquad \cos\theta = \frac{\text{adjacent}}{\text{hypotenuse}} \qquad \tan\theta = \frac{\text{opposite}}{\text{adjacent}}$$

The remaining three trigonometric functions can be derived from $\sin\theta$, $\cos\theta$, and $\tan\theta$ using the *reciprocal identities*. Recall that the **reciprocal** of x is $\dfrac{1}{x}$ for $x \neq 0$.

STUDY TIP

You only need to memorize the three main trigonometric ratios: $\sin\theta$, $\cos\theta$, and $\tan\theta$. The remaining three can always be calculated as reciprocals of the main three for an acute angle θ by remembering the reciprocal identities.

RECIPROCAL IDENTITIES

$$\csc\theta = \frac{1}{\sin\theta} \qquad \sec\theta = \frac{1}{\cos\theta} \qquad \cot\theta = \frac{1}{\tan\theta}$$

It is important to note that the reciprocal identities only hold for values of θ that do not make the denominator equal to zero (i.e., when $\sin\theta$, $\cos\theta$, or $\tan\theta$ are not equal to 0). Using this terminology, we have an alternative definition that is easier to remember.

DEFINITION Trigonometric Functions (Alternate Form)

Let θ be an acute angle in a right triangle; then

$$\sin\theta = \frac{\text{opposite}}{\text{hypotenuse}}$$

$$\cos\theta = \frac{\text{adjacent}}{\text{hypotenuse}}$$

$$\tan\theta = \frac{\text{opposite}}{\text{adjacent}}$$

and their reciprocals

$$\csc\theta = \frac{1}{\sin\theta} \qquad \sec\theta = \frac{1}{\cos\theta} \qquad \cot\theta = \frac{1}{\tan\theta}$$

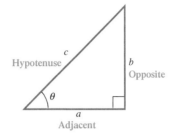

Notice that the tangent function can also be written as a quotient of the sine function and the cosine function:

$$\tan\theta = \frac{\sin\theta}{\cos\theta} = \frac{\dfrac{\text{opposite}}{\text{hypotenuse}}}{\dfrac{\text{adjacent}}{\text{hypotenuse}}} = \frac{\text{opposite}}{\text{adjacent}}$$

and similarly, $\cot\theta = \dfrac{\cos\theta}{\sin\theta}$. These relationships are called the *quotient identities*.

▶ **EXAMPLE 1** Finding Trigonometric Function Values of a General Angle θ in a Right Triangle

For the given right triangle, calculate the values of $\sin\theta$, $\tan\theta$, and $\csc\theta$.

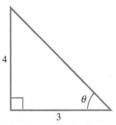

Solution:

STEP 1 **Find the length of the hypotenuse.**

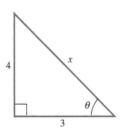

Apply the Pythagorean theorem.

$$4^2 + 3^2 = x^2$$
$$16 + 9 = x^2$$
$$x^2 = 25$$
$$x = \pm 5$$

Lengths of sides can be only positive. $\qquad x = 5$

STEP 2 **Label the sides of the triangle:**

- with numbers representing lengths.
- as hypotenuse or as opposite or adjacent with respect to θ.

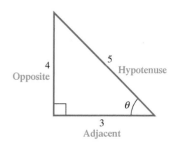

STEP 3 **Set up the trigonometric functions as ratios.**

Sine is Opposite over Hypotenuse (SOH). $\qquad \sin\theta = \dfrac{\text{opposite}}{\text{hypotenuse}} = \boxed{\dfrac{4}{5}}$

Tangent is Opposite over Adjacent (TOA). $\qquad \tan\theta = \dfrac{\text{opposite}}{\text{adjacent}} = \boxed{\dfrac{4}{3}}$

Cosecant is the reciprocal of sine. $\qquad \csc\theta = \dfrac{1}{\sin\theta} = \dfrac{1}{\dfrac{4}{5}} = \boxed{\dfrac{5}{4}}$

▼
ANSWER
$\cos\theta = \frac{3}{5}$, $\sec\theta = \frac{5}{3}$,
and $\cot\theta = \frac{3}{4}$

▼

YOUR TURN For the triangle in Example 1, calculate the values of $\cos\theta$, $\sec\theta$, and $\cot\theta$.

▶ **EXAMPLE 2** **Finding Trigonometric Function Values for a General Angle θ in a Right Triangle**

For the given right triangle, calculate $\cos\theta$, $\tan\theta$, and $\sec\theta$.

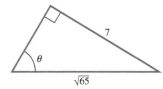

Solution:

STEP 1 **Find the length of the unknown leg.**

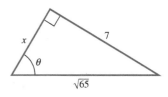

Apply the Pythagorean theorem.

$$x^2 + 7^2 = \left(\sqrt{65}\right)^2$$
$$x^2 + 49 = 65$$
$$x^2 = 16$$
$$x = \pm 4$$

Lengths of sides can be only positive. $\qquad x = 4$

STEP 2 **Label the sides of the triangle:**

- with numbers representing lengths.
- as hypotenuse or as opposite or adjacent with respect to θ.

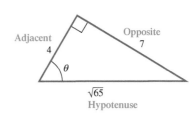

STEP 3 **Set up the trigonometric functions as ratios.**

Cosine is Adjacent over Hypotenuse (CAH). $\cos\theta = \dfrac{\text{adjacent}}{\text{hypotenuse}} = \boxed{\dfrac{4}{\sqrt{65}}}$

Tangent is Opposite over Adjacent (TOA). $\tan\theta = \dfrac{\text{opposite}}{\text{adjacent}} = \boxed{\dfrac{7}{4}}$

Secant is the reciprocal of cosine. $\sec\theta = \dfrac{1}{\cos\theta} = \dfrac{1}{\dfrac{4}{\sqrt{65}}} = \boxed{\dfrac{\sqrt{65}}{4}}$

STUDY TIP

Because we rationalize expressions containing radicals in the denominator, sometimes reciprocal ratios may not always look like reciprocal fractions:

$\cos\theta = \dfrac{4\sqrt{65}}{65}$; $\sec\theta = \dfrac{\sqrt{65}}{4}$.

Expressions that contain a radical in the denominator like $\dfrac{4}{\sqrt{65}}$ can be rationalized by multiplying both the numerator and the denominator by the radical, $\sqrt{65}$.

$$\dfrac{4}{\sqrt{65}} \cdot \underbrace{\left(\dfrac{\sqrt{65}}{\sqrt{65}}\right)}_{1} = \dfrac{4\sqrt{65}}{65}$$

In this example, the cosine function can now be written as $\boxed{\cos\theta = \dfrac{4\sqrt{65}}{65}}$.

▼

YOUR TURN For the triangle in Example 2, calculate $\sin\theta$, $\csc\theta$, and $\cot\theta$.

▼

ANSWER

$\sin\theta = \dfrac{7\sqrt{65}}{65}$,

$\csc\theta = \dfrac{\sqrt{65}}{7}$, and $\cot\theta = \dfrac{4}{7}$

1.3.2 Cofunctions

Notice the *co* in *co*sine, *co*secant, and *co*tangent functions. These *cofunctions* are based on the relationship of *co*mplementary angles. Let us look at a right triangle with labeled sides and angles.

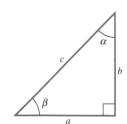

$\left.\begin{array}{l} \sin\beta = \dfrac{\text{opposite of }\beta}{\text{hypotenuse}} = \dfrac{b}{c} \\[2mm] \cos\alpha = \dfrac{\text{adjacent to }\alpha}{\text{hypotenuse}} = \dfrac{b}{c} \end{array}\right\} \sin\beta = \cos\alpha$

1.3.2 SKILL

Express trigonometric functions in terms of their cofunctions.

1.3.2 CONCEPTUAL

Understand that a trigonometric function of an angle is the cofunction of its complementary angle.

Recall that the sum of the measures of the three angles in a triangle is 180°. In a right triangle, one angle is 90°; therefore, the two acute angles are complementary angles (the measures sum to 90°). You can see in the triangle above that β and α are complementary angles. In other words, the sine of an angle is the same as the *co*sine of the *co*mplement of that angle. This is true for all *trigonometric cofunction* pairs.

COFUNCTION THEOREM

A trigonometric function of an angle is always equal to the cofunction of the complement of the angle. If $\alpha + \beta = 90°$, then

$$\sin\beta = \cos\alpha$$
$$\sec\beta = \csc\alpha$$
$$\tan\beta = \cot\alpha$$

[CONCEPT CHECK]

TRUE OR FALSE:

$\sin\theta = \cos(90° - \theta)$

▼

ANSWER True

COFUNCTION IDENTITIES

$$\sin\theta = \cos(90° - \theta) \qquad \cos\theta = \sin(90° - \theta)$$
$$\tan\theta = \cot(90° - \theta) \qquad \cot\theta = \tan(90° - \theta)$$
$$\sec\theta = \csc(90° - \theta) \qquad \csc\theta = \sec(90° - \theta)$$

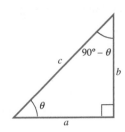

EXAMPLE 3 **Writing Trigonometric Function Values in Terms of Their Cofunctions**

Write each function or function value in terms of its cofunction.

a. $\sin 30°$ **b.** $\tan x$ **c.** $\csc 40°$

Solution (a):

Cosine is the cofunction of sine.	$\sin\theta = \cos(90° - \theta)$
Substitute $\theta = 30°$.	$\sin 30° = \cos(90° - 30°)$
Simplify.	$\boxed{\sin 30° = \cos 60°}$

Solution (b):

Cotangent is the cofunction of tangent.	$\tan\theta = \cot(90° - \theta)$
Substitute $\theta = x$.	$\boxed{\tan x = \cot(90° - x)}$

Solution (c):

Secant is the cofunction of cosecant.	$\csc\theta = \sec(90° - \theta)$
Substitute $\theta = 40°$.	$\csc 40° = \sec(90° - 40°)$
Simplify.	$\boxed{\csc 40° = \sec 50°}$

▼

ANSWER

a. $\sin 45°$ **b.** $\sec(90° - y)$

▶ **YOUR TURN** Write each function or function value in terms of its cofunction.

a. $\cos 45°$ **b.** $\csc y$

▶[SECTION 1.3] SUMMARY

In this section, we have defined trigonometric functions as ratios of the lengths of sides of right triangles. This approach is called *right triangle trigonometry*. This is the first of three definitions of trigonometric functions (others will follow in Chapters 2 and 3). We now can find trigonometric functions of an acute angle by taking ratios of the three sides of a right triangle: adjacent, opposite, and hypotenuse.

$$\sin\theta = \frac{\text{opposite}}{\text{hypotenuse}} \qquad \cos\theta = \frac{\text{adjacent}}{\text{hypotenuse}} \qquad \tan\theta = \frac{\text{opposite}}{\text{adjacent}}$$

It is important to remember that "adjacent" and "opposite" are with respect to one of the acute angles we are considering. We learned that the trigonometric functions of an angle are equal to the cofunctions of the complement to the angle.

[SECTION 1.3] EXERCISES

• SKILLS

In Exercises 1–6, refer to the triangle in the drawing to find the indicated trigonometric function values.

1. $\sin\theta$ **2.** $\cos\theta$ **3.** $\csc\theta$ **4.** $\sec\theta$ **5.** $\tan\theta$ **6.** $\cot\theta$

In Exercises 7–12, refer to the triangle in the drawing to find the indicated trigonometric function values. Rationalize any denominators containing radicals that you encounter in the answers.

7. $\cos\theta$ **8.** $\sin\theta$ **9.** $\sec\theta$ **10.** $\csc\theta$ **11.** $\tan\theta$ **12.** $\cot\theta$

In Exercises 13–18, refer to the triangle in the drawing to find the indicated trigonometric function values. Rationalize any denominators containing radicals that you encounter in the answers.

13. $\sin\theta$ **14.** $\cos\theta$ **15.** $\tan\theta$ **16.** $\csc\theta$ **17.** $\sec\theta$ **18.** $\cot\theta$

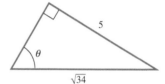

In Exercises 19–24, use the cofunction identities to fill in the blanks.

19. $\sin 60° = \cos$_____

20. $\sin 45° = \cos$_____

21. $\cos x = \sin$_____

22. $\cot A = \tan$_____

23. $\csc 30° = \sec$_____

24. $\sec B = \csc$_____

In Exercises 25–30, write the trigonometric function values in terms of its cofunction.

25. $\sin(x + y)$

26. $\sin(60° - x)$

27. $\cos(20° + A)$

28. $\cos(A + B)$

29. $\cot(45° - x)$

30. $\sec(30° - \theta)$

• APPLICATIONS

For Exercises 31 and 32, consider the following scenario:

A man lives in a house that borders a pasture. He decides to go to the grocery store to get some milk. He is trying to decide whether to drive along the roads in his car or take his all terrain vehicle (ATV) across the pasture. His car drives faster than the ATV, but the distance the ATV would travel is less than the distance he would travel in his car.

31. Shortcut. If $\sin\theta = \frac{3}{5}$ and $\cos\theta = \frac{4}{5}$ and if he drove his car along the streets, it would be 14 miles round trip. How far would he have to go on his ATV round trip? Round your answer to the nearest mile.

32. Shortcut. If $\tan\theta = 1$ and if he drove his car along the streets, it would be 200 yards round trip. How far would he have to go on his ATV round trip? Round your answer to the nearest yard.

33. Roofing. Bob is told that the pitch on the roof of his garage is 5-12, meaning that for every 5 feet the roof increases vertically, it increases 12 feet horizontally. If θ is defined as the angle at the corner of the roof formed by the pitch of the roof and a horizontal line, what is $\sin\theta$?

34. Roofing. Bob's roof has a 5-12 pitch, while his neighbor's roof has a 7-12 pitch. With θ defined as the angle formed at the corner of the roof by the pitch of the roof and a horizontal line, whose roof has a larger value for $\cos\theta$? Explain.

35. Automotive Design. The angle θ formed by a car's windshield and dashboard is such that $\tan\theta = \sqrt{3}$. What is the measure of angle θ?

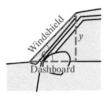

36. Automotive Design. Consider the information related to the interior of the car given in Exercise 35. If the vertical distance y between the top of the windshield and the horizontal plane of the dashboard is 3 feet, what is the length of the windshield?

37. Bookshelves. Juan is building a bookcase. To give each shelf extra strength, he is planning to add a triangle-shaped brace on each end. If each brace is a right triangle as shown below, such that $\cos A = \frac{3}{5}$, find $\sin B$.

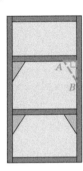

38. Bookshelves. If the height of the brace shown in Exercise 37 is 8 inches, find the width of the brace.

39. Hiking. Raja and Ariel are planning to hike to the top of a hill. If θ is the angle formed by the hill and the ground as shown below, such that $\sec\theta = 1.75$, find $\sin\theta$.

40. Hiking. Having recently had knee surgery, Lila is under doctor's orders not to hike anything steeper than a 45° incline. If θ is the angle formed by the hill and the ground such that $\tan\theta = 1.2$, is the hill too steep for Lila to hike? Explain.

For Exercises 41 and 42, refer to the following:

When traveling through air, a spherical drop of blood with diameter d maintains its spherical shape until hitting a flat surface. The drop of blood's direction of travel dictates the directionality of the blood splatter on the surface. For this reason, the diameter of the blood drop is equal to the width of the blood splatter on the surface. The angle at which a spherical drop of blood is deposited on a surface, called *angle of impact*, is related to the width w and the length l of the splatter by $\sin\theta = \dfrac{w}{l}$.

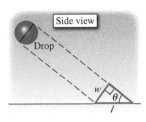
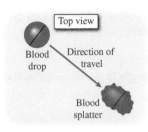

41. Forensic Science. If a drop of blood found at a crime scene has a width of 6 millimeters and a length of 12 millimeters, find the angle θ that represents the directionality.

42. Forensic Science. If a drop of blood found at a crime scene has a width of $\sqrt{2}$ millimeters and a length of 2 millimeters, find the angle θ that represents the directionality.

For Exercises 43 and 44, refer to the following:

The monthly profits of PizzaRia are a function of sales, that is, $p(s)$. A financial analysis has determined that the sales s in thousands of dollars of PizzaRia are also related to monthly profits p in thousands of dollars by the relationship

$$\tan\theta = \frac{p}{s} \quad \text{for} \quad 0 \le s \le 55 \quad \text{and} \quad 0 \le p \le 45$$

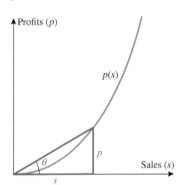

Based on sales and profits, it can be determined that the domain for angle θ is

$$0° \le \theta \le 40°$$

The ratio $\tan\theta$ represents the slope of the hypotenuse of the right triangle formed by sales s and profit p (see figure above). The angle θ can be interpreted as a measure of the relative size of s to p. The larger the angle θ is, the greater profit is relative to sales, and conversely, the smaller the angle θ is, the smaller profit is relative to sales.

43. Business. If PizzaRia's monthly sales are $25,000 and monthly profits are $10,000, find

    a. $\tan\theta$ **b.** $\cot\theta$

44. Business. If PizzaRia's monthly sales are s and monthly profits are p:

    a. Determine the hypotenuse in terms of s and p.

    b. Determine a formula for $\cos\theta$ in terms of s and p.

• **CATCH THE MISTAKE**

In Exercises 45–48, explain the mistake that is made.

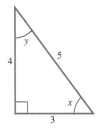

45. Calculate $\sin y$.

Solution:

Write the sine ratio. $\qquad \sin y = \dfrac{\text{opposite}}{\text{hypotenuse}}$

The opposite side is 4; the hypotenuse side is 5. $\qquad \sin y = \dfrac{4}{5}$

This is incorrect. What mistake was made?

46. Calculate $\tan x$.

Solution:

Write the tangent ratio. $\qquad \tan x = \dfrac{\text{adjacent}}{\text{opposite}}$

The adjacent side is 3; the opposite side is 4. $\qquad \tan x = \dfrac{3}{4}$

This is incorrect. What mistake was made?

47. Calculate $\sec x$.

Solution:

Write the sine ratio. $\qquad \sin x = \dfrac{\text{opposite}}{\text{hypotenuse}}$

The opposite side is 4; the hypotenuse side is 5. $\qquad \sin x = \dfrac{4}{5}$

Write secant as the reciprocal of sine. $\quad \sec x = \dfrac{1}{\sin x}$

Simplify. $\qquad \sec x = \dfrac{1}{4/5} = \dfrac{5}{4}$

This is incorrect. What mistake was made?

48. Calculate $\csc y$.

Solution:

Write the cosine ratio. $\qquad \cos y = \dfrac{\text{adjacent}}{\text{hypotenuse}}$

The adjacent side is 4; the hypotenuse side is 5. $\qquad \cos y = \dfrac{4}{5}$

Write cosecant as the reciprocal of cosine. $\qquad \csc y = \dfrac{1}{\cos y}$

Simplify. $\qquad \csc y = \dfrac{1}{4/5} = \dfrac{5}{4}$

This is incorrect. What mistake was made?

• **CONCEPTUAL**

In Exercises 49–52, determine whether each statement is true or false.

49. $\sin 45° = \cos 45°$ **50.** $\sin 60° = \cos 30°$ **51.** $\sec 60° = \csc 30°$ **52.** $\cot 45° = \tan 45°$

In Exercises 53–58, use the special triangles (30°-60°-90° and 45°-45°-90°) shown below:

53. Calculate $\sin 30°$ and $\cos 30°$.   **56.** Calculate $\sin 45°$, $\cos 45°$, and $\tan 45°$.

54. Calculate $\sin 60°$ and $\cos 60°$.   **57.** Calculate $\sec 45°$ and $\csc 45°$.

55. Calculate $\tan 30°$ and $\tan 60°$.   **58.** Calculate $\tan 60°$ and $\cot 30°$.

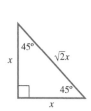
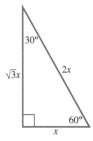

• **CHALLENGE**

59. What values can $\sin \theta$ and $\cos \theta$ take on?

60. What values can $\tan \theta$ and $\cot \theta$ take on?

61. What values can $\sec \theta$ and $\csc \theta$ take on?

62. As θ increases from 0° to 90°, how does $\sin \theta$ change?

63. As θ increases from 0° to 90°, how does the cofunction of $\sin \theta$ change?

64. As θ increases from 0° to 90°, how does $\csc \theta = \dfrac{1}{\sin \theta}$ change?

65. Calculate $\sec 70°$ the following two ways:

    a. Find $\cos 70°$ to three decimal places and then divide 1 by that number. Write that number to five decimal places.

    b. First find $\cos 70°$ and then find its reciprocal. Round the result to five decimal places.

66. Calculate $\csc 40°$ the following two ways:

    a. Find $\sin 40°$, write that down (round to three decimal places), and then divide 1 by that number. Write this last result to five decimal places.

    b. First find $\sin 40°$ and then find its reciprocal. Round the result to five decimal places.

67. Calculate $\cot(54.9°)$ the following two ways:

    a. Find $\tan(54.9°)$ to three decimal places and then divide 1 by that number. Write that number to five decimal places.

    b. First find $\tan(54.9°)$ and then find its reciprocal. Round the result to five decimal places.

68. Calculate $\sec(18.6°)$ the following two ways:

    a. Find $\cos(18.6°)$ to three decimal places and then divide 1 by that number. Write that number to five decimal places.

    b. First find $\cos(18.6°)$ and then find its reciprocal. Round the result to five decimal places.

1.4 EVALUATING TRIGONOMETRIC FUNCTIONS: EXACTLY AND WITH CALCULATORS

SKILLS OBJECTIVES	**CONCEPTUAL OBJECTIVES**
■ Evaluate trigonometric functions exactly for special angles.	■ Learn the trigonometric functions as ratios of sides of a right triangle.
■ Evaluate (approximate) trigonometric functions using a calculator.	■ Understand the difference between evaluating trigonometric functions exactly and using a calculator.
■ Represent partial degrees in either decimal degrees (DD) or degrees-minutes-seconds (DMS).	■ Understand that decimal degrees (DD) and degrees-minutes-seconds (DMS) are both ways of measuring parts of a degree and be able to recognize equivalent forms.

1.4.1 Evaluating Trigonometric Functions *Exactly* for Special Angle Measures: 30°, 45°, and 60°

1.4.1 SKILL

Evaluate trigonometric functions exactly for special angles.

1.4.1 CONCEPTUAL

Learn the trigonometric functions as ratios of sides of a right triangle.

We now turn our attention to evaluating trigonometric functions for known angles. In this section, we will distinguish between evaluating a trigonometric function *exactly* and *approximating* the value of a trigonometric function *with a calculator*. Throughout this text instructions will specify which is desired (exact or approximate), and the proper notation, $=$ or $\approx$, will be used.

    Three special acute angles are very important in trigonometry: $30°, 45°,$ and $60°$. In Section 1.1, we discussed two important triangles: $30°$-$60°$-$90°$ and $45°$-$45°$-$90°$. Recall the relationships between the lengths of the sides of these two right triangles.

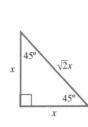
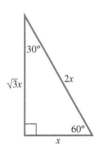

    We can combine these relationships with the trigonometric ratios developed in Section 1.3 to evaluate the trigonometric functions for the special angle measures of $30°, 45°,$ and $60°$.

▶ **EXAMPLE 1** **Evaluating the Trigonometric Functions Exactly for 30°**

Evaluate the six trigonometric functions for an angle that measures 30°.

Solution:

Label the sides (opposite, adjacent, and hypotenuse) of the 30°-60°-90° right triangle with respect to the **30°** angle.

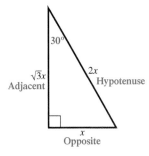

Use the right triangle ratio definitions of the sine, cosine, and tangent functions.

$$\sin 30° = \frac{\text{opposite}}{\text{hypotenuse}} = \frac{x}{2x} = \frac{1}{2}$$

$$\cos 30° = \frac{\text{adjacent}}{\text{hypotenuse}} = \frac{\sqrt{3}x}{2x} = \frac{\sqrt{3}}{2}$$

$$\tan 30° = \frac{\text{opposite}}{\text{adjacent}} = \frac{x}{\sqrt{3}x} = \frac{1}{\sqrt{3}} = \frac{1}{\sqrt{3}} \cdot \frac{\sqrt{3}}{\sqrt{3}} = \frac{\sqrt{3}}{3}$$

Use the reciprocal identities to obtain values for the cosecant, secant, and cotangent functions.

$$\csc 30° = \frac{1}{\sin 30°} = \frac{1}{\frac{1}{2}} = 2$$

$$\sec 30° = \frac{1}{\cos 30°} = \frac{1}{\frac{\sqrt{3}}{2}} = \frac{2}{\sqrt{3}} = \frac{2}{\sqrt{3}} \cdot \frac{\sqrt{3}}{\sqrt{3}} = \frac{2\sqrt{3}}{3}$$

$$\cot 30° = \frac{1}{\tan 30°} = \frac{1}{\frac{\sqrt{3}}{3}} = \frac{3}{\sqrt{3}} = \frac{3}{\sqrt{3}} \cdot \frac{\sqrt{3}}{\sqrt{3}} = \sqrt{3}$$

The six trigonometric functions evaluated for an angle measuring 30° are

$$\boxed{\sin 30° = \frac{1}{2}} \qquad \boxed{\cos 30° = \frac{\sqrt{3}}{2}} \qquad \boxed{\tan 30° = \frac{\sqrt{3}}{3}}$$

$$\boxed{\csc 30° = 2} \qquad \boxed{\sec 30° = \frac{2\sqrt{3}}{3}} \qquad \boxed{\cot 30° = \sqrt{3}}$$

▼ ....

YOUR TURN Evaluate the six trigonometric functions for an angle that measures 60°.

In comparing our answers in Example 1 and in Your Turn, we see that the following cofunction relationships are true:

$$\sin 30° = \cos 60° \qquad \sec 30° = \csc 60° \qquad \tan 30° = \cot 60°$$
$$\sin 60° = \cos 30° \qquad \sec 60° = \csc 30° \qquad \tan 60° = \cot 30°$$

which is expected, since 30° and 60° are complementary angles.

▼ ....

ANSWER

$$\sin 60° = \frac{\sqrt{3}}{2} \qquad \csc 60° = \frac{2\sqrt{3}}{3}$$

$$\cos 60° = \frac{1}{2} \qquad \sec 60° = 2$$

$$\tan 60° = \sqrt{3} \qquad \cot 60° = \frac{\sqrt{3}}{3}$$

EXAMPLE 2 **Evaluating the Trigonometric Functions Exactly for 45°**

Evaluate the six trigonometric functions for an angle that measures 45°.

Solution:

Label the sides of the 45°-45°-90° right triangle with respect to one of the **45°** angles.

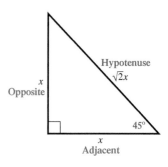

Use the right triangle ratio definitions of the sine, cosine, and tangent functions.

$$\sin 45° = \frac{\text{opposite}}{\text{hypotenuse}} = \frac{x}{\sqrt{2}x} = \frac{1}{\sqrt{2}} = \frac{1}{\sqrt{2}} \cdot \frac{\sqrt{2}}{\sqrt{2}} = \frac{\sqrt{2}}{2}$$

$$\cos 45° = \frac{\text{adjacent}}{\text{hypotenuse}} = \frac{x}{\sqrt{2}x} = \frac{1}{\sqrt{2}} = \frac{1}{\sqrt{2}} \cdot \frac{\sqrt{2}}{\sqrt{2}} = \frac{\sqrt{2}}{2}$$

$$\tan 45° = \frac{\text{opposite}}{\text{adjacent}} = \frac{x}{x} = 1$$

STUDY TIP

$\sin 45° = \frac{\sqrt{2}}{2}$ is exact, whereas if we evaluate with a calculator, we get an approximation:

$$\sin 45° \approx 0.7071$$

Use the reciprocal identities to obtain the values of the cosecant, secant, and cotangent functions.

$$\csc 45° = \frac{1}{\sin 45°} = \frac{1}{\frac{\sqrt{2}}{2}} = \frac{2}{\sqrt{2}} = \frac{2}{\sqrt{2}} \cdot \frac{\sqrt{2}}{\sqrt{2}} = \sqrt{2}$$

$$\sec 45° = \frac{1}{\cos 45°} = \frac{1}{\frac{\sqrt{2}}{2}} = \frac{2}{\sqrt{2}} = \frac{2}{\sqrt{2}} \cdot \frac{\sqrt{2}}{\sqrt{2}} = \sqrt{2}$$

$$\cot 45° = \frac{1}{\tan 45°} = \frac{1}{1} = 1$$

The six trigonometric functions evaluated for an angle measuring 45° are

$$\boxed{\sin 45° = \frac{\sqrt{2}}{2}} \qquad \boxed{\cos 45° = \frac{\sqrt{2}}{2}} \qquad \boxed{\tan 45° = 1}$$

$$\boxed{\csc 45° = \sqrt{2}} \qquad \boxed{\sec 45° = \sqrt{2}} \qquad \boxed{\cot 45° = 1}$$

We see that the following cofunction relationships are indeed true:

$$\sin 45° = \cos 45° \qquad \sec 45° = \csc 45° \qquad \tan 45° = \cot 45°$$

which is expected, since 45° and 45° are complementary angles.

The trigonometric function values for the three special angle measures (30°, 45°, and 60°) are summarized in the following table:

TRIGONOMETRIC FUNCTION VALUES FOR SPECIAL ANGLES (30°, 45°, AND 60°)						
θ	SIN θ	COS θ	TAN θ	COT θ	SEC θ	CSC θ
30°	$\dfrac{1}{2}$	$\dfrac{\sqrt{3}}{2}$	$\dfrac{\sqrt{3}}{3}$	$\sqrt{3}$	$\dfrac{2\sqrt{3}}{3}$	2
45°	$\dfrac{\sqrt{2}}{2}$	$\dfrac{\sqrt{2}}{2}$	1	1	$\sqrt{2}$	$\sqrt{2}$
60°	$\dfrac{\sqrt{3}}{2}$	$\dfrac{1}{2}$	$\sqrt{3}$	$\dfrac{\sqrt{3}}{3}$	2	$\dfrac{2\sqrt{3}}{3}$

It is important to **learn** the special values in red for the sine and cosine functions. All other values in the table can be found through reciprocals or quotients of these two functions. Remember that the tangent function is the ratio of the sine to cosine functions.

$$\sin\theta = \frac{\text{opposite}}{\text{hypotenuse}} \qquad \cos\theta = \frac{\text{adjacent}}{\text{hypotenuse}} \qquad \tan\theta = \frac{\sin\theta}{\cos\theta} = \frac{\dfrac{\text{opposite}}{\text{hypotenuse}}}{\dfrac{\text{adjacent}}{\text{hypotenuse}}} = \frac{\text{opposite}}{\text{adjacent}}$$

1.4.2 Using Calculators to Evaluate (Approximate) Trigonometric Function Values

We will now turn our attention to using calculators to evaluate trigonometric functions, which sometimes results in an approximation. Scientific and graphing calculators have buttons for the sine (sin), cosine (cos), and tangent (tan) functions. Let us start with what we already know and confirm it with our calculators.

EXAMPLE 3   **Evaluating Trigonometric Functions with a Calculator**

Use a calculator to find the values of
a. $\cos 30°$ **b.** $\sin 75°$ **c.** $\tan 67°$ **d.** $\sec 52°$
Round your answers to four decimal places.

Solution:

a. $\cos 30° \approx 0.866025403 \approx \boxed{0.8660}$

b. $\sin 75° \approx 0.965925826 \approx \boxed{0.9659}$

c. $\tan 67° \approx 2.355852366 \approx \boxed{2.3559}$

d. $\sec 52° = \dfrac{1}{\cos 52°} \approx 1.624269245 \approx \boxed{1.6243}$

Note: We know $\cos 30° = \dfrac{\sqrt{3}}{2} \approx 0.8660$.

▼

YOUR TURN   Use a calculator to find the values of
a. $\cos 22°$ **b.** $\tan 81°$ **c.** $\csc 37°$ **d.** $\sin 45°$
Round your answers to four decimal places.

STUDY TIP
If you memorize the values for sine and cosine for the angles given in the table, then the other trigonometric function values in the table can be found using the quotient and reciprocal identities.

STUDY TIP
SOHCAHTOA:
• Sine is Opposite over Hypotenuse
• Cosine is Adjacent over Hypotenuse
• Tangent is Opposite over Adjacent

1.4.2 SKILL
Evaluate (approximate) trigonometric functions using a calculator.

1.4.2 CONCEPTUAL
Understand the difference between evaluating trigonometric functions exactly and using a calculator.

STUDY TIP
Make sure your calculator is set in degrees (DEG) mode.

[CONCEPT CHECK]
TRUE OR FALSE $\sin(45°) = 0.07071$

▼

ANSWER False. That is an approximation. The exact value is $\sqrt{2}$.

▼

ANSWER
a. 0.9272 **b.** 6.3138
c. 1.6616 **d.** 0.7071

1.4.3 Representing Partial Degrees: DD or DMS

Since one revolution is equal to 360°, an angle of 1° seems very small. However, there are times when we want to break down a degree even further.

For example, if we are off even one-thousandth of a degree in pointing our antenna toward a geostationary satellite, we won't receive a signal. That is because the distance the signal travels (35,000 kilometers) is so much larger than the size of the satellite (5 meters).

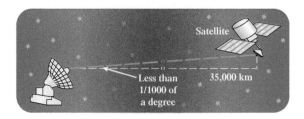

There are two traditional ways of representing part of a degree: *degrees-minutes-seconds* (DMS) and *decimal degrees* (DD). Most calculators allow you to enter angles in either DD or DMS format, and some even make the conversion between them. However, we will illustrate a manual conversion technique.

The **degrees-minutes-seconds** way of breaking down degrees is similar to how we break down time in hours-minutes-seconds. There are 60 minutes in an hour and 60 seconds in a minute, which results in an hour being broken down into 3,600 seconds. Similarly, we can think of degrees like hours. We can divide 1° into 60 equal parts. Each part is called a **minute** and is denoted $1'$. One minute is therefore $\frac{1}{60}$ of a degree.

$$1' = \left(\frac{1}{60}\right)^° \text{ or } 60' = 1°$$

We can then break down each minute into 60 seconds. Therefore, 1 **second**, denoted $1''$, is $\frac{1}{60}$ of a minute or $\frac{1}{3600}$ of a degree.

$$1'' = \left(\frac{1}{60}\right)' = \left(\frac{1}{3600}\right)^° \text{ or } 60'' = 1'$$

The following examples represent how DMS expressions are stated:

WORDS	MATH
22 degrees, 5 minutes	22° 5′
49 degrees, 21 minutes, 17 seconds	49° 21′ 17″

We add and subtract fractional values in DMS form similar to how we add and subtract time. For example, if Carol and Donna both run the Boston Marathon and Carol crosses the finish line in 5 hours, 42 minutes, and 10 seconds and Donna crosses the finish line in 6 hours, 55 minutes, and 22 seconds, how much time elapsed between the two women crossing the finish line? The answer is 1 hour, 13 minutes, and 12 seconds. We combine hours with hours, minutes with minutes, and seconds with seconds.

EXAMPLE 4 **Adding Degrees in DMS Form**

Add $27°5'17''$ and $35°17'52''$.

Solution:

Align degrees with degrees, minutes with minutes, and seconds with seconds.

Add degrees, minutes, and seconds, respectively.

Note that 69 seconds is equal to 1 minute, 9 seconds and simplify.

$$
\begin{array}{rccc}
 & 27° & 5' & 17'' \\
+ & 35° & 17' & 52'' \\
\hline
 & 62° & 22' & 69'' \\
 & & & \underbrace{\quad}_{1'+9''}
\end{array}
$$

$$\boxed{\;62° \quad 23' \quad 9''\;}$$

▼ YOUR TURN Add $35°\,21'\,42''$ and $7°\,5'\,30''$.

▼ ANSWER
$42°\,27'\,12''$

EXAMPLE 5 **Subtracting Degrees in DMS Form**

Subtract $15°\,28'$ from $90°$.

Solution:

Align degrees with degrees and minutes with minutes.

$$
\begin{array}{rcc}
 & 90° & 0' \\
- & 15° & 28' \\
\hline
\end{array}
$$

Borrow $1°$ from $90°$ and write it as 60 minutes.

$$
\begin{array}{rcc}
 & 89° & 60' \\
- & 15° & 28' \\
\hline
\end{array}
$$

Subtract degrees and minutes, respectively.

$$\boxed{\;74° \quad 32'\;}$$

▼ YOUR TURN Subtract $23°8'$ from $90°$.

▼ ANSWER
$66°\,52'$

An alternate way of representing parts of degrees is with *decimal degrees*. For example, $33.4°$ and $91.725°$ are measures of angles in **decimal degrees**. To convert from DD to DMS, multiply the decimal part once by 60 to get the minutes and—if necessary—multiply the resulting decimal part by 60 to get the seconds. A similar two-stage reverse process is necessary for the opposite conversion of DMS to DD.

▶ **EXAMPLE 6** **Converting from Degrees-Minutes-Seconds to Decimal Degrees**

Convert $17°39'22''$ to decimal degrees. Round to the nearest thousandth.

Solution:

Write the number of minutes in decimal form, where $1' = \left(\frac{1}{60}\right)°$.

$$39' = \left(\frac{39}{60}\right)° = 0.65°$$

Write the number of seconds in decimal form, where $1'' = \left(\frac{1}{3600}\right)°$.

$$22'' = \left(\frac{22}{3600}\right)° \approx 0.0061°$$

Write the expression as a sum.

$$17°39'22'' \approx 17° + 0.65° + 0.0061°$$

Add and round to the nearest thousandth.

$$17°39'22'' \approx \boxed{17.656°}$$

▼ YOUR TURN Convert $62°8'15''$ to decimal degrees. Round to the nearest thousandth.

▼ ANSWER
$62.138°$

EXAMPLE 7 **Converting from Decimal Degrees to Degrees-Minutes-Seconds**

Convert $29.538°$ to degrees, minutes, and seconds. Round to the nearest second.

Solution:

Write as a sum.	$29.538° = 29° + 0.538°$
To find the number of minutes, multiply the decimal part by 60, since $1° = 60'$.	$29.538° = 29° + 0.538°\left(\dfrac{60'}{1°}\right)$
Simplify.	$29.538° = 29° + 32.28'$
Write as a sum.	$29.538° = 29° + 32' + 0.28'$
To find the number of seconds, multiply the decimal by 60, since $1' = 60''$.	$29.538° = 29° + 32' + 0.28'\left(\dfrac{60''}{1'}\right)$
Simplify.	$29.538° = 29°\,32'\,16.8''$
Round to the nearest second.	$\boxed{29.538° = 29°\,32'\,17''}$

▼

ANSWER
$35°\,25'\,34''$

▼

YOUR TURN Convert $35.426°$ to degrees, minutes, and seconds. Round to the nearest second.

In this text, we will primarily use decimal degrees for angle measure, and we will also use decimal approximations for trigonometric function values. A common question that arises is how to round the decimals. There is a difference between specifying that a number be rounded to a particular *decimal place* and specifying rounding to a certain number of *significant digits*. More discussion on that topic will follow in the next section when solving right triangles. For now, we will round angle measures to the nearest minute in DMS or the nearest hundredth in DD, and we will round trigonometric function values to four decimal places for convenience.

EXAMPLE 8 **Evaluating Trigonometric Functions with Calculators**

Evaluate the following trigonometric functions for the specified angle measurements. Round your answers to four decimal places.

a. $\sin(18°9')$ **b.** $\sec(29.524°)$

Solution (a):

Write $18°9'$ in decimal degrees.	$18°9' = 18° + \left(\dfrac{9}{60}\right)°$
	$= 18° + 0.15°$
	$= 18.15°$
Use a calculator to evaluate the sine function.	$\sin(18.15°) = 0.311505793$
Round to four decimal places.	$\boxed{\approx 0.3115}$

Solution (b):

Write secant as the reciprocal of cosine.	$\sec(29.524°) = \dfrac{1}{\cos(29.524°)}$
Use a calculator to evaluate the expression.	$\sec(29.524°) \approx 1.149227998$
Round to four decimal places.	$\boxed{\sec(29.524°) \approx 1.1492}$

▶[SECTION 1.4] SUMMARY

In this section, you have learned the *exact* values of the sine, cosine, and tangent functions for the special angle measures: 30°, 45°, and 60°.

The values for each of the other trigonometric functions can be determined through reciprocal properties. Calculators can be used to *approximate* trigonometric values of any acute angle. Degrees can be broken down into smaller parts using one of two systems: decimal degrees and degrees-minutes-seconds.

θ	SIN θ	COS θ	TAN θ
30°	$\dfrac{1}{2}$	$\dfrac{\sqrt{3}}{2}$	$\dfrac{\sqrt{3}}{3}$
45°	$\dfrac{\sqrt{2}}{2}$	$\dfrac{\sqrt{2}}{2}$	1
60°	$\dfrac{\sqrt{3}}{2}$	$\dfrac{1}{2}$	$\sqrt{3}$

[SECTION 1.4] EXERCISES

• **SKILLS**

In Exercises 1–6, label each trigonometric function value with its correct value in (a), (b), and (c).

a. $\dfrac{1}{2}$ b. $\dfrac{\sqrt{3}}{2}$ c. $\dfrac{\sqrt{2}}{2}$

1. $\sin 30°$ **2.** $\sin 60°$ **3.** $\cos 30°$ **4.** $\cos 60°$ **5.** $\sin 45°$ **6.** $\cos 45°$

For Exercises 7–9, use the results in Exercises 1–6 and the trigonometric quotient identity, $\tan\theta = \dfrac{\sin\theta}{\cos\theta}$, to calculate the following values.

7. $\tan 30°$ **8.** $\tan 45°$ **9.** $\tan 60°$

For Exercises 10–18, use the results in Exercises 1–9 and the reciprocal identities, $\csc\theta = \dfrac{1}{\sin\theta}$, $\sec\theta = \dfrac{1}{\cos\theta}$, and $\cot\theta = \dfrac{1}{\tan\theta}$, to calculate the following values.

10. $\csc 30°$ **11.** $\sec 30°$ **12.** $\cot 30°$ **13.** $\csc 60°$ **14.** $\sec 60°$ **15.** $\cot 60°$

16. $\csc 45°$ **17.** $\sec 45°$ **18.** $\cot 45°$

In Exercises 19–30, use a calculator to evaluate the trigonometric functions for the indicated values. Round your answers to four decimal places.

19. $\sin 37°$ **20.** $\sin(17.8°)$ **21.** $\cos 82°$ **22.** $\cos(21.9°)$ **23.** $\tan 54°$ **24.** $\tan(43.2°)$

25. $\sec 8°$ **26.** $\sec 75°$ **27.** $\csc 89°$ **28.** $\csc 51°$ **29.** $\cot 55°$ **30.** $\cot 29°$

In Exercises 31–38, perform the indicated operations using the following angles:

$$\angle A = 5°17'29'' \quad \angle B = 63°28'35'' \quad \angle C = 16°11'30''$$

31. $\angle A + \angle B$ **32.** $\angle A + \angle C$ **33.** $\angle B + \angle C$ **34.** $\angle B - \angle A$

35. $\angle B - \angle C$ **36.** $\angle C - \angle A$ **37.** $90° - \angle A$ **38.** $90° - \angle B$

In Exercises 39–46, convert from degrees-minutes-seconds to decimal degrees. Round to the nearest hundredth if only minutes are given and to the nearest thousandth if seconds are given.

39. $33°20'$ **40.** $89°45'$ **41.** $59°27'$ **42.** $72°13'$

43. $27°45'15''$ **44.** $36°5'30''$ **45.** $42°28'12''$ **46.** $63°10'9''$

In Exercises 47–54, convert from decimal degrees to degrees-minutes-seconds. In Exercises 47–50, round to the nearest minute. In Exercises 51–54, round to the nearest second.

47. $15.75°$ **48.** $15.50°$ **49.** $22.35°$ **50.** $80.47°$

51. $30.175°$ **52.** $25.258°$ **53.** $77.535°$ **54.** $5.995°$

In Exercises 55–60, use a calculator to evaluate the trigonometric functions for the indicated values. **Round your answers to four decimal places.**

55. $\sin(10°\,25')$ **56.** $\cos(75°\,13')$ **57.** $\tan(22°\,15')$ **58.** $\sec(68°\,22')$ **59.** $\csc(28°\,25'\,35'')$ **60.** $\sec(50°\,20'\,19'')$

• APPLICATIONS

For Exercises 61 and 62, refer to the following:

X-ray crystallography is a method of determining the arrangement of atoms within a crystal. This method has revealed the structure and functioning of many biological molecules including vitamins, drugs, proteins, and nucleic acids (including DNA). The structure of a crystal can be determined experimentally using Bragg's law:

$$n\lambda = 2d\,\sin\theta$$

where λ is the wavelength of the x-ray (measured in angstroms), d is the distance between atomic planes (measured in angstroms), θ is the angle of reflection (in degrees), and n is the order of Bragg reflection (a positive integer).

61. Physics/Life Sciences. A diffractometer was used to make a diffraction pattern for a protein crystal from which it was determined experimentally that x-rays of wavelength 1.54 angstroms produced an angle of reflection of 45° corresponding to a Bragg reflection of order 1. Find the distance between atomic planes for the protein crystal to the nearest hundredth of an angstrom.

62. Physics/Life Sciences. A diffractometer was used to make a diffraction pattern for a salt crystal from which it was determined experimentally that x-rays of wavelength 1.67 angstroms produced an angle of reflection of 71.3° corresponding to a Bragg reflection of order 4. Find the distance between atomic planes for the salt crystal to the nearest hundredth of an angstrom.

For Exercises 63–66, refer to the following:

Have you ever noticed that if you put a stick in the water, it looks bent? We know the stick didn't bend. Instead, the light rays bent, which made the image appear to bend. Light rays propagating from one medium (like air) to another medium (like water) experience refraction, or "bending," with respect to the surface. Light bends according to Snell's law, which states:

$$n_i\sin(\theta_i) = n_r\sin(\theta_r)$$

where

- n_i is the refractive index of the medium the light is leaving, the incident medium.
- θ_i is the incident angle between the light ray and the normal (perpendicular) to the interface between mediums.
- n_r is the refractive index of the medium the light is entering.
- θ_r is the refractive angle between the light ray and the normal (perpendicular) to the interface between mediums.

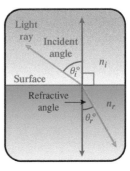

Calculate the index of refraction n_r of the indicated refractive medium given the following assumptions. Round answers to three decimal places.

- The incident medium is air.
- Air has an index of refraction of $n_i = 1.00$.
- The incidence angle is $\theta_i = 30°$.

63. Optics. Diamond, $\theta_r = 12°$
64. Optics. Emerald, $\theta_r = 18.5°$
65. Optics. Water, $\theta_r = 22°$
66. Optics. Plastic, $\theta_r = 20°$

67. Sprinkler. A water sprinkler is placed in the corner of the yard. If it disseminates water through an angle of 85°28′, how much of the corner is it missing?

68. Sprinkler. Sprinklers are set alongside a circle drive. If each sprinkler disseminates water through an angle of 36°19′, what angle is covered by both sprinklers?

69. Obstacle Course. As part of an obstacle course, participants are required to ascend to the top of a ladder placed against a building and then use a rope to climb the rest of the way to the roof. The distance traveled can be calculated using the formula $d = 15\sin\theta + 4\sqrt{3}$, where θ is the angle the ladder makes with the ground and d is the distance traveled, measured in feet. Find the exact distance traveled by the participants if $\theta = 60°$.

70. Obstacle Course. If the ladder in Exercise 69 is placed closer to the wall, the formula for distance traveled becomes $d = \dfrac{15}{\csc\theta} + 4$. Approximate the distance traveled by the participants if $\theta = 65°$.

71. Hot-Air Balloon. A hot-air balloon is tethered by ropes on two sides that form a 45° angle with the ground. If the height of the balloon can be determined by multiplying the length of one tether by $\sin 45°$, find the exact height of the balloon when 100-foot ropes are used.

72. Hot-Air Balloon. A hot-air balloon is tethered by ropes on two sides that form a 60° angle with the ground. If the height of the balloon can be determined by multiplying the length

of one tether by $\sin 60°$, find the exact height of the balloon when 100-foot ropes are used.

73. Staircase. The pitch of a staircase is given as $40°18'27''$. Write the pitch in decimal degrees.

74. Staircase. The height, measured in feet, of a certain staircase is given by the formula $h = 15\tan\theta$, where θ is the pitch of the staircase. What is the height of a staircase with a pitch of $39°28'37''$?

• **CATCH THE MISTAKE**

In Exercises 75 and 76, explain the mistake that is made.

75. $\sec 60°$

Solution:

Write secant as the reciprocal of cosine.	$\sec\theta = \dfrac{1}{\cos\theta}$
Substitute $\theta = 60°$.	$\sec 60° = \dfrac{1}{\cos 60°}$
Recall $\cos 60° = \dfrac{\sqrt{3}}{2}$.	$\sec 60° = \dfrac{1}{\frac{\sqrt{3}}{2}}$
Simplify.	$\sec 60° = \dfrac{2}{\sqrt{3}}$
Rationalize the denominator.	$\sec 60° = \dfrac{2\sqrt{3}}{3}$

This is incorrect. What mistake was made?

76. $\sec(36°25')$

Solution:

Convert $36°25'$ to decimal degrees.	$\dfrac{25}{100} = 0.25$
	$36°25' = 36.25°$
Use the reciprocal identity, $\sec\theta = \dfrac{1}{\cos\theta}$.	$\sec(36.25°) = \dfrac{1}{\cos(36.25°)}$
Approximate with a calculator.	$\sec(36.25°) \approx 1.2400$

This is incorrect. What mistake was made?

• **CONCEPTUAL**

In Exercises 77–80, determine whether each statement is true or false.

77. $\cos 30° = \sec\left(\dfrac{1}{30°}\right)$

78. $\sin 50° = 0.77$

79. When approximating values of $\sin 10°$ and $\cos 10°$ with a calculator, it is important for your calculator to be in degree mode.

80. $\tan(20°50') = \cot(70°10')$

For Exercises 81–84, refer to the following:

Thus far in this text, we have only discussed trigonometric values of acute angles, $0° < \theta < 90°$. What about when θ is approximately $0°$ or $90°$? We will formally consider these cases in the next chapter, but for now, draw and label a right triangle that has one angle very close to $0°$ so that the opposite side is very small compared to the adjacent side. Then the hypotenuse and the adjacent side are very close in size.

Use trigonometric ratios and the assumption that a is much larger than b to approximate the values without using a calculator.

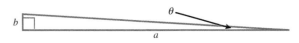

81. $\sin 0°$ **82.** $\cos 0°$ **83.** $\cos 90°$ **84.** $\sin 90°$

• **CHALLENGE**

85. Find the exact value of $\dfrac{\sec 45° + \tan 30°}{\cos 30°}$.

86. Find the exact value of $\dfrac{(\sin 45°)(\cot 30°)}{\csc^2 60°}$.

87. Find the exact value of $\cos 75° - (\csc 45°)(\cos 30°)$, given that $\sin 15° = \dfrac{\sqrt{6} - \sqrt{2}}{4}$.

88. Find the exact value of $\cot 75°(\cos 45° + \sin 30°)$, given that $\tan 15° = 2 - \sqrt{3}$.

89. Calculate $\dfrac{\tan(25°10'15'') + \sec(46°14'26'')}{\csc(23°17'23'')}$.

90. Calculate $\dfrac{\cos(33°38'1'')}{\csc(50°33'2'')} - \cot(36°58'6'')$.

• **TECHNOLOGY**

In Exercises 91 and 92, perform the indicated operations. Which gives you a more accurate value?

91. Calculate $\sec 70°$ the following two ways:
   a. Write down $\cos 70°$ (round to three decimal places), and then divide 1 by that number. Write the number to five decimal places.
   b. First find $\cos 70°$ and then find its reciprocal. Round the result to five decimal places.

92. Calculate $\csc 40°$ the following two ways:
   a. Find $\sin 40°$ (round to three decimal places), and then divide 1 by that number. Write this last result to five decimal places.
   b. First find $\sin 40°$ and then find its reciprocal. Round the result to five decimal places.

In Exercises 93 and 94, illustrate calculator procedures for converting between DMS and DD.

93. Convert $3°14'25''$ to decimal degrees. Round to three decimal places.

94. Convert $27.683°$ to degrees-minutes-seconds. Round to three decimal places.

1.5 SOLVING RIGHT TRIANGLES

SKILLS OBJECTIVES	CONCEPTUAL OBJECTIVES
▪ Identify the number of significant digits to express the lengths of sides and measures of angles when solving right triangles. ▪ Solve right triangles given the measure of an acute angle and the length of a side. ▪ Solve right triangles given two side lengths.	▪ Understand that the least accurate number in your calculation determines the accuracy of your result. ▪ Understand that when an acute angle is given, then the third angle can be found exactly. The remaining unknown side lengths can be found using right triangle trigonometry and the Pythagorean theorem. ▪ Understand that the trigonometric inverse keys on a calculator can be used to approximate the measure of an angle, given its trigonometric function value.

1.5.1 SKILL

Identify the number of significant digits to express the lengths of sides and measures of angles when solving right triangles.

1.5.1 CONCEPTUAL

Understand that the least accurate number in your calculation determines the accuracy of your result.

To **solve a triangle** means to find the measure of the three angles and three side lengths of the triangle. In this section, we will discuss only right triangles. Therefore, we know one angle has a measure of 90°. We will know some information (the lengths of two sides or the length of a side and the measure of an acute angle), and we will determine the measures of the unknown angles and the lengths of the unknown sides. However, before we start solving right triangles, we must first discuss accuracy and significant digits.

1.5.1 Accuracy and Significant Digits

If we are upgrading our flooring and quickly measure a room as 10 feet by 12 feet and we want to calculate the diagonal length of the room, we use the Pythagorean theorem.

WORDS	MATH
Apply the Pythagorean theorem.	$10^2 + 12^2 = d^2$
Simplify.	$d^2 = 244$
Use the square root property.	$d = \pm\sqrt{244}$
Be sure the length of the diagonal is positive.	$d = \sqrt{244}$
Approximate the radical with a calculator.	$d \approx 15.62049935$

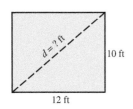

Would you say that the 10-foot by 12-foot room has a diagonal of 15.62049935 feet? No, because the known room measurements were given only with an accuracy of a foot and the diagonal above is calculated to eight decimal places. *Your results are no more accurate than the least accurate measure in your calculation.* In this example we round to the nearest foot, and hence we say that the diagonal of the 10-foot by 12-foot room is about 16 feet.

Significant digits are used to determine the precision of a measurement.

> **DEFINITION** Significant Digits
>
> The number of significant digits in a number is found by counting all of the digits from left to right starting with the first nonzero digit.

We have placed a question mark next to 8000 simply because we don't know the answer. If 8000 is a result from rounding to the nearest thousand, then it has one significant digit. If 8000 is the result of rounding to the nearest ten, then it has three significant digits, and if there are exactly 8000 people surveyed, then 8000 is an exact value and it has four significant digits. In this text, we will assume that integers have the greatest number of significant digits. Therefore, 8000 has four significant digits and can be expressed in scientific notation as 8.000×10^3.

In solving right triangles, we first determine which of the given measurements has the *least* number of significant digits so we can round our final answers to the same number of significant digits.

NUMBER	SIGNIFICANT DIGITS
0.04	1
0.276	3
0.2076	4
1.23	3
17	2
17.00	4
17.000	5
6.25	3
8000	?

EXAMPLE 1 Identifying the Least Number of Significant Digits

Determine the number of significant digits corresponding to the given information in the following triangle: measure of an acute angle and a side length. In solving this right triangle, what number of significant digits should be used to express the remaining angle measure and side lengths?

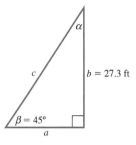

Solution:

Determine the significant digits corresponding to $\beta = 45°$.   Two significant digits

Determine the significant digits corresponding to $b = 27.3$ feet. Three significant digits

In solving this triangle, the remaining side lengths of a and c and the measure of α should be expressed to two significant digits.

1.5.2 Solving a Right Triangle Given an Acute Angle Measure and a Side Length

When solving a right triangle, we already know that one angle has measure 90°. Let us now consider the case when the measure of an acute angle and a side length are given. Since the measure of one of the acute angles is given, the remaining acute angle can be found using the fact that the sum of three angles in a triangle is 180°. Right triangle trigonometry is then used to find the remaining side lengths.

EXAMPLE 2   **Solving a Right Triangle Given an Angle and a Side**

Solve the right triangle—find a, b, and α.

Solution:

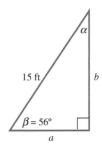

STEP 1 **Determine accuracy.**

Since the given quantities (15 feet and 56°) both are expressed to two significant digits, we will round final calculated values to two significant digits.

STEP 2 **Solve for α.**

Two acute angles in a right triangle are complementary.

$$\alpha + 56° = 90°$$

Solve for α.

$$\boxed{\alpha = 34°}$$

> **STUDY TIP**
> Make sure your calculator is in "degrees" mode.

STEP 3 **Solve for a.**

The cosine of an angle is equal to the adjacent side over the hypotenuse.

$$\cos 56° = \frac{a}{15}$$

Solve for a.

$$a = 15 \cos 56°$$

Evaluate the right side of the expression using a calculator.

$$a \approx 8.38789$$

Round a to two significant digits.

$$\boxed{a \approx 8.4 \text{ ft}}$$

STEP 4 **Solve for b.**

Notice that there are two ways to solve for b: trigonometric functions or the Pythagorean theorem. Although it is tempting to use the Pythagorean theorem, it is better to use the given information with trigonometric functions than to use a value that has already been rounded, which could make results less accurate.

The sine of an angle is equal to the opposite side over the hypotenuse.

$$\sin 56° = \frac{b}{15}$$

Solve for b.

$$b = 15 \sin 56°$$

Evaluate the right side of the expression using a calculator.

$$b \approx 12.43556$$

Round b to two significant digits.

$$\boxed{b \approx 12 \text{ ft}}$$

STEP 5 **Check the solution.**

Angles and sides are rounded to two significant digits.

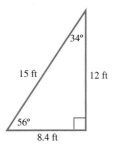

Check the trigonometric values of the specific angles by calculating the trigonometric ratios.

$$\sin 34° \overset{?}{=} \frac{8.4}{15} \qquad \cos 34° \overset{?}{=} \frac{12}{15} \qquad \tan 34° \overset{?}{=} \frac{8.4}{12}$$

$$0.5592 \approx 0.56 \qquad 0.8290 \approx 0.80 \qquad 0.6745 \approx 0.70$$

[CONCEPT CHECK]

In Your Turn, the angle θ can be calculated exactly. Can the side lengths be calculated exactly?

▼ ANSWER No, they must be approximated.

▼ **YOUR TURN** Given the triangle shown here, solve the right triangle—find a, b, and θ.

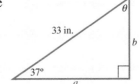

▼ **ANSWER**

$\theta = 53°$, $a \approx 26$ in., and $b \approx 20$ in.

1.5.3 Solving a Right Triangle Given the Lengths of Two Sides

When solving a right triangle, we already know that one angle has measure 90°. Let us now consider the case when the lengths of two sides are given. In this case, the third side can be found using the Pythagorean theorem. If we can determine the measure of one of the acute angles, then we can find the measure of the third acute angle using the fact that the sum of the three angle measures in a triangle is 180°. How do we find the measure of one of the acute angles? Since we know the side lengths, we can use right triangle ratios to determine the trigonometric function (sine, cosine, or tangent) values and then ask ourselves: What angle corresponds to that value?

Sometimes, we may know the answer exactly. For example, if we determine that $\sin\theta = \frac{1}{2}$, then we know that the acute angle θ is 30° because $\sin 30° = \frac{1}{2}$. Other times we may not know the corresponding angle, such as $\sin\theta = 0.9511$. Calculators have three keys ($\sin^{-1}$, $\cos^{-1}$, and $\tan^{-1}$) that help us determine the unknown angle. For example, a calculator can be used to assist us in finding what angle θ corresponds to $\sin\theta = 0.9511$.

$$\sin^{-1}(0.9511) = 72.00806419$$

At first glance, these three keys might appear to yield the reciprocal; however, the -1 superscript corresponds to an inverse function. We will learn more about inverse trigonometric functions in Chapter 6, but for now we will use these three calculator keys to help us solve right triangles.

1.5.3 SKILL

Solve right triangles given two side lengths.

1.5.3 CONCEPTUAL

Understand that the trigonometric inverse keys on a calculator can be used to approximate the measure of an angle, given its trigonometric function value.

EXAMPLE 3 **Using a Calculator to Determine an Acute Angle Measure**

Use a calculator to find θ. Round answers to the nearest degree.

a. $\cos\theta = 0.8734$

b. $\tan\theta = 2.752$

Solution (a):

Use a calculator to evaluate the inverse cosine function.

$$\theta = \cos^{-1}(0.8734) = 29.14382196°$$

Round to the nearest degree.

$$\theta \approx 29°$$

Solution (b):

Use a calculator to evaluate the inverse tangent function.

$$\theta = \tan^{-1}(2.752) = 70.03026784°$$

Round to the nearest degree.

$$\theta \approx 70°$$

[CONCEPT CHECK]

When two side lengths are given but the two acute angles in a right triangle are unknown, how do we find an approximation of the acute angles?

▼ ANSWER Inverse trigonometric function

▼ **YOUR TURN** Use a calculator to find θ, given $\sin\theta = 0.7739$. Round the answer to the nearest degree.

▼ **ANSWER**

51°

EXAMPLE 4 **Solving a Right Triangle Given Two Sides**

Solve the right triangle—find a, α, and β.

Solution:

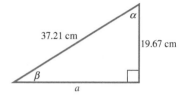

STEP 1 **Determine accuracy.**

The given sides have four significant digits; therefore, round final calculated values to four significant digits.

STEP 2 **Solve for α.**

The cosine of an angle is equal to the adjacent side over the hypotenuse.

$$\cos\alpha = \frac{19.67 \text{ cm}}{37.21 \text{ cm}}$$

Evaluate the right side with a calculator. $\cos\alpha \approx 0.528621338$

Write the angle α in terms of the inverse cosine function.

$$\alpha \approx \cos^{-1}(0.528621338)$$

Use a calculator to evaluate the inverse cosine function.

$$\alpha \approx 58.08764855°$$

Round α to four significant digits. $\boxed{\alpha \approx 58.09°}$

STEP 3 **Solve for β.**

The two acute angles in a right triangle are complementary. $\alpha + \beta = 90°$

Substitute $\alpha \approx 58.09°$. $58.09 + \beta \approx 90°$

Solve for β. $\boxed{\beta \approx 31.91°}$

The answer is already rounded to four significant digits.

STEP 4 **Solve for a.**

Use the Pythagorean theorem since the lengths of two sides are given. $a^2 + b^2 = c^2$

Substitute given values for b and c. $a^2 + 19.67^2 = 37.21^2$

Solve for a. $a \approx 31.5859969$

Round a to four significant digits. $\boxed{a \approx 31.59 \text{ cm}}$

STEP 5 **Check the solution.**

Angles are rounded to the nearest hundredth degree, and sides are rounded to four significant digits of accuracy.

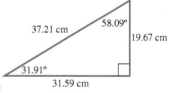

Check the trigonometric values of the specific angles by calculating the trigonometric ratios.

$$\sin(31.91°) \overset{?}{=} \frac{19.67}{37.21} \qquad \sin(58.09°) \overset{?}{=} \frac{31.59}{37.21}$$

$$0.5286 \approx 0.5286 \qquad\quad 0.8489 \approx 0.8490$$

▼
ANSWER

$a \approx 16.0$ mi,
$\alpha \approx 43.0°$, and $\beta \approx 47.0°$

▼
YOUR TURN Solve the right triangle—find a, α, and β.

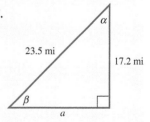

Applications

In many applications of solving right triangles, you are given the length of a side and the measure of an acute angle and asked to find one of the other sides. Two common examples involve an observer (or point of reference) located on the horizontal and an object that is either above or below the horizontal. If the object is above the horizontal, then the angle made is called the **angle of elevation,** and if the object is below the horizontal, then the angle made is called the **angle of depression**.

For example, if a race car driver is looking straight ahead (in a horizontal line of sight), then looking up is elevation and looking down is depression.

If the angle is a physical one (like a skateboard ramp), then the appropriate name is the **angle of inclination**.

Angle of inclination

EXAMPLE 5 **Angle of Depression (NASCAR)**

In this picture, car 19 is behind the leader car 2. If the angle of depression is 18° from the car 19's driver's eyes to the bottom of the 3-foot high back end of car 2 (side opposite the angle of depression), how far apart are their bumpers? Assume that the horizontal distance from the car 19's driver's eyes to the front of his car is 5 feet.

Solution:

Draw an appropriate right triangle and label the known quantities.

Because the sides of interest are adjacent and opposite to the known angle, identify the tangent ratio.

$$\tan 18° = \frac{3 \text{ ft}}{x}$$

Solve for x.

$$x = \frac{3 \text{ ft}}{\tan 18°}$$

Evaluate the right side.

$$x \approx 9.233 \text{ ft}$$

Round to the nearest foot.

$$x \approx 9 \text{ ft}$$

Subtract 5 feet from x.

$$9 - 5 = \boxed{4 \text{ ft}}$$

Their bumpers are approximately $\boxed{4 \text{ feet apart}}$.

Suppose NASA wants to talk with astronauts on the International Space Station (ISS), which is traveling at a speed of 17,700 mph, 400 kilometers (250 miles) above the surface of the Earth. If the antennas at the ground station in Houston have a pointing error of even 1 minute, that is $1' = \left(\dfrac{1}{60}\right)^{\circ} = 0.01667°$, they will miss the chance to talk with the astronauts.

EXAMPLE 6 **Pointing Error**

Assume that the ISS (which is 108 meters long and 73 meters wide) is in a 400-kilometer low Earth orbit. If the communications antennas have a 1-minute pointing error, how many meters "off" will the communications link be?

Solution:

Draw a right triangle that depicts this scenario.

Because the sides of interest are adjacent and opposite to the known angle, identify the tangent ratio.

$$\tan(1') = \frac{x}{400 \text{ km}}$$

Solve for x.

$$x = (400 \text{ km}) \tan(1')$$

Convert $1'$ to decimal degrees.

$$1' = \left(\frac{1}{60}\right)^{\circ} \approx 0.016666667°$$

Convert the equation for x in terms of decimal degrees.

$$x \approx (400 \text{ km}) \tan(0.016666667°)$$

Evaluate the expression on the right.

$$x \approx 0.116355 \text{ km} \approx 116 \text{ m}$$

400 kilometers is accurate to three significant digits. So we express the answer to three significant digits.

The pointing error causes the signal to be off by approximately $\boxed{116 \text{ meters}}$. Since the ISS is only 108 meters long, it is expected that the signal will be missed by the astronaut crew.

In navigation, the word **bearing** means the direction to which a vessel is pointed. **Heading** is the direction in which the vessel is actually traveling. Heading and bearing are only synonyms when there is no wind on land. Direction is often given as a bearing, which is the measure of an acute angle with respect to the north-south vertical line. "The plane has a bearing N 20° E" means that the plane is pointed 20° to the east of due north.

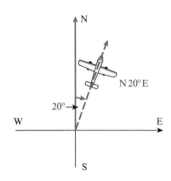

▶ **EXAMPLE 7**   **Bearing (Navigation)**

A jet takes off bearing N 28° E and flies 5 miles, and then makes a left (90°) turn and flies 12 miles farther. If the control tower operator wanted to locate the plane, what bearing would she use? Round to the nearest degree.

Solution:

Draw a picture that represents this scenario.

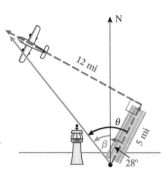

Identify the tangent ratio.

$$\tan\theta = \frac{12}{5}$$

Use the inverse tangent function to solve for θ.

$$\theta = \tan^{-1}\left(\frac{12}{5}\right) \approx 67.4°$$

Subtract 28° from θ to find the bearing, β.

$$\beta \approx 67.4° - 28° \approx 39.4°$$

Round to the nearest degree.

$$\boxed{\beta \approx N\,39°\,W}$$

▶[SECTION 1.5]   **SUMMARY**

In this section, we have solved right triangles. When either a side length and an acute angle measure are given or two side lengths are given, it is possible to solve the right triangle (find all unknown side lengths and angle measures). The least accurate number used in your calculations determines the appropriate number of significant digits for your results.

[SECTION 1.5] EXERCISES

• SKILLS

In Exercises 1–4, determine the number of significant digits corresponding to each of the given angle measures and side lengths.

1. $\alpha = 37.5°$　　**2.** $\beta = 49.76°$　　**3.** $a = 0.37$ km　　**4.** $b = 0.2$ mi

In Exercises 5–10, use a calculator to find the measure of angle θ. Round answer to the nearest degree.

5. $\sin\theta = 0.7264$　　**6.** $\sin\theta = 0.1798$　　**7.** $\cos\theta = 0.5674$　　**8.** $\cos\theta = 0.8866$　　**9.** $\tan\theta = 8.235$　　**10.** $\tan\theta = 3.563$

In Exercises 11–30, refer to the right triangle diagram and the given information to find the indicated measure. Write your answers for angle measures in decimal degrees.

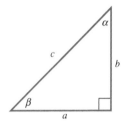

11. $\beta = 35°$, $c = 17$ in.; find a.
12. $\beta = 35°$, $c = 17$ in.; find b.
13. $\alpha = 55°$, $c = 22$ ft; find a.
14. $\alpha = 55°$, $c = 22$ ft; find b.
15. $\alpha = 20.5°$, $b = 14.7$ mi; find a.
16. $\beta = 69.3°$, $a = 0.752$ mi; find b.
17. $\beta = 25°$, $a = 11$ km; find c.
18. $\beta = 75°$, $b = 26$ km; find c.
19. $\alpha = 48.25°$, $a = 15.37$ cm; find c.
20. $\alpha = 29.80°$, $b = 16.79$ cm; find c.
21. $a = 29$ mm, $c = 38$ mm; find α.
22. $a = 89$ mm, $c = 99$ mm; find β.
23. $b = 2.3$ m, $c = 4.9$ m; find α.
24. $b = 7.8$ m, $c = 13$ m; find β
25. $\alpha = 21°17'$, $b = 210.8$ yd; find a.
26. $\beta = 27°21'$, $a = 117.0$ yd; find b.
27. $\beta = 15°20'$, $a = 10.2$ km; find c.
28. $\beta = 65°30'$, $b = 18.6$ km; find c.
29. $\alpha = 40°28'10''$, $a = 12,522$ km; find c.
30. $\alpha = 28°32'50''$, $b = 17,986$ km; find c.

In Exercises 31–48, refer to the right triangle diagram and the given information to solve the right triangle. Write your answers for angle measures in decimal degrees.

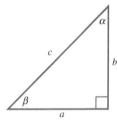

31. $\alpha = 32°$ and $c = 12$ ft
32. $\alpha = 65°$ and $c = 37$ ft
33. $\alpha = 44°$ and $b = 2.6$ cm
34. $\alpha = 12.0°$ and $b = 10.0$ m
35. $\alpha = 60°$ and $c = 5$ in.
36. $\alpha = 9.67°$ and $c = 5.38$ in.
37. $\beta = 72°$ and $c = 9.7$ mm
38. $\beta = 45°$ and $c = 7.8$ mm
39. $\alpha = 54.2°$ and $a = 111$ mi
40. $\beta = 47.2°$ and $a = 9.75$ mi
41. $\beta = 45°$, $b = 10.2$ km
42. $\beta = 85.5°$, $b = 14.3$ ft
43. $\alpha = 28°23'$ and $b = 1734$ ft
44. $\alpha = 72°59'$ and $a = 2175$ ft
45. $a = 42.5$ ft and $b = 28.7$ ft
46. $a = 19.8$ ft and $c = 48.7$ ft
47. $a = 35,236$ km and $c = 42,766$ km
48. $b = 0.1245$ mm and $c = 0.8763$ mm

• APPLICATIONS

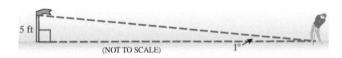

49. Golf. If the flagpole that a golfer aims at on a green measures 5 feet from the ground to the top of the flag and a golfer measures a 1° angle from top to bottom of the pole, how far (in horizontal distance) is the golfer from the flag? Round to the nearest foot.

50. Golf. If the flagpole that a golfer aims at on a green measures 5 feet from the ground to the top of the flag and a golfer measures a 3° angle from top to bottom of the pole, how far (in horizontal distance) is the golfer from the flag? Round to the nearest foot.

Exercises 51 and 52 illustrate a midair refueling scenario that military aircraft often enact. Assume the elevation angle that the hose makes with the plane being fueled is $\theta = 36°$.

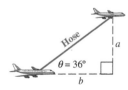

51. Midair Refueling. If the hose is 150 feet long, what should be the altitude difference a between the two planes? Round to the nearest foot.

52. Midair Refueling. If the smallest acceptable altitude difference *a* between the two planes is 100 feet, how long should the hose be? Round to the nearest foot.

Exercises 53–56 are based on the idea of a glide slope (the angle the flight path makes with the ground).

Precision Approach Path Indicator (PAPI) lights are used as a visual approach slope aid for pilots landing aircraft. The typical glide path for commercial jet airliners is 3°. A Navy fighter jet has an outer glide approach of 18°–20°. PAPI lights are typically configured as a row of four lights. All four lights are on, but in different combinations of red or white. If all four lights are white, then the angle of descent is too high; if all four lights are red, then the angle of descent is too low; and if there are two white and two red, then the approach is perfect.

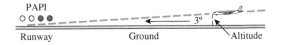

53. Glide Path of a Commercial Jet Airliner. If a commercial jetliner is 5000 feet (about 1 mile ground distance) from the runway, what should be the altitude of the plane to achieve 2 red/2 white PAPI lights? (Assume that this corresponds to a 3° glide path.)

54. Glide Path of a Commercial Jet Airliner. If a commercial jetliner is at an altitude of 450 feet when it is 5200 feet from the runway (approximately 1 mile ground distance), what is the glide slope angle? Will the pilot see white lights, red lights, or both?

55. Glide Path of a Navy Fighter Jet. If the pilot of a Navy fighter jet is at an altitude of 3000 feet when she is 15,500 feet (approximately 3 miles ground distance) from the landing facility, what is her glide slope angle (round to the nearest degree)? Is she too high or too low?

56. Glide Path of a Navy Fighter Jet. If the same pilot in Exercise 55 raises the nose of a Navy fighter jet so that she drops only 500 feet by the time she is 7800 feet from the landing strip (ground distance), what is her glide angle at that time (round to the nearest degree)? Is she within the specs (18°−20°) to land the jet?

In Exercises 57 and 58, refer to the illustration below that shows a search and rescue helicopter with a 30° field of view with a search light.

57. Search and Rescue. If the search and rescue helicopter is flying at an altitude of 150 feet above sea level, what is the diameter of the circle illuminated on the surface of the water?

58. Search and Rescue. If the search and rescue helicopter is flying at an altitude of 500 feet above sea level, what is the diameter of the circle illuminated on the surface of the water?

For Exercises 59–62, refer to the following:

Geostatic orbits are useful because they cause a satellite to appear stationary with respect to a fixed point on the rotating Earth. As a result, an antenna (dish TV) can point in a fixed direction and maintain a link with the satellite. The satellite orbits in the direction of the Earth's rotation at an altitude of approximately 35,000 kilometers.

59. Dish TV. If your dish TV antenna has a pointing error of 1″ (1 second), how long would the satellite have to be to maintain a link? Round your answer to the nearest meter.

60. Dish TV. If your dish TV antenna has a pointing error of $\frac{1}{2}''$ (half a second), how long would the satellite have to be to maintain a link? Round your answer to the nearest meter.

61. Dish TV. If the satellite in a geostationary orbit (at 35,000 kilometers) was only 10 meters long, about how accurately pointed would the dish have to be? Give the answer in degrees to two significant digits.

62. Dish TV. If the satellite in a geostationary orbit (at 35,000 kilometers) was only 30 meters long, about how accurately pointed would the dish have to be? Give the answer in degrees to two significant digits.

63. Angle of Elevation (Traffic). A person driving in a sedan is driving too close to the back of an 18 wheeler on an interstate highway. He decides to back off until he can see the entire truck (to the top). If the height of the trailer is 15 feet and the sedan driver's angle of elevation (to the top of the trailer from the horizontal line with the bottom of the trailer) is roughly 30°, how far is he sitting from the end of the trailer?

64. Angle of Depression (Opera). The balcony seats at the opera house have an angle of depression of 55° to center stage. If the horizontal (ground) distance to the center of the stage is 50 feet, how far are the patrons in the balcony from the singer at center stage?

65. Angle of Inclination (Skiing). The angle of inclination of a mountain with triple black diamond ski paths is 65°. If a skier at the top of the mountain is at an elevation of 4000 feet, how long is the ski run from the top to the base of the mountain?

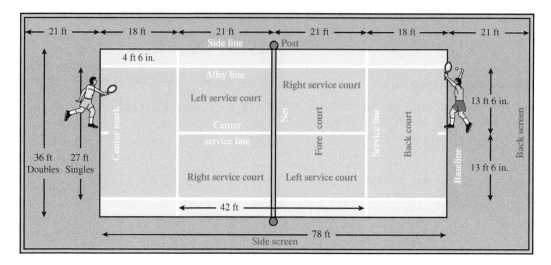

66. Bearing (Navigation). If a plane takes off bearing N33°W and flies 6 miles and then makes a right turn (90°) and flies 10 miles farther, what bearing will the traffic controller use to locate the plane?

67. Bearing (Navigation). If a plane takes off bearing N35°E and flies 3 miles and then makes a left turn (90°) and flies 8 miles farther, what bearing will the traffic controller use to locate the plane?

68. Bearing (Navigation). If a plane takes off bearing N48°W and flies 6 miles and then makes a right turn (90°) and flies 17 miles farther, what bearing will the traffic controller use to locate the plane?

For Exercises 69 and 70, refer to the following:

With the advent of new technology, tennis racquets can now be constructed to permit a player to serve at speeds in excess of 120 mph. One of the most effective serves in tennis is a power serve that is hit at top speed directly at the top left corner of the

For Exercises 71 and 72, refer to the following:

The structure of molecules is critical to the study of materials science and organic chemistry, and has countless applications to a variety of interesting phenomena. Trigonometry plays a critical role in determining bonding angles of molecules. For instance, the structure of the $(FeCl_4Br_2)^{-3}$ ion (dibromatetrachlorideferrate III) is shown in the figure below.

71. Chemistry. Determine the angle θ [i.e., the angle between the axis containing the apical bromide atom (Br) and the segment connecting Br to Cl].

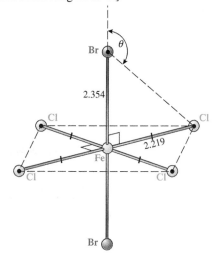

right service court (or top right corner of the left service court). When attempting this serve, a player will toss the ball rather high into the air, bring the racquet back, and then make contact with the ball at the precise moment when the position of the ball in the air coincides with the top of the netted part of the racquet when the player's arm is fully stretched over his or her head.

69. Tennis. Assume that the player is serving into the right service court and stands just *2 inches* to the right of the center line behind the baseline. If, at the moment the racquet strikes the ball, both are *72 inches from the ground* and the serve actually hits the top left corner of the right service court, determine the angle at which the ball meets the ground in the right service court. Round to the nearest degree.

70. Tennis. Assume that the player is serving into the right service court and stands just *2 inches* to the right of the center line behind the baseline. If the ball hits the top left corner of the right service court at an angle of 44°, at what height above the ground must the ball be struck?

72. Chemistry. Now, suppose one of the chlorides (Cl) is removed. The resulting structure is triagonal in nature, resulting in the figure below. Does the angle θ change? If so, what is its new value?

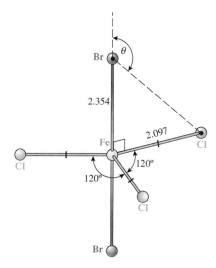

73. Navigation. A plane takes off headed due north. With the wind, the airplane actually travels on a heading of N 15° E. After traveling for 100 miles, how far north is the plane from its starting position?

74. Navigation. A boat must cross a 150-foot river. While the boat is pointed due east, perpendicular to the river, the current causes it to land 25 feet down river. What is the heading of the boat?

For Exercises 75 and 76, refer to the following:

A canal constructed by a water-users association can be approximated by an isosceles triangle (see the figure below). When the canal was originally constructed, the depth of the canal was 5.0 feet and the angle defining the shape of the canal was 60°.

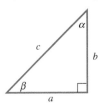

75. Environmental Science. If the width of the water surface today is 4.0 feet, find the depth of the water running through the canal.

76. Environmental Science. One year later, a survey is performed to measure the effects of erosion on the canal. It is determined that when the water depth is 4.0 feet, the width of the water surface is 5.0 feet. Find the angle θ defining the shape of the canal to the nearest degree. Has erosion affected the shape of the canal? Explain.

For Exercises 77 and 78, refer to the following:

After breaking a femur, a patient is placed in traction. The end of a femur of length l is lifted to an elevation forming an angle θ with the horizontal (angle of elevation).

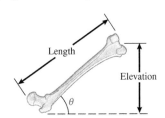

77. Health/Medicine. A femur 18 inches long is placed into traction, forming an angle of 15° with the horizontal. Find the height of elevation at the end of the femur.

78. Health/Medicine. A femur 18 inches long is placed in traction with an elevation of 6.2 inches. What is the angle of elevation of the femur?

● **CATCH THE MISTAKE**

For Exercises 79 and 80, refer to the right triangle diagram below and explain the mistake that is made.

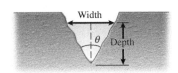

79. If $b = 800$ feet and $a = 10$ feet, find β.

Solution:

Represent tangent as the opposite side over the adjacent side.	$\tan\beta = \dfrac{b}{a}$
Substitute $b = 800$ feet and $a = 10$ feet.	$\tan\beta = \dfrac{800}{10} = 80$
Use a calculator to evaluate β.	$\beta = \tan 80° \approx 5.67°$

This is incorrect. What mistake was made?

80. If $\beta = 56°$ and $c = 15$ feet, find b and then find a.

Solution:

Write sine as the opposite side over the hypotenuse.	$\sin 56° = \dfrac{b}{15}$
Solve for b.	$b = 15 \sin 56°$
Use a calculator to approximate b.	$b \approx 12.4356$
Round the answer to two significant digits.	$b \approx 12$ ft
Use the Pythagorean theorem to find a.	$a^2 + b^2 = c^2$
Substitute $b = 12$ feet and $c = 15$ feet.	$a^2 + 12^2 = 15^2$
Solve for a.	$a = 9$
Round the answer to two significant digits.	$a = 9.0$ ft

Compare this with the results from Example 1. Why did we get a different value for a here?

● **CONCEPTUAL**

In Exercises 81–88, determine whether each statement is true or false.

81. If you are given the lengths of two sides of a right triangle, you can solve the right triangle.

82. If you are given the length of one side and the measure of one acute angle of a right triangle, you can solve the right triangle.

83. If you are given the measures of the two acute angles of a right triangle, you can solve the right triangle.

84. If you are given the length of the hypotenuse of a right triangle and the measures of the angle opposite the hypotenuse, you can solve the right triangle.

85. The number 0.0123 has five significant digits.

86. If the measurement 700 feet has been rounded to the nearest whole number, it has three significant digits.

87. If you are given the length of one side and the measure of one acute angle, you can solve the right triangle using either the sine or cosine function.

88. If you are given the lengths of two sides of a right triangle, you will need to use an inverse trigonometric function to find the third side length.

• CHALLENGE

89. Use the information in the picture below to determine the height of the mountain.

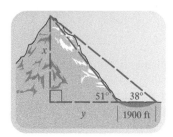

90. Two friends who are engineers at Kennedy Space Center (KSC) watched the shuttle launch. Carolyn was at the Vehicle Assembly Building (VAB) 3 miles from the launch pad and Jackie was across the Banana River, which is 8 miles from the launch pad. They called each other at liftoff, and after 10 seconds they each estimated the angle of elevation with respect to the ground. Carolyn thought the angle of elevation was approximately 40° and Jackie thought the angle of elevation was approximately 15°. Approximately how high was the shuttle after 10 seconds? (Average their estimates and round to the nearest mile.)

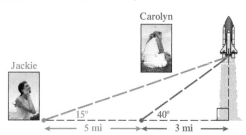

Image Bank/Getty Images, Inc.; Photonica/Getty Images

91. From the top of a 12-foot ladder, the angle of depression to the far side of a sidewalk is 45°, while the angle of depression to the near side of the sidewalk is 65°. How wide is the sidewalk?

92. Find the perimeter of triangle *A*.

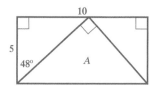

93. Solve for *x*.

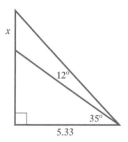

94. An electric line is strung from a 20-foot pole to a point 12 foot up on the side of a house. If the pole is 250 feet from the house, what angle does the electric line make with the pole?

• TECHNOLOGY

95. Use a calculator to find $\sin^{-1}(\sin 40°)$.

96. Use a calculator to find $\cos^{-1}(\cos 17°)$.

97. Use a calculator to find $\cos(\cos^{-1} 0.8)$.

98. Use a calculator to find $\sin(\sin^{-1} 0.3)$.

99. Based on the result from Exercise 95, what would $\sin^{-1}(\sin\theta)$ be for an acute angle θ?

100. Based on the result from Exercise 96, what would $\cos^{-1}(\cos\theta)$ be for an acute angle θ?

▶[CHAPTER 1 REVIEW]

SECTION	CONCEPT	KEY IDEAS/FORMULAS
1.1	**Angles, degrees, and triangles**	
	Angles and degree measure	Terminal side Positive angle Initial side Initial side Negative angle Terminal side One complete rotation = 360° $\theta = 90°$ Right angle: Quarter rotation θ Acute angle $0° < \theta < 90°$ θ Obtuse angle $90° < \theta < 180°$ β α Complementary angles $\alpha + \beta = 90°$ β α Supplementary angles $\alpha + \beta = 180°$
	Triangles	γ α β $\alpha + \beta + \gamma = 180°$ a c b $a^2 + b^2 = c^2$
	Special right triangles	$45°$ x $\sqrt{2}x$ $45°$ x $45°\text{-}45°\text{-}90°$ $30°$ $\sqrt{3}x$ $2x$ $60°$ x $30°\text{-}60°\text{-}90°$

SECTION	CONCEPT	KEY IDEAS/FORMULAS
1.2	**Similar triangles**	
	Finding angle measures using geometry	
	Classification of triangles	Similar triangles: Same shape $$\frac{a}{a'} = \frac{b}{b'} = \frac{c}{c'}$$ Congruent triangles: Same shape and size
1.3	**Definition 1 of trigonometric functions: Right triangle ratios**	Definition 1 defines trigonometric functions of acute angles as ratios of the length of sides in a right triangle.
	Trigonometric functions: Right triangle ratios	SOH $\quad \sin\theta = \dfrac{\text{opposite}}{\text{hypotenuse}}$ CAH $\quad \cos\theta = \dfrac{\text{adjacent}}{\text{hypotenuse}}$ TOA $\quad \tan\theta = \dfrac{\text{opposite}}{\text{adjacent}}$ Reciprocal identities: $$\cot\theta = \frac{1}{\tan\theta} \quad \csc\theta = \frac{1}{\sin\theta} \quad \sec\theta = \frac{1}{\cos\theta}$$
	Cofunctions	If $\alpha + \beta = 90°$, then: $\sin\alpha = \cos\beta$ $$\sec\alpha = \csc\beta$$ $$\tan\alpha = \cot\beta$$

SECTION	CONCEPT	KEY IDEAS/FORMULAS
1.4	**Evaluating trigonometric functions: Exactly and with calculators**	
	Evaluating trigonometric functions *exactly* for special angle measures: 30°, 45°, and 60°	<table><tr><td>θ</td><td>**SIN** θ</td><td>**COS** θ</td></tr><tr><td>30°</td><td>$\dfrac{1}{2}$</td><td>$\dfrac{\sqrt{3}}{2}$</td></tr><tr><td>45°</td><td>$\dfrac{\sqrt{2}}{2}$</td><td>$\dfrac{\sqrt{2}}{2}$</td></tr><tr><td>60°</td><td>$\dfrac{\sqrt{3}}{2}$</td><td>$\dfrac{1}{2}$</td></tr></table> The other trigonometric function values can be found for these angles using $\tan\theta = \dfrac{\sin\theta}{\cos\theta}$ and reciprocal identities.
	Using calculators to evaluate (approximate) trigonometric function values	Make sure the calculator is in degree mode. The sin, cos, and tan buttons can be combined with the reciprocal button, $1/x$ or x^{-1}, to get csc, sec, and cot, respectively.
	Representing partial degrees: DD or DMS	$1' = \left(\frac{1}{60}\right)^\circ$ or $60' = 1^\circ$ $1'' = \left(\frac{1}{60}\right)' = \left(\frac{1}{3600}\right)^\circ$ or $60'' = 1'$ DMS to DD: Divide by multiples of 60. DD to DMS: Multiply by multiples of 60.
1.5	**Solving right triangles**	
	Accuracy and significant digits	Always use given measurements if possible for accurate results. Pay attention to proper significant digits for accuracy of answer.
	Solving a right triangle given an acute angle measure and a side length	■ The third angle measure can be found exactly using $\alpha + \beta + \gamma = 180^\circ$. ■ Right triangle trigonometry is used to find the lengths of the remaining sides.
	Solving a right triangle given the lengths of two sides	■ The length of the third side can be found using the Pythagorean theorem. ■ The measures of the acute angles can be found using right triangle trigonometry.

[CHAPTER 1 REVIEW EXERCISES]

1.1 Angles, Degrees, and Triangles

Find (a) the complement and (b) the supplement of the given angles.

1. $28°$ 2. $17°$ 3. $35°$ 4. $78°$

Refer to the following triangle.

5. If $\alpha = 120°$ and $\beta = 35°$, find γ.

6. If $\alpha = 105°$ and $\beta = 25°$, find γ.

7. If $\gamma = \beta$ and $\alpha = 7\beta$, find all three angles.

8. If $\gamma = \beta$ and $\alpha = 6\beta$, find all three angles.

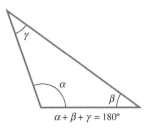

Refer to the following right triangle.

9. If $a = 4$ and $c = 12$, find b.

10. If $b = 9$ and $c = 15$, find a.

11. If $a = 7$ and $b = 4$, find c.

12. If $a = 10$ and $b = 8$, find c.

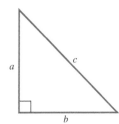

Refer to the following 45°-45°-90° triangle.

13. If the two legs have length 12 yards, how long is the hypotenuse?

14. If the hypotenuse has length $2\sqrt{2}$ feet, how long are the legs?

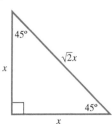

Refer to the following 30°-60°-90° triangle.

15. If the shorter leg has length 3 feet, what are the lengths of the other leg and the hypotenuse?

16. If the hypotenuse has length 12 kilometers, what are the lengths of the two legs?

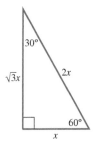

Applications

17. **Clock.** What is the measure (in degrees) of the angle that the minute hand sweeps in exactly 25 minutes?

18. **Clock.** What is the measure (in degrees) of the angle that the second hand sweeps in exactly 15 seconds?

1.2 Similar Triangles

Find the measure of the indicated angle.

19. $\angle F = 75°$, find $\angle G$.

20. $\angle F = 75°$, find $\angle D$.

21. $\angle F = 75°$, find $\angle C$.

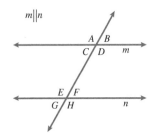

22. $\angle F = 75°$, find $\angle E$.

23. $\angle F = 75°$, find $\angle B$.

24. $\angle F = 75°$, find $\angle A$.

Calculate the specified lengths given that the two triangles are similar.

25. $A = 10$, $B = 8$, $D = 5$, $E = ?$

26. $A = 15$, $B = 12$, $E = 4$, $D = ?$

27. $D = 4.5$ m, $F = 8.2$ m, $A = 81$ km, $C = ?$

28. $E = 8$ cm, $F = 14$ cm, $C = 8$ m, $B = ?$

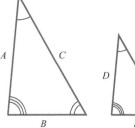

Applications

29. **Height of a Tree.** The shadow of a tree measures 9.6 meters. At the same time of day, the shadow of a 4-meter basketball backboard measures 1.2 meters. How tall is the tree?

30. **Height of a Man.** If an NBA center casts a 1 foot 9 inch shadow, and his 4-foot son casts a 1-foot shadow, how tall is the NBA center?

For Exercises 31 and 32, refer to the following:

In a home remodeling project, your architect gives you plans that have an indicated distance of 3 feet, measuring 1 inch with a ruler.

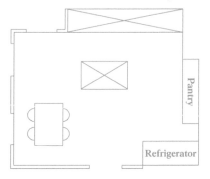

31. **Home Renovation.** How wide is the built-in refrigerator if it measures $1\frac{1}{3}$ inches with a ruler?

32. **Home Renovation.** How wide is the pantry if it measures $1\frac{1}{4}$ inches with a ruler?

1.3 Definition 1 of Trigonometric Functions: Right Triangle Ratios

Use the following triangle to find the indicated trigonometric function values. Rationalize any denominators that you encounter in your answers, but leave answers exact.

33. $\cos\theta$

34. $\sin\theta$

35. $\sec\theta$

36. $\csc\theta$

37. $\tan\theta$

38. $\cot\theta$

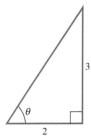

Use the cofunction identities to fill in the blanks.

39. $\sin 30° = \cos$ _____

40. $\cos A = \sin$ _____

41. $\tan 45° = \cot$ _____

42. $\csc 60° = \sec$ _____

Write the trigonometric function in terms of its cofunction.

43. $\sin(30° - x)$ **44.** $\cos(55° + A)$

45. $\csc(45° - x)$ **46.** $\sec(60° - \theta)$

1.4 Evaluating Trigonometric Functions: Exactly and with Calculators

Label each trigonometric function value with the corresponding value (a)–(c).

$$\text{a. } \frac{\sqrt{3}}{2} \quad \text{b. } \frac{1}{2} \quad \text{c. } \frac{\sqrt{2}}{2}$$

47. $\sin 30°$ **48.** $\cos 30°$ **49.** $\cos 60°$

50. $\sin 60°$ **51.** $\sin 45°$ **52.** $\cos 45°$

Use the results in Exercises 47–52 and the trigonometric quotient identity, $\tan \theta = \dfrac{\sin \theta}{\cos \theta}$, to calculate the following values.

53. $\tan 30°$ **54.** $\tan 45°$ **55.** $\tan 60°$

Use the results in Exercises 47–55 and the reciprocal identities to calculate the following values.

56. $\csc 30°$ **57.** $\csc 45°$ **58.** $\csc 60°$

59. $\sec 30°$ **60.** $\sec 45°$ **61.** $\sec 60°$

62. $\cot 30°$ **63.** $\cot 45°$ **64.** $\cot 60°$

Use a calculator to approximate the following trigonometric function values. Round answers to four decimal places.

65. $\sin 42°$ **66.** $\cos 57°$ **67.** $\cos(17.3°)$

68. $\tan(25.2°)$ **69.** $\cot 33°$ **70.** $\sec(16.8°)$

71. $\csc(40.25°)$ **72.** $\cot(19.76°)$

Convert from degrees-minutes-seconds to decimal degrees. Round to the nearest hundredth if only minutes are given and to the nearest thousandth if seconds are given.

73. $39°17'$ **74.** $68°15'$

75. $29°30'25''$ **76.** $25°45'15''$

Convert from decimal degrees to degrees-minutes-seconds.

77. $42.25°$ round to the nearest minute

78. $60.45°$ round to the nearest minute

79. $30.175°$ round to the nearest second

80. $25.258°$ round to the nearest second

Use a calculator to approximate the following trigonometric function values. Round answers to four decimal places.

81. $\sin(37°15')$ **82.** $\cos(42°35')$

83. $\cos(61°48')$ **84.** $\sin(20°17')$

Applications

Light bends according to Snell's law, which states

$$n_i \sin(\theta_i) = n_r \sin(\theta_r)$$

■ n_i is the refractive index of the medium the light is leaving.

■ θ_i is the incident angle between the light ray and the normal (perpendicular) to the interface between mediums.

■ n_r is the refractive index of the medium the light is entering.

■ θ_r is the refractive angle between the light ray and the normal (perpendicular) to the interface between mediums.

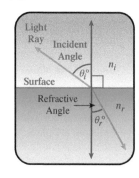

Calculate the index of refraction n_r of the indicated refractive medium given the following assumptions:

■ The incident medium is air.

■ Air has an index of refraction value of $n_i = 1.00$.

■ The incidence angle is $\theta_i = 60°$.

85. Optics. Glass, $\theta_r = 35.26°$

86. Optics. Glycerin, $\theta_r = 36.09°$

1.5 Solving Right Triangles

Use the right triangle diagram below and the information given to find the indicated measure. Write your answers for angle measures in decimal degrees.

87. $\beta = 25°$, $c = 15$ in., find a.

88. $\alpha = 50°$, $c = 27$ ft, find a.

89. $\alpha = 33.5°$, $b = 21.9$ mi, find a.

90. $\alpha = 47.45°$, $a = 19.22$ cm, find c.

91. $\beta = 37°45'$, $a = 120.0$ yd, find b.

92. $\beta = 75°10'$, $b = 96.5$ km, find c.

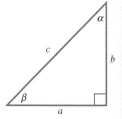

Use the right triangle diagram below and the information given to solve the right triangle. Write your answers for angle measures in decimal degrees.

93. $\alpha = 30°$ and $c = 21$ ft

94. $\beta = 65°$ and $c = 8.5$ mm

95. $\alpha = 48.5°$ and $a = 215$ mi

96. $\alpha = 30°15'$ and $b = 2154$ ft

97. $a = 30.5$ ft and $b = 45.7$ ft

98. $a = 11,798$ km and $c = 32,525$ km

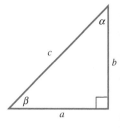

Applications

The illustration below shows a midair refueling scenario that our military aircraft often use. Assume the elevation angle that the hose makes with the plane being fueled is $\theta = 30°$.

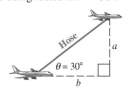

99. Midair Refueling. If the hose is 150 feet long, what should the altitude difference *a* be between the two planes?

100. Midair Refueling. If the smallest acceptable altitude difference *a* between the two planes is 100 feet, how long should the hose be?

101. Bearing (Navigation). If a plane takes off bearing N 15° W and flies 3 miles and then makes a right turn (90°) and flies 10 miles farther, what bearing will the traffic controller use to locate the plane?

102. Bearing (Navigation). If a plane takes off bearing N 20° E and flies 5 miles and then makes a left turn (90°) and flies 9 miles farther, what bearing will the traffic controller use to locate the plane?

Technology Exercises

Section 1.1

Assume a 30°-60°-90° triangle. Round your answers to two decimal places.

103. If the shorter leg has length 41.32 feet, what are the lengths of the other leg and hypotenuse?

104. If the longer leg has length 87.65 centimeters, what are the lengths of the other leg and the hypotenuse?

Section 1.4

105. Calculate csc(78.4°) in the following two ways:

  a. Find sin(78.4°) to three decimal places and then divide 1 by that number. Write that number to five decimal places.

  b. First find sin(78.4°) and then find its reciprocal. Round the result to five decimal places.

106. Calculate cot(34.8°) in the following two ways:

  a. Find tan(34.8°) to three decimal places and then divide 1 by that number. Write that number to five decimal places.

  b. First find tan(34.8°) and then find its reciprocal. Round the result to five decimal places.

107. Use a calculator to find $\tan[\tan^{-1}(2.612)]$.

108. Use a calculator to find $\cos[\cos^{-1}(0.125)]$.

Use a calculator to evaluate the following expressions. If you get an error, explain why.

109. sec180°   **110.** csc180°

[CHAPTER 1 PRACTICE TEST]

1. Calculate the measure of three angles in a triangle if the following are true:

   ■ The measure of the largest angle is five times the measure of the smallest angle.

   and

   ■ The larger of the two acute angles is three times the measure of the smallest angle.

2. In a right triangle, if the side opposite a 30° angle has a length of 5 centimeters, what is the length of the other leg and the hypotenuse?

3. A 5-foot girl is standing *in* the Grand Canyon, and she wants to estimate the height (depth) of the canyon. The sun casts her shadow 6 inches along the ground. To measure the shadow cast by the top of the canyon, she walks the length of the shadow. She takes 200 steps and estimates that each step is roughly 3 feet. Approximately how deep is the Grand Canyon?

For Exercises 4 and 5, use the triangle below:

4. Find the exact values for the indicated functions.
   a. $\sin\theta$ b. $\cos\theta$ c. $\tan\theta$
   d. $\sec\theta$ e. $\csc\theta$ f. $\cot\theta$

5. Find the exact values for the indicated functions.
   a. $\sin(90° - \theta)$
   b. $\sec(90° - \theta)$
   c. $\cot(90° - \theta)$

6. Fill in the exact values in the table below.

θ	SIN θ	COS θ	TAN θ	COT θ	SEC θ	CSC θ
30°						
45°						
60°						

7. Use a calculator to approximate $\sec(42.8°)$. Round your answer to four decimal places.

8. What is the difference between $\cos\theta = \frac{2}{3}$ and $\cos\theta \approx 0.66$?

9. Convert $33°45'20''$ to decimal degrees. Round to the appropriate decimal place.

For Exercises 10–15, refer to the triangle below:

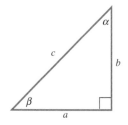

10. If $\alpha = 20°$ and $a = 10$ cm, find b.

11. If $a = 9.2$ km and $c = 23$ km, find β.

12. If $\beta = 50°$ and $a = 12$ mm, find c.

13. If $b = 13.3$ ft and $c = 14.0$ ft, find β.

14. If $\alpha = 33°10'$ and $c = 47$ m, find a.

15. If $a = 3.45$ and $b = 6.78$, find α.

16. What is the measure (in degrees) of the angle that a second hand sweeps in 5 seconds?

17. Light going from air to quartz crystal appears to bend according to Snell's law: $n_i \sin(\theta_i) = n_r \sin(\theta_r)$. Air has an index of refraction value of $n_i = 1.00$. If the incidence angle is 25°, $\theta_i = 25°$, and the refraction angle in the quartz is $\theta_r = 16°$, what is the index of refraction of quartz crystal?

18. If the search and rescue helicopter has a field of view of 40° and is flying at an altitude of 150 feet above sea level, what is the diameter of the circle that is illuminated on the surface of the water?

19. Convert 44.27° from decimal degrees to degrees-minutes-seconds.

20. Convert $22°10'23''$ from degrees-minutes-seconds to decimal degrees. Round to the nearest hundredth of a degree.

21. If $\sin 75° = \dfrac{\sqrt{6} + \sqrt{2}}{4}$, find the exact value of $\csc 75°$ and $\cos 15°$.

22. Find the exact value of $\sin 60° + \dfrac{\cos 45°}{\cot 30°}$.

23. Perform the indicated operation: $12°40' + 55°49'$.

24. Perform the indicated operation: $82°27' - 35°39'$.

25. The angle of elevation to the top of a building is 72° from a point 200 yards from the base of the building. How tall is the building?

Trigonometric Functions

Courtesy MGM Home Entertainment Distribution Corp.

Manny Millan/Getty Images, Inc.

One of the greatest basketball stories in history is immortalized in the movie *Hoosiers*. Gene Hackman stars as a coach of a men's basketball team from a tiny high school in the farmlands of Indiana in the 1950s that rose above all odds to win the state basketball championship. Their hometown gymnasium was very small. When the players arrived at the monstrous Hinkle Fieldhouse to play the state championship, they were in awe of the size of the arena. The coach made them measure the dimensions of the court and height of the rim. To their amazement, the measurements were the same as those of their court back home. It was as if their high school court had been picked up and placed in the large arena.

We will do the same thing with right triangles. We will superimpose them onto the Cartesian plane, which will allow us to arrive at a second definition of trigonometric functions in terms of ratios of *x*- and *y*-coordinates and distances. This new definition will allow us to find values of trigonometric functions for nonacute angles. Although the Cartesian plane might seem like the Hinkle Fieldhouse, remember that the hometown court (right triangles) lies within it.

LEARNING OBJECTIVES

- Sketch angles in the Cartesian plane.
- Define trigonometric functions as ratios of *x*- and *y*-coordinates and distances in the Cartesian plane.
- Evaluate trigonometric functions for nonacute angles.
- Apply basic trigonometric identities.

[IN THIS CHAPTER]

We will superimpose right triangles onto the Cartesian plane, which will allow us to define trigonometric functions as ratios of x- and y-coordinates. We will then be able to find values of trigonometric functions for both positive and negative angles as well as angles greater than or equal to $90°$ (nonacute angles). Finally, we will review some basic trigonometric identities and develop some new ones.

TRIGONOMETRIC FUNCTIONS

2.1 ANGLES IN THE CARTESIAN PLANE	**2.2** DEFINITION 2 OF TRIGONOMETRIC FUNCTIONS: THE CARTESIAN PLANE	**2.3** EVALUATING TRIGONOMETRIC FUNCTIONS FOR NONACUTE ANGLES	**2.4** BASIC TRIGONOMETRIC IDENTITIES
• Angles in Standard Position • Coterminal Angles	• Trigonometric Functions: The Cartesian Plane	• Algebraic Signs of the Trigonometric Functions • Ranges of the Trigonometric Functions • Reference Angles and Reference Right Triangles • Evaluating Trigonometric Functions for Nonacute Angles	• Reciprocal Identities • Quotient Identities • Pythagorean Identities

2.1 ANGLES IN THE CARTESIAN PLANE

SKILLS OBJECTIVES	CONCEPTUAL OBJECTIVES
■ Sketch angles in standard position. ■ Identify coterminal angles.	■ Understand that rotation in the counterclockwise direction corresponds to a positive angle measure, whereas rotation in a clockwise direction corresponds to a negative angle measure. ■ Understand that the measures of coterminal angles must differ by an integer multiple of 360°.

2.1.1 Angles in Standard Position

2.1.1 SKILL

Sketch angles in standard position.

2.1.1 CONCEPTUAL

Understand that rotation in the counterclockwise direction corresponds to a positive angle measure, whereas rotation in a clockwise direction corresponds to a negative angle measure.

In Section 1.1, we introduced angles. A common unit of measure for angles is degrees. We discussed triangles and the fact that the sum of the measures of the three interior angles of any triangle is always 180°. We also discussed the Pythagorean theorem, which relates the lengths of the three sides of a *right* triangle. Based on what we learned in Section 1.1, the following two angles, which correspond to a right angle or quarter rotation, both measure 90°.

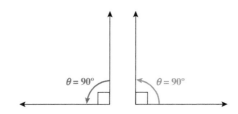

> **STUDY TIP**
>
> For a review of the rectangular (Cartesian) coordinate system, see Appendix A.2.

> **STUDY TIP**
>
> The initial side of an angle is a ray (initial ray), and the terminal side of an angle is also a ray (terminal ray).

We need a frame of reference. In this section, we use the Cartesian plane as our frame of reference by superimposing angles onto the *Cartesian coordinate system*, or *rectangular coordinate system* (Appendix A.2).

Let us now bridge concepts with which you are already familiar. First, recall that an angle is generated when a ray (the angle's *initial side*) is rotated around an endpoint (which becomes the angle's *vertex*). The ray in its new position after it is rotated is called the *terminal side* of the angle. Also recall the Cartesian (rectangular) coordinate system with the *x*-axis, *y*-axis, and origin. Using the *positive x*-axis combined with the origin as a frame of reference, we can graph angles in the Cartesian plane. If the *initial side* of the angle is aligned along the *positive x-axis* and the *vertex* of the angle is positioned at the *origin*, then the angle is said to be in *standard position*.

DEFINITION **Standard Position**

An angle is said to be in **standard position** if its initial side is along the positive *x*-axis and its vertex is at the origin.

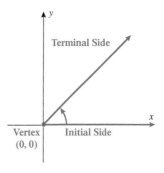

> **[CONCEPT CHECK]**
>
> Match the following: when the angle is formed using a clockwise/counterclockwise rotation that corresponds to an angle with positive/negative measure.
>
> ▼ ·
>
> **ANSWER** Clockwise—Negative
> Counterclockwise—Positive

We say that an angle lies in the quadrant in which its terminal side lies. For example, an acute angle $(0° < \theta < 90°)$ lies in **quadrant I**, whereas an obtuse angle $(90° < \theta < 180°)$ lies in quadrant II. Similarly, angles with measure $180° < \theta < 270°$ lie in quadrant III, and angles with measure $270° < \theta < 360°$ lie in **quadrant IV**. Angles in standard position with terminal sides along the x-axis or y-axis ($90°, 180°, 270°, 360°,$ and so on) are called **quadrantal angles**. An abbreviated way to represent an angle θ that lies in quadrant I is $\theta \in$ QI. Similarly, if an angle lies in quadrant II, we say $\theta \in$ QII, and so forth.

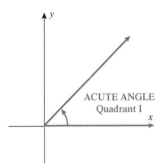

ACUTE ANGLE
Quadrant I

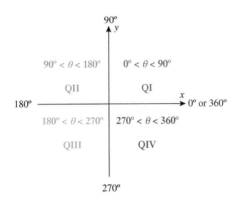

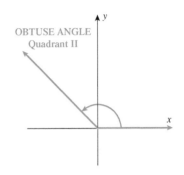

OBTUSE ANGLE
Quadrant II

Recall that rotation in a counterclockwise direction corresponds to a *positive angle*, whereas rotation in a clockwise direction corresponds to a *negative angle*. For example, an angle of measure $-92°$ lies in quadrant III and an angle of measure $-300°$ lies in quadrant I.

STUDY TIP

Counterclockwise rotation corresponds to a **positive angle**, and clockwise rotation corresponds to a **negative angle**.

EXAMPLE 1   **Sketching Angles in Standard Position**

Sketch the following angles in standard position, and state the quadrant in which (or the axis on which) the terminal side lies.

a. $-90°$ **b.** $210°$

Solution (a):

The initial side lies on the positive x-axis.

The negative angle indicates clockwise rotation.

$90°$ is a right angle.

The terminal side lies on the negative y-axis.

Solution (b):

The initial side lies on the positive x-axis.

The positive angle indicates counterclockwise rotation.

$180°$ represents a straight angle; an additional $30°$ yields a $210°$ angle.

The terminal side lies in quadrant III.

ANSWER

a.

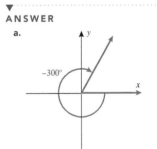

The terminal side lies in QI.

b.

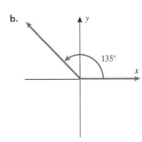

The terminal side lies in QII.

YOUR TURN   Sketch the following angles in standard position, and state the quadrant in which the terminal side lies.

a. $-300°$ **b.** $135°$

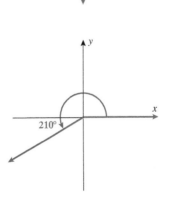

Common Angles in Standard Position

The common angles for which we determined the exact trigonometric function values in Chapter 1 are 30°, 45°, and 60°. Recall the relationships between the sides of 30°-60°-90° and 45°-45°-90° triangles.

Let us assume the hypotenuse is equal to 1. Then we have the following triangles:

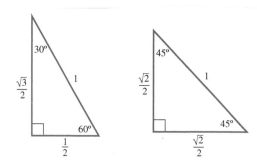

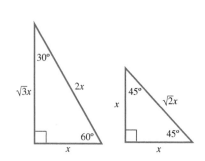

We can position these triangles on the Cartesian plane with one leg along the positive x-axis, so that we have three angles (30°, 45°, and 60°) in standard position. Remember that we are assuming that the hypotenuse is equal to 1. Notice that the x- and y-coordinates of the labeled points shown correspond to the side lengths.

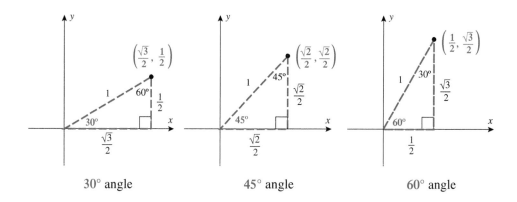

| 30° angle | 45° angle | 60° angle |

If we graph the three angles 30°, 45°, and 60° in the same Cartesian coordinate system, we get the following:

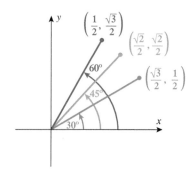

Using symmetry about the axes and the angles and coordinates in quadrant I (QI), we get the following angles and coordinates in quadrants II, III, and IV. Notice that all of these coordinate pairs satisfy the equation of the unit circle (radius equal to 1

whose center is at the origin): $x^2 + y^2 = 1$. In Chapter 3, we will define the values of the sine and cosine functions as the y- and x-coordinates along the unit circle, respectively.

STUDY TIP

Study the angles and point relationships in the diagram. Pay attention to the sign of the x- and y-coordinates in the respective quadrants.

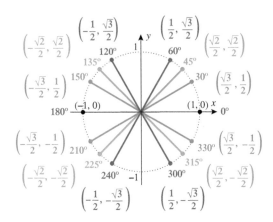

2.1.2 Coterminal Angles

DEFINITION	Coterminal Angles

Two angles in standard position with the same terminal side are called **coterminal angles**.

For example, $-40°$ and $320°$ are coterminal; their terminal rays are identical even though they are formed by rotation in opposite directions. The angles $60°$ and $420°$ are also coterminal; angles larger than $360°$ or less than $-360°$ are generated by continuing the rotation beyond a full rotation. All coterminal angles have the same initial side (positive x-axis) and the same terminal side, just different rotations.

2.1.2 SKILL

Identify coterminal angles.

2.1.2 CONCEPTUAL

Understand that the measures of coterminal angles must differ by an integer multiple of $360°$.

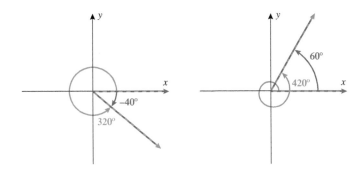

EXAMPLE 2	Recognizing Coterminal Angles

Determine whether the following pairs of angles are coterminal.

a. $\alpha = 120°, \beta = -180°$ **b.** $\alpha = 20°, \beta = 740°$

Solution (a):

The terminal side of α is in quadrant II.

The terminal side of β is along the negative x-axis.

Angles α and β are $\boxed{\text{not coterminal angles}}$.

[CONCEPT CHECK]

TRUE OR FALSE If an angle α has measure 50°, then the angle β that has measure 310° is coterminal with angle α.

ANSWER True

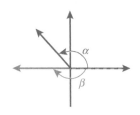

Solution (b):

Since 360° represents one rotation, 720° represents two rotations. Therefore, after 720° of rotation the angle is again along the positive *x*-axis. An additional 20° of rotation in the positive direction achieves a 740° angle that lies in quadrant I. Since $\alpha = 20°$ and $\beta = 740°$ have the same terminal side, they are ⎡coterminal angles⎤.

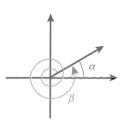

▼

────────────────────────────

▼

ANSWER

a. yes b. no

YOUR TURN Determine whether the following pairs of angles are coterminal.

a. $\alpha = 240°, \beta = -120°$ **b.** $\alpha = 20°, \beta = -380°$

⎡**STUDY TIP**

The measures of coterminal angles must differ by an integer multiple of 360°.⎤

The measures of coterminal angles must differ by an integer multiple of 360°. To find measures of the smallest positive coterminal angles, if the given angle is positive and greater than 360°, subtract 360° repeatedly until the result is a positive angle less than or equal to 360°. If the given angle is nonpositive, add 360° repeatedly until the result is a positive angle less than or equal to 360°.

▶ **EXAMPLE 3** **Finding Measures of Coterminal Angles**

Determine the angle of the smallest possible positive measure that is coterminal with each of the following angles.

a. 830° **b.** −520°

Solution (a):

Since 830° is positive, subtract 360°. $830° - 360° = 470°$

Subtract 360° again. $470° - 360° = 110°$

The angle with measure ⎡110°⎤ is the angle with the smallest positive measure that is coterminal with the angle of measure 830°.

Solution (b):

Since −520° is negative, add 360°. $-520° + 360° = -160°$

Add 360° again. $-160° + 360° = 200°$

The angle with measure ⎡200°⎤ is the angle with the smallest positive measure that is coterminal with the angle of measure −520°.

▼

────────────────────────────

▼

ANSWER

a. 180° b. 290°

YOUR TURN Determine the angle of the smallest possible positive measure that is coterminal with each of the following angles.

a. 900° **b.** −430°

▶ **[SECTION 2.1]** **SUMMARY**

An angle in the Cartesian plane is in standard position if its initial side lies along the positive *x*-axis and its vertex is located at the origin. Angles in standard position have terminal sides that lie either in one of the four quadrants or along one of the two axes. The special triangles, 30°-60°-90° and 45°-45°-90°, with hypotenuses having measure 1, were used to develop the coordinates in quadrant I for the special angles 30°, 45°, and 60°.

Symmetry was then used to locate similar pairs of coordinates in the other quadrants. All right triangles with one vertex (not the right angle vertex) located at the origin and hypotenuse equal to 1 have the other nonright angle vertex located along the unit circle $(x^2 + y^2 = 1)$. Coterminal angles are angles in standard position that have the same terminal side.

[SECTION 2.1] EXERCISES

• **SKILLS**

In Exercises 1–20, state in which quadrant or on which axis each of the following angles with given measure in standard position would lie.

1. 89°	**2.** 91°	**3.** 145°	**4.** 175°	**5.** 310°	**6.** 355°
7. 270°	**8.** 180°	**9.** −540°	**10.** −450°	**11.** 210.5°	**12.** 270.5°
13. 12.34°	**14.** 100.001°	**15.** 595°	**16.** 620°	**17.** 525°	**18.** 1085°
19. −905°	**20.** −640°				

In Exercises 21–32, sketch the angles with given measure in standard position.

21. 135°	**22.** 225°	**23.** −405°	**24.** −450°	**25.** −225°	**26.** −330°
27. 330°	**28.** −150°	**29.** 510°	**30.** −720°	**31.** 840°	**32.** −380°

In Exercises 33–38, match each of the angles (33–38) with its coterminal angle (a)–(f).

a. 30° **b.** −95° **c.** 185° **d.** −560° **e.** 780° **f.** 75°

33. −535°	**34.** −690°	**35.** 60°	**36.** 265°	**37.** −645°	**38.** 160°

In Exercises 39–50, determine the angle of the smallest possible positive measure that is coterminal with each of the following angles.

39. 412°	**40.** 379°	**41.** −92°	**42.** −187°	**43.** −390°	**44.** 945°
45. 510°	**46.** 1395°	**47.** 154°	**48.** 360°	**49.** −1050°	**50.** 2631°

• **APPLICATIONS**

51. Clock. What is the measure of the angle swept out by the second hand if it starts on the 3 and continues for 3 minutes and 20 seconds?

52. Clock. What is the measure of the angle swept out by the hour hand if it starts at 3 P.M. on Wednesday and continues until 5 P.M. on Thursday.

53. Tetherball. Joe and Alexandria are playing a game of tetherball. Alexandria begins the game and serves the ball counterclockwise. After traveling $3\frac{1}{2}$ revolutions, the ball is struck by Joe in a clockwise direction. If the path of the ball is modeled on a Cartesian plane with the initial position of the ball at 0°, at what angle is the ball 2 revolutions after Joe hits it?

54. Tetherball. If the game of tetherball described in Exercise 53 is won when one player hits the ball in his or her direction through 6 revolutions, through what angle must the ball be hit to win the game?

55. Track. Don and Ron both started running around a circular track, starting at the same point. But Don ran counterclockwise and Ron ran clockwise. The paths they ran swept through angles of 900° and −900°, respectively. Did they end up in the same spot when they finished?

56. Track. Dan and Stan both started running around a circular track, starting at the same point. The paths they ran swept through angles of 3640° and 1890°, respectively. Did they end up in the same spot when they finished?

For Exercises 57 and 58, refer to the following:

A common school locker combination lock is shown. The lock has a dial with forty calibration marks numbered 0 to 39. A combination

consists of three of these numbers (i.e., 5-35-20). To open the lock, the following steps are taken:

- Turn the dial clockwise two full turns.
- Continue turning clockwise until the first number of the combination is reached.
- Turn the dial counterclockwise one full turn.
- Continue turning counterclockwise until the second number is reached.
- Turn the dial clockwise again until the third number is reached.
- Pull the shank and the lock will open.

iStockphoto

57. Combination Lock. Given that the initial position of the dial is at zero (shown in the picture), how many degrees is the dial rotated in total (sum of clockwise and counterclockwise rotations) to open the lock if the combination is 35-5-20?

58. Combination Lock. Given that the initial position of the dial is at zero (shown in the picture), how many degrees is the dial rotated in total (sum of clockwise and counterclockwise rotations) to open the lock if the combination is 20-15-5?

59. Kite. Henri is flying a kite on the beach. He lets out 100 feet of string and has it flying at an angle of 60° to the ground. How far is the kite extended horizontally and vertically from Henri?

60. Kite. Camille flies her kite at an angle of 45° to the ground. If she has used 75 feet of string, how far is the kite extended horizontally and vertically from Camille?

61. Dartboard. If a dartboard is superimposed on a Cartesian plane, in what quadrant or on what axis does a dart land if its position is given by the point $(-3, 5)$?

62. Dartboard. If a dartboard is superimposed on a Cartesian plane, in what quadrant or on what axis does a dart land if its position is given by the point $(0, -2)$?

63. Ferris Wheel. The position of each car on a Ferris wheel, 200 feet in diameter, can be given in terms of its position on a Cartesian plane. If the Ferris wheel is centered at the origin and travels in a counterclockwise direction, through what angle has a car gone if it starts at $(100, 0)$ and stops at $(0, -100)$ and rotates through five full revolutions?

64. Ferris Wheel. A car on the Ferris wheel described in Exercise 63 starts at $(100, 0)$ and before completing one revolution is stopped at $(50, -50\sqrt{3})$. Through what angle has the car rotated?

• **CATCH THE MISTAKE**

In Exercises 65 and 66, explain the mistake that is made.

65. Find the angle with smallest positive measure that is coterminal with the angle with measure $-45°$ Assume that both angles are in standard position.

Solution:

| Coterminal angles are supplementary angles. | $-45° + \alpha = 180°$ |
| Add 45° to both sides. | $\alpha = 225°$ |

This is incorrect. What mistake was made?

66. In which quadrant (or axis) does the terminal side of an angle in standard position lie if the angle has measure $-1950°$?

Solution:

Keep adding 360°.	$-1950° + 360° = -1590°$
	$-1590° + 360° = -1230°$
	$-1230° + 360° = -870°$
	$-870° + 360° = -510°$
	$-510° + 360° = -150°$

The terminal side of the angle $-150°$ lies in quadrant II.

This is incorrect. What mistake was made?

• **CONCEPTUAL**

In Exercises 67–70, determine whether each statement is true or false.

67. The terminal sides of two coterminal angles must lie in the same quadrant or on the same axes.

68. An acute angle in standard position and an obtuse angle in standard position cannot be coterminal.

69. If the measures of two angles are $n°$ and $(-n)°$, then the angles are coterminal.

70. The difference in measure between two positive coterminal angles must be an integer multiple of 360°.

71. Write an expression for all angles coterminal with 30°.

72. Find the measure of the angle in standard position with all of the following characteristics:

■ Negative measure

■ Coterminal with the supplement of an angle with measure 130°

■ Less than one rotation

• **CHALLENGE**

73. Write an expression that represents all angles with negative measure that are coterminal with an angle that has measure 30°.

74. How many angles α that are coterminal to $-60°$ exist such that $-2000° < \alpha < 2000°$?

75. Use a right triangle to develop a formula for the distance between (x_1, y_1) and (x_2, y_2) on a Cartesian plane. *Hint:* Use the Pythagorean theorem.

76. In what quadrant is angle θ if $\theta = 4.5(360°) + 45°$?

2.2 DEFINITION 2 OF TRIGONOMETRIC FUNCTIONS: THE CARTESIAN PLANE

SKILLS OBJECTIVES	CONCEPTUAL OBJECTIVES
▪ Calculate trigonometric function values for nonacute and quadrantal angles.	▪ Understand that right triangle definitions of trigonometric functions for acute angles are consistent with definitions of trigonometric functions for all angles in the Cartesian plane.

2.2.1 Trigonometric Functions: The Cartesian Plane

In Chapter 1, we defined trigonometric functions as ratios of side lengths of right triangles. This definition holds only for acute $(0° < \theta < 90°)$ angles, since the two angles in a right triangle other than the right angle must be acute. In this chapter, we define trigonometric functions as ratios of x- and y-coordinates and distances in the Cartesian plane, which is consistent with right triangle trigonometry for acute angles. However, this second approach enables us to formulate trigonometric functions for quadrantal angles (whose terminal side lies along an axis) and for nonacute angles.

To define the trigonometric functions in the Cartesian plane, let us start with an acute angle θ in standard position. Choose any point (x, y) on the terminal side of the angle as long as it is not the vertex (the origin). A right triangle can be drawn so that the right angle is made when a perpendicular segment connects the point (x, y) to the x-axis. Notice that the side opposite θ has length y and the other leg of the right triangle has length x.

2.2.1 SKILL

Calculate trigonometric function values for nonacute and quadrantal angles.

2.2.1 CONCEPTUAL

Understand that right triangle definitions of trigonometric functions for acute angles are consistent with definitions of trigonometric functions for all angles in the Cartesian plane.

WORDS	MATH
The distance r from the origin $(0, 0)$ to the point (x, y) can be found using the distance formula:	$r = \sqrt{(x - 0)^2 + (y - 0)^2}$ $r = \sqrt{x^2 + y^2}$
Since r is a distance, it is always positive.	$r > 0$

Using our first definition of trigonometric functions in terms of right triangle ratios (Section 1.3), we say that $\sin\theta = \dfrac{\text{opposite}}{\text{hypotenuse}}$. From this picture we see that the sine function can also be defined by the relation $\sin\theta = \dfrac{y}{r}$. Similar reasoning holds for all six trigonometric functions and leads us to the second definition of the trigonometric functions, in terms of ratios of coordinates and distances in the Cartesian plane.

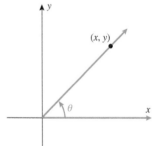

DEFINITION 2 Trigonometric Functions

Let (x, y) be any point, other than the origin, on the terminal side of an angle θ in standard position. Let r be the distance from the point (x, y) to the origin; then the six trigonometric functions are defined as

$$\sin\theta = \frac{y}{r} \qquad \cos\theta = \frac{x}{r} \qquad \tan\theta = \frac{y}{x}\,(x \neq 0)$$

$$\csc\theta = \frac{r}{y}\,(y \neq 0) \qquad \sec\theta = \frac{r}{x}\,(x \neq 0) \qquad \cot\theta = \frac{x}{y}\,(y \neq 0)$$

where $r = \sqrt{x^2 + y^2}$, or $x^2 + y^2 = r^2$.
The distance r is positive: $r > 0$.

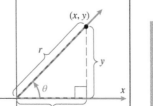

STUDY TIP

There is no need to memorize the definitions for secant, cosecant, and cotangent functions, since their values can be derived from the reciprocals of the sine, cosine, and tangent function values.

EXAMPLE 1 **Calculating Trigonometric Function Values for Acute Angles**

The terminal side of an angle θ in standard position passes through the point $(2, 5)$. Calculate the values of the six trigonometric functions for angle θ.

Solution:

STEP 1 **Draw the angle and label the point $(2, 5)$.**

STEP 2 **Calculate the distance r.**

$$r = \sqrt{2^2 + 5^2} = \sqrt{29}$$

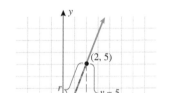

STEP 3 **Formulate the trigonometric functions in terms of x, y, and r.**

Let $x = 2$, $y = 5$, and $r = \sqrt{29}$.

$$\sin\theta = \frac{y}{r} = \frac{5}{\sqrt{29}} \qquad \cos\theta = \frac{x}{r} = \frac{2}{\sqrt{29}} \qquad \tan\theta = \frac{y}{x} = \frac{5}{2}$$

$$\csc\theta = \frac{r}{y} = \frac{\sqrt{29}}{5} \qquad \sec\theta = \frac{r}{x} = \frac{\sqrt{29}}{2} \qquad \cot\theta = \frac{x}{y} = \frac{2}{5}$$

STEP 4 **Rationalize any denominators containing a radical in the sine and cosine functions.**

$$\sin\theta = \frac{5}{\sqrt{29}} \cdot \frac{\sqrt{29}}{\sqrt{29}} = \frac{5\sqrt{29}}{29} \qquad \cos\theta = \frac{2}{\sqrt{29}} \cdot \frac{\sqrt{29}}{\sqrt{29}} = \frac{2\sqrt{29}}{29}$$

STEP 5 **Write the values of the six trigonometric functions for θ.**

$$\boxed{\sin\theta = \frac{5\sqrt{29}}{29}} \qquad \boxed{\cos\theta = \frac{2\sqrt{29}}{29}} \qquad \boxed{\tan\theta = \frac{5}{2}}$$

$$\boxed{\csc\theta = \frac{\sqrt{29}}{5}} \qquad \boxed{\sec\theta = \frac{\sqrt{29}}{2}} \qquad \boxed{\cot\theta = \frac{2}{5}}$$

Note: In Example 1, we could have used the values of the sine, cosine, and tangent functions along with the reciprocal identities to calculate the cosecant, secant, and cotangent function values.

▼
ANSWER

$$\sin\theta = \frac{7\sqrt{58}}{58} \qquad \csc\theta = \frac{\sqrt{58}}{7}$$

$$\cos\theta = \frac{3\sqrt{58}}{58} \qquad \sec\theta = \frac{\sqrt{58}}{3}$$

$$\tan\theta = \frac{7}{3} \qquad \cot\theta = \frac{3}{7}$$

▼
YOUR TURN The terminal side of an angle θ in standard position passes through the point $(3, 7)$. Calculate the values of the six trigonometric functions for angle θ.

We can now use this second definition of trigonometric functions to find values for any angle (acute or nonacute as well as negative angles).

▶ **EXAMPLE 2** **Calculating Trigonometric Function Values for Nonacute Angles**

The terminal side of an angle θ in standard position passes through the point $(-4, -7)$. Calculate the values of the six trigonometric functions for angle θ.

Solution:

STEP 1 **Draw the angle and label the point $(-4, -7)$.**

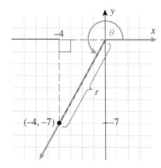

STEP 2 **Calculate the distance r.** $r = \sqrt{(-4)^2 + (-7)^2} = \sqrt{65}$

STEP 3 **Formulate the trigonometric functions in terms of x, y, and r.**

Let $x = -4$, $y = -7$, and $r = \sqrt{65}$.

$$\sin\theta = \frac{y}{r} = \frac{-7}{\sqrt{65}} \qquad \cos\theta = \frac{x}{r} = \frac{-4}{\sqrt{65}} \qquad \tan\theta = \frac{y}{x} = \frac{-7}{-4} = \frac{7}{4}$$

$$\csc\theta = \frac{r}{y} = \frac{\sqrt{65}}{-7} \qquad \sec\theta = \frac{r}{x} = \frac{\sqrt{65}}{-4} \qquad \cot\theta = \frac{x}{y} = \frac{-4}{-7} = \frac{4}{7}$$

STEP 4 **Rationalize any denominators that contain a radical in the sine and cosine functions.**

$$\sin\theta = \frac{y}{r} = \frac{-7}{\sqrt{65}} \cdot \frac{\sqrt{65}}{\sqrt{65}} = -\frac{7\sqrt{65}}{65}$$

$$\cos\theta = \frac{x}{r} = \frac{-4}{\sqrt{65}} \cdot \frac{\sqrt{65}}{\sqrt{65}} = -\frac{4\sqrt{65}}{65}$$

STEP 5 **Write the values of the six trigonometric functions for θ.**

$$\boxed{\sin\theta = -\frac{7\sqrt{65}}{65}} \qquad \boxed{\cos\theta = -\frac{4\sqrt{65}}{65}} \qquad \boxed{\tan\theta = \frac{7}{4}}$$

$$\boxed{\csc\theta = -\frac{\sqrt{65}}{7}} \qquad \boxed{\sec\theta = -\frac{\sqrt{65}}{4}} \qquad \boxed{\cot\theta = \frac{4}{7}}$$

Note: We could have used the values of the sine, cosine, and tangent functions along with the reciprocal identities to calculate the cosecant, secant, and cotangent functions.

▼

YOUR TURN   The terminal side of an angle θ in standard position passes through the point $(-3, 5)$. Calculate the values of the six trigonometric functions for angle θ.

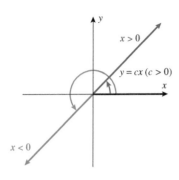

If we say that the terminal side of an angle lies on a line that passes through the origin, we must specify which part of the line contains the terminal ray to determine the angle. For example, for the line drawn in the diagram on the right, if we specify $x > 0$, then we know that the terminal side lies in quadrant I. Alternatively, if we specify $x < 0$, then we know that the terminal side lies in quadrant III.

Once we know which part of the line represents the terminal side of the angle, it does not matter which point on the line we use to formulate the trigonometric function values because corresponding sides of similar triangles are proportional.

EXAMPLE 3 **Calculating Trigonometric Function Values for Nonacute Angles**

Calculate the values for the six trigonometric functions of the angle θ, given in standard position, if the terminal side of θ lies on the line $y = 3x$, $x \le 0$.

Solution:

STEP 1 **Draw the line and label a point on the terminal side ($y = 3x$, $x \le 0$).**

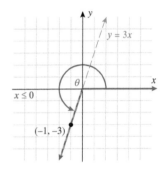

STEP 2 **Calculate the distance r.**
$$r = \sqrt{(-1)^2 + (-3)^2} = \sqrt{10}$$

STEP 3 **Formulate the trigonometric functions in terms of x, y, and r.**
Let $x = -1$, $y = -3$, and $r = \sqrt{10}$.

$$\sin\theta = \frac{y}{r} = \frac{-3}{\sqrt{10}} \qquad \cos\theta = \frac{x}{r} = \frac{-1}{\sqrt{10}} \qquad \tan\theta = \frac{y}{x} = \frac{-3}{-1} = 3$$

$$\csc\theta = \frac{r}{y} = \frac{\sqrt{10}}{-3} \qquad \sec\theta = \frac{r}{x} = \frac{\sqrt{10}}{-1} \qquad \cot\theta = \frac{x}{y} = \frac{-1}{-3} = \frac{1}{3}$$

STEP 4 **Rationalize radical denominators in the sine and cosine functions.**

$$\sin\theta = \frac{-3}{\sqrt{10}} \cdot \frac{\sqrt{10}}{\sqrt{10}} = -\frac{3\sqrt{10}}{10} \qquad \cos\theta = \frac{-1}{\sqrt{10}} \cdot \frac{\sqrt{10}}{\sqrt{10}} = -\frac{\sqrt{10}}{10}$$

STEP 5 **Write the values of the six trigonometric functions for θ.**

$$\boxed{\sin\theta = -\frac{3\sqrt{10}}{10}} \qquad \boxed{\cos\theta = -\frac{\sqrt{10}}{10}} \qquad \boxed{\tan\theta = 3}$$

$$\boxed{\csc\theta = -\frac{\sqrt{10}}{3}} \qquad \boxed{\sec\theta = -\sqrt{10}} \qquad \boxed{\cot\theta = \frac{1}{3}}$$

▼
ANSWER

$$\sin\theta = -\frac{2\sqrt{5}}{5} \qquad \csc\theta = -\frac{\sqrt{5}}{2}$$

$$\cos\theta = -\frac{\sqrt{5}}{5} \qquad \sec\theta = -\sqrt{5}$$

$$\tan\theta = 2 \qquad \cot\theta = \frac{1}{2}$$

▼

YOUR TURN Calculate the values for the six trigonometric functions of the angle θ given in standard position, if the terminal side of θ lies on the line $y = 2x$, $x \le 0$.

In Chapter 1, we avoided evaluating expressions such as $\sin 90°$ because $90°$ is not an acute angle. However, with our second definition of trigonometric functions in the Cartesian plane, we now are able to evaluate the trigonometric functions for *quadrantal angles* (angles in standard position whose terminal sides coincide with an axis) such as $90°$, $180°$, $270°$, and $360°$. Notice that $90°$ and $270°$ lie along the *y*-axis and therefore have an *x*-coordinate value equal to 0. Similarly, $180°$ and $360°$ lie along the *x*-axis and have a *y*-coordinate value equal to 0. Some of the trigonometric functions are defined with *x*- or *y*-coordinates in the denominator, and since dividing by 0 is undefined in mathematics, not all trigonometric functions are defined for quadrantal angles.

EXAMPLE 4 **Calculating Trigonometric Function Values for Quadrantal Angles**

Calculate the values for the six trigonometric functions when $\theta = 90°$.

Solution:

STEP 1 **Draw the angle and label a point on the terminal side.**

Note: A convenient point on the terminal side is $(0, 1)$.

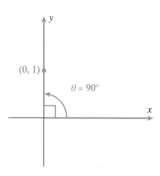

STEP 2 **Calculate the distance, *r*** $r = \sqrt{(0)^2 + (1)^2} = \sqrt{1} = 1$

STEP 3 **Formulate the trigonometric functions in terms of *x*, *y*, and *r*.**

Let $x = 0$, $y = 1$, and $r = 1$.

$$\sin\theta = \frac{y}{r} = \frac{1}{1} \qquad \cos\theta = \frac{x}{r} = \frac{0}{1} \qquad \tan\theta = \frac{y}{x} = \frac{1}{0}$$

$$\csc\theta = \frac{r}{y} = \frac{1}{1} \qquad \sec\theta = \frac{r}{x} = \frac{1}{0} \qquad \cot\theta = \frac{x}{y} = \frac{0}{1}$$

STEP 4 **Write the values of the six trigonometric functions for θ.**

| $\sin\theta = 1$ | $\cos\theta = 0$ | $\tan\theta$ is undefined |
| $\csc\theta = 1$ | $\sec\theta$ is undefined | $\cot\theta = 0$ |

Notice that $\csc\theta$, $\sec\theta$, and $\cot\theta$ could be found using the reciprocal identities, remembering that 0 has no reciprocal.

▼

YOUR TURN Calculate the values for the six trigonometric functions when $\theta = 270°$.

▼
ANSWER
$\sin\theta = -1$
$\cos\theta = 0$
$\tan\theta$ is undefined
$\csc\theta = -1$
$\sec\theta$ is undefined
$\cot\theta = 0$

EXAMPLE 5   **Calculating Trigonometric Function Values for Quadrantal Angles**

Calculate the values for the six trigonometric functions when $\theta = 180°$.

Solution:

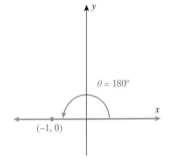

STEP 1 **Draw the angle and label a point on the terminal side.**

Note: A convenient point on the terminal side is $(-1, 0)$.

STEP 2 **Calculate the distance *r*.**

$$r = \sqrt{(-1)^2 + (0)^2} = \sqrt{1} = 1$$

STEP 3 **Formulate the trigonometric functions in terms of *x*, *y*, and *r*.**

Let $x = -1, y = 0,$ and $r = 1$.

$$\sin\theta = \frac{y}{r} = \frac{0}{1} \qquad \cos\theta = \frac{x}{r} = \frac{-1}{1} \qquad \tan\theta = \frac{y}{x} = \frac{0}{-1}$$

$$\csc\theta = \frac{r}{y} = \frac{1}{0} \qquad \sec\theta = \frac{r}{x} = \frac{1}{-1} \qquad \cot\theta = \frac{x}{y} = \frac{-1}{0}$$

STEP 4 **Write the values of the six trigonometric functions for θ.**

$\sin\theta = 0$	$\cos\theta = -1$	$\tan\theta = 0$
$\csc\theta$ is undefined	$\sec\theta = -1$	$\cot\theta$ is undefined

ANSWER

$\sin\theta = 0$
$\cos\theta = 1$
$\tan\theta = 0$
$\csc\theta$ is undefined
$\sec\theta = 1$
$\cot\theta$ is undefined

YOUR TURN Calculate the values for the six trigonometric functions when $\theta = 360°$.

The following table summarizes the trigonometric function values for the quadrantal angles $0°, 90°, 180°, 270°,$ and $360°$.

θ	SIN θ	COS θ	TAN θ	COT θ	SEC θ	CSC θ
0°	0	1	0	undefined	1	undefined
90°	1	0	undefined	0	undefined	1
180°	0	-1	0	undefined	-1	undefined
270°	-1	0	undefined	0	undefined	-1
360°	0	1	0	undefined	1	undefined

▶[SECTION 2.2]   **SUMMARY**

Trigonometric functions are defined in the Cartesian plane as ratios of coordinates and distances.

$$\sin\theta = \frac{y}{r} \qquad \cos\theta = \frac{x}{r} \qquad r = \sqrt{x^2 + y^2}$$

Right triangle trigonometric definitions learned in Chapter 1 are consistent with these definitions. We now have the ability to evaluate trigonometric functions for nonacute angles. Trigonometric functions are not always defined for quadrantal angles.

[SECTION 2.2]   EXERCISES

• **SKILLS**

In Exercises 1–16, the terminal side of an angle θ in standard position passes through the indicated point. Calculate the values of the six trigonometric functions for angle θ.

1. $(1, 2)$ **2.** $(2, 3)$ **3.** $(3, 6)$ **4.** $(8, 4)$

5. $\left(\frac{1}{2}, \frac{2}{5}\right)$ **6.** $\left(\frac{4}{7}, \frac{2}{3}\right)$ **7.** $(-2, 4)$ **8.** $(-1, 3)$

9. $(-4, -7)$ **10.** $(-9, -5)$ **11.** $(-\sqrt{2}, \sqrt{3})$ **12.** $(-\sqrt{3}, \sqrt{2})$

13. $(-\sqrt{5}, -\sqrt{3})$ **14.** $(-\sqrt{6}, -\sqrt{5})$ **15.** $\left(-\frac{10}{3}, \frac{4}{3}\right)$ **16.** $\left(-\frac{2}{9}, -\frac{1}{3}\right)$

In Exercises 17–24, calculate the values for the six trigonometric functions of the angle θ given in standard position, if the terminal side of θ lies on the given line.

17. $y = 2x \quad x \geq 0$ **18.** $y = 3x \quad x \geq 0$ **19.** $y = \frac{1}{2}x \quad x \geq 0$ **20.** $y = \frac{1}{2}x \quad x \leq 0$

21. $y = -\frac{1}{3}x \quad x \geq 0$ **22.** $y = -\frac{1}{3}x \quad x \leq 0$ **23.** $2x + 3y = 0 \quad x \leq 0$ **24.** $2x + 3y = 0 \quad x \geq 0$

In Exercises 25–40, calculate (if possible) the values for the six trigonometric functions of the angle θ given in standard position.

25. $\theta = 450°$ **26.** $\theta = 540°$ **27.** $\theta = 630°$ **28.** $\theta = 720°$

29. $\theta = -270°$ **30.** $\theta = -180°$ **31.** $\theta = -90°$ **32.** $\theta = -360°$

33. $\theta = -450°$ **34.** $\theta = -540°$ **35.** $\theta = -630°$ **36.** $\theta = -720°$

37. $\theta = 810°$ **38.** $\theta = 900°$ **39.** $\theta = -810°$ **40.** $\theta = -900°$

• **APPLICATIONS**

41. Geometry. A right triangle is drawn in quadrant I with one leg on the x-axis and its hypotenuse on the terminal side of $\angle\theta$ drawn in standard position. If $\sin\theta = \frac{7}{25}$, then what is the value of $\tan\theta$?

42. Geometry. A right triangle is drawn in quadrant I with one leg on the x-axis and its hypotenuse on the terminal side of $\angle\theta$ drawn in standard position. If $\tan\theta = \frac{84}{13}$, then what is the value of $\cos\theta$?

In Exercises 43 and 44, refer to the following figure.

43. Angle of Elevation. Let $\angle\theta$ be the angle of elevation from a point on the ground to the top of a tree. If $\cos\theta = \frac{11}{61}$ and the distance from the point on the ground to the base of the tree is 22 feet, then how high is the tree?

44. Angle of Elevation. Let $\angle\theta$ be the angle of elevation from a point on the ground to the top of a tree. If $\sin\theta = \frac{40}{41}$ and the tree is 20 feet high, then how far from the base of the tree is the point on the ground?

45. Flight. An airplane takes off and flies toward the north and east such that for each mile it travels east, it travels 4 miles north. If θ is the angle formed by east and the path of the plane, find $\sin\theta$, $\cos\theta$, and $\tan\theta$.

46. Flight. An airplane takes off and flies toward the north and west such that for each mile it travels west, it travels $\frac{1}{2}$ mile north. If θ is the angle formed by due west and the path of the plane, find $\sin\theta$, $\cos\theta$, and $\tan\theta$.

47. Volleyball. Angelina spikes a volleyball such that if θ is the angle of depression for the path of the ball (the angle the path of the ball makes with the ground), then $\cos\theta = \frac{9}{10}$. If the ball is hit from a height of 9.1 feet, how far does the ball travel (along the path of the ball) before hitting the ground?

48. Volleyball. Angelina spikes a volleyball such that if θ is the angle of depression for the path of the ball (the angle the path of the ball makes with the ground), then $\sec\theta = \frac{6}{5}$. If the ball is hit from a height of 8.9 feet, how far does the ball travel (along the path of the ball) before hitting the ground?

For Exercises 49 and 50, refer to the following:

The monthly revenues (measured in thousands of dollars) of PizzaRia are a function of monthly costs (measured in thousands of dollars), that is, $R(c)$. Recall that the angle θ can be interpreted as a measure of the sizes of cost c and revenue R relative to each other; that is, $\tan\theta = \dfrac{\text{revenue}}{\text{costs}}$. The larger the angle θ is, the greater revenue is relative to cost, and conversely; the smaller the angle θ is, the smaller revenue is relative to cost.

49. Business. An analysis of a month's revenue and costs indicates $\sin\theta = \frac{8}{10}$. Determine whether the company experiences a loss or profit for that month.

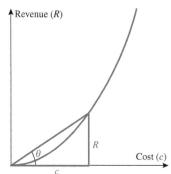

50. Business. An analysis of a month's revenue and costs indicates $\cos\theta = \frac{3}{4}$. Determine whether the company experiences a loss or profit for that month.

For Exercises 51 and 52, refer to the following:

A bunion is a progressive medical disorder of the foot in which the big toe gradually begins to lean toward the second toe. The condition can lead to a misalignment of the metatarsal and phalanges of the big toe when compared to the metatarsal and phalanges of the second toe. The angle θ formed between the metatarsal of the big toe and the metatarsal of the second toe is called the first intermetatarsal angle and is normally less than 9°. In the presence of a bunion, the intermetatarsal angle exceeds 9°, which may be caused by the length of the first metatarsal bone. Exercises 51 and 52 both correspond to the presence of a bunion.

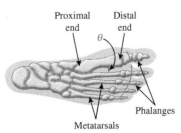

(*Source*: http://emedicine.medscape.com/article/1235796-media.)

51. Health/Medicine. If $\cos\theta = \dfrac{4.8 \text{ cm}}{5.0 \text{ cm}}$, find the distance between the distal ends of the two metatarsals. Round to the nearest millimeter.

52. Health/Medicine. If $\sec\theta = \dfrac{5.0 \text{ cm}}{4.9 \text{ cm}}$, find the distance between the distal ends of the two metatarsals. Round to the nearest millimeter.

• CATCH THE MISTAKE

In Exercises 53 and 54, explain the mistake that is made.

53. The terminal side of an angle θ, in standard position, passes through the point $(1, 2)$. Calculate $\sin\theta$.

Solution:

Label the coordinates.	$x = 1, y = 2$
Calculate r.	$r = 1^2 + 2^2 = 5$
Use the definition of sine.	$\sin\theta = \dfrac{y}{r}$
Substitute $y = 2, r = 5$.	$\sin\theta = \dfrac{2}{5}$

This is incorrect. What mistake was made?

54. Evaluate $\sec 810°$ exactly.

Solution:

Find a coterminal angle that lies between 0° and 360°.	$810° - 360° = 450°$
810° has the same terminal side as 90°.	$450° - 360° = 90°$
Evaluate cosine for 90°.	$\cos 90° = 0$

Evaluate secant for 90°. $\quad \sec 90° = \dfrac{1}{\cos 90°} = \dfrac{1}{0} = 0$

This is incorrect. What mistake was made?

• CONCEPTUAL

In Exercises 55–58, determine whether each statement is true or false.

55. $\sin 30° + \sin 60° = \sin 90°$

56. $\tan 0° + \tan 90° = \tan 90°$

57. $\cos\theta = \cos(\theta + 360°n)$, where n is an integer

58. $\sin\theta = \sin(\theta + 360°n)$, where n is an integer

59. If the terminal side of angle θ passes through the point $(-3a, 4a)$, find $\cos\theta$.

60. If the terminal side of angle θ passes through the point $(-3a, 4a)$, find $\sin\theta$.

• CHALLENGE

61. If the line $y = mx$ makes an angle θ with the x-axis, find the slope m in terms of a single trigonometric function.

62. If $(x, -2)$ is a point on the terminal side of angle θ and $\csc\theta = -\dfrac{\sqrt{29}}{2}$, find x.

63. Find the equation of the line with *positive* slope that passes through the point $(a, 0)$ and makes an acute angle θ with the x-axis. The equation of the line will be in terms of x, a, and a trigonometric function of θ.

64. Find the equation of the line with *negative* slope that passes through the point $(a, 0)$ and makes an acute angle θ with the x-axis. The equation of the line will be in terms of x, a, and a trigonometric function of θ.

65. Write an expression for all values of x for which $\sec x$ is undefined.

66. Given that $\csc x$ is undefined, what is the value of $\sin x$? Explain.

In Exercises 67–76, use a calculator to evaluate the following expressions. If you get an error, explain why.

67. $\sin 270°$ **68.** $\cos 270°$ **69.** $\tan 270°$ **70.** $\cot 270°$ **71.** $\cos(-270°)$

72. $\sin(-270°)$ **73.** $\sec(-270°)$ **74.** $\sec 270°$ **75.** $\csc 270°$ **76.** $\csc(-270°)$

2.3 EVALUATING TRIGONOMETRIC FUNCTIONS FOR NONACUTE ANGLES

SKILLS OBJECTIVES	CONCEPTUAL OBJECTIVES
▪ Determine the reference angle of a nonacute angle. ▪ Determine the ranges of the six trigonometric functions. ▪ Determine the reference angle for any angle whose terminal side lies in one of the four quadrants. ▪ Evaluate trigonometric functions for nonacute and quadrantal angles.	▪ Understand which trigonometric functions are positive or negative in each of the four quadrants. ▪ Understand why sine and cosine range between −1 and 1. ▪ Understand that reference angles are acute and have positive measure. ▪ Understand that the value of a trigonometric function at an angle is the same as the trigonometric value of its reference angle, except there may be a difference in its algebraic sign (+/−).

In Section 2.2, we defined trigonometric functions in the Cartesian plane as ratios of x, y, and r. We calculated the trigonometric function values given a point on the terminal side of the angle in standard position. We also discussed the special case of a quadrantal angle. In this section, we now evaluate trigonometric functions for common angles (angles with reference angles of $30°$, $45°$, and $60°$ and quadrantal angles) exactly. Additionally, we will approximate trigonometric functions for nonacute angles using an acute reference angle and knowledge of algebraic signs of trigonometric functions in particular quadrants.

2.3.1 Algebraic Signs of the Trigonometric Functions

In Section 2.2, we defined trigonometric functions as ratios of x, y, and r in the Cartesian plane. Since r is the distance from the origin to the point (x, y) and distance is never negative, r is always taken as the positive solution to $r^2 = x^2 + y^2$, so $r = \sqrt{x^2 + y^2}$.

The x-coordinate is positive in quadrants **I** and **IV** and negative in quadrants **II** and **III**.

The y-coordinate is positive in quadrants **I** and **II** and negative in quadrants **III** and **IV**.

Recall the definition of the six trigonometric functions in the Cartesian plane:

$$\sin\theta = \frac{y}{r} \qquad \cos\theta = \frac{x}{r} \qquad \tan\theta = \frac{y}{x}\,(x \neq 0)$$

$$\csc\theta = \frac{r}{y}\,(y \neq 0) \qquad \sec\theta = \frac{r}{x}\,(x \neq 0) \qquad \cot\theta = \frac{x}{y}\,(y \neq 0)$$

Therefore, the algebraic sign, + or −, of each trigonometric function will depend on the respective signs of x and y, which is determined by which quadrant contains the terminal side of angle θ. Let us look at the three main trigonometric functions: sine, cosine, and tangent. In quadrant I, all three functions are positive since x, y, and r are all positive. However, in quadrant II, only sine is positive since y and r are both

2.3.1 SKILL

Determine the reference angle of a nonacute angle.

2.3.1 CONCEPTUAL

Understand which trigonometric functions are positive or negative in each of the four quadrants.

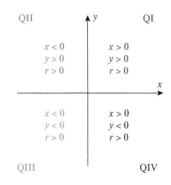

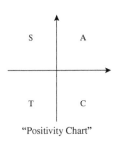

"Positivity Chart"

CONCEPT CHECK

In what quadrant are sine and cosine functions both negative?

▼ ....

ANSWER III

positive. In quadrant III, only tangent is positive, and in quadrant IV, only cosine is positive. The phrase "**A**ll **S**tudents **T**ake **C**alculus" helps us remember which of the main three trigonometric functions are positive in each quadrant. (A—all, S—sine, T—tangent, C—cosine corresponding to quadrants I through IV, respectively.)

PHRASE	QUADRANT	POSITIVE TRIGONOMETRIC FUNCTION
All	I	All three: sine, cosine, and tangent
Students	II	Sine
Take	III	Tangent
Calculus	IV	Cosine

The following table indicates the algebraic sign of all six trigonometric functions according to the quadrant in which the terminal side of an angle θ lies. Remember that reciprocal function pairs will have the same sign.

TERMINAL SIDE OF θ IN QUADRANT	SIN θ	COS θ	TAN θ	COT θ	SEC θ	CSC θ
I	+	+	+	+	+	+
II	+	−	−	−	−	+
III	−	−	+	+	−	−
IV	−	+	−	−	+	−

EXAMPLE 1 **Evaluating a Trigonometric Function When the Quadrant of the Terminal Side Is Known**

If $\cos\theta = -\frac{3}{5}$ and the terminal side of θ lies in quadrant III, find $\sin\theta$.

Solution:

STEP 1 **Draw some angle θ in quadrant III.**

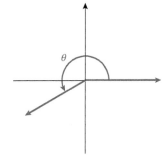

STEP 2 **Identify known quantities from the information given.**

Recall that $\cos\theta = \frac{x}{r}$ and $r > 0$. $\cos\theta = -\frac{3}{5} = \frac{-3}{5} = \frac{x}{r}$

Identify x and r. $x = -3$ and $r = 5$

STEP 3 **Since x and r are known, find y.**

Substitute $x = -3$ and $r = 5$ into $x^2 + y^2 = r^2$.
$$(-3)^2 + y^2 = 5^2$$

Solve for y.
$$9 + y^2 = 25$$
$$y^2 = 16$$
$$y = \pm 4$$

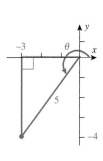

STEP 4 **Select the sign of y based on quadrant information.**

Since the terminal side of θ lies
in quadrant III, $y < 0$. $\qquad\qquad\qquad y = -4$

STEP 5 **Find $\sin\theta$.** $\qquad\qquad\qquad \sin\theta = \dfrac{y}{r} = \dfrac{-4}{5}$

$$\boxed{\sin\theta = -\dfrac{4}{5}}$$

▼

YOUR TURN If $\sin\theta = -\frac{3}{4}$ and the terminal side of θ lies in quadrant III, find $\cos\theta$.

▼ ANSWER

$$\cos\theta = -\dfrac{\sqrt{7}}{4}$$

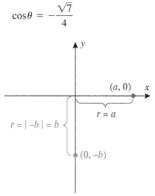

We can also make a table showing the values of the trigonometric functions when the terminal side of θ lies along each of the axes (i.e., when θ is any of the quadrantal angles). We calculated this table for specific angles in Section 2.2, and we will develop it again here using the algebraic signs of the trigonometric functions in the different quadrants.

When the terminal side lies along the x-axis, then $y = 0$. When $y = 0$, notice that $r = \sqrt{x^2 + y^2} = \sqrt{x^2} = |x|$. When the terminal side lies along the positive x-axis, $x > 0$, then $r = x$, which results in the cosine function being equal to 1. When the terminal side lies along the negative x-axis, $x < 0$, then $r = -x$, which results in the cosine function being equal to -1. A similar argument can be made for the y-axis that results in the sine function being equal to ± 1.

TERMINAL SIDE OF θ LIES ALONG THE . . .	SIN θ	COS θ	TAN θ	COT θ	SEC θ	CSC θ
Positive x-axis ($0°$ or $360°$)	0	1	0	undefined	1	undefined
Positive y-axis ($90°$)	1	0	undefined	0	undefined	1
Negative x-axis ($180°$)	0	-1	0	undefined	-1	undefined
Negative y-axis ($270°$)	-1	0	undefined	0	undefined	-1

EXAMPLE 2 **Working with Values of the Trigonometric Functions for Quadrantal Angles**

Evaluate the following expressions, if possible.
a. $\cos 540° + \sin 270°$ **b.** $\cot 90° + \tan(-90°)$

Solution (a):

The terminal side of an angle with measure $540°$ lies along the negative x-axis. $\qquad\qquad 540° - 360° = 180°$

Note: A convenient point on the terminal side is $(-1, 0)$.

Evaluate cosine of an angle whose terminal side lies along the negative x-axis. $\qquad\qquad \cos 540° = -1$

Note: $r = \sqrt{(-1)^2 + 0^2} = 1$. $\qquad\qquad \dfrac{x}{r} = \dfrac{-1}{1} = -1$

Evaluate sine of an angle whose terminal side lies along the negative y-axis. $\qquad\qquad \sin 270° = -1$

Note: A convenient point on the terminal side is $(0, -1)$. $\qquad\qquad \dfrac{y}{r} = \dfrac{-1}{1} = -1$

Sum the cosine and sine terms. $\qquad\qquad \cos 540° + \sin 270° = -1 + (-1)$

$$\cos 540° + \sin 270° = \boxed{-2}$$

Solution (b):

Evaluate cotangent of an angle whose terminal
side lies along the positive y-axis.

$$\cot 90° = 0$$

Note: A convenient point on the terminal side is $(0, 1)$.

$$\frac{x}{y} = \frac{0}{1} = 0$$

The terminal side of an angle with measure
$-90°$ lies along the negative y-axis.

$$\tan(-90°) = \tan 270°$$

Note: A convenient point on the teminal side is $(0, -1)$.

$$\frac{y}{x} = \frac{-1}{0}$$

The tangent function is undefined for an angle
whose terminal side lies along the negative y-axis.

$\tan(-90°)$ is $\boxed{\text{undefined}}$

Even though $\cot 90°$ is defined, since $\tan(-90°)$ is undefined, the sum of the two
expressions is also undefined.

▼
ANSWER
a. 0   b. undefined

▼
YOUR TURN   Evaluate the following expressions, if possible.

a. $\csc 90° + \sec 180°$ **b.** $\csc(-630°) + \sec(-630°)$

2.3.2 Ranges of the Trigonometric Functions

2.3.2 SKILL

Determine the ranges of the six
trigonometric functions.

2.3.2 CONCEPTUAL

Understand why sine and cosine
range between -1 and 1.

Thus far, we have discussed what the algebraic sign of a trigonometric function value is
for an angle in a particular quadrant, but we haven't discussed how to find actual values
of the trigonometric functions for nonacute angles. We will need to define reference
angles and reference right triangles. However, before we proceed, let's get a feel for the
ranges (set of values) of the functions we will expect.

Let us start with an angle θ in quadrant I and the sine function defined as the ratio:

$$\sin\theta = \frac{y}{r}$$

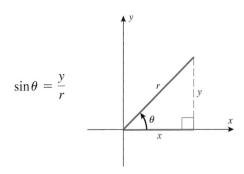

We keep the value of r constant; then as the measure of θ increases toward $90°$, y
increases. Notice that the value of y approaches the value of r until they are equal when
$\theta = 90°$, and y can never be larger than r.

[CONCEPT CHECK]

The value of r is always (greater
than/less than) the values of both
x and y.

▼
ANSWER Greater than

A similar analysis can be conducted in quadrant IV as θ approaches $-90°$ from $0°$ (note that y is negative in quadrant IV). A result that is valid in all four quadrants is $|y| \leq r$.

WORDS	MATH
Write $|y| \leq r$ as a double inequality.	$-r \leq y \leq r$
Divide the entire inequality by r.	$-1 \leq \dfrac{y}{r} \leq 1$
Substitute $\sin\theta = \dfrac{y}{r}$.	$-1 \leq \sin\theta \leq 1$

Similarly, by allowing θ to approach $0°$ from $-90°$ and θ to approach $180°$ from $90°$, we can show that $|x| \leq r$, which leads to the range of the cosine function: $-1 \leq \cos\theta \leq 1$ using a similar mathematical argument as for $\sin\theta$. Sine and cosine values range between -1 and 1, and since secant and cosecant are reciprocals of the cosine and sine functions, respectively, their ranges are stated as

$$\sec\theta \leq -1 \text{ or } \sec\theta \geq 1 \qquad \csc\theta \leq -1 \text{ or } \csc\theta \geq 1$$

Since $\tan\theta = \dfrac{y}{x}$ and $\cot\theta = \dfrac{x}{y}$ and since $x < y, x = y,$ or $x > y$ are all possible, the values of the tangent and cotangent functions can be any real numbers (positive or negative or zero) and very large or very small in absolute value. The following box summarizes the ranges of the trigonometric functions.

RANGES OF THE TRIGONOMETRIC FUNCTIONS

For any angle θ where the trigonometric functions are defined, the six trigonometric functions have the following ranges:

- $\sin\theta$: $[-1, 1]$
- $\csc\theta$: $(-\infty, -1] \cup [1, \infty)$
- $\cos\theta$: $[-1, 1]$
- $\sec\theta$: $(-\infty, -1] \cup [1, \infty)$
- $\tan\theta$: $(-\infty, \infty)$
- $\cot\theta$: $(-\infty, \infty)$

Notice that the secant and cosecant functions have the range $(-\infty, -1] \cup [1, \infty)$. It is not possible for $\sec\theta = \dfrac{1}{\cos\theta}$ and $\csc\theta = \dfrac{1}{\sin\theta}$ to correspond to values strictly between -1 and 1 because $\sin\theta$ and $\cos\theta$ cannot be less than -1 or greater than 1.

▶ **EXAMPLE 3** **Determining Whether a Value Is Within the Range of a Trigonometric Function**

Determine whether each statement is possible or not possible.

a. $\cos\theta = 1.001$ **b.** $\cot\theta = 0$ **c.** $\sec\theta = \dfrac{\sqrt{3}}{2}$

Solution (a): It is not possible because $-1 \leq \cos\theta \leq 1$.

Solution (b): It is possible because $\cot 90° = 0$.

Solution (c): It is not possible because $-1 < \dfrac{\sqrt{3}}{2} \approx 0.866 < 1$ and $\sec\theta \leq -1$ or $\sec\theta \geq 1$.

▼

YOUR TURN Determine whether each statement is possible or not possible.

a. $\sin\theta = -1.1$ **b.** $\tan\theta = 2$ **c.** $\csc\theta = \sqrt{3}$

▼
ANSWER
a. not possible
b. possible
c. possible

2.3.3 Reference Angles and Reference Right Triangles

Now that we know the trigonometric function ranges and their algebraic signs in each of the four quadrants, we can evaluate the trigonometric functions for nonacute angles. Before we do that, however, we must first discuss *reference angles* and *reference right triangles*.

Recall the story in the chapter opener about the small-town basketball team that played the state championship in the enormous arena. As it turned out, their small-town gymnasium court had the same measurements as the court in the large arena. It was as if their hometown court had been picked up and placed inside the arena. Right triangle trigonometry (Chapter 1) is our hometown court, and the Cartesian plane with nonacute angles is the large arena.

Every nonquadrantal angle in standard position has a corresponding *reference angle* and *reference right triangle*. We have already calculated the trigonometric function values for quadrantal angles.

DEFINITION | **Reference Angle**

For any angle θ, where $0° < \theta < 360°$, in standard position whose terminal side lies in one of the four quadrants, there exists a **reference angle** α that is the acute angle with positive measure formed by the terminal side of θ and the *x*-axis.

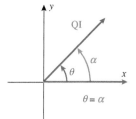

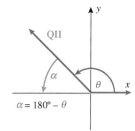

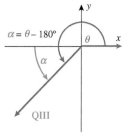

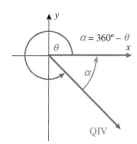

In the above formulas for α, if θ is less than 0°, or greater than 360°, find a corresponding coterminal angle that lies strictly between 0° and 360°.

The reference angle is the positive acute angle that the terminal side makes with the *x*-axis.

DEFINITION | Reference Right Triangle

To form a **reference right triangle** for angle θ, where $0° < \theta < 360°$, drop a perpendicular from the terminal side of the angle to the x-axis. The right triangle now has the reference angle α as one of its angles.

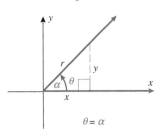

$\theta = \alpha$

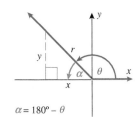

$\alpha = 180° - \theta$

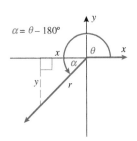

$\alpha = \theta - 180°$

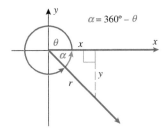

$\alpha = 360° - \theta$

EXAMPLE 4 | **Finding Reference Angles**

Find the reference angle for each angle given.

a. 210°   **b.** 135°   **c.** 422°

Solution (a):

The terminal side of θ lies in quadrant III.

The reference angle is made by the terminal side and the negative x-axis.

$210° - 180° = \boxed{30°}$

[CONCEPT CHECK]

Which of the following is true about the reference angle?
(A) It is the angle made by the terminal side and the x-axis (or negative x-axis);
(B) it is positive;
(C) it is an acute angle; or
(D) all of the above.

▼ ....

ANSWER (D) all of the above

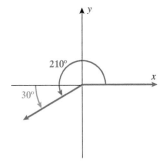

Solution (b):

The terminal side of θ lies in quadrant II.

The reference angle is made by the terminal side and the negative x-axis.

$180° - 135° = \boxed{45°}$

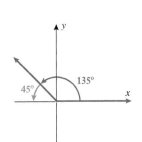

Solution (c):

The terminal side of θ lies in quadrant I.

The reference angle is made by the terminal side and the positive x-axis.

$422° - 360° = \boxed{62°}$

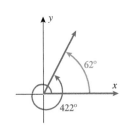

▼

YOUR TURN Find the reference angle for each angle given.

a. 160°   **b.** 285°   **c.** 600°

▼
ANSWER

a. 20°   **b.** 75°   **c.** 60°

In Section 1.3, we first defined the trigonometric functions of an acute angle as ratios of lengths of sides of a right triangle. For example, $\sin\theta = \dfrac{\text{opposite}}{\text{hypotenuse}}$. The lengths of the sides of triangles are always positive.

In Section 2.2, we defined the sine function of any angle as $\sin\theta = \dfrac{y}{r}$. Notice in the definition of a reference right triangle that for a nonacute angle θ, $\sin\theta = \dfrac{y}{r}$; and for the acute reference angle α, $\sin\alpha = \dfrac{|y|}{r}$. The only difference between these two expressions is the algebraic sign of y.

Therefore, to calculate the trigonometric function values for a nonacute angle, simply find the trigonometric function values for the reference angle and determine the correct algebraic sign according to the quadrant in which the terminal side lies.

2.3.4 Evaluating Trigonometric Functions for Nonacute Angles

2.3.4 SKILL

Evaluate trigonometric functions for nonacute and quadrantal angles

2.3.4 CONCEPTUAL

Understand that the value of a trigonometric function at an angle is the same as the trigonometric value of its reference angle, except there may be a difference in its algebraic sign $(+/-)$.

Let's look at a specific example before we generalize a procedure for evaluating trigonometric function values for nonacute angles. Suppose we have the angles in standard position with measures $60°$, $120°$, $240°$, and $300°$. Notice that the reference angle for all of these angles is $60°$.

If we draw reference triangles and let the shortest leg have length 1, we find that the other leg has length $\sqrt{3}$ and the hypotenuse has length 2. Notice that the legs of the triangles have lengths (always positive) 1 and $\sqrt{3}$; however, the coordinates are $(\pm 1, \pm\sqrt{3})$. Therefore, when we calculate the trigonometric functions for any of the angles $60°$, $120°$, $240°$, and $300°$, we can simply calculate the trigonometric function values for the reference angle $60°$, and determine the algebraic sign $(+ \text{ or } -)$ for the particular trigonometric function and quadrant.

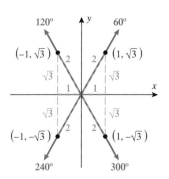

To find the value of $\cos 120°$, we first recognize that the terminal side of an angle with $120°$ measure lies in quadrant II. We also know that cosine is negative in quadrant II. We then calculate the cosine of the reference angle, $60°$.

$$\cos 60° = \frac{\text{adjacent}}{\text{hypotenuse}} = \frac{1}{2}$$

Since we know that $\cos 120°$ is negative because it lies in quadrant II, we know that

$$\cos 120° = -\frac{1}{2}$$

Similarly, we know that $\cos 240° = -\frac{1}{2}$ and $\cos 300° = \frac{1}{2}$.

For any angle whose terminal side lies along one of the axes, we consult the table in this section for the values of the trigonometric functions for *quadrantal angles*. An alternative to memorizing the table is to take the points $(1, 0)$, $(0, 1)$, $(-1, 0)$, and $(0, -1)$ for convenience on the axes and use Definition 2 in the Cartesian plane to determine the trigonometric function values for quadrantal angles. If the terminal side lies in one of the four quadrants, the angle is said to be nonquadrantal, and the following procedure can be used.

PROCEDURE FOR EVALUATING TRIGONOMETRIC FUNCTION VALUES FOR ANY NONQUADRANTAL ANGLE θ

Step 1: Find the smallest possible positive coterminal angle to θ:
- If $0° < \theta < 360°$, proceed to Step 2.
- If $\theta < 0°$, then add $360°$ as many times as needed to get a coterminal angle with measure between $0°$ and $360°$.
- If $\theta > 360°$, then subtract $360°$ as many times as needed to get a coterminal angle with measure between $0°$ and $360°$.

Step 2: Find the quadrant in which the terminal side of the angle in Step 1 lies.

Step 3: Find the reference angle α of the coterminal angle found in Step 1.

Step 4: Find the trigonometric function values for the reference angle α.

Step 5: Determine the correct algebraic signs $(+$ or $-)$ for the trigonometric function values based on the quadrant identified in Step 2.

Step 6: Combine the trigonometric values found in Step 4 with the algebraic signs in Step 5 to give the trigonometric function values of θ.

EXAMPLE 5 **Evaluating the Cosine Function of a Special Angle Exactly**

Find the exact value of $\cos 210°$.

Solution:

The terminal side of $\theta = 210°$ lies in quadrant III.

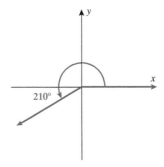

Find the reference angle for $\theta = 210°$. $210° - 180° = 30°$

Find the value of the cosine function of the reference angle. $\cos 30° = \dfrac{\sqrt{3}}{2}$

Determine the algebraic sign for the cosine function in quadrant III.

Negative $(-)$

Combine the algebraic sign of the cosine function in quadrant III with the value of the cosine function of the reference angle.

$$\cos 210° = \boxed{-\dfrac{\sqrt{3}}{2}}$$

▼

ANSWER
$-\frac{1}{2}$.

▼

YOUR TURN Find the exact value of $\sin 330°$.

EXAMPLE 6 **Evaluating the Tangent Function of a Special Angle Exactly**

Find the exact value of $\tan 495°$.

Solution:

Find the smallest possible coterminal angle.

Subtract $360°$ to get a coterminal angle between $0°$ and $360°$.

$$495° - 360° = 135°$$

The terminal side of the angle lies in quadrant II.

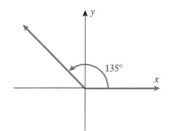

Find the reference angle for $135°$.

$$180° - 135° = 45°$$

Find the value of the tangent function of the reference angle.

$$\tan 45° = 1$$

Determine the algebraic sign for the tangent function in quadrant II.

Negative $(-)$

Combine the algebraic sign of the tangent function in quadrant II with the value of the tangent function of the reference angle.

$$\tan 495° = \boxed{-1}$$

▼

ANSWER
$-\sqrt{3}$

▼

YOUR TURN Find the exact value of $\tan 660°$.

[CONCEPT CHECK]

If the terminal side of an angle lies in a quadrant other than I, and if cosine is positive and sine is the negative of the reference angle, then the terminal side of the angle lies in which quadrant?

▼

ANSWER IV

▶ **EXAMPLE 7** **Evaluating the Cosecant Function of a Special Angle Exactly**

Find the exact value of $\csc(-210°)$.

Solution:

Find the smallest possible coterminal angle.

Add $360°$ to get a coterminal angle between $0°$ and $360°$.

$$-210° + 360° = 150°$$

The terminal side of the angle lies in quadrant II.

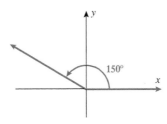

Find the reference angle for 150°.

$$180° - 150° = 30°$$

Find the value of the cosecant function of the reference angle.

$$\csc 30° = \frac{1}{\sin 30°} = \frac{1}{\frac{1}{2}} = 2$$

Determine the algebraic sign for the cosecant function in quadrant II.

Positive $(+)$

Combine the algebraic sign of the cosecant function in quadrant II with the value of the cosecant function of the reference angle.

$$\csc(-210°) = \boxed{2}$$

▼

YOUR TURN Find the exact value of $\sec(-330°)$.

▼
ANSWER

$$\frac{2\sqrt{3}}{3}$$

EXAMPLE 8 **Finding Exact Angle Measures Given Trigonometric Function Values**

Find all values of θ where $0° < \theta \le 360°$, when $\sin\theta = -\dfrac{\sqrt{3}}{2}$.

Solution:

Determine in what quadrants sine is negative.

QIII and QIV

Since the absolute value of $\sin\theta$ is $\dfrac{\sqrt{3}}{2}$, the reference angle is 60°.

$$\sin 60° = \frac{\sqrt{3}}{2}$$

Determine angles between 180° and 360° in quadrant III and quadrant IV with reference angle 60°.

Quadrant III:

$$180° + 60° = 240°$$

Quadrant IV:

$$360° - 60° = 300°$$

The two angles are $\boxed{240°}$ and $\boxed{300°}$.

▼

YOUR TURN Find all values of θ where $0° < \theta \le 360°$, when $\cos\theta = -\dfrac{\sqrt{3}}{2}$.

▼
ANSWER
150° and 210°

We can evaluate trigonometric functions exactly for quadrantal angles and any angle whose reference angle is a special angle (30°, 45°, or 60°). For all other angles, we can approximate the trigonometric function value with a calculator.

In the following examples, approximations will be rounded to the nearest ten thousandth (four decimal places).

EXAMPLE 9 **Using a Calculator to Evaluate Trigonometric Values of Angles**

Use a calculator to evaluate the following:

a. $\tan 466°$ **b.** $\csc(-313°)$

Solution (a): $\tan 466° \approx \boxed{-3.4874}$

Solution (b): $\sin(-313°) \approx 0.7314; \quad \dfrac{1}{\sin(-313°)} \approx \boxed{1.3673}$

YOUR TURN   Use a calculator to evaluate the following:

a. $\sec(-118°)$ **b.** $\cot 226°$

▶ **EXAMPLE 10** **Finding Approximate Angle Measures Given Trigonometric Function Values**

Find the smallest possible positive measure of θ (rounded to the nearest degree) if $\sin\theta = -0.6293$ and the terminal side of θ (in standard position) lies in quadrant III, where $0° \le \theta \le 360°$.

Solution:

Sine of the reference angle is 0.6293.	$\sin\alpha = 0.6293$
Find the reference angle.	$\alpha = \sin^{-1}(0.6293) \approx 38.998°$
Round the reference angle to the nearest degree.	$\alpha \approx 39°$
Find θ, which lies in quadrant III.	$180° + 39° \approx 219°$
	$\boxed{\theta \approx 219°}$
Check with a calculator.	$\sin 219° \approx -0.6293$

YOUR TURN   Find the smallest possible positive measure of θ (rounded to the nearest degree) if $\cos\theta = -0.5299$ and the terminal side of θ (in standard position) lies in quadrant II.

▶[SECTION 2.3]   **SUMMARY**

In this section, the algebraic signs of the trigonometric functions in each quadrant were identified.

θ	SIN θ	COS θ	TAN θ
QI	+	+	+
QII	+	−	−
QIII	−	−	+
QIV	−	+	−

The trigonometric functions can be evaluated for nonquadrantal angles using reference angles and knowledge of algebraic signs in each quadrant. We evaluated the trigonometric functions for quadrantal angles using the terminal side approach. When reference angles are 30°, 45°, or 60°, nonacute angles can be evaluated exactly; otherwise, we use a calculator to approximate trigonometric function values for any angle.

[SECTION 2.3] EXERCISES

• **SKILLS**

In Exercises 1–6, indicate the quadrant in which the terminal side of θ must lie in order for each of the following to be true.

1. $\cos\theta$ is positive and $\sin\theta$ is negative.

2. $\cos\theta$ is negative and $\sin\theta$ is positive.

3. $\tan\theta$ is negative and $\sin\theta$ is positive.

4. $\tan\theta$ is positive and $\cos\theta$ is negative.

5. $\sec\theta$ and $\csc\theta$ are both positive.

6. $\sec\theta$ and $\csc\theta$ are both negative.

In Exercises 7–20, find the indicated trigonometric function values if possible.

7. If $\cos\theta = -\dfrac{3}{5}$ and the terminal side of θ lies in quadrant III, find $\sin\theta$.

8. If $\tan\theta = -\dfrac{5}{12}$ and the terminal side of θ lies in quadrant II, find $\cos\theta$.

9. If $\sin\theta = \dfrac{60}{61}$ and the terminal side of θ lies in quadrant II, find $\tan\theta$.

10. If $\cos\theta = \dfrac{40}{41}$ and the terminal side of θ lies in quadrant IV, find $\tan\theta$.

11. If $\sin\theta = -\dfrac{3}{4}$ and the terminal side of θ lies in quadrant IV, find $\sec\theta$.

12. If $\cos\theta = -\dfrac{8}{5}$ and the terminal side of θ lies in quadrant III, find $\sin\theta$.

13. If $\tan\theta = \dfrac{84}{13}$ and the terminal side of θ lies in quadrant III, find $\sin\theta$.

14. If $\sin\theta = -\dfrac{7}{25}$ and the terminal side of θ lies in quadrant IV, find $\cos\theta$.

15. If $\csc\theta = -\sqrt{5}$ and the terminal side of θ lies in quadrant III, find $\cot\theta$.

16. If $\cos\theta = \dfrac{\sqrt{5}}{10}$ and the terminal side of θ lies in quadrant IV, find $\sin\theta$.

17. If $\sec\theta = -2$ and the terminal side of θ lies in quadrant III, find $\tan\theta$.

18. If $\cot\theta = 1$ and the terminal side of θ lies in quadrant I, find $\sin\theta$.

19. If $\sin\theta = \dfrac{\sqrt{11}}{6}$ and the terminal side of θ lies in quadrant II, find $\tan\theta$.

20. If $\tan\theta = -\dfrac{1}{2}$ and the terminal side of θ lies in quadrant II, find $\cos\theta$.

In Exercises 21–30, evaluate each expression if possible.

21. $\cos(-270°) + \sin 450°$

22. $\sin(-270°) + \cos 450°$

23. $\sin 630° + \tan(-540°)$

24. $\cos(-720°) + \tan 720°$

25. $\cos 540° - \sec(-540°)$

26. $\sin(-450°) + \csc 270°$

27. $\csc(-630°) - \cot 630°$

28. $\sec(-540°) + \tan 540°$

29. $\tan 720° + \sec 720°$

30. $\cot 450° - \cos(-450°)$

In Exercises 31–38, determine whether each statement is possible or not possible.

31. $\sin\theta = -0.999$

32. $\cos\theta = 1.0001$

33. $\cos\theta = \dfrac{2\sqrt{6}}{3}$

34. $\sin\theta = \dfrac{\sqrt{2}}{10}$

35. $\tan\theta = 4\sqrt{5}$

36. $\cot\theta = -\dfrac{\sqrt{6}}{7}$

37. $\sec\theta = -\dfrac{4}{\sqrt{7}}$

38. $\csc\theta = \dfrac{\pi}{2}$

In Exercises 39–52, evaluate the following expressions *exactly* by using a reference angle.

39. $\cos 240°$ **40.** $\cos 120°$ **41.** $\sin 300°$ **42.** $\sin 315°$

43. $\cot(-150°)$ **44.** $\sin(-135°)$ **45.** $\cos(-30°)$ **46.** $\sec(-135°)$

47. $\tan 210°$ **48.** $\sec 135°$ **49.** $\tan(-315°)$ **50.** $\sec(-330°)$

51. $\csc 330°$ **52.** $\csc(-240°)$

In Exercises 53–60, find all possible values of θ, where $0° < \theta \leq 360°$, when each of the following is true.

53. $\cos\theta = \dfrac{\sqrt{3}}{2}$ **54.** $\sin\theta = \dfrac{\sqrt{3}}{2}$ **55.** $\sin\theta = -\dfrac{1}{2}$ **56.** $\cos\theta = -\dfrac{1}{2}$

57. $\cos\theta = 0$ **58.** $\sin\theta = 0$ **59.** $\sin\theta = -1$ **60.** $\cos\theta = -1$

In Exercises 61–70, evaluate the trigonometric expressions with a calculator. Round your answer to four decimal places.

61. $\sin 237°$ **62.** $\cos 317°$ **63.** $\tan(-265°)$ **64.** $\tan 622°$

65. $\sec 421°$ **66.** $\sec(-222°)$ **67.** $\csc(-111°)$ **68.** $\csc 211°$

69. $\cot 159°$ **70.** $\cot(-82°)$

In Exercises 71–80, find the smallest possible positive measure of θ (rounded to the nearest degree) if the indicated information is true.

71. $\sin\theta = 0.9397$ and the terminal side of θ lies in quadrant II.

72. $\cos\theta = 0.7071$ and the terminal side of θ lies in quadrant IV.

73. $\cos\theta = -0.7986$ and the terminal side of θ lies in quadrant II.

74. $\sin\theta = -0.1746$ and the terminal side of θ lies in quadrant III.

75. $\tan\theta = -0.7813$ and the terminal side of θ lies in quadrant IV.

76. $\cos\theta = -0.3420$ and the terminal side of θ lies in quadrant III.

77. $\tan\theta = -0.8391$ and the terminal side of θ lies in quadrant II.

78. $\tan\theta = 11.4301$ and the terminal side of θ lies in quadrant III.

79. $\sin\theta = -0.3420$ and the terminal side of θ lies in quadrant IV.

80. $\sin\theta = -0.4226$ and the terminal side of θ lies in quadrant III.

• **APPLICATIONS** ....

In Exercises 81–84, refer to the following:

When light passes from one substance to another, such as from air to water, its path bends. This is called *refraction* and is what is seen in eyeglass lenses, camera lenses, and gems. The rule governing the change in the path is called *Snell's law*, named after a Dutch astronomer, Willebrord Snellius (1580–1626): $n_1 \sin\theta_1 = n_2 \sin\theta_2$, where the n_1 and n_2 are the indices of refraction of the different substances and the θ_1 and θ_2 are the respective angles that light makes with a line perpendicular to the surface at the boundary between substances. The figure shows the path of light rays going from air to water. Assume that the index of refraction in air is 1.

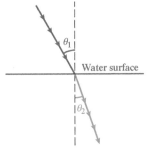
Water surface

81. **Optics.** If light rays hit the water's surface at an angle 30° from the perpendicular and are refracted to an angle of 22° from the perpendicular, then what is the refraction index for water?

82. **Optics.** If light rays hit a glass surface at an angle 30° from the perpendicular and are refracted to an angle of 18° from the perpendicular, then what is the refraction index for glass?

83. **Optics.** If the refraction index for a diamond is 2.4, then to what angle is light refracted if it enters the diamond at an angle of 30°?

84. **Optics.** If the refraction index for a rhinestone is 1.9, then to what angle is light refracted if it enters the rhinestone at an angle of 30°?

85. **Merry-Go-Round.** Penelope loves to ride merry-go-rounds. Ben models her path on a Cartesian plane, with the pole of the merry-go-round at the origin and θ being the angle between the positive x-axis and the ray from the origin through her current position. When Penelope gets on, her position makes $\theta = 30°$. If Penelope continues counterclockwise around the merry-go-round for $3\frac{1}{4}$ revolutions, what is the value for the sine of the new value of θ?

86. **Merry-Go-Round.** Given the model of Penelope on the merry-go-round in Exercise 85, consider the following. When Penelope gets on the merry-go-round, $\theta = 120°$. She rides clockwise for $\frac{2}{3}$ of a revolution and then stops and rides in the opposite direction for $\frac{3}{4}$ revolution before getting off. What is the cosine of the angle given by her new position?

87. **Clock.** Let α be the angle formed by a ray from the center of a clock through the 3 and the clock's minute hand. If $\tan\alpha = 0$, at what number is the minute hand pointing?

88. **Clock.** Consider α as given in Exercise 87. If $\sin\alpha = \frac{1}{2}$, at what number is the minute hand pointing?

For Exercises 89 and 90, refer to the following:

An orthotic knee brace can be used to treat knee injuries by locking the knee at an angle θ chosen to facilitate healing. The angle θ is measured from the metal bar on the side of the brace on the thigh to the metal bar on the side of the brace on the calf (see the figure on the left below). To make working with the brace more convenient, rotate the image such that the thigh aligns with the positive x-axis (see the figure on the right below).

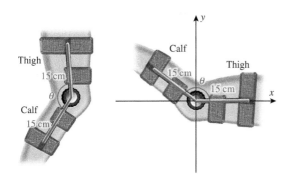

89. Health/Medicine. If $\theta = 165°$, find the measure of the reference angle. What is the physical meaning of the reference angle?

90. Health/Medicine. If $\theta = 160°$, find the measure of the reference angle. What would an angle greater than $180°$ represent?

For Exercises 91 and 92, refer to the following:

Water covers two-thirds of the Earth's surface, and every living thing is dependent on it. For example, the human body is made up of over 70% water. The water molecule is composed of one oxygen atom and two hydrogen atoms and exhibits a bent shape with the oxygen molecule at the center. The angle θ between the O-H bonds is $104.5°$.

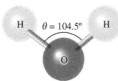

(*Source*: http://www.wiley.com/college/boyer/0470003790/reviews/pH/ph_water.htm.)

91. Chemistry. Sketch the water molecule in the xy-coordinate system in a convenient manner for illustrating angles. Find the reference angle. Illustrate both the angle θ and the reference angle on the sketch.

92. Chemistry. Find $\cos(104.5°)$ using the reference angle found in Exercise 91.

● **CATCH THE MISTAKE**

In Exercises 93 and 94, explain the mistake that is made.

93. Evaluate the expression $\sec 120°$ exactly.

Solution:

$120°$ lies in quadrant II.
The reference angle is $30°$.

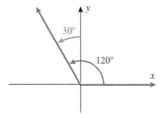

Find the cosine of the reference angle.	$\cos 30° = \dfrac{\sqrt{3}}{2}$
Cosine is negative in quadrant II.	$\cos 120° = -\dfrac{\sqrt{3}}{2}$
Secant is the reciprocal of cosine.	$\sec 120° = -\dfrac{2}{\sqrt{3}} = -\dfrac{2\sqrt{3}}{3}$

This is incorrect. What mistake was made?

94. Find the smallest possible positive measure of θ (rounded to the nearest degree) if $\cos\theta = -0.2388$ and the terminal side of θ (in standard position) lies in quadrant III.

Solution:

| Evaluate with a calculator. | $\theta = \cos^{-1}(-0.2388) = 103.8157°$ |
| Approximate to the nearest degree. | $\theta \approx 104°$ |

This is incorrect. What mistake was made?

● **CONCEPTUAL**

In Exercises 95–98, determine whether each statement is true or false.

95. It is possible for all six trigonometric functions of the same angle to have positive values.

96. It is possible for all six trigonometric functions of the same angle to have negative values.

97. The trigonometric function value for any angle with negative measure must be negative.

98. The trigonometric function value for any angle with positive measure must be positive.

99. If $\tan\theta = \dfrac{a}{b}$, where a and b are positive, and θ lies in quadrant III, find $\sin\theta$.

100. If $\tan\theta = -\dfrac{a}{b}$, where a and b are positive, and θ lies in quadrant II, find $\cos\theta$.

101. Name the reference angle for θ, given that $\cos\theta = -\dfrac{\sqrt{3}}{2}$.

102. Find $\tan x$ if $x = (45 + 180k)°$ for odd integer k.

103. If the reference angle for $\beta = 60°$, what are the possible values for $\cos\beta$?

104. Find $\csc(3x + 30°)$ for $x = -210°$.

105. Use a calculator to evaluate $\sin 80°$ and $\sin 100°$. Now use the calculator to evaluate $\sin^{-1}(0.9848)$. When sine is positive, in which of the quadrants, I or II, does the calculator assume the terminal side of the angle lies?

106. Use a calculator to evaluate $\sin 260°$ and $\sin 280°$. Now use the calculator to evaluate $\sin^{-1}(-0.9848)$. When sine is negative, in which of the quadrants, III or IV, does the calculator assume the terminal side of the angle lies?

107. Use a calculator to evaluate $\cos 75°$ and $\cos(-75°)$. Now use the calculator to evaluate $\cos^{-1}(0.2588)$. When cosine is positive, in which of the quadrants, I or IV, does the calculator assume the terminal side of the angle lies?

108. Use a calculator to evaluate $\cos 105°$ and $\cos 255°$. Now use the calculator to evaluate $\cos^{-1}(-0.2588)$. When cosine is negative, in which of the quadrants, II or III, does the calculator assume the terminal side of the angle lies?

109. Use a calculator to evaluate $\tan 65°$ and $\tan 245°$. Now use the calculator to evaluate $\tan^{-1}(2.1445)$. When tangent is positive, in which of the quadrants, I or III, does the calculator assume the terminal side of the angle lies?

110. Use a calculator to evaluate $\tan 136°$ and $\tan 316°$. Now use the calculator to evaluate $\tan^{-1}(-0.9657)$. When tangent is negative, in which of the quadrants, II or IV, does the calculator assume the terminal side of the angle lies?

2.4 BASIC TRIGONOMETRIC IDENTITIES

SKILLS OBJECTIVES	CONCEPTUAL OBJECTIVES
■ Apply the reciprocal identities. ■ Apply the quotient identities. ■ Apply the pythagorean identities.	■ Understand that trigonometric reciprocal identities are not always defined. ■ Understand that trigonometric quotient identities are not always defined. ■ Understand that the second and third Pythagorean identities are not always defined.

2.4.1 Reciprocal Identities

2.4.1 SKILL

Apply the reciprocal identities.

2.4.1 CONCEPTUAL

Understand that trigonometric reciprocal identities are not always defined.

In this section, we revisit the *reciprocal* and *quotient identities* and develop the Pythagorean identities. These identities will allow us to simplify expressions. In mathematics, an **identity** is a statement that is true for all values of the variable (for which the expressions are defined). For example, $x^2 = x \cdot x$ is an identity because the statement is true no matter what values are selected for x. Similarly, $\dfrac{2x}{5x} = \dfrac{2}{5}$ is an identity, but it is important to note that this identity holds for all x except 0 ($x \neq 0$), since the expression on the left is not defined when $x = 0$. In our discussions on identities, it is important to note that the identities hold for all values of θ for which the expressions are defined.

Recall that the **reciprocal** of a nonzero number x is $\dfrac{1}{x}$. For example, the reciprocal of 5 is $\dfrac{1}{5}$ and the reciprocal of $\dfrac{9}{17}$ is $\dfrac{17}{9}$. There is no reciprocal for 0. In mathematics, the expression $\dfrac{1}{0}$ is undefined. Calculators have a reciprocal key typically labeled $\dfrac{1}{x}$ or x^{-1}. This can be used to find the reciprocal for any number (except 0).

In Section 2.2, the trigonometric functions were defined as

$$\sin\theta = \frac{y}{r} \qquad \cos\theta = \frac{x}{r} \qquad \tan\theta = \frac{y}{x} \ (x \neq 0)$$

$$\csc\theta = \frac{r}{y} \ (y \neq 0) \qquad \sec\theta = \frac{r}{x} \ (x \neq 0) \qquad \cot\theta = \frac{x}{y} \ (y \neq 0)$$

Notice that $\csc\theta = \dfrac{r}{y} = \dfrac{1}{\frac{y}{r}} = \dfrac{1}{\sin\theta}$.

The following box summarizes the *reciprocal identities* that are true for all values of θ that do not correspond to the trigonometric function in the denominator having a zero value.

RECIPROCAL IDENTITIES

RECIPROCAL IDENTITIES		EQUIVALENT FORMS	
$\csc\theta = \dfrac{1}{\sin\theta}$	$\sin\theta \neq 0$	$\sin\theta = \dfrac{1}{\csc\theta}$	
$\sec\theta = \dfrac{1}{\cos\theta}$	$\cos\theta \neq 0$	$\cos\theta = \dfrac{1}{\sec\theta}$	
$\cot\theta = \dfrac{1}{\tan\theta}$	$\tan\theta \neq 0$	$\tan\theta = \dfrac{1}{\cot\theta}$	$\cot\theta \neq 0$

[CONCEPT CHECK]

For what values of "θ" is cosecant (θ) not defined?

▼ ·······

ANSWER "θ" = $n\cdot\pi$ for n = 0, 1, 2, 3 . . .

Reciprocals always have the same algebraic sign. Therefore, if the sine function is positive, then the cosecant function is positive. If the cosine function is negative, then the secant function is negative.

STUDY TIP
Reciprocals always have the same algebraic sign (+ or −).

▶ **EXAMPLE 1** **Using Reciprocal Identities**

a. If $\cos\theta = -\dfrac{1}{2}$, find $\sec\theta$.

b. If $\sin\theta = \dfrac{\sqrt{3}}{2}$, find $\csc\theta$.

c. If $\tan\theta = c$, find $\cot\theta$.

Solution (a):

Secant is the reciprocal of cosine. $\qquad \sec\theta = \dfrac{1}{\cos\theta}$

Substitute $\cos\theta = -\dfrac{1}{2}$ into the secant expression. $\qquad \sec\theta = \dfrac{1}{-\frac{1}{2}}$

Simplify. $\qquad \boxed{\sec\theta = -2}$

Solution (b):

Cosecant is the reciprocal of sine. $\qquad \csc\theta = \dfrac{1}{\sin\theta}$

Substitute $\sin\theta = \dfrac{\sqrt{3}}{2}$ into the cosecant expression. $\qquad \csc\theta = \dfrac{1}{\frac{\sqrt{3}}{2}}$

Simplify.

$$\csc\theta = \frac{2}{\sqrt{3}}$$

Rationalize the radical in the denominator.

$$\csc\theta = \frac{2}{\sqrt{3}} \cdot \frac{\sqrt{3}}{\sqrt{3}}$$

$$\boxed{\csc\theta = \frac{2\sqrt{3}}{3}}$$

Solution (c):

Cotangent is the reciprocal of tangent.

$$\cot\theta = \frac{1}{\tan\theta}$$

Substitute $\tan\theta = c$ into the cotangent expression.

$$\cot\theta = \frac{1}{c}$$

State any restrictions on c.

$$\boxed{\cot\theta = \frac{1}{c} \quad c \neq 0}$$

▼

ANSWER

a. $\sec\theta = -1$

b. $\csc\theta = \dfrac{1}{x} \quad x \neq 0$

▼ YOUR TURN **a.** If $\cos\theta = -1$, find $\sec\theta$.

b. If $\sin\theta = x$, find $\csc\theta$ and state any restrictions.

2.4.2 Quotient Identities

2.4.2 SKILL

Apply the quotient identities.

2.4.2 CONCEPTUAL

Understand that trigonometric quotient identities are not always defined.

Let us now use the trigonometric definitions to derive the *quotient identities*.

WORDS	MATH
Write the definition of the tangent function.	$\tan\theta = \dfrac{y}{x} \quad (x \neq 0)$
Multiply both the numerator and the denominator by $\dfrac{1}{r}$.	$\tan\theta = \dfrac{y/r}{x/r} \quad (x \neq 0)$
Substitute $\sin\theta = \dfrac{y}{r}$ and $\cos\theta = \dfrac{x}{r}$.	$\tan\theta = \dfrac{\sin\theta}{\cos\theta}$
Write the $x \neq 0$ in terms of θ.	$\cos\theta \neq 0$
State the quotient identity.	$\boxed{\tan\theta = \dfrac{\sin\theta}{\cos\theta} \quad \cos\theta \neq 0}$

The following box summarizes the quotient identities that are true for all values of θ that do not correspond to the trigonometric function in the denominator having a zero value.

[CONCEPT CHECK]

For what values of "☺" is tangent (☺) not defined?

▼

ANSWER "☺" $= \pm\dfrac{(2n+1)\pi}{2}$ for

$n = 0, 1, 2, 3 \dots$

QUOTIENT IDENTITIES

$$\tan\theta = \frac{\sin\theta}{\cos\theta} \quad \cos\theta \neq 0 \qquad \cot\theta = \frac{\cos\theta}{\sin\theta} \quad \sin\theta \neq 0$$

▶ **EXAMPLE 2**   **Using the Quotient Identities**

If $\sin\theta = \frac{3}{5}$ and $\cos\theta = -\frac{4}{5}$, find $\tan\theta$ and $\cot\theta$.

Solution:

Write the quotient identity involving the tangent function.

$$\tan\theta = \frac{\sin\theta}{\cos\theta}$$

Write the quotient identity involving the cotangent function.

$$\cot\theta = \frac{\cos\theta}{\sin\theta}$$

Substitute $\sin\theta = \frac{3}{5}$ and $\cos\theta = -\frac{4}{5}$.

$$\tan\theta = \frac{3/5}{-4/5} \quad\text{and}\quad \cot\theta = \frac{-4/5}{3/5}$$

Simplify.

$$\boxed{\tan\theta = -\frac{3}{4}} \quad\text{and}\quad \boxed{\cot\theta = -\frac{4}{3}}$$

Note: We could have found $\tan\theta$ and then used the reciprocal identity, $\cot\theta = \dfrac{1}{\tan\theta}$, to find the cotangent function value.

▼

YOUR TURN   If $\sin\theta = \frac{4}{5}$ and $\cos\theta = \frac{3}{5}$, find $\tan\theta$ and $\cot\theta$.

▼
ANSWER

$\tan\theta = \frac{4}{3}$ and $\cot\theta = \frac{3}{4}$

2.4.3 Pythagorean Identities

Before we develop the *Pythagorean identities*, we first need to discuss notation. We need to distinguish between $(\sin\theta)^2$, which implies that we first find $\sin\theta$ and then square the result, and $\sin(\theta^2)$, which implies that we first square the argument θ and then find the value of the sine function. We use the following notation to imply that the function output is raised to a power.

2.4.3 SKILL

Apply the Pythagorean identities.

2.4.3 CONCEPTUAL

Understand that the second and third Pythagorean identities are not always defined.

OPERATION	SHORTHAND NOTATION	WORDS	EXAMPLE
$(\sin\theta)^2$	$\sin^2\theta$	Sine squared of theta	$\sin\theta = \dfrac{\sqrt{3}}{2}$, then $\sin^2\theta = \left(\dfrac{\sqrt{3}}{2}\right)^2 = \dfrac{3}{4}$
$(\cos\theta)^3$	$\cos^3\theta$	Cosine cubed of theta	$\cos\theta = -\dfrac{1}{2}$, then $\cos^3\theta = \left(-\dfrac{1}{2}\right)^3 = -\dfrac{1}{8}$

Let us now derive our first *Pythagorean identity*. Recall that in the Cartesian plane, the relationship between the coordinates x and y and the distance r from the origin to the point (x, y) is given by $x^2 + y^2 = r^2$.

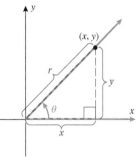

WORDS	MATH
Write the distance relationship in the Cartesian plane.	$x^2 + y^2 = r^2$
Divide the equation by r^2.	$\dfrac{x^2 + y^2}{r^2} = \dfrac{r^2}{r^2}$
Simplify.	$\dfrac{x^2}{r^2} + \dfrac{y^2}{r^2} = 1$

WORDS	MATH
Use the power rule of exponents, $\dfrac{a^2}{b^2} = \left(\dfrac{a}{b}\right)^2$.	$\left(\dfrac{x}{r}\right)^2 + \left(\dfrac{y}{r}\right)^2 = 1$
Substitute the sine and cosine definitions, $\sin\theta = \dfrac{y}{r}$ and $\cos\theta = \dfrac{x}{r}$, into the left side of the equation.	$(\cos\theta)^2 + (\sin\theta)^2 = 1$
Use shorthand notation. The result is the first *Pythagorean identity*.	$\boxed{\cos^2\theta + \sin^2\theta = 1}$

Note: $\cos^2\theta + \sin^2\theta = 1$ and $\sin^2\theta + \cos^2\theta = 1$ are equivalent identities since addition is a commutative operation.

Two important things to note at this time are:

■ The angle θ must be the same for this identity to hold.

■ This identity states that the sum of the squares of sine and cosine of the same angle θ is always 1.

Let's illustrate the Pythagorean identity with both a special angle and a calculator.

θ	SIN θ	COS θ	SIN² θ + COS² θ = 1
30°	$\dfrac{1}{2}$	$\dfrac{\sqrt{3}}{2}$	$\left(\dfrac{1}{2}\right)^2 + \left(\dfrac{\sqrt{3}}{2}\right)^2 = \dfrac{1}{4} + \dfrac{3}{4} = 1$
27°	≈ 0.4540	≈ 0.8910	$(0.4540)^2 + (0.8910)^2 \approx 0.999997 \approx 1$

The *second Pythagorean identity* can be derived from the first.

STUDY TIP

The second and third identities hold only for angles for which the functions are defined.

WORDS	MATH
Write the first Pythagorean identity.	$\sin^2\theta + \cos^2\theta = 1$
Divide both sides by $\cos^2\theta$.	$\dfrac{\sin^2\theta}{\cos^2\theta} + \dfrac{\cos^2\theta}{\cos^2\theta} = \dfrac{1}{\cos^2\theta}$
Use the property of exponents.	$\left(\dfrac{\sin\theta}{\cos\theta}\right)^2 + \left(\dfrac{\cos\theta}{\cos\theta}\right)^2 = \left(\dfrac{1}{\cos\theta}\right)^2$
Use the quotient and reciprocal identities. The result is the second *Pythagorean identity*.	$\boxed{\tan^2\theta + 1 = \sec^2\theta}$

In this derivation, had we divided by $\sin^2\theta$ instead of $\cos^2\theta$, we would have arrived at the *third Pythagorean identity*, $1 + \cot^2\theta = \csc^2\theta$.

[CONCEPT CHECK]

TRUE OR FALSE
$\sin^2(x) + \cos^2(y) = 1$, for all values x and y

▼

ANSWER False

PYTHAGOREAN IDENTITIES

$$\sin^2\theta + \cos^2\theta = 1 \qquad \tan^2\theta + 1 = \sec^2\theta \qquad 1 + \cot^2\theta = \csc^2\theta$$

▶ **EXAMPLE 3** **Using the Identities to Find Trigonometric Function Values**

Find $\cos\theta$ and $\tan\theta$ if $\sin\theta = \frac{4}{5}$ and the terminal side of θ lies in quadrant II.

Solution:

Write the first Pythagorean identity.

$$\sin^2\theta + \cos^2\theta = 1$$

Substitute $\sin\theta = \frac{4}{5}$ into the equation.

$$\left(\frac{4}{5}\right)^2 + \cos^2\theta = 1$$

Eliminate the parentheses.

$$\frac{16}{25} + \cos^2\theta = 1$$

Subtract $\frac{16}{25}$ from both sides of the equation.

$$\cos^2\theta = \frac{9}{25}$$

Use the square root property.

$$\cos\theta = \pm\sqrt{\frac{9}{25}}$$

Simplify.

$$\cos\theta = \pm\frac{3}{5}$$

Cosine is negative in quadrant II.

$$\boxed{\cos\theta = -\frac{3}{5}}$$

Write the quotient identity involving tangent.

$$\tan\theta = \frac{\sin\theta}{\cos\theta}$$

Substitute $\sin\theta = \frac{4}{5}$ and $\cos\theta = -\frac{3}{5}$.

$$\tan\theta = \frac{4/5}{-3/5}$$

Simplify.

$$\boxed{\tan\theta = -\frac{4}{3}}$$

▼ ....

YOUR TURN Find $\sin\theta$ and $\tan\theta$ if $\cos\theta = -\frac{4}{5}$ and the terminal side of θ lies in quadrant III.

▼ ....
ANSWER

$\sin\theta = -\frac{3}{5}$ and $\tan\theta = \frac{3}{4}$

▶ **EXAMPLE 4** **Using the Identities to Find Trigonometric Function Values**

Find $\sin\theta$ and $\cos\theta$ if $\tan\theta = \frac{3}{4}$ and if the terminal side of θ lies in quadrant III.

Solution:

Write the second Pythagorean identity.

$$\tan^2\theta + 1 = \sec^2\theta$$

Substitute $\tan\theta = \frac{3}{4}$ into the equation.

$$\left(\frac{3}{4}\right)^2 + 1 = \sec^2\theta$$

Eliminate the parentheses.

$$\frac{9}{16} + 1 = \sec^2\theta$$

Simplify.

$$\sec^2\theta = \frac{25}{16}$$

Apply the square root property.

$$\sec\theta = \pm\sqrt{\frac{25}{16}} = \pm\frac{5}{4}$$

Secant is negative in quadrant III.

$$\sec\theta = -\frac{5}{4}$$

Substitute the reciprocal identity, $\sec\theta = \dfrac{1}{\cos\theta}$.

$$\frac{1}{\cos\theta} = -\frac{5}{4}$$

▼ ....
CAUTION

A common mistake when given the tangent function is assuming that the numerator is the value of sine and the denominator is the value of cosine.

Don't assume that if $\tan\theta = \frac{3}{4}$, then since $\tan\theta = \dfrac{\sin\theta}{\cos\theta}$, $\sin\theta = 3$ and $\cos\theta = 4$. Recall that the ranges of the sine and cosine functions are $[-1, 1]$.

Solve for $\cos\theta$.

$$\boxed{\cos\theta = -\frac{4}{5}}$$

Write the first Pythagorean identity.

$$\sin^2\theta + \cos^2\theta = 1$$

Substitute $\cos\theta = -\frac{4}{5}$.

$$\sin^2\theta + \left(-\frac{4}{5}\right)^2 = 1$$

Simplify.

$$\sin^2\theta = \frac{9}{25}$$

Apply the square root property.

$$\sin\theta = \pm\sqrt{\frac{9}{25}} = \pm\frac{3}{5}$$

Sine is negative in quadrant III.

$$\boxed{\sin\theta = -\frac{3}{5}}$$

▶ **EXAMPLE 5** **Using Identities to Simplify Expressions**

Multiply $(1 - \cos\theta)(1 + \cos\theta)$ and simplify.

Solution:

Multiply using the FOIL (First-Outer-Inner-Last) method.
Combine like terms.

$$(1 - \cos\theta)(1 + \cos\theta) = 1 + \cos\theta - \cos\theta - \cos^2\theta$$
$$= 1 + \cos\theta - \cos\theta - \cos^2\theta$$
$$= 1 - \cos^2\theta$$

Rewrite the Pythagorean identity, $\sin^2\theta + \cos^2\theta = 1$ as $\sin^2\theta = 1 - \cos^2\theta$.

$$= \underbrace{1 - \cos^2\theta}_{\sin^2\theta}$$

$$\boxed{(1 - \cos\theta)(1 + \cos\theta) = \sin^2\theta}$$

▼ **ANSWER**

$\cos^2\theta$

YOUR TURN Multiply $(1 - \sin\theta)(1 + \sin\theta)$ and simplify.

▶ **[SECTION 2.4]** **SUMMARY**

In this section, basic trigonometric identities (reciprocal, quotient, and Pythagorean) were developed. These identities are only valid when the denominators are nonzero.

RECIPROCAL IDENTITIES

$$\csc\theta = \frac{1}{\sin\theta} \qquad \sec\theta = \frac{1}{\cos\theta} \qquad \cot\theta = \frac{1}{\tan\theta}$$

QUOTIENT IDENTITIES

$$\tan\theta = \frac{\sin\theta}{\cos\theta} \qquad \cot\theta = \frac{\cos\theta}{\sin\theta}$$

PYTHAGOREAN IDENTITIES

$$\sin^2\theta + \cos^2\theta = 1 \qquad \tan^2\theta + 1 = \sec^2\theta \qquad 1 + \cot^2\theta = \csc^2\theta$$

The identities were used to calculate trigonometric function values given other trigonometric function values and information regarding in which quadrant the terminal side of the angle lies. Basic trigonometric identities also can be used to simplify expressions.

[SECTION 2.4] EXERCISES

• **SKILLS**

In Exercises 1–10, use a reciprocal identity to find the function value indicated. Rationalize denominators if necessary.

1. If $\cos\theta = \dfrac{7}{8}$, find $\sec\theta$.

2. If $\sin\theta = -\dfrac{3}{7}$, find $\csc\theta$.

3. If $\sin\theta = -0.6$, find $\csc\theta$.

4. If $\cos\theta = 0.8$, find $\sec\theta$.

5. If $\tan\theta = -5$, find $\cot\theta$.

6. If $\tan\theta = 0.5$, find $\cot\theta$.

7. If $\csc\theta = \dfrac{\sqrt{5}}{2}$, find $\sin\theta$.

8. If $\sec\theta = \dfrac{\sqrt{11}}{2}$, find $\cos\theta$.

9. If $\cot\theta = -\dfrac{\sqrt{7}}{5}$, find $\tan\theta$.

10. If $\cot\theta = 3.5$, find $\tan\theta$.

In Exercises 11–18, use a quotient identity to find the function value indicated. Rationalize denominators if necessary.

11. If $\sin\theta = -\dfrac{1}{2}$ and $\cos\theta = \dfrac{\sqrt{3}}{2}$, find $\tan\theta$.

12. If $\sin\theta = -\dfrac{1}{2}$ and $\cos\theta = \dfrac{\sqrt{3}}{2}$, find $\cot\theta$.

13. If $\sin\theta = \dfrac{4}{5}$ and $\cos\theta = -\dfrac{3}{5}$, find $\tan\theta$.

14. If $\sin\theta = \dfrac{4}{5}$ and $\cos\theta = -\dfrac{3}{5}$, find $\cot\theta$.

15. If $\sin\theta = -0.6$ and $\cos\theta = -0.8$, find $\cot\theta$.

16. If $\sin\theta = -0.6$ and $\cos\theta = -0.8$, find $\tan\theta$.

17. If $\sin\theta = -\dfrac{\sqrt{11}}{6}$ and $\cos\theta = -\dfrac{5}{6}$, find $\tan\theta$.

18. If $\sin\theta = -\dfrac{\sqrt{11}}{6}$ and $\cos\theta = -\dfrac{5}{6}$, find $\cot\theta$.

In Exercises 19–22, find the indicated expression.

19. If $\sin\theta = \dfrac{\sqrt{5}}{8}$, find $\sin^2\theta$.

20. If $\cos\theta = 0.1$, find $\cos^3\theta$.

21. If $\csc\theta = -2$, find $\csc^3\theta$.

22. If $\tan\theta = -5$, find $\tan^2\theta$.

In Exercises 23–36, use a Pythagorean identity to find the function value indicated. Rationalize denominators if necessary.

23. If $\sin\theta = -\dfrac{1}{2}$ and the terminal side of θ lies in quadrant III, find $\cos\theta$.

24. If $\sin\theta = -\dfrac{3}{5}$ and the terminal side of θ lies in quadrant III, find $\cos\theta$.

25. If $\cos\theta = \dfrac{2}{5}$ and the terminal side of θ lies in quadrant IV, find $\sin\theta$.

26. If $\cos\theta = \dfrac{2}{7}$ and the terminal side of θ lies in quadrant IV, find $\sin\theta$.

27. If $\tan\theta = 4$ and the terminal side of θ lies in quadrant III, find $\sec\theta$.

28. If $\tan\theta = -5$ and the terminal side of θ lies in quadrant II, find $\sec\theta$.

29. If $\cot\theta = 2$ and the terminal side of θ lies in quadrant III, find $\csc\theta$.

30. If $\cot\theta = -3$ and the terminal side of θ lies in quadrant II, find $\csc\theta$.

31. If $\sin\theta = \dfrac{8}{15}$ and the terminal side of θ lies in quadrant II, find $\tan\theta$.

32. If $\sin\theta = -\dfrac{7}{15}$ and the terminal side of θ lies in quadrant III, find $\tan\theta$.

33. If $\cos\theta = -\dfrac{7}{15}$ and the terminal side of θ lies in quadrant III, find $\csc\theta$.

34. If $\cos\theta = -\dfrac{8}{15}$ and the terminal side of θ lies in quadrant II, find $\csc\theta$.

35. If $\sec\theta = \dfrac{\sqrt{13}}{3}$ and the terminal side of θ lies in quadrant IV, find $\sin\theta$.

36. If $\sec\theta = -\dfrac{\sqrt{61}}{5}$ and the terminal side of θ lies in quadrant II, find $\sin\theta$.

In Exercises 37–44, use appropriate identities to find the function value indicated. Rationalize denominators if necessary.

37. Find $\sin\theta$ and $\cos\theta$ if $\tan\theta = -\dfrac{4}{3}$ and the terminal side of θ lies in quadrant II.

38. Find $\sin\theta$ and $\cos\theta$ if $\tan\theta = -\dfrac{3}{4}$ and the terminal side of θ lies in quadrant II.

39. Find $\sin\theta$ and $\cos\theta$ if $\tan\theta = -\dfrac{1}{5}$ and the terminal side of θ lies in quadrant IV.

40. Find $\sin\theta$ and $\cos\theta$ if $\tan\theta = -\dfrac{2}{3}$ and the terminal side of θ lies in quadrant IV.

41. Find $\sin\theta$ and $\cos\theta$ if $\tan\theta = 2$ and the terminal side of θ lies in quadrant III.

42. Find $\sin\theta$ and $\cos\theta$ if $\tan\theta = 5$ and the terminal side of θ lies in quadrant III.

43. Find $\sin\theta$ and $\cos\theta$ if $\tan\theta = 0.6$ and the terminal side of θ lies in quadrant III.

44. Find $\sin\theta$ and $\cos\theta$ if $\tan\theta = 0.8$ and the terminal side of θ lies in quadrant III.

In Exercises 45–54, simplify each of the following expressions if possible. Leave all answers in terms of $\sin\theta$ and $\cos\theta$.

45. $\sec\theta\cot\theta$

46. $\csc\theta\tan\theta$

47. $\tan^2\theta - \sec^2\theta$

48. $\cot^2\theta - \csc^2\theta$

49. $\csc\theta - \sin\theta$

50. $\sec\theta - \cos\theta$

51. $\dfrac{\sec\theta}{\tan\theta}$

52. $\dfrac{\csc\theta}{\cot\theta}$

53. $(\sin\theta + \cos\theta)^2$

54. $(\sin\theta - \cos\theta)^2$

• **APPLICATIONS**

55. **Mountain Grade.** From the top of a mountain road looking down, the angle of depression is given by the angle θ. If the grade of the road is given by tangent of θ written as a percentage, find the grade of a road given

   that $\sin\theta = \dfrac{\sqrt{101}}{101}$ and $\cos\theta = \dfrac{10\sqrt{101}}{101}$.

56. **Mountain Grade.** Consider the grade of a road as defined in Exercise 55. If the angle of depression of a certain highway

   is such that $\sec\theta = \dfrac{\sqrt{409}}{20}$, find the grade of the highway.

57. **Geometry.** The slope of a line passing through the origin can be given as $m = \tan\alpha$, where α is the positive angle formed by the line and the positive x-axis. Find the slope of a line such that $\sin\theta = \frac{5}{13}$ and $\cos\theta = \frac{12}{13}$.

58. **Geometry.** The slope of a line passing through the origin can be given as $m = \tan\alpha$, where α is the positive angle formed by the line and the positive x-axis. Find the slope of a line such that $\sec\theta = \frac{25}{7}$.

For Exercises 59 and 60, refer to the following:

The *bifolium* is a curve that can be drawn using either an algebraic equation or an equation using trigonometric functions. Even though the trigonometric equation uses polar coordinates, it's much easier to solve the trigonometric equation for function values than to solve the algebraic equation. The graph of $r = 8\sin\theta\cos^2\theta$ is shown below. You'll see more on polar coordinates and graphs of polar equations in Chapter 8. The following graph is in polar coordinates.

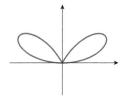

59. **Bifolium.** The equation for the bifolium above is $r = 8\sin\theta\cos^2\theta$. Use a Pythagorean identity to rewrite the equation using just the function $\sin\theta$. Then find r if $\theta = 30°, 60°$, and $90°$.

60. **Bifolium.** The equation for the bifolium above is $r = 8\sin\theta\cos^2\theta$. Use a Pythagorean identity to rewrite the equation using just the function $\sin\theta$. Then find r if $\theta = -30°, -60°$, and $-90°$.

For Exercises 61 and 62, refer to the following:

The monthly revenues (measured in thousands of dollars) of PizzaRia are a function of cost (measured in thousands of dollars), that is, $R(c)$. Recall that the angle θ can be interpreted as a measure of the sizes of c (cost) and R (revenue) relative to each other. The larger the angle θ is, the greater revenue is relative to cost, and conversely, the smaller the angle θ is, the smaller revenue is relative to cost. The ratio of revenue to costs can be approximated by the tangent function, $\tan\theta \approx \dfrac{R}{c}$.

61. **Business.** An analysis of the revenue and the costs of one month determined that $\cos\theta = \frac{3}{5}$ and $\sin\theta = \frac{4}{5}$. Find $\tan\theta$ and interpret its meaning.

62. **Business.** An analysis of the revenue and the costs of one

   month determined that $\cos\theta = \dfrac{2\sqrt{5}}{5}$ and $\sin\theta = \dfrac{\sqrt{5}}{5}$. Find $\tan\theta$ and interpret its meaning.

• **CATCH THE MISTAKE**

In Exercises 63 and 64, explain the mistake that is made.

63. If $\sin\theta = -\frac{1}{3}$ and the terminal side of θ lies in quadrant III, find $\cos\theta$.

Solution:

Write the Pythagorean identity. $\sin^2\theta + \cos^2\theta = 1$

Substitute $\sin\theta = -\frac{1}{3}$ into the identity. $\left(-\frac{1}{3}\right)^2 + \cos^2\theta = 1$

Solve for $\cos\theta$. $\frac{1}{9} + \cos^2\theta = 1$

$$\cos^2\theta = \frac{8}{9}$$

$$\cos\theta = \sqrt{\frac{8}{9}}$$

$$\cos\theta = \frac{2\sqrt{2}}{3}$$

This is incorrect. What mistake was made?

64. Find $\sin\theta$ and $\cos\theta$ if $\tan\theta = -\frac{1}{4}$ and the terminal side of θ lies in quadrant II.

Solution:

Write the quotient identity. $\tan\theta = \dfrac{\sin\theta}{\cos\theta}$

Substitute $\tan\theta = -\frac{1}{4}$ into the identity. $-\dfrac{1}{4} = \dfrac{\sin\theta}{\cos\theta}$

In quadrant II, sine is positive and cosine is negative. $\dfrac{1}{-4} = \dfrac{\sin\theta}{\cos\theta}$

Identify sine and cosine. $\sin\theta = 1$ and $\cos\theta = -4$

This is incorrect. What mistake was made?

• **CONCEPTUAL**

In Exercises 65 and 66, determine whether each statement is true or false.

65. It is possible for $\sin\theta > 0$ and $\csc\theta < 0$.

66. It is possible for $\cos\theta < 0$ and $\sec\theta > 0$.

67. If $\sin\theta = a$ and $\cos\theta = b$, find $\cot\theta$ and state any restrictions on a or b.

68. If $\sin\theta = a$ and $\cos\theta = b$, find $\tan\theta$ and state any restrictions on a or b.

69. Find the measure of an angle θ, $0° < \theta \le 180°$, that satisfies $\sin\theta = \csc\theta$.

70. Find the measure of an angle θ, $0° < \theta \le 180°$, that satisfies $\cos\theta = \sec\theta$.

71. Write $\cot\theta$ in terms of only $\sin\theta$.

72. Write $\csc\theta$ in terms of only $\cos\theta$.

• **CHALLENGE**

73. Let $x = 8\sin\theta$ in the expression $\sqrt{64 - x^2}$ and simplify.

74. Let $x = 6\cos\theta$ in the expression $\sqrt{36 - x^2}$ and simplify.

75. If $\cos\theta = \dfrac{1}{x}$ and the terminal side of θ lies in quadrant IV, find $\tan\theta$.

76. Write in terms of $\sin\theta$: $\dfrac{\cos\theta}{\tan\theta(1 - \sin\theta)}$.

• **TECHNOLOGY**

77. Verify the quotient $\tan\theta = \dfrac{\sin\theta}{\cos\theta}$ numerically for $\theta = 22°$. Use a calculator to find

   a. $\sin 22°$

   b. $\cos 22°$

   c. $\dfrac{\sin 22°}{\cos 22°}$

   d. $\tan 22°$

Are the results in (c) and (d) the same?

78. Is $\tan 10° = \dfrac{\sin 160°}{\cos 16°}$? Use a calculator to find

   a. $\sin 160°$

   b. $\cos 16°$

   c. $\dfrac{\sin 160°}{\cos 16°}$

   d. $\tan 10°$

Are the results in (c) and (d) the same?

79. Does $\sin 35° = \sqrt{1 - \cos^2 35°}$? Use a calculator to find

   a. $\sin 35°$

   b. $1 - \cos 35°$

   c. $\sqrt{1 - \cos^2 35°}$

Which results are the same?

80. Does $\sec 68° = \sqrt{1 + \tan^2 68°}$? Use a calculator to find

   a. $\tan 68°$

   b. $1 + \tan 68°$

   c. $\sqrt{1 + \tan^2 68°}$

Which results are the same?

▶[CHAPTER 2 REVIEW]

SECTION	CONCEPT	KEY IDEAS/FORMULAS
2.1	**Angles in the Cartesian plane**	
	Angles in standard position	
	Coterminal angles	Two angles are coterminal if they are in standard form and share the same terminal side. The measures of coterminal angles differ by multiples of 360°.
2.2	**Definition 2 of trigonometric functions: The Cartesian plane**	
	Trigonometric functions: The Cartesian plane	$\sin\theta = \dfrac{y}{r}$ $\cos\theta = \dfrac{x}{r}$ $\tan\theta = \dfrac{y}{x}$ $\csc\theta = \dfrac{r}{y}$ $\sec\theta = \dfrac{r}{x}$ $\cot\theta = \dfrac{x}{y}$ where $x^2 + y^2 = r^2$ or $r = \sqrt{x^2 + y^2}$. The distance r is positive $(r > 0)$. *Quadrantal Angles*

θ	0°	90°	180°	270°
$\sin\theta$	0	1	0	−1
$\cos\theta$	1	0	−1	0
$\tan\theta$	0	undefined	0	undefined
$\cot\theta$	undefined	0	undefined	0
$\sec\theta$	1	undefined	−1	undefined
$\csc\theta$	undefined	1	undefined	−1

SECTION	CONCEPT	KEY IDEAS/FORMULAS
2.3	**Evaluating trigonometric functions for nonacute angles**	
	Algebraic signs of the trigonometric functions	(see table below)

θ	QI	QII	QIII	QIV
$\sin\theta$	+	+	−	−
$\cos\theta$	+	−	−	+
$\tan\theta$	+	−	+	−

When $x > 0$, cosine is positive

When $x < 0$, cosine is negative

When $y > 0$, sine is positive

When $y < 0$, sine is negative

Note: The reciprocal function pairs have the same algebraic sign.

	Ranges of the trigonometric functions	$-1 \leq \sin\theta \leq 1$ and $-1 \leq \cos\theta \leq 1$ $\sec\theta \leq -1$ or $\sec\theta \geq 1$ $\csc\theta \leq -1$ or $\csc\theta \geq 1$ $-\infty < \tan\theta < \infty$ $-\infty < \cot\theta < \infty$
	Reference angles and reference right triangles	The reference angle α for angle θ (where θ is a nonquadrantal angle with positive measure $0° < \theta < 360°$) is given by ■ QI: $\alpha = \theta$ ■ QII: $\alpha = 180° - \theta$ ■ QIII: $\alpha = \theta - 180°$ ■ QIV: $\alpha = 360° - \theta$
	Evaluating trigonometric functions for nonacute angles	Step 1: Find the quadrant in which the terminal side of the angle lies. Step 2: Find the reference angle. Step 3: Find the trigonometric value of the reference angle. Step 4: Determine the correct algebraic sign $(+$ or $-)$ based on the quadrant found in Step 1. Step 5: Combine Steps 3 and 4.
2.4	**Basic trigonometric identities**	
	Reciprocal identities	$\csc\theta = \dfrac{1}{\sin\theta} \qquad \sec\theta = \dfrac{1}{\cos\theta} \qquad \cot\theta = \dfrac{1}{\tan\theta}$
	Quotient identities	$\tan\theta = \dfrac{\sin\theta}{\cos\theta} \qquad \cot\theta = \dfrac{\cos\theta}{\sin\theta}$
	Pythagorean identities	$\sin^2\theta + \cos^2\theta = 1$ $\tan^2\theta + 1 = \sec^2\theta$ $1 + \cot^2\theta = \csc^2\theta$

[CHAPTER 2 REVIEW EXERCISES]

2.1 Angles in the Cartesian Plane

State in which quadrant or on which axis the following angles with given measure in standard position would lie.

1. $300°$ **2.** $280°$ **3.** $150°$ **4.** $-180°$

Sketch the angles with given measure in standard position.

5. $120°$ **6.** $-30°$ **7.** $450°$ **8.** $-185°$

Determine the angle of the smallest possible positive measure that is coterminal with each of the following angles.

9. $1000°$ **10.** $-800°$ **11.** $480°$ **12.** $-10°$

2.2 Definition 2 of Trigonometric Functions: The Cartesian Plane

The terminal side of an angle θ in standard position passes through the indicated point. Calculate the values of the six trigonometric functions for angle θ.

13. $(6, -8)$ **14.** $(-24, -7)$

15. $(-6, 2)$ **16.** $(-40, 9)$

17. $(\sqrt{3}, 1)$ **18.** $(-9, -9)$

Calculate the values for the six trigonometric functions of the angle θ, given in standard position, if the terminal side of θ lies on the following line.

19. $y = x,\ x \le 0$ **20.** $y = -2x,\ x \le 0$

21. $3x + 4y = 0,\ x \ge 0$ **22.** $5x - 12y = 0,\ x \ge 0$

Calculate the values for the six trigonometric functions of the angle θ if possible.

23. $\theta = 270°$ **24.** $\theta = 1080°$

25. $\theta = -1080°$ **26.** $\theta = -360n°$, n is an integer

2.3 Evaluating Trigonometric Functions for Nonacute Angles

Indicate the quadrant in which the terminal side of θ must lie for each of the following to be true.

27. $\sin\theta$ is negative and $\tan\theta$ is positive.

28. $\cos\theta$ is negative and $\csc\theta$ is positive.

29. $\tan\theta$ is negative and $\sec\theta$ is positive.

30. $\tan\theta$ and $\sin\theta$ are both negative.

Find the indicated trigonometric function values.

31. If $\tan\theta = -\frac{4}{3}$ and the terminal side of θ lies in quadrant II, find $\sin\theta$.

32. If $\sin\theta = -\frac{8}{17}$ and the terminal side of θ lies in quadrant IV, find $\cos\theta$.

33. If $\cos\theta = \frac{1}{3}$ and the terminal side of θ lies in quadrant IV, find $\tan\theta$.

34. If $\cot\theta = \sqrt{3}$ and the terminal side of θ lies in quadrant III, find $\sin\theta$.

Evaluate the following expressions exactly if possible. Do not use a calculator.

35. $\cos 270° + \sin(-270°)$

36. $\tan 450° - \cot 540°$

37. $\sec(-720°) + \csc(-450°)$

38. $\tan 1080° - \sin(-1080°)$

Determine whether each statement is possible or impossible.

39. $\sin\theta = -\sqrt{2}$ **40.** $\cos\theta = 0.0004$

41. $\tan\theta = -5.4321$ **42.** $\csc\theta = \frac{15}{17}$

Evaluate the following trigonometric function values exactly.

43. $\sin 330°$ **44.** $\cos(-300°)$

45. $\tan 150°$ **46.** $\cot 315°$

47. $\sec(-150°)$ **48.** $\csc 210°$

Evaluate each of the trigonometric function values with a calculator. Round to four decimal places.

49. $\sin(-14°)$ **50.** $\cos 275°$

51. $\tan 46°$ **52.** $\cot 500°$

53. $\sec 111°$ **54.** $\csc 215°$

55. $\cot(-57°)$ **56.** $\csc(-59°)$

2.4 Basic Trigonometric Identities

Use trigonometric identities to find the indicated value(s).

57. If $\sin\theta = -\dfrac{7}{11}$, find $\csc\theta$.

58. If $\cot\theta = \dfrac{3\sqrt{5}}{5}$, find $\tan\theta$.

59. If $\sin\theta = \dfrac{8}{17}$ and $\cos\theta = -\dfrac{15}{17}$, find $\tan\theta$.

60. If $\sin\theta = -\dfrac{12}{13}$ and $\cos\theta = -\dfrac{5}{13}$, find $\cot\theta$.

61. If $\sin\theta = \dfrac{2\sqrt{3}}{9}$, find $\sin^2\theta$.

62. If $\cos\theta = -0.45$, find $\cos^2\theta$.

63. If $\tan\theta = \sqrt{3}$, find $\tan^3\theta$.

64. If $\sin\theta = \dfrac{11}{61}$ and the terminal side of θ lies in quadrant II, find $\tan\theta$.

65. If $\cos\theta = \dfrac{7}{25}$ and the terminal side of θ lies in quadrant IV, find $\cot\theta$.

66. If $\tan\theta = -1$ and the terminal side of θ lies in quadrant IV, find $\sec\theta$.

67. If $\csc\theta = \dfrac{13}{12}$ and the terminal side of θ lies in quadrant II, find $\cot\theta$.

68. Find $\sin\theta$ and $\cos\theta$ if $\tan\theta = -\dfrac{24}{7}$ and the terminal side of θ lies in quadrant II.

69. Find $\sin\theta$ and $\cos\theta$ if $\cot\theta = \dfrac{\sqrt{3}}{3}$ and the terminal side of θ lies in quadrant III.

70. Find $\sin\theta$ and $\cos\theta$ if $\cot\theta = \sqrt{3}$ and the terminal side of θ lies in quadrant III.

Simplify the following expressions. Write answers in terms of $\sin\theta$ and $\cos\theta$.

71. $\sec\theta \cot^2\theta$

72. $\sin\theta \cot\theta$

73. $\dfrac{\tan\theta}{\sec\theta}$

74. $(\cos\theta + \sec\theta)^2$

75. $\cot\theta(\sec\theta + \sin\theta)$

76. $\dfrac{\sin^2\theta}{\csc\theta}$

Technology Exercises

Section 2.2

Use a calculator to evaluate the following expressions. If you get an error, explain why.

77. $\sec 180°$

78. $\csc 180°$

Section 2.3

79. Use a calculator to evaluate $\sin 218°$ and $\sin 322°$. Now use the calculator to evaluate $\sin^{-1}(-0.6157)$. When sine is negative, in which of the quadrants, III or IV, does the calculator assume the terminal side of the angle lies?

80. Use a calculator to evaluate $\cos 28°$ and $\cos 332°$. Now use the calculator to evaluate $\cos^{-1}(0.8829)$. When cosine is positive, in which of the quadrants, I or IV, does the calculator assume the terminal side of the angle lies?

Section 2.4

81. Is $\cos 73° = \sqrt{1 - \sin^2 73°}$? Use a calculator to find each of the following:

  a. $\cos 73°$

  b. $1 - \sin 73°$

  c. $\sqrt{1 - \sin^2 73°}$

Which results are the same?

82. Is $\csc 28° = \sqrt{1 + \cot^2 28°}$? Use a calculator to find each of the following:

  a. $\csc 28°$

  b. $1 + \cot 28°$

  c. $\sqrt{1 + \cot^2 28°}$

Which results are the same?

REVIEW EXERCISES

[CHAPTER 2 PRACTICE TEST]

1. Fill in the table with exact values for the quadrantal angles and algebraic signs for the nonquadrantal angles that lie in the indicated quadrants.

	0°	QI	90°	QII	180°	QIII	270°	QIV	360°
$\sin\theta$									
$\cos\theta$									

2. If $\cot\theta < 0$ and $\sec\theta > 0$, in which quadrant does the terminal side of θ lie?

3. Find the values of θ, $0° < \theta \le 360°$, for which $\tan\theta$ is undefined.

4. Are the two angles with measures $-170°$ and $890°$ coterminal?

5. Find the smallest angle with positive measure that is coterminal with an angle which has measure $-500°$.

6. Find $\cos\theta$ if the terminal side of angle θ passes through the point $(-2, -5)$.

7. Find the value of $\sin\theta$ if the terminal side of angle θ lies along the graph of $y = |x|$.

8. Evaluate $\sin 210°$ exactly.

9. Evaluate $\cos 315°$ exactly.

10. Evaluate $\tan 150°$ exactly.

11. Evaluate $\tan(-315°)$ exactly.

12. Evaluate $\sec 135°$ exactly.

13. Evaluate $\cot 222°$ with a calculator. Round to four decimal places.

14. Evaluate $\sec(-165°)$ with a calculator. Round to four decimal places.

15. Evaluate $\cos(110°11')$ with a calculator. Round to four decimal places.

16. Find $\sin\theta$ and $\cos\theta$ if $\tan\theta = \frac{4}{3}$ and the terminal side of angle θ lies in quadrant III.

17. Find $\sin\theta$ and $\tan\theta$ if $\cos\theta = \frac{1}{6}$ and the terminal side of angle θ lies in quadrant IV.

18. Simplify the expression $\dfrac{(1 - \cos\theta)(1 + \cos\theta)}{\tan^2\theta}$ and write the answer only in terms of $\cos\theta$.

19. Which of the following is not possible?
    a. $\cos\theta = -\sqrt{3}$
    b. $\tan\theta = -\sqrt{3}$

20. If $\sin\theta = \dfrac{1}{4}$ and $\cos\theta = \dfrac{\sqrt{15}}{4}$, use identities to find $\cot\theta$.

21. If $\csc^2\theta = \dfrac{4}{3}$, use identities to find $\cot\theta$.

22. If $\sin\theta = \dfrac{3}{7}$, use identities to find $\cos\theta$.

23. A man is taking a picture of the top of a tall building. If the angle of elevation to the top of the building from a point 100 yards away from the base of the building is such that $\tan\theta = \frac{5}{2}$, how tall is the building?

24. Find the positive measure of θ (rounded to the nearest degree) if $\cos\theta = 0.1125$ and the terminal side of θ lies in quadrant IV.

25. Evaluate $\sec 870°$ exactly.

26. Identify which of the following are undefined:
    a. $\tan 270°$
    b. $\sin 180°$
    c. $\sec 90°$

[CHAPTERS 1–2 CUMULATIVE TEST]

1. For the angle 37°, find

   a. its complement

   b. its supplement

2. In the right triangle with legs a and b and hypotenuse c, if $a = 24$ and $c = 25$, find b.

3. **Clock.** What is the measure (in degrees) of the angle the minute hand sweeps in 35 minutes?

4. In a 45°-45°-90° triangle, if the two legs have length 15 feet, how long is the hypotenuse?

5. **Height of a Tree.** The shadow of a tree measures $15\frac{1}{3}$ feet. At the same time of day, the shadow of a 6-foot pole measures 2.3 feet. How tall is the tree?

6. If $B = 85°$, find the measure of angle E.

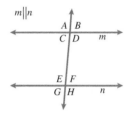

7. Write $\tan 60°$ in terms of its cofunction.

8. Write $\cos(30° - x)$ in terms of its cofunction.

9. Give the exact value of $\cos 30°$.

10. Use a calculator to evaluate $\cos 62°$. Round to four decimal places.

11. Convert 17.525° from decimal degrees to degrees-minutes-seconds. Round to the nearest second.

12. Given $\beta = 58.32°$ and $a = 11.6$ miles, use the right triangle diagram shown below to find c.

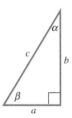

13. Given $a = 39$ m and $c = 48$ m, use the right triangle diagram shown above to find α.

14. **Angle of Inclination (Skiing).** The angle of inclination of a mountain with triple black diamond ski paths is 63°. If a skier at the top of the mountain is at an elevation of 4200 feet, how long is the ski run from the top to the base of the mountain? Round to the nearest foot.

15. State the quadrant in which the terminal side of the angle 210° lies.

16. Determine the angle of smallest possible positive measure that is coterminal with the angle $-170°$.

17. **Dartboard.** If a dartboard is superimposed on a Cartesian plane, in what quadrant does a dart land if its position is given by the point $(2, -5)$?

18. The terminal side of angle θ in standard position passes through the point $(-5, 2)$. Calculate the exact values of the six trigonometric functions for angle θ.

19. The angle $\theta = -1170°$ is in standard position. Calculate the values for the six trigonometric functions of θ if possible.

20. An airplane takes off and flies toward the north and west such that for each mile it travels west it travels 3 miles north. If θ is the angle formed by east and the path of the plane, find $\sin\theta$, $\cos\theta$, and $\tan\theta$.

21. If $\tan\theta = -\dfrac{7}{24}$ and the terminal side of θ lies in quadrant IV, find $\sec\theta$.

22. Determine if the statement $\csc\theta = \dfrac{2}{\pi}$ is possible or not possible.

23. Give the exact value $\cos 210°$.

24. If $\cos\theta = 0.2$, find $\cos^3\theta$.

25. Simplify $\tan\theta(\csc\theta + \cos\theta)$. Leave answer in terms of $\sin\theta$ and $\cos\theta$.

[3] Radian Measure and the Unit Circle Approach

SHANNON STAPLETON/REUTERS / Newscom

ONCE YOU RX
THERE'S NO GOING BACK

How does an odometer or speedometer on an automobile work? The transmission counts how many times the tires rotate (how many full revolutions take place) per second. A computer then calculates how far the car has traveled in that second by multiplying the number of revolutions by the tire circumference. Distance is given by the odometer, and the speedometer takes the distance per second and converts to miles per hour (or km/h). The computer chip is programmed to the tire size designed for the vehicle. If a person were to change the tire size (smaller or larger than the original specifications), then the odometer and speedometer would need to be adjusted.

Suppose you bought a 2017 Lexus RX 450h, which comes standard with 18-inch rims (corresponding to a tire with 28-inch diameter), and you decide to later upgrade these tires for 20-inch rims (corresponding to a tire with 30-inch diameter). If the onboard computer is not adjusted, is the actual speed faster or slower than the speedometer indicator?*

In this case, the speedometer would read too low. For example, if your speedometer read 75 mph, your actual speed would be 80 mph. In this chapter, you will see how the *angular speed* (rotations of tires per second), *radius* (of the tires), and *linear speed* (speed of the automobile) are related.

*Section 3.3, Example 3 and Exercises 53 and 54.

LEARNING OBJECTIVES

- Convert between degrees and radians.
- Calculate arc length and the area of a circular sector.

- Relate angular and linear speeds.

- Draw the unit circle and label the sine and cosine values for special angles (in both degrees and radians).

You will learn a second way to measure angles using radians. You will convert between degrees and radians. You will calculate arc lengths, areas of circular sectors, and angular and linear speeds. Finally, the third definition of trigonometric functions using the unit circle approach will be given. You will work with the trigonometric functions in the context of a

RADIAN MEASURE AND THE UNIT CIRCLE APPROACH

3.1 RADIAN MEASURE	**3.2** ARC LENGTH AND AREA OF A CIRCULAR SECTOR	**3.3** LINEAR AND ANGULAR SPEEDS	**3.4** DEFINITION 3 OF TRIGONOMETRIC FUNCTIONS: UNIT CIRCLE APPROACH
• The Radian Measure of an Angle • Converting Between Degrees and Radians	• Arc Length • Area of a Circular Sector	• Linear Speed • Angular Speed • Relationship Between Linear and Angular Speeds	• Trigonometric Functions and the Unit Circle (Circular Functions) • Properties of Circular Functions

3.1 RADIAN MEASURE

SKILLS OBJECTIVES	CONCEPTUAL OBJECTIVES
■ Calculate the radian measure of an angle. ■ Convert between degrees and radians.	■ Realize that radian measure allows us to write trigonometric functions as functions of real numbers. ■ Understand that degrees and radians are both units for measuring angles.

3.1.1 The Radian Measure of an Angle

3.1.1 SKILL

Calculate the radian measure of an angle.

3.1.1 CONCEPTUAL

Realize that radian measure allows us to write trigonometric functions as functions of real numbers.

In geometry and most everyday applications, angles are measured in degrees. However, *radian measure* is another way to measure angles. Using radian measure allows us to write trigonometric functions as functions not only of angles but also of real numbers in general.

Recall that in Section 1.1 we defined one full rotation as an angle having measure 360°. Now we think of the angle in the context of a circle. A **central angle** is an angle that has its vertex at the center of a circle.

When the intercepted arc's length is equal to the radius, the measure of the central angle is 1 **radian**. From geometry, we know that the ratio of the measures of two angles is equal to the ratio of the lengths of the arcs subtended by those angles (along the same circle).

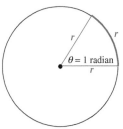

$$\frac{\theta_1}{\theta_2} = \frac{s_1}{s_2}$$

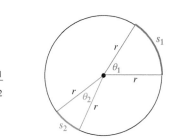

If $\theta_1 = 1$ radian, then the length of the subtended arc is equal to the radius, $s_1 = r$. This leads to a general definition of *radian measure*.

▼ . . . .
CAUTION

To correctly calculate radians from the formula $\theta = \dfrac{s}{r}$, the radius and arc length must be expressed in the same units.

DEFINITION Radian Measure

If a central angle θ in a circle with radius r intercepts an arc on the circle of length s, then the measure of θ, in **radians**, is given by

$$\theta \text{ (in radians)} = \frac{s}{r}$$

Note: The formula is valid only if s (arc length) and r (radius) are expressed in the same units.

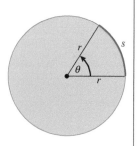

Note that both s and r are measured in units of length. When both are given in the same units, the units cancel, giving the number of radians as a *dimensionless* (unitless) real number.

One full rotation corresponds to an arc length equal to the circumference $2\pi r$ of the circle with radius r. We see then that one full rotation is equal to 2π radians.

$$\theta_{\text{full rotation}} = \frac{2\pi r}{r} = 2\pi$$

EXAMPLE 1   Finding the Radian Measure of an Angle

What is the measure (in radians) of a central angle θ that intercepts an arc of length 4 feet on a circle with radius 10 feet?

Solution:

Write the formula relating radian measure to arc length and radius.

$$\theta = \frac{s}{r}$$

Let $s = 4$ feet and $r = 10$ feet.

$$\theta = \frac{4\text{ ft}}{10\text{ ft}} = \boxed{0.4\text{ rad}}$$

▼

YOUR TURN   What is the measure (in radians) of a central angle θ that intercepts an arc of length 3 inches on a circle with radius 50 inches?

▶ EXAMPLE 2   Finding the Radian Measure of an Angle

What is the measure (in radians) of a central angle θ that intercepts an arc of length 6 centimeters on a circle with radius 2 meters?

common mistake

A common mistake is forgetting to first put the radius and arc length in the same units.

✓CORRECT

Write the formula relating radian measure to arc length and radius.

$$\theta \text{ (in radians)} = \frac{s}{r}$$

Substitute $s = 6$ centimeters and $r = 2$ meters into the radian expression.

$$\theta = \frac{6\text{ cm}}{2\text{ m}}$$

Convert the radius (2) meters to centimeters: 2 meters = 200 centimeters

$$\theta = \frac{6\text{ cm}}{200\text{ cm}}$$

The units, centimeters, cancel and the result is a unitless real number.

$$\boxed{\theta = 0.03\text{ rad}}$$

✗INCORRECT

Substitute $s = 6$ centimeters and $r = 2$ meters into the radian expression.

$$\theta = \frac{6\text{ cm}}{2\text{ m}}$$
$$= 3$$

ERROR: (not converting both numerator and denominator to the same units)

▼

YOUR TURN   What is the measure (in radians) of a central angle θ that intercepts an arc of length 12 millimeters on a circle with radius 4 centimeters?

Because radians are unitless, the word *radians* (or *rad*) is often omitted. If an angle measure is given simply as a real number, then radians are implied.

WORDS	MATH
The measure of θ is 4 degrees.	$\theta = 4°$
The measure of θ is 4 radians.	$\theta = 4$

3.1.2 Converting Between Degrees and Radians

3.1.2 **SKILL**

Convert between degrees and radians.

3.1.2 **CONCEPTUAL**

Understand that degrees and radians are both units for measuring angles.

To convert between degrees and radians, we must first look for a relationship between them. We start by considering one full rotation around the circle. An angle corresponding to one full rotation is said to have measure 360°, and we saw previously that one full rotation corresponds to $\theta = 2\pi$ rad.

WORDS	MATH
Write the angle measure (in degrees) that corresponds to one full rotation.	$\theta = 360°$
Write the angle measure (in radians) that corresponds to one full rotation.	
Arc length is the circumference of the circle.	$s = 2\pi r$
Substitute $s = 2\pi r$ into θ (in radians) $= \dfrac{s}{r}$.	$\theta = \dfrac{2\pi r}{r} = 2\pi$ rad
Equate the measures corresponding to one full rotation.	$360° = 2\pi$ rad
Divide by 2.	$180° = \pi$ rad
Divide by 180° or π.	$\boxed{1 = \dfrac{\pi}{180°} \text{ or } 1 = \dfrac{180°}{\pi}}$

This leads us to formulas $\left(\text{unit conversations, like } \dfrac{1 \text{ hr}}{60 \text{ min}}\right)$ that convert between degrees and radians. Let θ_d represent an angle measure given in degrees and θ_r represent the corresponding angle measure given in radians.

CONVERTING DEGREES TO RADIANS

To convert degrees to radians, multiply the degree measure by $\dfrac{\pi}{180°}$.

$$\theta_r = \theta_d\left(\frac{\pi}{180°}\right)$$

[CONCEPT CHECK]

Which angle would have greater measure: *B* radians or *B* degrees?

▼

ANSWER *B* rad

CONVERTING RADIANS TO DEGREES

To convert radians to degrees, multiply the radian measure by $\dfrac{180°}{\pi}$.

$$\theta_d = \theta_r\left(\frac{180°}{\pi}\right)$$

Before we begin converting between degrees and radians, let's first get a feel for radians. How many degrees is 1 radian?

WORDS	MATH
Multiply 1 radian by $\dfrac{180°}{\pi}$.	$1\left(\dfrac{180°}{\pi}\right)$
Approximate π by 3.14.	$1\left(\dfrac{180°}{3.14}\right)$
Use a calculator to evaluate and round to the nearest degree.	$\approx 57°$
	$\boxed{1 \text{ rad} \approx 57°}$

A radian is much larger than a degree (almost 57 times larger). Let's compare two angles, one measuring 30 radians and the other measuring 30°. Note that $\dfrac{30 \text{ rad}}{2\pi \text{ rad/rev}} \approx 4.77$ revolutions, whereas $30° = \frac{1}{12}$ revolution.

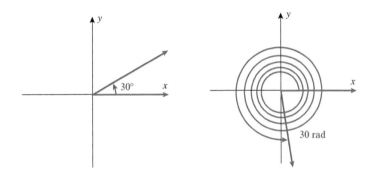

EXAMPLE 3   Converting Degrees to Radians

Convert 45° to radians.

Solution:

Multiply 45° by $\dfrac{\pi}{180°}$. $(45°)\left(\dfrac{\pi}{180°}\right) = \dfrac{45°\pi}{180°}$

Simplify. $= \dfrac{\pi}{4} \text{ rad}$

Note: $\dfrac{\pi}{4}$ is the exact value. A calculator can be used to approximate this expression. Scientific and graphing calculators have a π button. The decimal approximation of $\dfrac{\pi}{4}$ rounded to three decimal places is 0.785.

Exact Value: $\boxed{\dfrac{\pi}{4}}$

Approximate Value: $\boxed{0.785}$

▼

YOUR TURN   Convert 60° to radians.

▼
ANSWER
$\dfrac{\pi}{3}$ or   1.047

EXAMPLE 4 **Converting Degrees to Radians**

Convert 472° to radians.

Solution:

Multiply 472° by $\dfrac{\pi}{180°}$.

$$472°\left(\dfrac{\pi}{180°}\right)$$

Simplify (factor out the common 4).

$$\boxed{= \dfrac{118}{45}\pi}$$

Use a calculator to approximate.

$$\boxed{\approx 8.238 \text{ rad}}$$

▼ **ANSWER**

$\dfrac{23}{9}\pi$ or 8.029

▼ **YOUR TURN** Convert 460° to radians.

EXAMPLE 5 **Converting Radians to Degrees**

Convert $\dfrac{2\pi}{3}$ to degrees.

Solution:

Multiply $\dfrac{2\pi}{3}$ by $\dfrac{180°}{\pi}$.

$$\dfrac{2\pi}{3}\cdot\dfrac{180°}{\pi}$$

Simplify.

$$\boxed{= 120°}$$

▼ **ANSWER**
270°

▼ **YOUR TURN** Convert $\dfrac{3\pi}{2}$ to degrees.

EXAMPLE 6 **Converting Radians to Degrees**

Convert 10 radians to degrees.

Solution:

Multiply 10 radians by $\dfrac{180°}{\pi}$.

$$10\cdot\dfrac{180°}{\pi}$$

Simplify.

$$= \dfrac{1800°}{\pi} \approx \boxed{573°}$$

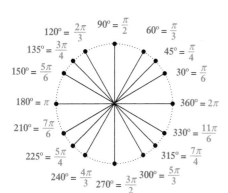

Since $\pi = 180°$, we know the following special angles:

$$\dfrac{\pi}{2} = 90° \quad \dfrac{\pi}{3} = 60° \quad \dfrac{\pi}{4} = 45° \quad \dfrac{\pi}{6} = 30°$$

and we can now draw the unit circle with the special angles in both **degrees** and **radians**.

The following table lists sine and cosine values for special angles in both degrees and radians. Tangent, secant, cosecant, and cotangent values can all be found from sine and cosine values using quotient and reciprocal identities. The table only lists special angles in quadrant I and quadrantal angles ($0° \leq \theta \leq 360°$ or $0 \leq \theta \leq 2\pi$). Values in quadrants II, III, and IV can be found using reference angles and knowledge of the algebraic sign ($+$ or $-$) of the sine and cosine functions in each quadrant.

ANGLE, θ		VALUE OF TRIGONOMETRIC FUNCTION	
RADIANS	DEGREES	$\sin \theta$	$\cos \theta$
0	0°	0	1
$\dfrac{\pi}{6}$	30°	$\dfrac{1}{2}$	$\dfrac{\sqrt{3}}{2}$
$\dfrac{\pi}{4}$	45°	$\dfrac{\sqrt{2}}{2}$	$\dfrac{\sqrt{2}}{2}$
$\dfrac{\pi}{3}$	60°	$\dfrac{\sqrt{3}}{2}$	$\dfrac{1}{2}$
$\dfrac{\pi}{2}$	90°	1	0
π	180°	0	-1
$\dfrac{3\pi}{2}$	270°	-1	0
2π	360°	0	1

EXAMPLE 7   **Evaluating Trigonometric Functions for Angles in Radian Measure**

Evaluate $\sin\left(\dfrac{\pi}{3}\right)$ exactly.

Solution:

Recognize that $\dfrac{\pi}{3} = 60°$ or convert $\dfrac{\pi}{3}$ to degrees. $\dfrac{\pi}{3} \cdot \dfrac{180°}{\pi} = 60°$

Find the value of $\sin 60°$. $\sin 60° = \dfrac{\sqrt{3}}{2}$

Equate $\sin 60°$ and $\sin\left(\dfrac{\pi}{3}\right)$. $\boxed{\sin\left(\dfrac{\pi}{3}\right) = \dfrac{\sqrt{3}}{2}}$

▼ YOUR TURN   Evaluate $\cos\left(\dfrac{\pi}{3}\right)$ exactly.

▼ ANSWER
$\dfrac{1}{2}$

If the angle of the trigonometric function to be evaluated has its terminal side in quadrants II, III, or IV, then we use reference angles and knowledge of the algebraic sign ($+$ or $-$) in that quadrant. We know how to find reference angles in degrees. Now we will find reference angles in radians.

TERMINAL SIDE LIES IN . . .	DEGREES	RADIANS
QI	$\alpha = \theta$	$\alpha = \theta$
QII	$\alpha = 180° - \theta$	$\alpha = \pi - \theta$
QIII	$\alpha = \theta - 180°$	$\alpha = \theta - \pi$
QIV	$\alpha = 360° - \theta$	$\alpha = 2\pi - \theta$

EXAMPLE 8   **Finding Reference Angles in Radians**

Find the reference angle for each angle given.

a. $\dfrac{3\pi}{4}$ b. $\dfrac{11\pi}{6}$

Solution (a):

The terminal side of θ lies in quadrant II.

Recall that π radians is $\frac{1}{2}$ of a full revolution, so $\frac{3}{4}\pi$ is $\frac{3}{4}$ of a half of revolution.

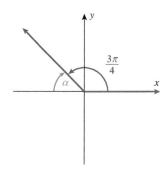

The reference angle is made with the terminal side and the negative x-axis.

$$\pi - \frac{3\pi}{4} = \frac{4\pi}{4} - \frac{3\pi}{4} = \boxed{\frac{\pi}{4}}$$

Solution (b):

The terminal side of θ lies in quadrant IV.

Recall that 2π is a complete revolution. Note that $\dfrac{11\pi}{6}$ is not quite $2\pi\left(\text{or } \dfrac{12\pi}{6}\right)$.

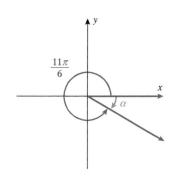

The reference angle is made with the terminal side and the positive x-axis.

$$2\pi - \frac{11\pi}{6} = \frac{12\pi}{6} - \frac{11\pi}{6} = \boxed{\frac{\pi}{6}}$$

▼
ANSWER
$\dfrac{\pi}{3}$

▼
YOUR TURN   Find the reference angle for $\dfrac{5\pi}{3}$.

▶ **EXAMPLE 9**   **Evaluating Trigonometric Functions for Angles in Radian Measure Using Reference Angles**

Evaluate $\cos\left(\dfrac{5\pi}{4}\right)$ exactly.

Solution:

The terminal side of angle $\dfrac{5\pi}{4}$ lies in quadrant III since $\dfrac{5\pi}{4} = \pi + \dfrac{\pi}{4}$.

The reference angle is $\dfrac{\pi}{4} = 45°$.

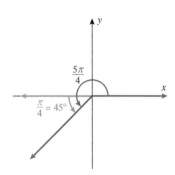

Find the cosine value for the reference angle.

$$\cos\left(\frac{\pi}{4}\right) = \cos 45° = \frac{\sqrt{2}}{2}$$

Determine the algebraic sign for the cosine function in quadrant III.

Negative $(-)$

Combine the algebraic sign of the cosine function in quadrant III with the value of the cosine function of the reference angle.

$$\boxed{\cos\left(\frac{5\pi}{4}\right) = -\frac{\sqrt{2}}{2}}$$

Confirm with a calculator.

$-0.707 \approx -0.707$

▼

YOUR TURN   Evaluate $\sin\left(\dfrac{7\pi}{4}\right)$ exactly.

▼
ANSWER
$-\dfrac{\sqrt{2}}{2}$

◉ **[SECTION 3.1]   SUMMARY**

In this section, a second measure of angles was introduced, which allows us to write trigonometric functions as functions of real numbers. A central angle of a circle has radian measure equal to the ratio of the arc length intercepted by the angle to the radius of the circle, $\theta = \dfrac{s}{r}$.

Radians and degrees are related by the relation that $\pi = 180°$.

■ To convert from radians to degrees, multiply the radian measure by $\dfrac{180°}{\pi}$.

■ To convert from degrees to radians, multiply the degree measure by $\dfrac{\pi}{180°}$.

One radian is approximately equal to $57°$. Careful attention must be paid to what mode (degrees or radians) calculators are set when evaluating trigonometric functions. To evaluate a trigonometric function for nonacute angles in radians, we use reference angles (in radians) and knowledge of the algebraic sign of the trigonometric function.

[SECTION 3.1] EXERCISES

• SKILLS

In Exercises 1–10, find the measure (in radians) of a central angle θ that intercepts an arc on a circle of radius r with indicated arc length s.

1. $r = 10\,\text{cm}$, $s = 2\,\text{cm}$
2. $r = 20\,\text{cm}$, $s = 2\,\text{cm}$
3. $r = 22\,\text{in.}$, $s = 4\,\text{in.}$
4. $r = 6\,\text{in.}$, $s = 1\,\text{in.}$
5. $r = 100\,\text{cm}$, $s = 20\,\text{mm}$
6. $r = 1\,\text{m}$, $s = 2\,\text{cm}$
7. $r = \frac{1}{4}\,\text{in.}$, $s = \frac{1}{32}\,\text{in.}$
8. $r = \frac{3}{4}\,\text{cm}$, $s = \frac{3}{14}\,\text{cm}$
9. $r = 2.5\,\text{cm}$, $s = 5\,\text{mm}$
10. $r = 1.6\,\text{cm}$, $s = 0.2\,\text{mm}$

In Exercises 11–24, convert each angle measure from degrees to radians. Leave answers in terms of π.

11. $30°$
12. $60°$
13. $45°$
14. $90°$
15. $315°$
16. $270°$
17. $75°$
18. $100°$
19. $170°$
20. $340°$
21. $780°$
22. $540°$
23. $-210°$
24. $-320°$

In Exercises 25–38, convert each angle measure from radians to degrees.

25. $\dfrac{\pi}{6}$
26. $\dfrac{\pi}{4}$
27. $\dfrac{3\pi}{4}$
28. $\dfrac{7\pi}{6}$
29. $\dfrac{3\pi}{8}$
30. $\dfrac{11\pi}{9}$
31. $\dfrac{5\pi}{12}$
32. $\dfrac{7\pi}{3}$
33. 9π
34. -6π
35. $\dfrac{19\pi}{20}$
36. $\dfrac{13\pi}{36}$
37. $-\dfrac{7\pi}{15}$
38. $-\dfrac{8\pi}{9}$

In Exercises 39–44, convert each angle measure from radians to degrees. Round answers to the nearest hundredth of a degree.

39. 4
40. 3
41. 0.85
42. 3.27
43. -2.7989
44. -5.9841

In Exercises 45–50, convert each angle measure from degrees to radians. Round answers to three significant digits.

45. $47°$
46. $65°$
47. $112°$
48. $172°$
49. $56.5°$
50. $298.7°$

In Exercises 51–58, find the reference angle for each of the following angles in terms of both radians and degrees.

51. $\dfrac{2\pi}{3}$
52. $\dfrac{3\pi}{4}$
53. $\dfrac{7\pi}{4}$
54. $\dfrac{5\pi}{4}$
55. $\dfrac{5\pi}{12}$
56. $\dfrac{7\pi}{12}$
57. $\dfrac{4\pi}{3}$
58. $\dfrac{9\pi}{4}$

In Exercises 59–84, find the *exact* value of the following expressions. Do not use a calculator.

59. $\sin\left(\dfrac{\pi}{4}\right)$
60. $\cos\left(\dfrac{\pi}{6}\right)$
61. $\sin\left(\dfrac{7\pi}{4}\right)$
62. $\cos\left(\dfrac{2\pi}{3}\right)$

63. $\sin\left(-\dfrac{3\pi}{4}\right)$
64. $\cos\left(-\dfrac{7\pi}{6}\right)$
65. $\sin\left(\dfrac{4\pi}{3}\right)$
66. $\cos\left(\dfrac{11\pi}{6}\right)$

67. $\sin\left(-\dfrac{\pi}{6}\right)$
68. $\cos\left(-\dfrac{\pi}{4}\right)$
69. $\cos\left(-\dfrac{5\pi}{3}\right)$
70. $\sin\left(\dfrac{5\pi}{6}\right)$

71. $\tan\left(\dfrac{11\pi}{6}\right)$
72. $\tan\left(\dfrac{5\pi}{3}\right)$
73. $\tan\left(\dfrac{\pi}{6}\right)$
74. $\tan\left(\dfrac{5\pi}{6}\right)$

75. $\tan\left(-\dfrac{5\pi}{6}\right)$
76. $\tan\left(-\dfrac{3\pi}{4}\right)$
77. $\cot\left(\dfrac{3\pi}{2}\right)$
78. $\csc\left(-\dfrac{\pi}{2}\right)$

79. $\sec(5\pi)$
80. $\cot\left(-\dfrac{3\pi}{2}\right)$
81. $\sin\left(\dfrac{13\pi}{4}\right)$
82. $\cos\left(\dfrac{11\pi}{3}\right)$

83. $\cos\left(-\dfrac{17\pi}{6}\right)$
84. $\sin\left(-\dfrac{8\pi}{3}\right)$

• **APPLICATIONS**

For Exercises 85 and 86, refer to the following:

Two electronic signals that are not co-phased are called out of phase. Two signals that cancel each other out are said to be 180° out of phase, or the difference in their phases is 180°.

85. Electronic Signals. How many radians out of phase are two signals whose phase difference is 270°?

86. Electronic Signals. How many radians out of phase are two signals whose phase difference is 110°?

87. Construction. In China, you find circular clan homes called *tulou*. Some tulou are three or four stories high and exceed 70 meters in diameter. If a wedge or section on the third floor of such a building has a central angle measuring 36°, how many radians is this?

Kin Cheung/Reuters

88. Construction. In China, you find circular clan homes called *tulou*. Some tulou are three or four stories high and exceed 70 meters in diameter. If a wedge or section on the third floor of such a building has a central angle measuring 72°, how many radians is this?

89. Clock. How many radians does the second hand of a clock turn in $2\frac{1}{2}$ minutes?

90. Clock. How many radians does the second hand of a clock turn in 3 minutes and 15 seconds?

91. London Eye. The London Eye has 32 capsules (each capable of holding 25 passengers with an unobstructed view of London). What is the radian measure of the angle made between the center of the wheel and the spokes aligning with each capsule?

92. Space Needle. The space needle in Seattle has a restaurant that offers views of Mount Rainier and Puget Sound. The restaurant completes one full rotation in approximately 45 minutes. How many radians will the restaurant have rotated in 25 minutes?

93. Sprinkler. A water sprinkler can reach an arc of 15 feet, 20 feet from the sprinkler as shown. Through how many radians does the sprinkler rotate?

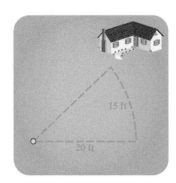

94. Sprinkler. A sprinkler is set to reach an arc of 35 feet, 15 feet from the sprinkler. Through how many radians does the sprinkler rotate?

95. Engine. If a car engine is said to be running at 1500 RPMs (revolutions per minute), through how many radians is the engine turning every second?

96. Engine. If a car engine is said to rotate 15,000° per second, through how many radians does the engine turn each second?

For Exercises 97 and 98, refer to the following:

A traction splint is commonly used to treat complete long-bone fractures of the leg. The angle between the leg and torso is an oblique angle θ. The reference angle α is the acute angle between the leg in traction and the bed.

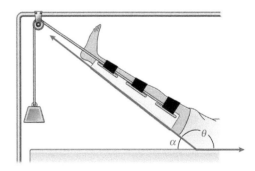

97. Health/Medicine. If $\theta = \dfrac{3\pi}{4}$, find the measure of the reference angle in both radians and degrees.

98. Health/Medicine. If $\theta = \dfrac{2\pi}{3}$, find the measure of the reference angle in both radians and degrees.

For Exercises 99–102, refer to the following:

A water molecule is composed of one oxygen atom and two hydrogen atoms and exhibits a bent shape with the oxygen atom at the center.

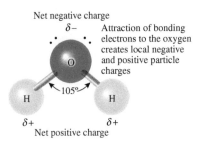

Net negative charge

$\delta-$ Attraction of bonding electrons to the oxygen creates local negative and positive particle charges

105°

H   H

$\delta+$ $\delta+$

Net positive charge

99. **Chemistry.** The angle between the O–H bonds in a water molecule is approximately 105°. Find the angle between the O–H bonds of a water molecule in radians.

100. **Chemistry.** The angle between the S–O bonds in sulfur dioxide (SO_2) is approximately 120°. Find the angle between the S–O bonds of sulfur dioxide in radians.

101. **Chemistry/Environment.** Nitrogen dioxide (NO_2) is a toxic gas and prominent air pollutant. The angle between the N–O bond is 134.3°. Find the angle between the N–O bonds in radians.

119.7 pm

N

O   O

134.3°

102. **Chemistry/Environment.** Methane (CH_4) is a chemical compound and potent greenhouse gas. The angle between the C–H bonds is 109.5°. Find the angle between the C–H bonds in radians.

H   108.70 pm

C   H

H

109.5°

● **CATCH THE MISTAKE**

In Exercises 103–106, explain the mistake that is made.

103. What is the measure (in radians) of a central angle θ that intercepts an arc of length 6 centimeters on a circle with radius 2 meters?

Solution:

Write the formula for radians.

Substitute $s = 6$, $r = 2$. $\theta = \dfrac{6}{2}$

Write the angle in terms of radians. $\theta = 3$ rad

This is incorrect. What mistake was made?

104. What is the measure (in radians) of a central angle θ that intercepts an arc of length 2 inches on a circle with radius 1 foot?

Solution:

Write the formula for radians. $\theta = \dfrac{s}{r}$

Substitute $s = 2$, $r = 1$. $\theta = \dfrac{2}{1}$

Write the angle in terms of radians. $\theta = 2$ rad

This is incorrect. What mistake was made?

105. Evaluate $6\tan(45) + 5\sec\left(\dfrac{\pi}{3}\right)$.

Solution:

Evaluate $\tan(45)$ and $\sec\left(\dfrac{\pi}{3}\right)$. $\tan(45) = 1$ $\sec\left(\dfrac{\pi}{3}\right) = 2$

Substitute the values of the trigonometric functions. $6\tan(45) + 5\sec\left(\dfrac{\pi}{3}\right) = 6(1) + 5(2)$

Simplify. $6\tan(45) + 5\sec\left(\dfrac{\pi}{3}\right) = 16$

This is incorrect. What mistake was made?

106. Approximate with a calculator $\cos(42) + \tan(65) - \sin(12)$. Round to three decimal places.

Solution:

Evaluate the trigonometric functions individually.

$\cos(42) \approx 0.743$ $\tan(65) \approx 2.145$ $\sin(12) \approx 0.208$

Substitute the values into the expression.

$\cos(42) + \tan(65) - \sin(12) \approx 0.743 + 2.145 - 0.208$

Simplify.

$\cos(42) + \tan(65) - \sin(12) \approx 2.680$

This is incorrect. What mistake was made?

• CONCEPTUAL

In Exercises 107–110, determine whether each statement is true or false.

107. An angle with measure 4 radians is a quadrant II angle.

108. Angles expressed exactly in radian measure are always given in terms of π.

109. For an angle with positive measure, it is possible for the numerical values of the degree and radian measures to be equal.

110. The sum of the angles with radian measure in a triangle is π.

111. Find the sum of complementary angles in radian measure.

112. How many complete revolutions does an angle with measure 92 radians make?

• CHALLENGE

113. The distance between Atlanta, Georgia, and Boston, Massachusetts, is approximately 900 miles along the curved surface of the Earth. The radius of the Earth is approximately 4000 miles. What is the central angle with vertex at the center of the Earth and sides of the angles intersecting the surface of the Earth in Atlanta and Boston?

114. The radius of the Earth is approximately 6400 kilometers. If a central angle, with vertex at the center of the Earth, intersects the surface of the Earth in London (UK) and Rome (Italy) with a central angle of 0.22 radians, what is the distance along the Earth's surface between London and Rome? Round to the nearest hundred kilometers.

115. At 8:20, what is the radian measure of the smaller angle between the hour hand and minute hand?

116. At 9:05, what is the radian measure of the larger angle between the hour hand and minute hand?

117. Find the exact value for
$$5\cos\left(3x + \frac{\pi}{2}\right) - 2\sin(2x) + 5 \text{ for } x = \frac{\pi}{3}.$$

118. Find the exact value for
$$-2\cos\left(3x + \frac{\pi}{3}\right) - 2\sin\left(\frac{x}{6}\right) + 5 \text{ for } x = -\pi.$$

• TECHNOLOGY

119. With a calculator set in radian mode, find $\sin 42$. With a calculator set in degree mode, find $\sin\left(42\dfrac{180°}{\pi}\right)$. Why do your results make sense?

120. With a calculator set in radian mode, find $\cos 5$. With a calculator set in degree mode, find $\cos\left(5\dfrac{180°}{\pi}\right)$. Why do your results make sense?

3.2 ARC LENGTH AND AREA OF A CIRCULAR SECTOR

SKILLS OBJECTIVES	CONCEPTUAL OBJECTIVES
▪ Calculate the length of an arc along a circle. ▪ Calculate the area of a circular sector.	▪ Understand why the angle in the arc length formula must be in radian measure. ▪ Understand that the central angle should be given in radians to use the area of a circular sector formula.

In Section 3.1, radian measure was defined in terms of the ratio of a circular arc of length s and length of the circle's radius r.

$$\theta \text{ (in radians)} = \frac{s}{r}$$

In this section (3.2) and the next (3.3), we look at applications of radian measure that involve calculating *arc lengths* and *areas of circular sectors* and calculating *angular and linear speeds*.

3.2.1 Arc Length

From geometry we know the length of an arc of a circle is proportional to its central angle. In Section 3.1, we learned that for the special case when the arc length is equal to the circumference of the circle, the angle measure in radians corresponding to one full rotation is 2π. Let us now assume that we are given the central angle and we want to find the arc length.

WORDS	MATH
Write the definition of radian measure.	$\theta = \dfrac{s}{r}$
Multiply both sides of the equation by r.	$r \cdot \theta = \dfrac{s}{r} \cdot r$
Simplify.	$r\theta = s$

The formula $s = r\theta$ is true only when θ is in radians. To develop a formula when θ is in degrees, we multiply θ by $\dfrac{\pi}{180°}$ to convert the angle measure to radians.

DEFINITION Arc Length

If a central angle θ in a circle with radius r intercepts an arc on the circle of length s, then the **arc length** s is given by

$s = r\theta_r$	θ_r is in radians.
$s = r\theta_d\left(\dfrac{\pi}{180°}\right)$	θ_d is in degrees.

EXAMPLE 1 Finding Arc Length When the Angle Has Radian Measure

In a circle with radius 10 centimeters, an arc is intercepted by a central angle with measure $\dfrac{7\pi}{4}$. Find the arc length.

Solution:

Write the formula for arc length when the angle has radian measure. $s = r\theta_r$

Substitute $r = 10$ centimeters and $\theta_r = \dfrac{7\pi}{4}$. $s = (10\,\text{cm})\left(\dfrac{7\pi}{4}\right)$

Simplify. $s = \boxed{\dfrac{35\pi}{2}\,\text{cm}}$

YOUR TURN In a circle with radius 15 inches, an arc is intercepted by a central angle with measure $\dfrac{\pi}{3}$. Find the arc length.

EXAMPLE 2 **Finding Arc Length When the Angle Has Degree Measure**

In a circle with radius 7.5 centimeters, an arc is intercepted by a central angle with measure 76°. Find the arc length. Approximate the arc length to the nearest centimeter.

Solution:

Write the formula for arc length when the angle has degree measure.

$$s = r\theta_d\left(\frac{\pi}{180°}\right)$$

Substitute $r = 7.5$ centimeters and $\theta_d = 76°$.

$$s = (7.5\text{ cm})(76°)\left(\frac{\pi}{180°}\right)$$

Evaluate the result with a calculator.

$$s \approx 9.948\text{ cm}$$

Round to the nearest centimeter.

$$\boxed{s \approx 10\text{ cm}}$$

▼

YOUR TURN In a circle with radius 20 meters, an arc is intercepted by a central angle with measure 113°. Find the arc length. Approximate the arc length to the nearest meter.

▼ ANSWER

39 m

EXAMPLE 3 **Path of International Space Station**

The International Space Station (ISS) is in an approximate circular orbit 400 kilometers above the surface of the Earth. If the ground station tracks the space station when it is within a 45° central angle of this circular orbit about the center of the Earth above the tracking antenna, how many kilometers does the ISS cover while it is being tracked by the ground station? Assume that the radius of the Earth is 6400 kilometers. Round to the nearest kilometer.

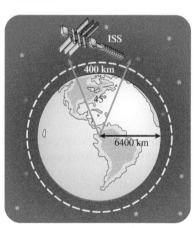

Solution:

Write the formula for arc length when the angle has degree measure.

$$s = r\theta_d\left(\frac{\pi}{180°}\right)$$

Recognize that the radius of the orbit is $r = 6400 + 400 = 6800$ kilometers and that $\theta_d = 45°$.

$$s = (6800\text{ km})(45°)\left(\frac{\pi}{180°}\right)$$

Evaluate with a calculator.

$$s \approx 5340.708\text{ km}$$

Round to the nearest kilometer.

$$\boxed{s \approx 5341\text{ km}}$$

The ISS travels approximately 5341 kilometers during the ground station tracking.

▼

YOUR TURN If the ground station in Example 3 could track the ISS within a 60° central angle of its circular orbit about the center of the Earth, how far would the ISS travel during the ground station tracking?

▼ ANSWER

7121 km

▶ **EXAMPLE 4** **Gears**

Gears are inside many devices like automobiles and power meters. When the smaller gear drives the larger gear, then typically the driving gear is rotated faster than a larger gear would be if it were the drive gear. In general, smaller ratios of radius of the driving gear to the driven gear are called for when machines are expected to yield more power. The smaller gear has a radius of 3 centimeters, and the larger gear has a radius of 6.4 centimeters. If the smaller gear rotates 170°, how many degrees has the larger gear rotated? Round the answer to the nearest degree.

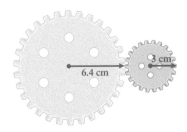

Solution:

Recognize that the small gear arc length = the large gear arc length.

Smaller Gear

Write the formula for arc length when the angle has degree measure.

$$s = r\theta_d \left(\frac{\pi}{180°} \right)$$

Substitute the values for the smaller gear: $r = 3$ centimeters and $\theta_d = 170°$.

$$s_{\text{smaller}} = (3 \text{ cm})(170°) \left(\frac{\pi}{180°} \right)$$

Simplify.

$$s_{\text{smaller}} = \left(\frac{17\pi}{6} \right) \text{ cm}$$

Larger Gear

Remember that the larger gear's arc length is equal to the smaller gear's arc length.

$$s = \left(\frac{17\pi}{6} \right) \text{ cm}$$

Write the formula for arc length when the angle has degree measure.

$$s = r\theta_d \left(\frac{\pi}{180°} \right)$$

Substitute $s = \left(\frac{17\pi}{6} \right)$ centimeter and $r = 6.4$ centimeters

$$\left(\frac{17\pi}{6} \text{ cm} \right) = (6.4 \text{ cm})\theta_d \left(\frac{\pi}{180°} \right)$$

Solve for θ_d.

$$\theta_d = \frac{180°}{\pi} \cdot \frac{17\pi \text{ cm}}{6(6.4 \text{ cm})}$$

Simplify.

$$\theta_d \approx 79.6875°$$

Round to the nearest degree.

$$\boxed{\theta_d = 80°}$$

The larger gear rotates approximately 80°.

3.2.2 **SKILL**

Calculate the area of a circular sector.

3.2.2 **CONCEPTUAL**

Understand that the central angle should be given in radians to use the area of a circular sector formula.

3.2.2 Area of a Circular Sector

A restaurant lists a piece of French silk pie as having 400 calories. How does the chef arrive at that number? She calculates the calories of all the ingredients that went into making the entire pie and then divides by the number of slices the pie yields. For example, if an entire pie has 3200 calories and it is sliced into eight equal pieces, then

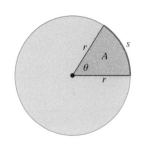

each piece has 400 calories. Although that example involves volume, the idea is the same with areas of sectors of circles. *Circular sectors* can be thought of as "pieces of a pie."

Recall that arc lengths of a circle are proportional to the central angle (in radians) and the radius. Similarly, a circular sector is a portion of the entire circle. Let A represent the area of the *sector of the circle* and θ_r represent the central angle (in radians) that forms the sector. Then, let us consider the entire circle whose area is πr^2 and the angle that represents one full rotation has measure 2π (radians).

WORDS	MATH
Write the ratio of the area of the sector to the area of the entire circle.	$\dfrac{A}{\pi r^2}$
Write the ratio of the central angle θ_r to the measure of one full rotation.	$\dfrac{\theta_r}{2\pi}$
The ratios must be equal (proportionality of sector to circle).	$\dfrac{A}{\pi r^2} = \dfrac{\theta_r}{2\pi}$
Multiply both sides of the equation by πr^2.	$\pi r^2 \cdot \dfrac{A}{\pi r^2} = \dfrac{\theta_r}{2\pi} \cdot \pi r^2$
Simplify.	$A = \dfrac{1}{2} r^2 \theta_r$

DEFINITION Area of a Circular Sector

The **area of a sector of a circle** with radius r and central angle θ is given by

$A = \dfrac{1}{2} r^2 \theta_r$	θ_r is in radians.
$A = \dfrac{1}{2} r^2 \theta_d \left(\dfrac{\pi}{180°} \right)$	θ_d is in degrees.

[CONCEPT CHECK]

If the area of a circular sector A is given by $A = \dfrac{1}{2} r^2 \theta_r$ and r is given in feet, then what are the units of A?

▼

ANSWER Square feet

EXAMPLE 5 **Finding the Area of a Circular Sector When the Angle Has Radian Measure**

Find the area of the sector associated with a single slice of pizza if the entire pizza has a 14-inch diameter and the pizza is cut into eight equal pieces.

Solution:

The radius is half the diameter.	$r = \dfrac{14}{2} = 7 \text{ in.}$
Find the angle of each slice if the pizza is cut into eight pieces ($\tfrac{1}{8}$ of the complete 2π revolution).	$\theta_r = \dfrac{2\pi}{8} = \dfrac{\pi}{4}$
Write the formula for circular sector area in radians.	$A = \dfrac{1}{2} r^2 \theta_r$

Substitute $r = 7$ inches and $\theta_r = \dfrac{\pi}{4}$ into the area equation.

$$A = \frac{1}{2}(7 \text{ in.})^2\left(\frac{\pi}{4}\right)$$

Simplify.

$$\boxed{A = \frac{49\pi}{8} \text{ in.}^2}$$

Approximate the area with a calculator.

$$\boxed{A \approx 19 \text{ in.}^2}$$

▼
ANSWER

8π in.$^2 \approx 25$ in.^{2}

▼
YOUR TURN Find the area of a slice of pizza (cut into eight equal pieces) if the entire pizza has a 16-inch diameter.

▶ **EXAMPLE 6** **Finding the Area of a Circular Sector When the Angle Has Degree Measure**

Sprinkler heads come in all different sizes depending on the angle of rotation desired. If a sprinkler head rotates 90° and has enough pressure to keep a constant 25-foot spray, what is the area of the sector of the lawn that gets watered? Round to the nearest square foot.

Solution:

Write the formula for circular sector area in degrees.

$$A = \frac{1}{2}r^2\theta_d\left(\frac{\pi}{180°}\right)$$

Substitute $r = 25$ feet and $\theta_d = 90°$ into the area equation.

$$A = \frac{1}{2}(25 \text{ ft})^2(90°)\left(\frac{\pi}{180°}\right)$$

Simplify.

$$A = \left(\frac{625\pi}{4}\right) \text{ ft}^2 \approx 490.87 \text{ ft}^2$$

Round to the nearest square foot.

$$\boxed{A \approx 491 \text{ ft}^2}$$

▼
ANSWER

450π ft$^2 \approx 1414$ ft^2

▼
YOUR TURN If a sprinkler head rotates 180° and has enough pressure to keep a constant 30-foot spray, what is the area of the sector of the lawn it can water? Round to the nearest square foot.

▶[SECTION 3.2] **SUMMARY**

In this section, we used the proportionality concept (both the arc length and area of a sector are proportional to the central angle of a circle). The definition of radian measure was used to develop formulas for the **arc length** of a circle when the central angle is given in either radians or degrees.

The formula for the **area of a sector of a circle** was also developed for the cases in which the central angle is given in either radians or degrees.

$s = r\theta_r$	θ_r is in radians.
$s = r\theta_d\left(\dfrac{\pi}{180°}\right)$	θ_d is in degrees.

$A = \dfrac{1}{2}r^2\theta_r$	θ_r is in radians.
$A = \dfrac{1}{2}r^2\theta_d\left(\dfrac{\pi}{180°}\right)$	θ_d is in degrees.

[SECTION 3.2] EXERCISES

• SKILLS

In Exercises 1–12, find the exact length of each arc made by the indicated central angle and radius of each circle.

1. $\theta = 3, r = 4$ mm

2. $\theta = 4, r = 5$ cm

3. $\theta = \dfrac{\pi}{12}, r = 8$ ft

4. $\theta = \dfrac{\pi}{8}, r = 6$ yd

5. $\theta = \dfrac{2\pi}{7}, r = 3.5$ m

6. $\theta = \dfrac{\pi}{4}, r = 10$ in.

7. $\theta = 22°, r = 18\,\mu$m

8. $\theta = 14°, r = 15\,\mu$m

9. $\theta = 8°, r = 1500$ km

10. $\theta = 3°, r = 1800$ km

11. $\theta = 48°, r = 24$ cm

12. $\theta = 30°, r = 120$ cm

In Exercises 13–24, find the exact length of each radius given the arc length and central angle of each circle.

13. $s = \dfrac{5\pi}{2}$ ft, $\theta = \dfrac{\pi}{10}$

14. $s = \dfrac{5\pi}{6}$ m, $\theta = \dfrac{\pi}{12}$

15. $s = \dfrac{24\pi}{5}$ in., $\theta = \dfrac{3\pi}{5}$

16. $s = \dfrac{5\pi}{9}$ km, $\theta = \dfrac{\pi}{180}$

17. $s = \dfrac{12\pi}{5}$ yd, $\theta = \dfrac{4\pi}{5}$

18. $s = 4\pi$ in., $\theta = \dfrac{3\pi}{2}$

19. $s = \dfrac{4\pi}{9}$ yd, $\theta = 20°$

20. $s = \dfrac{11\pi}{6}$ cm, $\theta = 15°$

21. $s = \dfrac{8\pi}{3}$ mi, $\theta = 40°$

22. $s = \dfrac{\pi}{4}\,\mu$m, $\theta = 30°$

23. $s = \dfrac{2\pi}{11}$ km, $\theta = 45°$

24. $s = \dfrac{3\pi}{16}$ ft, $\theta = 35°$

In Exercises 25–36, use a calculator to approximate the length of each arc made by the indicated central angle and radius of each circle. Round answers to two significant digits.

25. $\theta = 3.3, r = 0.4$ mm

26. $\theta = 2.4, r = 5.5$ cm

27. $\theta = \dfrac{\pi}{15}, r = 8$ yd

28. $\theta = \dfrac{\pi}{10}, r = 6$ ft

29. $\theta = 4.95, r = 30$ mi

30. $\theta = \dfrac{7\pi}{8}, r = 17$ mm

31. $\theta = 79.5°, r = 1.55\,\mu$m

32. $\theta = 19.7°, r = 0.63\,\mu$m

33. $\theta = 29°, r = 2500$ km

34. $\theta = 11°, r = 2200$ km

35. $\theta = 57°, r = 22$ ft

36. $\theta = 127°, r = 58$ in.

In Exercises 37–48, find the area of the circular sector given the indicated radius and central angle. Round answers to three significant digits.

37. $\theta = \dfrac{\pi}{6}, r = 7$ ft

38. $\theta = \dfrac{\pi}{5}, r = 3$ in.

39. $\theta = \dfrac{3\pi}{8}, r = 2.2$ km

40. $\theta = \dfrac{5\pi}{6}, r = 13$ mi

41. $\theta = \dfrac{3\pi}{11}, r = 10$ cm

42. $\theta = \dfrac{2\pi}{3}, r = 33$ m

43. $\theta = 56°, r = 4.2$ cm

44. $\theta = 27°, r = 2.5$ mm

45. $\theta = 1.2°, r = 1.5$ ft

46. $\theta = 14°, r = 3.0$ ft

47. $\theta = 22.8°, r = 2.6$ mi

48. $\theta = 60°, r = 15$ km

● **APPLICATIONS**

49. Low Earth Orbit Satellites. A low Earth orbit (LEO) satellite is in an approximate circular orbit 300 kilometers above the surface of the Earth. If the ground station tracks the satellite when it is within a 45° cone above the tracking antenna (directly overhead), how many kilometers does the satellite cover during the ground station track? Assume the radius of the Earth is 6400 kilometers. Round your answer to the nearest kilometer.

50. Low Earth Orbit Satellites. A low Earth orbit (LEO) satellite is in an approximate circular orbit 250 kilometers above the surface of the Earth. If the ground station tracks the satellite when it is within a 30° cone above the tracking antenna (directly overhead), how many kilometers does the satellite cover during the ground station track? Assume the radius of the Earth is 6400 kilometers. Round your answer to the nearest kilometer.

51. Big Ben. The famous clock tower in London has a minute hand that is 14 feet long. How far does the tip of the minute hand of Big Ben travel in 25 minutes? Round your answer to the nearest foot.

Peter Adams/Digital Vision/Getty Images, Inc.

52. Big Ben. The famous clock tower in London has a minute hand that is 14 feet long. How far does the tip of the minute hand of Big Ben travel in 35 minutes? Round your answer to two decimal places.

53. London Eye. The London Eye is a wheel that has 32 capsules and a diameter of 400 feet. What is the distance someone has traveled once they reach the highest point for the first time?

David Ball/Getty Images, Inc.

54. London Eye. Assuming the wheel stops at each capsule in Exercise 53, what is the distance someone has traveled from the point he or she first gets in the capsule to the point at which the Eye stops for the sixth time during the ride?

55. Gears. The smaller gear shown below has a radius of 5 centimeters, and the larger gear has a radius of 12.1 centimeters. If the smaller gear rotates 120°, how many degrees has the larger gear rotated? Round the answer to the nearest degree.

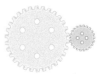

56. Gears. The smaller gear has a radius of 3 inches, and the larger gear has a radius of 15 inches (see the figure above). If the smaller gear rotates 420°, how many degrees has the larger gear rotated? Round the answer to the nearest degree.

57. Bicycle Low Gear. If a bicycle has 26-inch diameter wheels, the front chain drive has a radius of 2.2 inches, and the back drive has a radius of 3 inches, how far does the bicycle travel for every one rotation of the cranks (pedals)?

58. Bicycle High Gear. If a bicycle has 26-inch diameter wheels, the front chain drive has a radius of 4 inches, and the back drive has a radius of 1 inch, how far does the bicycle travel for every one rotation of the cranks (pedals)?

Getty Images, Inc.

59. Odometer. A Ford Expedition comes standard with -inch rims (which corresponds to a tire with -inch diameter). Suppose you decide to later upgrade these tires for -inch rims (corresponding to a tire with -inch diameter). If you do not get your onboard computer reset for the new tires, the odometer will not be accurate. After your new tires have actually driven 1000 miles, how many miles will the odometer report the Expedition has been driven? Round to the nearest mile.

60. Odometer. For the same Ford Expedition in Exercise 59, after you have driven 50,000 miles, how many miles will the odometer report the Expedition has been driven if the computer is not reset to account for the new oversized tires? Round to the nearest mile.

61. Sprinkler Coverage. A sprinkler has a 20-foot spray and covers an angle of 45°. What is the area that the sprinkler waters?

62. Sprinkler Coverage. A sprinkler has a 22-foot spray and covers an angle of 60°. What is the area that the sprinkler waters?

63. Windshield Wiper. A windshield wiper that is 12 inches long (blade and arm) rotates 70°. If the rubber part is 8 inches long, what is the area cleared by the wiper? Round to the nearest square inch.

64. Windshield Wiper. A windshield wiper that is 11 inches long (blade and arm) rotates 65°. If the rubber part is 7 inches long, what is the area cleared by the wiper? Round to the nearest square inch.

65. Bicycle Wheel. A bicycle wheel 26 inches in diameter travels 45° in 0.05 seconds. Through how many revolutions does the wheel turn in 30 seconds?

66. Bicycle Wheel. A bicycle wheel 26 inches in diameter travels $\dfrac{2\pi}{3}$ in 0.075 seconds. Through how many revolutions does the wheel turn in 30 seconds?

67. Bicycle Wheel. A bicycle wheel 26 inches in diameter travels 20 inches in 0.10 seconds. What is the speed of the wheel in revolutions per second?

68. Bicycle Wheel. A bicycle wheel 26 inches in diameter travels at four revolutions per second. Through how many radians does the wheel turn in 0.5 seconds?

For Exercises 69 and 70, refer to the following:

Sniffers outside a chemical munitions disposal site monitor the atmosphere surrounding the site to detect any toxic gases. In the event that there is an accidental release of toxic fumes, the data provided by the sniffers make it possible to determine both the distance d that the fumes reach and the angle of spread θ that sweep out a circular sector.

69. Environment. If the maximum angle of spread is 105° and the maximum distance at which the toxic fumes were detected was 9 miles from the site, find the area of the circular sector affected by the accidental release.

70. Environment. To protect the public from the fumes, officials must secure the perimeter of this area. Find the perimeter of the circular sector in Exercise 69.

For Exercises 71 and 72, refer to the following:

The structure of human DNA is a *linear* double helix formed of nucleotide base pairs (two nucleotides) that are stacked with spacing of 3.4 angstroms (3.4×10^{-12} m). Each base pair is rotated 36° with respect to an adjacent pair and has 10 base pairs per helical turn. The DNA of a virus or a bacterium, however, is a *circular* double helix (see the figure below) with the structure varying among species.

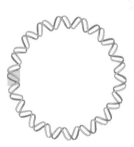

(*Source*: Karp, *Cell and Molecular Biology*, 8e, Figure 13.5, p. 517.)

71. Biology. If the circular DNA of a virus has 10 twists (or turns) per circle and an inner diameter of 4.5 nanometers, find the arc length between consecutive twists of the DNA.

72. Biology. If the circular DNA of a virus has 40 twists (or turns) per circle and an inner diameter of 2.0 nanometers, find the arc length between consecutive twists of the DNA.

● **CATCH THE MISTAKE**

In Exercises 73 and 74, explain the mistake that is made.

73. A circle with radius 5 centimeters has an arc that is made from a central angle with measure 65°. Approximate the arc length to the nearest millimeter.

Solution:

Write the formula for arc length. $\qquad s = r\theta$

Substitute $r = 5$ centimeters and $\theta = 65°$ into the formula. $\qquad s = (5 \text{ cm})(65)$

Simplify. $\qquad s = 325 \text{ cm}$

This is incorrect. What mistake was made?

74. For a circle with radius $r = 2.2$ centimeters, find the area of the circular sector with central angle measuring $\theta = 25°$. Round the answer to three significant digits.

Solution:

Write the formula for area of a circular sector. $\qquad A = \dfrac{1}{2}r^2\theta_r$

Substitute $r = 2.2$ centimeters and $\theta = 25°$ into the formula. $\qquad A = \dfrac{1}{2}(2.2 \text{ cm})^2(25°)$

Simplify. $\qquad A = 60.5 \text{ cm}^2$

This is incorrect. What mistake was made?

● **CONCEPTUAL**

In Exercises 75–78, determine whether each statement is true or false.

75. The length of an arc with central angle 45° in a unit circle is 45.

76. The length of an arc with central angle $\dfrac{\pi}{3}$ in a unit circle is $\dfrac{\pi}{3}$.

77. If the radius of a circle doubles, then the arc length (associated with a fixed central angle) doubles.

78. If the radius of a circle doubles, then the area of the sector (associated with a fixed central angle) doubles.

79. If a smaller gear has radius r_1, a larger gear has radius r_2, and the smaller gear rotates $\theta°_1$, what is the degree measure of the angle the larger gear rotates?

80. If a circle with radius r_1 has an arc length s_1 associated with a particular central angle, write the formula for the area of the sector of the circle formed by that central angle, in terms of the radius and arc length.

● **CHALLENGE**

For Exercises 81–84, refer to the following:

You may think that a baseball field is a circular sector, but it is not. If it were, the distances from home plate to left field, center field, and right field would all be the same (the radius). Where the infield dirt meets the outfield grass and along the fence in the outfield are arc lengths associated with a circle of radius 95 feet and with a vertex located at the pitcher's mound (not home plate).

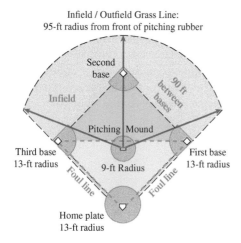

Infield / Outfield Grass Line:
95-ft radius from front of pitching rubber

Second base

90 ft between bases

Infield

Pitching Mound

Third base
13-ft radius

9-ft Radius

First base
13-ft radius

Foul line

Foul line

Home plate
13-ft radius

81. What is the area enclosed in the circular sector with radius 95 feet and central angle 150°? Round to the nearest hundred square feet.

82. Approximate the area of the infield by adding the area in blue to the result in Exercise 81. Neglect the area near first and third bases and the foul line. Round to the nearest hundred square feet.

83. If a batter wants to bunt a ball so that it is fair (in front of home plate and between the foul lines) but keep it in the dirt (in the sector in front of home plate), within how small of an area is the batter trying to keep his bunt? Round to the nearest square foot.

84. Most bunts would fall within the blue triangle in the diagram on the left. Assume the catcher only fields bunts that fall in the sector described in Exercise 83 and the pitcher only fields bunts that fall on the pitcher's mound. Approximately how much area do the first baseman and third baseman *each* need to cover? Round to the nearest square foot.

3.3 LINEAR AND ANGULAR SPEEDS

SKILLS OBJECTIVES	CONCEPTUAL OBJECTIVES
▪ Calculate linear speed. ▪ Calculate angular speed. ▪ Solve application problems involving both angular and linear speeds.	▪ Understand that linear speed has units of length/time. ▪ Understand that angular speed has units of radians/time. ▪ Understand that angular speed and linear speed are related through the radius.

In the chapter opener about a 2016 Lexus RS with standard 18-inch rims, we learned that the onboard computer that determines distance (odometer reading) and speed (speedometer) combines the number of tire rotations and the size of the tire. Because the onboard computer is set for 18-inch rims (which corresponds to a tire with 30-inch diameter), if the owner decided to upgrade to 20-inch rims (corresponding to a tire with 30.2-inch diameter), the computer would have to be updated with this new information. If the computer is not updated with the new tire size, both the odometer and speedometer readings will be incorrect.

You will see in this section that the *angular speed* (rotations of tires per second), *radius* (of the tires), and *linear speed* (speed of the automobile) are related. In the context of a circle, we will first define *linear speed*, then *angular speed*, and finally relate them using the *radius*.

3.3.1 Linear Speed

Although **velocity** and **speed** are often used as synonyms, speed is how fast you are traveling, whereas velocity is the speed *and* the direction in which you are traveling. In physics the difference between speed and velocity is that velocity has direction and is written as a vector (Chapter 7), and speed is the *magnitude* of the velocity vector, which results in a real number. In this chapter, *speed* will be used.

Recall the relationship between distance, rate, and time: $d = rt$. Rate is speed, and in words this formula can be rewritten as

$$\text{distance} = \text{speed} \cdot \text{time} \quad \text{or} \quad \text{speed} = \frac{\text{distance}}{\text{time}}$$

It is important to note that we assume speed is constant. If we think of a car driving around a circular track, the distance it travels is the arc length s, and if we let v represent speed and t represent time, we have the formula for speed around a circle (*linear speed*):

$$v = \frac{s}{t}$$

3.3.1 SKILL

Calculate linear speed.

3.3.1 CONCEPTUAL

Understand that linear speed has units of length/time.

DEFINITION Linear Speed

If a point P moves along the circumference of a circle at a constant speed, then the **linear speed** v is given by

$$v = \frac{s}{t}$$

where s is the arc length and t is the time.

Which of the following is not a measure of linear speed?
(A) Miles per hour
(B) Feet per second
(C) Degrees per second
(D) Meters per minute

▼ ···

ANSWER Degrees per second

▼ ···

ANSWER

105 mph

3.3.2 **SKILL**

Calculate angular speed.

3.3.2 **CONCEPTUAL**

Understand that angular speed has units of radians/time.

[**STUDY TIP**

The units of angular speed will be in radians per unit time (e.g., radians per minute).
]

[CONCEPT CHECK]

If we know that it takes a beacon, A minutes to complete one full rotation then what is its angular speed?

▼ ···

ANSWER $\frac{2\pi}{A}$ rad/min

EXAMPLE 1 **Linear Speed**

A car travels at a constant speed around a circular track with circumference equal to 2 miles. If the car records a time of 15 minutes for 9 laps, what is the linear speed of the car in miles per hour?

Solution:

Calculate the distance traveled around the circular track.

$$s = (9\ \text{laps})\left(\frac{2\ \text{mi}}{\text{lap}}\right) = 18\ \text{mi}$$

Substitute $t = 15$ minutes and $s = 18$ miles into $v = \dfrac{s}{t}$.

$$v = \frac{18\ \text{mi}}{15\ \text{min}}$$

Convert the linear speed from miles per minute to miles per hour.

$$v = \left(\frac{18\ \text{mi}}{15\ \text{min}}\right)\left(\frac{60\ \text{min}}{1\ \text{hr}}\right)$$

Simplify.

$$\boxed{v = 72\ \text{mph}}$$

▼

YOUR TURN A car travels at a constant speed around a circular track with circumference equal to 3 miles. If the car records a time of 12 minutes for 7 laps, what is the linear speed of the car in miles per hour?

3.3.2 Angular Speed

To calculate linear speed, we find how fast a position along the circumference of a circle is changing. To calculate **angular speed**, we find how fast the central angle is changing.

DEFINITION **Angular Speed**
If a point P moves along the circumference of a circle at a constant speed, then the central angle θ that is formed with the terminal side passing through point P also changes over some time t at a constant speed. The **angular speed** ω (omega) is given by $$\omega = \frac{\theta}{t} \quad \text{where } \theta \text{ is given in radians}$$

▶ **EXAMPLE 2** **Angular Speed**

A lighthouse in the middle of a channel rotates its light in a circular motion with constant speed. If the beacon of light completes one rotation every 10 seconds, what is the angular speed of the beacon in radians per minute?

Solution:

Calculate the angle measure in radians associated with one rotation.

$$\theta = 2\pi$$

Substitute $\theta = 2\pi$ and $t = 10$ seconds into $\omega = \dfrac{\theta}{t}$.

$$\omega = \frac{2\pi\ (\text{rad})}{10\ \text{sec}}$$

Convert the angular speed from radians per second to radians per minute.

$$\omega = \frac{2\pi \text{ (rad)}}{10 \text{ sec}} \cdot \frac{60 \text{ sec}}{1 \text{ min}}$$

Simplify.

$$\boxed{\omega = 12\pi \text{ rad/min}}$$

▼

YOUR TURN If the lighthouse in Example 2 is adjusted so that the beacon rotates one time every 40 seconds, what is the angular speed of the beacon in radians per minute?

▼
ANSWER

$\omega = 3\pi$ rad/min

3.3.3 Relationship Between Linear and Angular Speeds

In the chapter opener, we discussed the 2016 Lexus RX with 18-inch standard rims that would have odometer and speedometer errors if the owner decided to upgrade to 20-inch rims without updating the onboard computer. That is because *angular speed* (rotations of tires per second), *radius* (of the tires), and *linear speed* (speed of the automobile) are related. To see how, let us start with the definition of arc length (Section 3.2), which comes from the definition of radian measure (Section 3.1).

3.3.3 SKILL

Solve application problems involving both angular and linear speeds.

3.3.3 CONCEPTUAL

Understand that angular speed and linear speed are related through the radius.

WORDS	MATH
Write the definition of radian measure.	$\theta = \dfrac{s}{r}$
Write the definition of arc length (θ in radians).	$s = r\theta$
Divide both sides by t.	$\dfrac{s}{t} = \dfrac{r\theta}{t}$
Rewrite the right side of the equation.	$\dfrac{s}{t} = r\dfrac{\theta}{t}$
Recall the definitions of **linear** and **angular** speeds.	$v = \dfrac{s}{t}$ and $\omega = \dfrac{\theta}{t}$
Substitute $v = \dfrac{s}{t}$ and $\omega = \dfrac{\theta}{t}$ into $\dfrac{s}{t} = r\dfrac{\theta}{t}$.	$v = r\omega$

RELATING LINEAR AND ANGULAR SPEEDS

If a point *P* moves at a constant speed along the circumference of a circle with radius *r*, then the **linear speed** *v* and the **angular speed** ω are related by

$$v = r\omega \quad \text{or} \quad \omega = \frac{v}{r}$$

Note: This relationship is true only when θ is given in radians.

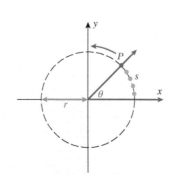

STUDY TIP

This relationship between linear speed and angular speed assumes the angle is given in radians.

[CONCEPT CHECK]

For a fixed angular speed, if the radius is increased, then the linear speed (increases/decreases)?

▼

ANSWER Increases

We now will investigate the Lexus scenario with upgraded tires. Notice that tires of two different radii with the same angular speed have different linear speeds since $v = r\omega$. The larger tire (larger r) has the faster linear speed.

▶ **EXAMPLE 3** **Relating Linear and Angular Speeds**

A 2016 Lexus RX 450 h comes standard with wheels that have a rim diameter of 18 inches and tire diameter of 28 inches. If the owner decided to upgrade the wheels with a rim diameter of 20 inches (tire diameter of 30 inches) without having the onboard computer updated, how fast will the truck *actually* be traveling when the speedometer reads 75 miles per hour?

Solution:

The computer in the 2016 Lexus RX "thinks" the tires are 28 inches in diameter and knows the angular speed. Use the programmed tire diameter and speedometer reading to calculate the angular speed. Then use that angular speed and the upgraded tire diameter to get the actual speed (linear speed).

STEP 1 **Calculate the angular speed of the tires.**

Write the formula for the angular speed. $\quad \omega = \dfrac{v}{r}$

Substitute $v = 75$ miles per hour and
$r = \dfrac{28}{2} = 14$ inches into the formula. $\quad \omega = \dfrac{75 \text{ mi/hr}}{14 \text{ in.}}$

1 mile $= 5280$ feet $= 63{,}360$ inches. $\quad \omega = \dfrac{75(63{,}360) \text{ in./hr}}{14 \text{ in.}}$

Simplify. $\quad \omega \approx 339{,}429 \dfrac{\text{rad}}{\text{hr}}$

STEP 2 **Calculate the actual linear speed of the car.**

Write the linear speed formula. $\quad v = r\omega$

Substitute $r = \dfrac{30}{2} = 15$ inches
and $\omega \approx 339{,}429$ radians per hour. $\quad v = (15 \text{ in.})\left(339{,}429 \dfrac{\text{rad}}{\text{hr}}\right)$

Simplify. $\quad v \approx 5{,}091{,}435 \dfrac{\text{in.}}{\text{hr}}$

1 mile $= 5280$ feet $= 63{,}360$ inches. $\quad v \approx 5{,}091{,}435 \dfrac{\text{in.}}{\text{hr}} \cdot \dfrac{1 \text{ mi}}{63{,}360 \text{ in.}}$

$\boxed{v \approx 80 \text{ mph}}$

Although the speedometer indicates a speed of 75 miles per hour, the actual speed is approximately $\boxed{80 \text{ miles per hour}}$.

STUDY TIP

We could have solved Example 3 the following way:

$\dfrac{75 \text{ mph}}{14 \text{ in.}} = \dfrac{x}{15 \text{ in.}}$

$x = \dfrac{15 \text{ in.}}{14 \text{ in.}} \times 75 \text{ mph}$

$\approx 80 \text{ mph}$

▼
ANSWER

Approximately 63 mph

YOUR TURN Suppose the owner of the 2016 Lexus RX in Example 3 decides to downsize the tires from their original 18-inch rims (28-in tires) to a wheel with 17-inch rims (27-in tires). If the speedometer indicates a speed of 65 miles per hour, what is the actual speed of the car?

▶[SECTION 3.3]   SUMMARY

In this section, circular motion was defined in terms of linear speed (speed along the circumference of a circle) v and angular speed (speed of angle rotation) ω.

Linear speed: $v = \dfrac{s}{t}$

Angular speed: $\omega = \dfrac{\theta}{t}$, where θ is given in radians.

Linear and angular speeds associated with circular motion are related through the radius r of the circle.

$$v = r\omega \qquad \text{or} \qquad \omega = \frac{v}{r}$$

These formulas hold true only when angular speed is given in radians per unit of time.

[SECTION 3.3]   EXERCISES

• SKILLS

In Exercises 1–10, find the linear speed of a point that moves with constant speed in a circular motion if the point travels along the circle of arc length s in time t. Label your answer with correct units.

1. $s = 2\,\text{m}, t = 5\,\text{sec}$
2. $s = 12\,\text{ft}, t = 3\,\text{min}$
3. $s = 68{,}000\,\text{km}, t = 250\,\text{hr}$
4. $s = 7{,}524\,\text{mi}, t = 12\,\text{days}$
5. $s = 1.75\,\text{nm (nanometers)}, t = 0.25\,\text{ms (milliseconds)}$
6. $s = 3.6\,\mu\text{m (microns)}, t = 9\,\text{ns (nanoseconds)}$
7. $s = \frac{1}{16}\,\text{in.}, t = 4\,\text{min}$
8. $s = \frac{2}{5}\,\text{cm}, t = 8\,\text{hr}$
9. $s = \frac{3}{10}\,\text{m}, t = 5.2\,\text{sec}$
10. $s = 12.2\,\text{mm}, t = 3.4\,\text{min}$

In Exercises 11–20, find the distance traveled (arc length) of a point that moves with constant speed v along a circle in time t.

11. $v = 2.8\,\text{m/sec}, t = 3.5\,\text{sec}$
12. $v = 6.2\,\text{km/hr}, t = 4.5\,\text{hr}$
13. $v = 4.5\,\text{mi/hr}, t = 20\,\text{min}$
14. $v = 5.6\,\text{ft/sec}, t = 2\,\text{min}$
15. $v = 60\,\text{mi/hr}, t = 15\,\text{min}$
16. $v = 72\,\text{km/hr}, t = 10\,\text{min}$
17. $v = 750\,\text{km/min}, t = 4\,\text{days}$
18. $v = 120\,\text{ft/sec}, t = 27\,\text{min}$
19. $v = 23\,\text{ft/s}, t = 3\,\text{min}$
20. $v = 46\,\text{km/hr}, t = 20\,\text{min}$

In Exercises 21–32, find the angular speed associated with rotating a central angle θ in time t.

21. $\theta = 25\pi, t = 10\,\text{sec}$
22. $\theta = \dfrac{3\pi}{4}, t = \dfrac{1}{6}\,\text{sec}$
23. $\theta = 100\pi, t = 5\,\text{min}$
24. $\theta = \dfrac{\pi}{2}, t = \dfrac{1}{10}\,\text{min}$

25. $\theta = \dfrac{7\pi}{2}, t = 12\,\text{hr}$
26. $\theta = 18.3, t = 30.45\,\text{hr}$
27. $\theta = 200°, t = 5\,\text{sec}$
28. $\theta = 60°, t = 0.2\,\text{sec}$

29. $\theta = 780°, t = 3\,\text{min}$
30. $\theta = 420°, t = 6\,\text{min}$
31. $\theta = 900°, t = 3.5\,\text{sec}$
32. $\theta = 350°, t = 5.6\,\text{sec}$

In Exercises 33–42, find the linear speed of a point traveling at a constant speed along the circumference of a circle with radius r and angular speed ω.

33. $\omega = \dfrac{2\pi\,\text{rad}}{3\,\text{sec}}, r = 9\,\text{in.}$

34. $\omega = \dfrac{3\pi\,\text{rad}}{4\,\text{sec}}, r = 8\,\text{cm}$

35. $\omega = \dfrac{\pi\,\text{rad}}{20\,\text{sec}}, r = 5\,\text{mm}$

36. $\omega = \dfrac{5\pi\,\text{rad}}{16\,\text{sec}}, r = 24\,\text{ft}$

37. $\omega = \dfrac{4\pi\,\text{rad}}{15\,\text{sec}}, r = 2.5\,\text{in.}$

38. $\omega = \dfrac{8\pi\,\text{rad}}{15\,\text{sec}}, r = 4.5\,\text{cm}$

39. $\omega = \dfrac{16\pi\,\text{rad}}{3\,\text{sec}}, r = \dfrac{7}{3}\,\text{yd}$

40. $\omega = \dfrac{\pi\,\text{rad}}{8\,\text{min}}, r = 10.2\,\text{in.}$

41. $\omega = 10\pi\dfrac{\text{rad}}{\text{sec}}, r = 40\,\text{cm}$

42. $\omega = 27.3\dfrac{\text{rad}}{\text{sec}}, r = 22.6\,\text{mm}$

In Exercises 43–52, find the distance a point travels along a circle s, over a time t, given the angular speed ω, and radius of the circle r. Round to three significant digits.

43. $r = 5\,\text{cm}, \omega = \dfrac{\pi\,\text{rad}}{6\,\text{sec}}, t = 10\,\text{sec}$

44. $r = 2\,\text{mm}, \omega = 6\pi\dfrac{\text{rad}}{\text{sec}}, t = 11\,\text{sec}$

45. $r = 5.2\,\text{in.}, \omega = \dfrac{\pi\,\text{rad}}{15\,\text{sec}}, t = 10\,\text{min}$

46. $r = 3.2\,\text{ft}, \omega = \dfrac{\pi\,\text{rad}}{4\,\text{sec}}, t = 3\,\text{min}$

47. $r = 12\,\text{m}, \omega = \dfrac{3\pi\,\text{rad}}{2\,\text{sec}}, t = 100\,\text{sec}$

48. $r = 6.5\,\text{cm}, \omega = \dfrac{2\pi\,\text{rad}}{15\,\text{sec}}, t = 50.5\,\text{min}$

49. $r = 30\,\text{cm}, \omega = \dfrac{\pi\,\text{rad}}{10\,\text{sec}}, t = 25\,\text{sec}$

50. $r = 5\,\text{cm}, \omega = \dfrac{5\pi\,\text{rad}}{3\,\text{sec}}, t = 9\,\text{min}$

51. $r = 15\,\text{in.}, \omega = 5$ rotations per second, $t = 15\,\text{min}$ (express distance in miles*)

52. $r = 17\,\text{in.}, \omega = 6$ rotations per second, $t = 10\,\text{min}$ (express distance in miles*)

*1 mi = 5280 ft

- **APPLICATIONS**

53. **Tires.** A car owner decides to upgrade from tires with a diameter of 24.3 inches to tires with a diameter of 26.1 inches. If she doesn't update the onboard computer, how fast will she actually be traveling when the speedometer reads 65 mph?

54. **Tires.** A car owner decides to upgrade from tires with a diameter of 24.8 inches to tires with a diameter of 27.0 inches. If she doesn't update the onboard computer, how fast will she actually be traveling when the speedometer reads 70 mph?

55. **Planets.** The Earth rotates every 24 hours (actually 23 hours, 56 minutes, and 4 seconds) and has a diameter of 7926 miles. If you're standing on the equator, how fast are you traveling in miles per hour (how fast is the Earth spinning)? Compute this using 24 hours and then with 23 hours, 56 minutes, 4 seconds as time of rotation.

56. **Planets.** The planet Jupiter rotates every 9.9 hours and has a diameter of 88,846 miles. If you're standing on its equator, how fast are you traveling in miles per hour?

57. **Carousel.** A boy wants to jump onto a moving carousel that is spinning at the rate of five revolutions per minute. If the carousel is 60 feet in diameter, how fast must the boy run, in feet per second, to match the speed of the carousel and jump on?

58. **Carousel.** A boy wants to jump onto a playground carousel that is spinning at the rate of 30 revolutions per minute. If the carousel is 6 feet in diameter, how fast must the boy run, in feet per second, to match the speed of the carousel and jump on?

59. **Music.** Some people still have their phonograph collections and play the records on turntables. A phonograph record is a vinyl disc that rotates on the turntable. If a 12-inch-diameter record rotates at $33\frac{1}{3}$ revolutions per minute, what is the angular speed in radians per minute?

60. **Music.** Some people still have their phonograph collections and play the records on turntables. A phonograph record is a vinyl disc that rotates on the turntable. If a 12-inch-diameter record rotates at $33\frac{1}{3}$ revolutions per minute, what is the linear speed of a point on the outer edge in inches per minute?

61. **Bicycle.** How fast is a bicyclist traveling in miles per hour if his tires are 27 inches in diameter and his angular speed is 5π radians per second?

62. **Bicycle.** How fast is a bicyclist traveling in miles per hour if his tires are 22 inches in diameter and his angular speed is 5π radians per second?

63. **Electric Motor.** If a 2-inch-diameter pulley that's being driven by an electric motor and running at 1600 revolutions per minute is connected by a belt to a 5-inch-diameter pulley to drive a saw, what is the speed of the saw in revolutions per minute?

64. **Electric Motor.** If a 2.5-inch-diameter pulley that's being driven by an electric motor and running at 1800 revolutions per minute is connected by a belt to a 4-inch-diameter pulley to drive a saw, what is the speed of the saw in revolutions per minute?

For Exercises 65 and 66, refer to the following:

NASA explores artificial gravity as a way to counter the physiologic effects of extended weightlessness for future space exploration. NASA's centrifuge has a 58-foot-diameter arm.

65. **NASA.** If two humans are on opposite (red and blue) ends of the centrifuge and their linear speed is 200 miles per hour, how fast is the arm rotating?

66. **NASA.** If two humans are on opposite (red and blue) ends of the centrifuge and they rotate one full rotation every second, what is their linear speed in feet per second?

For Exercises 67 and 68, refer to the following:

To achieve similar weightlessness as that on NASA's centrifuge, ride the *Gravitron* at a carnival or fair. The Gravitron has a diameter of 14 meters, and in the first 20 seconds it achieves zero gravity and the floor drops.

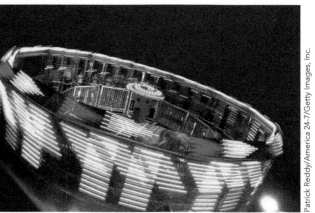

67. **Gravitron.** If the Gravitron rotates 24 times per minute, find the linear speed of the people riding it in meters per second.

68. **Gravitron.** If the Gravitron rotates 30 times per minute, find the linear speed of the people riding it in kilometers per hour.

69. **Clock.** What is the linear speed of a point on the end of a 10-centimeter second hand given in meters per second?

70. **Clock.** What is the angular speed of a point on the end of a 10-centimeter second hand given in radians per second?

• CATCH THE MISTAKE

In Exercises 71 and 72, explain the mistake that is made.

71. If the radius of a set of tires on a car is 15 inches and the tires rotate 180° per second, how fast is the car traveling (linear speed) in miles per hour?

Solution:

Write the formula for
linear speed. $v = r\omega$

Let $r = 15$ inches and
$\omega = 180°$ per second. $v = (15\ \text{in.})(180°/\text{sec})$

Simplify. $v = 2700\ \text{in./sec}$

Let 1 mile = 5280 feet
= 63,360 inches and $v = \left(\dfrac{2700 \cdot 3600}{63{,}360}\right)\text{mph}$
1 hour = 3600 seconds.

Simplify. $v \approx 153.4\ \text{mph}$

This is incorrect. The correct answer is approximately 2.7 miles per hour. What mistake was made?

72. If a bicycle has tires with radius 10 inches and the tires rotate 90° per $\frac{1}{2}$ second, how fast is the bicycle traveling (linear speed) in miles per hour?

Solution:

Write the formula for
linear speed. $v = r\omega$

Let $r = 10$ inches and
$\omega = 180°$ per second. $v = (10\ \text{in.})(180°/\text{sec})$

Simplify. $v = 1800\ \text{in./sec}$

Let 1 mile = 5280 feet
= 63,360 inches and $v = \left(\dfrac{1800 \cdot 3600}{63{,}360}\right)\text{mph}$
1 hour = 3600 seconds.

Simplify. $v \approx 102.3\ \text{mph}$

This is incorrect. The correct answer is approximately 1.8 miles per hour. What mistake was made?

• CONCEPTUAL

In Exercises 73 and 74, determine whether each statement is true or false.

73. Angular and linear speed are inversely proportional.

74. Angular and linear speed are directly proportional.

75. In the chapter opener about the 2016 Lexus RX, if the standard tires have radius r_1 and the upgraded tires have radius r_2, assuming the owner does not get the onboard computer adjusted, find the actual speed the Lexus RX is traveling, v_2, in terms of the indicated speed on the speedometer, v_1.

76. For the Lexus RX in Exercise 75, find the actual mileage the Lexus RX has traveled, s_2, in terms of the indicated mileage on the odometer, s_1.

In Exercises 77 and 78, use the diagram below:

The large gear has a radius of 6 centimeters, the medium gear has a radius of 3 centimeters, and the small gear has a radius of 1 centimeter.

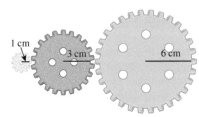

77. If the small gear rotates 1 revolution per second, what is the linear speed of a point traveling along the circumference of the large gear?

78. If the small gear rotates 1.5 revolutions per second, what is the linear speed of a point traveling along the circumference of the large gear?

• CHALLENGE

79. A boy swings a red ball attached to a 10-foot string around his head as fast as he can. He then picks up a blue ball attached to a 5-foot string and swings it at the same angular speed. How does the linear velocity of the blue ball compare to that of the red ball.

80. One of the cars on a Ferris wheel, 100 feet in diameter, goes all of the way around in 35 seconds. What is the linear speed of a point halfway between the car and the hub?

3.4 DEFINITION 3 OF TRIGONOMETRIC FUNCTIONS: UNIT CIRCLE APPROACH

SKILLS OBJECTIVES	CONCEPTUAL OBJECTIVES
■ Draw the unit circle illustrating the special angles and label cosine and sine values.	■ Relate *x*-coordinates and *y*-coordinates of points on the unit circle to the values of cosine and sine functions.
■ Classify circular functions as even or odd.	■ Visualize the periodic properties of circular functions.

Recall that the first definition of trigonometric functions we developed was in terms of ratios of sides of right triangles (Section 1.3). Then, in Section 2.2, we superimposed right triangles on the Cartesian plane, which led to a second definition of trigonometric functions (for any angle) in terms of ratios of *x*- and *y*-coordinates of a point and the distance from the origin to that point. In this section, we inscribe right triangles into the unit circle in the Cartesian plane, which will yield a third definition of trigonometric functions. It is important to note that all three definitions are consistent with one another.

3.4.1 Trigonometric Functions and the Unit Circle (Circular Functions)

Recall that the equation for the **unit circle** (radius of 1 centered at the origin) is given by $x^2 + y^2 = 1$. We will use the term *circular function* later in this section, but it is important to note that a circle is not a function (it does not pass the vertical line test).

   If we form a central angle θ in the unit circle such that the terminal side lies in quadrant I, we can use the previous two definitions of the sine and cosine functions when $r = 1$ (i.e., on the unit circle) and note that we can form a right triangle with legs of lengths *x* and *y* and hypotenuse $r = 1$.

3.4.1 SKILL

Draw the unit circle illustrating the special angles and label cosine and sine values.

3.4.1 CONCEPTUAL

Relate *x*-coordinates and *y*-coordinates of points on the unit circle to the values of cosine and sine functions.

TRIGONOMETRIC FUNCTION	RIGHT TRIANGLE TRIGONOMETRY	CARTESIAN PLANE
$\sin\theta$	$\dfrac{\text{opposite}}{\text{hypotenuse}} = \dfrac{y}{1} = y$	$\dfrac{y}{r} = \dfrac{y}{1} = y$
$\cos\theta$	$\dfrac{\text{adjacent}}{\text{hypotenuse}} = \dfrac{x}{1} = x$	$\dfrac{x}{r} = \dfrac{x}{1} = x$

   Notice that any point (x, y) on the unit circle can be written as $(\cos\theta, \sin\theta)$, where θ is the measure of a trigonometric angle defined in Chapter 2. If we recall the unit circle coordinate values for special angles (Section 2.1), we can now summarize the exact values for the **sine** and **cosine** functions in the illustration below.

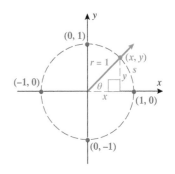

[CONCEPT CHECK]

If the point $(-A, -B)$ lies in quadrant III, then what must be true so that $\cos\theta = -A$ and $\sin\theta = -B$?

ANSWER $A^2 + B^2 = 1$

[STUDY TIP]

$(\cos\theta, \sin\theta)$ represents a point (x, y) on the unit circle.

$(x, y) = (\cos\theta, \sin\theta)$

(unit circle diagram showing special angles with coordinates:)

$\left(-\dfrac{1}{2}, \dfrac{\sqrt{3}}{2}\right)$ $\left(\dfrac{1}{2}, \dfrac{\sqrt{3}}{2}\right)$

$\left(-\dfrac{\sqrt{2}}{2}, \dfrac{\sqrt{2}}{2}\right)$ (0, 1) 90° $\left(\dfrac{\sqrt{2}}{2}, \dfrac{\sqrt{2}}{2}\right)$

$\left(-\dfrac{\sqrt{3}}{2}, \dfrac{1}{2}\right)$ $\dfrac{2\pi}{3}$ $\dfrac{3\pi}{4}$ 120° $\dfrac{\pi}{2}$ $\dfrac{\pi}{3}$ $\dfrac{\pi}{4}$ 60° $\left(\dfrac{\sqrt{3}}{2}, \dfrac{1}{2}\right)$

$\dfrac{5\pi}{6}$ 135° 45° $\dfrac{\pi}{6}$ 30°

(−1, 0)   150°   0° 0   (1, 0)

π 180°   360° 2π

$\left(-\dfrac{\sqrt{3}}{2}, -\dfrac{1}{2}\right)$ $\dfrac{7\pi}{6}$ 210° 330° $\dfrac{11\pi}{6}$ $\left(\dfrac{\sqrt{3}}{2}, -\dfrac{1}{2}\right)$

225° 315° $\dfrac{7\pi}{4}$

$\left(-\dfrac{\sqrt{2}}{2}, -\dfrac{\sqrt{2}}{2}\right)$ $\dfrac{5\pi}{4}$ $\dfrac{4\pi}{3}$ 240° 300° $\dfrac{5\pi}{3}$ $\left(\dfrac{\sqrt{2}}{2}, -\dfrac{\sqrt{2}}{2}\right)$

270° $\dfrac{3\pi}{2}$ (0, −1)

$\left(-\dfrac{1}{2}, -\dfrac{\sqrt{3}}{2}\right)$ $\left(\dfrac{1}{2}, -\dfrac{\sqrt{3}}{2}\right)$

The following observations are consistent with properties of trigonometric functions we've studied already:

- $\sin\theta > 0$ in quadrant I and quadrant II, where $y > 0$.
- $\cos\theta > 0$ in quadrant I and quadrant IV, where $x > 0$.
- The equation of the unit circle $x^2 + y^2 = 1$ leads also to the Pythagorean identity $\cos^2\theta + \sin^2\theta = 1$ that we derived in Section 2.4.

Circular Functions

Using the unit circle relationship $(x, y) = (\cos\theta, \sin\theta)$, where θ is the central angle whose terminal side intersects the unit circle at the point (x, y), we can now define the remaining trigonometric functions using this unit circle approach and the quotient and reciprocal identities. Because the trigonometric functions are defined in terms of the unit *circle*, the trigonometric functions are often called *circular* functions. Recall that $\theta = \dfrac{s}{r}$, and since $r = 1$, we know that $\theta = s$.

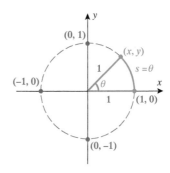

DEFINITION 3 | **Trigonometric Functions: Unit Circle Approach**

Let (x, y) be any point on the unit circle $(x^2 + y^2 = 1)$. If θ is the real number that represents the distance from the point $(1, 0)$ along the circumference of the circle to the point (x, y), then

$$\sin\theta = y \qquad\qquad \cos\theta = x \qquad\qquad \tan\theta = \frac{y}{x} \quad x \neq 0$$

$$\csc\theta = \frac{1}{y} \quad y \neq 0 \qquad \sec\theta = \frac{1}{x} \quad x \neq 0 \qquad \cot\theta = \frac{x}{y} \quad y \neq 0$$

The coordinates of the points on the unit circle can be written as $(\cos\theta, \sin\theta)$, and since θ is a real number, the **trigonometric functions** are often called **circular functions**.

▶ **EXAMPLE 1**   **Finding Exact Trigonometric (Circular) Function Values**

Find the exact values for each of the following using the unit circle definition.

a. $\sin\left(\dfrac{7\pi}{4}\right)$ **b.** $\cos\left(\dfrac{5\pi}{6}\right)$ **c.** $\tan\left(\dfrac{3\pi}{2}\right)$

Solution (a):

The angle $\dfrac{7\pi}{4}$ corresponds to the coordinates $\left(\dfrac{\sqrt{2}}{2}, -\dfrac{\sqrt{2}}{2}\right)$ on the unit circle.

The value of the sine function is the y-coordinate. $\boxed{\sin\left(\dfrac{7\pi}{4}\right) = -\dfrac{\sqrt{2}}{2}}$

Solution (b):

The angle $\dfrac{5\pi}{6}$ corresponds to the coordinates $\left(-\dfrac{\sqrt{3}}{2}, \dfrac{1}{2}\right)$ on the unit circle.

The value of the cosine function is the x-coordinate. $\boxed{\cos\left(\dfrac{5\pi}{6}\right) = -\dfrac{\sqrt{3}}{2}}$

Solution (c):

The angle $\dfrac{3\pi}{2}$ corresponds to the coordinates $(0, -1)$ on the unit circle.

The value of the cosine function is the x-coordinate. $\cos\left(\dfrac{3\pi}{2}\right) = 0$

The value of the sine function is the y-coordinate. $\sin\left(\dfrac{3\pi}{2}\right) = -1$

The tangent function is the ratio of the sine to cosine functions. $\tan\left(\dfrac{3\pi}{2}\right) = \dfrac{\sin(3\pi/2)}{\cos(3\pi/2)}$

Substitute $\cos\left(\dfrac{3\pi}{2}\right) = 0$ and $\sin\left(\dfrac{3\pi}{2}\right) = -1$. $\tan\left(\dfrac{3\pi}{2}\right) = \dfrac{-1}{0}$

$\boxed{\tan\left(\dfrac{3\pi}{2}\right) \text{ is undefined}}$

▼ **YOUR TURN**   Find the exact values for each of the following using the unit circle definition.

a. $\sin\left(\dfrac{5\pi}{6}\right)$ **b.** $\cos\left(\dfrac{7\pi}{4}\right)$ **c.** $\tan\left(\dfrac{2\pi}{3}\right)$

▼ **ANSWER**

a. $\dfrac{1}{2}$ **b.** $\dfrac{\sqrt{2}}{2}$ **c.** $-\sqrt{3}$

EXAMPLE 2	**Solving Equations Involving Trigonometric (Circular) Functions**

Use the unit circle to find all values of θ, $0 \leq \theta \leq 2\pi$, for which $\sin\theta = -\frac{1}{2}$.

Solution:

Since the value of the sine function is negative, θ must lie in quadrants III or IV.

The value of sine is the y-coordinate. The angles corresponding to

$$\sin\theta = -\frac{1}{2} \text{ are}$$

$$\frac{7\pi}{6} \text{ and } \frac{11\pi}{6}.$$

There are two values for θ that are greater than or equal to zero and less than or equal to 2π that satisfy the equation $\sin\theta = -\frac{1}{2}$.

$$\boxed{\theta = \frac{7\pi}{6}, \frac{11\pi}{6}}$$

▼
ANSWER
$\theta = \dfrac{2\pi}{3}, \dfrac{4\pi}{3}$

▼
YOUR TURN Find all values of θ, $0 \leq \theta \leq 2\pi$, for which $\cos\theta = -\frac{1}{2}$.

3.4.2 Properties of Circular Functions

3.4.2 SKILL

Classify circular functions as even or odd.

3.4.2 CONCEPTUAL

Visualize the periodic properties of circular functions.

WORDS	MATH
The coordinates of any point (x, y) that lies on the unit circle satisfies the equation $x^2 + y^2 = 1$.	$-1 \leq x \leq 1$ and $-1 \leq y \leq 1$
Since $x = \cos\theta$ and $y = \sin\theta$, the following trigonometric inequalities hold.	$-1 \leq \cos\theta \leq 1$ and $-1 \leq \sin\theta \leq 1$
State the **domain and range of the cosine and sine functions**.	Domain: $(-\infty, \infty)$ Range: $[-1, 1]$
Since $\cot\theta = \dfrac{\cos\theta}{\sin\theta}$ and $\csc\theta = \dfrac{1}{\sin\theta}$, the values for θ that make $\sin\theta = 0$ must be eliminated from the **domain of the cotangent and cosecant functions**.	Domain: $\theta \neq n\pi$, where n is an integer
Since $\tan\theta = \dfrac{\sin\theta}{\cos\theta}$ and $\sec\theta = \dfrac{1}{\cos\theta}$, the values for θ that make $\cos\theta = 0$ must be eliminated from the **domain of the tangent and secant functions**.	Domain: $\theta \neq \dfrac{(2n+1)\pi}{2} = \dfrac{\pi}{2} + n\pi$, where n is an integer

[CONCEPT CHECK]

Which of the following is NOT equal to the others:

(A) $\sin(2\pi + A)$

(B) $\sin(A - 2\pi)$

(C) $\sin(\pi + A)$

(D) $\sin(A + 8\pi)$

▼
ANSWER C

The following box summarizes the domains and ranges of the trigonometric (circular) functions.

DOMAINS AND RANGES OF THE TRIGONOMETRIC (CIRCULAR) FUNCTIONS

For any real number θ and integer n,

FUNCTION	DOMAIN	RANGE
$\sin\theta$	$(-\infty, \infty)$	$[-1, 1]$
$\cos\theta$	$(-\infty, \infty)$	$[-1, 1]$
$\tan\theta$	all real numbers such that $\theta \neq \dfrac{(2n+1)\pi}{2} = \dfrac{\pi}{2} + n\pi$	$(-\infty, \infty)$
$\cot\theta$	all real numbers such that $\theta \neq n\pi$	$(-\infty, \infty)$
$\sec\theta$	all real numbers such that $\theta \neq \dfrac{(2n+1)\pi}{2} = \dfrac{\pi}{2} + n\pi$	$(-\infty, -1] \cup [1, \infty)$
$\csc\theta$	all real numbers such that $\theta \neq n\pi$	$(-\infty, -1] \cup [1, \infty)$

Recall from algebra that even and odd functions have both an algebraic and a graphical interpretation. **Even functions** are functions for which $f(-x) = f(x)$ for all x in the domain of f, and the graph of an even function is symmetric about the y-axis. **Odd functions** are functions for which $f(-x) = -f(x)$ for all x in the domain of f, and the graph of an odd function is symmetric about the origin.

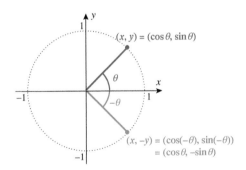

The cosine function is an even function. $\cos\theta = \cos(-\theta)$

The sine function is an odd function. $\sin(-\theta) = -\sin\theta$

EXAMPLE 3 Using Properties of Trigonometric (Circular) Functions

Evaluate $\cos\left(-\dfrac{5\pi}{6}\right)$.

Solution:

The cosine function is an even function.
$$\cos\left(-\frac{5\pi}{6}\right) = \cos\left(\frac{5\pi}{6}\right)$$

Use the unit circle to evaluate cosine.
$$\cos\left(\frac{5\pi}{6}\right) = -\frac{\sqrt{3}}{2}$$

$$\boxed{\cos\left(-\frac{5\pi}{6}\right) = -\frac{\sqrt{3}}{2}}$$

▼ **ANSWER**
$-\frac{1}{2}$

YOUR TURN Evaluate $\sin\left(-\dfrac{5\pi}{6}\right)$.

> **STUDY TIP**
> Set the calculator to radian mode before evaluating circular functions in radians. Alternatively, convert the radian measure to degrees before evaluating the trigonometric function value.

Although trigonometric (circular) functions can be evaluated exactly for some special angles, a calculator can be used to approximate trigonometric (circular) functions for any value.

▶ **EXAMPLE 4** Evaluating Trigonometric (Circular) Functions with a Calculator

Use a calculator to evaluate $\sin\left(\dfrac{7\pi}{12}\right)$. Round the answer to four decimal places.

---

common mistake

✓CORRECT	✗INCORRECT
Evaluate with a calculator.	Evaluate with a calculator.
0.965925826	0.031979376 **ERROR**
Round to four decimal places.	(calculator in degree mode)
$\boxed{\sin\left(\dfrac{7\pi}{12}\right) \approx 0.9659}$	

Many calculators automatically reset to degree mode after every calculation, so be sure to always check what mode the calculator indicates.

▼ **ANSWER**
-0.7265

YOUR TURN Use a calculator to evaluate $\tan\left(\dfrac{9\pi}{5}\right)$. Round the answer to four decimal places.

EXAMPLE 5 **Even and Odd Trigonometric (Circular) Functions**

Show that the secant function is an even function.

Solution: Show that $\sec(-\theta) = \sec\theta$.

Secant is the reciprocal of cosine.
$$\sec(-\theta) = \frac{1}{\cos(-\theta)}$$

Cosine is an even function, so $\cos(-\theta) = \cos\theta$.
$$\sec(-\theta) = \frac{1}{\cos\theta}$$

Secant is the reciprocal of cosine, $\sec\theta = \frac{1}{\cos\theta}$.
$$\sec(-\theta) = \frac{1}{\cos\theta} = \sec\theta$$

Since $\sec(-\theta) = \sec\theta$, the secant function is an even function.

▶[SECTION 3.4] SUMMARY

In this section, we have defined trigonometric functions in terms of the unit circle. The coordinates of any point (x, y) that lies on the unit circle satisfy the equation $x^2 + y^2 = 1$. The Pythagorean identity $\cos^2\theta + \sin^2\theta = 1$ follows immediately from the unit circle equation if $(x, y) = (\cos\theta, \sin\theta)$, where θ is the central angle whose terminal side intersects the unit circle at the point (x, y). The cosine function is an even function, $\cos(-\theta) = \cos\theta$, and the sine function is an odd function, $\sin(-\theta) = -\sin\theta$.

[SECTION 3.4] EXERCISES

• **SKILLS**

In Exercises 1–14, find the *exact* values of the indicated trigonometric functions using the unit circle.

1. $\sin\left(\dfrac{5\pi}{3}\right)$

2. $\cos\left(\dfrac{5\pi}{3}\right)$

3. $\cos\left(\dfrac{7\pi}{6}\right)$

4. $\sin\left(\dfrac{7\pi}{6}\right)$

5. $\sin\left(\dfrac{3\pi}{4}\right)$

6. $\cos\left(\dfrac{3\pi}{4}\right)$

7. $\tan\left(\dfrac{7\pi}{4}\right)$

8. $\cot\left(\dfrac{7\pi}{4}\right)$

9. $\sec\left(\dfrac{5\pi}{6}\right)$

10. $\csc\left(\dfrac{11\pi}{6}\right)$

11. $\sec 225°$

12. $\csc 300°$

13. $\tan 240°$

14. $\cot 330°$

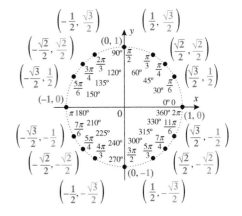

In Exercises 15–30, use the unit circle and the fact that sine is an odd function and cosine is an even function to find the *exact* values of the indicated functions.

15. $\sin\left(-\dfrac{2\pi}{3}\right)$ **16.** $\sin\left(-\dfrac{5\pi}{4}\right)$ **17.** $\sin\left(-\dfrac{\pi}{3}\right)$ **18.** $\sin\left(-\dfrac{7\pi}{6}\right)$

19. $\cos\left(-\dfrac{3\pi}{4}\right)$ **20.** $\cos\left(-\dfrac{5\pi}{3}\right)$ **21.** $\cos\left(-\dfrac{5\pi}{6}\right)$ **22.** $\cos\left(-\dfrac{7\pi}{4}\right)$

23. $\sin(-225°)$ **24.** $\sin(-180°)$ **25.** $\sin(-270°)$ **26.** $\sin(-60°)$

27. $\cos(-45°)$ **28.** $\cos(-135°)$ **29.** $\cos(-90°)$ **30.** $\cos(-210°)$

In Exercises 31–50, use the unit circle to find all of the exact values of θ that make the equation true in the indicated interval.

31. $\cos\theta = \dfrac{\sqrt{3}}{2}, 0 \le \theta \le 2\pi$ **32.** $\cos\theta = -\dfrac{\sqrt{3}}{2}, 0 \le \theta \le 2\pi$

33. $\sin\theta = -\dfrac{\sqrt{3}}{2}, 0 \le \theta \le 2\pi$ **34.** $\sin\theta = \dfrac{\sqrt{3}}{2}, 0 \le \theta \le 2\pi$

35. $\cos\theta = \dfrac{1}{2}, 0 \le \theta \le 2\pi$ **36.** $\sin\theta = -\dfrac{1}{2}, 0 \le \theta \le 2\pi$

37. $\cos\theta = -\dfrac{\sqrt{2}}{2}, 0 \le \theta \le 2\pi$ **38.** $\sin\theta = \dfrac{\sqrt{2}}{2}, 0 \le \theta \le 2\pi$

39. $\sin\theta = 0, 0 \le \theta \le 4\pi$ **40.** $\sin\theta = -1, 0 \le \theta \le 4\pi$

41. $\cos\theta = -1, 0 \le \theta \le 4\pi$ **42.** $\cos\theta = 0, 0 \le \theta \le 4\pi$

43. $\tan\theta = -1, 0 \le \theta \le 2\pi$ **44.** $\cot\theta = 1, 0 \le \theta \le 2\pi$

45. $\sec\theta = -\sqrt{2}, 0 \le \theta \le 2\pi$ **46.** $\csc\theta = \sqrt{2}, 0 \le \theta \le 2\pi$

47. $\csc\theta$ is undefined, $0 \le \theta \le 2\pi$ **48.** $\sec\theta$ is undefined, $0 \le \theta \le 2\pi$

49. $\tan\theta$ is undefined, $0 \le \theta \le 2\pi$ **50.** $\cot\theta$ is undefined, $0 \le \theta \le 2\pi$

In Exercises 51–58, approximate the trigonometric function values. Round answers to four decimal places.

51. $\cos\left(\dfrac{7\pi}{11}\right)$ **52.** $\sin\left(\dfrac{5\pi}{9}\right)$ **53.** $\cot\left(\dfrac{11\pi}{5}\right)$ **54.** $\tan\left(\dfrac{12\pi}{7}\right)$

55. $\sin 4$ **56.** $\cos 7$ **57.** $\tan(2.5)$ **58.** $\csc 1$

• APPLICATIONS

For Exercises 59 and 60, refer to the following:

The average daily temperature in Peoria, Illinois, can be predicted by the formula $T = 50 - 28\cos\dfrac{2\pi(x - 31)}{365}$, where x is the number of the day in a nonleap year (January 1 = 1, February 1 = 32, etc.) and T is in degrees Fahrenheit.

59. Atmospheric Temperature. What is the expected temperature on February 15?

60. Atmospheric Temperature. What is the expected temperature on August 15?

For Exercises 61 and 62, refer to the following:

The human body temperature normally fluctuates during the day. Assume a person's body temperature can be predicted by the formula $T = 99.1 - 0.5\sin\left(x + \dfrac{\pi}{12}\right)$, where x is the number of hours since midnight and T is in degrees Fahrenheit.

61. Body Temperature. What is the person's temperature at 6:00 A.M.?

62. Body Temperature. What is the person's temperature at 9:00 P.M.?

For Exercises 63 and 64, refer to the following:

The height of the water in a harbor changes with the tides. On a particular day, it can be determined by the formula $h(x) = 5 + 4.8 \sin\left[\dfrac{\pi}{6}(x + 4)\right]$, where x is the number of hours since midnight and h is the height of the tide in feet.

Bill Brooks/Alamy

63. **Tides.** What is the height of the tide at 3:00 P.M.?

64. **Tides.** What is the height of the tide at 5:00 A.M.?

65. **Yo-Yo Dieting.** A woman has been yo-yo dieting for years. Her weight changes throughout the year as she gains and loses weight. Her weight in a particular month can be determined by the formula $w(x) = 145 + 10 \cos\left(\dfrac{\pi}{6}x\right)$, where x is the month and w is in pounds. If $x = 1$ corresponds to January, how much does she weigh in June?

66. **Yo-Yo Dieting.** How much does the woman in Exercise 65 weigh in December?

67. **Seasonal Sales.** The average number of guests visiting the Magic Kingdom at Walt Disney World per day is given by $n(x) = 30{,}000 + 20{,}000 \sin\left[\dfrac{\pi}{2}(x + 1)\right]$, where n is the number of guests and x is the month. If January corresponds to $x = 1$, how many people, on average, are visiting the Magic Kingdom per day in February?

68. **Seasonal Sales.** How many guests are visiting the Magic Kingdom in Exercise 67 in December?

69. **Temperature.** The average high temperature for a certain city is given by the equation $T = 60 - 20 \cos\left(\dfrac{\pi}{6}t\right)$, where T is degrees Fahrenheit and t is time in months. What is the average temperature in June ($t = 6$)?

70. **Temperature.** The average high temperature for a certain city is given by the equation $T = 65 - 25 \cos\left(\dfrac{\pi}{6}t\right)$, where T is degrees Fahrenheit and t is time in months. What is the average temperature in October ($t = 10$)?

71. **Gear.** The vertical position in centimeters of a tooth on a gear is given by the function $y = 3 \sin(10t)$, where t is time in seconds. Find the vertical position after 2.5 seconds.

72. **Gear.** The vertical position in centimeters of a tooth on a gear is given by the equation $y = 5 \sin(3.6t)$, where t is time in seconds. Find the vertical position after 10 seconds.

73. **Oscillating Spring.** A weight is attached to a spring and then pulled down and let go to begin a vertical motion. The position of the weight in inches from equilibrium is given by the equation $y = -15 \sin\left(\dfrac{7}{2}t + \dfrac{7\pi}{2}\right)$, where t is time in seconds after the spring is let go. Find the position of the weight 3.5 seconds after being let go.

74. **Oscillating Spring.** A weight is attached to a spring and then pulled down and let go to begin a vertical motion. The position of the weight in inches from equilibrium is given by the equation $y = -15 \sin\left(4.6t - \dfrac{\pi}{2}\right)$, where t is time in seconds after the spring is let go. Find the position of the weight 5 seconds after being let go.

For Exercises 75 and 76, refer to the following:

During the course of treatment of an illness, the concentration of a drug in the bloodstream in micrograms per microliter fluctuates during the dosing period of 8 hours according to the model

$$C(t) = 15.4 - 4.7 \sin\left(\dfrac{\pi}{4}t + \dfrac{\pi}{2}\right), \quad 0 \le t \le 8$$

Note: This model does not apply to the first dose of the medication.

75. **Health/Medicine.** Find the concentration of the drug in the bloodstream at the beginning of a dosing period.

76. **Health/Medicine.** Find the concentration of the drug in the bloodstream 6 hours after taking a dose of the drug.

In Exercises 77 and 78, refer to the following:

By analyzing available empirical data, it has been determined that the body temperature of a particular species fluctuates during a 24-hour day according to the model

$$T(t) = 36.3 - 1.4 \cos\left[\frac{\pi}{12}(t - 2)\right], \quad 0 \le t \le 24$$

where T represents temperature in degrees Celsius and t represents time in hours measured from 12:00 A.M. (midnight).

77. Biology. Find the approximate body temperature at midnight. Round your answer to the nearest degree.

78. Biology. Find the approximate body temperature at 2:45 P.M. Round your answer to the nearest degree.

• CATCH THE MISTAKE

In Exercises 79 and 80, explain the mistake that is made.

79. Use the unit circle to evaluate $\tan\left(\dfrac{5\pi}{6}\right)$ exactly.

Solution:

Tangent is the ratio of sine to cosine.
$$\tan\left(\frac{5\pi}{6}\right) = \frac{\sin\left(\dfrac{5\pi}{6}\right)}{\cos\left(\dfrac{5\pi}{6}\right)}$$

Use the unit circle to identify sine and cosine.
$$\sin\left(\frac{5\pi}{6}\right) = -\frac{\sqrt{3}}{2} \text{ and } \cos\left(\frac{5\pi}{6}\right) = \frac{1}{2}$$

Substitute values for sine and cosine.
$$\tan\left(\frac{5\pi}{6}\right) = \frac{-\dfrac{\sqrt{3}}{2}}{\dfrac{1}{2}}$$

Simplify.
$$\tan\left(\frac{5\pi}{6}\right) = -\sqrt{3}$$

This is incorrect. What mistake was made?

80. Use the unit circle to evaluate $\sec\left(\dfrac{11\pi}{6}\right)$ exactly.

Solution:

Secant is the reciprocal of cosine.
$$\sec\left(\frac{11\pi}{6}\right) = \frac{1}{\cos\left(\dfrac{11\pi}{6}\right)}$$

Use the unit circle to evaluate cosine.
$$\cos\left(\frac{11\pi}{6}\right) = -\frac{1}{2}$$

Substitute the value for cosine.
$$\sec\left(\frac{11\pi}{6}\right) = \frac{1}{-\dfrac{1}{2}}$$

Simplify.
$$\sec\left(\frac{11\pi}{6}\right) = -2$$

This is incorrect. What mistake was made?

• CONCEPTUAL

In Exercises 81–84, determine whether each statement is true or false.

81. $\sin(2n\pi + \theta) = \sin\theta$, for n an integer.

82. $\cos(2n\pi + \theta) = \cos\theta$, for n an integer.

83. $\sin\theta = 1$ when $\theta = \dfrac{(2n + 1)\pi}{2}$, for n an integer.

84. $\cos\theta = 1$ when $\theta = n\pi$, for n an integer.

85. Is $y = \csc x$ an even or an odd function? Justify your answer.

86. Is $y = \tan x$ an even or an odd function? Justify your answer.

87. Find all the values of θ, $0 \le \theta \le 2\pi$, for which the equation is true: $\sin\theta = \cos\theta$.

88. Find all the values of θ (θ is any real number) for which the equation is true: $\sin\theta = \cos\theta$.

• **CHALLENGE**

89. How many times is the expression $\left|\cos(2\pi t)\right| = 1$ true for $0 \le t \le 12$?

90. How many times is the expression $\left|\sin\left(\frac{\pi}{2}t\right)\right| = 1$ true for $0 \le t \le 10$?

91. For what values of x, such that $0 \le x < 2\pi$, is the expression $\left|\cos t\right| = \left|\sin t\right|$ true?

92. For what values of x, such that $0 \le x < 2\pi$, is the expression $\left|\sec t\right| = \left|\cos t\right|$ true?

93. Find values of x such that $0 \le x < 2\pi$ and both of the following are true: $\sin x < \frac{1}{2}$ and $\cos x < \frac{1}{2}$.

94. Find values of x such that $0 \le x < 2\pi$ and both of the following are true: $\tan x < 1$ and $\sec x < 0$.

• **TECHNOLOGY**

95. Use a calculator to approximate $\sin 423°$. What do you expect $\sin(-423°)$ to be? Verify your answer with a calculator.

96. Use a calculator to approximate $\cos 227°$. What do you expect $\cos(-227°)$ to be? Verify your answer with a calculator.

For Exercises 97 and 98, refer to the following:

A graphing calculator can be used to graph the unit circle with parametric equations (these will be covered in more detail in Section 8.5). For now, set the calculator in parametric and radian modes and let

$$X_1 = \cos T$$
$$Y_1 = \sin T$$

Set the window so that $0 \le t \le 2\pi$, step $= \frac{\pi}{15}$, $-2 \le X \le 2$, and $-2 \le Y \le 2$.

97. To approximate $\cos\left(\frac{\pi}{3}\right)$, use the trace function to move 5 steps $\left(\text{of } \frac{\pi}{15} \text{ each}\right)$ to the right of $t = 0$ and read the x-coordinate.

98. To approximate $\sin\left(\frac{\pi}{3}\right)$, use the trace function to move 5 steps $\left(\text{of } \frac{\pi}{15} \text{ each}\right)$ to the right of $t = 0$ and read the y-coordinate.

▶[CHAPTER 3 REVIEW]

SECTION	CONCEPT	KEY IDEAS/FORMULAS
3.1	**Radian measure**	
	The radian measure of an angle	θ (in radians) $= \dfrac{s}{r}$ s (arc length) and r (radius) must have the same units.
	Converting between degrees and radians	Degrees to radians: $\theta_r = \theta_d\left(\dfrac{\pi}{180°}\right)$ Radians to degrees: $\theta_d = \theta_r\left(\dfrac{180°}{\pi}\right)$
3.2	**Arc length and area of a circular sector**	
	Arc length	$s = r\theta_r$, where θ_r is in radians, or $s = r\theta_d\left(\dfrac{\pi}{180°}\right)$, where θ_d is in degrees
	Area of circular sector	$A = \dfrac{1}{2}r^2\theta_r$, where θ_r is in radians, or $A = \dfrac{1}{2}r^2\theta_d\left(\dfrac{\pi}{180°}\right)$, where θ_d is in degrees
3.3	**Linear and angular speeds**	Uniform circular motion ▪ Linear speed: speed around the circumference of a circle ▪ Angular speed: rotation speed of angle
	Linear speed	Linear speed v is given by $$v = \frac{s}{t}$$ where s is the arc length and t is time.
	Angular speed	Angular speed ω is given by $$\omega = \frac{\theta}{t}$$ where θ is given in radians.
	Relationship between linear and angular speeds	$v = r\omega$ or $\omega = \dfrac{v}{r}$ It is important to note that these formulas hold true only when angular speed is given in *radians* per unit of time.

SECTION	CONCEPT	KEY IDEAS/FORMULAS
3.4	**Definition 3 of trigonometric functions: Unit circle approach**	
	Trigonometric functions and the unit circle (circular functions)	
	Properties of circular functions	Cosine is an even function: $\cos(-\theta) = \cos\theta$ Sine is an odd function: $\sin(-\theta) = -\sin\theta$

FUNCTION	DOMAIN	RANGE
$\sin\theta$	$(-\infty, \infty)$	$[-1, 1]$
$\cos\theta$	$(-\infty, \infty)$	$[-1, 1]$
$\tan\theta$	$\theta \neq \dfrac{(2n+1)\pi}{2} = \dfrac{\pi}{2} + n\pi$	$(-\infty, \infty)$
$\cot\theta$	$\theta \neq n\pi$	$(-\infty, \infty)$
$\sec\theta$	$\theta \neq \dfrac{(2n+1)\pi}{2} = \dfrac{\pi}{2} + n\pi$	$(-\infty, -1] \cup [1, \infty)$
$\csc\theta$	$\theta \neq n\pi$	$(-\infty, -1] \cup [1, \infty)$

*n is an integer.

[CHAPTER 3 REVIEW EXERCISES]

3.1 Radian Measure

Convert from degrees to radians. Leave your answers exact in terms of π.

1. $135°$ 2. $240°$ 3. $330°$ 4. $180°$

5. $216°$ 6. $108°$ 7. $504°$ 8. $600°$

9. $-150°$ 10. $-15°$

Convert from radians to degrees.

11. $\dfrac{\pi}{3}$ 12. $\dfrac{11\pi}{6}$ 13. $\dfrac{5\pi}{4}$ 14. $\dfrac{2\pi}{3}$

15. $\dfrac{5\pi}{9}$ 16. $\dfrac{17\pi}{10}$ 17. $\dfrac{13\pi}{4}$ 18. $\dfrac{11\pi}{3}$

19. $-\dfrac{5\pi}{18}$ 20. $-\dfrac{13\pi}{6}$

Find the reference angle of each angle given (in radians).

21. $\dfrac{7\pi}{4}$ 22. $\dfrac{5\pi}{6}$ 23. $\dfrac{7\pi}{6}$ 24. $\dfrac{2\pi}{3}$

3.2 Arc Length and Area of a Circular Sector

Find the arc length intercepted by the indicated central angle of the circle with the given radius. Round to two decimal places.

25. $\theta = \dfrac{\pi}{3}, r = 5$ cm   26. $\theta = \dfrac{5\pi}{6}, r = 10$ in.

27. $\theta = 100°, r = 5$ in.   28. $\theta = 36°, r = 12$ ft

Find the measure of the angle whose intercepted arc and radius of a circle are given.

29. $r = 12$ in., $s = 6$ in.   30. $r = 10$ ft, $s = 27$ in.

31. $r = 6$ ft, $s = 4\pi$ ft   32. $r = 8$ m, $s = 2\pi$ m

33. $r = 5$ ft, $s = 10$ in.   34. $r = 4$ km, $s = 4$ m

Find the measure of each radius given the arc length and central angle of each circle.

35. $\theta = \dfrac{5\pi}{8}, s = \pi$ in.   36. $\theta = \dfrac{2\pi}{3}, s = 3\pi$ km

37. $\theta = 150°, s = 14\pi$ m   38. $\theta = 63°, s = 14\pi$ in.

39. $\theta = 10°, s = \dfrac{5\pi}{18}$ yd   40. $\theta = 80°, s = \dfrac{5\pi}{3}$ ft

Find the area of the circular sector given the indicated radius and central angle.

41. $\theta = \dfrac{\pi}{3}, r = 24$ mi   42. $\theta = \dfrac{5\pi}{12}, r = 9$ in.

43. $\theta = 60°, r = 60$ m   44. $\theta = 81°, r = 36$ cm

3.3 Linear and Angular Speeds

Find the linear speed of a point that moves with constant speed in a circular motion if the point travels arc length s in time t.

45. $s = 3$ ft, $t = 9$ sec   46. $s = 5280$ ft, $t = 4$ min

47. $s = 15$ mi, $t = 3$ min   48. $s = 12$ cm, $t = 0.25$ sec

Find the distance traveled by a point that moves with constant speed v along a circle in time t.

49. $v = 15$ mi/hr, $t = 1$ day   50. $v = 16$ ft/sec, $t = 1$ min

51. $v = 80$ mi/hr, $t = 15$ min   52. $v = 1.5$ cm/hr, $t = 6$ sec

Find the angular speed (radians/second) associated with rotating a central angle θ in time t.

53. $\theta = 6\pi, t = 9$ sec   54. $\theta = \pi, t = 0.05$ sec

55. $\theta = 225°, t = 20$ sec   56. $\theta = 330°, t = 22$ sec

Find the linear speed of a point traveling at a constant speed along the circumference of a circle with radius r and angular speed ω.

57. $\omega = \dfrac{5\pi \, \text{rad}}{6 \, \text{sec}}, r = 12$ m   58. $\omega = \dfrac{\pi \, \text{rad}}{20 \, \text{sec}}, r = 30$ in.

Find the distance s a point travels along a circle over a time t, given the angular speed ω and radius of the circle r.

59. $r = 10$ ft, $\omega = \dfrac{\pi \, \text{rad}}{4 \, \text{sec}}, t = 30$ sec

60. $r = 6$ in., $\omega = \dfrac{3\pi \, \text{rad}}{4 \, \text{sec}}, t = 6$ sec

61. $r = 12$ yd, $\omega = \dfrac{2\pi \, \text{rad}}{3 \, \text{sec}}, t = 30$ sec

62. $r = 100$ in., $\omega = \dfrac{\pi \, \text{rad}}{18 \, \text{sec}}, t = 3$ min

Applications

63. A ladybug is clinging to the outer edge of a child's spinning disk. The disk is 4 inches in diameter and is spinning at 60 revolutions per minute. How fast is the ladybug traveling?

64. How fast is a motorcyclist traveling in miles per hour if his tires are 30 inches in diameter and his angular speed is 10π radians per second?

3.4 Definition 3 of Trigonometric Functions: Unit Circle Approach

Find the exact values of the indicated trigonometric functions.

65. $\tan\left(\dfrac{5\pi}{6}\right)$

66. $\cos\left(\dfrac{5\pi}{6}\right)$

67. $\sin\left(\dfrac{11\pi}{6}\right)$

68. $\sec\left(\dfrac{11\pi}{6}\right)$

69. $\cot\left(\dfrac{5\pi}{4}\right)$

70. $\csc\left(\dfrac{5\pi}{4}\right)$

71. $\sin\left(\dfrac{3\pi}{2}\right)$

72. $\cos\left(\dfrac{3\pi}{2}\right)$

73. $\cos\pi$

74. $\tan 315°$

75. $\cos 60°$

76. $\sin 330°$

77. $\sin\left(-\dfrac{5\pi}{6}\right)$

78. $\cos\left(-\dfrac{5\pi}{4}\right)$

79. $\cos(-240°)$

80. $\sin(-135°)$

Find all of the exact values of θ that make the equation true in the indicated interval.

81. $\sin\theta = -\frac{1}{2}, 0 \le \theta \le 2\pi$

82. $\cos\theta = -\frac{1}{2}, 0 \le \theta \le 2\pi$

83. $\tan\theta = 0, 0 \le \theta \le 4\pi$

84. $\sin\theta = -1, 0 \le \theta \le 4\pi$

Technology Exercises

Section 3.1

Find the measure (in degrees, minutes, and nearest seconds) of a central angle θ that intercepts an arc on a circle with radius r with indicated arc length s. Use the TI calculator commands ANGLE **and** DMS **to change to degrees, minutes, and seconds.**

85. $r = 11.2$ ft, $s = 19.7$ ft

86. $r = 56.9$ cm, $s = 139.2$ cm

Section 3.4

For Exercises 87 and 88, refer to the following:

A graphing calculator can be used to graph the unit circle with parametric equations (these will be covered in more detail in Section 8.3). For now, set the calculator in parametric and radian modes and let

$$X_1 = \cos T$$
$$Y_1 = \sin T$$

Set the window so that $0 \le T \le 2\pi$, step $= \dfrac{\pi}{15}$, $-2 \le X \le 2$, and $-2 \le Y \le 2$. To approximate the sine or cosines of a T value, use the TRACE key, enter the T value, and read the corresponding coordinates from the screen.

87. Use the above steps to approximate $\cos\left(\dfrac{13\pi}{12}\right)$ to four decimal places.

88. Use the above steps to approximate $\sin\left(\dfrac{5\pi}{6}\right)$ to four decimal places.

[CHAPTER 3 PRACTICE TEST]

1. Find the measure (in radians) of a central angle θ that intercepts an arc on a circle with radius $r = 20$ centimeters and arc length $s = 4$ millimeters.

2. Convert $\dfrac{13\pi}{4}$ to degree measure.

3. Convert $260°$ to radian measure. Leave the answer exact in terms of π.

4. Convert $217°$ to radian measure. Round to two decimal places.

5. What is the reference angle to $\theta = \dfrac{7\pi}{12}$?

6. Find the radius of the minute hand on a clock if a point on the end travels 10 centimeters in 20 minutes.

7. Betty is walking around a circular walking path. If the radius of the path is 0.50 miles and she has walked through an angle of 120°, how far has she walked?

8. Calculate the arc length on a circle with central angle $\theta = \dfrac{\pi}{15}$ and radius $r = 8$ yards.

9. A sprinkler has a 25-foot spray and covers an angle of 30°. What is the area that the sprinkler waters? Round to the nearest square foot.

10. A bicycle with tires of radius $r = 15$ inches is being ridden by a boy at a constant speed—the tires are making five rotations per second. How many miles will he ride in 15 minutes? (1 mi = 5280 ft)

11. The smaller gear in the diagram below has a radius of 2 centimeters, and the larger gear has a radius of 5.2 centimeters. If the smaller gear rotates 135°, how many degrees has the larger gear rotated? Round answer to the nearest degree.

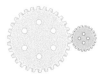

12. Samuel rides 55 feet on a merry-go-round that is 10 feet in diameter in a clockwise direction. Through what angle has Samuel rotated?

13. Layla is building an ornamental wall that is in the shape of a piece of a circle 12 feet in diameter. If the central angle of the circle is 40°, how long is the rock wall?

14. A blueberry pie is made in a 9-inch-diameter pie pan. If a 1-inch-radius circle is cut out of the middle for decoration, what is the area of each piece of pie if the pie is cut into eight equal pieces?

15. Tom's hands go in a 9-inch-radius circular pattern as he rows his boat across a lake. If his hands make a complete rotation every 1.5 seconds, what are the angular speed and linear speed of his hands?

In Exercises 16–20, if possible, find the exact value of the indicated trigonometric function using the unit circle.

16. $\sin\left(-\dfrac{7\pi}{6}\right)$

17. $\tan\left(\dfrac{7\pi}{4}\right)$

18. $\csc\left(-\dfrac{3\pi}{4}\right)$

19. $\cot\left(-\dfrac{3\pi}{2}\right)$

20. $\sec\left(-\dfrac{7\pi}{2}\right)$

21. What is the measure in radians of the smaller angle between the hour and minute hands at 10:10?

22. Find all of the exact values of θ that make the equation
$$\sin\theta = -\dfrac{\sqrt{3}}{2}$$ true in the interval $0 \le \theta \le 2\pi$.

23. Find all of the exact values of θ that make the equation
$$\tan\theta = \dfrac{\sqrt{3}}{3}$$ true in the interval $0 \le \theta \le 2\pi$.

24. Sales of a seasonal product s vary according to the time of year sold given as t. If the equation that models sales is
$$s = 500 - 125\cos\left(\dfrac{\pi t}{6}\right),$$ what were the sales in March $(t = 3)$?

25. The manager of a 24-hour plant tracks productivity throughout the day and finds that the equation
$$p = 50 - 12\cos\left(\dfrac{\pi}{12}t - \dfrac{\pi}{4}\right)$$ accurately models output p from his workers at time t, where p is the number of units produced by the workers and t is the time in hours after midnight. What is the plant's output at 5:00 in the evening?

[CHAPTERS 1-3 CUMULATIVE TEST]

1. In a triangle with angles α, β, and γ, and $\alpha + \beta + \gamma = 180°$, if $\alpha = 115°$, $\beta = 25°$, find γ.

2. In a 30°-60°-90° triangle, if the hypotenuse is 24 meters, what are the lengths of the two legs?

3. If $D = (9x + 6)°$ and $G = (7x - 2)°$ in the diagram below, find the measures D and G.

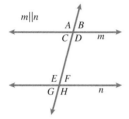

4. **Height of a Woman.** If a 6-foot volleyball player has a 1-foot 4-inch shadow, how long a shadow will her 4-foot 6-inch daughter cast?

5. Use the triangle below to find $\cos\theta$.

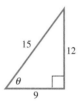

6. Write $\csc 30°$ in terms of its cofunction.

7. Perform the operation $\angle B - \angle A$, where $\angle A = 9°24'15''$ and $\angle B = 74°13'29''$.

8. Use a calculator to approximate $\sec(78°25')$. Round the answer to four decimal places.

9. Given $\alpha = 37.4°$ and $a = 132$ miles, use the right triangle diagram to solve the right triangle. Write the answer for angle measures in decimal degrees.

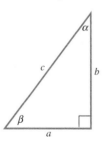

10. Given the angle 99.99° in standard position, state the quadrant of this angle.

11. Given the angle $-270°$ in standard position, find the axis of this angle.

12. The angle θ in standard position has the terminal side defined by the line $3x + 2y = 0$, $x \le 0$. Calculate the values for the six trigonometric functions of θ.

13. Given the angle $\theta = -900°$ in standard position, calculate, if possible, the values for the six trigonometric functions of θ.

14. If $\cos\theta = -\frac{9}{41}$, and the terminal side of θ lies in quadrant III, find $\csc\theta$.

15. Evaluate the expression $\sin 540° - \sec(-540°)$.

16. Find the positive measure of θ (rounded to the nearest degree) if $\tan\theta = 1.4285$ and the terminal side of θ lies in quadrant III.

17. Given $\cot\theta = -\dfrac{\sqrt{5}}{3}$, use the reciprocal identity to find $\tan\theta$.

18. If $\cos\theta = \frac{1}{6}$ and the terminal side of θ lies in quadrant IV, find $\sin\theta$.

19. Find $\sin\theta$ and $\cos\theta$ if $\tan\theta = 6$ and the terminal side of θ lies in quadrant III.

20. Find the measure (in radians) of a central angle θ that intercepts an arc on a circle of radius $r = 1.6$ centimeters with arc length $s = 4$ millimeters.

21. **Clock.** How many radians does the second hand of a clock turn in 1 minute, 45 seconds?

22. Find the exact length of the radius with arc length $s = \dfrac{9\pi}{7}$ meters and central angle $\theta = \dfrac{2\pi}{7}$.

23. Find the distance traveled (arc length) of a point that moves with constant speed $v = 2.6$ meters per second along a circle in 3.3 seconds.

24. **Bicycle.** How fast is a bicyclist traveling in miles per hour if his tires are 24 inches in diameter and his angular speed is 5π radians per second?

25. Find all of the exact values of θ, when $\tan\theta = 1$ and $0 \le \theta \le 2\pi$.

Graphing Trigonometric Functions

Sine and cosine functions are used to represent periodic phenomena. Orbits, tide levels, the biological clock in animals and plants, and radio signals are all periodic (repetitive).

When you are standing on the shore of a placid lake and a motor boat goes by, the waves lap up to the shore at regular intervals and for a while the height of each wave appears to be constant.

If we graph the height of the water as a function of time, the result is a *sine wave*. The duration of time between each wave hitting the shore is called the *period*, and the height of the wave is called the *amplitude*.

LEARNING OBJECTIVES

- Graph basic sine and cosine functions using amplitude and period.
- Graph general sine and cosine functions using translations.
- Graph tangent, cotangent, secant, and cosecant functions.

We will graph trigonometric functions. We will start with simple sine and cosine functions and then proceed to translations of sine and cosine functions. We will then use the basic graphs of the sine and cosine functions to determine graphs of the tangent, secant, cosecant, and cotangent functions. Many periodic phenomena can be represented with sine and cosine functions, such as orbits, tides, biological clocks, and electromagnetic waves (radio, cell phones, and lasers).

GRAPHING TRIGONOMETRIC FUNCTIONS

4.1
BASIC GRAPHS OF SINE AND COSINE FUNCTIONS: AMPLITUDE AND PERIOD

- The Graphs of Sinusoidal Functions
- Harmonic Motion

4.2
TRANSLATIONS OF THE SINE AND COSINE FUNCTIONS: ADDITION OF ORDINATES

- Reflections and Vertical Shifts of Sinusoidal Functions
- Horizontal Shifts: Phase Shift
- Graphing $y = k + A\sin(Bx + C)$ and $y = k + A\cos(Bx + C)$
- Graphing Sums of Functions: Addition of Ordinates

4.3
GRAPHS OF TANGENT, COTANGENT, SECANT, AND COSECANT FUNCTIONS

- Graphing the Tangent, Cotangent, Secant, and Cosecant Functions
- Translations of Tangent, Cotangent, Secant, and Cosecant Functions

4.1 BASIC GRAPHS OF SINE AND COSINE FUNCTIONS: AMPLITUDE AND PERIOD

SKILLS OBJECTIVES	CONCEPTUAL OBJECTIVES
■ Determine the amplitude and period of sinusoidal functions. ■ Solve harmonic motion problems.	■ Understand why the graphs of the sine and cosine functions are called sinusoidal graphs. ■ Visualize harmonic motion as a sinusoidal function.

4.1.1 The Graphs of Sinusoidal Functions

4.1.1 SKILL

Determine the amplitude and period of sinusoidal functions.

4.1.1 CONCEPTUAL

Understand why the graphs of the sine and cosine functions are called sinusoidal graphs.

The following are examples of things that repeat in a predictable way (are roughly periodic):

■ a heartbeat

■ tide levels

■ time of sunrise

■ average outdoor temperature for the time of year

The trigonometric functions are *strictly* periodic. In the unit circle, the value of any of the trigonometric functions is the same for any coterminal angle (same initial and terminal sides) no matter how many full rotations the angle makes. For example, if we add (or subtract) integer multiples of 2π to the angle θ, the values for sine and cosine are unchanged.

$$\sin(\theta + 2n\pi) = \sin\theta \quad \text{or} \quad \cos(\theta + 2n\pi) = \cos\theta \ (n \text{ is any integer})$$

DEFINITION | Periodic Function

A function f is called a **periodic function** if there is a positive number p such that

$$f(x + p) = f(x) \qquad \text{for all } x \text{ in the domain of } f$$

If p is the smallest such number for which this equation holds, then p is called the **fundamental period.**

The above equations confirm that the sine and cosine functions (and hence the secant and cosecant functions) are periodic with fundamental period 2π. We'll see in this chapter that the tangent and cotangent functions are periodic with fundamental period π.

The Graph of $f(x) = \sin x$

Let us start by point-plotting the sine function. We select values for the sine function corresponding to some of the "special angles" we covered in the previous chapters.

x	y = sin x	(x, y)
0	$\sin 0 = 0$	$(0, 0)$
$\dfrac{\pi}{4}$	$\sin\left(\dfrac{\pi}{4}\right) = \dfrac{\sqrt{2}}{2}$	$\left(\dfrac{\pi}{4}, \dfrac{\sqrt{2}}{2}\right)$
$\dfrac{\pi}{2}$	$\sin\left(\dfrac{\pi}{2}\right) = 1$	$\left(\dfrac{\pi}{2}, 1\right)$
$\dfrac{3\pi}{4}$	$\sin\left(\dfrac{3\pi}{4}\right) = \dfrac{\sqrt{2}}{2}$	$\left(\dfrac{3\pi}{4}, \dfrac{\sqrt{2}}{2}\right)$
π	$\sin \pi = 0$	$(\pi, 0)$
$\dfrac{5\pi}{4}$	$\sin\left(\dfrac{5\pi}{4}\right) = -\dfrac{\sqrt{2}}{2}$	$\left(\dfrac{5\pi}{4}, -\dfrac{\sqrt{2}}{2}\right)$
$\dfrac{3\pi}{2}$	$\sin\left(\dfrac{3\pi}{2}\right) = -1$	$\left(\dfrac{3\pi}{2}, -1\right)$
$\dfrac{7\pi}{4}$	$\sin\left(\dfrac{7\pi}{4}\right) = -\dfrac{\sqrt{2}}{2}$	$\left(\dfrac{7\pi}{4}, -\dfrac{\sqrt{2}}{2}\right)$
2π	$\sin(2\pi) = 0$	$(2\pi, 0)$

Plotting the above coordinates (x, y) in the Cartesian plane, we can obtain the graph of one **period**, or **cycle**, of the graph of $y = \sin x$. Note that $\dfrac{\sqrt{2}}{2} \approx 0.7$.

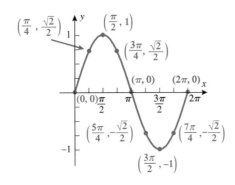

We can extend the graph horizontally in both directions (left and right) since the domain of the sine function is the set of all real numbers.

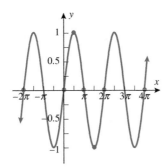

In this chapter, we are no longer showing angles on the unit circle but are now showing angles as *real numbers* in radians on the *x*-axis of the *Cartesian* graph. Therefore, we no longer illustrate a terminal side to an angle—the physical arcs and angles no longer exist; only their measures exist, as values of the *x*-coordinate.

If we graph the function $f(x) = \sin x$, the x-intercepts of the graph correspond to values of x at which the sine function is equal to zero.

x	$y = \sin x$	(x, y)
0	$\sin 0 = 0$	$(0, 0)$
π	$\sin \pi = 0$	$(\pi, 0)$
2π	$\sin(2\pi) = 0$	$(2\pi, 0)$
3π	$\sin(3\pi) = 0$	$(3\pi, 0)$
4π	$\sin(4\pi) = 0$	$(4\pi, 0)$
$\ldots$		
$n\pi$	$\sin(n\pi) = 0$	$(n\pi, 0)$

where n is an integer.

Notice that the point $(0, 0)$ is both a y-intercept and an x-intercept, but all x-intercepts have the form $(n\pi, 0)$, where n is an integer. The maximum value of the sine function is 1, and the minimum value of the sine function is -1; these values occur at odd integer multiples of $\dfrac{\pi}{2}$.

x	$y = \sin x$	(x, y)
$\dfrac{\pi}{2}$	$\sin\left(\dfrac{\pi}{2}\right) = 1$	$\left(\dfrac{\pi}{2}, 1\right)$
$\dfrac{3\pi}{2}$	$\sin\left(\dfrac{3\pi}{2}\right) = -1$	$\left(\dfrac{3\pi}{2}, -1\right)$
$\dfrac{5\pi}{2}$	$\sin\left(\dfrac{5\pi}{2}\right) = 1$	$\left(\dfrac{5\pi}{2}, 1\right)$
$\dfrac{7\pi}{2}$	$\sin\left(\dfrac{7\pi}{2}\right) = -1$	$\left(\dfrac{7\pi}{2}, -1\right)$
$\ldots$		
$\dfrac{(2n + 1)\pi}{2}$	$\sin\left(\dfrac{(2n + 1)\pi}{2}\right) = \pm 1$	$\left(\dfrac{(2n + 1)\pi}{2}, \pm 1\right)$

n is an integer.

The following box summarizes the sine function:

SINE FUNCTION $f(x) = \sin x$

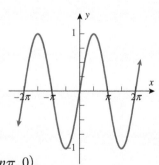

- Domain: $(-\infty, \infty)$ or $-\infty < x < \infty$
- Range: $[-1, 1]$ or $-1 \leq y \leq 1$
- The sine function is an odd function:
  - Symmetric about the origin
  - $\sin(-x) = -\sin x$
- The sine function is a periodic function with fundamental period 2π.
- The x-intercepts, $0, \pm\pi, \pm 2\pi, \ldots$, are integer multiples of π, $n\pi$, where n is an integer: $(n\pi, 0)$.
- The maximum (1) and minimum (-1) values of the sine function correspond to x values that are odd integer multiples of $\dfrac{\pi}{2}$, $\dfrac{(2n + 1)\pi}{2}$, such as $\pm\dfrac{\pi}{2}, \pm\dfrac{3\pi}{2}, \pm\dfrac{5\pi}{2}, \ldots$.

The Graph of $f(x) = \cos x$

Let us start by point-plotting the cosine function.

x	$y = \cos x$	(x, y)
0	$\cos 0 = 1$	$(0, 1)$
$\dfrac{\pi}{4}$	$\cos\left(\dfrac{\pi}{4}\right) = \dfrac{\sqrt{2}}{2}$	$\left(\dfrac{\pi}{4}, \dfrac{\sqrt{2}}{2}\right)$
$\dfrac{\pi}{2}$	$\cos\left(\dfrac{\pi}{2}\right) = 0$	$\left(\dfrac{\pi}{2}, 0\right)$
$\dfrac{3\pi}{4}$	$\cos\left(\dfrac{3\pi}{4}\right) = -\dfrac{\sqrt{2}}{2}$	$\left(\dfrac{3\pi}{4}, -\dfrac{\sqrt{2}}{2}\right)$
π	$\cos \pi = -1$	$(\pi, -1)$
$\dfrac{5\pi}{4}$	$\cos\left(\dfrac{5\pi}{4}\right) = -\dfrac{\sqrt{2}}{2}$	$\left(\dfrac{5\pi}{4}, -\dfrac{\sqrt{2}}{2}\right)$
$\dfrac{3\pi}{2}$	$\cos\left(\dfrac{3\pi}{2}\right) = 0$	$\left(\dfrac{3\pi}{2}, 0\right)$
$\dfrac{7\pi}{4}$	$\cos\left(\dfrac{7\pi}{4}\right) = \dfrac{\sqrt{2}}{2}$	$\left(\dfrac{7\pi}{4}, \dfrac{\sqrt{2}}{2}\right)$
2π	$\cos(2\pi) = 1$	$(2\pi, 1)$

Plotting the above coordinates (x, y), we obtain the graph of one period, or cycle, of the graph of $y = \cos x$. Note that $\dfrac{\sqrt{2}}{2} \approx 0.7$.

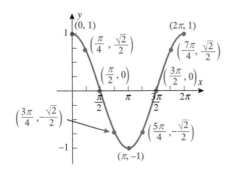

We can extend the graph horizontally in both directions (left and right) since the domain of the cosine function is the set of all real numbers.

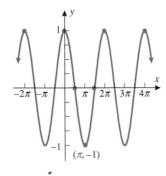

If we graph the function $y = \cos x$, the x-intercepts correspond to values of x at which the cosine function is equal to zero.

x	y = cos x	(x, y)
$\dfrac{\pi}{2}$	$\cos\left(\dfrac{\pi}{2}\right) = 0$	$\left(\dfrac{\pi}{2}, 0\right)$
$\dfrac{3\pi}{2}$	$\cos\left(\dfrac{3\pi}{2}\right) = 0$	$\left(\dfrac{3\pi}{2}, 0\right)$
$\dfrac{5\pi}{2}$	$\cos\left(\dfrac{5\pi}{2}\right) = 0$	$\left(\dfrac{5\pi}{2}, 0\right)$
$\dfrac{7\pi}{2}$	$\cos\left(\dfrac{7\pi}{2}\right) = 0$	$\left(\dfrac{7\pi}{2}, 0\right)$
...		
$\dfrac{(2n+1)\pi}{2}$	$\cos\left(\dfrac{(2n+1)\pi}{2}\right) = 0$	$\left(\dfrac{(2n+1)\pi}{2}, 0\right)$

n is an integer.

The point $(0, 1)$ is the y-intercept, since $\cos 0 = 1$, and there are several x-intercepts of the form $\left(\dfrac{(2n+1)\pi}{2}, 0\right)$, where n is an integer. The maximum value of the cosine function is 1, and the minimum value of the cosine function is -1; these values occur at integer multiples of π, $n\pi$.

x	y = cos x	(x, y)
0	$\cos 0 = 1$	$(0, 1)$
π	$\cos \pi = -1$	$(\pi, -1)$
2π	$\cos(2\pi) = 1$	$(2\pi, 1)$
3π	$\cos(3\pi) = -1$	$(3\pi, -1)$
4π	$\cos(4\pi) = 1$	$(4\pi, 1)$
...		
$n\pi$	$\cos(n\pi) = \pm 1$	$(n\pi, \pm 1)$

n is an integer.

The following box summarizes the cosine function:

COSINE FUNCTION $f(x) = \cos x$

- Domain: $(-\infty, \infty)$ or $-\infty < x < \infty$
- Range: $[-1, 1]$ or $-1 \leq y \leq 1$
- The cosine function is an even function:
  - Symmetric about the y-axis
  - $\cos(-x) = \cos x$
- The cosine function is a periodic function with fundamental period 2π.

- The x-intercepts, $\pm\dfrac{\pi}{2}, \pm\dfrac{3\pi}{2}, \pm\dfrac{5\pi}{2}, \ldots$, are odd integer multiples of $\dfrac{\pi}{2}$, which have the form $\dfrac{(2n+1)\pi}{2}$, where n is an integer: $\left(\dfrac{(2n+1)\pi}{2}, 0\right)$.

- The maximum (1) and minimum (-1) values of the cosine function correspond to x-values that are integer multiples of π, $n\pi$, such as $0, \pm\pi, \pm2\pi, \ldots$.

The Amplitude and Period of Sinusoidal Graphs

In mathematics, the word **sinusoidal** means "resembling the sine function." Let us start by graphing $f(x) = \sin x$ and $f(x) = \cos x$ on the same graph. Notice that they have similar characteristics (domain, range, period, and shape).

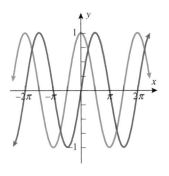

In fact, if we were to shift the cosine graph to the right $\dfrac{\pi}{2}$ units, the two graphs would be identical. For that reason, we refer to any graphs of the form $y = \cos x$ or $y = \sin x$ as **sinusoidal functions**.

We now turn our attention to graphs of the form $y = A\sin(Bx)$ and $y = A\cos(Bx)$, which are graphs like $y = \sin x$ and $y = \cos x$ that have been stretched or compressed vertically and horizontally. (See the Appendix for a review of translations of functions.)

EXAMPLE 1 **Vertical Stretching and Compressing**

Plot the functions $y = 2\sin x$ and $y = \frac{1}{2}\sin x$ on the same graph with $y = \sin x$ on the interval $-4\pi \le x \le 4\pi$.

Solution:

STEP 1 Make a table with key values of the three functions.

x	0	$\dfrac{\pi}{2}$	π	$\dfrac{3\pi}{2}$	2π
$\sin x$	0	1	0	-1	0
$2\sin x$	0	2	0	-2	0
$\frac{1}{2}\sin x$	0	$\frac{1}{2}$	0	$-\frac{1}{2}$	0

STEP 2 Label the points on the graph and connect with a smooth curve over one period, $0 \le x \le 2\pi$.

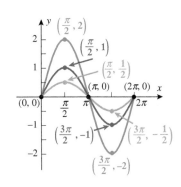

STEP 3 Extend the graph in both directions
(repeats every 2π).

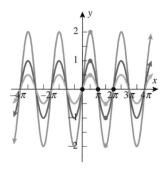

▼
ANSWER

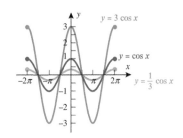

▼ YOUR TURN Plot the functions $y = 3\cos x$ and $y = \frac{1}{3}\cos x$ on the same graph
with $y = \cos x$ on the interval $-2\pi \le x \le 2\pi$.

Notice in Example 1 and the corresponding Your Turn that:

■ $y = 2\sin x$ has the shape and period of $y = \sin x$ but stretched vertically.

■ $y = \frac{1}{2}\sin x$ has the shape and period of $y = \sin x$ but compressed vertically.

■ $y = 3\cos x$ has the shape and period of $y = \cos x$ but stretched vertically.

■ $y = \frac{1}{3}\cos x$ has the shape and period of $y = \cos x$ but compressed vertically.

In general, functions of the form $y = A\sin x$ and $y = A\cos x$ are stretched vertically
when $|A| > 1$ and compressed vertically when $|A| < 1$.

The **amplitude** of a periodic function is half the distance between the maximum
value of the function and the minimum value of the function. For the functions $y = \sin x$
and $y = \cos x$, the maximum value is 1 and the minimum value is -1. Therefore, the
amplitude of each of these two functions is $|A| = \frac{1}{2}|1 - (-1)| = 1$.

AMPLITUDE OF SINUSOIDAL FUNCTIONS

For sinusoidal functions of the form $y = A\sin x$ and $y = A\cos x$, the **amplitude**
is $|A|$. When $|A| < 1$, the graph is compressed vertically and when $|A| > 1$, the
graph is stretched vertically.

EXAMPLE 2 **Finding the Amplitude of Sinusoidal Functions**

State the amplitude of
a. $f(x) = -4\cos x$
b. $g(x) = \frac{1}{5}\sin x$

Solution (a): $|A| = |-4| = \boxed{4}$

Solution (b): $|A| = |\frac{1}{5}| = \boxed{\frac{1}{5}}$

EXAMPLE 3 **Horizontal Stretching and Compressing**

Plot the functions $y = \cos(2x)$ and $y = \cos\left(\frac{1}{2}x\right)$ on the same graph with $y = \cos x$ on the interval $-2\pi \leq x \leq 2\pi$.

Solution:

STEP 1 Make a table with the coordinate values of the graphs. It is necessary only to select the points that correspond to x-intercepts, $(y = 0)$, and maximum and minimum points, $(y = \pm 1)$. Usually, the period is divided into four subintervals (which you will see in Examples 5 to 7).

x	O	$\frac{\pi}{4}$	$\frac{\pi}{2}$	$\frac{3\pi}{4}$	π	$\frac{5\pi}{4}$	$\frac{3\pi}{2}$	$\frac{7\pi}{4}$	2π
$\cos x$	1		0		-1		0		1
$\cos(2x)$	1	0	-1	0	1	0	-1	0	1
$\cos\left(\frac{1}{2}x\right)$	1				0				-1

STEP 2 Label the points on the graph and connect with a smooth curve.

$y = \cos x$
$y = \cos(2x)$
$y = \cos\left(\frac{1}{2}x\right)$

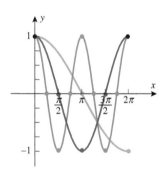
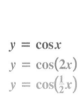

STEP 3 Extend the graph to cover the interval: $-2\pi \leq x \leq 2\pi$.

$y = \cos x$
$y = \cos(2x)$
$y = \cos\left(\frac{1}{2}x\right)$

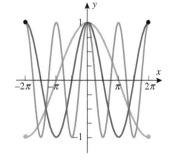

ANSWER

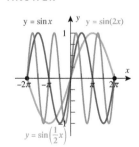

YOUR TURN Plot the functions $y = \sin(2x)$ and $y = \sin\left(\frac{1}{2}x\right)$ on the same graph with $y = \sin x$ on the interval $-2\pi \leq x \leq 2\pi$.

Notice in Example 3 and the corresponding Your Turn that:

- $y = \cos(2x)$ has the shape and amplitude of $y = \cos x$ but compressed horizontally.
- $y = \cos(\frac{1}{2}x)$ has the shape and amplitude of $y = \cos x$ but stretched horizontally.
- $y = \sin(2x)$ has the shape and amplitude of $y = \sin x$ but compressed horizontally.
- $y = \sin(\frac{1}{2}x)$ has the shape and amplitude of $y = \sin x$ but stretched horizontally.

In general, functions of the form $y = \sin(Bx)$ and $y = \cos(Bx)$, with $B > 0$, are *compressed* horizontally when $B > 1$ and *stretched* horizontally when $0 < B < 1$. We will discuss negative arguments $(B < 0)$ in the next section in the context of *reflections*.

The period of the functions $y = \sin x$ and $y = \cos x$ is 2π. To find the period of a function of the form $y = A \sin(Bx)$ or $y = A \cos(Bx)$, set Bx equal to 2π and solve for x.

$$Bx = 2\pi$$

$$x = \frac{2\pi}{B}$$

PERIOD OF SINUSOIDAL FUNCTIONS

For sinusoidal functions of the form $y = A \sin(Bx)$ and $y = A \cos(Bx)$, with $B > 0$, the **period** is $\frac{2\pi}{B}$. When $0 < B < 1$, the graph is stretched horizontally since the period is larger than 2π, and when $B > 1$, the graph is compressed horizontally since the period is smaller than 2π.

EXAMPLE 4 **Finding the Period of a Sinusoidal Function**

State the period of
a. $f(x) = \cos(4x)$
b. $g(x) = \sin(\frac{1}{3}x)$

Solution (a):

Compare $\cos(4x)$ with $\cos(Bx)$ to identify B. $\qquad\qquad B = 4$

Calculate the period of $\cos(4x)$, using $p = \dfrac{2\pi}{B}$. $\qquad p = \dfrac{2\pi}{4} = \dfrac{\pi}{2}$

The period of $\cos(4x)$ is $\boxed{p = \dfrac{\pi}{2}}$.

Solution (b):

Compare $\sin(\frac{1}{3}x)$ with $\sin(Bx)$ to identify B. $\qquad\qquad B = \dfrac{1}{3}$

Calculate the period of $\sin(\frac{1}{3}x)$, using $p = \dfrac{2\pi}{B}$. $\qquad p = \dfrac{2\pi}{\frac{1}{3}} = 6\pi$

The period of $\sin(\frac{1}{3}x)$ is $\boxed{p = 6\pi}$.

▼
ANSWER

a. $p = \dfrac{2\pi}{3}$ **b.** $p = 4\pi$

▼
YOUR TURN State the period of
a. $f(x) = \sin(3x)$ **b.** $g(x) = \cos(\frac{1}{2}x)$

Now that you know the basic graphs of $y = \sin x$ and $y = \cos x$, you can sketch one cycle (period) of these graphs with the following x values: $0, \dfrac{\pi}{2}, \pi, \dfrac{3\pi}{2}, 2\pi$. For a period of 2π, we use steps of $\dfrac{\pi}{2}$. Therefore, for functions of the form $y = A\sin(Bx)$ and $y = A\cos(Bx)$, we start at the origin, and as long as we include these five basic values (corresponding to four equal intervals) during one period, we are able to sketch the graphs.

STUDY TIP

Divide the period by 4 to get the key values along the x-axis for graphing.

STRATEGY FOR SKETCHING GRAPHS OF SINUSOIDAL FUNCTIONS

To graph $y = A\sin(Bx)$ or $y = A\cos(Bx)$ with $B > 0$:

Step 1: Find the amplitude $|A|$ and period $\dfrac{2\pi}{B}$.

Step 2: Divide the period into four equal parts (steps).

Step 3: Make a table and evaluate the function for the x values from Step 2 (starting at $x = 0$).

Step 4: Draw the xy-plane (label the y-axis up to $\pm A$) and plot the points found in Step 3.

Step 5: Connect the points with a sinusoidal curve (with amplitude $|A|$).

Step 6: Extend the graph over one or two additional periods in both directions (left and right).

EXAMPLE 5 **Graphing Sinusoidal Functions of the Form $y = A\sin(Bx)$**

Use the strategy for graphing sinusoidal functions to graph $y = 3\sin(2x)$.

Solution:

STEP 1 Find the amplitude and period for $A = 3$ and $B = 2$.

$|A| = |3| = 3$

$$p = \frac{2\pi}{B} = \frac{2\pi}{2} = \pi$$

STEP 2 Divide the period π into four equal steps.

$$\frac{\pi}{4}$$

STEP 3 Make a table, starting at $x = 0$ to the period $x = \pi$ in steps of $\dfrac{\pi}{4}$.

x	$y = 3\sin(2x)$	(x, y)
0	$3[\sin 0] = 3[0] = 0$	$(0, 0)$
$\dfrac{\pi}{4}$	$3\left[\sin\left(\dfrac{\pi}{2}\right)\right] = 3[1] = 3$	$\left(\dfrac{\pi}{4}, 3\right)$
$\dfrac{\pi}{2}$	$3[\sin \pi] = 3[0] = 0$	$\left(\dfrac{\pi}{2}, 0\right)$
$\dfrac{3\pi}{4}$	$3\left[\sin\left(\dfrac{3\pi}{2}\right)\right] = 3[-1] = -3$	$\left(\dfrac{3\pi}{4}, -3\right)$
π	$3[\sin(2\pi)] = 3[0] = 0$	$(\pi, 0)$

STEP 4 Draw the *xy*-plane and label the points in the table.

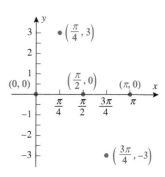

STEP 5 Connect the points with a sinusoidal curve.

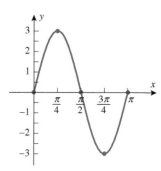

▼
ANSWER

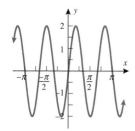

x	**y**
0	0
$\frac{\pi}{6}$	2
$\frac{\pi}{3}$	0
$\frac{\pi}{2}$	−2
$\frac{2\pi}{3}$	0

STEP 6 Repeat over several periods (to the left and right).

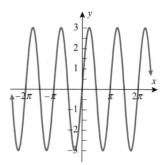

▼
YOUR TURN   Use the strategy for graphing sinusoidal functions to graph $y = 2\sin(3x)$.

EXAMPLE 6   **Graphing Sinusoidal Functions of the Form $y = A\cos(Bx)$**

Use the strategy for graphing sinusoidal functions to graph $y = -2\cos\left(\frac{1}{3}x\right)$.

Solution:

STEP 1 Find the amplitude and period for $A = -2$ and $B = \frac{1}{3}$.

$$|A| = |-2| = 2$$

$$p = \frac{2\pi}{B} = \frac{2\pi}{\frac{1}{3}} = 6\pi$$

STEP 2 Divide the period 6π into four equal steps.

$$\frac{6\pi}{4} = \frac{3\pi}{2}$$

STEP 3 Make a table starting at $x = 0$ and completing one period of 6π in steps of $\dfrac{3\pi}{2}$.

x	$y = -2\cos\left(\frac{1}{3}x\right)$	(x, y)
0	$-2[\cos 0] = -2[1] = -2$	$(0, -2)$
$\dfrac{3\pi}{2}$	$-2\left[\cos\left(\dfrac{\pi}{2}\right)\right] = -2[0] = 0$	$\left(\dfrac{3\pi}{2}, 0\right)$
3π	$-2[\cos \pi] = -2[-1] = 2$	$(3\pi, 2)$
$\dfrac{9\pi}{2}$	$-2\left[\cos\left(\dfrac{3\pi}{2}\right)\right] = -2[0] = 0$	$\left(\dfrac{9\pi}{2}, 0\right)$
6π	$-2[\cos(2\pi)] = -2[1] = -2$	$(6\pi, -2)$

STEP 4 Draw the *xy*-plane and label the points in the table.

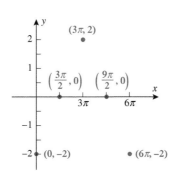

STEP 5 Connect the points with a sinusoidal curve.

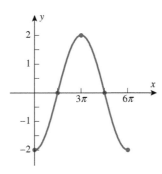

STEP 6 Repeat over several periods (to the left and right).

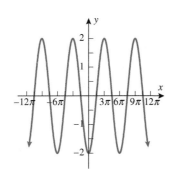

▼
ANSWER

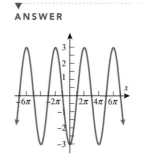

YOUR TURN Use the strategy for graphing sinusoidal functions to graph $y = -3\cos\left(\frac{1}{2}x\right)$.

EXAMPLE 7 **Finding an Equation for a Sinusoidal Graph**

Find an equation for the graph.

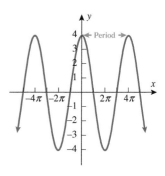

Solution:

This graph represents a cosine function.	$y = A\cos(Bx)$
The amplitude is 4 (half the maximum vertical spread).	$\|A\| = 4$ so $A = \pm 4$
Since the point $(0, 4)$ lies on the graph, we know that $4 = A\cos 0$, so we select $A = 4$.	
The period $\dfrac{2\pi}{B}$ is equal to 4π.	$\dfrac{2\pi}{B} = 4\pi$
Solve for B.	$B = \dfrac{1}{2}$
Substitute $A = 4$ and $B = \frac{1}{2}$ into $y = A\cos(Bx)$.	$\boxed{y = 4\cos\left(\dfrac{1}{2}x\right)}$

▼
ANSWER

$y = 6\sin(2x)$

▼
YOUR TURN Find an equation of the graph.

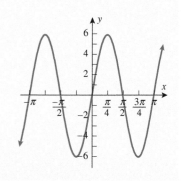

4.1.2 Harmonic Motion

4.1.2 SKILL

Solve harmonic motion problems.

4.1.2 CONCEPTUAL

Visualize harmonic motion as a sinusoidal function.

One of the most important applications of sinusoidal functions is in describing *harmonic motion*, which we define as the symmetric periodic movement of an object or quantity about a center (equilibrium) position or value. The oscillation of a pendulum is a form of harmonic motion. Other examples are the recoil of a spring balance scale when a weight is placed on the tray, or the variation of current or voltage within an AC circuit.

There are three types of harmonic motion: *simple harmonic motion*, *damped harmonic motion*, and *resonance*.

Simple Harmonic Motion

Simple harmonic motion is the kind of *unvarying* periodic motion that would occur in an ideal situation in which no resistive forces, such as friction, cause the amplitude of oscillation to decrease over time. This type of motion will also occur if energy is being supplied at the correct rate to overcome resistive forces. Simple harmonic motion occurs, for example, in an AC electric circuit when a power source is consistently supplying energy. When you are swinging on a swing and "pumping" energy into the swing to keep it in motion at a constant period and amplitude, you are sustaining simple harmonic motion.

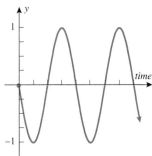

Damped Harmonic Motion

In damped harmonic motion, the amplitude of the periodic motion decreases as time increases. If you are on a moving swing and stop "pumping" new energy into the swing, the swing will continue moving with a constant period, but the amplitude, the height to which the swing will rise, will diminish with each cycle as the swing is slowed down by friction with the air or between its own moving parts.

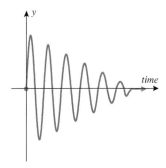

Resonance

Resonance is what occurs when the amplitude of periodic motion increases as time increases. It is caused when the energy applied to an oscillating object or system is more than what is needed to oppose friction or other forces and sustain simple harmonic motion; instead, the applied energy *increases* the amplitude of harmonic motion with each cycle. With resonance, eventually, the amplitude becomes unbounded and the result is disastrous. Bridges have collapsed because of resonance. Military soldiers know that when they march across a bridge, they must break cadence to prevent resonance. Below are two pictures of the Tacoma Narrows Bridge (near Seattle, Washington) that opened to traffic on July 1, 1940, and collapsed into Puget Sound on November 7, 1940. The collapsing of the Tacoma Narrows Bridge is often presented as an example of resonance (the wind providing an external periodic frequency that coincided with the bridge's natural frequency), but many believe the actual angle of failure was torsional motion that twisted cubes.

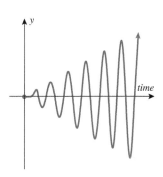

© Keystone/Stringer/Getty Images News and Sports Services

Topham/The Image Works

Examples of Harmonic Motion

If we hang a weight from a spring, then while the resulting system is at rest, we say it is in the equilibrium position.

If we then pull down on the weight and release it, the elasticity in the spring pulls the weight up and causes it to start oscillating up and down.

If we neglect friction and air resistance, we can imagine that the combination of the weight and the spring will oscillate indefinitely; the height of the weight with respect to the equilibrium position can be modeled by a simple sinusoidal function. This is an example of **simple harmonic motion**.

SIMPLE HARMONIC MOTION

The position of a point oscillating around an equilibrium position at time t can be modeled by the sinusoidal function

$$y = A \sin(\omega t) \qquad \text{or} \qquad y = A \cos(\omega t)$$

where $|A|$ is the amplitude and the period is $\dfrac{2\pi}{\omega}$, where $\omega > 0$.

Note: The symbol ω (Greek lowercase Omega) represents the angular frequency.

EXAMPLE 8 **Simple Harmonic Motion**

Let the height of the seat of a swing be equal to zero when the swing is at rest. Assume that a child starts swinging until she reaches the highest she can swing and keeps her effort constant. The height $h(t)$ of the seat is given by

$$h(t) = 8 \sin\left(\frac{\pi}{2}t\right)$$

where t is time in seconds and h is the height in feet. Note that positive h indicates height reached swinging forward, and negative h indicates height reached swinging backward. Assume that $t = 0$ when the child passes through the equilibrium position swinging forward.

a. What is the maximum height above the resting level reached by the seat of the swing?

b. What is the period of the swinging child?

c. Graph the height function $h(t)$ for $0 \le t \le 4$.

Solution (a):

The amplitude is 8. $|A| = |8| = \boxed{8 \text{ ft}}$

Solution (b):

The period is $\dfrac{2\pi}{B}$, where $B = \dfrac{\pi}{2}$. $p = \dfrac{2\pi}{\dfrac{\pi}{2}} = \boxed{4 \text{ sec}}$

Solution (c):

Divide the period 4 into four equal parts (steps of 1).

t (SECONDS)	$h(t) = 8\sin\left(\frac{\pi}{2}t\right) = y$	(t, y)
0	$8\sin 0 = 8(0) = 0$	$(0, 0)$
1	$8\sin\left(\dfrac{\pi}{2}\right) = 8(1) = 8$	$(1, 8)$
2	$8\sin \pi = 8(0) = 0$	$(2, 0)$
3	$8\sin\left(\dfrac{3\pi}{2}\right) = 8(-1) = -8$	$(3, -8)$
4	$8\sin(2\pi) = 8(0) = 0$	$(4, 0)$

From the graph, we see that the maximum height is 8 feet and the period is 4 seconds.

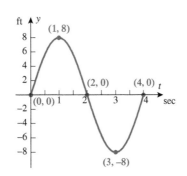

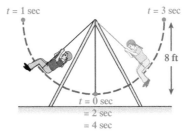

Damped harmonic motion is any sinusoidal function whose amplitude decreases as time increases. If we again hang a weight from a spring so that it is suspended at rest, and then pull down on the weight and release, the weight will oscillate about the equilibrium point. This time we will not neglect friction and air resistance: The weight will oscillate closer and closer to the equilibrium point over time until the weight eventually comes to rest at the equilibrium point. This is an example of damped harmonic motion.

The product of any decreasing function and the original periodic function will describe damped oscillatory motion. Here are two examples of functions that describe damped harmonic motion:

$$y = \frac{1}{t}\sin(\omega t) \qquad y = e^{-t}\cos(\omega t)$$

where e^{-t} is a decreasing exponential function (exponential decay).

EXAMPLE 9 **Damped Harmonic Motion**

Assume that the child in Example 8 decides to stop pumping and allows the swing to continue moving until she eventually comes to rest. If the height is given by

$$h(t) = \frac{8}{t} \cos\left(\frac{\pi}{2}t\right)$$

where t is time in seconds and h is the height in feet above the resting position. Note that positive h indicates height reached swinging forward and negative h indicates height reached swinging backward, assuming that $t = 1$ when the child initially passes through the equilibrium position swinging backward and stops "pumping."

a. Graph the height function $h(t)$ for $1 \le t \le 8$.

b. What is the height above the resting level at 4 seconds? At 8 seconds? After 1 minute?

Solution (a):

Make a table with integer values of t. $\qquad\qquad 1 \le t \le 8$

t (SECONDS)	$h(t) = \frac{8}{t}\cos\left(\frac{\pi}{2}t\right) = y$	(t, y)
1	$\frac{8}{1}\cos\left(\frac{\pi}{2}\right) = 8(0) = 0$	$(1, 0)$
2	$\frac{8}{2}\cos\pi = 4(-1) = -4$	$(2, -4)$
3	$\frac{8}{3}\cos\left(\frac{3\pi}{2}\right) = \frac{8}{3}(0) = 0$	$(3, 0)$
4	$\frac{8}{4}\cos(2\pi) = 2(1) = 2$	$(4, 2)$
5	$\frac{8}{5}\cos\left(\frac{5\pi}{2}\right) = \frac{8}{5}(0) = 0$	$(5, 0)$
6	$\frac{8}{6}\cos(3\pi) = -\frac{4}{3}(-1) = -\frac{4}{3}$	$\left(6, -\frac{4}{3}\right)$
7	$\frac{8}{7}\cos\left(\frac{7\pi}{2}\right) = \frac{8}{7}(0) = 0$	$(7, 0)$
8	$\frac{8}{8}\cos(4\pi) = 1(1) = 1$	$(8, 1)$

Plot the points in the above table and connect with a smooth curve.

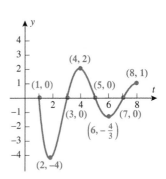

Solution (b):

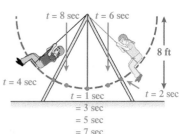

The height is 2 feet when t is 4 seconds.	$\dfrac{8}{4}\cos(2\pi) = 2(1) = 2$
The height is 1 foot when t is 8 seconds.	$\dfrac{8}{8}\cos(4\pi) = 1(1) = 1$
The height is 0.13 feet when t is 1 minute (60 seconds).	$\dfrac{8}{60}\cos(30\pi) \approx 0.1333$

Resonance can be represented by the product of any increasing function and the original sinusoidal function. Here are two examples of functions that result in resonance as time increases.

$$y = t\cos(\omega t) \qquad y = e^t\sin(\omega t)$$

▶[SECTION 4.1] SUMMARY

The sine function is an odd function and its graph is symmetric about the origin.

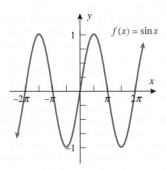

The cosine function is an even function, and its graph is symmetric about the y-axis.

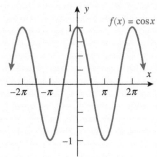

Graphs of the form $y = A\sin(Bx)$ and $y = A\cos(Bx)$ have amplitude $|A|$ and period $\dfrac{2\pi}{B}$.

Point-plotting can be used to graph sinusoidal functions. A more efficient way is to first determine the amplitude and period. Divide the period into four equal parts and choose those values for x. Make a table of those four points and graph (this is the graph of one period). Extend the graph to the left and right.

To find an equation of a sinusoidal function, given its graph, start by first finding the amplitude (half the distance between the maximum and minimum values). Then determine the period. Use these values to determine A and B in the expressions $f(x) = A\sin(Bx)$ or $f(x) = A\cos(Bx)$. Harmonic motion is one of the primary applications of sinusoidal functions.

[SECTION 4.1] EXERCISES

• SKILLS

In Exercises 1–10, match each sinusoidal function with its graph (a)–(j).

1. $y = -\sin x$ **2.** $y = \sin x$ **3.** $y = \cos x$ **4.** $y = -\cos x$

5. $y = 2\sin x$ **6.** $y = 2\cos x$ **7.** $y = \sin\left(\frac{1}{2}x\right)$ **8.** $y = \cos\left(\frac{1}{2}x\right)$

9. $y = -2\cos\left(\frac{1}{2}x\right)$ **10.** $y = -2\sin\left(\frac{1}{2}x\right)$

a.

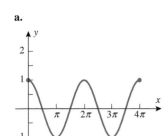

b.

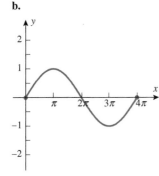

c.

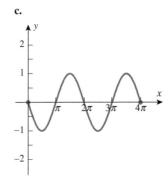

d.

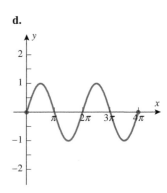

e.

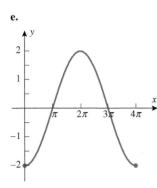

f.

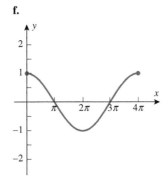

g.

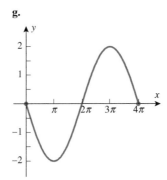

h.

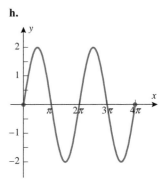

i.

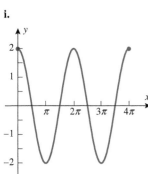

j.

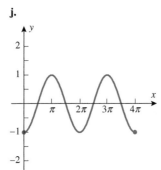

In Exercises 11–24, state the amplitude and period of each sinusoidal function.

11. $y = \dfrac{3}{2}\cos(3x)$

12. $y = \dfrac{2}{3}\sin(4x)$

13. $y = -\sin(5x)$

14. $y = -\cos(7x)$

15. $y = \dfrac{2}{3}\cos\left(\dfrac{3}{2}x\right)$

16. $y = \dfrac{3}{2}\sin\left(\dfrac{2}{3}x\right)$

17. $y = -3\cos(\pi x)$

18. $y = -2\sin(\pi x)$

19. $y = 5\sin\left(\dfrac{\pi}{3}x\right)$

20. $y = 4\cos\left(\dfrac{\pi}{4}x\right)$

21. $y = -\dfrac{4}{5}\cos\left(\dfrac{x}{2}\right)$

22. $y = 3\sin\left(\dfrac{x}{\pi}\right)$

23. $y = -\dfrac{1}{3}\sin\left(\dfrac{1}{4}x\right)$

24. $y = \cos\left(\dfrac{\pi x}{3}\right)$

In Exercises 25–40, graph the given sinusoidal functions over one period.

25. $y = 8\cos x$

26. $y = 7\sin x$

27. $y = \sin(4x)$

28. $y = \cos(3x)$

29. $y = -2\cos x$

30. $y = -\frac{1}{2}\sin(2x)$

31. $y = -4\cos(2x)$

32. $y = \sin(0.5x)$

33. $y = -3\cos\left(\dfrac{1}{2}x\right)$

34. $y = -2\sin\left(\dfrac{1}{4}x\right)$

35. $y = -3\sin(\pi x)$

36. $y = -2\cos(\pi x)$

37. $y = 5\cos(2\pi x)$

38. $y = 4\sin(2\pi x)$

39. $y = -3\sin\left(\dfrac{\pi}{4}x\right)$

40. $y = -4\sin\left(\dfrac{\pi}{2}x\right)$

In Exercises 41–52, graph the given sinusoidal function over the interval $[-2p, 2p]$, where p is the period of the function.

41. $y = 2\cos\left(\dfrac{\pi}{2}x\right)$

42. $y = -3\sin\left(\dfrac{\pi}{2}x\right)$

43. $y = 6\sin(3x)$

44. $y = -2\cos\left(\dfrac{\pi}{3}x\right)$

45. $y = -4\cos\left(\dfrac{1}{2}x\right)$

46. $y = -5\sin\left(\dfrac{1}{2}x\right)$

47. $y = -\sin(6x)$

48. $y = -\cos(4x)$

49. $y = 3\cos\left(\dfrac{\pi}{4}x\right)$

50. $y = 4\sin\left(\dfrac{\pi}{4}x\right)$

51. $y = \sin(4\pi x)$

52. $y = \cos(6\pi x)$

In Exercises 53–60, find an equation for each graph.

53.

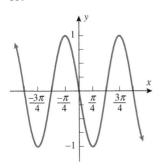

54.

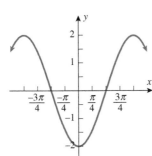

55.

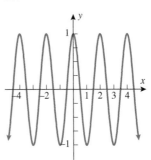

56.

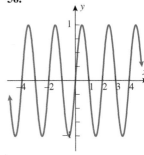

57.

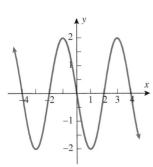

58.

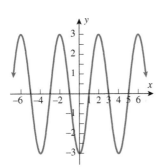

59.

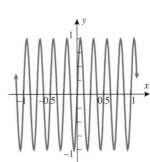

60.

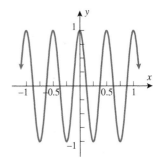

● **APPLICATIONS**

For Exercises 61 and 62, refer to the following:

An analysis of demand d for widgets manufactured by WidgetsRUs (measured in thousands of units per week) indicates that demand can be modeled by the graph below, where t is time in months since January 2010 (note that $t = 0$ corresponds to January 2010).

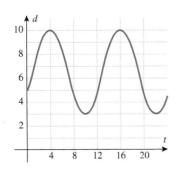

61. Business. Find the amplitude of the graph.

62. Business. Find the period of the graph.

For Exercises 63 and 64, refer to the following:

Researchers have been monitoring oxygen levels (milligrams per liter) in the water of a lake and have found that the oxygen levels fluctuate with an eight-week period. The following tables illustrate data from eight weeks.

63. Environment. Find the amplitude of the oxygen level fluctuations.

Week: t	0 (initial measurement)	1	2	3	4	5	6	7	8
Oxygen levels: mg/L	7	7.7	8	7.7	7	6.3	6	6.3	7

64. Environment. Find the amplitude of the oxygen level fluctuations.

Week: t	0 (initial measurement)	1	2	3	4	5	6	7	8
Oxygen levels: mg/L	7	8.4	9	8.4	7	5.6	5	5.6	7

For Exercises 65–68, refer to the following:

A weight hanging on a spring will oscillate up and down about its equilibrium position after it's pulled down and released. This is an example of simple harmonic motion. This motion would continue forever if there were not any friction or air resistance. Simple harmonic motion can be described with the function $y = A\cos\left(t\sqrt{\dfrac{k}{m}}\right)$, where A is the amplitude, t is the time in seconds, m is the mass, and k is a constant particular to that spring.

65. Simple Harmonic Motion. If the distance a spring is displaced y is measured in centimeters and the weight in grams, then what are the amplitude and mass if
$$y = 4\cos\left(\frac{t\sqrt{k}}{2}\right)?$$

66. Simple Harmonic Motion. Find the amplitude and mass if $y = 3\cos(3t\sqrt{k})$.

67. Frequency of Oscillations. The frequency of oscillation f is given by $f = \dfrac{1}{p}$, where p is the period. What is the frequency of oscillation modeled by $y = 3\cos\left(\dfrac{t}{2}\right)$?

68. Frequency of Oscillations. What is the frequency of oscillation modeled by $y = 3.5\cos(3t)$?

In Exercises 69–72, the term _frequency_ is used. Frequency is the reciprocal of the period, $f = \dfrac{1}{p}$.

69. Sound Waves. A pure tone created by a vibrating tuning fork shows up as a sine wave on an oscilloscope's screen. A tuning fork vibrating at 256 hertz gives the tone middle C and can have the equation $y = 0.005\sin[2\pi(256t)]$, where the amplitude is in centimeters and the time in seconds. What are the amplitude and frequency of the wave?

70. Sound Waves. A pure tone created by a vibrating tuning fork shows up as a sine wave on an oscilloscope's screen. A tuning fork vibrating at 288 hertz gives the tone D and can have the equation $y = 0.005\sin[2\pi(288t)]$, where the amplitude is in centimeters and the time in seconds. What are the amplitude and frequency of the wave?

71. Sound Waves. If a sound wave is represented by $y = 0.008\sin(750\pi t)$ cm, what are its amplitude and frequency?

72. Sound Waves. If a sound wave is represented by $y = 0.006\cos(1000\pi t)$ cm, what are its amplitude and frequency?

For Exercises 73–76, refer to the following:

When an airplane flies faster than the speed of sound, the sound waves that are formed take on a cone shape, and where the cone hits the ground, a sonic boom is heard. If θ is the angle of the vertex of the cone, then $\sin\left(\dfrac{\theta}{2}\right) = \dfrac{330\,\text{m/sec}}{V} = \dfrac{1}{M}$, where V is the speed of the plane and M is the mach number.

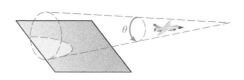

73. Sonic Booms. What is the speed of the plane if the plane is flying at mach 2?

74. Sonic Booms. What is the mach number if the plane is flying at 990 m/sec?

75. Sonic Booms. What is the speed of the plane if the cone angle is 60°?

76. Sonic Booms. What is the speed of the plane if the cone angle is 30°?

• **CATCH THE MISTAKE**

In Exercises 77 and 78, explain the mistake that is made.

77. Graph the function $y = -2\cos x$.

Solution:

Find the amplitude. $\qquad\qquad A = |-2| = 2$

The graph of $y = -2\cos x$ is similar to the graph of $y = \cos x$ with amplitude 2.

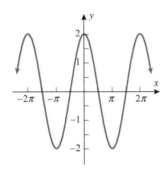

This is incorrect. What mistake was made?

78. Graph the function $y = -\sin(2x)$.

Solution:

Make a table with values.

x	$y = -\sin(2x)$	(x, y)
0	$y = -\sin 0 = 0$	$(0, 0)$
$\dfrac{\pi}{2}$	$y = -\sin \pi = 0$	$\left(\dfrac{\pi}{2}, 0\right)$
π	$y = -\sin(2\pi) = 0$	$(\pi, 0)$
$\dfrac{3\pi}{2}$	$y = -\sin(3\pi) = 0$	$\left(\dfrac{3\pi}{2}, 0\right)$
2π	$y = -\sin(4\pi) = 0$	$(2\pi, 0)$

Graph the function by plotting these points and connecting with a sinusoidal curve.

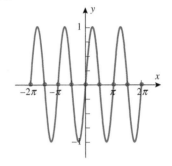

This is incorrect. What mistake was made?

• **CONCEPTUAL**

In Exercises 79–82, determine whether each statement is true or false. Assume A and B are positive real numbers.

79. The graph of $y = -A\cos(Bx)$ is the same as the graph of $y = A\cos(Bx)$ reflected about the x-axis.

80. The graph of $y = A\sin(-Bx)$ is the same as the graph of $y = A\sin(Bx)$ reflected about the x-axis.

81. The graph of $y = -A\cos(-Bx)$ is the same as the graph of $y = A\cos(Bx)$.

82. The graph of $y = -A\sin(-Bx)$ is the same as the graph of $y = A\sin(Bx)$.

In Exercises 83–86, assume A and B are positive real numbers.

83. Find the y-intercept of the function $y = A\cos(Bx)$.

84. Find the y-intercept of the function $y = A\sin(Bx)$.

85. Find the x-intercepts of the function $y = A\sin(Bx)$.

86. Find the x-intercepts of the function $y = A\cos(Bx)$.

• **CHALLENGE**

87. Sketch the graph of each of the following on the same set of axes over the interval $0 \le x \le 2\pi$: $y = 2\sin x$ and $y = \cos(2x)$. Then sketch the graph of the equation $y = 2\sin x + \cos(2x)$ by combining the y-coordinates of the two original graphs.

88. Sketch the graph of each of the following on the same set of axes over the interval $0 \le x \le 2\pi$: $y = 2\cos x$ and $y = 2\sin\left(\frac{1}{2}x\right)$. Then sketch the graph of the equation $y = 2\cos x + 2\sin\left(\frac{1}{2}x\right)$ by combining the y-coordinates of the two original graphs.

89. Sketch the graph of each of the following on the same set of axes over the interval $0 \le x \le 2\pi$: $y = x$ and $y = 2\sin x$. Then sketch the graph of the equation $y = x + 2\sin x$ by combining the y-coordinates of the two original graphs.

90. Sketch the graph of each of the following on the same set of axes over the interval $0 \le x \le 2\pi$: $y = \frac{1}{2}x$ and $y = -\cos x$. Then sketch the graph of the equation $y = \frac{1}{2}x - \cos x$ by combining the y-coordinates of the two original graphs.

• **TECHNOLOGY**

91. Use a graphing calculator to graph $Y_1 = 5\sin x$ and $Y_2 = \sin(5x)$. Is the following statement true based on what you see? $y = \sin(cx)$ has the same graph as $y = c\sin x$.

92. Use a graphing calculator to graph $Y_1 = 3\cos x$ and $Y_2 = \cos(3x)$. Is the following statement true based on what you see? $y = \cos(cx)$ has the same graph as $y = c\cos x$.

93. Use a graphing calculator to graph $Y_1 = \sin x$ and $Y_2 = \cos\left(x - \dfrac{\pi}{2}\right)$. What do you notice?

94. Use a graphing calculator to graph $Y_1 = \cos x$ and $Y_2 = \sin\left(x + \dfrac{\pi}{2}\right)$. What do you notice?

95. Use a graphing calculator to graph $Y_1 = \cos x$ and $Y_2 = \cos(x + c)$, where

a. $c = \dfrac{\pi}{3}$, and explain the relationship between Y_2 and Y_1.

b. $c = -\dfrac{\pi}{3}$, and explain the relationship between Y_2 and Y_1.

96. Use a graphing calculator to graph $Y_1 = \sin x$ and $Y_2 = \sin(x + c)$, where

a. $c = \dfrac{\pi}{3}$, and explain the relationship between Y_2 and Y_1.

b. $c = -\dfrac{\pi}{3}$, and explain the relationship between Y_2 and Y_1.

For Exercises 97 and 98, refer to the following:

Damped oscillatory motion, or **damped oscillation,** occurs when things in oscillatory motion experience friction or resistance. The friction causes the amplitude to decrease as a function of time. Mathematically, we use a negative exponential to damp the oscillations in the form of

$$f(t) = e^{-t}\sin t$$

97. Damped Oscillation. Graph the functions $Y_1 = e^{-t}$, $Y_2 = \sin t$, and $Y_3 = e^{-t}\sin t$ in the same viewing window (let t range from 0 to 2π). What happens as t increases?

98. Damped Oscillation. Graph $Y_1 = e^{-t}\sin t$, $Y_2 = e^{-2t}\sin t$, and $Y_3 = e^{-4t}\sin t$ in the same viewing window. What happens to $Y = e^{-kt}\sin t$ as k increases?

4.2 TRANSLATIONS OF THE SINE AND COSINE FUNCTIONS: ADDITION OF ORDINATES

SKILLS OBJECTIVES	CONCEPTUAL OBJECTIVES
▪ Graph reflections and vertical shifts of sine and cosine functions.	▪ For $y = k + A\sin(Bx)$ or $y = k + A\cos(Bx)$ recognize which constants indicate whether the sine or cosine function is reflected or shifted vertically and why.
▪ Determine the phase shift of a sinusoidal function.	
▪ Graph a function of the form $y = k + A\sin(Bx + C)$ or $y = k + A\cos(Bx + C)$.	▪ Understand that rewriting the sinusoidal function in standard form makes identifying the phase shift easier.
▪ Graph sums of trigonometric and other algebraic functions.	▪ Understand how to pull together the concepts of reflections, vertical shifts, and horizontal shifts to graph a sine or cosine function that has all of these properties.
	▪ Understand that the y-coordinates of the combined function are found by adding the y-coordinates of the individual functions.

> **STUDY TIP**
>
> Recall that even functions satisfy $f(-x) = f(x)$ and odd functions satisfy $f(-x) = -f(x)$. The sine function is odd and the cosine function is even.

Appendix A.5 includes a discussion on translations of functions. A shift inside the function $f(x \pm c)$ results in a horizontal shift of the graph of f opposite the sign of c, where $c > 0$. A shift outside the function $f(x) \pm c$ results in a vertical shift of the graph of f with the sign of c. A negative sign inside the function $f(-x)$ results in the reflection of the graph of f about the y-axis, and a negative sign outside the function $-f(x)$ results in the reflection of the graph of f about the x-axis. These rules are for all functions; therefore, they apply to trigonometric functions.

In Section 4.1, we graphed functions of the form $y = A\sin(Bx)$ and $y = A\cos(Bx)$. Using the properties of translations and reflections of functions in this chapter, we will graph functions of the form $y = k + A\sin(Bx + C)$ or $y = k + A\cos(Bx + C)$.

4.2.1 Reflections and Vertical Shifts of Sinusoidal Functions

Reflections

In Section 4.1, when graphing the functions $f(x) = A \sin(Bx)$ and $f(x) = A \cos(Bx)$, we determined the period to be $\dfrac{2\pi}{B}$, but we discussed only cases when B was positive, $B > 0$. Realize that if the argument is negative, we can use the properties of even and odd functions to rewrite the expression with a positive argument.

"Odd Function"   "Even Function"

$A \sin(-Bx) = -A \sin(Bx)$ or $A \cos(-Bx) = A \cos(Bx)$

Since we do not need to consider negative values of B (because of the properties of even and odd functions), there is no need to consider reflection about the y-axis. We do, however, need to consider reflection about the x-axis.

Recall that a negative sign outside the function $-f(x)$ results in the reflection of the graph of f about the x-axis. The following are graphs of $y = \sin x$ and $y = -\sin x$:

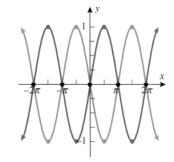

The following is a table of values for one period:

x	$y = \sin x$	(x, y)	$y = -\sin x$	(x, y)
0	$\sin 0 = 0$	$(0, 0)$	$-\sin 0 = 0$	$(0, 0)$
$\dfrac{\pi}{2}$	$\sin\left(\dfrac{\pi}{2}\right) = 1$	$\left(\dfrac{\pi}{2}, 1\right)$	$-\sin\left(\dfrac{\pi}{2}\right) = -1$	$\left(\dfrac{\pi}{2}, -1\right)$
π	$\sin \pi = 0$	$(\pi, 0)$	$-\sin \pi = 0$	$(\pi, 0)$
$\dfrac{3\pi}{2}$	$\sin\left(\dfrac{3\pi}{2}\right) = -1$	$\left(\dfrac{3\pi}{2}, -1\right)$	$-\sin\left(\dfrac{3\pi}{2}\right) = 1$	$\left(\dfrac{3\pi}{2}, 1\right)$
2π	$\sin(2\pi) = 0$	$(2\pi, 0)$	$-\sin(2\pi) = 0$	$(2\pi, 0)$

Suppose we are asked to graph $y = -2 \cos(\pi x)$. We can proceed in one of two ways:

- Make a table for the values of $y = -2 \cos(\pi x)$ and point-plot, or
- Graph the function $y = 2 \cos(\pi x)$ and reflect that graph about the x-axis to arrive at the graph of $y = -2 \cos(\pi x)$.

We will proceed with the second option.

WORDS

Determine the amplitude and period of $y = 2 \cos(\pi x)$.

Divide the period 2 into four equal parts and make a table.

MATH

$|A| = |2|, \; p = \dfrac{2\pi}{\pi} = 2$

x	0	$\frac{1}{2}$	1	$\frac{3}{2}$	2
$y = 2\cos(\pi x)$	2	0	-2	0	2

4.2.1 SKILL

Graph reflections and vertical shifts of sine and cosine functions.

4.2.1 CONCEPTUAL

For $y = k + A \sin(Bx)$, or $y = k + A \cos(Bx)$, recognize which constants indicate whether the sine or cosine function is reflected or shifted vertically and why.

STUDY TIP

The graph of $-f(x)$ is the graph of $f(x)$ reflected about the x-axis.

Graph $y = 2\cos(\pi x)$.

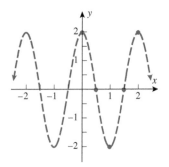

Reflect around the x-axis.

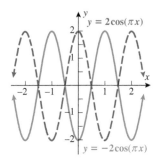

Vertical Shifts

Recall that graphing functions using vertical shifts occur in the following way ($k > 0$):

- To graph $f(x) + k$, shift the graph of $f(x)$ **up** k units.
- To graph $f(x) - k$, shift the graph of $f(x)$ **down** k units.

Therefore, functions like $y = k + A\sin(Bx)$ or $y = k + A\cos(Bx)$ are graphed by shifting the graphs of $y = A\sin(Bx)$ or $y = A\cos(Bx)$ vertically (up or down) k units. Although we found with reflection that it was just as easy to plot points and graph as opposed to first graphing one function and then reflecting about the x-axis, in this case it is much easier to first graph the simpler function and then perform a vertical shift.

VERTICAL TRANSLATIONS (SHIFTS) OF SINUSOIDAL FUNCTIONS

To graph sinusoidal functions of the form $y = A\sin(Bx) \pm k$ or $y = A\cos(Bx) \pm k$, where $k > 0$, start with the graphs of $y = A\sin(Bx)$ or $y = A\cos(Bx)$ and shift them up (+) or down (−) k units.

EXAMPLE 1 **Graphing Functions of the Form $y = k + A\sin(Bx)$**

Graph $y = -3 + 2\sin(\pi x)$, $-2 \leq x \leq 2$.

Solution:

STEP 1 Graph $y = 2\sin(\pi x)$ over one period.

The amplitude is 2, the period is $\dfrac{2\pi}{\pi} = 2$.

Divide 2 into four equal parts. The step size is $\frac{1}{2}$.

x	y = 2 sin (πx)	(x, y)
0	$y = 2\sin 0 = 0$	$(0, 0)$
$\frac{1}{2}$	$y = 2\sin\left(\frac{\pi}{2}\right) = 2(1) = 2$	$\left(\frac{1}{2}, 2\right)$
1	$y = 2\sin \pi = 0$	$(1, 0)$
$\frac{3}{2}$	$y = 2\sin\left(\frac{3\pi}{2}\right) = 2(-1) = -2$	$\left(\frac{3}{2}, -2\right)$
2	$y = 2\sin(2\pi) = 0$	$(2, 0)$

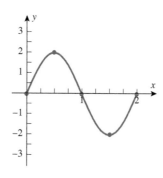

STEP 2 Extend the graph over two periods, $-2 \le x \le 2$.

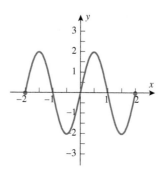

STEP 3 The graph of $y = -3 + 2\sin(\pi x)$ is the graph of $y = 2\sin(\pi x)$ shifted down three units.

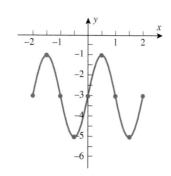

▼
ANSWER

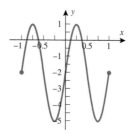

▼

YOUR TURN Graph $y = -2 + 3\sin(2\pi x)$, $-1 \le x \le 1$.

▶ **EXAMPLE 2** **Graphing Functions of the Form** $y = k + A\cos(Bx)$

Graph $y = 1 - \cos\left(\frac{1}{2}x\right)$, $-4\pi \leq x \leq 4\pi$.

Solution:

STEP 1 Graph $y = -\cos\left(\frac{1}{2}x\right)$ over one period.

The amplitude is 1, the period is $\dfrac{2\pi}{\dfrac{1}{2}} = 4\pi$.

Divide 4π into four equal parts. The step size is π.

x	$y = -\cos\left(\frac{1}{2}x\right)$	(x, y)
0	$y = -\cos 0 = -1$	$(0, -1)$
π	$y = -\cos\left(\dfrac{\pi}{2}\right) = 0$	$(\pi, 0)$
2π	$y = -\cos \pi = -(-1) = 1$	$(2\pi, 1)$
3π	$y = -\cos\left(\dfrac{3\pi}{2}\right) = 0$	$(3\pi, 0)$
4π	$y = -\cos(2\pi) = -1$	$(4\pi, -1)$

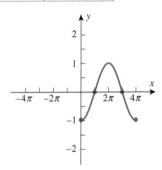

STEP 2 Extend the graph over two periods, $-4\pi \leq x \leq 4\pi$.

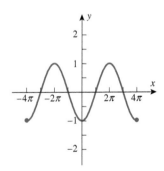

STEP 3 The graph of $y = 1 - \cos\left(\frac{1}{2}x\right)$ is the graph of $y = -\cos\left(\frac{1}{2}x\right)$ shifted up one unit.

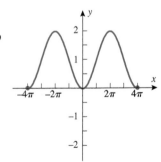

▼
ANSWER

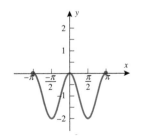

▼

YOUR TURN Graph $y = -1 + \cos(2x)$, $-\pi \leq x \leq \pi$.

4.2.2 Horizontal Shifts: Phase Shift

Recall that graphing functions using horizontal shifts occur in the following way ($C > 0$):

- To graph $f(x + C)$, shift the graph of $f(x)$ to the **left** C units.
- To graph $f(x - C)$, shift the graph of $f(x)$ to the **right** C units.

Therefore, to graph functions such as $y = A\sin(x \pm C)$ or $y = A\cos(x \pm C)$, $C > 0$, we simply shift the graphs of $y = A\sin x$ and $y = A\cos x$ horizontally (left or right) C units.

For example, to graph $y = \cos\left(x - \dfrac{\pi}{2}\right)$ we first graph $y = \cos x$ and then shift that graph $\dfrac{\pi}{2}$ units to the right.

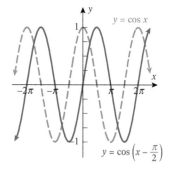

Notice that the cosine function shifted to the right $\dfrac{\pi}{2}$ units coincides with the graph of $y = \sin x$. This leads to the statement $\cos\left(x - \dfrac{\pi}{2}\right) = \sin x$. The shift of $\dfrac{\pi}{2}$ units is often called the *phase shift*. For this reason, we say that the sine and cosine functions are $90°$, or $\dfrac{\pi}{2}$, out of phase.

If we want to graph functions such as $y = A\sin(Bx \pm C)$ or $y = A\cos(Bx \pm C)$, we must first factor the common B term into **standard form** $y = A\sin\left[B\left(x \pm \dfrac{C}{B}\right)\right]$ or $y = A\cos\left[B\left(x \pm \dfrac{C}{B}\right)\right]$. The period of functions such as $y = A\sin(Bx \pm C)$ or $y = A\cos(Bx \pm C)$ is still $\dfrac{2\pi}{B}$; only now there is a phase shift. The graphs of these functions can be obtained by shifting the functions $y = A\sin(Bx)$ or $y = A\cos(Bx)$ horizontally $\dfrac{C}{B}$ units. This horizontal shift $\dfrac{C}{B}$ is called the **phase shift**. The sign of the phase shift determines if the shift is to the left or to the right.

HORIZONTAL TRANSLATIONS (PHASE SHIFTS) OF SINUSOIDAL FUNCTIONS

To graph sinusoidal functions of the form $y = A\sin(Bx \pm C)$ or $y = A\cos(Bx \pm C)$, where $B > 0$ and $C > 0$, start by rewriting the sinusoidal functions in the forms

$y = A\sin\left[B\left(x \pm \dfrac{C}{B}\right)\right]$ and $y = A\cos\left[B\left(x \pm \dfrac{C}{B}\right)\right]$ and then shift the graphs

of $y = A\sin(Bx)$ or $y = A\cos(Bx)$, $\dfrac{C}{B}$ units to the left ($+$) or right ($-$).

Note: When $B = 1$, the graphs of $y = A\sin(x \pm C)$ or $y = A\cos(x \pm C)$ can be obtained by shifting the graphs of $y = A\sin x$ or $y = A\cos x$ left ($+$) or right ($-$) C units.

A second way of interpreting this is as follows. Since the graph of $y = A\sin x$ or $y = A\cos x$ sketched over one period goes from $x = 0$ to $x = 2\pi$, then the graph of $y = A\sin(Bx + C)$ or $y = A\cos(Bx + C)$ sketched over one period goes from

$$Bx + C = 0 \qquad \text{to} \qquad Bx + C = 2\pi$$

$$x = -\frac{C}{B} \qquad \text{to} \qquad x = -\frac{C}{B} + \frac{2\pi}{B}$$

The phase shift $-\dfrac{C}{B}$ is $\dfrac{C}{B}$ units to the *left* for $C > 0$ and the period is $\dfrac{2\pi}{B}$.

Note: Had we used the function $y = A\sin(Bx - C)$ or $y = A\cos(Bx - C)$, the phase shift would be $\dfrac{C}{B}$, which is $\dfrac{C}{B}$ units to the *right* for $C > 0$ and the period is still $\dfrac{2\pi}{B}$.

EXAMPLE 3   **Graphing Functions of the Form $y = A\cos(Bx \pm C)$**

Graph $y = 5\cos(4x + \pi)$ over one period.

Solution:

STEP 1 State the amplitude. $|A| = |5| = 5$

STEP 2 Calculate the period and phase shift.

The interval for one period is from 0 to 2π. $4x + \pi = 0$ to $4x + \pi = 2\pi$

Solve for x. $x = -\dfrac{\pi}{4}$ to $x = -\dfrac{\pi}{4} + \dfrac{\pi}{2}$

Identify the **phase shift**. $-\dfrac{C}{B} = -\dfrac{\pi}{4}$

Identify the **period** $\dfrac{2\pi}{B}$. $p = \dfrac{2\pi}{B} = \dfrac{2\pi}{4} = \dfrac{\pi}{2}$

STEP 3 Graph.

Draw a cosine function starting at $x = -\dfrac{\pi}{4}$ with period $p = \dfrac{\pi}{2}$ and amplitude 5. Note that key points are $\dfrac{p}{4} = \dfrac{\pi/2}{4} = \dfrac{\pi}{8}$ units from

$$x = -\frac{\pi}{4}\left(-\frac{\pi}{4}, -\frac{\pi}{8}, 0, \frac{\pi}{8}, \text{ and } \frac{\pi}{4}\right).$$

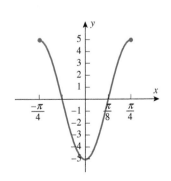

▼

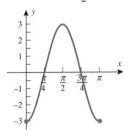

YOUR TURN   Graph $y = 3\cos(2x - \pi)$ over one period.

EXAMPLE 4 **Graphing Functions of the Form** $y = A\sin(Bx \pm C)$

Graph $y = 3\sin\left[-\frac{1}{2}(x - \pi)\right]$.

Solution:

STEP 1 Use the property of odd functions:

$$f(-x) = -f(x).$$
$$y = -3\sin\left[\frac{1}{2}(x - \pi)\right]$$

STEP 2 State the amplitude.
$$|A| = |-3| = 3$$

The graph of $y = -3\sin\left[\frac{1}{2}(x - \pi)\right]$ will be a reflection of
$y = 3\sin\left[\frac{1}{2}(x - \pi)\right]$ about the x-axis.

STEP 3 Calculate the period and phase shift.

The interval for one period
is from 0 to 2π.
$$\frac{1}{2}(x - \pi) = 0 \quad \text{to} \quad \frac{1}{2}(x - \pi) = 2\pi$$

Solve for x.
$$x = \pi \quad \text{to} \quad x = \pi + 4\pi$$

Identify the **phase shift**.
$$-\frac{C}{B} = \pi$$

Identify the **period**.
$$\frac{2\pi}{B} = 4\pi$$

STEP 4 Graph.

Draw a sine function starting at
$x = \pi$ with period $p = 4\pi$ and
amplitude 3 with a dashed curve.
Then reflect this curve about the
x-axis (solid). Note the four key
x-values are every π.

$$\frac{p}{4} = \frac{4\pi}{4} = \pi.$$

Extend the graph in both
directions (left and right).

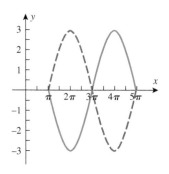

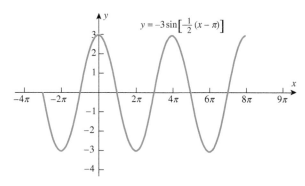

$$y = -3\sin\left[-\frac{1}{2}(x - \pi)\right]$$

YOUR TURN Graph $y = -4\sin(-2x + \pi)$.

STUDY TIP

An alternative method for finding
the period and phase shift is to
first write the function in standard
form.

$$3\sin\left[-\frac{1}{2}(x - \pi)\right]$$
$$= -3\sin\left[\frac{1}{2}(x - \pi)\right]$$

$$B = \frac{1}{2}$$

$$\text{Period} = \frac{2\pi}{B} = \frac{2\pi}{1/2} = 4\pi$$

Phase shift $= \pi$ units to the right.

▼ ANSWER

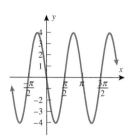

4.2.3 SKILL

Graph a function of the form
$y = k + A\sin(Bx + C)$ or
$y = k + A\cos(Bx + C)$.

4.2.3 CONCEPTUAL

Understand how to pull together
the concepts of reflections,
vertical shifts, and horizontal
shifts to graph a sine or cosine
function that has all of these
properties.

4.2.3 Graphing $y = k + A\sin(Bx + C)$ and $y = k + A\cos(Bx + C)$

We now put everything together (reflections, vertical shifts, and horizontal shifts) to graph functions of the form $y = k + A\sin(Bx + C)$ and $y = k + A\cos(Bx + C)$. We summarize with the following strategy.

Strategy for Graphing $y = k + A\sin(Bx + C)$ AND $y = k + A\cos(Bx + C)$

A strategy for graphing $y = k + A\sin(Bx + C)$ is outlined below. The same strategy can be used to graph $y = k + A\cos(Bx + C)$, $C > 0$.

Step 1: Find the amplitude $|A|$.

Step 2: Find the period $\dfrac{2\pi}{B}$ and phase shift $-\dfrac{C}{B}$.

Step 3: Graph $y = A\sin(Bx + C)$ over one period $\left(\text{from } -\dfrac{C}{B} \text{ to } -\dfrac{C}{B} + \dfrac{2\pi}{B}\right)$, dividing the period into four equal parts.

Step 4: Extend the graph over several periods.

Step 5: Shift the graph of $y = A\sin(Bx + C)$ vertically k units.

Note: If we rewrite the function in **standard form,** we get

$$y = k + A\sin\left[B\left(x + \frac{C}{B}\right)\right]$$

which makes it easier to identify the phase shift.

If $B < 0$, we can use properties of even and odd functions

$$\sin(-x) = -\sin x \qquad \cos(-x) = \cos x$$

to rewrite the function with $B > 0$.

EXAMPLE 5 **Graphing Sinusoidal Functions**

Graph $y = -3 + 2\cos(2x - \pi)$.

STEP 1 Find the amplitude. $\qquad |A| = |2| = 2$

STEP 2 Find the phase shift and period.

Identify the phase shift and period with $B = 2$ and $C = -\pi$.

$$-\frac{C}{B} = -\left(-\frac{\pi}{2}\right) = \frac{\pi}{2} \text{ (Phase shift)}$$

$$p = \frac{2\pi}{B} = \frac{2\pi}{2} = \pi \text{ (Period)}$$

STEP 3 Graph $y = 2\cos(2x - \pi)$ starting at $x = \dfrac{\pi}{2}$ over one period π using steps of $\dfrac{\pi}{4}$ $\left(\dfrac{\pi}{2}, \dfrac{3\pi}{4}, \pi, \dfrac{5\pi}{4}, \text{ and } \dfrac{3\pi}{2}\right)$.

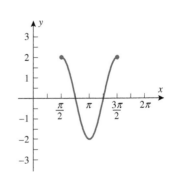

STEP 4 Extend the graph of $y = 2\cos(2x - \pi)$ over several periods.

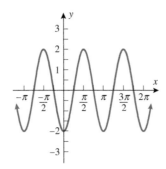

STEP 5 Shift the graph of $y = 2\cos(2x - \pi)$ down three units to arrive at the graph of $y = -3 + 2\cos(2x - \pi)$.

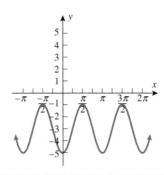

YOUR TURN Graph $y = 2 - 3\sin(2x + \pi)$.

ANSWER

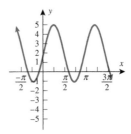

Finding Sinusoidal Functions from Data

Often things that are periodic in nature can be modeled with sinusoidal functions—for example, the temperature in a particular city. It is said that Orlando, Florida, has an average temperature of 72.3°. If we look at the average temperature by month, we see that it is warmest in the summer (June, July, and August) and cooler in the winter (December, January, and February). We expect this pattern to repeat every year (12 months).

Average Temperature by Month (°F)
Orlando, FL

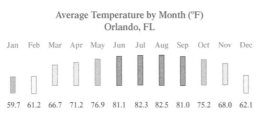

Jan	Feb	Mar	Apr	May	Jun	Jul	Aug	Sep	Oct	Nov	Dec
59.7	61.2	66.7	71.2	76.9	81.1	82.3	82.5	81.0	75.2	68.0	62.1

If we let the x-axis represent the months ($x = 1$ corresponds to January and $x = 12$ corresponds to December) and the y-axis represent the temperature in degrees Fahrenheit, we can find a sinusoidal function that models the data.

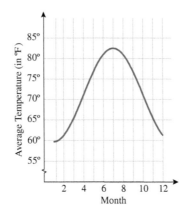

The most general sinusoidal function is given by $y = k + A\sin(Bx + C)$. Since the graphs of the sine and cosine functions differ only by a horizontal shift, we can assume the general form of the sine function without loss of generality.

The following procedure summarizes how to find a sinusoidal function that models (fits) the data we are given.

PROCEDURE FOR FINDING A SINUSOIDAL FUNCTION GIVEN DATA

Assume a general sinusoidal function: $y = k + A\sin(Bx + C)$.

Step 1: Calculate A (Amplitude $= |A|$).

$$|A| = \left| \frac{\text{Largest data value} - \text{smallest data value}}{2} \right|$$

Step 2: Calculate the vertical shift k.

$$k = \frac{\text{Largest data value} + \text{smallest data value}}{2}$$

Step 3: Calculate B. Assume that the period $p = \dfrac{2\pi}{B}$ is equal to the time it takes for the data to repeat.

Step 4: Select a point (x, y) from the data and solve $y = k + A\sin(Bx + C)$ for C. Answers will vary depending on which point is selected.

Example 6 illustrates a sinusoidal function with a vertical (but no horizontal) shift. In Example 7, we will return to the temperature in Orlando data and illustrate a sinusoidal function with both horizontal and vertical translations.

Circadian rhythms are the many different patterns that repeat "daily" in several living organisms arising from the so-called biological clock. Some of these cycles last 24 hours, while others are somewhat longer or shorter. In fact, experimentation has shown that a given Circadian rhythm can be altered if one controls the environment correctly. This was often done with Space Shuttle experiments related to germination to speed up the process.

▶ **EXAMPLE 6** **Circadian Rhythms in Photosynthesis**

It is known that photosynthesis (the process by which plants convert carbon dioxide and water into sugar and water) is a Circadian process exhibited by some plants. Assuming that a normal cycle takes 24 hours, the following is a typical graph of photosynthesis for a given plant. Here, photosynthesis is measured in terms of carbon assimilation, the units of which are micromoles of carbon per square meter per second. Find its equation.

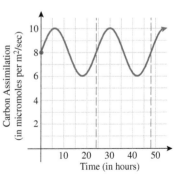

Solution:

Let $y = k + A\sin(Bt)$, where y represents the carbon assimilation and t represents time in hours. *Note*: The graph shows that there is no phase shift for the sine function.

STEP 1 Find A.

$$|A| = \left| \frac{10 - 6}{2} \right| = |2|$$

Select $A = 2$ since data are positive.

$$A = \pm 2$$

STEP 2 Find the vertical shift.

$$k = \frac{10 + 6}{2} = 8$$

STEP 3 Calculate B.

The period is 24 hours.

$$24 = \frac{2\pi}{B}$$

$$B = \frac{2\pi}{24} = \frac{\pi}{12}$$

$\boxed{y = 2\sin\left(\frac{\pi}{12}t\right) + 8}$ micromoles of carbon per square meter per second, where t is in hours.

EXAMPLE 7 **Finding a Sinusoidal Function from Data**

Find a sinusoidal function that represents the monthly average temperature in Orlando, Florida.

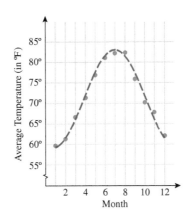

Average Temperature by Month (°F)
Orlando, FL

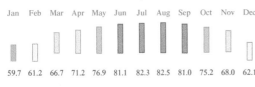

Jan Feb Mar Apr May Jun Jul Aug Sep Oct Nov Dec

59.7 61.2 66.7 71.2 76.9 81.1 82.3 82.5 81.0 75.2 68.0 62.1

Solution:

Assume a general sine function: $y = k + A\sin(Bx + C)$, where x represents the month ($x = 1$ corresponds to January and $x = 12$ corresponds to December).

STEP 1 Find A.

$$|A| = \left|\frac{\text{Largest data value} - \text{smallest data value}}{2}\right| = \left|\frac{82.5 - 59.7}{2}\right|$$

$$= |11.4| \Rightarrow A = \pm 11.4$$

Select $A = 11.4$ since the temperature in this graph is positive.

STEP 2 Find the vertical shift.

$$k = \frac{\text{Largest data value} + \text{smallest data value}}{2} = \frac{82.5 + 59.7}{2} = 71.1$$

STEP 3 Calculate B.

The period is 12 months.

$$12 = \frac{2\pi}{B}$$

$$B = \frac{\pi}{6}$$

STEP 4 Solve for C.

Substitute $A = 11.4, k = 71.1$, and $B = \frac{\pi}{6}$.

$$y = 71.1 + 11.4\sin\left(\frac{\pi}{6}x + C\right)$$

Select the point $x = 1$ and $y = 59.7$. $\quad 59.7 = 71.1 + 11.4 \sin\left(\dfrac{\pi}{6} + C\right)$

Subtract 71.1. $\qquad\qquad\qquad\qquad -11.4 = 11.4 \sin\left(\dfrac{\pi}{6} + C\right)$

Divide by 11.4. $\qquad\qquad\qquad\qquad\quad -1 = \sin\left(\dfrac{\pi}{6} + C\right)$

$\sin\theta = -1$ when $\theta = \dfrac{3\pi}{2}$. $\qquad\quad \dfrac{\pi}{6} + C = \dfrac{3\pi}{2}$

Subtract $\dfrac{\pi}{6}$. $\qquad\qquad\qquad\qquad\qquad\quad C = \dfrac{4\pi}{3}$

The function $y = 71.1 + 11.4 \sin\left(\dfrac{\pi}{6}x + \dfrac{4\pi}{3}\right)$ models the average monthly temperature in Orlando when $x = 1$ corresponds to January.

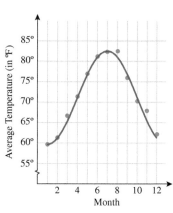

4.2.4 Graphing Sums of Functions: Addition of Ordinates

4.2.4 SKILL

Graph sums of trigonometric and other algebraic functions.

4.2.4 CONCEPTUAL

Understand that the y-coordinates of the combined function are found by adding the y-coordinates of the individual functions.

[CONCEPT CHECK]

For addition of ordinates divide the smallest period of the periodic functions by ___ in order to get the corresponding ordinates, to add.

▼
ANSWER 4

Since you have the ability to graph sinusoidal functions, let us now consider graphing sums of functions such as

$$y = x - \sin\left(\dfrac{\pi x}{2}\right) \qquad y = \sin x + \cos x \qquad y = 3\sin x + \cos(2x)$$

The method for graphing these sums is called the **addition of ordinates** because we add the corresponding y-values (ordinates). The following table illustrates the ordinates (y-values) of the two sinusoidal functions $\sin x$ and $\cos x$; adding the corresponding ordinates leads to the y-values of $y = \sin x + \cos x$.

x	$\sin x$	$\cos x$	$y = \sin x + \cos x$
0	0	1	1
$\dfrac{\pi}{4}$	$\dfrac{\sqrt{2}}{2}$	$\dfrac{\sqrt{2}}{2}$	$\sqrt{2}$
$\dfrac{\pi}{2}$	1	0	1
$\dfrac{3\pi}{4}$	$\dfrac{\sqrt{2}}{2}$	$-\dfrac{\sqrt{2}}{2}$	0
π	0	-1	-1
$\dfrac{5\pi}{4}$	$-\dfrac{\sqrt{2}}{2}$	$-\dfrac{\sqrt{2}}{2}$	$-\sqrt{2}$
$\dfrac{3\pi}{2}$	-1	0	-1
$\dfrac{7\pi}{4}$	$-\dfrac{\sqrt{2}}{2}$	$\dfrac{\sqrt{2}}{2}$	0
2π	0	1	1

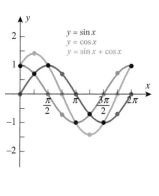

Using a graphing utility, we can graph $y_1 = \sin x$, $y_2 = \cos x$, and $y_3 = y_1 + y_2$.

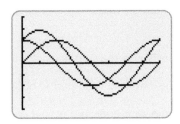

EXAMPLE 8   **Graphing Sums of Functions**

Graph $y = x - \sin\left(\dfrac{\pi x}{2}\right)$ on the interval $0 \le x \le 4$.

Solution:

Notice that $y = x - \sin\left(\dfrac{\pi x}{2}\right)$ is the sum of $y_1 = x$ and $y_2 = -\sin\left(\dfrac{\pi x}{2}\right)$.

State the amplitude and period for the graph of y_2. $A = 1,\ p = 4$

Make a table of x- and y-values of y_1, y_2, and $y = y_1 + y_2$.

x	$y_1 = x$	$y_2 = -\sin\left(\dfrac{\pi x}{2}\right)$	$y = y_1 + y_2 = x + \left[-\sin\left(\dfrac{\pi x}{2}\right)\right]$
0	0	0	0
1	1	-1	0
2	2	0	2
3	3	1	4
4	4	0	4

Graph.

$y_1 = x$

$y_2 = -\sin\left(\dfrac{\pi x}{2}\right)$

$y = x - \sin\left(\dfrac{\pi x}{2}\right)$

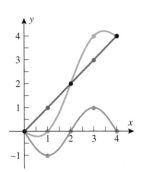

Using a graphing utility, we can graph $y_1 = x$, $y_2 = -\sin\left(\dfrac{\pi x}{2}\right)$, and $y_3 = y_1 + y_2$.

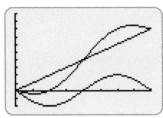

EXAMPLE 9 **Graphing Sums of Sine and Cosine Functions**

Graph $y = 3\sin x + \cos(2x)$ on the interval $0 \leq x \leq 2\pi$.

Solution:

Notice that $y = 3\sin x + \cos(2x)$ is the sum of $y_1 = 3\sin x$ and $y_2 = \cos(2x)$.

Let $y_1 = 3\sin x$ and state the amplitude and period of its graph. $|A| = 3,\ p = 2\pi$

Let $y_2 = \cos(2x)$ and state the amplitude and period of its graph. $|A| = 1,\ p = \pi$

Make a table of x- and y-values of y_1, y_2, and $y = y_1 + y_2$.

x	$y_1 = 3\sin x$	$y_2 = \cos(2x)$	$y = y_1 + y_2 = 3\sin x + \cos(2x)$
0	0	1	1
$\dfrac{\pi}{4}$	$\dfrac{3\sqrt{2}}{2}$	0	$\dfrac{3\sqrt{2}}{2}$
$\dfrac{\pi}{2}$	3	-1	2
$\dfrac{3\pi}{4}$	$\dfrac{3\sqrt{2}}{2}$	0	$\dfrac{3\sqrt{2}}{2}$
π	0	1	1
$\dfrac{5\pi}{4}$	$-\dfrac{3\sqrt{2}}{2}$	0	$-\dfrac{3\sqrt{2}}{2}$
$\dfrac{3\pi}{2}$	-3	-1	-4
$\dfrac{7\pi}{4}$	$-\dfrac{3\sqrt{2}}{2}$	0	$-\dfrac{3\sqrt{2}}{2}$
2π	0	1	1

Graph.

$y_1 = 3\sin x$

$y_2 = \cos(2x)$

$y = 3\sin x + \cos(2x)$

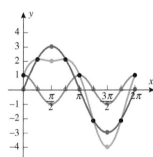

EXAMPLE 10 **Graphing Sums of Cosine Functions**

Graph $y = \cos\left(\dfrac{x}{2}\right) - \cos x$ on the interval $0 \le x \le 4\pi$.

Solution:

Notice that $y = \cos\left(\dfrac{x}{2}\right) - \cos x$ is the sum of $y_1 = \cos\left(\dfrac{x}{2}\right)$ and $y_2 = -\cos x$.

Let $y_1 = \cos\left(\dfrac{x}{2}\right)$ and state the amplitude and
period of its graph. $\quad |A| = 1,\ p = 4\pi$

Let $y_2 = -\cos x$ and state the amplitude and
period of its graph. $\quad |A| = |-1| = 1,\ p = 2\pi$

Make a table of x- and y-values of y_1, y_2, and $y = y_1 + y_2$.

x	$y_1 = \cos\left(\dfrac{x}{2}\right)$	$y_2 = -\cos x$	$y = y_1 + y_2 = \cos\left(\dfrac{x}{2}\right) + (-\cos x)$
0	1	-1	0
$\dfrac{\pi}{2}$	$\dfrac{\sqrt{2}}{2}$	0	$\dfrac{\sqrt{2}}{2}$
π	0	1	1
$\dfrac{3\pi}{2}$	$-\dfrac{\sqrt{2}}{2}$	0	$-\dfrac{\sqrt{2}}{2}$
2π	-1	-1	-2
$\dfrac{5\pi}{2}$	$-\dfrac{\sqrt{2}}{2}$	0	$-\dfrac{\sqrt{2}}{2}$
3π	0	1	1
$\dfrac{7\pi}{2}$	$\dfrac{\sqrt{2}}{2}$	0	$\dfrac{\sqrt{2}}{2}$
4π	1	-1	0

Graph.

$y_1 = \cos\left(\dfrac{x}{2}\right)$

$y_2 = -\cos x$

$y = \cos\left(\dfrac{x}{2}\right) + (-\cos x)$

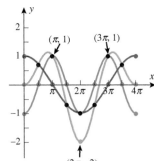

▶[SECTION 4.2] SUMMARY

Graphs of the form $y = -A\sin(Bx)$ are reflections of graphs of $y = A\sin(Bx)$ about the x-axis. Graphs of the form $y = A\sin(Bx) \pm k$ are obtained by first graphing $y = A\sin(Bx)$ and shifting up $(+)$ or down $(-)$ k units. Graphs of the form $y = A\sin(Bx \pm C)$ have the shape of a sine graph but with period $\dfrac{2\pi}{B}$ and horizontal shift to the left $(+)$ or the right

$(-) \dfrac{C}{B}$ units. To graph $y = k + A\sin(Bx + C)$, we first graph $y = A\sin(Bx + C)$ and then shift up or down k units. Many physical phenomena can be modeled by sinusoidal graphs. To graph sums of functions, add the corresponding y-values of each individual function.

[SECTION 4.2] EXERCISES

• SKILLS

In Exercises 1–8, match the function with its graph (a)–(h).

1. $\sin\left(x - \dfrac{\pi}{2}\right)$ 2. $\cos\left(x + \dfrac{\pi}{2}\right)$ 3. $-\cos\left(x + \dfrac{\pi}{2}\right)$ 4. $-\sin\left(x - \dfrac{\pi}{2}\right)$

5. $\cos x + 1$ 6. $1 - \sin x$ 7. $-1 + \sin x$ 8. $1 - \cos x$

a.
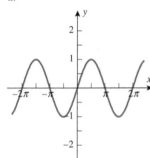

b.
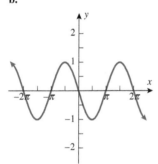

c.
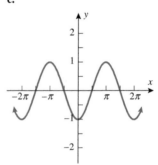

d.
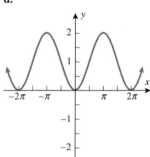

e.
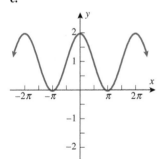

f.
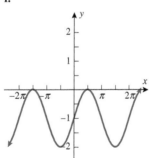

g.
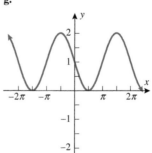

h.
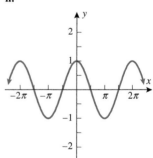

In Exercises 9–24, sketch the graph of each sinusoidal function over one period.

9. $y = -3 + \sin x$ 10. $y = 3 - \cos x$ 11. $y = \sin\left(x + \dfrac{\pi}{2}\right)$ 12. $y = \cos\left(x - \dfrac{\pi}{4}\right)$

13. $y = 1 - \sin(x - \pi)$ 14. $y = -1 + \cos(x + \pi)$ 15. $y = 2 + \cos(2x)$ 16. $y = 3 + \sin(\pi x)$

17. $y = 2 - \sin(4x)$ 18. $y = 1 - \cos(2x)$ 19. $y = 3 - 2\cos(\frac{1}{2}x)$ 20. $y = 2 - 3\sin(\frac{1}{2}x)$

21. $y = -1 + 2\sin\left(\dfrac{\pi}{2}x\right)$ 22. $y = -2 + \cos\left(\dfrac{\pi}{2}x\right)$ 23. $y = 2 + \sin\left(x + \dfrac{\pi}{4}\right)$ 24. $y = -1 + \cos(x - \pi)$

In Exercises 25–36, state the amplitude, period, and phase shift of each sinusoidal function.

25. $y = -\dfrac{1}{3}\cos\left(x - \dfrac{\pi}{2}\right)$ **26.** $y = 10\sin\left(x + \dfrac{3\pi}{4}\right)$ **27.** $y = 2\sin(\pi x - 1)$ **28.** $y = 4\cos(x + \pi)$

29. $y = -5\cos(3x + 2)$ **30.** $y = -7\sin(4x - 3)$ **31.** $y = 6\sin[-\pi(x + 2)]$ **32.** $y = 3\sin\left[-\dfrac{\pi}{2}(x - 1)\right]$

33. $y = -3\sin\left(2x - \dfrac{3\pi}{4}\right)$ **34.** $y = -\dfrac{1}{2}\cos\left(\pi x - \dfrac{\pi}{2}\right)$ **35.** $y = 2\sin(3x + \pi)$ **36.** $y = -\cos(3x - \pi)$

In Exercises 37–46, sketch the graph of each sinusoidal function over the indicated interval.

37. $y = \dfrac{1}{2} + \dfrac{3}{2}\cos(2x + \pi), \left[-\dfrac{3\pi}{2}, \dfrac{3\pi}{2}\right]$ **38.** $y = \dfrac{1}{3} + \dfrac{2}{3}\sin(2x - \pi), \left[-\dfrac{3\pi}{2}, \dfrac{3\pi}{2}\right]$

39. $y = \dfrac{1}{2} - \dfrac{1}{2}\sin\left(\dfrac{1}{2}x - \dfrac{\pi}{4}\right), \left[-\dfrac{7\pi}{2}, \dfrac{9\pi}{2}\right]$ **40.** $y = -\dfrac{1}{2} + \dfrac{1}{2}\cos\left(\dfrac{1}{2}x + \dfrac{\pi}{4}\right), \left[-\dfrac{9\pi}{2}, \dfrac{7\pi}{2}\right]$

41. $y = -3 + 4\sin[\pi(x - 2)], |0, 4|$ **42.** $y = 4 - 3\cos[\pi(x + 1)], [-1, 3]$

43. $y = 2 - 3\cos\left(3x - \dfrac{\pi}{2}\right), \left[-\dfrac{\pi}{2}, \dfrac{5\pi}{6}\right]$ **44.** $y = 1 - 2\sin\left(3x + \dfrac{\pi}{2}\right), \left[-\dfrac{5\pi}{6}, \dfrac{\pi}{2}\right]$

45. $y = 2 - \sin\left[-\dfrac{\pi}{2}\left(x - \dfrac{1}{2}\right)\right], \left[-\dfrac{7}{2}, \dfrac{9}{2}\right]$ **46.** $y = 1 - \cos\left[-\dfrac{\pi}{2}\left(x + \dfrac{1}{2}\right)\right], \left[-\dfrac{9}{2}, \dfrac{7}{2}\right]$

In Exercises 47–50, find an equation that corresponds to each graph.

47. 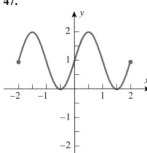   **48.** 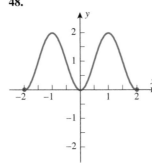   **49.** 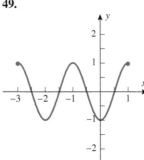   **50.**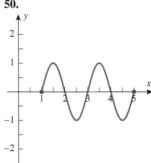

In Exercises 51–62, add the ordinates of the indicated functions to graph each summed function on the indicated interval.

51. $y = 2x - \cos(\pi x)$, on $0 \le x \le 4$ **52.** $y = 3x - 2\cos(\pi x)$, on $0 \le x \le 4$

53. $y = \dfrac{1}{3}x + 2\cos(2x)$, on $0 \le x \le 2\pi$ **54.** $y = \dfrac{1}{4}x + 3\cos\left(\dfrac{x}{2}\right)$, on $0 \le x \le 4\pi$

55. $y = \sin x - \cos x$, on $0 \le x \le 2\pi$ **56.** $y = \cos x - \sin x$, on $0 \le x \le 2\pi$

57. $y = 3\cos x + \sin x$, on $0 \le x \le 2\pi$ **58.** $y = 3\sin x - \cos x$, on $0 \le x \le 2\pi$

59. $y = \cos\left(\dfrac{x}{2}\right) + \cos(2x)$, on $0 \le x \le 4\pi$ **60.** $y = \sin\left(\dfrac{x}{2}\right) + \sin(2x)$, on $0 \le x \le 4\pi$

61. $y = 2\sin\left(\dfrac{x}{2}\right) - \cos(2x)$, on $0 \le x \le 4\pi$ **62.** $y = 2\cos\left(\dfrac{x}{2}\right) + \sin(2x)$, on $0 \le x \le 4\pi$

• **APPLICATIONS**

For Exercises 63 and 64, refer to the following:

If water is polluted by organic material, the levels of dissolved oxygen generally decrease. Oxygen depletion (oxygen levels *below* 4 milligrams per liter) can be detrimental to fish and other aquatic life. Oxygen levels from samples taken from the same location in a lake were studied over a period of 10 years. It was found that over the course of a year, the oxygen levels in the lake could be approximated by the function

$$f(t) = 3.1\cos t + 7.4 \qquad 1 \le t \le 365$$

where t is the day of the year and $t = 1$ represents January 1st. Assume that it is not a leap year.

63. **Environment/Life Sciences.** Find the oxygen level of the lake (to two decimal places) on November 20th.

64. **Environment/Life Sciences.** Find the oxygen level of the lake (to two decimal places) on June 30th.

For Exercises 65–68, refer to the following:

Computer sales are generally subject to seasonal fluctuations. An analysis of the sales of QualComp computers during 2017–2019 is approximated by the function

$$s(t) = 0.098\sin(0.79t + 2.37) + 0.387 \qquad 1 \le t \le 12$$

where t represents time in quarters ($t = 1$ represents the end of the first quarter of 2017), and s(t) represents computer sales (quarterly revenue) in millions of dollars.

65. Business/Economics. Find the amplitude. Interpret its meaning.

66. Business/Economics. Find the period. Interpret its meaning.

67. Business/Economics. Find the vertical shift. Interpret its meaning.

68. Business/Economics. Find the phase shift. Interpret its meaning.

69. Electrical Current. The current, in amperes, flowing through an alternating current circuit at time t is

$$I = 220\sin\left[20\pi\left(t - \frac{1}{100}\right)\right] \qquad t \ge 0$$

What is the maximum current? The minimum current? The period? The phase shift?

70. Electrical Current. The current, in amperes, flowing through an alternating current circuit at time t is

$$I = 120\sin\left[30\pi\left(t - \frac{1}{200}\right)\right] \qquad t \ge 0$$

What is the maximum current? The minimum current? The period? The phase shift?

For Exercises 71 and 72, find a sinusoidal function of the form $y = k + A\sin(Bx + C)$ that fits the data, where y is the temperature in degrees Fahrenheit and x is the number of the month with $x = 1$ corresponding to January. Assume the period is 12 months.

71. Temperature. The average temperature in Charlotte, North Carolina, is 60.1°F. The average monthly temperatures are given by

Average Temperature by Month (°F)
Charlotte, NC

Jan Feb Mar Apr May Jun Jul Aug Sep Oct Nov Dec

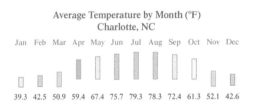

39.3 42.5 50.9 59.4 67.4 75.7 79.3 78.3 72.4 61.3 52.1 42.6

72. Temperature. The average temperature in Houston, Texas, is 67.9°F. The average monthly temperatures are given by

Average Temperature by Month (°F)
Houston, TX

Jan Feb Mar Apr May Jun Jul Aug Sep Oct Nov Dec

50.4 53.9 60.6 68.3 74.5 80.4 82.6 82.3 78.2 69.6 61.0 53.5

73. Roller Coaster. If a roller coaster at an amusement park is built using the sine curve determined by

$$y = 20\sin\left(\frac{\pi}{400}x\right) + 30, \text{ where } x \text{ is the horizontal}$$

(ground) distance from the beginning of the roller coaster in feet and y is the height of the track, then how high does the roller coaster go, and what distance does the roller coaster travel if it goes through three complete sine cycles?

74. Roller Coaster. If a roller coaster at an amusement park is built using the sine curve determined by

$$y = 25\sin\left(\frac{\pi}{500}x\right) + 30, \text{ where } x \text{ is the horizontal}$$

(ground) distance from the beginning of the roller coaster in feet and y is the height of the track, then how high does the roller coaster go, and what distance does the roller coaster travel if it goes through four complete sine cycles?

75. Biology. The number of deer on an island varies over time because of the amount of available food on the island. If the number of deer is determined by $y = 100\sin\left(\frac{\pi t}{4}\right) + 500$, where t is in years, then what are the highest and lowest numbers of deer on the island, and how long is the cycle?

76. Biology. The number of deer on an island varies over time because of the amount of available food on the island. If the number of deer is determined by

$$y = 500\sin\left(\frac{\pi t}{2}\right) + 1000, \text{ where } t \text{ is in years, then what}$$

are the highest and lowest numbers of deer on the island, and how long is the cycle?

77. Life Sciences. Suppose the cycle of a Circadian rhythm in photosynthesis is 22 hours and the maximum and minimum values obtained are 12 and 6 micromoles of carbon per square meter per second. Find an equation that models this photosynthesis.

78. Life Sciences. For the Circadian rhythm given in Exercise 77, find the carbon assimilation at

a. 46 hours

b. 68 hours

For Exercises 79 and 80, refer to the following:

With the advent of summer come fireflies. They are intriguing because they emit a flashing luminescence that beckons their mate to them. It is known that the speed and intensity of the flashing are related to the temperature—the higher the temperature, the quicker and more intense the flashing becomes. If you ever watch a single firefly, you will see that the intensity of the flashing is periodic with time. The intensity of light emitted is measured in *candelas per square meter (of firefly)*. To give an idea of this unit of measure, the intensity of a picture on a typical TV screen is about 450 candelas. The measurement for the intensity of the light emitted by a typical firefly at its brightest moment is about 50 candelas. Assume that a typical cycle of this flashing is 4 seconds and that the intensity is essentially zero candelas at the beginning and ending of a cycle.

79. Biology: Bioluminescence in Fireflies. Find an equation that describes this flashing. What is the intensity of the flashing at 4 minutes?

80. Biology: Bioluminescence in Fireflies. Graph the equation from Exercise 79 for a period of 30 seconds.

● CATCH THE MISTAKE

In Exercises 81 and 82, explain the mistake that is made.

81. Find the amplitude, phase shift, and period of
$y = -\frac{2}{3}\sin(2x - \pi)$.

Solution:

Find the amplitude. $|A| = \left| -\frac{2}{3} \right| = \frac{2}{3}$

Find the phase shift. π units to the right

Find the period. $\frac{2\pi}{2} = \pi$

One of these is incorrect. Which is it, and what mistake was made?

82. Graph $y = 1 + \sin(2x)$.

Solution:

Step 1: Graph $y = \sin(2x)$.
Find the amplitude. $|A| = |-1| = 1$
Make a table.

x	$y = -\sin(2x)$	(x, y)
0	$y = -\sin 0 = 0$	$(0, 0)$
π	$y = -\sin(2\pi) = 0$	$(\pi, 0)$
2π	$y = -\sin(4\pi) = 0$	$(2\pi, 0)$

Plot the points.

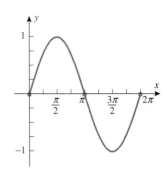

Step 2: Shift the graph of $y = \sin(2x)$ up one unit to arrive at the graph of $y = 1 + \sin(2x)$.

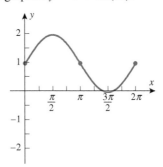

This is incorrect. What mistake was made?

● CONCEPTUAL

In Exercises 83–86, determine whether each statement is true or false.

83. The period of $y = k + A\sin(\omega t - \phi)$ increases as ω increases.

84. The phase shift of $y = k + A\sin(\omega t - \phi)$ is ϕ units.

85. $A\sin\left(\omega t + \frac{\omega\pi}{2}\right) = A\cos(\omega t)$

86. $A\sin\left(\omega t - \frac{\omega\pi}{2}\right) = A\cos(\omega t)$

87. Find the x-intercepts for the graph of the equation $y = 1 + \cos(\omega x)$.

88. Find the x-intercepts for the graph of the equation $y = \sin\left(\omega x - \frac{\pi}{2}\right)$.

89. If asked to graph $y = k + A\cos(\omega t - \phi)$ over one period, state the interval for x over which the graph could be drawn [phase shift, phase shift + period].

90. If asked to graph $y = k + A\cos(\omega t + \phi)$ over one period, state the interval for x over which the graph could be drawn [phase shift, phase shift + period].

91. Write the equation for a sine graph that coincides with the graph of $y = \cos\left(2x - \frac{\pi}{3}\right)$.

92. Solve the equation $\sin\left(4x + \frac{\pi}{2}\right) = 1$.

● TECHNOLOGY

93. Using a graphing calculator, graph $Y_1 = x - \frac{x^3}{6}$ and $Y_2 = \sin x$ for x in $[0, 0.2]$. Is the polynomial in Y_1 a good approximation of the sine function?

94. Using a graphing calculator, graph $Y_1 = 1 - \frac{x^2}{2}$ and $Y_2 = \cos x$ for x in $[0, 0.2]$. Is the polynomial in Y_1 a good approximation of the cosine function?

95. Use a graphing calculator and the $\boxed{\text{SIN}}$ $\boxed{\text{REG}}$ program to find the sine function that best fits the data in Exercise 71. Graph the function found by the graphing calculator and the function you found in Exercise 71 in the same viewing rectangle.

96. Use a graphing calculator and the $\boxed{\text{SIN}}$ $\boxed{\text{REG}}$ program to find the sine function that best fits the data in Exercise 72. Graph the function found by the graphing calculator and the function you found in Exercise 72 in the same viewing rectangle.

4.3 GRAPHS OF TANGENT, COTANGENT, SECANT, AND COSECANT FUNCTIONS

SKILLS OBJECTIVES	CONCEPTUAL OBJECTIVES
▪ Graph basic tangent, cotangent, secant, and cosecant functions. ▪ Graph translated tangent, cotangent, secant, and cosecant functions.	▪ Understand the relationships between the graphs of cosine and secant functions and the sine and cosecant functions. ▪ Understand that graph-shifting techniques for tangent and cotangent are consistent with translations used for sinusoidal functions, but for secant and cosecant functions we first graph the horizontally translated sinusoidal functions and then shift up or down depending on the vertical translations.

4.3.1 Graphing the Tangent, Cotangent, Secant, and Cosecant Functions

4.3.1 SKILL

Graph basic tangent, cotangent, secant, and cosecant functions.

4.3.1 CONCEPTUAL

Understand the relationships between the graphs of cosine and secant functions and the sine and cosecant functions.

[CONCEPT CHECK]

If the range of the graph of a sinusoidal function is $[-A, A]$, what is the range of the corresponding secant or cosecant functions?

▼ ....

ANSWER $(-\infty, -A) \cup (A, \infty)$

The first two sections of this chapter focused on graphing sinusoidal functions (sine and cosine). We now turn our attention to graphing the other trigonometric functions: tangent, cotangent, secant, and cosecant. From the graphs of the sine and cosine functions, we can get the graphs of the other trigonometric functions. Recall the reciprocal and quotient identities:

$$\tan x = \frac{\sin x}{\cos x} \qquad \cot x = \frac{\cos x}{\sin x} \qquad \sec x = \frac{1}{\cos x} \qquad \csc x = \frac{1}{\sin x}$$

Recall that when graphing rational functions, a *vertical asymptote* corresponds to a denominator equal to zero (as long as the numerator and denominator have no common factors). As you will see in this section, the tangent and secant functions have graphs with vertical asymptotes at the x-values where the cosine function is equal to zero, and the cotangent and cosecant functions have graphs with vertical asymptotes at the x-values where the sine function is equal to zero.

One important difference between the sinusoidal functions ($y = \sin x$ and $y = \cos x$) and the other four ($y = \tan x$, $y = \sec x$, $y = \csc x$, $y = \cot x$) is that the sinusoidal functions have defined amplitudes, whereas the other four trigonometric functions do not (since they are unbounded vertically).

The Tangent Function

Since the tangent function is a quotient that relies on the sine and cosine functions, let us start with a table of values for the quadrantal angles, the only angles for which the sine or cosine functions are zero.

x	$\sin x$	$\cos x$	$\tan x = \frac{\sin x}{\cos x}$	(x, y) OR ASYMPTOTE
0	0	1	0	$(0, 0)$ (x-intercept)
$\dfrac{\pi}{2}$	1	0	undefined	vertical asymptote: $x = \dfrac{\pi}{2}$
π	0	-1	0	$(\pi, 0)$ (x-intercept)
$\dfrac{3\pi}{2}$	-1	0	undefined	vertical asymptote: $x = \dfrac{3\pi}{2}$
2π	0	1	0	$(2\pi, 0)$ (x-intercept)

Notice that the x-intercepts correspond to integer multiples of π where $\sin x = 0$ and the vertical asymptotes correspond to odd integer multiples of $\dfrac{\pi}{2}$ where $\cos x = 0$.

We know that the graph of the tangent function is undefined at the odd integer multiples of $\dfrac{\pi}{2}$, so its graph cannot cross the vertical asymptotes. What happens between the asymptotes? We know the x-intercepts; let us now make a table for special values of x.

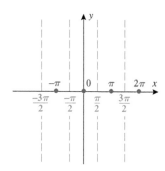

x	$\sin x$	$\cos x$	$y = \tan x = \frac{\sin x}{\cos x}$	(x, y)
$\dfrac{\pi}{6}$	$\dfrac{1}{2}$	$\dfrac{\sqrt{3}}{2}$	$\dfrac{1}{\sqrt{3}} = \dfrac{\sqrt{3}}{3} \approx 0.577$	$\approx \left(\dfrac{\pi}{6}, 0.577\right)$
$\dfrac{\pi}{4}$	$\dfrac{\sqrt{2}}{2}$	$\dfrac{\sqrt{2}}{2}$	1	$\left(\dfrac{\pi}{4}, 1\right)$
$\dfrac{\pi}{3}$	$\dfrac{\sqrt{3}}{2}$	$\dfrac{1}{2}$	$\sqrt{3} \approx 1.732$	$\approx \left(\dfrac{\pi}{3}, 1.732\right)$
$\dfrac{2\pi}{3}$	$\dfrac{\sqrt{3}}{2}$	$-\dfrac{1}{2}$	$-\sqrt{3} \approx -1.732$	$\approx \left(\dfrac{2\pi}{3}, -1.732\right)$
$\dfrac{3\pi}{4}$	$\dfrac{\sqrt{2}}{2}$	$-\dfrac{\sqrt{2}}{2}$	-1	$\left(\dfrac{3\pi}{4}, -1\right)$
$\dfrac{5\pi}{6}$	$\dfrac{1}{2}$	$-\dfrac{\sqrt{3}}{2}$	$\dfrac{1}{-\sqrt{3}} = -\dfrac{\sqrt{3}}{3} \approx -0.577$	$\approx \left(\dfrac{5\pi}{6}, -0.577\right)$

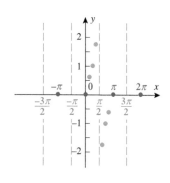

What happens to $\tan x$ as x approaches $\dfrac{\pi}{2}$? We know $\tan x$ is undefined at $x = \dfrac{\pi}{2}$ but we must consider x-values both larger and smaller than $\dfrac{\pi}{2} \approx 1.571$.

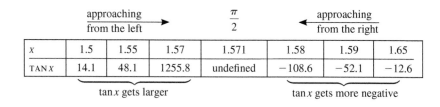

	approaching from the left $\longrightarrow$		$\dfrac{\pi}{2}$	$\longleftarrow$ approaching from the right		
x	1.5	1.55	1.57	1.571	1.58	1.59	1.65
TAN x	14.1	48.1	1255.8	undefined	-108.6	-52.1	-12.6

$\underbrace{\qquad\qquad}_{\tan x \text{ gets larger}}$ $\underbrace{\qquad\qquad}_{\tan x \text{ gets more negative}}$

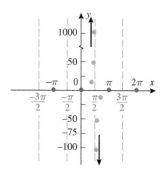

The arrows on the graph on the right indicate increasing without bound (in the positive and negative directions).

GRAPH OF $y = \tan x$

1. The x-intercepts occur at integer multiples of π. $(n\pi, 0)$
2. Vertical asymptotes occur at odd integer multiples of $\dfrac{\pi}{2}$. $x = \dfrac{(2n + 1)\pi}{2}$
3. The domain is the set of all real numbers except odd integer multiples of $\dfrac{\pi}{2}$. $x \neq \dfrac{(2n + 1)\pi}{2}$

4. The range is the set of all real numbers. $(-\infty, \infty)$

5. $y = \tan x$ has period π.

6. $y = \tan x$ is an odd function (graph is symmetric about the origin). $\tan(-x) = -\tan x$

7. The graph has no defined amplitude, since the function is unbounded.

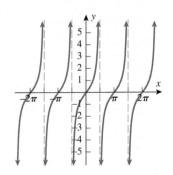

Note: n is an integer.

The Cotangent Function

The cotangent function is similar to the tangent function in that it is a quotient involving the sine and cosine functions. The difference is that the cotangent function has the cosine function in the numerator and the sine function in the denominator: $\cot x = \dfrac{\cos x}{\sin x}$. The graph of $y = \tan x$ has x-intercepts corresponding to integer multiples of π and vertical asymptotes corresponding to odd integer multiples of $\dfrac{\pi}{2}$. The graph of the cotangent function is the opposite in that it has x-intercepts corresponding to odd integer multiples of $\dfrac{\pi}{2}$ and vertical asymptotes corresponding to integer multiples of π. This is because the x-intercepts occur when the numerator, $\cos x$, is equal to 0 and the vertical asymptotes occur when the denominator, $\sin x$, is equal to 0.

GRAPH OF $y = \cot x$

1. The x-intercepts occur at odd integer multiples of $\dfrac{\pi}{2}$. $\left(\dfrac{(2n + 1)\pi}{2}, 0 \right)$

2. Vertical asymptotes occur at integer multiples of π. $x = n\pi$

3. The domain is the set of all real numbers except integer multiples of π. $x \neq n\pi$

4. The range is the set of all real numbers. $(-\infty, \infty)$

5. $y = \cot x$ has period π.

6. $y = \cot x$ is an odd function (graph is symmetric about the origin). $\cot(-x) = -\cot x$

7. The graph has no defined amplitude, since the function is unbounded.

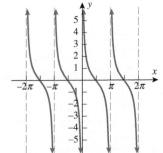

Note: n is an integer.

The Secant Function

Since $y = \cos x$ has period 2π, the secant function, which is the reciprocal of the cosine function, $\sec x = \dfrac{1}{\cos x}$, also has period 2π. We now illustrate values of the secant function with a table.

x	$\cos x$	$y = \sec x = \frac{1}{\cos x}$	(x, y) OR ASYMPTOTE
0	1	1	$(0, 1)$ (y-intercept)
$\dfrac{\pi}{2}$	0	undefined	vertical asymptote: $x = \dfrac{\pi}{2}$
π	-1	-1	$(\pi, -1)$
$\dfrac{3\pi}{2}$	0	undefined	vertical asymptote: $x = \dfrac{3\pi}{2}$
2π	1	1	$(2\pi, 1)$

Again, we ask the same question: What happens as x approaches the vertical asymptotes? The secant function grows without bound in either the positive or negative direction.

If we graph $y = \cos x$ (the "guide" function) and $y = \sec x$ on the same graph, we notice the following:

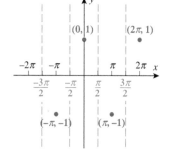

- x-intercepts of $y = \cos x$ correspond to the vertical asymptotes of $y = \sec x$.
- The range of cosine is $[-1, 1]$ and the range of secant is $(-\infty, -1] \cup [1, \infty)$.
- When cosine is positive, secant is positive, and when one is negative, the other is negative.

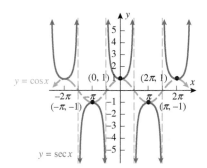

The cosine function is used as the "guide" function to graph the secant function.

GRAPH OF $y = \sec x$

1. There are no x-intercepts.

   $\dfrac{1}{\cos x} \neq 0$

2. Vertical asymptotes occur at odd integer multiples of $\dfrac{\pi}{2}$.

   $x = \dfrac{(2n + 1)\pi}{2}$

3. The domain is the set of all real numbers except odd integer multiples of $\dfrac{\pi}{2}$.

   $x \neq \dfrac{(2n + 1)\pi}{2}$

4. The range is $(-\infty, -1] \cup [1, \infty)$.

5. $y = \sec x$ has period 2π.

6. $y = \sec x$ is an even function (graph is symmetric about the y-axis).

$$\sec(-x) = \sec x$$

7. The graph has no defined amplitude, since the function is unbounded.

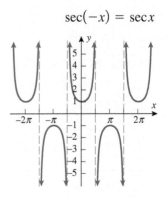

Note: n is an integer.

The Cosecant Function

Since $y = \sin x$ has period 2π, the cosecant function, which is the reciprocal of the sine function, $\csc x = \dfrac{1}{\sin x}$, also has period 2π. We now illustrate values of cosecant with a table.

x	$\sin x$	$y = \csc x = \frac{1}{\sin x}$	(x, y) OR ASYMPTOTE
0	0	undefined	vertical asymptote: $x = 0$
$\dfrac{\pi}{2}$	1	1	$\left(\dfrac{\pi}{2}, 1\right)$
π	0	undefined	vertical asymptote: $x = \pi$
$\dfrac{3\pi}{2}$	-1	-1	$\left(\dfrac{3\pi}{2}, -1\right)$
2π	0	undefined	vertical asymptote: $x = 2\pi$

Again, we ask the same question: What happens as x approaches the vertical asymptotes? The cosecant function grows without bound in either the positive or negative direction.

If we graph $y = \sin x$ (the "guide" function) and $y = \csc x$ on the same graph, we notice the following:

- x-intercepts of $y = \sin x$ correspond to the vertical asymptotes of $y = \csc x$.
- The range of sine is $[-1, 1]$, and the range of cosecant is $(-\infty, -1] \cup [1, \infty)$.
- When sine is positive, cosecant is positive, and when one is negative, the other is negative.

The sine function is used as the "guide" function to graph the cosecant function.

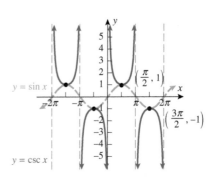

GRAPH OF $y = \csc x$

1. There are no x-intercepts. $\dfrac{1}{\sin x} \neq 0$

2. Vertical asymptotes occur at integer multiples of π. $x = n\pi$

3. The domain is the set of all real numbers
   except integer multiples of π. $x \neq n\pi$

4. The range is $(-\infty, -1] \cup [1, \infty)$.

5. $y = \csc x$ has period 2π.

6. $y = \csc x$ is an odd function (symmetric
   about the origin). $\csc(-x) = -\csc x$

7. The graph has no defined amplitude, since
   the function is unbounded.

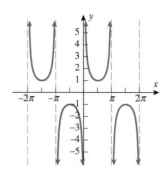

Note: n is an integer.

Graphing More General Tangent, Cotangent, Secant, and Cosecant Functions

FUNCTION	$y = \sin x$	$y = \cos x$	$y = \tan x$	$y = \cot x$	$y = \sec x$	$y = \csc x$
Graph						
Domain	$(-\infty, \infty)$	$(-\infty, \infty)$	$x \neq \dfrac{(2n+1)\pi}{2}$	$x \neq n\pi$	$x \neq \dfrac{(2n+1)\pi}{2}$	$x \neq n\pi$
Range	$[-1, 1]$	$[-1, 1]$	$(-\infty, \infty)$	$(-\infty, \infty)$	$(-\infty, -1] \cup [1, \infty)$	$(-\infty, -1] \cup [1, \infty)$
Amplitude	1	1	none	none	none	none
Period	2π	2π	π	π	2π	2π
x-intercepts	$(n\pi, 0)$	$\left(\dfrac{(2n+1)\pi}{2}, 0\right)$	$(n\pi, 0)$	$\left(\dfrac{(2n+1)\pi}{2}, 0\right)$	none	none
Vertical Asymptotes	none	none	$x = \dfrac{(2n+1)\pi}{2}$	$x = n\pi$	$x = \dfrac{(2n+1)\pi}{2}$	$x = n\pi$

Note: n is an integer.

We use these basic functions as the starting point for graphing general tangent, cotangent, secant, and cosecant functions.

GRAPHING TANGENT AND COTANGENT FUNCTIONS

Graphs of $y = A\tan(Bx)$ and $y = A\cot(Bx)$ can be obtained using the following steps (assume $B > 0$):

Step 1: Calculate the period: $\dfrac{\pi}{B}$.

Step 2: Find two neighboring vertical asymptotes.

For $y = A\tan(Bx)$: $Bx = -\dfrac{\pi}{2}$ and $Bx = \dfrac{\pi}{2}$ so $x = -\dfrac{\pi}{2B}$ and $x = \dfrac{\pi}{2B}$

For $y = A\cot(Bx)$: $Bx = 0$ and $Bx = \pi$ so $x = 0$ and $x = \dfrac{\pi}{B}$

Step 3: Find the x-intercept between the two asymptotes.

For $y = A\tan(Bx)$: $Bx = 0$ so $x = 0$

For $y = A\cot(Bx)$: $Bx = \dfrac{\pi}{2}$ so $x = \dfrac{\pi}{2B}$

Step 4: Draw the vertical asymptotes and label the x-intercept.

Step 5: Divide the interval between the asymptotes into four equal parts. Set up a table with coordinates corresponding to points in the interval.

Step 6: Connect the points with a smooth curve. Use arrows to indicate the behavior toward the asymptotes.

- If $A > 0$,
  - $y = A\tan(Bx)$ increases between vertical asymptotes from left to right.
  - $y = A\cot(Bx)$ decreases between vertical asymptotes from left to right.
- If $A < 0$,
  - $y = A\tan(Bx)$ decreases between vertical asymptotes from left to right.
  - $y = A\cot(Bx)$ increases between vertical asymptotes from left to right.

> **STUDY TIP**
>
> Notice that the tangent and cotangent functions have period π (not 2π like the other four trigonometric functions).

▶ **EXAMPLE 1** Graphing $y = A\tan(Bx)$

Graph $y = -3\tan(2x)$ on the interval $-\dfrac{\pi}{2} \le x \le \dfrac{\pi}{2}$.

Solution: $A = -3, B = 2$

STEP 1 Calculate the period. $\qquad p = \dfrac{\pi}{B} = \dfrac{\pi}{2}$

STEP 2 Find two vertical asymptotes. $\qquad Bx = -\dfrac{\pi}{2}$ and $Bx = \dfrac{\pi}{2}$

Substitute $B = 2$ and solve for x. $\qquad x = -\dfrac{\pi}{4}$ and $x = \dfrac{\pi}{4}$

STEP 3 Find the x-intercept between $\qquad Bx = 0$
the asymptotes. $\qquad x = 0$

STEP 4 Draw the vertical asymptotes with dashed lines, $x = -\dfrac{\pi}{4}$ and $x = \dfrac{\pi}{4}$, and label the x-intercept, $(0, 0)$.

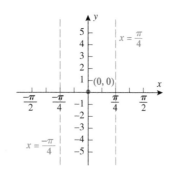

STEP 5 Divide the period $\dfrac{\pi}{2}$ into four equal parts in steps of $\dfrac{\pi}{8}$. Set up a table with coordinates corresponding to values of $y = -3\tan(2x)$.

x	$y = -3\tan(2x)$	(x, y)
$-\dfrac{\pi}{4}$	undefined	vertical asymptote, $x = -\dfrac{\pi}{4}$
$-\dfrac{\pi}{8}$	3	$\left(-\dfrac{\pi}{8}, 3\right)$
0	0	$(0, 0)$ $(x\text{-intercept})$
$\dfrac{\pi}{8}$	-3	$\left(\dfrac{\pi}{8}, -3\right)$
$\dfrac{\pi}{4}$	undefined	vertical asymptote, $x = \dfrac{\pi}{4}$

STEP 6 Graph the points from the table and connect with a smooth curve. Repeat to the right and left until reaching the interval endpoints. Notice the distance between the vertical asymptotes is the period length $\dfrac{\pi}{2}$.

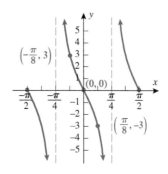

YOUR TURN Graph $y = \frac{1}{3}\tan\left(\frac{1}{2}x\right)$ on the interval $-\pi \le x \le \pi$.

ANSWER

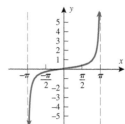

EXAMPLE 2   **Graphing $y = A\cot(Bx)$**

Graph $y = 4\cot\left(\frac{1}{2}x\right)$ on the interval $-2\pi \le x \le 2\pi$.

Solution: $A = 4$, $B = \frac{1}{2}$

STEP 1 Calculate the period.
$$p = \frac{\pi}{B} = 2\pi$$

STEP 2 Find two vertical asymptotes.
$$Bx = 0 \quad \text{and} \quad Bx = \pi$$
Substitute $B = \frac{1}{2}$ and solve for x.
$$x = 0 \quad \text{and} \quad x = 2\pi$$

STEP 3 Find the x-intercept between the asymptotes.
$$Bx = \frac{\pi}{2}$$
$$x = \pi$$

STEP 4 Draw the vertical asymptotes, $x = 0$ and $x = 2\pi$, and label the x-intercept, $(\pi, 0)$. Notice the distance between the vertical asymptotes is the period length 2π.

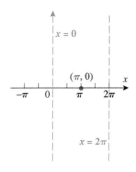

STEP 5 Divide the period 2π into four equal parts in steps of $\dfrac{\pi}{2}$. Set up a table with coordinates corresponding to values of $y = 4\cot\left(\frac{1}{2}x\right)$.

x	$y = 4\cot\left(\frac{1}{2}x\right)$	(x, y)
0	undefined	vertical asymptote, $x = 0$
$\dfrac{\pi}{2}$	4	$\left(\dfrac{\pi}{2}, 4\right)$
π	0	$(\pi, 0)$
$\dfrac{3\pi}{2}$	-4	$\left(\dfrac{3\pi}{2}, -4\right)$
2π	undefined	vertical asymptote, $x = 2\pi$

STEP 6 Graph the points from the table and connect with a smooth curve. Repeat to the right and left until reaching the interval endpoints.

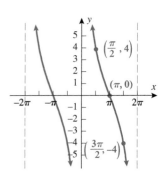

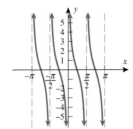

YOUR TURN Graph $y = 2\cot(2x)$ on the interval $-\pi \le x \le \pi$.

GRAPHING SECANT AND COSECANT FUNCTIONS

Graphs of $y = A\sec(Bx)$ and $y = A\csc(Bx)$ can be obtained using the following steps:

Step 1: Graph the corresponding guide function with a dashed curve.

For $y = A\sec(Bx)$, use $y = A\cos(Bx)$ as a guide.

For $y = A\csc(Bx)$, use $y = A\sin(Bx)$ as a guide.

Step 2: Draw the vertical asymptotes that correspond to the x-intercepts of the guide function.

Step 3: Draw the U and ∩ shapes, between the asymptotes. If the guide function has a positive value between the asymptotes, the U opens upward; and if the guide function has a negative value, the U opens downward.

▶ **EXAMPLE 3**   **Graphing $y = A\sec(Bx)$**

Graph $y = 2\sec(\pi x)$ on the interval $-2 \le x \le 2$.

STEP 1 Graph the corresponding guide function
with a dashed curve.

For $y = 2\sec(\pi x)$, use $y = 2\cos(\pi x)$
as a guide $(A = 2; \text{ period} = 2)$.

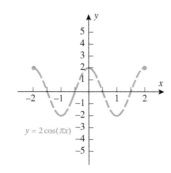

STEP 2 Draw the asymptotes that correspond to
the x-intercepts of the guide function.

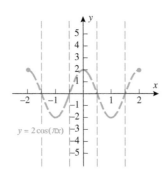

STEP 3 Draw the U shape between the asymptotes.
If the guide function is positive, the U
opens upward, and if the guide function is
negative, the U opens downward.

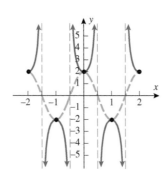

▼

YOUR TURN   Graph $y = -\sec(2\pi x)$ on the interval $-1 \le x \le 1$.

▼
ANSWER

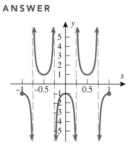

EXAMPLE 4   **Graphing $y = A\csc(Bx)$**

Graph $y = -3\csc(2\pi x)$ on the interval $-1 \le x \le 1$.

STEP 1 Graph the corresponding guide function
with a dashed curve.

For $y = -3\csc(2\pi x)$, use
$y = -3\sin(2\pi x)$ as a guide
$(A = 3; \text{ period} = 1)$.

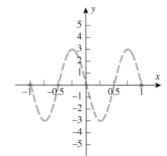

STEP 2 Draw the asymptotes that correspond to the *x*-intercepts of the guide function.

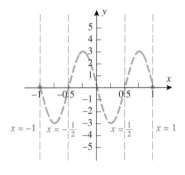

STEP 3 Draw the U shape between the asymptotes. If the guide function is positive, the U opens upward, and if the guide function is negative, the U opens downward.

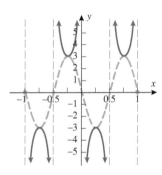

▼
ANSWER

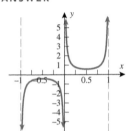

▼
YOUR TURN Graph $y = \frac{1}{2}\csc(\pi x)$ on the interval $-1 \le x \le 1$.

4.3.2 Translations of Tangent, Cotangent, Secant, and Cosecant Functions

4.3.2 SKILL

Graph translated tangent, cotangent, secant, and cosecant functions.

4.3.2 CONCEPTUAL

Understand that graph-shifting techniques for tangent and cotangent are consistent with translations used for sinusoidal functions, but for secant and cosecant functions we first graph the horizontally translated sinusoidal functions and then shift up or down depending on the vertical translations.

Vertical translations and horizontal translations (phase shifts) of the tangent, cotangent, secant, and cosecant functions are graphed the same way as vertical and horizontal translations of sinusoidal graphs. For tangent and cotangent functions, we follow the same procedure as we did with sinusoidal functions. For secant and cosecant functions, we graph the guide function first and then translate up or down depending on the sign of the vertical shift.

EXAMPLE 5 **Graphing $y = k + A\tan(Bx \pm C)$**

Graph $y = 1 - \tan\left(x - \dfrac{\pi}{2}\right)$ on $-\pi \le x \le \pi$. State the domain and range on the interval. There are two ways to approach graphing this function. Both will be illustrated.

Solution (1):

Plot $y = \tan x$, and then do the following:

- Shift the curve to the right $\dfrac{\pi}{2}$ units. $y = \tan\left(x - \dfrac{\pi}{2}\right)$

- Reflect the curve about the *x*-axis (because *A* is negative). $y = -\tan\left(x - \dfrac{\pi}{2}\right)$

- Shift the entire graph up one unit.

$$y = 1 - \tan\left(x - \frac{\pi}{2}\right)$$

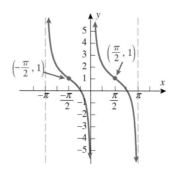

Solution (2):

Graph $y = -\tan\left(x - \frac{\pi}{2}\right)$, and then shift the entire graph up one unit because $k = 1$.

STEP 1 Calculate the period.

$$\frac{\pi}{B} = \pi$$

STEP 2 Find two vertical asymptotes. Solve for x.

$$x - \frac{\pi}{2} = -\frac{\pi}{2} \quad \text{and} \quad x - \frac{\pi}{2} = \frac{\pi}{2}$$

$$x = 0 \quad \text{and} \quad x = \pi$$

are vertical asymptotes

STEP 3 Find the x-intercept between the asymptotes.

$$x - \frac{\pi}{2} = 0$$

$$x = \frac{\pi}{2}$$

STEP 4 Draw the vertical asymptotes, $x = 0$ and $x = \pi$, and label the x-intercept, $\left(\frac{\pi}{2}, 0\right)$.

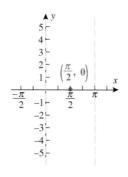

STEP 5 Divide the period π into four equal parts in steps of $\frac{\pi}{4}$. Set up a table with coordinates corresponding to values of $y = -\tan\left(x - \frac{\pi}{2}\right)$ between the two asymptotes.

x	$y = -\tan\left(x - \frac{\pi}{2}\right)$	(x, y)
$x = 0$	undefined	vertical asymptote, $x = 0$
$\dfrac{\pi}{4}$	1	$\left(\dfrac{\pi}{4}, 1\right)$
$\dfrac{\pi}{2}$	0	$\left(\dfrac{\pi}{2}, 0\right)$ (x-intercept)
$\dfrac{3\pi}{4}$	-1	$\left(\dfrac{3\pi}{4}, -1\right)$
$x = \pi$	undefined	vertical asymptote, $x = \pi$

STEP 6 Graph the points from the table and connect with a smooth curve. Repeat to the right and left until reaching the interval endpoints.

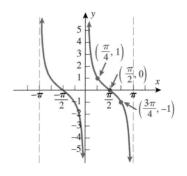

STEP 7 Shift the entire graph up one unit to arrive at the graph of $y = 1 - \tan\left(x - \frac{\pi}{2}\right)$.

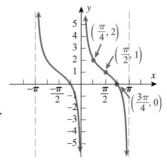

STEP 8 State the domain and range on the interval.

Domain: $(-\pi, 0) \cup (0, \pi)$
Range: $(-\infty, \infty)$

▼

YOUR TURN Graph $y = -1 + \cot\left(x + \frac{\pi}{2}\right)$ on $-\pi \leq x \leq \pi$. State the domain and range on the interval.

▼ · · · · · · · · · · · · ·
ANSWER

Domain:

$$\left[-\pi, -\frac{\pi}{2}\right) \cup \left(-\frac{\pi}{2}, \frac{\pi}{2}\right) \cup \left(\frac{\pi}{2}, \pi\right]$$

Range: $(-\infty, \infty)$

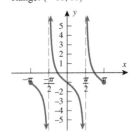

$\left[\text{CONCEPT CHECK}\right]$

If $k > 0$, what is the last step in graphing the function $y = -k + \sec(\pi x)$?

▼ · · · · · · · · · · · · · · · · ·

ANSWER Shift the graph of $\sec(\pi x)$ down k units.

EXAMPLE 6 **Graphing** $y = k + A\csc(Bx + C)$

Graph $y = 1 - \csc(2x - \pi)$ on $-\pi \leq x \leq \pi$. State the domain and range on the interval.

Solution:

Graph $y = -\csc(2x - \pi)$ and shift the entire graph up one unit to arrive at the graph of $y = 1 - \csc(2x - \pi)$.

STEP 1 Draw the guide function, $y = -\sin(2x - \pi)$.

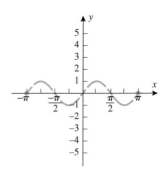

STEP 2 Draw the vertical asymptotes of $y = -\csc(2x - \pi)$ that correspond to the x-intercepts of $y = -\sin(2x - \pi)$.

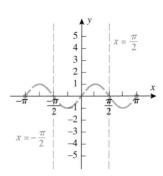

STEP 3 Draw the U shape between the asymptotes. If the guide function is positive, the U opens upward, and if the guide function is negative, the U opens downward.

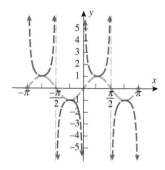

STEP 4 Shift the entire graph up one unit to arrive at the graph of $y = 1 - \csc(2x - \pi)$.

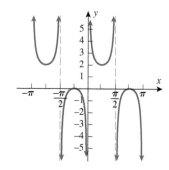

STEP 5 State the domain and range on the interval.

$$\text{Domain: } \left(-\pi, -\frac{\pi}{2}\right) \cup \left(-\frac{\pi}{2}, 0\right) \cup \left(0, \frac{\pi}{2}\right) \cup \left(\frac{\pi}{2}, \pi\right)$$

$$\text{Range: } (-\infty, 0] \cup [2, \infty)$$

▼
ANSWER

Domain:

$$\left[-1, -\frac{1}{2}\right) \cup \left(-\frac{1}{2}, \frac{1}{2}\right) \cup \left(\frac{1}{2}, 1\right]$$

Range: $(-\infty, -3] \cup [-1, \infty)$

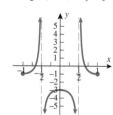

▼
YOUR TURN Graph $y = -2 + \sec(\pi x - \pi)$ on $-1 \le x \le 1$. State the domain and range on the interval.

▶[SECTION 4.3] SUMMARY

The tangent and cotangent functions have period π, whereas the secant and cosecant functions have period 2π. The tangent and cotangent functions are graphed by first identifying the vertical asymptotes and x-intercepts, then by finding values of the function within a period (i.e., between the asymptotes). Graphs of the secant and cosecant functions are found by first graphing their guide function (cosine and sine, respectively) and then labeling vertical asymptotes that correspond to x-intercepts of the guide function. The graphs of secant and cosecant functions resemble the letter U opening up or down. The secant and cosecant functions are positive when their guide function is positive and negative when their guide function is negative because of their reciprocal relationships.

[SECTION 4.3] EXERCISES

● **SKILLS**

In Exercises 1–8, match each function to its graph (a)–(h).

1. $y = -\tan x$ **2.** $y = -\csc x$ **3.** $y = \sec(2x)$ **4.** $y = \csc(2x)$

5. $y = \cot(\pi x)$ **6.** $y = -\cot(\pi x)$ **7.** $y = 3\sec x$ **8.** $y = 3\csc x$

a.
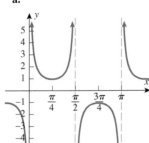

b.
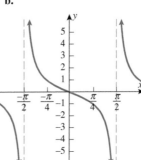

c.
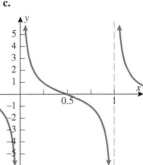

d.
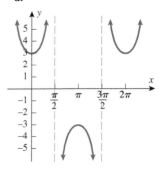

e.
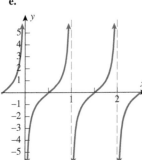

f.
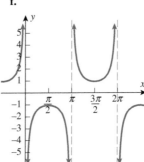

g.
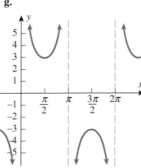

h.
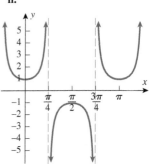

In Exercises 9–18, determine the period and phase shift (if there is one) for each function.

9. $y = \tan(x - \pi)$

10. $y = 3\tan(3x)$

11. $y = -\sec\left(2x - \dfrac{\pi}{2}\right)$

12. $y = \sec\left(\dfrac{1}{2}x + \dfrac{\pi}{4}\right)$

13. $y = \csc\left[3\left(x - \dfrac{\pi}{6}\right)\right]$

14. $y = -4\csc\left[\dfrac{1}{4}\left(x + \dfrac{3}{2}\right)\right]$

15. $y = \tan(0.5x + 2\pi)$

16. $y = \sec\left(0.1x + \dfrac{\pi}{10}\right)$

17. $y = -3\cot\left(\pi x - \dfrac{\pi}{2}\right)$

18. $y = \cot(2x + 2)$

In Exercises 19–30, graph the functions over the indicated intervals.

19. $y = \tan(\frac{1}{2}x),\ -2\pi \le x \le 2\pi$

20. $y = \cot(\frac{1}{2}x),\ -2\pi \le x \le 2\pi$

21. $y = -\cot(2\pi x),\ -1 \le x \le 1$

22. $y = -\tan(2\pi x),\ -1 \le x \le 1$

23. $y = 2\tan(3x),\ -\pi \le x \le \pi$

24. $y = 2\tan\left(\frac{1}{3}x\right),\ -3\pi \le x \le 3\pi$

25. $y = -\tan\left(x - \dfrac{\pi}{2}\right),\ -\pi \le x \le \pi$

26. $y = \tan\left(x + \dfrac{\pi}{4}\right),\ -\pi \le x \le \pi$

27. $y = \cot\left(x - \dfrac{\pi}{4}\right), -\pi \le x \le \pi$

28. $y = -\cot\left(x + \dfrac{\pi}{2}\right), -\pi \le x \le \pi$

29. $y = \tan(2x - \pi), -2\pi \le x \le 2\pi$

30. $y = \cot(2x - \pi), -2\pi \le x \le 2\pi$

In Exercises 31–42, graph the functions over the indicated intervals.

31. $y = \sec\left(\tfrac{1}{2}x\right), -2\pi \le x \le 2\pi$

32. $y = \csc\left(\tfrac{1}{2}x\right), -2\pi \le x \le 2\pi$

33. $y = -\csc(2\pi x), -1 \le x \le 1$

34. $y = -\sec(2\pi x), -1 \le x \le 1$

35. $y = 2\sec(3x), 0 \le x \le 2\pi$

36. $y = 2\csc\left(\tfrac{1}{3}x\right), -3\pi \le x \le 3\pi$

37. $y = -3\csc\left(x - \dfrac{\pi}{2}\right)$, over at least one period

38. $y = 5\sec\left(x + \dfrac{\pi}{4}\right)$, over at least one period

39. $y = \tfrac{1}{2}\sec(x - \pi)$, over at least one period

40. $y = -4\csc(x + \pi)$, over at least one period

41. $y = 2\sec(2x - \pi), -2\pi \le x \le 2\pi$

42. $y = 2\csc(2x + \pi), -2\pi \le x \le 2\pi$

In Exercises 43–50, graph the functions over at least one period.

43. $y = 3 - 2\sec\left(x - \dfrac{\pi}{2}\right)$

44. $y = -3 + 2\csc\left(x + \dfrac{\pi}{2}\right)$

45. $y = \dfrac{1}{2} + \dfrac{1}{2}\tan\left(x - \dfrac{\pi}{2}\right)$

46. $y = \dfrac{3}{4} - \dfrac{1}{4}\cot\left(x + \dfrac{\pi}{2}\right)$

47. $y = -2 + 3\csc(2x - \pi)$

48. $y = -1 + 4\sec(2x + \pi)$

49. $y = -1 - \sec\left(\dfrac{1}{2}x - \dfrac{\pi}{4}\right)$

50. $y = -2 + \csc\left(\dfrac{1}{2}x + \dfrac{\pi}{4}\right)$

In Exercises 51–56, state the domain and range of the functions.

51. $y = \tan\left(\pi x - \dfrac{\pi}{2}\right)$

52. $y = \cot\left(x - \dfrac{\pi}{2}\right)$

53. $y = 2\sec(5x)$

54. $y = -4\sec(3x)$

55. $y = 2 - \csc\left(\tfrac{1}{2}x - \pi\right)$

56. $y = 1 - 2\sec\left(\tfrac{1}{2}x + \pi\right)$

● **APPLICATIONS**

57. Gaming. A video game developer has a monster that moves on a path given by the equation $y = 2 + \tfrac{3}{2}\csc x$ on the interval $0 \le x < \pi$. If a Cartesian plane is overlaid on the video screen such that $0 \le x \le 3$ and $0 \le y \le 8$, sketch the path of the monster on the screen.

58. Gaming. A video game developer has a monster that moves on a path given by the equation $y = 3 + \cot\left(\tfrac{2}{3}x\right)$ on the interval $0 \le x < \dfrac{3\pi}{2}$. If a Cartesian plane is overlaid on the video screen such that $0 \le x \le 5$ and $0 \le y \le 8$, sketch the path of the monster on the screen.

59. Diving. When viewing a person diving into a swimming pool from the side, the equation $y = 4 + \sec\left(2x + \dfrac{\pi}{2}\right)$ can be used to model the motion, where $\dfrac{\pi}{8} \le x < \dfrac{\pi}{2}$. Sketch the path of the diver as modeled by this equation.

60. Diving. When viewing a person diving into a swimming pool from the side, the equation $y = 4 + \csc(3x + \pi)$ can be used to model the motion, where $\dfrac{\pi}{12} \le x < \dfrac{\pi}{3}$. Sketch the path of the diver as modeled by this equation.

● **CATCH THE MISTAKE**

In Exercises 61 and 62, explain the mistake that is made.

61. Graph $y = 3\csc(2x)$.

Solution:

Graph the guide function, $y = \sin(2x)$.

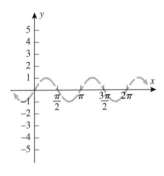

Draw vertical asymptotes at x-values that correspond to x-intercepts of the guide function.

Draw the cosecant function.

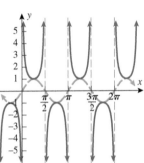

This is incorrect. What mistake was made?

62. Graph $y = \tan(4x)$.

Solution:

Step 1: Calculate the period. $\quad p = \dfrac{\pi}{B} = \dfrac{\pi}{4}$

Step 2: Find two vertical asymptotes. Solve for x.
$$4x = 0 \quad \text{and} \quad 4x = \pi$$
$$x = 0 \quad \text{and} \quad x = \dfrac{\pi}{4}$$

Step 3: Find the x-intercept between the asymptotes.
$$4x = \dfrac{\pi}{2}$$
$$x = \dfrac{\pi}{8}$$

Step 4: Draw the vertical asymptotes, $x = 0$ and $x = \dfrac{\pi}{4}$, and label the x-intercept, $\left(\dfrac{\pi}{8}, 0\right)$.

Step 5: Graph.

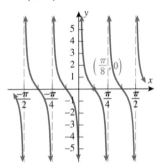

This is incorrect. What mistake was made?

● **CONCEPTUAL**

In Exercises 63–66, determine whether each statement is true or false.

63. $\sec\left(x - \dfrac{\pi}{2}\right) = \csc x$

64. $\csc\left(x - \dfrac{\pi}{2}\right) = \sec x$

65. The range for the function $y = A\csc\left(2x - \dfrac{\pi}{3}\right)$ is $(-\infty, -A] \cup [A, \infty)$.

66. The range for the function is $y = k + A\csc\left(2x - \dfrac{\pi}{3}\right)$ is $(-\infty, -A - k] \cup [A + k, \infty)$.

67. For what values of n do $y = \tan x$ and $y = \tan(x - n\pi)$ have the same graph?

68. For what values of n do $y = \csc x$ and $y = \csc(x - n\pi)$ have the same graph?

69. Solve the equation $\tan(2x - \pi) = 0$ for x in the interval $[-\pi, \pi]$ by graphing.

70. Solve the equation $\csc(2x + \pi) = 0$ for x in the interval $[-\pi, \pi]$ by graphing.

• CHALLENGE

71. Find the domain for the function $y = \csc\left(2x + \dfrac{\pi}{2}\right)$.

72. Find the domain for the function $y = \cot(4x - \pi)$.

73. Sketch the graph of $y = \sec x + \csc x$ over the interval $0 \le x \le 2\pi$ using the addition of ordinates method.

74. Sketch the graph of $y = \frac{1}{2}x + \csc(2x)$ over the interval $0 \le x \le 2\pi$ using the addition of ordinates method.

• TECHNOLOGY

75. What is the amplitude of the function $y = \cos x + \sin x$? Use a graphing calculator to graph $Y_1 = \cos x$, $Y_2 = \sin x$, and $Y_3 = \cos x + \sin x$ in the same viewing window.

76. Graph $Y_1 = \cos x + \sin x$ and $Y_2 = \sec x + \csc x$ in the same viewing window. Based on what you see, is $Y_1 = \cos x + \sin x$ the guide function for $Y_2 = \sec x + \csc x$?

▶ [**CHAPTER 4 REVIEW**]

SECTION	CONCEPT	KEY IDEAS/FORMULAS
4.1	**Basic graphs of sine and cosine functions: Amplitude and period**	
	The graphs of sinusoidal functions	Odd function: $\sin(-x) = -\sin x$ Period: 2π Amplitude: 1 Even function: $\cos(-x) = \cos x$ Period: 2π Amplitude: 1 **The amplitude and period of sinusoidal graphs:** Amplitude $= \lvert A \rvert$ ▪ $\lvert A \rvert > 1$ stretch vertically ▪ $\lvert A \rvert < 1$ compress vertically Period $= \dfrac{2\pi}{B}$ $B > 0$ ▪ $B > 1$, compress horizontally (period shorter) ▪ $B < 1$, stretch horizontally (period longer)
	Harmonic motion	▪ simple ▪ damped ▪ resonance
4.2	**Translations of the sine and cosine functions: Addition of ordinates**	$y = k + A\sin(Bx \pm C)$ or $y = k + A\cos(Bx \pm C)$
	Reflections and vertical shifts of sinusoidal functions	**Reflections** ▪ $y = -A\sin(Bx)$ is the reflection of $y = A\sin(Bx)$ about the x-axis. ▪ $y = -A\cos(Bx)$ is the reflection of $y = A\cos(Bx)$ about the x-axis. **Vertical shifts** ▪ $y = k + A\sin(Bx)$ is found by shifting $y = A\sin(Bx)$ up or down k units. ▪ $y = k + A\cos(Bx)$ is found by shifting $y = A\cos(Bx)$ up or down k units.

SECTION	CONCEPT	KEY IDEAS/FORMULAS
	Horizontal shifts: Phase shift	■ $y = A\sin(Bx \pm C) = A\sin\left[B\left(x \pm \dfrac{C}{B}\right)\right]$ has period $\dfrac{2\pi}{B}$ and a phase shift of $\dfrac{C}{B}$ units to the left $(+)$ or the right $(-)$. ■ $y = A\cos(Bx \pm C) = A\cos\left[B\left(x \pm \dfrac{C}{B}\right)\right]$ has period $\dfrac{2\pi}{B}$ and a phase shift of $\dfrac{C}{B}$ units to the left $(+)$ or the right $(-)$.
	Graphing $y = k + A\sin(Bx + C)$ and $y = k + A\cos(Bx + C)$	To graph $y = k + A\sin(Bx + C)$ or $y = k + A\cos(Bx + C)$, start with the graph of $y = A\sin(Bx + C)$ or $y = A\cos(Bx + C)$ and shift up or down k units.
	Graphing sums of functions: Addition of ordinates	Sum the y-coordinates (ordinates) of the functions that are summed.
4.3	**Graphs of tangent, cotangent, secant, and cosecant functions**	
	Graphing the tangent, cotangent, secant, and cosecant functions	**The tangent function** $\quad f(x) = \tan x$ x-intercepts: $(n\pi, 0)$ Asymptotes: $x = \dfrac{(2n+1)}{2}\pi$ Period: π Amplitude: none **The cotangent function** $\quad f(x) = \cot x$ Asymptotes: $x = n\pi$ x-intercepts: $\left(\dfrac{(2n+1)}{2}\pi, 0\right)$ Period: π Amplitude: none

SECTION	CONCEPT	KEY IDEAS/FORMULAS

The secant function

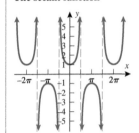

$f(x) = \sec x$

Asymptotes: $x = \dfrac{(2n + 1)}{2}\pi$

Period: 2π

Amplitude: none

The cosecant function

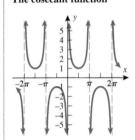

$f(x) = \csc x$

Asymptotes: $x = n\pi$

Period: 2π

Amplitude: none

■ These four trigonometric functions have no amplitude.

■ Tangent and cotangent functions have period π, whereas secant and cosecant functions have period 2π.

■ To determine asymptotes, find the values of the argument that correspond to the trigonometric ratios having a zero denominator.

Translations of tangent, cotangent, secant, and cosecant functions

$y = A\tan(Bx + C)$ and $y = A\cot(Bx + C)$

To find asymptotes, set $Bx + C$ equal to

■ $-\dfrac{\pi}{2}$ and $\dfrac{\pi}{2}$ for tangent and solve for x.

■ 0 and π for cotangent and solve for x.

To find the x-intercept, set $Bx + C$ equal to

■ 0 for tangent and solve for x.

■ $\dfrac{\pi}{2}$ for cotangent and solve for x.

$y = A\sec(Bx + C)$ and $y = A\csc(Bx + C)$

■ To graph $y = A\sec(Bx + C)$, use $y = A\cos(Bx + C)$ as the guide.

■ To graph $y = A\csc(Bx + C)$, use $y = A\sin(Bx + C)$ as the guide.

Intercepts on the guide function correspond to vertical asymptotes of secant or cosecant functions.

[CHAPTER 4 REVIEW EXERCISES]

4.1 Basic Graphs of Sine and Cosine Functions: Amplitude and Period

Refer to the graph of the sinusoidal function to answer the questions.

1. Determine the period of the function.

2. Determine the amplitude of the function.

3. Write an equation for the sinusoidal function.

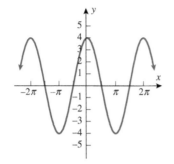

Refer to the graph of the sinusoidal function to answer the questions.

4. Determine the period of the function.

5. Determine the amplitude of the function.

6. Write an equation for the sinusoidal function.

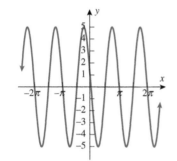

Refer to the graph of the sinusoidal function to answer the questions.

7. Determine the period of the function.

8. Determine the amplitude of the function.

9. Write an equation for the sinusoidal function.

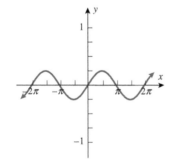

Refer to the graph of the cosine function to answer the questions.

10. Determine the period of the function.

11. Determine the amplitude of the function.

12. Write an equation for the cosine function.

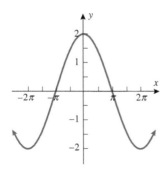

Determine the amplitude and period of each function.

13. $y = -2\cos(2\pi x)$

14. $y = \dfrac{1}{3}\sin\left(\dfrac{\pi}{2}x\right)$

15. $y = \dfrac{1}{5}\sin(3x)$

16. $y = -\dfrac{7}{6}\cos(6x)$

Graph each function from -2π to 2π.

17. $y = -2\sin\left(\dfrac{x}{2}\right)$

18. $y = 3\sin(3x)$

19. $y = \dfrac{1}{2}\cos(2x)$

20. $y = -\dfrac{1}{4}\cos\left(\dfrac{x}{2}\right)$

4.2 Translations of the Sine and Cosine Functions: Addition of Ordinates

State the amplitude, period, phase shift, and vertical shift of each function.

21. $y = 2 + 3\sin\left(x - \dfrac{\pi}{2}\right)$

22. $y = 3 - \dfrac{1}{2}\sin\left(x + \dfrac{\pi}{4}\right)$

23. $y = -2 - 4\cos\left[3\left(x + \dfrac{\pi}{4}\right)\right]$

24. $y = -1 + 2\cos\left[2\left(x - \dfrac{\pi}{3}\right)\right]$

25. $y = \dfrac{1}{2} - \cos(x - \pi)$

26. $y = \dfrac{3}{2} + \sin(x + \pi)$

27. $y = -\sin(\dfrac{1}{2}x + \pi)$

28. $y = \cos(\dfrac{1}{4}x - \pi)$

29. $y = 4 + 5\sin(4x - \pi)$

30. $y = -6\cos(3x + \pi)$

Sketch the graph of the function from -2π to 2π.

31. $y = 3 + \sin\left(x + \dfrac{\pi}{2}\right)$

32. $y = -2 + \sin\left(x - \dfrac{\pi}{2}\right)$

33. $y = -2 + \cos(x + \pi)$

34. $y = 3 + \cos(x - \pi)$

35. $y = 3 - \sin\left(\dfrac{\pi}{2}x\right)$

36. $y = 2\cos\left(\dfrac{\pi}{2}x\right)$

Match the graphs with their function equations.

37. $y = 2\sin\left(x - \dfrac{\pi}{4}\right)$

38. $y = -2 + 3\sin x$

39. $y = 3 - 3\cos x$

40. $y = 3\cos\left(x + \dfrac{\pi}{4}\right)$

a.

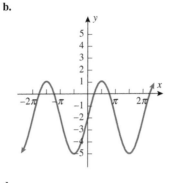

b.

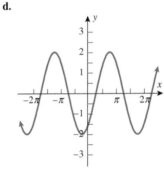

c.

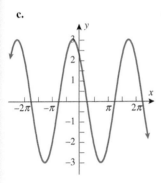

d.

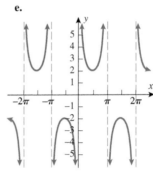

c.

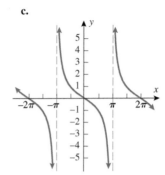

d.

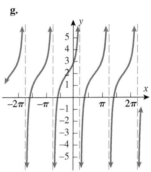

e.

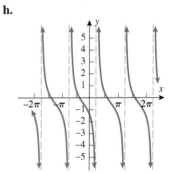

f.

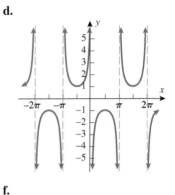

Use addition of ordinates to graph each of the following on the indicated interval.

41. $y = \sin(2x) + 2\cos x$ on the interval $0 \le x \le 2\pi$.

42. $y = \sin(\pi x) + \cos\left(\dfrac{\pi}{2}x\right)$ on the interval $0 \le x \le 2$.

43. $y = -2\sin x + \cos\left(x + \dfrac{\pi}{2}\right)$ on the interval $0 \le x \le 2\pi$.

44. $y = \sin(2x) - 3\cos(4x)$ on the interval $0 \le x \le \pi$.

4.3 Graphs of Tangent, Cotangent, Secant, and Cosecant Functions

Match each function with its graph (a)–(j).

45. $y = \tfrac{1}{2}\tan x$

46. $y = -\tan\left(\tfrac{1}{2}x\right)$

47. $y = \cot(2x)$

48. $y = 2\cot x$

49. $y = 2 + \tan\left(x + \dfrac{\pi}{4}\right)$

50. $y = -\dfrac{1}{2} + \cot\left(x - \dfrac{\pi}{4}\right)$

51. $y = 2 + 2\sec x$

52. $y = -\sec\left(x - \dfrac{\pi}{2}\right)$

53. $y = 2 + \csc\left(x + \dfrac{\pi}{4}\right)$

54. $y = 2\csc x$

g.

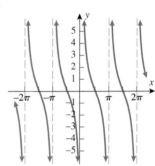

h.

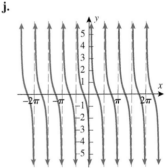

i.

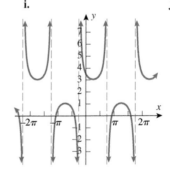

j.

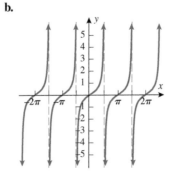

State the domain and range of each function.

55. $y = 4\tan\left(x + \dfrac{\pi}{2}\right)$

56. $y = \cot\left[2\left(x - \dfrac{\pi}{2}\right)\right]$

57. $y = 3\sec(2x)$

58. $y = 1 + 2\csc x$

a.

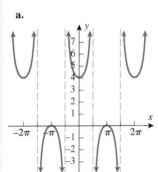

b.

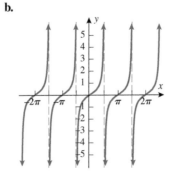

Graph each function on $[-2\pi, 2\pi]$.

59. $y = -\tan\left(x - \dfrac{\pi}{4}\right)$ **60.** $y = 1 + \cot(2x)$

61. $y = 2 + \sec(x - \pi)$ **62.** $y = -\csc\left(x + \dfrac{\pi}{4}\right)$

63. $y = 2 + 3\sec\left(x - \dfrac{\pi}{3}\right)$ **64.** $y = 1 + \tan(2x + \pi)$

Technology Exercises

Section 4.2

65. Use a graphing calculator to graph $Y_1 = \cos x$ and $Y_2 = \cos(x + c)$, where

a. $c = \dfrac{\pi}{6}$, and explain the relationship between Y_2 and Y_1.

b. $c = -\dfrac{\pi}{6}$, and explain the relationship between Y_2 and Y_1.

66. Use a graphing calculator to graph $Y_1 = \sin x$ and $Y_2 = c + \sin x$, where

a. $c = \dfrac{1}{2}$, and explain the relationship between Y_2 and Y_1.

b. $c = -\dfrac{1}{2}$, and explain the relationship between Y_2 and Y_1.

67. What is the amplitude of the function $y = 4\cos x - 3\sin x$? Use a graphing calculator to graph $Y_1 = 4\cos x$, $Y_2 = 3\sin x$, and $Y_3 = 4\cos x - 3\sin x$ in the same viewing window.

68. What is the amplitude of the function $y = \sqrt{3}\sin x + \cos x$? Use a graphing calculator to graph $Y_1 = \sqrt{3}\sin x$, $Y_2 = \cos x$, and $Y_3 = \sqrt{3}\sin x + \cos x$ in the same viewing window.

[CHAPTER 4 PRACTICE TEST]

1. State the amplitude and period of $y = -5\sin(3x)$.

2. Graph $y = \sin(3x)$ over $0 \le x \le \dfrac{4\pi}{3}$.

3. Graph $y = \cos\left(x - \dfrac{\pi}{4}\right)$ over the interval $-\dfrac{3\pi}{4} \le x \le \dfrac{3\pi}{4}$.

4. Graph $y = -4\sin x$ over the interval $-3\pi \le x \le 3\pi$.

5. Graph $y = \cos\left(\dfrac{x}{2}\right)$ over the interval $-4\pi \le x \le 4\pi$.

6. Graph $y = \sin\left(x + \dfrac{\pi}{6}\right)$ over the interval $-\dfrac{\pi}{6} \le x \le \dfrac{11\pi}{6}$.

7. Graph $y = -2 + \cos x$ over the interval $-2\pi \le x \le 2\pi$.

8. Graph $y = -2\cos\left(\tfrac{1}{2}x\right)$ over $-4\pi \le x \le 4\pi$.

9. Graph $y = \tan\left(\pi x - \dfrac{\pi}{2}\right)$ over two periods.

10. State the vertical asymptotes of $y = \csc(2\pi x)$ for all x.

11. Graph $y = -2\sec(\pi x)$ over two periods.

12. State the amplitude, period, and phase shift (if possible) of $y = -3\cot(4x - 2\pi)$.

13. Graph $y = -2 + 3\sin(2x - \pi)$ over $0 \le x \le 2\pi$.

14. Graph $y = 1 - \cos\left(\pi x - \dfrac{\pi}{2}\right)$ over at least one period.

15. Graph $y = 2 - \csc\left(\pi x - \dfrac{\pi}{2}\right)$ over at least one period.

16. Graph $y = \tfrac{1}{2} + \tfrac{1}{2}\cot(2x)$ over $0 \le x \le 2\pi$.

17. An electromagnetic signal has the form $E(t) = A\cos(\omega t + \phi)$. What are the period and the phase shift?

18. The table given shows the average high temperatures for each month during the year. Find a trigonometric function that expresses the average high temperature T, in Fahrenheit, and the number of months since the beginning of the year in terms of t.

t	1	2	3	4	5	6	7	8	9	10	11	12
T	34°	43°	51°	62°	73°	84°	88°	86°	72°	65°	49°	38°

19. The table given shows the average sales on a seasonal product for each month during the year. Find a trigonometric function that expresses the average sales S, in thousands of dollars, and the number of months since the beginning of the year in terms of t.

t	1	2	3	4	5	6	7	8	9	10	11	12
S	45	39	36	30	24	22	17	21	25	32	37	43

20. The population of rabbits in a certain county varies over time due to various environmental factors. If the equation $y = 450 - 25\sin\left(\dfrac{2\pi}{5}t - \dfrac{8\pi}{5}\right)$ represents the population of rabbits (in thousands) for year t, where t is the number of years since 2015, by how much does the rabbit population fluctuate over time? What is the first year after 2015 in which the rabbit population is at a max?

21. House values have experienced tremendous highs and lows over the last several years. If the average housing prices in thousands of dollars for a specific subdivision are modeled by the function $y = 200 - 35\sin\left(\dfrac{\pi}{3}t\right)$ for year t, where t is the number of years since 2015, by how much do home prices fluctuate over time? In what year do home prices max?

22. The vertical asymptotes of $y = 2\csc(3x - \pi)$ correspond to the _____ of $y = 2\sin(3x - \pi)$.

23. State the x-intercepts of $y = \tan(2x)$ for all x.

24. Determine the limit, $\lim\limits_{x \to \frac{\pi}{4}}\left[-\tan(2x)\right]$. That is, determine what $\tan(2x)$ approaches as x approaches $\dfrac{\pi}{4}$ from the left.

25. Select all of the following that are true statements for all x:

   a. $\cos\left(x - \dfrac{\pi}{2}\right) = \sin x$

   b. $\cos\left(x + \dfrac{\pi}{2}\right) = \sin x$

   c. $\csc\left(x + \dfrac{\pi}{2}\right) = \sec x$

   d. $\sec\left(x + \dfrac{\pi}{2}\right) = \csc x$

   e. $\tan(x - \pi) = \tan x$

26. Use the addition of ordinates to graph $y = 2\cos x - \sin(2x)$ on the interval $0 \le x \le 2\pi$.

[CHAPTERS 1–4 CUMULATIVE TEST]

1. In a 30°-60°-90° triangle, if the shortest leg has length 8 inches, what are the lengths of the other leg and the hypotenuse?

2. Given $F = 72°$, use the diagram below to find C.

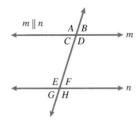

3. Use the triangle below to find $\sec\theta$.

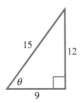

4. Convert $33°\ 44'\ 10''$ from degrees–minutes–seconds to decimal degrees. Round to the nearest thousandth.

5. Given $\alpha = 64.3°$ and $a = 132$ feet, use the right triangle below to find b.

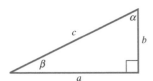

6. Determine the angle of smallest possible positive measure that is coterminal with the angle 1579°.

7. The angle θ, in standard position, has the terminal side defined by the line $y = 5x$, $x \geq 0$. Calculate the values for the six trigonometric functions of θ.

8. If both $\tan\theta$ and $\sec\theta$ are negative, find the quadrant in which the terminal side of θ must lie.

9. Determine if the statement $\sec\theta = \frac{24}{25}$ is possible or impossible.

10. Find all values of θ, where $0° \leq \theta \leq 360°$, when
$$\cos\theta = -\frac{\sqrt{3}}{2}.$$

11. Given $\sin\theta = -\frac{2}{3}$ and $\cos\theta = \frac{\sqrt{5}}{3}$, find $\tan\theta$.

12. If $\csc\theta = -\frac{\sqrt{41}}{4}$ and the terminal side of θ lies in quadrant III, find $\cos\theta$.

13. **Geometry.** The slope of a line passing through the origin can be given as $m = \tan\theta$, where θ is the positive angle formed by the line and the positive x-axis. Find the slope of a line such that $\sin\theta = \frac{3}{5}$ and $\cos\theta = \frac{4}{5}$.

14. Convert $-105°$ to radians. Leave the answer in terms of π.

15. Find the reference angle for $\frac{7\pi}{6}$ in terms of both radians and degrees.

16. Find the exact value of $\sin\left(-\frac{7\pi}{4}\right)$.

17. Find the exact length of the arc with central angle $\theta = \frac{\pi}{3}$ and radius $r = 6$ centimeters.

18. Find the area of a circular sector with radius 3.4 meters and central angle $\theta = 35°$. Round the answer to three significant digits.

19. Find the linear speed of a point that moves with constant speed in a circular motion if the point travels along the circle of arc length $s = 3$ meters in 4 seconds.

20. Find the angular speed (radians/second) associated with rotating a central angle $\theta = 24\pi$ in 16 seconds.

21. **Yo-Yo Dieting.** A woman has been yo-yo dieting for years. Her weight changes throughout the year as she gains and loses weight. Her weight in a particular month can be determined with $w(x) = 145 + 10\cos\left(\frac{\pi}{6}x\right)$, where x is the month and w is in pounds. If $x = 1$ corresponds to January, how much does she weigh in May?

22. Graph $y = 3\cos(2x)$ over one period.

23. **Sound Waves.** If a sound wave is represented by $y = 0.007\sin(850\pi t)$ centimeters, what are its amplitude and frequency?

24. State the amplitude, period, and phase shift of the function
$$y = 2\sin\left(x - \frac{\pi}{4}\right).$$

25. State the domain and range of the function
$$y = 5\tan\left(x - \frac{\pi}{2}\right).$$

[5]

Trigonometric Identities

laurentiu iordache/Alamy StockPhotoE7JRC2

When you dial a phone number on your iPhone®, how does the smart phone know which key you have pressed? Dual Tone Multi-Frequency (DTMF), also known as touch-tone dialing, was developed by Bell Labs in the 1960s. The Touch-Tone® system also introduced a standardized keypad layout. After testing 18 different layouts, Bell Labs eventually chose the one familiar to us today, with 1 in the upper-left and 0 at the bottom between the star and the pound keys.

The keypad is laid out in a 4 × 3 matrix, with each row representing a low frequency and each column representing a high frequency.

FREQUENCY	1209 Hz	1336 Hz	1477 Hz
697 Hz	1	2	3
770 Hz	4	5	6
852 Hz	7	8	9
941 Hz	*	0	#

When you press the number 8, the phone sends a sinusoidal tone that combines a low-frequency tone of 852 hertz and a high-frequency tone of 1336 hertz. The result can be found using sum-to-product *trigonometric identities*.

LEARNING OBJECTIVES

- Verify a trigonometric identity.
- Apply the sum and difference identities.
- Apply the double-angle identities.
- Apply the half-angle identities.
- Apply the product-to-sum and sum-to-product identities.

[IN THIS CHAPTER]

We will review basic identities and use them to simplify trigonometric expressions. We will verify trigonometric identities. Specific identities that will be discussed are sum and difference, double-angle and half-angle, and product-to-sum and sum-to-product. Music and touch-tone keypads are applications of trigonometric identities. The trigonometric identities have useful applications, and they are used most frequently in calculus.

TRIGONOMETRIC IDENTITIES

5.1 TRIGONOMETRIC IDENTITIES	**5.2** SUM AND DIFFERENCE IDENTITIES	**5.3** DOUBLE-ANGLE IDENTITIES	**5.4** HALF-ANGLE IDENTITIES	**5.5** PRODUCT-TO-SUM AND SUM-TO-PRODUCT IDENTITIES
• Verifying Trigonometric Identities	• Sum and Difference Identities for the Cosine Function • Sum and Difference Identities for the Sine Function • Sum and Difference Identities for the Tangent Function	• Applying Double-Angle Identities	• Applying Half-Angle Identities	• Product-to-Sum Identities • Sum-to-Product Identities

5.1 TRIGONOMETRIC IDENTITIES

SKILLS OBJECTIVE	CONCEPTUAL OBJECTIVE
■ Simplify trigonometric expressions using identities.	■ Understand that there is more than one way to verify an identity.

5.1.1 Verifying Trigonometric Identities

Basic Identities

In Section 2.4, we discussed the *basic* (fundamental) trigonometric identities: reciprocal, quotient, and Pythagorean. In mathematics, an **identity** is an equation that is true for *all* values for which the expressions in the equation are defined. If an equation is true only for *some* values of the variable, it is a **conditional equation**. If an equation is true for no values of the variable, then it is a **contradiction**.

The following are **identities** (true for all x for which the expressions are defined):

IDENTITY	TRUE FOR THESE VALUES OF x
$x^2 + 3x + 2 = (x + 2)(x + 1)$	All real numbers
$\tan x = \dfrac{\sin x}{\cos x}$	All real numbers except $x = \dfrac{(2n + 1)\pi}{2}$, where n is an integer $\left(x = \dfrac{n\pi}{2}, \text{ where } n \text{ is an odd integer}\right)$
$\sin^2 x + \cos^2 x = 1$	All real numbers

The following are **conditional equations** (true only for particular values of x):

EQUATION	TRUE FOR THESE VALUES OF x
$x^2 + 3x + 2 = 0$	$x = -2$ and $x = -1$
$\tan x = 0$	$x = n\pi$, where n is an integer
$\sin^2 x - \cos^2 x = 1$	$x = \dfrac{(2n + 1)\pi}{2}$, where n is an integer $\left(x = \dfrac{n\pi}{2}, \text{ where } n \text{ is an odd integer}\right)$

The following are **contradictions** (*not* true for *any* values of x):

EQUATION	TRUE FOR THESE VALUES OF x
$x + 5 = x + 7$	none
$\sin^2 x + \cos^2 x = 5$	none

The following boxes summarize the identities that were discussed in Section 2.4:

RECIPROCAL IDENTITIES

RECIPROCAL IDENTITIES	DOMAIN RESTRICTIONS
$\csc x = \dfrac{1}{\sin x}$	$x \neq n\pi$ n is an integer
$\sec x = \dfrac{1}{\cos x}$	$x \neq \dfrac{n\pi}{2}$ n is an odd integer
$\cot x = \dfrac{1}{\tan x}$	$x \neq \dfrac{n\pi}{2}$ n is an integer

STUDY TIP
• The reciprocal identity $\cot x = \dfrac{1}{\tan x}$ is only valid if *both* $\tan x$ and $\cot x$ are defined.
• $x \neq \dfrac{n\pi}{2}$ includes both integer multiples of π and integer multiples of $\dfrac{\pi}{2}$.

QUOTIENT IDENTITIES

QUOTIENT IDENTITIES	DOMAIN RESTRICTIONS
$\tan x = \dfrac{\sin x}{\cos x}$	$\cos x \neq 0$ or $x \neq \dfrac{(2n+1)\pi}{2}$ n is an integer
$\cot x = \dfrac{\cos x}{\sin x}$	$\sin x \neq 0$ or $x \neq n\pi$ n is an integer

PYTHAGOREAN IDENTITIES

PYTHAGOREAN IDENTITIES	DOMAIN RESTRICTIONS
$\sin^2 x + \cos^2 x = 1$	
$\tan^2 x + 1 = \sec^2 x$	$\cos x \neq 0$ or $x \neq \dfrac{(2n+1)\pi}{2}$ n is an integer
$1 + \cot^2 x = \csc^2 x$	$\sin x \neq 0$ or $x \neq n\pi$ n is an integer

In the previous chapters, we have discussed even and odd functions that have these respective properties:

TYPE OF FUNCTION	ALGEBRAIC IDENTITY	GRAPH
Even	$f(-x) = f(x)$	Symmetry about the y-axis
Odd	$f(-x) = -f(x)$	Symmetry about the origin

We have already pointed out in previous chapters that the sine function is an odd function and the cosine function is an even function. Combining this knowledge with the reciprocal and quotient identities, we arrive at the *even–odd identities*.

EVEN–ODD IDENTITIES

Odd:
- $\sin(-x) = -\sin x$
- $\csc(-x) = -\csc x$
- $\tan(-x) = -\tan x$
- $\cot(-x) = -\cot x$

Even:
- $\cos(-x) = \cos x$
- $\sec(-x) = \sec x$

Simplifying Trigonometric Expressions Using Identities

In Section 2.4, we used the basic trigonometric identities to find values for trigonometric functions, and we simplified trigonometric expressions using the identities. We now will use the basic identities and algebraic manipulation to simplify more complicated trigonometric expressions. In simplifying trigonometric expressions, one approach is to first convert all expressions into sines and cosines and then simplify. We will try that approach here.

▶ **EXAMPLE 1**   **Simplifying Trigonometric Expressions**

Simplify the expression $\tan x \sin x + \cos x$.

Solution:

$$\tan x \sin x + \cos x$$

Write the tangent function in terms of the sine and cosine functions, $\tan x = \dfrac{\sin x}{\cos x}$.

$$= \left(\overbrace{\dfrac{\sin x}{\cos x}}^{\tan x}\right)\sin x + \cos x$$

Simplify.

$$= \dfrac{\sin^2 x}{\cos x} + \cos x$$

$$= \dfrac{\sin^2 x}{\cos x} + \dfrac{\cos^2 x}{\cos x}$$

Write as a fraction with a single quotient by finding a common denominator, $\cos x$.

$$= \dfrac{\sin^2 x + \cos^2 x}{\cos x}$$

Recognize the Pythagorean identity: $\sin^2 x + \cos^2 x = 1$.

$$= \dfrac{1}{\cos x}$$

Use the reciprocal identity, $\sec x = \dfrac{1}{\cos x}$.

$$= \boxed{\sec x}$$

▼
ANSWER

$\csc x$

YOUR TURN   Simplify the expression $\cot x \cos x + \sin x$.

In Example 1, $\tan x$ and $\sec x$ are not defined for odd integer multiples of $\dfrac{\pi}{2}$. In Your Turn, $\cot x$ and $\csc x$ are not defined for integer multiples of π. Both the original expression and the simplified form are governed by the same restrictions. There are times when the original expression is subject to more domain restrictions than the simplified form, and thus special attention must be given to note all domain restrictions.

For example, the algebraic expression $\dfrac{x^2 - 1}{x + 1}$ has the domain restriction $x \neq -1$ because that value for x makes the value of the denominator equal to zero. If we forget to state the domain restrictions, we might simplify the algebraic expressions $\dfrac{x^2 - 1}{x + 1} = \dfrac{(x - 1)(x + 1)}{(x + 1)} = x - 1$ and assume this is true for all values of x. The correct result is $\dfrac{x^2 - 1}{x + 1} = x - 1$ for $x \neq -1$. In fact, if we were to graph both the original expression $y = \dfrac{x^2 - 1}{x + 1}$ and the line $y = x - 1$, they would coincide, except the graph of the original expression would have a "hole" or discontinuity at $x = -1$. In this chapter, it is assumed that the domain of the simplified expression is the same as the domain of the original expression.

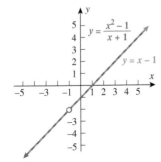

EXAMPLE 2 **Simplifying Trigonometric Expressions**

Simplify the expression $\dfrac{1}{\csc^2 x} + \dfrac{1}{\sec^2 x}$.

Solution:

Rewrite the expression in terms
of quotients squared.

$$\frac{1}{\csc^2 x} + \frac{1}{\sec^2 x} = \left(\frac{1}{\csc x}\right)^2 + \left(\frac{1}{\sec x}\right)^2$$

Use the reciprocal identities to write
the cosecant and secant functions
in terms of sines and cosines:

$$\sin x = \frac{1}{\csc x} \text{ and } \cos x = \frac{1}{\sec x}.$$

$$= \sin^2 x + \cos^2 x$$

Recognize the Pythagorean identity:
$\sin^2 x + \cos^2 x = 1$.

$$= \boxed{1}$$

▼
YOUR TURN Simplify the expression $\dfrac{1}{\cos^2 x} - 1$.

▼
ANSWER
$\tan^2 x$

Verifying Identities

We will now use the trigonometric identities to verify, or prove, other trigonometric
identities for all values for which the expressions in the equation are defined. For
example,

$$(\sin x - \cos x)^2 - 1 = -2\sin x \cos x$$

The good news is that we will know we are done when we get there, since we know the
left side is supposed to equal the right side. But how do we get there? How do we verify
that the identity is true? Remember that it must be true for *all x*, not just some *x*. Therefore,
it is not enough to simply select values for *x* and show it is true for those specific values.

WORDS	MATH
Start with one side of the equation (the more complicated side).	$(\sin x - \cos x)^2 - 1$
Remember that $(a - b)^2 = a^2 - 2ab + b^2$ and expand $(\sin x - \cos x)^2$.	$= \sin^2 x - 2\sin x \cos x + \cos^2 x - 1$
Group the $\sin^2 x$ and $\cos^2 x$ terms and use the Pythagorean identity.	$= -2\sin x \cos x + \underbrace{(\sin^2 x + \cos^2 x)}_{1} - 1$
Simplify.	$= -2\sin x \cos x$

When we arrive at the right side of the equation, then we have succeeded in verifying
the identity. In verifying trigonometric identities, there is no one procedure that works
for all identities. You can manipulate one side of the equation until it looks like the other
side. Looking at "where you want to be" will help you make proper decisions on "how"
to manipulate the one side. Here are two suggestions that are generally helpful:

1. Convert all trigonometric expressions to sines and cosines.
2. Write sums or differences of fractions (quotients) as a single fraction (quotient).

The following suggestions help guide the way to verifying trigonometric identities:

GUIDELINES FOR VERIFYING TRIGONOMETRIC IDENTITIES

- Start with the more complicated side of the equation.
- Combine sums and differences of quotients into a single quotient.
- Use basic trigonometric identities.
- Use algebraic techniques to manipulate one side of the equation until it looks like the other side.
- Sometimes it is helpful to convert all trigonometric functions into sines and cosines.

It is important to note that trigonometric identities must be valid for all values of the independent variable (usually x or θ) for which the expressions in the equation are defined (domain of the equation).

EXAMPLE 3 **Verifying Trigonometric Identities**

Verify the identity $(1 + \sin x)[1 + \sin(-x)] = \cos^2 x$.

Solution:

Start with the more complicated side (left side).	$(1 + \sin x)[1 + \sin(-x)]$
The sine function is odd: $\sin(-x) = \sin x$.	$= (1 + \sin x)(1 - \sin x)$
Eliminate the parentheses.	$= 1 - \sin^2 x$
Apply the Pythagorean identity, $\sin^2 x + \cos^2 x = 1$.	$= \cos^2 x$

EXAMPLE 4 **Verifying Trigonometric Identities**

Verify the identity $\dfrac{\tan x - \cot x}{\tan x + \cot x} = \sin^2 x - \cos^2 x$.

Solution:

Start with the more complicated side of the equation.

$$\frac{\tan x - \cot x}{\tan x + \cot x}$$

Use the quotient identities to write the tangent and cotangent functions in terms of the sine and cosine functions.

$$= \frac{\dfrac{\sin x}{\cos x} - \dfrac{\cos x}{\sin x}}{\dfrac{\sin x}{\cos x} + \dfrac{\cos x}{\sin x}}$$

Find the LCD of the numerator and the denominator.

$$\text{LCD} = \sin x \cos x$$

Multiply the numerator and the denominator by the $\sin x \cos x$.

$$= \frac{\left(\dfrac{\sin x}{\cos x} - \dfrac{\cos x}{\sin x}\right)}{\dfrac{\sin x}{\cos x} + \dfrac{\cos x}{\sin x}}\left(\dfrac{\sin x \cos x}{\sin x \cos x}\right)$$

Distribute.

$$\frac{\sin x \cdot \cos x \left(\dfrac{\sin x}{\cos x} - \dfrac{\cos x}{\sin x}\right)}{\sin x \cdot \cos x \left(\dfrac{\sin x}{\cos x} + \dfrac{\cos x}{\sin x}\right)} = \frac{\dfrac{\sin^2 x \cos x}{\cos x} - \dfrac{\sin x \cos^2 x}{\sin x}}{\dfrac{\sin^2 x \cos x}{\cos x} + \dfrac{\sin x \cos^2 x}{\sin x}}$$

Simplify.

$$= \frac{\sin^2 x - \cos^2 x}{\sin^2 x + \cos^2 x}$$

Recognize the Pythagorean identity in the denominator, $\sin^2 x + \cos^2 x = 1$.

$$= \boxed{\sin^2 x - \cos^2 x}$$

EXAMPLE 5 **Determining Whether a Trigonometric Equation Is an Identity**

Determine whether $(1 - \cos^2 x)(1 + \cot^2 x) = 0$ is an identity, a conditional equation, or a contradiction.

Solution:

Approach 1: Start with the left side of the equation and apply two Pythagorean identities. Apply the reciprocal identity, $\csc x = \dfrac{1}{\sin x}$.

$$\underbrace{(1 - \cos^2 x)}_{\sin^2 x}\underbrace{(1 + \cot^2 x)}_{\csc^2 x} = \sin^2 x \cdot \csc^2 x$$

$$= \sin^2 x \cdot \frac{1}{\sin^2 x} = \boxed{1}$$

Approach 2: Use the quotient identity to write the cotangent function in terms of the sine and cosine functions.

$$(1 - \cos^2 x)(1 + \cot^2 x) = (1 - \cos^2 x)\left(1 + \frac{\cos^2 x}{\sin^2 x}\right)$$

Combine the expression in the second parentheses into a single fraction.

$$= (1 - \cos^2 x)\left(\frac{\sin^2 x + \cos^2 x}{\sin^2 x}\right)$$

Use the Pythagorean identity.

$$= \underbrace{(1 - \cos^2 x)}_{\sin^2 x}\left(\frac{\overbrace{\sin^2 x + \cos^2 x}^{1}}{\sin^2 x}\right)$$

Simplify.

$$= \frac{\sin^2 x}{\sin^2 x} = 1$$

Since $1 \neq 0$, this is $\boxed{\text{not an identity}}$, but rather a contradiction.

▶ **EXAMPLE 6** **Verifying Trigonometric Identities**

Verify that $\dfrac{\sin(-x)}{\cos(-x)\tan(-x)} = 1$.

Solution:

Start with the left side of the equation.

$$\frac{\sin(-x)}{\cos(-x)\tan(-x)}$$

Use the quotient identity to write the tangent function in terms of the sine and cosine functions.

$$= \frac{\sin(-x)}{[\cos(-x)]\left[\dfrac{\sin(-x)}{\cos(-x)}\right]}$$

Simplify the product in the denominator.

$$= \frac{\sin(-x)}{\sin(-x)}$$

Simplify.

$$= \boxed{1}$$

Note: In the first step, we could have used the properties of even and odd functions: $\sin(-x) = -\sin x$, $\cos(-x) = \cos x$, and $\tan(-x) = -\tan x$.

EXAMPLE 7 **Verifying a More Complicated Identity**

Verify that $-\dfrac{\cot^2 x}{1 - \csc x} = \dfrac{\sin x + 1}{\sin x}$.

Solution:

Start with the left side of the equation.

$$-\frac{\cot^2 x}{1 - \csc x}$$

Use the Pythagorean identity, $\cot^2 x = \csc^2 x - 1$.

$$= -\frac{(\csc^2 x - 1)}{1 - \csc x}$$

Distribute the negative throughout the numerator.

$$= \frac{1 - \csc^2 x}{1 - \csc x}$$

Factor the numerator (difference of two squares).

$$= \frac{(1 - \csc x)(1 + \csc x)}{1 - \csc x}$$

Divide out the $1 - \csc x$ in both the numerator and denominator.

$$= 1 + \csc x$$

Use the reciprocal identity.

$$= 1 + \frac{1}{\sin x}$$

Combine the two expressions into a single fraction.

$$= \frac{\sin x + 1}{\sin x}$$

▶[SECTION 5.1] SUMMARY

In this section, we combined the basic trigonometric identities (reciprocal, quotient, Pythagorean, and even–odd) with algebraic techniques to simplify trigonometric expressions and verify more complex trigonometric identities. Two steps that are often used in both simplifying trigonometric expressions and verifying trigonometric identities are (1) writing all of the trigonometric functions in terms of the sine and cosine functions, and (2) combining sums or differences of fractions (quotients) into a single fraction (quotient). When verifying trigonometric identities, we work with the more complicated side (keeping the other side in mind as our goal).

[SECTION 5.1] EXERCISES

• **SKILLS**

In Exercises 1–20, simplify each of the following trigonometric expressions.

1. $\sin x \csc x$

2. $\tan x \cot x$

3. $\sec(-x)\cot x$

4. $\tan(-x)\cos(-x)$

5. $\sin^2 x(\cot^2 x + 1)$

6. $\cos^2 x(\tan^2 x + 1)$

7. $(\sin x - \cos x)(\sin x + \cos x)$

8. $(\sin x + \cos x)^2$

9. $\dfrac{\csc x}{\cot x}$

10. $\dfrac{\sec x}{\tan x}$

11. $\dfrac{1 - \cos^4 x}{1 + \cos^2 x}$

12. $\dfrac{1 - \sin^4 x}{1 + \sin^2 x}$

13. $1 - \dfrac{\sin^2 x}{1 - \cos x}$

14. $1 - \dfrac{\cos^2 x}{1 + \sin x}$

15. $\dfrac{\tan x - \cot x}{\tan x + \cot x} + 2\cos^2 x$

16. $\dfrac{\tan x - \cot x}{\tan x + \cot x} + \cos^2 x$

17. $\dfrac{\cot^2 x + 1}{\csc x} - \csc x$

18. $\dfrac{\cos^2 x}{1 + \cos x} + \dfrac{\cos^2 x}{1 - \cos x}$

19. $\dfrac{1 + \sin(-x)}{1 + \cos(-x)} - \dfrac{1 - \sin(-x)}{1 - \cos(-x)}$

20. $\dfrac{\cot^2 x}{\csc x} - \csc x$

In Exercises 21–46, verify each of the trigonometric identities.

21. $(\sin x + \cos x)^2 + (\sin x - \cos x)^2 = 2$

22. $(1 - \sin x)(1 + \sin x) = \cos^2 x$

23. $(\csc x + 1)(\csc x - 1) = \cot^2 x$

24. $(\sec x + 1)(\sec x - 1) = \tan^2 x$

25. $\tan x + \cot x = \csc x \sec x$

26. $\csc x - \sin x = \cot x \cos x$

27. $\dfrac{2 - \sin^2 x}{\cos x} = \sec x + \cos x$

28. $\dfrac{2 - \cos^2 x}{\sin x} = \csc x + \sin x$

29. $\dfrac{1}{\csc^2 x} + \dfrac{1}{\sec^2 x} = 1$

30. $\dfrac{1}{\cot^2 x} - \dfrac{1}{\tan^2 x} = \sec^2 x - \csc^2 x$

31. $\dfrac{1}{1 - \sin x} + \dfrac{1}{1 + \sin x} = 2\sec^2 x$

32. $\dfrac{1}{1 - \cos x} + \dfrac{1}{1 + \cos x} = 2\csc^2 x$

33. $\dfrac{\sin^2 x}{1 - \cos x} = 1 + \cos x$

34. $\dfrac{\cos^2 x}{1 - \sin x} = 1 + \sin x$

35. $\sec x + \tan x = \dfrac{1}{\sec x - \tan x}$

36. $\csc x + \cot x = \dfrac{1}{\csc x - \cot x}$

37. $\dfrac{\csc x - \tan x}{\sec x + \cot x} = \dfrac{\cos x - \sin^2 x}{\sin x + \cos^2 x}$

38. $\dfrac{\sec x + \tan x}{\csc x + 1} = \tan x$

39. $\dfrac{\cos^2 x + 1 + \sin x}{\cos^2 x + 3} = \dfrac{1 + \sin x}{2 + \sin x}$

40. $\dfrac{\sin x + 1 - \cos^2 x}{\cos^2 x} = \dfrac{\sin x}{1 - \sin x}$

41. $\sec x(\tan x + \cot x) = \dfrac{\csc x}{\cos^2 x}$

42. $\tan x(\csc x - \sin x) = \cos x$

43. $\dfrac{1 + \cos^2(-x)}{1 - \csc^2(-x)} = -\sin^2 x - \tan^2 x$

44. $\cos x(1 + \cos x) + \dfrac{\sin x}{\csc x} = 1 + \cos x$

45. $\sin x\left(\tan x + \dfrac{1}{\tan x}\right) = \sec x$

46. $\dfrac{1 + \sin x}{\cos x} = \dfrac{\cos x}{1 - \sin x}$

In Exercises 47–58, determine whether each equation is an identity, a conditional equation, or a contradiction.

47. $\cos^2 x(\tan x - \sec x)(\tan x + \sec x) = 1$

48. $\cos^2 x(\tan x - \sec x)(\tan x + \sec x) = \sin^2 x - 1$

49. $\dfrac{\csc x \cot x}{\sec x \tan x} = \cot^3 x$

50. $\sin x \cos x = 0$

51. $\sin x + \cos x = \sqrt{2}$

52. $\sin^2 x + \cos^2 x = 1$

53. $\tan^2 x - \sec^2 x = 1$

54. $\sec^2 x - \tan^2 x = 1$

55. $\sin x = \sqrt{1 - \cos^2 x}$

56. $\csc x = \sqrt{1 + \cot^2 x}$

57. $\sqrt{\sin^2 x + \cos^2 x} = 1$

58. $\sqrt{\sin^2 x + \cos^2 x} = \sin x + \cos x$

• APPLICATIONS

For Exercises 59 and 60, refer to the following:

In calculus, when integrating expressions such as $\sqrt{a^2 - x^2}$, $\sqrt{a^2 + x^2}$, and $\sqrt{x^2 - a^2}$, trigonometric functions are used as "dummy" functions to eliminate the radical. Once the integration is performed, the trigonometric function is "un-substituted." The following trigonometric substitutions (and corresponding trigonometric identities) are used to simplify these types of expressions.

EXPRESSION	SUBSTITUTION	TRIGONOMETRIC IDENTITY
$\sqrt{a^2 - x^2}$	$x = a\sin\theta \quad -\dfrac{\pi}{2} \le \theta \le \dfrac{\pi}{2}$	$1 - \sin^2\theta = \cos^2\theta$
$\sqrt{a^2 + x^2}$	$x = a\tan\theta \quad -\dfrac{\pi}{2} \le \theta \le \dfrac{\pi}{2}$	$1 + \tan^2\theta = \sec^2\theta$
$\sqrt{x^2 - a^2}$	$x = a\sec\theta \quad 0 \le \theta < \dfrac{\pi}{2} \text{ or } \pi \le \theta < \dfrac{3\pi}{2}$	$\sec^2\theta - 1 = \tan^2\theta$

When simplifying, it is important to remember that

$$|x| = \begin{cases} x & \text{if } x > 0 \\ -x & \text{if } x < 0 \end{cases}$$

59. Calculus (Trigonometric Substitution). Start with the expression $\sqrt{a^2 - x^2}$ and let $x = a\sin\theta$. Assuming $-\dfrac{\pi}{2} \leq \theta \leq \dfrac{\pi}{2}$, simplify the original expression so that it contains no radicals.

60. Calculus (Trigonometric Substitution). Start with the expression $\sqrt{a^2 + x^2}$ and let $x = a\tan\theta$. Assuming $-\dfrac{\pi}{2} < \theta < \dfrac{\pi}{2}$, simplify the original expression so that it contains no radicals.

61. Harmonic Motion. A weight is tied to the end of a spring and then set into motion. The displacement of the weight from equilibrium is given by the equation

$$y = -3\left[\dfrac{\sin(2t)\cos(2t) + 1}{\sin(2t) + \sec(2t)}\right], \text{ where } t \text{ is time in seconds.}$$

Simplify the equation and then describe how long it takes for the weight to go from its minimum to its maximum displacement.

62. Harmonic Motion. A weight is tied to the end of a spring and then set into motion. The displacement of the weight from equilibrium is given by the equation

$$y = 4\left[\dfrac{\sin\left(2t - \dfrac{\pi}{2}\right) + 1}{1 + \csc\left(2t - \dfrac{\pi}{2}\right)}\right], \text{ where } t \text{ is time in seconds.}$$

Simplify the equation and then describe the maximum displacement with respect to the equilibrium position, and how long it takes for the weight to first achieve its maximum starting at $t = 0$.

• **CATCH THE MISTAKE**

In Exercises 63–66, explain the mistake that is made.

63. Verify the identity $\dfrac{\cos x}{1 - \tan x} + \dfrac{\sin x}{1 - \cot x} = \sin x + \cos x$.

Solution:

Start with the left side of the equation.	$\dfrac{\cos x}{1 - \tan x} + \dfrac{\sin x}{1 - \cot x}$
Write the tangent and cotangent functions in terms of sines and cosines.	$\dfrac{\cos x}{1 - \dfrac{\sin x}{\cos x}} + \dfrac{\sin x}{1 - \dfrac{\cos x}{\sin x}}$
Cancel the common cosine in the first term and sine in the second term.	$\dfrac{1}{1 - \sin x} + \dfrac{1}{1 - \cos x}$

This is incorrect. What mistake was made?

64. Verify the identity $\dfrac{\cos^3 x \sec x}{1 - \sin x} = 1 + \sin x$.

Solution:

Start with the left side of the equation.	$\dfrac{\cos^3 x \sec x}{1 - \sin x}$
Rewrite secant in terms of sine.	$\dfrac{\cos^3 x \dfrac{1}{\sin x}}{1 - \sin x}$
Simplify.	$\dfrac{\cos^3 x}{\sin x - \sin^2 x}$

This is incorrect. What mistake was made?

65. Determine whether the equation is a conditional equation or an identity: $\dfrac{\tan x}{\cot x} = 1$.

Solution:

Start with the left side of the equation.	$\dfrac{\tan x}{\cot x}$
Rewrite the tangent and cotangent functions in terms of sines and cosines.	$= \dfrac{\dfrac{\sin x}{\cos x}}{\dfrac{\cos x}{\sin x}}$
Simplify.	$= \dfrac{\sin^2 x}{\cos^2 x} = \tan^2 x$
Let $x = \dfrac{\pi}{4}$. *Note*: $\tan\left(\dfrac{\pi}{4}\right) = 1$.	$= 1$

Since $\dfrac{\tan x}{\cot x} = 1$, this equation is an identity.
This is incorrect. What mistake was made?

66. Determine whether the equation is a conditional equation or an identity: $|\sin x| - \cos x = 1$.

Solution:

Start with the left side of the equation.	$|\sin x| - \cos x$
Let $x = \dfrac{(2n + 1)\pi}{2}$, where n is an integer.	$\left|\sin\left(\dfrac{(2n + 1)\pi}{2}\right)\right| - \cos\left(\dfrac{(2n + 1)\pi}{2}\right)$
Simplify.	$|\pm 1| - 0 = 1$

Since $|\sin x| - \cos x = 1$, this is an identity.
This is incorrect. What mistake was made?

• **CONCEPTUAL**

In Exercises 67–70, determine whether each statement is true or false.

67. If an equation is true for some values (but not all values), then it is still an identity.

68. If an equation has an infinite number of solutions, then it is an identity.

69. The following is an identity true for all values in the domain of the functions: $\tan^2 x - \sec^2 x = 1$.

70. The following is an identity true for all values in the domain of the functions: $\csc^2 x - \cot^2 x = 1$.

71. In what quadrants is the equation $\cos\theta = \sqrt{1 - \sin^2\theta}$ true?

72. In what quadrants is the equation $-\cos\theta = \sqrt{1 - \sin^2\theta}$ true?

73. Do you think $\sin(A + B) = \sin A + \sin B$? Why?

74. Do you think $\cos\left(\frac{1}{2}A\right) = \frac{1}{2}\cos A$? Why?

• **CHALLENGE**

75. Simplify $(a\sin x + b\cos x)^2 + (b\sin x - a\cos x)^2$.

76. Simplify $\dfrac{1 + \cot^3 x}{1 + \cot x} + \cot x$.

77. Verify the trigonometric identity $\dfrac{1 - \sin^4 x}{2\cos^2 x - \cos^4 x} = 1$.

78. Verify the trigonometric identity $\dfrac{1 - 2\sin x - \sin^2 x + 2\sin^3 x}{\cos^2 x} = 1 - 2\sin x$.

79. Simplify the expression $\dfrac{a + 2c - d}{2 - b}$ if $a = \sin x$, $b = \cos x$, $c = \cot x$, and $d = \csc x$.

80. Simplify the expression $\dfrac{a + b^2}{ab}$ if $a = \sin x$ and $b = \cos x$.

• **TECHNOLOGY**

In the next section, you will learn the sum and difference identities. In Exercises 81–84, we illustrate these identities with graphing calculators.

81. Determine the correct sign $(+ \text{ or } -)$ for
$\cos(A + B) = \cos A \cos B \underset{?}{\pm} \sin A \sin B$ by graphing

$Y_1 = \cos(A + B)$, $Y_2 = \cos A \cos B + \sin A \sin B$, and $Y_2 = \cos A \cos B - \sin A \sin B$ in the same viewing rectangle for chosen values for A and B, with $A \neq B$.

82. Determine the correct sign $(+ \text{ or } -)$ for
$\cos(A - B) = \cos A \cos B \underset{?}{\pm} \sin A \sin B$ by graphing

$Y_1 = \cos(A - B)$, $Y_2 = \cos A \cos B + \sin A \sin B$, and $Y_2 = \cos A \cos B - \sin A \sin B$ in the same viewing rectangle for chosen values for A and B with $A \neq B$.

83. Determine the correct sign $(+ \text{ or } -)$ for
$\sin(A + B) = \sin A \cos B \underset{?}{\pm} \cos A \sin B$ by graphing

$Y_1 = \sin(A + B)$, $Y_2 = \sin A \cos B + \cos A \sin B$, and $Y_2 = \sin A \cos B - \cos A \sin B$ in the same viewing rectangle for chosen values for A and B, with $A \neq B$.

84. Determine the correct sign $(+ \text{ or } -)$ for
$\sin(A - B) = \sin A \cos B \underset{?}{\pm} \cos A \sin B$ by graphing

$Y_1 = \sin(A - B)$, $Y_2 = \sin A \cos B + \cos A \sin B$, and $Y_2 = \sin A \cos B - \cos A \sin B$ in the same viewing rectangle for chosen values for A and B, with $A \neq B$.

5.2 SUM AND DIFFERENCE IDENTITIES

SKILLS OBJECTIVES

- Find exact values for the cosine function using sum and difference identities.
- Find exact values for the sine function using sum and difference identities.
- Find exact values for the tangent function using the sum or difference identities for the tangent function.

CONCEPTUAL OBJECTIVES

- Understand that some sums or differences of trigonometric functions can be written as a single cosine expression.
- Understand that some sums or differences of trigonometric functions can be written as a single sine expression.
- Understand that the sum and difference identities for sine and cosine functions are used in the derivation of the sum and difference identities for the tangent function.

In this section, we will consider trigonometric functions with arguments that are sums and differences. In general, $f(A + B) \neq f(A) + f(B)$. First, it is important to note that function notation is not distributive:

$$\cos(A + B) \neq \cos A + \cos B$$

This principle is easy to prove. Let $A = \pi$ and $B = 0$; then

$$\cos(A + B) = \cos(\pi + 0) = \cos \pi = -1$$
$$\cos A + \cos B = \cos \pi + \cos 0 = -1 + 1 = 0$$

Since $-1 \neq 0$, we know that $\cos(A + B) \neq \cos A + \cos B$.

In this section, we will derive some new and important identities (sum and difference identities for the cosine, sine, and tangent functions) and revisit cofunction identities. We begin with the familiar distance formula, from which we can derive the sum and difference identities for the cosine function. From there we can derive the sum and difference formulas for the sine and tangent functions.

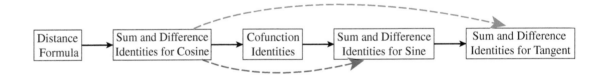

Before we start deriving and working with trigonometric sum and difference identities, let us first discuss why these are important. Sum and difference identities (and later product-to-sum and sum-to-product identities) are important because they allow the calculation of trigonometric function values in functional (analytic) form and often lead to evaluating expressions *exactly* (as opposed to approximating them with calculators). The identities developed in this chapter are useful in such applications as musical sound where they allow the determination of the "beat" frequency. In calculus, these identities will simplify the integration and differentiation processes.

5.2.1 SKILL

Find exact values for the cosine function using sum and difference identities.

5.2.1 CONCEPTUAL

Understand that sums or differences of trigonometric functions can be written as a single cosine expression.

5.2.1 Sum and Difference Identities for the Cosine Function

Derivation of the Sum and Difference Identities for the Cosine Function

Recall from Section 3.4 that the unit circle approach for defining trigonometric functions gave the relationship between the coordinates along the unit circle and the sine and cosine functions. Specifically, the x-coordinate corresponds to the value of the cosine function, and the y-coordinate corresponds to the value of the sine function for a given angle θ and the point (x, y) where the terminal side of the angle θ intersects the unit circle.

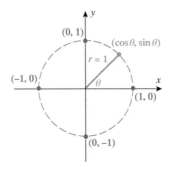

Let us now draw the unit circle with two angles, α and β, realizing that the two terminal sides of these angles form a third angle, $\alpha - \beta$.

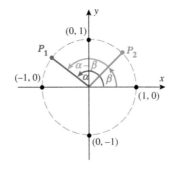

If we label the points $P_1 = (\cos\alpha, \sin\alpha)$ and $P_2 = (\cos\beta, \sin\beta)$, we can then draw a **segment P_1P_2** connecting points P_1 and P_2.

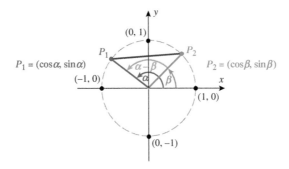

If we rotate the angle clockwise so that the central angle $\alpha - \beta$ is in standard position, then the two points where the initial and terminal sides intersect the unit circle are $P_4 = (1, 0)$ and $P_3 = (\cos(\alpha - \beta), \sin(\alpha - \beta))$, respectively.

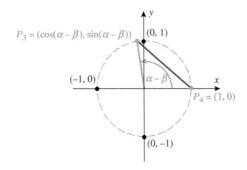

The distance from P_1 to P_2 is equal to the length of the **segment**. Similarly, the distance from P_3 to P_4 is equal to the length of the **segment**. Since the lengths of the **segments** are equal, we say that the distances are equal.

WORDS	MATH
Set the distances (segment lengths) equal by applying the distance formulas.	$\sqrt{(x_2 - x_1)^2 + (y_2 - y_1)^2} = \sqrt{(x_4 - x_3)^2 + (y_4 - y_3)^2}$

Substitute $(x_1, y_1) = (\cos\alpha, \sin\alpha)$ and $(x_2, y_2) = (\cos\beta, \sin\beta)$ into the left side of the equation and $(x_3, y_3) = (\cos(\alpha - \beta), \sin(\alpha - \beta))$ and $(x_4, y_4) = (1, 0)$ into the right side of the equation.

$$\sqrt{(\cos\beta - \cos\alpha)^2 + (\sin\beta - \sin\alpha)^2} = \sqrt{(1 - \cos(\alpha - \beta))^2 + (0 - \sin(\alpha - \beta))^2}$$

WORDS	MATH
Square both sides of the equation.	$(\cos\beta - \cos\alpha)^2 + (\sin\beta - \sin\alpha)^2 = (1 - \cos(\alpha - \beta))^2 + (0 - \sin(\alpha - \beta))^2$
Eliminate the brackets.	$\cos^2\beta - 2\cos\alpha\cos\beta + \cos^2\alpha + \sin^2\beta - 2\sin\alpha\sin\beta + \sin^2\alpha$ $= 1 - 2\cos(\alpha - \beta) + \cos^2(\alpha - \beta) + \sin^2(\alpha - \beta)$
Regroup terms on each side and use the Pythagorean identity.	$\underbrace{\cos^2\alpha + \sin^2\alpha}_{1} - 2\cos\alpha\cos\beta - 2\sin\alpha\sin\beta + \underbrace{\cos^2\beta + \sin^2\beta}_{1}$ $= 1 - 2\cos(\alpha - \beta) + \underbrace{\cos^2(\alpha - \beta) + \sin^2(\alpha - \beta)}_{1}$
Simplify.	$2 - 2\cos\alpha\cos\beta - 2\sin\alpha\sin\beta = 2 - 2\cos(\alpha - \beta)$
Subtract 2 from both sides.	$-2\cos\alpha\cos\beta - 2\sin\alpha\sin\beta = -2\cos(\alpha - \beta)$
Divide by -2.	$\cos\alpha\cos\beta + \sin\alpha\sin\beta = \cos(\alpha - \beta)$
We call the resulting identity the **difference identity for cosine**.	$\boxed{\cos(\alpha - \beta) = \cos\alpha\cos\beta + \sin\alpha\sin\beta}$

We can now derive the sum identity for the cosine function from the difference identity for the cosine function and the properties of even and odd functions.

WORDS	MATH
Replace β with $-\beta$ in the difference identity.	$\cos(\alpha - (-\beta)) = \cos\alpha\cos(-\beta) + \sin\alpha\sin(-\beta)$
Simplify the left side and use properties of even and odd functions on the right side.	$\cos(\alpha + \beta) = \cos\alpha(\cos\beta) + \sin\alpha(-\sin\beta)$
We call the resulting identity the **sum identity for cosine**.	$\boxed{\cos(\alpha + \beta) = \cos\alpha\cos\beta - \sin\alpha\sin\beta}$

SUM AND DIFFERENCE IDENTITIES FOR THE COSINE FUNCTION

$$\text{Sum:} \quad \cos(A + B) = \cos A \cos B - \sin A \sin B$$
$$\text{Difference:} \quad \cos(A - B) = \cos A \cos B + \sin A \sin B$$

EXAMPLE 1   Finding Exact Values for the Cosine Function

Evaluate each of the following cosine expressions exactly:

a. $\cos 15°$ **b.** $\cos\left(\dfrac{7\pi}{12}\right)$

Solution (a):

Write 15° as a difference of known "special" angles.
$$\cos 15° = \cos(45° - 30°)$$

Write the difference identity for the cosine function.
$$\cos(A - B) = \cos A \cos B + \sin A \sin B$$

Substitute $A = 45°$ and $B = 30°$.
$$\cos 15° = \cos 45° \cos 30° + \sin 45° \sin 30°$$

Evaluate the expressions on the right exactly.
$$\cos 15° = \left(\frac{\sqrt{2}}{2}\right)\left(\frac{\sqrt{3}}{2}\right) + \left(\frac{\sqrt{2}}{2}\right)\left(\frac{1}{2}\right)$$

Simplify.
$$\boxed{\cos 15° = \frac{\sqrt{6} + \sqrt{2}}{4}}$$

Solution (b):

Write $\dfrac{7\pi}{12}$ as a sum of known "special" angles.
$$\cos\left(\frac{7\pi}{12}\right) = \cos\left(\frac{4\pi}{12} + \frac{3\pi}{12}\right)$$

Simplify.
$$\cos\left(\frac{7\pi}{12}\right) = \cos\left(\frac{\pi}{3} + \frac{\pi}{4}\right)$$

Write the sum identity for the cosine function.
$$\cos(A + B) = \cos A \cos B - \sin A \sin B$$

Substitute $A = \dfrac{\pi}{3}$ and $B = \dfrac{\pi}{4}$.
$$\cos\left(\frac{7\pi}{12}\right) = \cos\left(\frac{\pi}{3}\right)\cos\left(\frac{\pi}{4}\right) - \sin\left(\frac{\pi}{3}\right)\sin\left(\frac{\pi}{4}\right)$$

Evaluate the expressions on the right exactly.
$$\cos\left(\frac{7\pi}{12}\right) = \left(\frac{1}{2}\right)\left(\frac{\sqrt{2}}{2}\right) - \left(\frac{\sqrt{3}}{2}\right)\left(\frac{\sqrt{2}}{2}\right)$$

Simplify.
$$\boxed{\cos\left(\frac{7\pi}{12}\right) = \frac{\sqrt{2} - \sqrt{6}}{4}}$$

Note: $\dfrac{7\pi}{12}$ can also be represented as 105°.

YOUR TURN Use the sum or difference identities for the cosine function to evaluate each cosine expression exactly.

a. $\cos 75°$ **b.** $\cos\left(\dfrac{5\pi}{12}\right)$

ANSWER
a. $\dfrac{\sqrt{6} - \sqrt{2}}{4}$
b. $\dfrac{\sqrt{6} - \sqrt{2}}{4}$

Example 1 illustrates an important characteristic of the sum and difference identities: that we can now find the exact trigonometric function value of angles that are multiples of 15° $\left(\text{or equivalently } \dfrac{\pi}{12}\right)$.

EXAMPLE 2	Writing a Sum or Difference as a Single Cosine Expression

Use the sum or the difference identity for the cosine function to write the expressions as a single cosine expression.

a. $\sin(5x)\sin(2x) + \cos(5x)\cos(2x)$

b. $\cos x \cos(3x) - \sin x \sin(3x)$

Solution (a):

Because of the positive sign, this will be a cosine of a difference.

Reverse the expression and write the formula.
$$\cos A \cos B + \sin A \sin B = \cos(A - B)$$

Identify A and B.
$$A = 5x \quad \text{and} \quad B = 2x$$

Substitute $A = 5x$ and $B = 2x$ into the difference identity.
$$\cos(5x)\cos(2x) + \sin(5x)\sin(2x) = \cos(5x - 2x)$$

Simplify.
$$\boxed{\cos(5x)\cos(2x) + \sin(5x)\sin(2x) = \cos(3x)}$$

Notice that if we had selected $A = 2x$ and $B = 5x$ instead, the result would have been $\cos(-3x)$, but since the cosine function is an even function, this would have simplified to $\cos(3x)$.

Solution (b):

Because of the negative sign, this will be a cosine of a sum.

Reverse the expression and write the formula.
$$\cos A \cos B - \sin A \sin B = \cos(A + B)$$

Identify A and B.
$$A = x \quad \text{and} \quad B = 3x$$

Substitute $A = x$ and $B = 3x$ into the sum identity.
$$\cos x \cos(3x) - \sin x \sin(3x) = \cos(x + 3x)$$

Simplify.
$$\boxed{\cos x \cos(3x) - \sin x \sin(3x) = \cos(4x)}$$

▼

ANSWER

$\cos(3x)$

▼

YOUR TURN Write $\cos(4x)\cos(7x) + \sin(4x)\sin(7x)$ as a single cosine expression.

Cofunction Identities

In Section 1.3, we discussed cofunction relationships for acute angles. Recall that a trigonometric function value of an angle is equal to the corresponding cofunction value of its complementary angle. Now we use the sum and difference identities for the cosine function to develop the cofunction identities for any angle θ.

WORDS	MATH
Write the difference identity for the cosine function.	$\cos(A - B) = \cos A \cos B + \sin A \sin B$
Let $A = \dfrac{\pi}{2}$ and $B = \theta$.	$\cos\left(\dfrac{\pi}{2} - \theta\right) = \cos\left(\dfrac{\pi}{2}\right)\cos\theta + \sin\left(\dfrac{\pi}{2}\right)\sin\theta$
Evaluate known values for the sine and cosine functions.	$\cos\left(\dfrac{\pi}{2} - \theta\right) = 0 \cdot \cos\theta + 1 \cdot \sin\theta$
Simplify.	$\boxed{\cos\left(\dfrac{\pi}{2} - \theta\right) = \sin\theta}$

Similarly, to determine the other corresponding cofunction identity:

WORDS	MATH

Write the difference identity for the cosine function.

$$\cos(A - B) = \cos A \cos B + \sin A \sin B$$

Let $A = \dfrac{\pi}{2}$ and $B = \dfrac{\pi}{2} - \theta$.

$$\cos\left[\frac{\pi}{2} - \left(\frac{\pi}{2} - \theta\right)\right] = \cos\left(\frac{\pi}{2}\right)\cos\left(\frac{\pi}{2} - \theta\right) + \sin\left(\frac{\pi}{2}\right)\sin\left(\frac{\pi}{2} - \theta\right)$$

Evaluate the known values for the sine and cosine functions.

$$\cos\theta = 0 \cdot \cos\left(\frac{\pi}{2} - \theta\right) + 1 \cdot \sin\left(\frac{\pi}{2} - \theta\right)$$

Simplify.

$$\boxed{\cos\theta = \sin\left(\frac{\pi}{2} - \theta\right)}$$

COFUNCTION IDENTITIES FOR THE SINE AND COSINE FUNCTIONS

$$\cos\left(\frac{\pi}{2} - \theta\right) = \sin\theta \qquad \sin\left(\frac{\pi}{2} - \theta\right) = \cos\theta$$

5.2.2 SKILL

Find exact values for the sine function using sum and difference identities.

5.2.2 Sum and Difference Identities for the Sine Function

5.2.2 CONCEPTUAL

Understand that some sums or differences of trigonometric functions can be written as a single sine expression.

We can now use the cofunction identities for the sine and cosine functions together with the sum and difference identities for the cosine function to develop the sum and difference identities for the sine function.

WORDS	MATH

Start with the cofunction identity.

$$\sin\theta = \cos\left(\frac{\pi}{2} - \theta\right)$$

Let $\theta = A + B$.

$$\sin(A + B) = \cos\left[\frac{\pi}{2} - (A + B)\right]$$

Regroup the terms in the cosine expression.

$$\sin(A + B) = \cos\left[\left(\frac{\pi}{2} - A\right) - B\right]$$

Use the difference identity for the cosine function.

$$\sin(A + B) = \cos\left(\frac{\pi}{2} - A\right)\cos B + \sin\left(\frac{\pi}{2} - A\right)\sin B$$

Use the cofunction identities.

$$\sin(A + B) = \underbrace{\cos\left(\frac{\pi}{2} - A\right)}_{\sin A}\cos B + \underbrace{\sin\left(\frac{\pi}{2} - A\right)}_{\cos A}\sin B$$

Simplify.

$$\boxed{\sin(A + B) = \sin A \cos B + \cos A \sin B}$$

Now we can derive the difference identity for the sine function using the sum identity for the sine function and the properties of even and odd functions.

WORDS	MATH
Replace B with $-B$ in the sum identity.	$\sin[A + (-B)] = \sin A \cos(-B) + \cos A \sin(-B)$
Simplify using even and odd identities.	$\boxed{\sin(A - B) = \sin A \cos B - \cos A \sin B}$

SUM AND DIFFERENCE IDENTITIES FOR THE SINE FUNCTION

Sum: $\quad\sin(A + B) = \sin A \cos B + \cos A \sin B$
Difference: $\quad\sin(A - B) = \sin A \cos B - \cos A \sin B$

[CONCEPT CHECK]
TRUE OR FALSE
$\sin(A - B) = \sin(B - A)$

▼
ANSWER False

▶ **EXAMPLE 3** **Finding Exact Values for the Sine Function**

Use the sum or the difference identity for the sine function to evaluate each sine expression exactly.

a. $\sin 75°$ **b.** $\sin\left(\dfrac{5\pi}{12}\right)$

Solution (a):

Write $75°$ as a sum of known "special" angles.
$$\sin 75° = \sin(45° + 30°)$$

Write the sum identity for the sine function.
$$\sin(A + B) = \sin A \cos B + \cos A \sin B$$

Substitute $A = 45°$ and $B = 30°$.
$$\sin 75° = \sin 45° \cos 30° + \cos 45° \sin 30°$$

Evaluate the expressions on the right exactly.
$$\sin 75° = \left(\frac{\sqrt{2}}{2}\right)\left(\frac{\sqrt{3}}{2}\right) + \left(\frac{\sqrt{2}}{2}\right)\left(\frac{1}{2}\right)$$

Simplify.
$$\boxed{\sin 75° = \frac{\sqrt{6} + \sqrt{2}}{4}}$$

Solution (b):

Write $\dfrac{5\pi}{12}$ as a sum of known "special" angles.
$$\sin\left(\frac{5\pi}{12}\right) = \sin\left(\frac{2\pi}{12} + \frac{3\pi}{12}\right)$$

Simplify.
$$\sin\left(\frac{5\pi}{12}\right) = \sin\left(\frac{\pi}{6} + \frac{\pi}{4}\right)$$

Write the sum identity for the sine function.
$$\sin(A + B) = \sin A \cos B + \cos A \sin B$$

Substitute $A = \dfrac{\pi}{6}$ and $B = \dfrac{\pi}{4}$.
$$\sin\left(\frac{5\pi}{12}\right) = \sin\left(\frac{\pi}{6}\right)\cos\left(\frac{\pi}{4}\right) + \cos\left(\frac{\pi}{6}\right)\sin\left(\frac{\pi}{4}\right)$$

Evaluate the expressions on the right exactly.
$$\sin\left(\frac{5\pi}{12}\right) = \left(\frac{1}{2}\right)\left(\frac{\sqrt{2}}{2}\right) + \left(\frac{\sqrt{3}}{2}\right)\left(\frac{\sqrt{2}}{2}\right)$$

Simplify.
$$\boxed{\sin\left(\frac{5\pi}{12}\right) = \frac{\sqrt{2} + \sqrt{6}}{4}}$$

Note: $\dfrac{5\pi}{12}$ can also be represented as $75°$.

▼
ANSWER

a. $\dfrac{\sqrt{6} - \sqrt{2}}{4}$

b. $\dfrac{\sqrt{6} + \sqrt{2}}{4}$

▼

YOUR TURN Use the sum or the difference identity for the sine function to evaluate the sine expressions exactly.

a. $\sin 15°$ **b.** $\sin\left(\dfrac{7\pi}{12}\right)$

We see in Example 3 that the sum and difference identities allow us to calculate exact values for trigonometric functions of angles that are multiples of 15° $\left(\text{or equivalently } \dfrac{\pi}{12}\right)$, as we saw with the cosine function.

EXAMPLE 4 **Writing a Sum or Difference as a Single Sine Expression**

Graph $y = 3\sin x\,\cos(3x) + 3\cos x\,\sin(3x)$.

Solution:

Use the sum identity for the sine function to write the expression as a single sine expression.

Factor out the common 3. $y = 3[\sin x\,\cos(3x) + \cos x\,\sin(3x)]$

Write the sum identity for the sine function. $\sin A\,\cos B + \cos A\,\sin B = \sin(A + B)$

Identify A and B. $A = x$ and $B = 3x$

Substitute $A = x$ and $B = 3x$ into the sum identity.
$$y = 3\,[\underbrace{\sin x\cos(3x) + \cos x\sin(3x)}_{\sin(x + 3x)\,=\,\sin(4x)}]$$

Graph $y = 3\sin(4x)$.

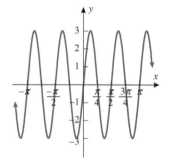

5.2.3 SKILL

Find exact values for the tangent function using the sum or difference identities for the tangent function.

5.2.3 CONCEPTUAL

Understand that the sum and difference identities for sine and cosine functions are used in the derivation of the sum and difference identities for the tangent function.

5.2.3 Sum and Difference Identities for the Tangent Function

We now develop the sum and difference identities for the tangent function.

WORDS	MATH
Start with the quotient identity.	$\tan x = \dfrac{\sin x}{\cos x}$
Let $x = A + B$.	$\tan(A + B) = \dfrac{\sin(A + B)}{\cos(A + B)}$
Use the sum identities for the sine and cosine functions.	$\tan(A + B) = \dfrac{\sin A\,\cos B + \cos A\,\sin B}{\cos A\,\cos B - \sin A\,\sin B}$
To be able to write the right-hand side in terms of tangents, we multiply the numerator and the denominator by $\dfrac{1}{\cos A\cos B}$.	$\tan(A + B) = \dfrac{\dfrac{\sin A\,\cos B + \cos A\,\sin B}{\cos A\,\cos B}}{\dfrac{\cos A\,\cos B - \sin A\,\sin B}{\cos A\,\cos B}}$

Divide out (cancel) common factors.

$$= \dfrac{\dfrac{\sin A \cos B}{\cos A \cos B} + \dfrac{\cos A \sin B}{\cos A \cos B}}{\dfrac{\cos A \cos B}{\cos A \cos B} - \dfrac{\sin A \sin B}{\cos A \cos B}}$$

Simplify.

$$\tan(A + B) = \dfrac{\left(\dfrac{\sin A}{\cos A}\right) + \left(\dfrac{\sin B}{\cos B}\right)}{1 - \left(\dfrac{\sin A}{\cos A}\right)\left(\dfrac{\sin B}{\cos B}\right)}$$

Write the expressions inside the parentheses in terms of the tangent function.

$$\boxed{\tan(A + B) = \dfrac{\tan A + \tan B}{1 - \tan A \tan B}}$$

Replace B with $-B$.

$$\tan[A + (-B)] = \tan(A - B) = \dfrac{\tan A + \tan(-B)}{1 - \tan A \tan(-B)}$$

Since the tangent function is an odd function, $\tan(-B) = -\tan B$.

$$\boxed{\tan(A - B) = \dfrac{\tan A - \tan B}{1 + \tan A \tan B}}$$

SUM AND DIFFERENCE IDENTITIES FOR THE TANGENT FUNCTION

Sum:
$$\tan(A + B) = \dfrac{\tan A + \tan B}{1 - \tan A \tan B}$$

Difference:
$$\tan(A - B) = \dfrac{\tan A - \tan B}{1 + \tan A \tan B}$$

[CONCEPT CHECK]

TRUE OR FALSE
$\tan(A - B) = -\tan(B - A)$

ANSWER True

▶ **EXAMPLE 5** **Finding Exact Values for the Tangent Function**

Find the exact value of $\tan(\alpha + \beta)$ if $\sin\alpha = -\frac{1}{3}$ and $\cos\beta = -\frac{1}{4}$, given that the terminal side of α lies in quadrant III and the terminal side of β lies in quadrant II.

Solution:

STEP 1 Write the sum identity for the tangent function.

$$\tan(\alpha + \beta) = \dfrac{\tan\alpha + \tan\beta}{1 - \tan\alpha \tan\beta}$$

STEP 2 Find $\tan\alpha$.

The terminal side of α lies in quadrant III.

$$\sin\alpha = \dfrac{y}{r} = -\dfrac{1}{3}$$

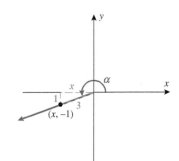

Solve for x. $(x^2 + y^2 = r^2)$ $x^2 + 1^2 = 3^2$

$$x = \pm\sqrt{8} = \pm 2\sqrt{2}$$

Take the negative sign since
we are in quadrant III. $x = -2\sqrt{2}$

Find $\tan\alpha$. $\tan\alpha = \dfrac{y}{x} = \dfrac{-1}{-2\sqrt{2}} = \dfrac{1}{2\sqrt{2}} \cdot \dfrac{\sqrt{2}}{\sqrt{2}} = \dfrac{\sqrt{2}}{4}$

STEP 3 Find $\tan\beta$.

The terminal side of β lies
in quadrant II.

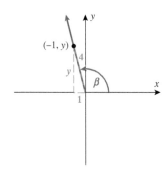

Solve for y. $(-1)^2 + y^2 = 4^2$
$(x^2 + y^2 = r^2)$ $y = \pm\sqrt{15}$

Take the positive sign
since we are in quadrant II. $y = \sqrt{15}$

Find $\tan\beta$. $\tan\beta = \dfrac{y}{x} = \dfrac{\sqrt{15}}{-1} = -\sqrt{15}$

STEP 4 Substitute $\tan\alpha = \dfrac{\sqrt{2}}{4}$
and $\tan\beta = -\sqrt{15}$
into the sum identity
for the tangent function. $\tan(\alpha + \beta) = \dfrac{\dfrac{\sqrt{2}}{4} - \sqrt{15}}{1 - \left(\dfrac{\sqrt{2}}{4}\right)\left(-\sqrt{15}\right)}$

Multiply the
numerator and
the denominator
by 4. $\tan(\alpha + \beta) = \dfrac{4\left(\dfrac{\sqrt{2}}{4} - \sqrt{15}\right)}{4\left(1 + \dfrac{\sqrt{30}}{4}\right)} = \dfrac{\sqrt{2} - 4\sqrt{15}}{4 + \sqrt{30}}$

The expression $\boxed{\tan(\alpha + \beta) = \dfrac{\sqrt{2} - 4\sqrt{15}}{4 + \sqrt{30}}}$ can be simplified further if we

rationalize the denominator.

It is important to note in Example 5 that right triangles have been superimposed in the Cartesian plane. The coordinate pair (x, y) can have positive or negative values, but the radius r is always positive. When right triangles are superimposed with one vertex at the point (x, y) and another vertex at the origin, it is important to understand that triangles have positive side *lengths*.

▶[SECTION 5.2] SUMMARY

In this section, we derived the sum and difference identities for the cosine function using the distance formula. We then used these identities to derive the cofunction identities. The cofunction identities and sum and difference identities for the cosine function were used to derive the sum and difference identities for the sine function. The sine and cosine sum and difference identities were combined to determine the tangent sum and difference identities. The sum and difference identities enabled us to evaluate trigonometric expressions exactly for arguments that are integer multiples of $15°$ $\left(\text{i.e., } \dfrac{\pi}{12}\right)$.

$$\cos(A + B) = \cos A \cos B - \sin A \sin B$$
$$\cos(A - B) = \cos A \cos B + \sin A \sin B$$
$$\sin(A + B) = \sin A \cos B + \cos A \sin B$$
$$\sin(A - B) = \sin A \cos B - \cos A \sin B$$
$$\tan(A + B) = \frac{\tan A + \tan B}{1 - \tan A \tan B}$$
$$\tan(A - B) = \frac{\tan A - \tan B}{1 + \tan A \tan B}$$

[SECTION 5.2] EXERCISES

• **SKILLS**

In Exercises 1–20, find exact values for each trigonometric expression.

1. $\sin\left(\dfrac{\pi}{12}\right)$
2. $\cos\left(\dfrac{\pi}{12}\right)$
3. $\cos\left(-\dfrac{5\pi}{12}\right)$
4. $\sin\left(-\dfrac{5\pi}{12}\right)$
5. $\sin\left(\dfrac{7\pi}{12}\right)$
6. $\cos\left(\dfrac{7\pi}{12}\right)$

7. $\tan\left(-\dfrac{\pi}{12}\right)$
8. $\tan\left(\dfrac{13\pi}{12}\right)$
9. $\sin 105°$
10. $\cos 195°$
11. $\tan(-105°)$
12. $\tan 165°$

13. $\cot\left(\dfrac{\pi}{12}\right)$
14. $\cot\left(-\dfrac{5\pi}{12}\right)$
15. $\sec\left(-\dfrac{11\pi}{12}\right)$
16. $\sec\left(-\dfrac{13\pi}{12}\right)$
17. $\csc\left(\dfrac{17\pi}{12}\right)$
18. $\csc\left(\dfrac{23\pi}{12}\right)$

19. $\sec(-195°)$
20. $\csc 285°$

In Exercises 21–34, write each expression as a single trigonometric function.

21. $\sin(2x)\sin(3x) + \cos(2x)\cos(3x)$
22. $\sin x \sin(2x) - \cos x \cos(2x)$
23. $\sin x \cos(2x) - \cos x \sin(2x)$
24. $\sin(2x)\cos(3x) + \cos(2x)\sin(3x)$
25. $(\sin A - \sin B)^2 + (\cos A - \cos B)^2 - 2$
26. $(\sin A + \sin B)^2 + (\cos A + \cos B)^2 - 2$
27. $\cos\left(\tfrac{1}{2}x\right)\sin\left(\tfrac{5}{2}x\right) + \cos\left(\tfrac{5}{2}x\right)\sin\left(\tfrac{1}{2}x\right)$
28. $\cos 50° \cos x + \sin 50° \sin x$
29. $2 - (\sin A + \cos B)^2 - (\cos A + \sin B)^2$
30. $2 - (\sin A - \cos B)^2 - (\cos A + \sin B)^2$

31. $\dfrac{\tan 49° - \tan 23°}{1 + \tan 49° \tan 23°}$
32. $\dfrac{\tan 49° + \tan 23°}{1 - \tan 49° \tan 23°}$
33. $\dfrac{\tan(\pi/8) - \tan(3\pi/8)}{1 + \tan(\pi/8)\tan(3\pi/8)}$
34. $\dfrac{\tan(7\pi/12) + \tan(\pi/6)}{1 - \tan(7\pi/12)\tan(\pi/6)}$

In Exercises 35–40, find the exact value of the indicated expression using the given information and identities.

35. Find the exact value of $\cos(\alpha + \beta)$ if $\cos\alpha = -\tfrac{1}{3}$ and $\cos\beta = -\tfrac{1}{4}$, if the terminal side of α lies in quadrant III and the terminal side of β lies in quadrant II.

36. Find the exact value of $\cos(\alpha - \beta)$ if $\cos\alpha = \tfrac{1}{3}$ and $\cos\beta = -\tfrac{1}{4}$, if the terminal side of α lies in quadrant IV and the terminal side of β lies in quadrant II.

37. Find the exact value of $\sin(\alpha - \beta)$ if $\sin\alpha = -\tfrac{3}{5}$ and $\sin\beta = \tfrac{1}{5}$, if the terminal side of α lies in quadrant III and the terminal side of β lies in quadrant I.

38. Find the exact value of $\sin(\alpha + \beta)$ if $\sin\alpha = -\tfrac{3}{5}$ and $\sin\beta = \tfrac{1}{5}$, if the terminal side of α lies in quadrant III and the terminal side of β lies in quadrant II.

39. Find the exact value of $\tan(\alpha + \beta)$ if $\sin\alpha = -\tfrac{3}{5}$ and $\cos\beta = -\tfrac{1}{4}$, if the terminal side of α lies in quadrant III and the terminal side of β lies in quadrant II.

40. Find the exact value of $\tan(\alpha - \beta)$ if $\sin\alpha = -\tfrac{3}{5}$ and $\cos\beta = -\tfrac{1}{4}$, if the terminal side of α lies in quadrant III and the terminal side of β lies in quadrant II.

In Exercises 41–50, determine whether each equation is a conditional equation or an identity.

41. $\sin(A + B) + \sin(A - B) = 2\sin A \cos B$

42. $\cos(A + B) + \cos(A - B) = 2\cos A \cos B$

43. $\sin\left(x - \dfrac{\pi}{2}\right) = \cos\left(x + \dfrac{\pi}{2}\right)$

44. $\sin\left(x + \dfrac{\pi}{2}\right) = \cos\left(x + \dfrac{\pi}{2}\right)$

45. $\sin(2x) = 2\sin x \cos x$

46. $\cos(2x) = \cos^2 x - \sin^2 x$

47. $\sin(A + B) = \sin A + \sin B$

48. $\cos(A + B) = \cos A + \cos B$

49. $\tan(\pi + B) = \tan B$

50. $\tan(A - \pi) = \tan A$

In Exercises 51–56, graph each of the functions by first rewriting it as a sine, cosine, or tangent of a difference or sum.

51. $y = \cos\left(\dfrac{\pi}{3}\right)\sin x + \cos x \sin\left(\dfrac{\pi}{3}\right)$

52. $y = \cos\left(\dfrac{\pi}{3}\right)\sin x - \cos x \sin\left(\dfrac{\pi}{3}\right)$

53. $y = \sin x \sin\left(\dfrac{\pi}{4}\right) + \cos x \cos\left(\dfrac{\pi}{4}\right)$

54. $y = \sin x \sin\left(\dfrac{\pi}{4}\right) - \cos x \cos\left(\dfrac{\pi}{4}\right)$

55. $y = -\sin x \cos(3x) - \cos x \sin(3x)$

56. $y = \sin x \sin(3x) + \cos x \cos(3x)$

• **APPLICATIONS**

For Exercises 57 and 58, refer to the following:

The difference quotient, $\dfrac{f(x + h) - f(x)}{h}$ is used to approximate the rate of change of the function f and will be used frequently in calculus.

57. Difference Quotient. Show that the difference quotient for
$$f(x) = \sin x \text{ is } \cos x\left(\dfrac{\sin h}{h}\right) - \sin x\left(\dfrac{1 - \cos h}{h}\right).$$

58. Difference Quotient. Show that the difference quotient for
$$f(x) = \cos x \text{ is } -\sin x\left(\dfrac{\sin h}{h}\right) - \cos x\left(\dfrac{1 - \cos h}{h}\right).$$

For Exercises 59 and 60, refer to the following:

A nonvertical line makes an angle with the x-axis. In the figure, we see that the line L_1 makes an acute angle θ_1 with the x-axis. Similarly, the line L_2 makes an acute angle θ_2 with the x-axis. In Exercises 59 and 60, use the following:

$\tan\theta_1 = $ slope of $L_1 = m_1$

$\tan\theta_2 = $ slope of $L_2 = m_2$

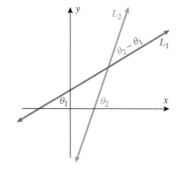

59. Relating Tangent and Scope. Show that
$$\tan(\theta_2 - \theta_1) = \dfrac{m_2 - m_1}{1 + m_1 m_2}.$$

60. Relating Tangent and Scope. Show that
$$\tan(\theta_1 - \theta_2) = \dfrac{m_1 - m_2}{1 + m_1 m_2}.$$

For Exercises 61 and 62, refer to the following:

An electric field E of a wave with constant amplitude A propagating a distance z is given by

$$E = A\cos(kz - ct)$$

where k is the propagation wave number, which is related to the wavelength λ by $k = \dfrac{2\pi}{\lambda}$, $c = 3.0 \times 10^8$ meters per second is the speed of light in a vacuum, and t is time in seconds.

61. Electromagnetic Wave Propagation. Use the cosine difference identity to express the electric field in terms of both sine and cosine functions. When the quotient of the propagation distance z and the wavelength λ are equal to an integer, what do you notice?

62. Electromagnetic Wave Propagation. Use the cosine difference identity to express the electric field in terms of both sine and cosine functions. When $t = 0$, what do you notice?

63. Functions. Consider a 15-foot ladder placed against a wall such that the distance from the top of the ladder to the floor is h feet and the angle between the floor and the ladder is θ.

   a. Write the height h as a function of angle θ.

   b. If the ladder is pushed toward the wall, increasing the angle θ by 10°, write a new function for the height as a function of $\theta + 10°$ and then express in terms of sines and cosines of θ and 10°.

64. Functions. Consider a 15-foot ladder placed against a wall such that the distance from the bottom of the ladder to the wall is x feet and the angle between the floor and the ladder is θ.

   a. Write the distance x as a function of angle θ.

   b. If the ladder is pushed toward the wall, increasing the angle θ by 10°, write a new function for the height as a function of $\theta + 10°$ and then express in terms of sines and cosines of θ and 10°.

65. Biology. By analyzing available empirical data, it has been determined that the body temperature of a species fluctuates according to the model

$$T(t) = 38 - 2.5 \cos\left[\frac{\pi}{6}(t - 3)\right], \qquad 0 \le t \le 24$$

where T represents temperature in degrees Celsius and t represents time (in hours) measured from 12:00 P.M. (noon). Use an identity to express $T(t)$ in terms of the sine function.

66. Health/Medicine. During the course of treatment of an illness, the concentration of a drug (in micrograms per milliliter) in the bloodstream fluctuates during the dosing period of 8 hours according to the model

$$C(t) = 15.4 - 4.7 \sin\left(\frac{\pi}{4}t + \frac{\pi}{2}\right), \qquad 0 \le t \le 8$$

Use an identity to express the concentration $C(t)$ in terms of the cosine function.

Note: This model does not apply to the first dose of the medication as there will be no medication in the bloodstream.

• CATCH THE MISTAKE

In Exercises 67 and 68, explain the mistake that is made.

67. Find the exact value of $\tan\left(\frac{5\pi}{12}\right)$.

Solution:

Write $\frac{5\pi}{12}$ as a sum. $\qquad \tan\left(\frac{5\pi}{12}\right) = \tan\left(\frac{\pi}{4} + \frac{\pi}{6}\right)$

Distribute. $\qquad\qquad\qquad = \tan\left(\frac{\pi}{4}\right) + \tan\left(\frac{\pi}{6}\right)$

Evaluate the tangent function for $\frac{\pi}{4}$ and $\frac{\pi}{6}$. $\qquad = 1 + \frac{\sqrt{3}}{3}$

This is incorrect. What mistake was made?

68. Find the exact value of $\tan\left(-\frac{7\pi}{6}\right)$.

Solution:

The tangent function is an even function. $\qquad \tan\left(-\frac{7\pi}{6}\right) = \tan\left(\frac{7\pi}{6}\right)$

Write $\frac{7\pi}{6}$ as a sum. $\qquad\qquad = \tan\left(\pi + \frac{\pi}{6}\right)$

Use the tangent sum identity, $\tan(A + B) = \frac{\tan A + \tan B}{1 - \tan A \tan B}$. $\qquad = \frac{\tan \pi + \tan\left(\frac{\pi}{6}\right)}{1 - \tan \pi \tan\left(\frac{\pi}{6}\right)}$

Evaluate the tangent functions on the right. $\qquad = \frac{0 + \frac{1}{\sqrt{3}}}{1 - 0}$

Simplify. $\qquad\qquad = \frac{\sqrt{3}}{3}$

This is incorrect. What mistake was made?

• CONCEPTUAL

In Exercises 69–72, determine whether each statement is true or false.

69. $\cos 15° = \cos 45° - \cos 30°$

70. $\sin\left(\frac{\pi}{2}\right) = \sin\left(\frac{\pi}{3}\right) + \sin\left(\frac{\pi}{6}\right)$

71. $\sin(90° + x) = \cos x$

72. $\cos(90° + x) = \sin x$

73. Simplify the expression $\cos(\alpha x)\cos(\beta x) + \sin(\alpha x)\sin(\beta x)$ to a single trigonometric function.

74. Simplify the expression $\cos(\alpha x)\sin(\beta x) + \sin(\alpha x)\cos(\beta x)$ to a single trigonometric function.

• **CHALLENGE**

75. Verify that $\sin(A + B + C) = \sin A \cos B \cos C + \cos A \sin B \cos C + \cos A \cos B \sin C - \sin A \sin B \sin C$.

76. Verify that $\cos(A + B + C) = \cos A \cos B \cos C - \sin A \sin B \cos C - \sin A \cos B \sin C - \cos A \sin B \sin C$.

77. Although, in general, the statement $\sin(A - B) = \sin A - \sin B$ is not true, it is true for some values. Determine some values of A and B that make this statement true.

78. Although, in general, the statement $\sin(A + B) = \sin A + \sin B$ is not true, it is true for some values. Determine some values of A and B that make this statement true.

79. Verify the identity $\dfrac{(\sin x + \cos y)^2}{\sin x \cos y} - 1 + \cot x \tan y = \dfrac{\sin(x + y)}{\sin x \cos y} + \dfrac{\sin x}{\cos y} + \dfrac{\cos y}{\sin x}$.

80. Verify the identity $\tan \alpha = \cot \beta - \cos(\alpha + \beta) \sec \alpha \csc \beta$.

• **TECHNOLOGY**

81. In Exercise 57, you showed that the difference quotient for $f(x) = \sin x$ is $\cos x \left(\dfrac{\sin h}{h} \right) - \sin x \left(\dfrac{1 - \cos h}{h} \right)$.

Plot $Y_1 = \cos x \left(\dfrac{\sin h}{h} \right) - \sin x \left(\dfrac{1 - \cos h}{h} \right)$ for

a. $h = 1$ **b.** $h = 0.1$ **c.** $h = 0.01$

What function does the difference quotient for $f(x) = \sin x$ resemble when h approaches zero?

82. In Exercise 58, you showed that the difference quotient for $f(x) = \cos x$ is $-\sin x \left(\dfrac{\sin h}{h} \right) - \cos x \left(\dfrac{1 - \cos h}{h} \right)$.

Plot $Y_1 = -\sin x \left(\dfrac{\sin h}{h} \right) - \cos x \left(\dfrac{1 - \cos h}{h} \right)$ for

a. $h = 1$ **b.** $h = 0.1$ **c.** $h = 0.01$

What function does the difference quotient for $f(x) = \cos x$ resemble when h approaches zero?

5.3 DOUBLE-ANGLE IDENTITIES

SKILLS OBJECTIVE	CONCEPTUAL OBJECTIVE
■ Use double-angle identities in simplifying some trigonometric expressions.	■ Understand that the double-angle identities are derived from the sum identities.

5.3.1 Applying Double-Angle Identities

Throughout this text, much attention has been given to distinguishing between evaluating trigonometric functions exactly (for special angles) or approximating values of trigonometric functions with a calculator. In previous chapters, we could only evaluate trigonometric functions exactly for reference angles of $30°$, $45°$, and $60°$ or $\dfrac{\pi}{6}, \dfrac{\pi}{4}$, and $\dfrac{\pi}{3}$, and note that as of the previous section, we can now include multiples of $\dfrac{\pi}{12}$ among these "special" angles. Now we can use *double-angle identities* to evaluate trigonometric function values for other angles that are even integer multiples of the special angles or to verify other trigonometric identities. One important distinction now is that we will be able to find exact values of many functions using the double-angle identities without needing to know the actual value of the angle.

5.3.1 SKILL

Use double-angle identities in simplifying some trigonometric expressions.

5.3.1 CONCEPTUAL

Understand that the double-angle identities are derived from the sum identities.

Derivation of Double-Angle Identities

To derive the *double-angle identities*, we let $A = B$ in the sum identities:

WORDS	MATH
Write the identity for the sine of a sum.	$\sin(A + B) = \sin A \cos B + \cos A \sin B$
Let $B = A$.	$\sin(A + A) = \sin A \cos A + \cos A \sin A$
Simplify.	$\boxed{\sin(2A) = 2\sin A \cos A}$
Write the identity for the cosine of a sum.	$\cos(A + B) = \cos A \cos B - \sin A \sin B$
Let $B = A$.	$\cos(A + A) = \cos A \cos A - \sin A \sin A$
Simplify.	$\boxed{\cos(2A) = \cos^2 A - \sin^2 A}$

The *double-angle identity* for the cosine function can be written two other ways if we use the Pythagorean identity:

WORDS	MATH
1. Write the identity for the cosine function of a double angle.	$\cos(2A) = \cos^2 A - \sin^2 A$
Use the Pythagorean identity for cosine.	$\cos(2A) = \underbrace{\cos^2 A}_{1 - \sin^2 A} - \sin^2 A$
Simplify.	$\boxed{\cos(2A) = 1 - 2\sin^2 A}$
2. Write the identity for the cosine function of a double angle.	$\cos(2A) = \cos^2 A - \underbrace{\sin^2 A}_{1 - \cos^2 A}$
Use the Pythagorean identity for the sine function.	$\cos(2A) = \cos^2 A - (1 - \cos^2 A)$ $= \cos^2 A - 1 + \cos^2 A$
Simplify.	$\boxed{\cos(2A) = 2\cos^2 A - 1}$

The tangent function can always be written as a quotient, $\tan(2A) = \dfrac{\sin(2A)}{\cos(2A)}$, if $\sin(2A)$ and $\cos(2A)$ are known. Here we write the double-angle identity for the tangent function in terms of only the tangent function.

WORDS	MATH
Write the identity for the tangent of a sum.	$\tan(A + B) = \dfrac{\tan A + \tan B}{1 - \tan A \tan B}$
Let $B = A$.	$\tan(A + A) = \dfrac{\tan A + \tan A}{1 - \tan A \tan A}$
Simplify.	$\boxed{\tan(2A) = \dfrac{2\tan A}{1 - \tan^2 A}}$

DOUBLE-ANGLE IDENTITIES FOR THE SINE, COSINE, AND TANGENT FUNCTIONS

SINE	COSINE	TANGENT
$\sin(2A) = 2\sin A\cos A$	$\cos(2A) = \cos^2 A - \sin^2 A$	$\tan(2A) = \dfrac{2\tan A}{1 - \tan^2 A}$
	$\cos(2A) = 1 - 2\sin^2 A$	
	$\cos(2A) = 2\cos^2 A - 1$	

Applying Double-Angle Identities

EXAMPLE 1 **Finding Exact Values of Trigonometric Functions Using Double-Angle Identities**

If $\cos x = \frac{2}{3}$, find $\sin(2x)$ given $\sin x < 0$.

Solution:

Find $\sin x$.

Use the Pythagorean identity. $\qquad\qquad \sin^2 x + \cos^2 x = 1$

Substitute $\cos x = \frac{2}{3}$. $\qquad\qquad \sin^2 x + \left(\dfrac{2}{3}\right)^2 = 1$

Solve for $\sin x$, which is negative. $\qquad \sin x = -\sqrt{1 - \dfrac{4}{9}}$

Simplify. $\qquad\qquad\qquad\qquad\qquad \sin x = -\sqrt{\dfrac{5}{9}} = -\dfrac{\sqrt{5}}{3}$

Find $\sin(2x)$.

Use the double-angle formula for the sine function. $\qquad\qquad \sin(2x) = 2\sin x\cos x$

Substitute $\sin x = -\dfrac{\sqrt{5}}{3}$ and $\cos x = \dfrac{2}{3}$. $\qquad \sin(2x) = 2\left(-\dfrac{\sqrt{5}}{3}\right)\left(\dfrac{2}{3}\right)$

Simplify. $\qquad\qquad\qquad\qquad \boxed{\sin(2x) = -\dfrac{4\sqrt{5}}{9}}$

▼

YOUR TURN If $\cos x = -\frac{1}{3}$, find $\sin(2x)$ given $\sin x < 0$.

▶ **EXAMPLE 2** **Finding Exact Values Using Double-Angle Identities**

If $\sin x = -\frac{4}{5}$ and $\cos x < 0$, find $\sin(2x)$, $\cos(2x)$, and $\tan(2x)$.

Solution:

Solve for $\cos x$.

Use the Pythagorean identity. $\qquad\qquad$ $\sin^2 x + \cos^2 x = 1$

Substitute $\sin x = -\frac{4}{5}$. $\qquad\qquad$ $\left(-\frac{4}{5}\right)^2 + \cos^2 x = 1$

Simplify. $\qquad\qquad$ $\cos^2 x = \frac{9}{25}$

Solve for $\cos x$, which is negative. $\qquad\qquad$ $\cos x = -\sqrt{\frac{9}{25}} = -\frac{3}{5}$

Find $\sin(2x)$.

Use the double-angle identity for the sine function. $\quad$ $\sin(2x) = 2\sin x \cos x$

Substitute $\sin x = -\frac{4}{5}$ and $\cos x = -\frac{3}{5}$. $\qquad$ $\sin(2x) = 2\left(-\frac{4}{5}\right)\left(-\frac{3}{5}\right)$

Simplify. $\qquad\qquad$ $\boxed{\sin(2x) = \dfrac{24}{25}}$

Find $\cos(2x)$.

Use the double-angle identity for the
cosine function. $\qquad\qquad$ $\cos(2x) = \cos^2 x - \sin^2 x$

Substitute $\sin x = -\frac{4}{5}$ and $\cos x = -\frac{3}{5}$. $\qquad$ $\cos(2x) = \left(-\frac{3}{5}\right)^2 - \left(-\frac{4}{5}\right)^2$

Simplify. $\qquad\qquad$ $\boxed{\cos(2x) = -\dfrac{7}{25}}$

Find $\tan(2x)$.

Use the quotient identity. $\qquad\qquad$ $\tan\theta = \dfrac{\sin\theta}{\cos\theta}$

Let $\theta = 2x$. $\qquad\qquad$ $\tan(2x) = \dfrac{\sin(2x)}{\cos(2x)}$

Substitute $\sin(2x) = \frac{24}{25}$ and $\cos(2x) = -\frac{7}{25}$. $\quad$ $\tan(2x) = \dfrac{24/25}{-7/25}$

Simplify. $\qquad\qquad$ $\boxed{\tan(2x) = -\dfrac{24}{7}}$

Note: $\tan(2x)$ could also have been found by first finding $\tan x = \dfrac{\sin x}{\cos x}$ and then

using the value for $\tan x$ in the double-angle identity, $\tan(2A) = \dfrac{2\tan A}{1 - \tan^2 A}$.

▼
ANSWER

$\sin(2x) = -\frac{24}{25}$
$\cos(2x) = -\frac{7}{25}$
$\tan(2x) = \frac{24}{7}$

▼
YOUR TURN If $\cos x = \frac{3}{5}$ and $\sin x < 0$, find $\sin(2x)$, $\cos(2x)$, and $\tan(2x)$.

EXAMPLE 3   **Verifying Trigonometric Identities Using Double-Angle Identities**

Verify the identity $(\sin x - \cos x)^2 = 1 - \sin(2x)$.

Solution:

Start with the left side of the equation.	$(\sin x - \cos x)^2$
Expand by squaring.	$= \sin^2 x - 2\sin x \cos x + \cos^2 x$
Group the $\sin^2 x$ and $\cos^2 x$ terms.	$= \sin^2 x + \cos^2 x - 2\sin x \cos x$
Recognize the Pythagorean identity.	$= \underbrace{\sin^2 x + \cos^2 x}_{1} - 2\sin x \cos x$
Recognize the sine double-angle identity.	$= 1 - \underbrace{2\sin x \cos x}_{\sin(2x)}$
Simplify.	$\boxed{(\sin x - \cos x)^2 = 1 - \sin(2x)}$

▶ **EXAMPLE 4**   **Verifying Multiple-Angle Identities Using Double-Angle Identities**

Verify the identity $\cos(3x) = (1 - 4\sin^2 x)\cos x$.

Solution:

Write the cosine of a sum identity.	$\cos(A + B) = \cos A \cos B - \sin A \sin B$
Let $A = 2x$ and $B = x$.	$\cos(2x + x) = \cos(2x)\cos x - \sin(2x)\sin x$
Recognize the double-angle identities.	$\cos(3x) = \underbrace{\cos(2x)}_{1-2\sin^2 x}\cos x - \underbrace{\sin(2x)}_{2\sin x \cos x}\sin x$
Simplify.	$\cos(3x) = \cos x - 2\sin^2 x \cos x - 2\sin^2 x \cos x$
	$\cos(3x) = \cos x - 4\sin^2 x \cos x$
Factor out the common cosine term.	$\boxed{\cos(3x) = (1 - 4\sin^2 x)\cos x}$

EXAMPLE 5 **Simplifying Trigonometric Expressions Using Double-Angle Identities**

Graph $y = \dfrac{\cot x - \tan x}{\cot x + \tan x}$.

Solution:

Simplify $y = \dfrac{\cot x - \tan x}{\cot x + \tan x}$ first.

Write the cotangent and tangent functions in terms of the sine and cosine functions.

$$y = \dfrac{\dfrac{\cos x}{\sin x} - \dfrac{\sin x}{\cos x}}{\dfrac{\cos x}{\sin x} + \dfrac{\sin x}{\cos x}}$$

Multiply the numerator and the denominator by $\sin x \cos x$.

$$y = \left(\dfrac{\dfrac{\cos x}{\sin x} - \dfrac{\sin x}{\cos x}}{\dfrac{\cos x}{\sin x} + \dfrac{\sin x}{\cos x}} \right) \left(\dfrac{\sin x \cos x}{\sin x \cos x} \right)$$

Simplify.

$$y = \dfrac{\cos^2 x - \sin^2 x}{\cos^2 x + \sin^2 x}$$

Recognize the double-angle and Pythagorean identities.

$$y = \dfrac{\overbrace{\cos^2 x - \sin^2 x}^{\cos(2x)}}{\underbrace{\cos^2 x + \sin^2 x}_{1}}$$

Simplify.

$$\boxed{y = \cos(2x)}$$

Graph $y = \cos(2x)$.

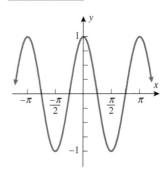

▶[SECTION 5.3] **SUMMARY**

In this section, we derived the double-angle identities from the sum identities. We then used the double-angle identities to find exact values of trigonometric functions, to verify other trigonometric identities and to simplify trigonometric expressions. There is no need to memorize the second and third forms of the cosine double-angle identity since they can be derived from the first using the Pythagorean identity.

$$\sin(2A) = 2\sin A \cos A$$
$$\cos(2A) = \cos^2 A - \sin^2 A$$
$$= 1 - 2\sin^2 A$$
$$= 2\cos^2 A - 1$$
$$\tan(2A) = \dfrac{2\tan A}{1 - \tan^2 A}$$

[SECTION 5.3] **EXERCISES**

• **SKILLS**

In Exercises 1–12, use the double-angle identities to find the indicated values.

1. If $\sin x = \dfrac{1}{\sqrt{5}}$ and $\cos x < 0$, find $\sin(2x)$.

2. If $\sin x = \dfrac{1}{\sqrt{5}}$ and $\cos x < 0$, find $\cos(2x)$.

3. If $\cos x = \dfrac{5}{13}$ and $\sin x < 0$, find $\tan(2x)$.

4. If $\cos x = -\dfrac{5}{13}$ and $\sin x < 0$, find $\tan(2x)$.

5. If $\tan x = \dfrac{12}{5}$ and $\pi < x < \dfrac{3\pi}{2}$, find $\sin(2x)$.

6. If $\tan x = \dfrac{12}{5}$ and $\pi < x < \dfrac{3\pi}{2}$, find $\cos(2x)$.

7. If $\sec x = \sqrt{5}$ and $\sin x > 0$, find $\tan(2x)$.

8. If $\sec x = \sqrt{3}$ and $\sin x < 0$, find $\tan(2x)$.

9. If $\csc x = -2\sqrt{5}$ and $\cos x < 0$, find $\sin(2x)$.

10. If $\csc x = -\sqrt{13}$ and $\cos x > 0$, find $\sin(2x)$.

11. If $\cos x = -\dfrac{12}{13}$ and $\csc x < 0$, find $\cot(2x)$.

12. If $\sin x = \dfrac{12}{13}$ and $\cot x < 0$, find $\csc(2x)$.

In Exercises 13–22, simplify each expression. Evaluate the resulting expression exactly, if possible.

13. $\dfrac{2\tan 15°}{1 - \tan^2 15°}$

14. $\dfrac{2\tan\left(\dfrac{\pi}{8}\right)}{1 - \tan^2\left(\dfrac{\pi}{8}\right)}$

15. $\sin\left(\dfrac{\pi}{8}\right)\cos\left(\dfrac{\pi}{8}\right)$

16. $\sin 15° \cos 15°$

17. $\cos^2(2x) - \sin^2(2x)$

18. $\cos^2(x + 2) - \sin^2(x + 2)$

19. $1 - 2\sin^2 105°$

20. $-4\sin\left(\dfrac{3\pi}{8}\right)\cos\left(\dfrac{3\pi}{8}\right)$

21. $\dfrac{\tan(4x)}{1 - \tan^2(4x)}$

22. $1 - 2\cos^2 195°$

In Exercises 23–42, verify each identity.

23. $\csc(2A) = \frac{1}{2}\csc A \sec A$

24. $\cot(2A) = \frac{1}{2}(\cot A - \tan A)$

25. $(\sin x - \cos x)(\cos x + \sin x) = -\cos(2x)$

26. $(\sin x + \cos x)^2 = 1 + \sin(2x)$

27. $\cos^2 x = \dfrac{1 + \cos(2x)}{2}$

28. $\sin^2 x = \dfrac{1 - \cos(2x)}{2}$

29. $\cos^4 x - \sin^4 x = \cos(2x)$

30. $\cos^4 x + \sin^4 x = 1 - \frac{1}{2}\sin^2(2x)$

31. $8\sin^2 x\cos^2 x = 1 - \cos(4x)$

32. $\sin^2(4x) = 2\cos^2(2x) - 2\cos^4(2x) + 2\cos^2(2x)\sin^2(2x)$

33. $-\frac{1}{2}\sec^2 x = -2\sin^2 x\csc^2(2x)$

34. $4\csc(4x) = \dfrac{\sec x \csc x}{\cos(2x)}$

35. $\sin(3x) = \sin x(4\cos^2 x - 1)$

36. $\tan(3x) = \dfrac{\tan x(3 - \tan^2 x)}{(1 - 3\tan^2 x)}$

37. $\frac{1}{2}\sin(4x) = 2\sin x\cos x - 4\sin^3 x\cos x$

38. $\cos(4x) = [\cos(2x) - \sin(2x)][\cos(2x) + \sin(2x)]$

39. $\sin(4x) = \sin(2x)(2 - 4\sin^2 x)$

40. $\cos(4x) = 1 - \sin^2(2x) - 4(\sin x\cos x)^2$

41. $\tan(4x) = \dfrac{4(\sin x)(\cos x)[\cos(2x)]}{1 - 2\sin^2(2x)}$

42. $\cos(6x) = 1 - 2[2\sin x\cos^2 x + \cos(2x)\sin x]^2$

In Exercises 43–46, graph the functions.

43. $y = \dfrac{\sin(2x)}{1 - \cos(2x)}$

44. $y = \dfrac{2\tan x}{2 - \sec^2 x}$

45. $y = \dfrac{\cot x + \tan x}{\cot x - \tan x}$

46. $y = \frac{1}{2}(\tan x)(\cot x)(\sec x)(\csc x)$

• **APPLICATIONS**

47. Business/Economics. Annual cash flow of a stock fund (measured as a percentage of total assets) has fluctuated in cycles. The highs were roughly +12% of total assets and lows were roughly −8% of total assets. This cash flow can be modeled by the function

$$C(t) = 12 - 20\sin^2 t$$

Use a double-angle identity to express $C(t)$ in terms of the cosine function.

48. Business. Computer sales are generally subject to seasonal fluctuations. An analysis of the sales of a computer manufacturer during 2016–2018 is approximated by the function

$$s(t) = 0.098\cos^2 t + 0.387 \qquad 1 \le t \le 12$$

where t represents time in quarters ($t = 1$ represents the end of the first quarter of 2016), and $s(t)$ represents computer sales (quarterly revenue) in millions of dollars.
Use a double-angle identity to express $s(t)$ in terms of the cosine function.

49. Hiking. Two hikers leave from the same campsite and walk in different directions. The distance d in miles between the hikers can be found using the function $d = \sqrt{13 - 12\cos\theta}$, where θ is the angle between the directions traveled by the hikers. Find a function for the distance between the hikers if θ is doubled and then use a double-angle formula to write the function in terms of the cosine of a single angle θ.

50. Hiking. Two hikers leave from the same campsite and walk in different directions. The distance d in miles between the hikers can be found using the function $d = \sqrt{41 - 40\cos\theta}$, where θ is the angle between the directions traveled by the hikers. Find a function for the distance between the hikers if θ is doubled and then use a double-angle formula to write the function in terms of the sine of a single angle θ.

51. Biology/Health. The rise and fall of a person's body temperature t days after contracting a certain virus can be modeled by the function $T = 98.6 + 4\sin^2 t$, where T is body temperature in degrees Fahrenheit and $0 \le t \le 3$. Write the function in terms of the cosine of a double angle and then sketch its graph.

52. Biology/Health. The rise and fall of a person's body temperature t days after contracting a certain virus can be modeled by the function $T = 98.6 + 6\sin t\cos t$, where T is body temperature in degrees Fahrenheit and $0 \le t \le 1.5$. Write the function in terms of the sine of a double angle and then sketch its graph.

For Exercises 53 and 54, refer to the following:

An ore crusher wheel consists of a heavy disk spinning on its axle. Its normal (crushing) force F in pounds between the wheel and the inclined track is determined by

$$F = W\sin\theta + \frac{1}{2}\psi^2\left[\frac{C}{R}(1 - \cos 2\theta) + \frac{A}{l}\sin 2\theta\right]$$

where W is the weight of the wheel, θ is the angle of the axis, C and A are moments of inertia, R is the radius of the wheel, l is the distance from the wheel to the pin where the axle is attached, and ψ is the speed in rpm that the wheel is spinning.
The optimum crushing force occurs when the angle θ is between 45° and 90°.

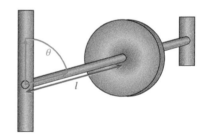

53. Ore-Crusher Wheel. Find F if the angle is 60°, W is 500 pounds, and ψ is 200 rpm, $\dfrac{C}{R} = 750$, and $\dfrac{A}{l} = 3.75$.

54. Ore-Crusher Wheel. Find F if the angle is 75°, W is 500 pounds, and ψ is 200 rpm, $\dfrac{C}{R} = 750$, and $\dfrac{A}{l} = 3.75$.

• **CATCH THE MISTAKE**

In Exercises 55 and 56, explain the mistake that is made.

55. If $\cos x = \frac{1}{3}$, find $\sin(2x)$ given $\sin x < 0$.

Solution:

Write the double-angle identity for the sine function. $\quad \sin(2x) = 2\sin x \cos x$

Solve for $\sin x$ using the Pythagorean identity. $\quad \sin^2 x + \left(\frac{1}{3}\right)^2 = 1$

$$\sin x = \frac{2\sqrt{2}}{3}$$

Substitute $\cos x = \frac{1}{3}$ and $\sin x = \frac{2\sqrt{2}}{3}$. $\quad \sin(2x) = 2\left(\frac{2\sqrt{2}}{3}\right)\left(\frac{1}{3}\right)$

Simplify. $\quad \sin(2x) = \frac{4\sqrt{2}}{9}$

This is incorrect. What mistake was made?

56. If $\sin x = \frac{1}{3}$, find $\tan(2x)$ given $\cos x < 0$.

Solution:

Use the quotient identity. $\quad \tan(2x) = \dfrac{\sin(2x)}{\cos x}$

Use the double-angle formula for the sine function. $\quad \tan(2x) = \dfrac{2\sin x \cos x}{\cos x}$

Cancel the common cosine factors. $\quad \tan(2x) = 2\sin x$

Substitute $\sin x = \frac{1}{3}$. $\quad \tan(2x) = \dfrac{2}{3}$

This is incorrect. What mistake was made?

• **CONCEPTUAL**

In Exercises 57–60, determine whether each statement is true or false.

57. $\sin(2A) + \sin(2A) = \sin(4A)$

58. $\cos(4A) - \cos(2A) = \cos(2A)$

59. If $\tan x > 0$, then $\tan(2x) > 0$.

60. If $\sin x > 0$, then $\sin(2x) > 0$.

61. Let n be a positive integer. Express $\sin\left(\dfrac{n\pi}{2}\right)\cos\left(\dfrac{n\pi}{2}\right)$ as a single trigonometric function, and then evaluate if possible.

62. Let n be a positive integer. Write the expression $\sin^2\left(\dfrac{n\pi}{2}\right)$ in terms of the cosine of a multiple angle, and then evaluate if possible.

• **CHALLENGE**

63. Express $\tan(4x)$ in terms of functions of $\tan x$.

64. Express $\tan(-4x)$ in terms of functions of $\tan x$.

65. Is the identity $2\csc(2x) = \dfrac{1 + \tan^2 x}{\tan x}$ true for $x = \dfrac{\pi}{2}$? Explain.

66. Is the identity $\tan(2x) = \dfrac{2\tan x}{1 - \tan^2 x}$ true for $x = \dfrac{\pi}{4}$? Explain.

67. Find all values for x where $0 \le x < 2\pi$ and $\sin(2x) = \sin x$.

68. Find all values for x where $0 \le x < 2\pi$ and $\cos(2x) = \cos x$.

• **TECHNOLOGY**

For Exercises 69–72, refer to the following:

We cannot *prove* that an equation is an identity using technology, but we can use technology as a first step to see whether or not the equation *seems* to be an identity.

69. Using a graphing calculator, plot

$$Y_1 = (2x) - \frac{(2x)^3}{3!} + \frac{(2x)^5}{5!} \text{ and } Y_2 = \sin(2x) \text{ for } x \text{ ranging}$$

$[-1, 1]$. Is Y_1 a good approximation to Y_2?

70. Using a graphing calculator, plot

$$Y_1 = 1 - \frac{(2x)^2}{2!} + \frac{(2x)^4}{4!} \text{ and } Y_2 = \cos(2x) \text{ for } x \text{ ranging}$$

$[-1, 1]$. Is Y_1 a good approximation to Y_2?

71. Using a graphing calculator, determine whether

$$\frac{\tan(4x) - \tan(3x)}{\tan x} = \frac{\csc(2x)}{1 - \sec(2x)} \text{ by plotting each side}$$

of the equation and seeing if the graphs coincide.

72. Using a graphing calculator, determine whether

$\csc(2x)\sec(2x)[\cos(2x) - \sin(2x)]$

$$= \frac{1 - 2\sin^2 x - 2\sin x \cos x}{2\sin x \cos x(\cos^2 x - \sin^2 x)} \text{ by plotting each side of the}$$

equation and seeing if the graphs coincide.

5.4 HALF-ANGLE IDENTITIES

SKILLS OBJECTIVE	CONCEPTUAL OBJECTIVE
▪ Use half-angle identities in simplifying some trigonometric expressions.	▪ Understand that the half-angle identities are derived from the double-angle identities.

5.4.1 Applying Half-Angle Identities

5.4.1 **SKILL**

Use half-angle identities in simplifying some trigonometric expressions.

5.4.1 **CONCEPTUAL**

Understand that the half-angle identities are derived from the double-angle identities.

We now use the *double-angle identities* from Section 5.3 to derive the *half-angle identities*. Like the double-angle identities, the half-angle *identities* will allow us to find certain exact values of trigonometric functions and to verify other trigonometric identities. We start by rewriting the second and third forms of the cosine double-angle identity to obtain identities for the square of the sine and cosine functions, $\sin^2 x$ and $\cos^2 x$, otherwise known as the *power reduction formulas*.

WORDS	MATH
Power reduction formula for the sine function	
Write the second form of the cosine double-angle identity.	$\cos(2A) = 1 - 2\sin^2 A$
Isolate the $2\sin^2 A$ term on one side of the equation.	$2\sin^2 A = 1 - \cos(2A)$
Divide both sides by 2.	$\sin^2 A = \dfrac{1 - \cos(2A)}{2}$
Power reduction formula for the cosine function	
Write the third form of the cosine double-angle identity.	$\cos(2A) = 2\cos^2 A - 1$
Isolate the $2\cos^2 A$ term on one side of the equation.	$2\cos^2 A = \cos(2A) + 1$
Divide both sides by 2.	$\cos^2 A = \dfrac{1 + \cos(2A)}{2}$
Power reduction formula for the tangent function	
Taking the quotient of these leads us to another identity.	$\tan^2 A = \dfrac{\sin^2 A}{\cos^2 A} = \dfrac{\dfrac{1 - \cos(2A)}{2}}{\dfrac{1 + \cos(2A)}{2}} = \dfrac{1 - \cos(2A)}{1 + \cos(2A)}$

These three identities for the squared functions, which are essentially alternative forms of the double-angle identities, are used in calculus as **power reduction formulas** (identities that allow us to reduce the power of the trigonometric function from 2 to 1):

$$\sin^2 A = \frac{1 - \cos(2A)}{2} \qquad \cos^2 A = \frac{1 + \cos(2A)}{2} \qquad \tan^2 A = \frac{1 - \cos(2A)}{1 + \cos(2A)}$$

Derivation of the Half-Angle Identities

We can now use these forms of the double-angle identities to derive the *half-angle identities*.

WORDS	MATH
Sine half-angle identity	
For the **sine half-angle identity**, start with the double-angle formula involving both the sine and cosine functions, $\cos(2x) = 1 - 2\sin^2 x$, and solve for $\sin^2 x$.	$\sin^2 x = \dfrac{1 - \cos(2x)}{2}$
Solve for $\sin x$.	$\sin x = \pm\sqrt{\dfrac{1 - \cos(2x)}{2}}$
Let $x = \dfrac{A}{2}$.	$\sin\left(\dfrac{A}{2}\right) = \pm\sqrt{\dfrac{1 - \cos\left(2 \cdot \dfrac{A}{2}\right)}{2}}$
Simplify.	$\boxed{\sin\left(\dfrac{A}{2}\right) = \pm\sqrt{\dfrac{1 - \cos A}{2}}}$
Cosine half-angle identity	
For the **cosine half-angle identity**, start with the double-angle formula involving only the cosine functions, $\cos(2x) = 2\cos^2 x - 1$, and solve for $\cos^2 x$.	$\cos^2 x = \dfrac{1 + \cos(2x)}{2}$
Solve for $\cos x$.	$\cos x = \pm\sqrt{\dfrac{1 + \cos(2x)}{2}}$
Let $x = \dfrac{A}{2}$.	$\cos\left(\dfrac{A}{2}\right) = \pm\sqrt{\dfrac{1 + \cos\left(2 \cdot \dfrac{A}{2}\right)}{2}}$
Simplify.	$\boxed{\cos\left(\dfrac{A}{2}\right) = \pm\sqrt{\dfrac{1 + \cos A}{2}}}$
Tangent half-angle identity	
For the **tangent half-angle identity**, start with the quotient identity.	$\tan\left(\dfrac{A}{2}\right) = \dfrac{\sin\left(\dfrac{A}{2}\right)}{\cos\left(\dfrac{A}{2}\right)}$
Substitute half-angle identities for the sine and cosine functions.	$\tan\left(\dfrac{A}{2}\right) = \dfrac{\pm\sqrt{\dfrac{1 - \cos A}{2}}}{\pm\sqrt{\dfrac{1 + \cos A}{2}}}$
Simplify.	$\boxed{\tan\left(\dfrac{A}{2}\right) = \pm\sqrt{\dfrac{1 - \cos A}{1 + \cos A}}}$

Note: $\tan\left(\dfrac{A}{2}\right)$ can also be found by starting with the identity, $\tan^2 x = \dfrac{1 - \cos(2x)}{1 + \cos(2x)}$, solving for $\tan x$, and letting $x = \dfrac{A}{2}$. The tangent function also has two other similar forms for $\tan\left(\dfrac{A}{2}\right)$ (see Exercises 63 and 64).

HALF-ANGLE IDENTITIES FOR THE SINE, COSINE, AND TANGENT FUNCTIONS

sine	cosine	tangent
$\sin\left(\dfrac{A}{2}\right) = \pm\sqrt{\dfrac{1 - \cos A}{2}}$	$\cos\left(\dfrac{A}{2}\right) = \pm\sqrt{\dfrac{1 + \cos A}{2}}$	$\tan\left(\dfrac{A}{2}\right) = \pm\sqrt{\dfrac{1 - \cos A}{1 + \cos A}}$
		$\tan\left(\dfrac{A}{2}\right) = \dfrac{\sin A}{1 + \cos A}$
		$\tan\left(\dfrac{A}{2}\right) = \dfrac{1 - \cos A}{\sin A}$

STUDY TIP

The sign (+ or −) is determined by which quadrant contains $\dfrac{A}{2}$ and the sign of the particular trigonometric function in that quadrant.

Applying Half-Angle Identities

These identities hold for any real number A or any angle with either degree measure or radian measure as long as both sides of the equation are defined. The sign (+ or −) is determined by the sign of the trigonometric function in the quadrant that contains $\dfrac{A}{2}$.

▶ **EXAMPLE 1** **Finding Exact Values Using Half-Angle Identities**

Use a half-angle identity to find $\cos 15°$.

Solution:

Write $\cos 15°$ in terms of a half-angle.

$$\cos 15° = \cos\left(\dfrac{30°}{2}\right)$$

Write the half-angle identity for the cosine function.

$$\cos\left(\dfrac{A}{2}\right) = \pm\sqrt{\dfrac{1 + \cos A}{2}}$$

Substitute $A = 30°$.

$$\cos\left(\dfrac{30°}{2}\right) = \pm\sqrt{\dfrac{1 + \cos 30°}{2}}$$

Simplify.

$$\cos 15° = \pm\sqrt{\dfrac{1 + \dfrac{\sqrt{3}}{2}}{2}}$$

$15°$ is in quadrant I where the cosine function is positive.

$$\cos 15° = \sqrt{\dfrac{2 + \sqrt{3}}{4}} = \boxed{\dfrac{\sqrt{2 + \sqrt{3}}}{2}}$$

▼

ANSWER

$$\dfrac{\sqrt{2 - \sqrt{2}}}{2}$$

▼

YOUR TURN Use a half-angle identity to find $\sin(22.5°)$.

EXAMPLE 2 **Finding Exact Values Using Half-Angle Identities**

Use a half-angle identity to find $\tan\left(\dfrac{11\pi}{12}\right)$.

Solution:

Write $\tan\left(\dfrac{11\pi}{12}\right)$ in terms of a half angle.

$$\tan\left(\frac{11\pi}{12}\right) = \tan\left(\frac{\frac{11\pi}{6}}{2}\right)$$

Write the half-angle identity for the tangent function.*

$$\tan\left(\frac{A}{2}\right) = \frac{1 - \cos A}{\sin A}$$

Substitute $A = \dfrac{11\pi}{6}$.

$$\tan\left(\frac{\frac{11\pi}{6}}{2}\right) = \frac{1 - \cos\left(\frac{11\pi}{6}\right)}{\sin\left(\frac{11\pi}{6}\right)}$$

Simplify.

$$\tan\left(\frac{11\pi}{12}\right) = \frac{1 - \frac{\sqrt{3}}{2}}{-\frac{1}{2}}$$

$$\tan\left(\frac{11\pi}{12}\right) = \boxed{\sqrt{3} - 2}$$

$\dfrac{11\pi}{12}$ is in quadrant **II** where the tangent function is negative. Notice that if we approximate $\tan\left(\dfrac{11\pi}{12}\right)$ with a calculator, we find that $\tan\left(\dfrac{11\pi}{12}\right) \approx -0.2679$, and $\sqrt{3} - 2 \approx -0.2679$.

*This form of the tangent half-angle identity was selected because of mathematical simplicity. If we had selected either of the other forms, we would have obtained an expression that had a square root of a square root or a radical in the denominator.

▼

YOUR TURN Use a half-angle identity to find $\tan\left(\dfrac{\pi}{8}\right)$.

▼
ANSWER
$\dfrac{2 - \sqrt{2}}{\sqrt{2}}$ or $\sqrt{2} - 1$

▶ **EXAMPLE 3** **Finding Exact Values Using Half-Angle Identities**

If $\cos x = \dfrac{3}{5}$ and $\dfrac{3\pi}{2} < x < 2\pi$, find $\sin\left(\dfrac{x}{2}\right)$, $\cos\left(\dfrac{x}{2}\right)$, and $\tan\left(\dfrac{x}{2}\right)$.

Solution:

Determine in which quadrant $\dfrac{x}{2}$ lies. Since $\dfrac{3\pi}{2} < x < 2\pi$, we divide the inequality by 2.

$$\dfrac{3\pi}{4} < \dfrac{x}{2} < \pi$$

$\dfrac{x}{2}$ lies in quadrant II; therefore, the sine function is positive, and the cosine and tangent functions are negative.

Write the half-angle identity for the sine function. $\qquad \sin\left(\dfrac{x}{2}\right) = \pm\sqrt{\dfrac{1 - \cos x}{2}}$

Substitute $\cos x = \dfrac{3}{5}$. $\qquad \sin\left(\dfrac{x}{2}\right) = \pm\sqrt{\dfrac{1 - \dfrac{3}{5}}{2}}$

Simplify. $\qquad \sin\left(\dfrac{x}{2}\right) = \pm\sqrt{\dfrac{1}{5}} = \pm\dfrac{\sqrt{5}}{5}$

Since $\dfrac{x}{2}$ lies in quadrant II, choose the positive value for the sine function. $\qquad \boxed{\sin\left(\dfrac{x}{2}\right) = \dfrac{\sqrt{5}}{5}}$

Write the half-angle identity for the cosine function. $\quad \cos\left(\dfrac{x}{2}\right) = \pm\sqrt{\dfrac{1 + \cos x}{2}}$

Substitute $\cos x = \dfrac{3}{5}$. $\qquad \cos\left(\dfrac{x}{2}\right) = \pm\sqrt{\dfrac{1 + \dfrac{3}{5}}{2}}$

Simplify. $\qquad \cos\left(\dfrac{x}{2}\right) = \pm\sqrt{\dfrac{4}{5}} = \pm\dfrac{2\sqrt{5}}{5}$

Since $\dfrac{x}{2}$ lies in quadrant II, choose the negative value for the cosine function. $\qquad \boxed{\cos\left(\dfrac{x}{2}\right) = -\dfrac{2\sqrt{5}}{5}}$

Use the quotient identity for tangent. $\qquad \tan\left(\dfrac{x}{2}\right) = \dfrac{\sin\left(\dfrac{x}{2}\right)}{\cos\left(\dfrac{x}{2}\right)}$

Substitute $\sin\left(\dfrac{x}{2}\right) = \dfrac{\sqrt{5}}{5}$ and $\cos\left(\dfrac{x}{2}\right) = -\dfrac{2\sqrt{5}}{5}$. $\qquad \tan\left(\dfrac{x}{2}\right) = \dfrac{\dfrac{\sqrt{5}}{5}}{-\dfrac{2\sqrt{5}}{5}}$

Simplify. $\qquad \boxed{\tan\left(\dfrac{x}{2}\right) = -\dfrac{1}{2}}$

▼

YOUR TURN If $\cos x = -\dfrac{3}{5}$ and $\pi < x < \dfrac{3\pi}{2}$, find $\sin\left(\dfrac{x}{2}\right)$, $\cos\left(\dfrac{x}{2}\right)$, and $\tan\left(\dfrac{x}{2}\right)$.

EXAMPLE 4 **Using Half-Angle Identities to Verify Other Trigonometric Identities**

Verify the identity $\cos^2\left(\dfrac{x}{2}\right) = \dfrac{\tan x + \sin x}{2\tan x}$.

Solution:

Start with the left side.

Write the cosine half-angle identity.
$$\cos\left(\frac{x}{2}\right) = \pm\sqrt{\frac{1 + \cos x}{2}}$$

Square both sides of the equation.
$$\cos^2\left(\frac{x}{2}\right) = \frac{1 + \cos x}{2}$$

Multiply the numerator and the denominator on the right side by $\tan x$.
$$\cos^2\left(\frac{x}{2}\right) = \left(\frac{1 + \cos x}{2}\right)\left(\frac{\tan x}{\tan x}\right)$$

Simplify.
$$\cos^2\left(\frac{x}{2}\right) = \frac{\tan x + \cos x \tan x}{2\tan x}$$

Note that $\cos x \tan x = \sin x$.
$$\boxed{\cos^2\left(\frac{x}{2}\right) = \frac{\tan x + \sin x}{2\tan x}}$$

An alternative solution is to start with the right-hand side.

Solution (alternative):

Start with the right-hand side.
$$\frac{\tan x + \sin x}{2\tan x}$$

Write this expression as the sum of two expressions.
$$= \frac{\tan x}{2\tan x} + \frac{\sin x}{2\tan x}$$

Simplify.
$$= \frac{1}{2} + \frac{1}{2}\cdot\frac{\sin x}{\tan x}$$

Write $\tan x = \dfrac{\sin x}{\cos x}$.
$$= \frac{1}{2} + \frac{1}{2}\cdot\frac{\sin x}{\dfrac{\sin x}{\cos x}}$$

Simplify.
$$= \frac{1}{2}(1 + \cos x)$$
$$= \cos^2\left(\frac{x}{2}\right)$$

EXAMPLE 5 **Using Half-Angle Identities to Verify Other Trigonometric Identities**

Verify the identity $\tan x = \csc(2x) - \cot(2x)$.

Solution:

Notice that x is $\frac{1}{2}$ of $2x$.

Write the third half-angle formula for the tangent function.
$$\tan\left(\frac{A}{2}\right) = \frac{1 - \cos A}{\sin A}$$

Write the right side as a difference of two expressions having the same denominator.
$$\tan\left(\frac{A}{2}\right) = \frac{1}{\sin A} - \frac{\cos A}{\sin A}$$

Substitute the reciprocal and quotient identities, respectively, on the right.
$$\tan\left(\frac{A}{2}\right) = \csc A - \cot A$$

Let $A = 2x$.
$$\boxed{\tan x = \csc(2x) - \cot(2x)}$$

Notice that in Example 5, we started with the third half-angle identity for the tangent function. In Example 6, we will start with the second half-angle identity for the tangent function. In general, you select the form that appears to lead to the desired expression.

EXAMPLE 6 **Using Half-Angle Identities to Simplify Trigonometric Expressions**

Graph $y = \dfrac{\sin(2\pi x)}{1 + \cos(2\pi x)}$ by first simplifying the trigonometric expression to a more recognizable form.

Solution:

Simplify the trigonometric expression using a half-angle identity for the tangent function.

Write the second half-angle identity for the tangent function.

$$\tan\left(\frac{A}{2}\right) = \frac{\sin A}{1 + \cos A}$$

Let $A = 2\pi x$.

$$\tan(\pi x) = \frac{\sin(2\pi x)}{1 + \cos(2\pi x)}$$

Graph $y = \tan(\pi x)$.

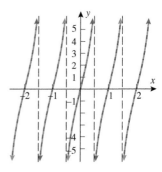

$$y = \tan(\pi x) \qquad y = \frac{\sin(2\pi x)}{1 + \cos(2\pi x)}$$

▼
ANSWER

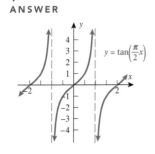

▼

YOUR TURN   Graph $y = \dfrac{1 - \cos(\pi x)}{\sin(\pi x)}$.

▶[SECTION 5.4]   SUMMARY

In this section, we used the double-angle identities to derive the half-angle identities. We then used the half-angle identities to find certain exact values of trigonometric functions, verify other trigonometric identities, and simplify trigonometric expressions.

$$\sin\left(\frac{A}{2}\right) = \pm\sqrt{\frac{1 - \cos A}{2}}$$

$$\cos\left(\frac{A}{2}\right) = \pm\sqrt{\frac{1 + \cos A}{2}}$$

$$\tan\left(\frac{A}{2}\right) = \pm\sqrt{\frac{1 - \cos A}{1 + \cos A}}$$

The sign ($+$ or $-$) is chosen by first determining which quadrant contains $\dfrac{A}{2}$ and then determining the sign of the indicated trigonometric function in that quadrant.

Recall that there are three forms of the tangent half-angle identity. There is no need to memorize the other forms of the tangent half-angle identity, since they can be derived by first using the Pythagorean identity and algebraic manipulation.

[SECTION 5.4] **EXERCISES**

• **SKILLS**

In Exercises 1–16, use the half-angle identities to find the exact values of the trigonometric expressions.

1. $\sin 15°$

2. $\cos(22.5°)$

3. $\cos\left(\dfrac{11\pi}{12}\right)$

4. $\sin\left(\dfrac{\pi}{8}\right)$

5. $\cos 75°$

6. $\sin 75°$

7. $\cos\left(\dfrac{23\pi}{12}\right)$

8. $\sin\left(\dfrac{9\pi}{8}\right)$

9. $\cos\left(\dfrac{3\pi}{8}\right)$

10. $\sin\left(\dfrac{7\pi}{12}\right)$

11. $\tan(67.5°)$

12. $\tan(202.5°)$

13. $\sec\left(-\dfrac{9\pi}{8}\right)$

14. $\csc\left(\dfrac{9\pi}{8}\right)$

15. $\cot\left(\dfrac{13\pi}{8}\right)$

16. $\cot\left(\dfrac{7\pi}{8}\right)$

In Exercises 17–26, use the half-angle identities to find the desired function values.

17. If $\cos x = \dfrac{5}{13}$ and $\dfrac{3\pi}{2} < x < 2\pi$, find $\sin\left(\dfrac{x}{2}\right)$.

18. If $\cos x = -\dfrac{5}{13}$ and $\pi < x < \dfrac{3\pi}{2}$, find $\cos\left(\dfrac{x}{2}\right)$.

19. If $\tan x = \dfrac{12}{5}$ and $\pi < x < \dfrac{3\pi}{2}$, find $\sin\left(\dfrac{x}{2}\right)$.

20. If $\tan x = \dfrac{12}{5}$ and $\pi < x < \dfrac{3\pi}{2}$, find $\cos\left(\dfrac{x}{2}\right)$.

21. If $\sec x = \sqrt{5}$ and $0 < x < \dfrac{\pi}{2}$, find $\tan\left(\dfrac{x}{2}\right)$.

22. If $\sec x = \sqrt{3}$ and $\dfrac{3\pi}{2} < x < 2\pi$, find $\tan\left(\dfrac{x}{2}\right)$.

23. If $\csc x = 3$ and $\dfrac{\pi}{2} < x < \pi$, find $\sin\left(\dfrac{x}{2}\right)$.

24. If $\csc x = -3$ and $\dfrac{3\pi}{2} < x < 2\pi$, find $\cos\left(\dfrac{x}{2}\right)$.

25. If $\cos x = -\dfrac{1}{4}$ and $\pi < x < \dfrac{3\pi}{2}$, find $\cot\left(\dfrac{x}{2}\right)$.

26. If $\cos x = \dfrac{1}{4}$ and $\dfrac{3\pi}{2} < x < 2\pi$, find $\csc\left(\dfrac{x}{2}\right)$.

In Exercises 27–30, simplify each expression using half-angle identities. Do not evaluate.

27. $\sqrt{\dfrac{1 + \cos\left(\dfrac{5\pi}{6}\right)}{2}}$

28. $\sqrt{\dfrac{1 - \cos\left(\dfrac{\pi}{4}\right)}{2}}$

29. $\dfrac{\sin 150°}{1 + \cos 150°}$

30. $\dfrac{1 - \cos 150°}{\sin 150°}$

In Exercises 31–44, verify the identities.

31. $\sin^2\left(\dfrac{x}{2}\right) + \cos^2\left(\dfrac{x}{2}\right) = 1$

32. $\cos^2\left(\dfrac{x}{2}\right) - \sin^2\left(\dfrac{x}{2}\right) = \cos x$

33. $\sin^2\left(\dfrac{x}{2}\right) = \dfrac{1 - \sin(90° - x)}{2}$

34. $2\sin\left(\dfrac{x}{2}\right)\cos\left(\dfrac{x}{2}\right) = \sin x$

35. $1 + \dfrac{\sin\left(\dfrac{x}{2}\right)}{\cos\left(\dfrac{x}{2}\right)} = \dfrac{1 - \cos x + \sin x}{\sin x}$

36. $2\cos^2\left(\dfrac{x}{2}\right) = 1 \pm \sqrt{1 - \sin^2 x}$

37. $\tan^2\left(\dfrac{x}{2}\right) = \dfrac{1 - \cos x}{1 + \cos x}$

38. $\tan^2\left(\dfrac{x}{2}\right) = (\csc x - \cot x)^2$

39. $\tan\left(\dfrac{A}{2}\right) + \cot\left(\dfrac{A}{2}\right) = 2\csc A$

40. $\cot\left(\dfrac{A}{2}\right) - \tan\left(\dfrac{A}{2}\right) = 2\cot A$

41. $\csc^2\left(\dfrac{A}{2}\right) = \dfrac{2(1 + \cos A)}{\sin^2 A}$

42. $\sec^2\left(\dfrac{A}{2}\right) = \dfrac{2(1 - \cos A)}{\sin^2 A}$

43. $\csc\left(\dfrac{A}{2}\right) = \pm|\csc A|\sqrt{2 + 2\cos A}$

44. $\sec\left(\dfrac{A}{2}\right) = \pm|\csc A|\sqrt{2 - 2\cos A}$

In Exercises 45–48, graph the functions.

45. $y = 4\cos^2\left(\dfrac{x}{2}\right)$

46. $y = -6\sin^2\left(\dfrac{x}{2}\right)$

47. $y = \dfrac{1 - \tan^2\left(\dfrac{x}{2}\right)}{1 + \tan^2\left(\dfrac{x}{2}\right)}$

48. $y = 1 - \left[\sin\left(\dfrac{x}{2}\right) + \cos\left(\dfrac{x}{2}\right)\right]^2$

• APPLICATIONS

49. Area of an Isosceles Triangle. Consider the triangle below, where the vertex angle measures θ, the equal sides measure a, the height is h, and half the base is b. (In an isosceles triangle, the perpendicular dropped from the vertex angle divides the triangle into two congruent triangles.) The two triangles formed are right triangles.

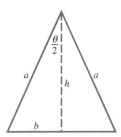

In the right triangles, $\sin\left(\dfrac{\theta}{2}\right) = \dfrac{b}{a}$ and $\cos\left(\dfrac{\theta}{2}\right) = \dfrac{h}{a}$. Multiply each side of each equation by a to get $b = a\sin\left(\dfrac{\theta}{2}\right)$, $h = a\cos\left(\dfrac{\theta}{2}\right)$. The area of the entire isosceles triangle is $A = \frac{1}{2}(2b)h = bh$. Substitute the values for b and h into the area formula. Show that the area is equivalent to $\dfrac{a^2}{2}\sin\theta$.

50. Area of an Isosceles Triangle. Use the results from Exercise 49 to find the area of an isosceles triangle whose equal sides measure 7 inches and whose base angles each measure 75°.

51. Biking. A bicycle ramp is made so that it can easily be raised and lowered for different levels of competition. For the advance division, the angle formed by the ramp and the ground is θ such that $\tan\theta$, the steepness of the ramp, is $\frac{5}{3}$. For the novice division, the angle θ is cut in half to lower the ramp. What is the steepness of the ramp for angle $\dfrac{\theta}{2}$?

52. Biking. A bicycle ramp is made so that it can easily be raised and lowered for different levels of competition. For the advance division, the angle formed by the ramp and the ground is θ such that $\sin\theta = \dfrac{2\sqrt{2}}{3}$. For the novice division, the angle θ is cut in half to lower the ramp. What is $\tan\left(\dfrac{\theta}{2}\right)$, the steepness of the ramp?

53. Farming. An auger used to deliver grain to a storage bin can be raised and lowered, thus allowing for different-size bins. Let α be the angle formed by the auger and the ground for bin A such that $\sin\alpha = \frac{40}{41}$. The angle formed by the auger and the ground for bin B is half of α. If the height h, in feet, of a bin can be found using the formula $h = 75\sin\theta$, where θ is the angle formed by the ground and the auger, find the height of bin B.

54. Farming. An auger used to deliver grain to a storage bin can be raised and lowered, thus allowing for different-size bins. Let α be the angle formed by the auger and the ground for bin A such that $\tan\alpha = \frac{24}{7}$. The angle formed by the auger and the ground for bin B is half of α. If the height h, in feet, of a bin can be found using the formula $h = 75\sin\theta$, where θ is the angle formed by the ground and the auger, find the height of bin B.

For Exercises 55 and 56, refer to the following:

Monthly profits can be expressed as a function of sales, that is, $p(s)$. A financial analysis of a company has determined that the sales s in thousands of dollars are also related to monthly profits p in thousands of dollars by the relationship:

$$\tan\theta = \dfrac{p}{s} \quad \text{for} \quad 0 \le s \le 50, 0 \le p < 40$$

Based on sales and profits, it can be determined that the domain for angle θ is $0 \le \theta \le 38°$.

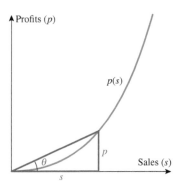

55. Business. If monthly profits are $3000 and monthly sales are $4000, find $\tan\left(\dfrac{\theta}{2}\right)$.

56. Business. If monthly profits are p and monthly sales are s (where $p < s$), find $\tan\left(\dfrac{\theta}{2}\right)$.

• **CATCH THE MISTAKE**

In Exercises 57 and 58, explain the mistake that is made.

57. If $\cos x = -\dfrac{1}{3}$, find $\sin\left(\dfrac{x}{2}\right)$ given $\pi < x < \dfrac{3\pi}{2}$.

Solution:

Write the half-angle identity for the sine function.	$\sin\left(\dfrac{x}{2}\right) = \pm\sqrt{\dfrac{1 - \cos x}{2}}$
Substitute $\cos x = -\dfrac{1}{3}$.	$\sin\left(\dfrac{x}{2}\right) = \pm\sqrt{\dfrac{1 + \dfrac{1}{3}}{2}}$
Simplify.	$\sin\left(\dfrac{x}{2}\right) = \pm\sqrt{\dfrac{2}{3}}$
The sine function is negative.	$\sin\left(\dfrac{x}{2}\right) = -\sqrt{\dfrac{2}{3}}$

This is incorrect. What mistake was made?

58. If $\cos x = \dfrac{1}{3}$, find $\tan^2\left(\dfrac{x}{2}\right)$.

Solution:

Use the quotient identity.	$\tan^2\left(\dfrac{x}{2}\right) = \dfrac{\sin^2\left(\dfrac{x}{2}\right)}{\cos^2 x}$
Use the half-angle identity for sine.	$\tan^2\left(\dfrac{x}{2}\right) = \dfrac{\dfrac{1 - \cos x}{2}}{\cos^2 x}$
Simplify.	$\tan^2\left(\dfrac{x}{2}\right) = \dfrac{1}{2}\left(\dfrac{1}{\cos^2 x} - \dfrac{\cos x}{\cos^2 x}\right)$
	$\tan^2\left(\dfrac{x}{2}\right) = \dfrac{1}{2}\left(\dfrac{1}{\cos^2 x} - \dfrac{1}{\cos x}\right)$
Substitute $\cos x = \dfrac{1}{3}$.	$\tan^2\left(\dfrac{x}{2}\right) = \dfrac{1}{2}\left(\dfrac{1}{\dfrac{1}{9}} - \dfrac{1}{\dfrac{1}{3}}\right)$
	$\tan^2\left(\dfrac{x}{2}\right) = 3$

This is incorrect. What mistake was made?

• **CONCEPTUAL**

In Exercises 59–62, determine whether each statement is true or false.

59. $\sin\left(\dfrac{A}{2}\right) + \sin\left(\dfrac{A}{2}\right) = \sin A$

60. $\cos\left(\dfrac{A}{2}\right) + \cos\left(\dfrac{A}{2}\right) = \cos A$

61. If $\tan x > 0$, then $\tan\left(\dfrac{x}{2}\right) > 0$.

62. If $\sin x > 0$, then $\sin\left(\dfrac{x}{2}\right) > 0$.

63. Given $\tan\left(\dfrac{A}{2}\right) = \pm\sqrt{\dfrac{1 - \cos A}{1 + \cos A}}$, verify that $\tan\left(\dfrac{A}{2}\right) = \dfrac{\sin A}{1 + \cos A}$. Substitute $A = \pi$ into the identity and explain your results.

64. Given $\tan\left(\dfrac{A}{2}\right) = \pm\sqrt{\dfrac{1 - \cos A}{1 + \cos A}}$, verify that $\tan\left(\dfrac{A}{2}\right) = \dfrac{1 - \cos A}{\sin A}$. Substitute $A = \pi$ into the identity and explain your results.

• CHALLENGE

65. Find the exact value of $\sin 15°$ in two ways, using sum and difference identities and half-angle identities; then show that they are equal.

66. Find the exact value of $\tan 15°$ in two ways, using sum and difference identities and half-angle identities; then show that they are equal.

67. Express $\tan\left(\dfrac{x}{4}\right)$ in terms of the cosine of a single angle.

68. Express $\sin\left(\dfrac{x}{4}\right)$ in terms of the cosine of a single angle.

• TECHNOLOGY

For Exercises 69–72, refer to the following:

One cannot *prove* that an equation is an identity using technology, but one can use it as a first step to see whether the equation *seems* to be an identity.

69. Using a graphing calculator, plot $Y_1 = \left(\dfrac{x}{2}\right) - \dfrac{\left(\dfrac{x}{2}\right)^3}{3!} + \dfrac{\left(\dfrac{x}{2}\right)^5}{5!}$ and $Y_2 = \sin\left(\dfrac{x}{2}\right)$ for x range $[-1, 1]$. Is Y_1 a good approximation to Y_2?

70. Using a graphing calculator, plot $Y_1 = 1 - \dfrac{\left(\dfrac{x}{2}\right)^2}{2!} + \dfrac{\left(\dfrac{x}{2}\right)^4}{4!}$ and $Y_2 = \cos\left(\dfrac{x}{2}\right)$ for x range $[-1, 1]$. Is Y_1 a good approximation to Y_2?

71. Using a graphing calculator, determine whether $\csc^2\left(\dfrac{x}{2}\right) + \sec^2\left(\dfrac{x}{2}\right) = 4\csc^2 x$ is an identity by plotting each side of the equation and seeing if the graphs coincide.

72. Using a graphing calculator, determine whether $\tan^2\left(\dfrac{x}{2}\right) + \cot^2\left(\dfrac{x}{2}\right) = -2\cot^2 x \sec x$ is an identity by plotting each side of the equation and seeing if the graphs coincide.

5.5 PRODUCT-TO-SUM AND SUM-TO-PRODUCT IDENTITIES

SKILLS OBJECTIVES	CONCEPTUAL OBJECTIVES
▪ Express products of trigonometric functions as sums of trigonometric functions. ▪ Express sums of trigonometric functions as products of trigonometric functions.	▪ Understand that the sum and difference identities are used to derive product-to-sum identities. ▪ Understand that the product-to-sum identities are used to derive the sum-to-product identities.

Often in calculus it is helpful to write products of trigonometric functions as sums of other trigonometric functions and vice versa. In this section, we discuss the *product-to-sum identities*, which convert products of trigonometric functions to sums of trigonometric functions, and *sum-to-product identities*, which convert sums of trigonometric functions to products of trigonometric functions.

5.5.1 Product-to-Sum Identities

The *product-to-sum identities* are derived from the sum and difference identities.

WORDS	MATH
Write the identity for the cosine of a sum.	$\cos A \cos B - \sin A \sin B = \cos(A + B)$
Write the identity for the cosine of a difference.	$\cos A \cos B + \sin A \sin B = \cos(A - B)$
Add the two identities.	$2\cos A \cos B = \cos(A + B) + \cos(A - B)$
Divide both sides by 2.	$\boxed{\cos A \cos B = \tfrac{1}{2}[\cos(A + B) + \cos(A - B)]}$
Subtract the sum identity from the difference identity.	$\cos A \cos B + \sin A \sin B = \cos(A - B)$ $-\cos A \cos B + \sin A \sin B = -\cos(A + B)$ $2\sin A \sin B = \cos(A - B) - \cos(A + B)$
Divide both sides by 2.	$\boxed{\sin A \sin B = \tfrac{1}{2}[\cos(A - B) - \cos(A + B)]}$
Write the identity for the sine of a sum.	$\sin A \cos B + \cos A \sin B = \sin(A + B)$
Write the identity for the sine of a difference.	$\sin A \cos B - \cos A \sin B = \sin(A - B)$
Add the two identities.	$2\sin A \cos B = \sin(A + B) + \sin(A - B)$
Divide both sides by 2.	$\boxed{\sin A \cos B = \tfrac{1}{2}[\sin(A + B) + \sin(A - B)]}$

5.5.1 SKILL

Express products of trigonometric functions as sums of trigonometric functions.

5.5.1 CONCEPTUAL

Understand that the sum and difference identities are used to derive product-to-sum identities.

PRODUCT-TO-SUM IDENTITIES

1. $\cos A \cos B = \tfrac{1}{2}[\cos(A + B) + \cos(A - B)]$
2. $\sin A \sin B = \tfrac{1}{2}[\cos(A - B) - \cos(A + B)]$
3. $\sin A \cos B = \tfrac{1}{2}[\sin(A + B) + \sin(A - B)]$

EXAMPLE 1   Illustrating a Product-to-Sum Identity for Specific Values

Show that product-to-sum identity (3) is true when $A = 30°$ and $B = 90°$.

Solution:

Write product-to-sum identity (3).

$$\sin A \cos B = \frac{1}{2}[\sin(A + B) + \sin(A - B)]$$

Let $A = 30°$ and $B = 90°$.

$$\sin 30° \cos 90° = \frac{1}{2}[\sin(30° + 90°) + \sin(30° - 90°)]$$

Simplify.

$$\sin 30° \cos 90° = \frac{1}{2}[\sin 120° + \sin(-60°)]$$

Evaluate the trigonometric functions.

$$\frac{1}{2} \cdot 0 = \frac{1}{2}\left[\frac{\sqrt{3}}{2} - \frac{\sqrt{3}}{2}\right]$$

Simplify.

$$0 = 0$$

EXAMPLE 2 **Converting a Product to a Sum**

Convert the product $\cos(4x)\cos(3x)$ to a sum.

Solution:

Write the product-to-sum
identity (1).

$$\cos A \cos B = \frac{1}{2}[\cos(A + B) + \cos(A - B)]$$

Let $A = 4x$ and $B = 3x$.

$$\cos(4x)\cos(3x) = \frac{1}{2}[\cos(4x + 3x) + \cos(4x - 3x)]$$

Simplify.

$$\boxed{\cos(4x)\cos(3x) = \frac{1}{2}[\cos(7x) + \cos x]}$$

▼

▼ **ANSWER**
$\frac{1}{2}[\cos(7x) + \cos(3x)]$

YOUR TURN Convert the product $\cos(2x)\cos(5x)$ to a sum.

▶ **EXAMPLE 3** **Converting Products to Sums**

Express $\sin(2x)\sin(3x)$ in terms of cosines.

Solution:

Write the product-to-sum
identity (2).

$$\sin A \sin B = \frac{1}{2}[\cos(A - B) - \cos(A + B)]$$

Let $A = 2x$ and $B = 3x$.

$$\sin(2x)\sin(3x) = \frac{1}{2}[\cos(2x - 3x) - \cos(2x + 3x)]$$

Simplify.

$$\sin(2x)\sin(3x) = \frac{1}{2}[\cos(-x) - \cos(5x)]$$

The cosine function is an
even function.

$$\boxed{\sin(2x)\sin(3x) = \frac{1}{2}[\cos x - \cos(5x)]}$$

▼

▼ **ANSWER**
$\frac{1}{2}[\cos(x) - \cos(3x)]$

YOUR TURN Express $\sin x \sin(2x)$ in terms of cosines.

5.5.2 Sum-to-Product Identities

The *sum-to-product identities* can be obtained from the product-to-sum identities.

WORDS	MATH
Write the identity for the product of the sine and cosine functions.	$\frac{1}{2}[\sin(x + y) + \sin(x - y)] = \sin x \cos y$
Let $x + y = A$ and $x - y = B$.	$\begin{array}{cc} x + y = A & x + y = A \\ \underline{x - y = B} & \underline{-x + y = -B} \\ 2x = A + B & 2y = A - B \end{array}$
Solve for x and y in term of A and B.	$x = \dfrac{A + B}{2} \qquad y = \dfrac{A - B}{2}$
Substitute these values into the identity.	$\frac{1}{2}(\sin A + \sin B) = \sin\left(\dfrac{A + B}{2}\right)\cos\left(\dfrac{A - B}{2}\right)$
Multiply by 2.	$\sin A + \sin B = 2\sin\left(\dfrac{A + B}{2}\right)\cos\left(\dfrac{A - B}{2}\right)$

The other four *sum-to-product* identities can be found similarly. All are summarized in the following box:

SUM-TO-PRODUCT IDENTITIES

4. $\sin A + \sin B = 2\sin\left(\dfrac{A + B}{2}\right)\cos\left(\dfrac{A - B}{2}\right)$

5. $\sin A - \sin B = 2\sin\left(\dfrac{A - B}{2}\right)\cos\left(\dfrac{A + B}{2}\right)$

6. $\cos A + \cos B = 2\cos\left(\dfrac{A + B}{2}\right)\cos\left(\dfrac{A - B}{2}\right)$

7. $\cos A - \cos B = -2\sin\left(\dfrac{A + B}{2}\right)\sin\left(\dfrac{A - B}{2}\right)$

EXAMPLE 4 **Illustrating a Sum-to-Product Identity for Specific Values**

Show the sum-to-product identity (7) is true when $A = 30°$ and $B = 90°$.

Solution:

Write the sum-to-product identity (7).
$$\cos A - \cos B = -2\sin\left(\dfrac{A + B}{2}\right)\sin\left(\dfrac{A - B}{2}\right)$$

Let $A = 30°$ and $B = 90°$.
$$\cos 30° - \cos 90° = -2\sin\left(\dfrac{30° + 90°}{2}\right)\sin\left(\dfrac{30° - 90°}{2}\right)$$

Simplify.
$$\cos 30° - \cos 90° = -2\sin 60° \sin(-30°)$$

The sine function is an odd function.
$$\cos 30° - \cos 90° = 2\sin 60° \sin 30°$$

Evaluate the trigonometric functions.
$$\dfrac{\sqrt{3}}{2} - 0 = 2 \cdot \dfrac{\sqrt{3}}{2} \cdot \dfrac{1}{2}$$

Simplify.
$$\dfrac{\sqrt{3}}{2} = \dfrac{\sqrt{3}}{2}$$

▶ **EXAMPLE 5** **Converting a Sum to a Product**

Convert $-9[\sin(2x) - \sin(10x)]$ to a product.

Solution:

The expression inside the brackets is in the form of identity (5).
$$\sin A - \sin B = 2\sin\left(\dfrac{A - B}{2}\right)\cos\left(\dfrac{A + B}{2}\right)$$

Let $A = 2x$ and $B = 10x$.
$$\sin(2x) - \sin(10x) = 2\sin\left(\dfrac{2x - 10x}{2}\right)\cos\left(\dfrac{2x + 10x}{2}\right)$$

Simplify.
$$\sin(2x) - \sin(10x) = 2\sin(-4x)\cos(6x)$$

The sine function is an odd function.
$$\sin(2x) - \sin(10x) = -2\sin(4x)\cos(6x)$$

Multiply both sides by -9.
$$\boxed{-9[\sin(2x) - \sin(10x)] = 18\sin(4x)\cos(6x)}$$

EXAMPLE 6 **Simplifying a Trigonometric Expression Using Sum-to-Product Identities**

Simplify the expression $\sin\left(\dfrac{x+y}{2}\right)\cos\left(\dfrac{x-y}{2}\right) + \sin\left(\dfrac{x-y}{2}\right)\cos\left(\dfrac{x+y}{2}\right)$.

Solution:

Use identities (4) and (5).

$$\underbrace{\sin\left(\frac{x+y}{2}\right)\cos\left(\frac{x-y}{2}\right)}_{\frac{1}{2}(\sin x + \sin y)} + \underbrace{\sin\left(\frac{x-y}{2}\right)\cos\left(\frac{x+y}{2}\right)}_{\frac{1}{2}(\sin x - \sin y)}$$

$$= \frac{1}{2}\sin x + \frac{1}{2}\sin y + \frac{1}{2}\sin x - \frac{1}{2}\sin y$$

Simplify.

$$= \boxed{\sin x}$$

Applications

In music, a note is a fixed pitch (frequency) that is given a name. If two notes are sounded simultaneously, then they combine to produce another note often called a "beat." The beat frequency is the difference of the two frequencies. The more rapid the beat, the further apart the two frequencies of the notes are. When musicians "tune" their instruments, they use a tuning fork to sound a note and then tune the instrument until the beat is eliminated; hence, the fork and instrument are in tune with each other. Mathematically, a note or tone is represented as $A\cos(2\pi ft)$, where A is the amplitude (loudness), f is the frequency in hertz, and t is time in seconds. The following figure summarizes common notes and frequencies:

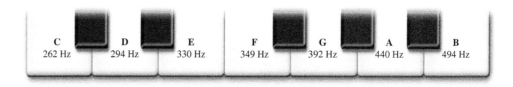

C	D	E	F	G	A	B
262 Hz	294 Hz	330 Hz	349 Hz	392 Hz	440 Hz	494 Hz

EXAMPLE 7 **Music**

Express the musical tone when a C and G are simultaneously struck (assume with the same loudness).

Find the beat frequency $f_2 - f_1$ Assume uniform loudness, $A = 1$.

Solution:

Write the mathematical description of a C note. $\cos(2\pi f_1 t)$, $f_1 = 262\,\text{Hz}$

Write the mathematical description of a G note. $\cos(2\pi f_2 t)$, $f_2 = 392\,\text{Hz}$

Add the two notes. $\cos(524\pi t) + \cos(784\pi t)$

Use a sum-to-product identity:
$\cos(524\pi t) + \cos(784\pi t)$.

$$= 2\cos\left(\frac{524\pi t + 784\pi t}{2}\right)\cos\left(\frac{524\pi t - 784\pi t}{2}\right)$$

Simplify.

$$= 2\cos(654\pi t)\cos(-130\pi t)$$

Cosine is an even function:
$\cos(-x) = \cos x$.

$$= 2\cos(654\pi t)\cos(130\pi t)$$

Identify average frequency and beat of the tone.

$$= 2\cos(2 \cdot \underbrace{327}_{\text{average frequency}}\pi t)\cos(\underbrace{130}_{\text{beats per second}}\pi t)$$

The beat frequency can also be found by subtracting f_1 from f_2.

$$f_2 - f_1 = 392 - 262 = 130\,\text{Hz}$$

Therefore, the tone of average frequency, 327 hertz, has a beat of 130 hertz (beats/per second).

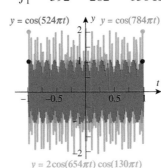

$$y = \cos(524\pi t) \qquad y = \cos(784\pi t)$$

$$y = 2\cos(654\pi t)\cos(130\pi t)$$

▶[SECTION 5.5] SUMMARY

In this section, we used the sum and difference identities (Section 5.2) to derive the product-to-sum identities. The product-to-sum identities allowed us to express products of trigonometric functions as sums of trigonometric functions.

$$\cos A \cos B = \tfrac{1}{2}[\cos(A + B) + \cos(A - B)]$$
$$\sin A \sin B = \tfrac{1}{2}[\cos(A - B) - \cos(A + B)]$$
$$\sin A \cos B = \tfrac{1}{2}[\sin(A + B) + \sin(A - B)]$$

We then used the product-to-sum identities to derive the sum-to-product identities. The sum-to-product identities allow us to express sums as products.

$$\sin A + \sin B = 2\sin\left(\frac{A + B}{2}\right)\cos\left(\frac{A - B}{2}\right)$$

$$\sin A - \sin B = 2\sin\left(\frac{A - B}{2}\right)\cos\left(\frac{A + B}{2}\right)$$

$$\cos A + \cos B = 2\cos\left(\frac{A + B}{2}\right)\cos\left(\frac{A - B}{2}\right)$$

$$\cos A - \cos B = -2\sin\left(\frac{A + B}{2}\right)\sin\left(\frac{A - B}{2}\right)$$

[SECTION 5.5] EXERCISES

• **SKILLS**

In Exercises 1–12, write each product as a sum or difference of sines and/or cosines.

1. $\sin(2x)\cos x$

2. $\cos(10x)\sin(5x)$

3. $5\sin(4x)\sin(6x)$

4. $-3\sin(2x)\sin(4x)$

5. $4\cos(-x)\cos(2x)$

6. $-8\cos(3x)\cos(5x)$

7. $\sin\left(\dfrac{3x}{2}\right)\sin\left(\dfrac{5x}{2}\right)$

8. $\sin\left(\dfrac{\pi x}{2}\right)\sin\left(\dfrac{5\pi x}{2}\right)$

9. $\cos\left(\dfrac{2x}{3}\right)\cos\left(\dfrac{4x}{3}\right)$

10. $\sin\left(-\dfrac{\pi}{4}x\right)\cos\left(-\dfrac{\pi}{2}x\right)$

11. $\sin(97.5°)\sin(22.5°)$

12. $\cos(85.5°)\cos(4.5°)$

In Exercises 13–24, write each expression as a product of sines and/or cosines.

13. $\cos(5x) + \cos(3x)$

14. $\cos(2x) - \cos(4x)$

15. $\sin(3x) - \sin x$

16. $\sin(10x) + \sin(5x)$

17. $\sin(8x) - \sin(6x)$

18. $\cos(2x) - \cos(6x)$

19. $\sin\left(\dfrac{x}{2}\right) - \sin\left(\dfrac{5x}{2}\right)$

20. $\cos\left(\dfrac{x}{2}\right) - \cos\left(\dfrac{5x}{2}\right)$

21. $\cos\left(\dfrac{2}{3}x\right) + \cos\left(\dfrac{7}{3}x\right)$

22. $\sin\left(\dfrac{2}{3}x\right) + \sin\left(\dfrac{7}{3}x\right)$

23. $\sin(0.4x) + \sin(0.6x)$

24. $\cos(0.3x) - \cos(0.5x)$

In Exercises 25–28, simplify the trigonometric expressions.

25. $\dfrac{\cos(3x) - \cos x}{\sin(3x) + \sin x}$

26. $\dfrac{\sin(4x) + \sin(2x)}{\cos(4x) - \cos(2x)}$

27. $\dfrac{\cos x - \cos(3x)}{\sin(3x) - \sin x}$

28. $\dfrac{\sin(4x) + \sin(2x)}{\cos(4x) + \cos(2x)}$

In Exercises 29–38, verify the identities.

29. $\dfrac{\sin A + \sin B}{\cos A + \cos B} = \tan\left(\dfrac{A + B}{2}\right)$

30. $\dfrac{\sin A - \sin B}{\cos A + \cos B} = \tan\left(\dfrac{A - B}{2}\right)$

31. $\dfrac{\cos A - \cos B}{\sin A + \sin B} = -\tan\left(\dfrac{A - B}{2}\right)$

32. $\dfrac{\cos A - \cos B}{\sin A - \sin B} = -\tan\left(\dfrac{A + B}{2}\right)$

33. $\dfrac{\sin A + \sin B}{\sin A - \sin B} = \tan\left(\dfrac{A + B}{2}\right)\cot\left(\dfrac{A - B}{2}\right)$

34. $\dfrac{\cos A - \cos B}{\cos A + \cos B} = -\tan\left(\dfrac{A + B}{2}\right)\tan\left(\dfrac{A - B}{2}\right)$

35. $\sin A + \sin B = \pm\sqrt{[1 - \cos(A + B)][1 + \cos(A - B)]}$

36. $\sin A \cos B = \cos A \sin B + \sin(A - B)$

37. $\dfrac{\cos^2 A - \cos^2 B}{\cos A + \cos B} = -2\left[\sin\left(\dfrac{A}{2}\right)\cos\left(\dfrac{B}{2}\right) + \cos\left(\dfrac{A}{2}\right)\sin\left(\dfrac{B}{2}\right)\right]\left[\sin\left(\dfrac{A}{2}\right)\cos\left(\dfrac{B}{2}\right) - \cos\left(\dfrac{A}{2}\right)\sin\left(\dfrac{B}{2}\right)\right]$

38. $\sin A = \sin\left(\dfrac{A + B}{2}\right)\cos\left(\dfrac{A - B}{2}\right) + \sin\left(\dfrac{A - B}{2}\right)\cos\left(\dfrac{A + B}{2}\right)$

• APPLICATIONS

39. **Business.** An analysis of the monthly costs and monthly revenues of a toy store indicates that monthly costs fluctuate (increase and decrease) according to the function

$$C(t) = \sin\left(\dfrac{\pi}{6}t + \pi\right)$$

and monthly revenues fluctuate (increase and decrease) according to the function

$$R(t) = \sin\left(\dfrac{\pi}{6}t + \dfrac{5\pi}{3}\right)$$

Find the function that describes how the monthly profits fluctuate: $P(t) = R(t) - C(t)$. Using identities in this section, express $P(t)$ in terms of a cosine function.

40. **Business.** An analysis of the monthly costs and monthly revenues of an electronics manufacturer indicates that monthly costs fluctuate (increase and decrease) according to the function

$$C(t) = \cos\left(\dfrac{\pi}{3}t + \dfrac{\pi}{3}\right)$$

and monthly revenues fluctuate (increase and decrease) according to the function

$$R(t) = \cos\left(\dfrac{\pi}{3}t\right)$$

Find the function that describes how the monthly profits fluctuate: $P(t) = R(t) - C(t)$. Using identities in this section, express $P(t)$ in terms of a sine function.

41. **Music.** Write a mathematical description of a tone that results from simultaneously playing a G and a B. What is the beat frequency? What is the average frequency?

42. **Music.** Write a mathematical description of a tone that results from simultaneously playing an F and an A. What is the beat frequency? What is the average frequency?

43. **Optics.** Two optical signals with uniform $(A = 1)$ intensities and wavelengths of 1.55 micrometer and 0.63 micrometer are "beat" together. What is the resulting sum if their individual signals are given by $\sin\left(\dfrac{2\pi tc}{1.55\mu m}\right)$ and $\sin\left(\dfrac{2\pi tc}{0.63\mu m}\right)$, where $c = 3.0 \times 10^8$ meters per second? (*Note*: 1 μm = 10^{-6} m.)

44. **Optics.** The two optical signals in Exercise 43 are beat together. What are the average frequency and the beat frequency?

For Exercises 45 and 46, refer to the following:

Touch-tone keypads have the following simultaneous low and high frequencies:

FREQUENCY	1209 Hz	1336 Hz	1477 Hz
697 Hz	1	2	3
770 Hz	4	5	6
852 Hz	7	8	9
941 Hz	*	0	#

The signal made when a key is pressed is $\sin(2\pi f_1 t) + \sin(2\pi f_2 t)$, where f_1 is the low frequency and f_2 is the high frequency.

45. **Touch-Tone Dialing.** What is the mathematical function that models the sound of dialing 4?

46. Touch-Tone Dialing. What is the mathematical function that models the sound of dialing 3?

47. Area of a Triangle. A formula for finding the area of a triangle when given the measures of the angles and one side is area $= \dfrac{a^2 \sin B \sin C}{2 \sin A}$, where a is the side opposite angle A. If the measures of angles B and C are 52.5° and 7.5°, respectively, and if $a = 10$ feet, use the appropriate product-to-sum identity to change the formula so that you can solve for the area of the triangle exactly.

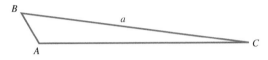

48. Area of a Triangle. If the measures of angles B and C in Exercise 47 are 75° and 45°, respectively, and if $a = 12$ inches, use the appropriate product-to-sum identity to change the formula so that you can solve for the area of the triangle exactly.

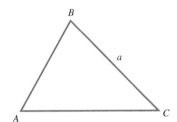

49. Calculus. In calculus, there is an operation called integration that serves a number of purposes. When performing integration on trigonometric functions, it is much easier if the expression contains a single trigonometric function or the sum of trigonometric functions instead of the product of trigonometric functions. Using identities, change the following expression so that it does not contain the product of trigonometric functions.

$$\sin A \sin B + \sin A \cos B + \cos A \cos B$$

50. Calculus. In calculus, there is an operation called integration that serves a number of purposes. When performing integration on trigonometric functions, it is much easier if the expression contains a single trigonometric function or the sum of trigonometric functions instead of the product of trigonometric functions. Using identities, change the following expression so that it does not contain the product of trigonometric functions.

$$2\cos\left(\frac{A-B}{2}\right)\left[\sin\left(\frac{A+B}{2}\right) + \cos\left(\frac{A+B}{2}\right)\right]$$

• **CATCH THE MISTAKE**

In Exercises 51 and 52, explain the mistake that is made.

51. Simplify the expression
$(\cos A - \cos B)^2 + (\sin A - \sin B)^2$.

Solution:

Expand by squaring.

$$= \cos^2 A - 2\cos A \cos B + \cos^2 B + \sin^2 A - 2\sin A \sin B + \sin^2 B$$

Regroup terms.

$$= \cos^2 A + \sin^2 A - 2\cos A \cos B - 2\sin A \sin B + \cos^2 B + \sin^2 B$$

Simplify using the Pythagorean identity.

$$= \underbrace{\cos^2 A + \sin^2 A}_{1} - 2\cos A \cos B - 2\sin A \sin B + \underbrace{\cos^2 B + \sin^2 B}_{1}$$

Factor the comon 2.

$$= 2(1 - \cos A \cos B - \sin A \sin B)$$

Simplify.

$$= 2[1 - \cos(AB) - \sin(AB)]$$

This is incorrect. What mistakes were made?

52. Simplify the expression $(\sin A - \sin B)(\cos A + \cos B)$.

Solution:

Multiply the expressions using the distributive property.

$$= \sin A \cos A + \sin A \cos B - \sin B \cos A - \sin B \cos B$$

Cancel the second and third terms.

$$= \sin A \cos A - \sin B \cos B$$

Use the product-to-sum identity.

$$= \underset{\frac{1}{2}[\sin(A+A)+\sin(A-A)]}{\underbrace{\sin A \cos A}} - \underset{\frac{1}{2}[\sin(B+B)+\sin(B-B)]}{\underbrace{\sin B \cos B}}$$

Simplify.

$$= \frac{1}{2}\sin(2A) - \frac{1}{2}\sin(2B)$$

This is incorrect. What mistake was made?

• **CONCEPTUAL**

In Exercises 53–56, determine whether each statement is true or false.

53. $\cos A \cos B = \cos(AB)$

54. $\sin A \sin B = \sin(AB)$

55. The product of two cosine functions is a sum of two other cosine functions.

56. The product of two sine functions is a difference of two cosine functions.

57. Write $\sin A \sin B \sin C$ as a sum or difference of sines and cosines.

58. Write $\cos A \cos B \cos C$ as a sum or difference of sines and cosines.

• **CHALLENGE**

59. Find all values of A and B such that $\cos A \cos B = \cos(A + B)$.

60. Find all values of A and B such that $\sin A \cos B = \sin(A + B)$.

61. Let $A = B$ and then verify that the identity for $\sin A \cos B$ still holds true.

62. Let $A = B$ and then verify that the identity for $\sin A \sin B$ still holds true.

63. Find an expression in terms of cosine for $\sin A + \sin B$ if A and B are complementary angles.

64. Find an expression in terms of cosine for $\cos A - \cos B$ if A and B are complementary angles.

• **TECHNOLOGY**

65. Suggest an identity $4\sin x \cos x \cos(2x) = $ _____ by graphing $Y_1 = 4\sin x \cos x \cos(2x)$ and determining the function based on the graph.

66. Suggest an identity $1 + \tan x \tan(2x) = $ _____ by graphing $Y_1 = 1 + \tan x \tan(2x)$ and determining the function based on the graph.

▶[CHAPTER 5 REVIEW]

SECTION	CONCEPT	KEY IDEAS/FORMULAS
5.1	Trigonometric identities	Identities must hold for *all* values of x (not just some values of x) for which both sides of the equation are defined.
	Verifying trigonometric identities	**Reciprocal identities** $$\csc x = \frac{1}{\sin x} \quad \sec x = \frac{1}{\cos x} \quad \cot x = \frac{1}{\tan x}$$ **Quotient identities** $$\tan x = \frac{\sin x}{\cos x} \quad \cot x = \frac{\cos x}{\sin x}$$ **Pythagorean identities** $$\sin^2 x + \cos^2 x = 1$$ $$\tan^2 x + 1 = \sec^2 x$$ $$1 + \cot^2 x = \csc^2 x$$ **Even–odd identities** Even: $f(-x) = f(x)$ symmetry about the y-axis Odd: $f(-x) = -f(x)$ symmetry about the origin Odd: $\quad\quad\quad\quad\quad$ Even: $\sin(-x) = -\sin x \quad\quad \cos(-x) = \cos x$ $\tan(-x) = -\tan x \quad\quad \sec(-x) = \sec x$ $\csc(-x) = -\csc x$ $\cot(-x) = -\cot x$ **Verifying identities** ▪ Convert all trigonometric functions to sines and cosines. ▪ Write all sums or differences of quotients as a single quotient.
5.2	Sum and difference identities	
	Sum and difference identities for the cosine function	$\cos(A + B) = \cos A \cos B - \sin A \sin B$ $\cos(A - B) = \cos A \cos B + \sin A \sin B$
	Sum and difference identities for the sine function	$\sin(A + B) = \sin A \cos B + \cos A \sin B$ $\sin(A - B) = \sin A \cos B - \cos A \sin B$
	Sum and difference identities for the tangent function	$\tan(A + B) = \dfrac{\tan A + \tan B}{1 - \tan A \tan B}$ $\tan(A - B) = \dfrac{\tan A - \tan B}{1 + \tan A \tan B}$
5.3	Double-angle identities	
	Applying double-angle identities	$\sin(2A) = 2\sin A \cos A$ $\cos(2A) = \cos^2 A - \sin^2 A$ $\quad\quad\quad = 1 - 2\sin^2 A = 2\cos^2 A - 1$ $\tan(2A) = \dfrac{2\tan A}{1 - \tan^2 A}$

SECTION	CONCEPT	KEY IDEAS/FORMULAS
5.4	**Half-angle identities**	
	Applying half-angle identities	$\sin\left(\dfrac{A}{2}\right) = \pm\sqrt{\dfrac{1 - \cos A}{2}}$ $\cos\left(\dfrac{A}{2}\right) = \pm\sqrt{\dfrac{1 + \cos A}{2}}$ $\tan\left(\dfrac{A}{2}\right) = \pm\sqrt{\dfrac{1 - \cos A}{1 + \cos A}} = \dfrac{\sin A}{1 + \cos A} = \dfrac{1 - \cos A}{\sin A}$
5.5	**Product-to-sum and sum-to-product identities**	
	Product-to-sum identities	$\cos A \cos B = \frac{1}{2}[\cos(A + B) + \cos(A - B)]$ $\sin A \sin B = \frac{1}{2}[\cos(A - B) - \cos(A + B)]$ $\sin A \cos B = \frac{1}{2}[\sin(A + B) + \sin(A - B)]$
	Sum-to-product identities	$\sin A + \sin B = 2\sin\left(\dfrac{A + B}{2}\right)\cos\left(\dfrac{A - B}{2}\right)$ $\sin A - \sin B = 2\sin\left(\dfrac{A - B}{2}\right)\cos\left(\dfrac{A + B}{2}\right)$ $\cos A + \cos B = 2\cos\left(\dfrac{A + B}{2}\right)\cos\left(\dfrac{A - B}{2}\right)$ $\cos A - \cos B = -2\sin\left(\dfrac{A + B}{2}\right)\sin\left(\dfrac{A - B}{2}\right)$

[CHAPTER 5 REVIEW EXERCISES]

5.1 Trigonometric Identities

Simplify the following trigonometric expressions.

1. $\tan x(\cot x + \tan x)$

2. $(\sec x + 1)(\sec x - 1)$

3. $\dfrac{\tan^4 x - 1}{\tan^2 x - 1}$

4. $\sec^2 x(\cot^2 x - \cos^2 x)$

5. $\cos x[\cos(-x) - \tan(-x)] - \sin x$

6. $\dfrac{\tan^2 x + 1}{2\sec^2 x}$

Verify the trigonometric identities.

7. $(\tan x + \cot x)^2 - 2 = \tan^2 x + \cot^2 x$

8. $\csc^2 x - \cot^2 x = 1$

9. $\dfrac{1}{\sin^2 x} - \dfrac{1}{\tan^2 x} = 1$

10. $\dfrac{1}{\csc x + 1} + \dfrac{1}{\csc x - 1} = \dfrac{2\tan x}{\cos x}$

11. $\dfrac{\tan^2 x - 1}{\sec^2 x + 3\tan x + 1} = \dfrac{\tan x - 1}{\tan x + 2}$

12. $\cot x(\sec x - \cos x) = \sin x$

Determine whether each of the following equations is an identity, a conditional equation, or a contradiction.

13. $2\tan^2 x + 1 = \dfrac{1 + \sin^2 x}{\cos^2 x}$

14. $\sin x - \cos x = 0$

15. $\cot^2 x - 1 = \tan^2 x$

16. $\cos^2 x(1 + \cot^2 x) = \cot^2 x$

5.2 Sum and Difference Identities

Find exact values for each trigonometric expression.

17. $\cos\left(\dfrac{7\pi}{12}\right)$

18. $\sin\left(\dfrac{\pi}{12}\right)$

19. $\tan(-15°)$

20. $\cot 105°$

Write each expression as a single trigonometric function.

21. $\sin(4x)\cos(3x) - \cos(4x)\sin(3x)$

22. $\sin(-x)\sin(-2x) + \cos(-x)\cos(-2x)$

23. $\dfrac{\tan(5x) - \tan(4x)}{1 + \tan(5x)\tan(4x)}$

24. $\dfrac{\tan\left(\dfrac{\pi}{4}\right) + \tan\left(\dfrac{\pi}{3}\right)}{1 - \tan\left(\dfrac{\pi}{4}\right)\tan\left(\dfrac{\pi}{3}\right)}$

Find the exact value of the indicated expression using the given information and identities.

25. Find the exact value of $\tan(\alpha - \beta)$ if $\sin\alpha = -\frac{3}{5}$ and $\sin\beta = -\frac{24}{25}$, if the terminal side of α lies in quadrant IV and the terminal side of β lies in quadrant III.

26. Find the exact value of $\cos(\alpha + \beta)$ if $\cos\alpha = -\frac{5}{13}$ and $\sin\beta = \frac{7}{25}$, if the terminal side of α lies in quadrant II and the terminal side of β also lies in quadrant II.

27. Find the exact value of $\cos(\alpha - \beta)$ if $\cos\alpha = \frac{9}{41}$ and $\cos\beta = \frac{7}{25}$, if the terminal side of α lies in quadrant IV and the terminal side of β lies in quadrant I.

28. Find the exact value of $\sin(\alpha - \beta)$ if $\sin\alpha = -\frac{5}{13}$ and $\cos\beta = -\frac{4}{5}$, if the terminal side of α lies in quadrant III and the terminal side of β lies in quadrant II.

Determine whether each equation is a conditional equation or an identity.

29. $2\cos A\cos B = \cos(A + B) + \cos(A - B)$

30. $2\sin A\sin B = \cos(A - B) - \cos(A + B)$

Graph the following functions.

31. $y = \cos\left(\dfrac{\pi}{2}\right)\cos x - \sin\left(\dfrac{\pi}{2}\right)\sin x$

32. $y = \sin\left(\dfrac{2\pi}{3}\right)\cos x + \cos\left(\dfrac{2\pi}{3}\right)\sin x$

5.3 Double-Angle Identities

Use double-angle identities to answer the following questions.

33. If $\sin x = \dfrac{3}{5}$ and $\dfrac{\pi}{2} < x < \pi$, find $\cos(2x)$.

34. If $\cos x = \dfrac{7}{25}$ and $\dfrac{3\pi}{2} < x < 2\pi$, find $\sin(2x)$.

35. If $\cot x = -\dfrac{11}{61}$ and $\dfrac{3\pi}{2} < x < 2\pi$, find $\tan(2x)$.

36. If $\tan x = -\dfrac{12}{5}$ and $\dfrac{\pi}{2} < x < \pi$, find $\cos(2x)$.

37. If $\sec x = \dfrac{25}{24}$ and $0 < x < \dfrac{\pi}{2}$, find $\sin(2x)$.

38. If $\csc x = \dfrac{5}{4}$ and $\dfrac{\pi}{2} < x < \pi$, find $\tan(2x)$.

Simplify each of the following. Evaluate exactly, if possible.

39. $\cos^2 15° - \sin^2 15°$

40. $\dfrac{2\tan\left(-\dfrac{\pi}{12}\right)}{1 - \tan^2\left(-\dfrac{\pi}{12}\right)}$

41. $6\sin\left(\dfrac{\pi}{12}\right)\cos\left(\dfrac{\pi}{12}\right)$

42. $1 - 2\sin^2\left(\dfrac{\pi}{8}\right)$

Verify the following identities.

43. $\sin^3 A - \cos^3 A = (\sin A - \cos A)\left[1 + \tfrac{1}{2}\sin(2A)\right]$

44. $2\sin A\cos^3 A - 2\sin^3 A\cos A = \cos(2A)\sin(2A)$

45. $\tan A = \dfrac{\sin(2A)}{1 + \cos(2A)}$

46. $\tan A = \dfrac{1 - \cos(2A)}{\sin(2A)}$

Applications

47. Launching a Missile. When launching a missile for a given range, the minimum velocity that's needed is related to the angle of the launch, and the velocity is determined by

$V = \dfrac{2\cos(2\theta)}{1 + \cos(2\theta)}$. Show that V is equivalent to $1 - \tan^2\theta$.

48. Launching a Missile. When launching a missile for a given range, the minimum velocity that's needed is related to the angle of the launch, and the velocity is determined by

$V = \dfrac{2\cos(2\theta)}{1 + \cos(2\theta)}$. Find the value of V when $\theta = \dfrac{\pi}{6}$.

5.4 Half-Angle Identities

Use half-angle identities to find the exact values of the trigonometric expressions.

49. $\sin(-22.5°)$

50. $\cos(67.5°)$

51. $\cot\left(\dfrac{3\pi}{8}\right)$

52. $\csc\left(-\dfrac{7\pi}{8}\right)$

Use half-angle identities to answer the following questions.

53. If $\sin x = -\dfrac{7}{25}$ and $\pi < x < \dfrac{3\pi}{2}$, find $\sin\left(\dfrac{x}{2}\right)$.

54. If $\cos x = -\dfrac{4}{5}$ and $\dfrac{\pi}{2} < x < \pi$, find $\cos\left(\dfrac{x}{2}\right)$.

55. If $\tan x = \dfrac{40}{9}$ and $\pi < x < \dfrac{3\pi}{2}$, find $\tan\left(\dfrac{x}{2}\right)$.

56. If $\sec x = \dfrac{17}{15}$ and $\dfrac{3\pi}{2} < x < 2\pi$, find $\sin\left(\dfrac{x}{2}\right)$.

Simplify each expression using half-angle identities. Do not evaluate.

57. $\sqrt{\dfrac{1 - \cos\left(\dfrac{\pi}{6}\right)}{2}}$

58. $\sqrt{\dfrac{1 - \cos\left(\dfrac{11\pi}{6}\right)}{1 + \cos\left(\dfrac{11\pi}{6}\right)}}$

Verify the following identities.

59. $\left[\sin\left(\dfrac{A}{2}\right) + \cos\left(\dfrac{A}{2}\right)\right]^2 = 1 + \sin A$

60. $\sec^2\left(\dfrac{A}{2}\right) + \tan^2\left(\dfrac{A}{2}\right) = \dfrac{3 - \cos A}{1 + \cos A}$

61. $\csc^2\left(\dfrac{A}{2}\right) + \cot^2\left(\dfrac{A}{2}\right) = \dfrac{3 + \cos A}{1 - \cos A}$

62. $\tan^2\left(\dfrac{A}{2}\right) + 1 = \sec^2\left(\dfrac{A}{2}\right)$

Graph the functions.

63. $y = \sqrt{\dfrac{1 - \cos\left(\dfrac{\pi}{12}x\right)}{2}}$

64. $y = \cos^2\left(\dfrac{x}{2}\right) - \sin^2\left(\dfrac{x}{2}\right)$

5.5 Product-to-Sum and Sum-to-Product Identities

Write each product as a sum or difference of sines and/or cosines.

65. $6\sin(5x)\cos(2x)$

66. $3\sin(4x)\sin(2x)$

Write each expression as a product of sines and/or cosines.

67. $\cos(5x) - \cos(3x)$

68. $\sin\left(\dfrac{5x}{2}\right) + \sin\left(\dfrac{3x}{2}\right)$

69. $\sin\left(\dfrac{4x}{3}\right) - \sin\left(\dfrac{2x}{3}\right)$

70. $\cos(7x) + \cos x$

Simplify the trigonometric expressions.

71. $\dfrac{\cos(8x) + \cos(2x)}{\sin(8x) - \sin(2x)}$

72. $\dfrac{\sin(5x) + \sin(3x)}{\cos(5x) + \cos(3x)}$

Verify the identities.

73. $\dfrac{\sin A + \sin B}{\cos A - \cos B} = -\cot\left(\dfrac{A - B}{2}\right)$

74. $\dfrac{\sin A - \sin B}{\cos A - \cos B} = -\cot\left(\dfrac{A + B}{2}\right)$

Technology Exercises

Section 5.1

75. Determine the correct sign $(+/-)$ for

$$\sin\left(2x + \frac{\pi}{4}\right) = \sin(2x)\cos\left(\frac{\pi}{4}\right) \underset{?}{\pm} \cos(2x)\sin\left(\frac{\pi}{4}\right)$$

by graphing $Y_1 = \sin\left(2x + \frac{\pi}{4}\right)$,

$Y_2 = \sin(2x)\cos\left(\frac{\pi}{4}\right) + \cos(2x)\sin\left(\frac{\pi}{4}\right)$, and

$Y_3 = \sin(2x)\cos\left(\frac{\pi}{4}\right) - \cos(2x)\sin\left(\frac{\pi}{4}\right)$ in the same viewing rectangle. Does $Y_1 = Y_2$ or Y_3?

76. Determine the correct sign $(+/-)$ for

$$\cos\left(2x - \frac{\pi}{3}\right) = \cos(2x)\cos\left(\frac{\pi}{3}\right) \underset{?}{\pm} \sin(2x)\sin\left(\frac{\pi}{3}\right)$$

by graphing $Y_1 = \cos\left(2x - \frac{\pi}{3}\right)$,

$Y_2 = \cos(2x)\cos\left(\frac{\pi}{3}\right) - \sin(2x)\sin\left(\frac{\pi}{3}\right)$, and

$Y_3 = \cos(2x)\cos\left(\frac{\pi}{3}\right) + \sin(2x)\sin\left(\frac{\pi}{3}\right)$ in the same viewing rectangle. Does $Y_1 = Y_2$ or Y_3?

Section 5.2

Recall that the difference quotient for a function f is given by $\dfrac{f(x + h) + f(x)}{h}$.

77. Show that the difference quotient for $f(x) = \sin(3x)$ is

$$[\cos(3x)]\left[\frac{\sin(3h)}{h}\right] - [\sin(3x)]\left[\frac{1 - \cos(3h)}{h}\right].$$

Plot $Y_1 = [\cos(3x)]\left[\dfrac{\sin(3h)}{h}\right] - [\sin(3x)]\left[\dfrac{1 - \cos(3h)}{h}\right]$

for

a. $h = 1$

b. $h = 0.1$

c. $h = 0.01$

What function does the difference quotient for $f(x) = \sin(3x)$ resemble when h approaches zero?

78. Show that the difference quotient for $f(x) = \cos(3x)$ is

$$[-\sin(3x)]\left[\frac{\sin(3h)}{h}\right] - [\cos(3x)]\left[\frac{1 - \cos(3h)}{h}\right].$$

Plot $Y_1 = [-\sin(3x)]\left[\dfrac{\sin(3h)}{h}\right] - [\cos(3x)]\left[\dfrac{1 - \cos(3h)}{h}\right]$

for

a. $h = 1$

b. $h = 0.1$

c. $h = 0.01$

What function does the difference quotient for $f(x) = \cos(3x)$ resemble when h approaches zero?

Section 5.3

79. With a graphing calculator, plot $Y_1 = \tan(2x)$, $Y_2 = 2\tan x$, and $Y_3 = \dfrac{2\tan x}{1 - \tan^2 x}$ in the same viewing rectangle $[-2\pi, 2\pi]$ by $[-10, 10]$. Which graphs are the same?

80. With a graphing calculator, plot $Y_1 = \cos(2x)$, $Y_2 = 2\cos x$, and $Y_3 = 1 - 2\sin^2 x$ in the same viewing rectangle $[-2\pi, 2\pi]$ by $[-2, 2]$. Which graphs are the same?

Section 5.4

81. With a graphing calculator, plot $Y_1 = \cos\left(\dfrac{x}{2}\right)$, $Y_2 = \dfrac{1}{2}\cos x$, and $Y_3 = -\sqrt{\dfrac{1 + \cos x}{2}}$ in the same viewing rectangle $[\pi, 2\pi]$ by $[-1, 1]$. Which graphs are the same?

82. With a graphing calculator, plot $Y_1 = \sin\left(\dfrac{x}{2}\right)$, $Y_2 = \dfrac{1}{2}\sin x$, and $Y_3 = -\sqrt{\dfrac{1 - \cos x}{2}}$ in the same viewing rectangle $[2\pi, 4\pi]$ by $[-1, 1]$. Which graphs are the same?

Section 5.5

83. With a graphing calculator, plot $Y_1 = \sin(5x)\cos(3x)$, $Y_2 = \sin(4x)$, and $Y_3 = \frac{1}{2}[\sin(8x) + \sin(2x)]$ in the same viewing rectangle $[0, 2\pi]$ by $[-1, 1]$. Which graphs are the same?

84. With a graphing calculator, plot $Y_1 = \sin(3x)\cos(5x)$, $Y_2 = \cos(4x)$, and $Y_3 = \frac{1}{2}[\sin(8x) - \sin(2x)]$ in the same viewing rectangle $[0, 2\pi]$ by $[-1, 1]$. Which graphs are the same?

[CHAPTER 5 PRACTICE TEST]

1. Verify the identity of $\sec x - \tan x = \dfrac{1}{\sec x + \tan x}$.

2. For what values of x does the quotient identity, $\tan x = \dfrac{\sin x}{\cos x}$, not hold?

3. Is the equation $\sqrt{\sin^2 x + \cos^2 x} = \sin x + \cos x$ conditional or an identity?

4. Evaluate $\sin\left(-\dfrac{\pi}{8}\right)$ exactly.

5. Evaluate $\cos\left(\dfrac{11\pi}{12}\right)$ exactly.

6. Evaluate $\sin\left(\dfrac{13\pi}{8}\right)$ exactly.

7. Evaluate $\tan\left(\dfrac{7\pi}{12}\right)$ exactly.

8. If $\cos x = \dfrac{2}{5}$ and $\dfrac{3\pi}{2} < x < 2\pi$, find $\sin\left(\dfrac{x}{2}\right)$.

9. If $\sin x = -\dfrac{1}{5}$ and $\pi < x < \dfrac{3\pi}{2}$, find $\cos(2x)$.

10. If $\tan x = -\dfrac{4}{5}$ and $\dfrac{\pi}{2} < x < \pi$, find $\sin(2x)$.

11. If $\sin x = -\dfrac{\sqrt{7}}{4}$ and $\dfrac{3\pi}{2} < x < 2\pi$, find $\tan(2x)$.

12. Write $\cos(7x)\cos(3x) - \sin(3x)\sin(7x)$ as a cosine or sine of a sum or difference.

13. Let $\sin A = \dfrac{2}{3}$ and $\cos B = -\dfrac{1}{3}$, where $\dfrac{\pi}{2} < A < \pi$ and $\pi < B < \dfrac{3\pi}{2}$. Find $\sin(A + B)$.

14. Let $\sin A = -\dfrac{3}{4}$ and $\cos B = -\dfrac{\sqrt{5}}{5}$, where $\pi < A < \dfrac{3\pi}{2}$ and $\dfrac{\pi}{2} < B < \pi$. Find $\sec(A + B)$.

15. Write $-\dfrac{2\tan x}{1 - \tan^2 x}$ as a single tangent function.

16. Write $\sqrt{\dfrac{1 + \cos(a + b)}{2}}$ as a single cosine function if $a + b$ is an angle in quadrant II.

17. Write $2\sin\left(\dfrac{x + 3}{2}\right)\cos\left(\dfrac{x - 3}{2}\right)$ as a sum of two sine functions.

18. Write $10\cos(3 - x) + 10\cos(x + 3)$ as a product of two cosine functions.

19. In the expression $\sqrt{9 - u^2}$, let $u = 3\sin x$. What is the resulting expression?

20. Verify the identity
$$\sin^2 x \sec(2x) = \dfrac{(1 + \cos x)(1 - \cos x)}{(\cos x + \sin x)(\cos x - \sin x)}.$$

21. If $\sin x = -\dfrac{\sqrt{3}}{6}$ and $\pi < x < \dfrac{3\pi}{2}$, find $\cos\left(\dfrac{x}{2}\right)$.

22. Write the expression $\dfrac{1}{2} - \sin^2 A$ in terms of a single trigonometric function.

23. Write the expression $\cos^2\left(\dfrac{5\pi}{8}\right) - \sin^2\left(\dfrac{5\pi}{8}\right)$ in terms of a single trigonometric function, and then simplify as much as possible.

24. Verify the identity $\cos(45° + x) = \dfrac{\sqrt{2}}{2}(\cos x - \sin x)$.

25. Simplify the expression
$$\dfrac{1}{2}\left[\cos\left(\dfrac{2\pi}{3} + \dfrac{\pi}{3}\right) + \cos\left(\dfrac{2\pi}{3} - \dfrac{\pi}{3}\right)\right]$$
to the product of two trigonometric functions, and then evaluate.

[CHAPTERS 1–5 CUMULATIVE TEST]

1. **Revolving Restaurant.** If a revolving restaurant overlooking a waterfall can rotate 240° in 30 minutes, how long does it take to make a complete revolution?

2. If $A = 110°$ in the diagram below, find H.

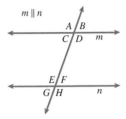

3. Use a calculator to evaluate $\sin(57°39')$.

4. Given $b = 54$ feet and $c = 73$ feet, use the right triangle diagram shown to find α.

5. Write an expression for all angles coterminal with 45°.

6. The terminal side of angle θ, in standard position, passes through the point $\sqrt{2}, -\sqrt{3}$. Calculate the exact values of the six trigonometric functions for angle θ.

7. Evaluate the expression $\cot(-450°) - \sin(-450°)$.

8. Find the positive measure of θ (rounded to the nearest degree) if $\sin\theta = -0.3424$ and the terminal side of θ lies in quadrant IV.

9. If $\tan\theta = -4$ and the terminal side of θ lies in quadrant II, find $\sec\theta$.

10. Convert $\dfrac{17\pi}{20}$ to degree measure.

11. Find the exact value of $\cos\left(\dfrac{11\pi}{6}\right)$.

12. Find the exact length to the radius with arc length $s = \dfrac{5\pi}{18}$ feet and central angle $\theta = 25°$.

13. **Sprinkler Coverage.** A sprinkler has a 21-foot spray and it covers an angle of 50°. What is the area that the sprinkler covers?

14. Find the linear speed of a point traveling at a constant speed along the circumference of a circle with radius 6 inches and angular speed $\dfrac{5\pi}{12}\dfrac{\text{rad}}{\text{sec}}$.

15. Find the distance a point travels along a circle s, over the time $t = 8$ seconds, with angular speed $\omega = \dfrac{\pi\,\text{rad}}{4\,\text{sec}}$, and radius of the circle $r = 6$ centimeters. Round to three significant digits.

16. State the amplitude and period of the function $y = \frac{7}{3}\cos(5x)$.

17. Graph $y = -2\cos(3x)$ over one period.

18. Sketch the graph of the function $y = 2\sin[\pi(x - 1)]$ over one period.

19. Find an equation of a cosine function that corresponds to the graph.

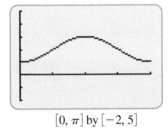

$[0, \pi]$ by $[-2, 5]$

20. Use addition of ordinates to graph the function $y = 2x - 3\cos(\pi x)$ over the interval $0 \le x \le 6$.

21. Graph the function $y = \tan(\frac{1}{4}x)$ over the interval $-2\pi \le x \le 2\pi$.

22. Simplify the trigonometric expression $\dfrac{\sec^4 x - 1}{\sec^2 x + 1}$.

23. Use the sum or difference identities to find the exact value of $\cos\left(-\dfrac{11\pi}{12}\right)$.

24. If $\tan x = \dfrac{7}{24}$ and $\pi < x < \dfrac{3\pi}{2}$, find $\cos(2x)$.

25. Use the half-angle identities to find the exact value of $\tan\left(-\dfrac{3\pi}{8}\right)$.

Solving Trigonometric Equations

© Prisma Bildagentur AG/Alamy Limited

In the summer, St. Petersburg, Russia, has many more daylight hours than St. Petersburg, Florida (United States), because of their locations with respect to the equator. From late May to early July, nights are bright in St. Petersburg, Russia. The nature of the *white nights* can be explained by the geographical location of St. Petersburg in Russia. It is one of the world's most northern cities: located at 59°57′ north (approximately the same latitude as Seward, Alaska). Due to such a high latitude, the sun does not go under the horizon deep enough for the sky to get dark.

Trigonometric equations can be used to represent the relationship between the number of daylight hours and the latitude of the city. *Inverse trigonometric functions* allow us to find the latitude given the number of hours of sunlight.

LEARNING OBJECTIVES

- Find the domain and range of inverse trigonometric functions.
- Solve trigonometric equations using algebraic techniques and inverse functions.
- Solve trigonometric equations using trigonometric identities.

[IN THIS CHAPTER]

We first develop the inverse trigonometric functions and then use these functions to assist us in solving trigonometric equations. As with all equations, the goal is to find the value(s) of the variable that make the equation true. You will find that trigonometric equations (like algebraic equations) can have no solution, one or more solutions, or even an infinite number of solutions.

SOLVING TRIGONOMETRIC EQUATIONS

6.1
INVERSE TRIGONOMETRIC FUNCTIONS

6.2
SOLVING TRIGONOMETRIC EQUATIONS THAT INVOLVE ONLY ONE TRIGONOMETRIC FUNCTION

6.3
SOLVING TRIGONOMETRIC EQUATIONS THAT INVOLVE MULTIPLE TRIGONOMETRIC FUNCTIONS

- Inverse Sine Function
- Inverse Cosine Function
- Inverse Tangent Function
- Remaining Inverse Trigonometric Functions
- Finding Exact Values for Expressions Involving Inverse Trigonometric Functions

- Solving Trigonometric Equations by Inspection
- Solving Trigonometric Equations Using Algebraic Techniques
- Solving Trigonometric Equations That Require the Use of Inverse Functions

- Solving Trigonometric Equations

6.1 INVERSE TRIGONOMETRIC FUNCTIONS

SKILLS OBJECTIVES	CONCEPTUAL OBJECTIVES
■ Find the exact values of an inverse sine function.	■ Understand that the domain of the sine function is restricted $\left[-\frac{\pi}{2}, \frac{\pi}{2}\right]$ in order for the inverse sine function to exist.
■ Find the exact values of an inverse cosine function.	
■ Find the exact values of an inverse tangent function.	■ Understand that the domain of the cosine function is restricted $[0, \pi]$ in order for the inverse cosine function to exist.
■ Find the exact values of the cotangent, cosecant, and secant inverse functions.	
■ Use identities to find exact values of trigonometric expressions involving inverse trigonometric functions.	■ Understand that the domain of the tangent function is restricted $\left(-\frac{\pi}{2}, \frac{\pi}{2}\right)$ in order for the inverse tangent function to exist.
	■ Understand that the cotangent, cosecant, and secant inverse functions are not found from the reciprocal of the tangent, sine, and cosine functions respectively, but rather from the inverse secant, inverse cosecant, and inverse cotangent identities.
	■ Visualize the quadrants in order to find exact values of trigonometric expressions involving inverse trigonometric functions.

In Appendix A.6, one-to-one functions and inverse functions are discussed. Here we present a summary of that section. A function is one-to-one if it passes the horizontal line test: No two *x*-values map to the same *y*-value.

Notice that the sine function does not pass the horizontal line test.

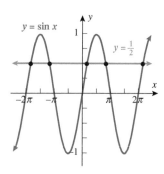

However, if we restrict the domain to $-\dfrac{\pi}{2} \le x \le \dfrac{\pi}{2}$, then the restricted function is one-to-one.

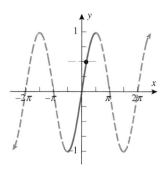

Recall that if $y = f(x)$, then $x = f^{-1}(y)$.

The following are the properties of inverse functions:

1. If f is a one-to-one function, then the inverse function f^{-1} exists.

2. The domain of f^{-1} = the range of f.

The range of f^{-1} = the domain of f.

3. $f^{-1}(f(x)) = x$ for all x in the domain of f.
   $f(f^{-1}(x)) = x$ for all x in the domain of f^{-1}.
4. The graph of f^{-1} is the reflection of the graph of f about the line $y = x$. If the point (a, b) lies on the graph of a function, then the point (b, a) lies on the graph of its inverse.

6.1.1 Inverse Sine Function

Let us start with the sine function with the restricted domain $\left[-\dfrac{\pi}{2}, \dfrac{\pi}{2}\right]$:

$$y = \sin x \quad \text{Domain: } \left[-\dfrac{\pi}{2}, \dfrac{\pi}{2}\right] \quad \text{Range: } [-1, 1]$$

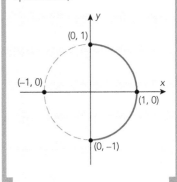

x	y
$-\dfrac{\pi}{2}$	-1
$-\dfrac{\pi}{4}$	$-\dfrac{\sqrt{2}}{2}$
0	0
$\dfrac{\pi}{4}$	$\dfrac{\sqrt{2}}{2}$
$\dfrac{\pi}{2}$	1

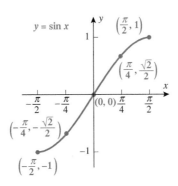

By the properties of inverse functions, the inverse sine function will have a domain of $[-1, 1]$ and a range of $\left[-\dfrac{\pi}{2}, \dfrac{\pi}{2}\right]$. To find the inverse sine function, the x- and y-values of $y = \sin x$ are interchanged.

$$y = \sin^{-1} x \quad \text{Domain: } [-1, 1] \quad \text{Range: } \left[-\dfrac{\pi}{2}, \dfrac{\pi}{2}\right]$$

x	y
-1	$-\dfrac{\pi}{2}$
$-\dfrac{\sqrt{2}}{2}$	$-\dfrac{\pi}{4}$
0	0
$\dfrac{\sqrt{2}}{2}$	$\dfrac{\pi}{4}$
1	$\dfrac{\pi}{2}$

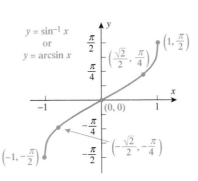

Notice that the inverse sine function, like the sine function, is an odd function (symmetric about the origin). Also notice that the graph of $y = \sin^{-1} x$ is the reflection of $y = \sin x$ about the line $y = x$.

If the sine of an angle is known, what is the measure of that angle? The inverse sine function determines that angle measure. Another notation for the inverse sine function is arcsin x.

INVERSE SINE FUNCTION

$$y = \sin^{-1} x \text{ or } y = \arcsin x \qquad \text{means} \qquad x = \sin y$$
"y is the inverse sine of x"   "y is the angle measure whose sine equals x"

$$\text{where} \quad -1 \le x \le 1 \quad \text{and} \quad -\dfrac{\pi}{2} \le y \le \dfrac{\pi}{2} \quad \text{or} \quad -90° \le y \le 90°$$

The -1 superscript indicates an inverse function. Therefore, the inverse sine function should not be interpreted as a reciprocal; the -1 is *not* an exponent.

$$\sin^{-1}x \neq \frac{1}{\sin x}$$

EXAMPLE 1 **Finding Exact Values of an Inverse Sine Function**

Find the exact value of each of the following expressions:

a. $\sin^{-1}\left(\dfrac{\sqrt{3}}{2}\right)$ **b.** $\arcsin\left(-\dfrac{1}{2}\right)$

Solution (a):

Let $\theta = \sin^{-1}\left(\dfrac{\sqrt{3}}{2}\right)$.　　　　　$\sin\theta = \dfrac{\sqrt{3}}{2}$ when $-\dfrac{\pi}{2} \leq \theta \leq \dfrac{\pi}{2}$

Which value of θ in the range $-\dfrac{\pi}{2} \leq \theta \leq \dfrac{\pi}{2}$ corresponds to a sine value of $\dfrac{\sqrt{3}}{2}$?

- The range $-\dfrac{\pi}{2} \leq \theta \leq \dfrac{\pi}{2}$ corresponds to quadrants I $\left(0 < \theta < \dfrac{\pi}{2}\right)$ and IV $\left(-\dfrac{\pi}{2} < \theta < 0\right)$.

- The sine function is positive in quadrant I.
- We look for a value of θ in quadrant I that has a sine value of $\dfrac{\sqrt{3}}{2}$.　　$\theta = \dfrac{\pi}{3}$

$\sin\left(\dfrac{\pi}{3}\right) = \dfrac{\sqrt{3}}{2}$ and $\dfrac{\pi}{3}$ is in the interval $\left[-\dfrac{\pi}{2}, \dfrac{\pi}{2}\right]$.　　$\boxed{\sin^{-1}\left(\dfrac{\sqrt{3}}{2}\right) = \dfrac{\pi}{3}}$

Solution (b):

Let $\theta = \arcsin\left(-\dfrac{1}{2}\right)$.　　　　　$\sin\theta = -\dfrac{1}{2}$ when $-\dfrac{\pi}{2} \leq \theta \leq \dfrac{\pi}{2}$

Which value of θ in the range $-\dfrac{\pi}{2} \leq \theta \leq \dfrac{\pi}{2}$ corresponds to a sine value of $-\dfrac{1}{2}$?

- The range $-\dfrac{\pi}{2} \leq \theta \leq \dfrac{\pi}{2}$ corresponds to quadrants I and IV.
- The sine function is negative in quadrant IV.
- We look for a value of θ in quadrant IV $\left(-\dfrac{\pi}{2} < \theta < 0\right)$ that has a sine value of $-\dfrac{1}{2}$.　　$\theta = -\dfrac{\pi}{6}$

$\sin\left(-\dfrac{\pi}{6}\right) = -\dfrac{1}{2}$ and $-\dfrac{\pi}{6}$ is in the interval $\left[-\dfrac{\pi}{2}, \dfrac{\pi}{2}\right]$.　　$\boxed{\arcsin\left(-\dfrac{1}{2}\right) = -\dfrac{\pi}{6}}$

STUDY TIP

In Example 1, note that the graphs help identify the desired angles.

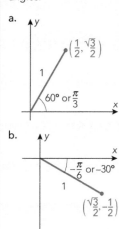

ANSWER

a. $-\dfrac{\pi}{3}$　**b.** $\dfrac{\pi}{6}$

YOUR TURN Find the exact value of the following expressions:

a. $\sin^{-1}\left(-\dfrac{\sqrt{3}}{2}\right)$ **b.** $\arcsin\left(\dfrac{1}{2}\right)$

It is important to note that the inverse sine function has a domain $[-1, 1]$. For example, $\sin^{-1} 3$ does not exist because 3 is not in the domain of the inverse sine function. Notice that the calculator evaluation of $\sin^{-1} 3$ says *error*. Calculators can be used to evaluate inverse sine functions when an exact evaluation is not possible. For example, $\sin^{-1} 0.3 \approx 17.46°$, or 0.305 radians.

We now state the properties relating the sine function and the inverse sine function that follow directly from properties of inverse functions.

SINE-INVERSE SINE IDENTITIES

$$\sin^{-1}(\sin x) = x \qquad \text{for} \qquad -\frac{\pi}{2} \le x \le \frac{\pi}{2} \text{ or } -90° \le x \le 90°$$

$$\sin(\sin^{-1} x) = x \qquad \text{for} \qquad -1 \le x \le 1$$

For example, $\sin^{-1}\left[\sin\left(\dfrac{\pi}{12}\right)\right] = \dfrac{\pi}{12}$, since $\dfrac{\pi}{12}$ is in the interval $\left[-\dfrac{\pi}{2}, \dfrac{\pi}{2}\right]$. However, you must be careful not to overlook the domain restriction for which these identities hold, as illustrated in Example 2.

▶ **EXAMPLE 2**   **Using Inverse Identities to Evaluate Expressions Involving Inverse Sine Functions**

Find the exact value of each of the following trigonometric expressions:

a. $\sin\left[\sin^{-1}\left(\dfrac{\sqrt{2}}{2}\right)\right]$

b. $\sin^{-1}\left[\sin\left(\dfrac{3\pi}{4}\right)\right]$

Solution (a):

Write the appropriate identity. $\sin(\sin^{-1} x) = x \quad \text{for} \quad -1 \le x \le 1$

Let $x = \dfrac{\sqrt{2}}{2}$, which is in the interval $[-1, 1]$.

Since the domain restriction is met, the identity can be used.

$$\boxed{\sin\left[\sin^{-1}\left(\dfrac{\sqrt{2}}{2}\right)\right] = \dfrac{\sqrt{2}}{2}}$$

$\big[$CONCEPT CHECK$\big]$

Evaluate (if possible) $\sin[\sin^{-1}(A)]$ when
(A) $A > 1$
(B) $A = 0$
(C) $A = -1$

▼

ANSWER (A) DNE (B) 0 (C) −1

Solution (b):

common mistake

Ignoring the domain restrictions on inverse identities.

✓CORRECT	✗INCORRECT
Write the appropriate identity.	
$\sin^{-1}(\sin x) = x$ for $-\dfrac{\pi}{2} \le x \le \dfrac{\pi}{2}$	$\sin^{-1}(\sin x) = x$ **ERROR**
Let $x = \dfrac{3\pi}{4}$, but $\dfrac{3\pi}{4}$ is *not* in the	Let $x = \dfrac{3\pi}{4}$. (Forgot domain restriction.)
interval $\left[-\dfrac{\pi}{2}, \dfrac{\pi}{2}\right]$.	$\sin^{-1}\left[\sin\left(\dfrac{3\pi}{4}\right)\right] \ne \dfrac{3\pi}{4}$ **INCORRECT**

Since the domain restriction is not met, the identity cannot be used. Instead, we look for a value in the domain that corresponds to the same value of sine.

$$\sin\left(\frac{3\pi}{4}\right) = \sin\left(\frac{\pi}{4}\right)$$

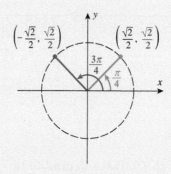

Substitute $\sin\left(\dfrac{3\pi}{4}\right) = \sin\left(\dfrac{\pi}{4}\right)$ into the expression.

$$\sin^{-1}\left[\sin\left(\frac{3\pi}{4}\right)\right] = \sin^{-1}\left[\sin\left(\frac{\pi}{4}\right)\right]$$

Since $\dfrac{\pi}{4}$ is in the interval $\left[-\dfrac{\pi}{2}, \dfrac{\pi}{2}\right]$, we can use the identity.

$$\sin^{-1}\left[\sin\left(\frac{3\pi}{4}\right)\right] = \sin^{-1}\left[\sin\left(\frac{\pi}{4}\right)\right] = \boxed{\dfrac{\pi}{4}}$$

▼
ANSWER
a. $-\dfrac{1}{2}$ b. $\dfrac{\pi}{6}$

▼
YOUR TURN Find the exact value of each of the following trigonometric expressions:

a. $\sin\left[\sin^{-1}\left(-\dfrac{1}{2}\right)\right]$ **b.** $\sin^{-1}\left[\sin\left(\dfrac{5\pi}{6}\right)\right]$

6.1.2 Inverse Cosine Function

The cosine function is not a one-to-one function, so we must restrict the domain to develop the inverse cosine function.

$$y = \cos x \quad \text{Domain: } [0, \pi] \quad \text{Range: } [-1, 1]$$

x	y
0	1
$\dfrac{\pi}{4}$	$\dfrac{\sqrt{2}}{2}$
$\dfrac{\pi}{2}$	0
$\dfrac{3\pi}{4}$	$-\dfrac{\sqrt{2}}{2}$
π	-1

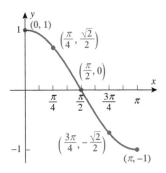

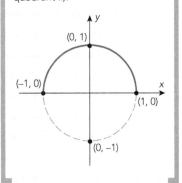
By properties of inverses, the inverse cosine function will have a domain of $[-1, 1]$ and a range of $[0, \pi]$. To find the inverse cosine function, the x- and y-values of $y = \cos x$ are interchanged.

$$y = \cos^{-1} x \quad \text{Domain: } [-1, 1] \quad \text{Range: } [0, \pi]$$

x	y
-1	π
$-\dfrac{\sqrt{2}}{2}$	$\dfrac{3\pi}{4}$
0	$\dfrac{\pi}{2}$
$\dfrac{\sqrt{2}}{2}$	$\dfrac{\pi}{4}$
1	0

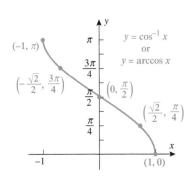

6.1.2 SKILL

Find the exact values of an inverse cosine function.

6.1.2 CONCEPTUAL

Understand that the domain of the cosine function is restricted $[0, \pi]$ in order for the inverse cosine function to exist.

Notice that the graph of $y = \cos^{-1} x$ is the reflection of the graph of $y = \cos x$ about the line $y = x$. However, the inverse cosine function, unlike the cosine function, is not symmetric about the y-axis, or the origin. Although the inverse sine and inverse cosine functions have the same domain, they behave differently. The inverse sine function increases on its domain (from left to right), whereas the inverse cosine function decreases on its domain (from left to right).

If the cosine of an angle is known, what is the measure of that angle? The inverse cosine function determines that angle measure. Another notation for the inverse cosine function is arccos x.

INVERSE COSINE FUNCTION

$$y = \cos^{-1} x \text{ or } y = \arccos x \qquad \text{means} \qquad x = \cos y$$
"y is the inverse cosine of x"   "y is the angle measure whose cosine equals x"

where $\quad -1 \leq x \leq 1 \quad$ and $\quad 0 \leq y \leq \pi \quad$ or $\quad 0° \leq y \leq 180°$

> **EXAMPLE 3** **Finding Exact Values of an Inverse Cosine Function**

Find the exact value of each of the following trigonometric expressions:

a. $\cos^{-1}\left(-\dfrac{\sqrt{2}}{2}\right)$ **b.** $\arccos 0$

Solution (a):

Let $\theta = \cos^{-1}\left(-\dfrac{\sqrt{2}}{2}\right)$. $\cos\theta = -\dfrac{\sqrt{2}}{2}$ when $0 \le \theta \le \pi$

Which value of θ in the range $0 \le \theta \le \pi$ corresponds to a cosine value of $-\dfrac{\sqrt{2}}{2}$?

■ The range $0 \le \theta \le \pi$ corresponds to quadrant I $\left(0 < \theta < \dfrac{\pi}{2}\right)$ and quadrant II $\left(\dfrac{\pi}{2} < \theta < \pi\right)$.

■ The cosine function is negative in quadrant II.

■ We look for a value of θ in quadrant II $\left(\dfrac{\pi}{2} < \theta < \pi\right)$ that has a cosine value of $-\dfrac{\sqrt{2}}{2}$. $\theta = \dfrac{3\pi}{4}$

$\cos\left(\dfrac{3\pi}{4}\right) = -\dfrac{\sqrt{2}}{2}$ and $\dfrac{3\pi}{4}$ is in the interval $[0, \pi]$. $\boxed{\cos^{-1}\left(-\dfrac{\sqrt{2}}{2}\right) = \dfrac{3\pi}{4}}$

Solution (b):

Let $\theta = \arccos 0$. $\cos\theta = 0$ when $0 \le \theta \le \pi$

Which value of θ in the range $0 \le \theta \le \pi$ corresponds to a cosine value of 0? $\theta = \dfrac{\pi}{2}$

$\cos\left(\dfrac{\pi}{2}\right) = 0$ and $\dfrac{\pi}{2}$ is in the interval $[0, \pi]$. $\boxed{\arccos 0 = \dfrac{\pi}{2}}$

▼
ANSWER

a. $\dfrac{\pi}{4}$ **b.** 0

▼ **YOUR TURN** Find the exact value of each of the following trignometric expressions:

a. $\cos^{-1}\left(\dfrac{\sqrt{2}}{2}\right)$ **b.** $\arccos 1$

We now state the properties relating the cosine function and the inverse cosine function that follow directly from properties of inverses.

COSINE–INVERSE COSINE IDENTITIES

$\cos^{-1}(\cos x) = x$ for $0 \le x \le \pi$ or $0° \le x \le 180°$

$\cos(\cos^{-1}x) = x$ for $-1 \le x \le 1$

As was the case with inverse identities for the sine function, you must be careful not to overlook the domain restrictions governing when each of these identities holds.

EXAMPLE 4   Using Inverse Identities to Evaluate Expressions Involving Inverse Cosine Functions

Find the exact value of each of the following trigonometric expressions:

a. $\cos\left[\cos^{-1}\left(-\dfrac{1}{2}\right)\right]$

b. $\cos^{-1}\left[\cos\left(\dfrac{7\pi}{4}\right)\right]$

Solution (a):

Write the appropriate identity. $\cos(\cos^{-1}x) = x$ for $-1 \le x \le 1$

Let $x = -\dfrac{1}{2}$, which is in the interval $[-1, 1]$.

Since the domain restriction is met, $\boxed{\cos\left[\cos^{-1}\left(-\dfrac{1}{2}\right)\right] = -\dfrac{1}{2}}$
the identity can be used.

Solution (b):

Write the appropriate identity. $\cos^{-1}(\cos x) = x$ for $0 \le x \le \pi$

Let $x = \dfrac{7\pi}{4}$, but $\dfrac{7\pi}{4}$ is *not* in the interval $[0, \pi]$.

Since the domain restriction is not met, the identity cannot be used. Instead, we look for a value in the interval that has the same cosine value.

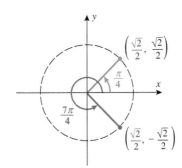

$\cos\left(\dfrac{7\pi}{4}\right) = \cos\left(\dfrac{\pi}{4}\right)$

Substitute $\cos\left(\dfrac{7\pi}{4}\right) = \cos\left(\dfrac{\pi}{4}\right)$
into the expression. $\cos^{-1}\left[\cos\left(\dfrac{7\pi}{4}\right)\right] = \cos^{-1}\left[\cos\left(\dfrac{\pi}{4}\right)\right]$

Since $\dfrac{\pi}{4}$ is in the interval $[0, \pi]$, $\boxed{\cos^{-1}\left[\cos\left(\dfrac{7\pi}{4}\right)\right] = \dfrac{\pi}{4}}$
we can use the identity.

▼ **YOUR TURN**   Find the exact value of each of the following trigonometric expressions:

a. $\cos\left[\cos^{-1}\left(\dfrac{1}{2}\right)\right]$ **b.** $\cos^{-1}\left[\cos\left(-\dfrac{\pi}{6}\right)\right]$

▼
ANSWER
a. $\dfrac{1}{2}$ b. $\dfrac{\pi}{6}$

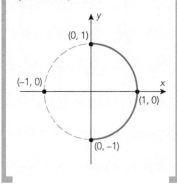

6.1.3 Inverse Tangent Function

The tangent function is not a one-to-one function (it fails the horizontal line test). Let us start with the tangent function with a restricted domain:

$$y = \tan x \quad \text{Domain: } \left(-\frac{\pi}{2}, \frac{\pi}{2}\right) \quad \text{Range: } (-\infty, \infty)$$

x	y
$-\dfrac{\pi}{2}$	$-\infty$
$-\dfrac{\pi}{4}$	-1
0	0
$\dfrac{\pi}{4}$	1
$\dfrac{\pi}{2}$	∞

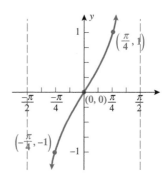

6.1.3 SKILL

Find the exact values of an inverse cosine function.

6.1.3 CONCEPTUAL

Understand that the domain of the tangent function is restricted $\left[-\dfrac{\pi}{2}, \dfrac{\pi}{2}\right]$ in order for the inverse tangent function to exist.

By the properties of inverse functions, the inverse tangent function will have a domain of $(-\infty, \infty)$ and a range of $\left(-\dfrac{\pi}{2}, \dfrac{\pi}{2}\right)$. To find the inverse tangent function, the x- and y-values of $y = \tan x$ are interchanged.

$$y = \tan^{-1}x \quad \text{Domain: } (-\infty, \infty) \quad \text{Range: } \left(-\frac{\pi}{2}, \frac{\pi}{2}\right)$$

x	y
$-\infty$	$-\dfrac{\pi}{2}$
-1	$-\dfrac{\pi}{4}$
0	0
1	$\dfrac{\pi}{4}$
∞	$\dfrac{\pi}{2}$

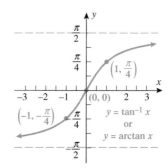

Notice that the inverse tangent function, like the tangent function, is an odd function (it is symmetric about the origin). Also note that the graph of $y = \tan^{-1}x$ is the reflection of $y = \tan x$ about the line $y = x$. The inverse tangent function allows us to answer the question: If the tangent of an angle is known, what is the measure of that angle? Another notation for the inverse tangent function is $\arctan x$.

INVERSE TANGENT FUNCTION

$$y = \tan^{-1}x \text{ or } y = \arctan x \qquad \text{means} \qquad x = \tan y$$

"y is the inverse tangent of x"   "y is the angle measure whose tangent equals x"

$$\text{where} \quad -\frac{\pi}{2} < y < \frac{\pi}{2} \quad \text{or} \quad -90° < y < 90°$$

▶ **EXAMPLE 5**   **Finding Exact Values of an Inverse Tangent Function**

Find the exact value of each of the following expressions:

a. $\tan^{-1}(\sqrt{3})$ **b.** $\arctan 0$

Solution (a):

Let $\theta = \tan^{-1}(\sqrt{3})$. $\tan\theta = \sqrt{3}$ when $-\dfrac{\pi}{2} < \theta < \dfrac{\pi}{2}$

Which value of θ in the range $-\dfrac{\pi}{2} < \theta < \dfrac{\pi}{2}$

corresponds to a tangent value of $\sqrt{3}$? $\theta = \dfrac{\pi}{3}$

$\tan\left(\dfrac{\pi}{3}\right) = \sqrt{3}$ and $\dfrac{\pi}{3}$ is in the interval $\left(-\dfrac{\pi}{2}, \dfrac{\pi}{2}\right)$. $\boxed{\tan^{-1}(\sqrt{3}) = \dfrac{\pi}{3}}$

Solution (b):

Let $\theta = \arctan 0$. $\tan\theta = 0$ when $-\dfrac{\pi}{2} < \theta < \dfrac{\pi}{2}$

Which value of θ in the range $-\dfrac{\pi}{2} < \theta < \dfrac{\pi}{2}$

corresponds to a tangent value of 0? $\theta = 0$

$\tan 0 = 0$ and 0 is in the interval $\left(-\dfrac{\pi}{2}, \dfrac{\pi}{2}\right)$. $\boxed{\arctan 0 = 0}$

     We now state the properties relating the tangent function and the inverse tangent function that follow directly from properties of inverses.

TANGENT-INVERSE TANGENT IDENTITIES

$$\tan^{-1}(\tan x) = x \qquad \text{for} \qquad -\dfrac{\pi}{2} < x < \dfrac{\pi}{2} \quad \text{or} \quad -90° < x < 90°$$

$$\tan(\tan^{-1} x) = x \qquad \text{for} \qquad -\infty < x < \infty$$

EXAMPLE 6   **Using Inverse Identities to Evaluate Expressions Involving Inverse Tangent Functions**

Find the exact value of each of the following trigonometric expressions:

a. $\tan(\tan^{-1} 17)$ **b.** $\tan^{-1}\left[\tan\left(\dfrac{2\pi}{3}\right)\right]$

Solution (a):

Write the appropriate identity. $\tan(\tan^{-1} x) = x$ for $-\infty < x < \infty$

Let $x = 17$, which is in the interval $(-\infty, \infty)$.

Since the domain restriction is met, the identity can be used. $\boxed{\tan(\tan^{-1} 17) = 17}$

Solution (b):

Write the appropriate identity. $\tan^{-1}(\tan x) = x$ for $-\dfrac{\pi}{2} < x < \dfrac{\pi}{2}$

Let $x = \dfrac{2\pi}{3}$, but $\dfrac{2\pi}{3}$ is *not* in the interval $\left(-\dfrac{\pi}{2}, \dfrac{\pi}{2}\right)$.

Since the domain restriction is not met, the identity cannot be used.

Instead, we look for another value in the interval that has the same tangent value. $\tan\left(\dfrac{2\pi}{3}\right) = \tan\left(-\dfrac{\pi}{3}\right)$

[**CONCEPT CHECK**]

Evaluate $\arctan\left[\tan\left(2n\pi + \dfrac{5\pi}{4}\right)\right]$, where n is an integer.

▼ ....

ANSWER $\pi/4$

302 CHAPTER 6 Solving Trigonometric Equations

Substitute $\tan\left(\dfrac{2\pi}{3}\right) = \tan\left(-\dfrac{\pi}{3}\right)$ into the expression.

$$\tan^{-1}\left[\tan\left(\dfrac{2\pi}{3}\right)\right] = \tan^{-1}\left[\tan\left(-\dfrac{\pi}{3}\right)\right]$$

Since $-\dfrac{\pi}{3}$ is in the interval $\left(-\dfrac{\pi}{2}, \dfrac{\pi}{2}\right)$,

$\tan^{-1}\left[\tan\left(-\dfrac{\pi}{3}\right)\right] = -\dfrac{\pi}{3}$.

$$\tan^{-1}\left[\tan\left(\dfrac{2\pi}{3}\right)\right] = \boxed{-\dfrac{\pi}{3}}$$

▼

YOUR TURN Find the exact value of $\tan^{-1}\left[\tan\left(\dfrac{7\pi}{6}\right)\right]$.

ANSWER
$\dfrac{\pi}{6}$

6.1.4 Remaining Inverse Trigonometric Functions

6.1.4 SKILL

Find the exact values of the cotangent, cosecant, and secant inverse functions.

6.1.4 CONCEPTUAL

Understand that the cotangent, cosecant, and secant inverse functions are not found from the reciprocal of the tangent, sine, and cosine functions respectively, but rather from the inverse secant, inverse cosecant, and inverse cotangent identities.

[CONCEPT CHECK]

Find the exact value (if possible) of $\csc^{-1}(A)$ for

(A) $A = 0$ and

(B) $A = 2$

▼

ANSWER (A) DNE   (B) $\dfrac{\pi}{6}$

The remaining three inverse trigonometric functions are defined similarly to the previous ones.

- Inverse cotangent function: $\cot^{-1}x$ or $\operatorname{arccot}x$
- Inverse secant function: $\sec^{-1}x$ or $\operatorname{arcsec}x$
- Inverse cosecant function: $\csc^{-1}x$ or $\operatorname{arccsc}x$

A table summarizing all six of the inverse trigonometric functions is given below:

INVERSE FUNCTION	DOMAIN	RANGE	GRAPH
$y = \sin^{-1}x$	$[-1, 1]$	$\left[-\dfrac{\pi}{2}, \dfrac{\pi}{2}\right]$ or $[-90°, 90°]$	
$y = \cos^{-1}x$	$[-1, 1]$	$[0, \pi]$ or $[0°, 180°]$	
$y = \tan^{-1}x$	$(-\infty, \infty)$	$\left(-\dfrac{\pi}{2}, \dfrac{\pi}{2}\right)$ or $(-90°, 90°)$	
$y = \cot^{-1}x$	$(-\infty, \infty)$	$(0, \pi)$ or $(0°, 180°)$	
$y = \sec^{-1}x$	$(-\infty, -1] \cup [1, \infty)$	$\left[0, \dfrac{\pi}{2}\right) \cup \left(\dfrac{\pi}{2}, \pi\right]$ or $[0°, 90°) \cup (90°, 180°]$	
$y = \csc^{-1}x$	$(-\infty, -1] \cup [1, \infty)$	$\left[-\dfrac{\pi}{2}, 0\right) \cup \left(0, \dfrac{\pi}{2}\right]$ or $[-90°, 0°) \cup (0°, 90°]$	

EXAMPLE 7 **Finding Exact Values of Inverse Trigonometric Functions**

Find the exact value of each of the following expressions:

a. $\cot^{-1}(\sqrt{3})$

b. $\csc^{-1}(\sqrt{2})$

c. $\sec^{-1}(-\sqrt{2})$

Solution (a):

Let $\theta = \cot^{-1}(\sqrt{3})$. $\cot\theta = \sqrt{3}$ when $0 < \theta < \pi$

Which value of θ in the range $0 < \theta < \pi$ $\theta = \dfrac{\pi}{6}$
corresponds to a cotangent value of $\sqrt{3}$?

$\cot\left(\dfrac{\pi}{6}\right) = \sqrt{3}$ and $\dfrac{\pi}{6}$ is in the

interval $(0, \pi)$. $\boxed{\cot^{-1}(\sqrt{3}) = \dfrac{\pi}{6}}$

Solution (b):

Let $\theta = \csc^{-1}(\sqrt{2})$. $\csc\theta = \sqrt{2}$

Which value of θ in the range $\left[-\dfrac{\pi}{2}, 0\right) \cup \left(0, \dfrac{\pi}{2}\right]$

corresponds to a cosecant value of $\sqrt{2}$? $\theta = \dfrac{\pi}{4}$

$\csc\left(\dfrac{\pi}{4}\right) = \sqrt{2}$ and $\dfrac{\pi}{4}$ is in the

interval $\left[-\dfrac{\pi}{2}, 0\right) \cup \left(0, \dfrac{\pi}{2}\right]$. $\boxed{\csc^{-1}(\sqrt{2}) = \dfrac{\pi}{4}}$

Solution (c):

Let $\theta = \sec^{-1}(-\sqrt{2})$. $\sec\theta = -\sqrt{2}$

Which value of θ in the range $\left[0, \dfrac{\pi}{2}\right) \cup \left(\dfrac{\pi}{2}, \pi\right]$

corresponds to a secant value of $-\sqrt{2}$? $\theta = \dfrac{3\pi}{4}$

$\sec\left(\dfrac{3\pi}{4}\right) = -\sqrt{2}$ and $\dfrac{3\pi}{4}$ is in the

interval $\left[0, \dfrac{\pi}{2}\right) \cup \left(\dfrac{\pi}{2}, \pi\right]$. $\boxed{\sec^{-1}(-\sqrt{2}) = \dfrac{3\pi}{4}}$

How do we approximate the inverse secant, inverse cosecant, and inverse cotangent functions with a calculator? Calculators have keys ($\sin^{-1}$, $\cos^{-1}$, and $\tan^{-1}$) for three of the inverse trigonometric functions but not for the other three. Recall that we find the cosecant, secant, and cotangent function values by first finding sine, cosine, or tangent, and then finding its reciprocal.

$$\csc x = \frac{1}{\sin x} \qquad \sec x = \frac{1}{\cos x} \qquad \cot x = \frac{1}{\tan x}$$

However, *the reciprocal approach cannot be used for inverse functions.* The three inverse trigonometric functions $\csc^{-1}x$, $\sec^{-1}x$, and $\cot^{-1}x$ cannot be found by finding the reciprocals of $\sin^{-1}x$, $\cos^{-1}x$, or $\tan^{-1}x$:

$$\csc^{-1}x \neq \frac{1}{\sin^{-1}x} \qquad \sec^{-1}x \neq \frac{1}{\cos^{-1}x} \qquad \cot^{-1}x \neq \frac{1}{\tan^{-1}x}$$

Instead, we seek the equivalent $\sin^{-1}x$, $\cos^{-1}x$, or $\tan^{-1}x$ values by algebraic means, always remembering to look within the correct domain and range.

WORDS	MATH
Start with the inverse secant function.	$y = \sec^{-1}x$ for $x \leq -1$ or $x \geq 1$
Write the equivalent secant expression.	$\sec y = x$ for $0 \leq y < \dfrac{\pi}{2}$ or $\dfrac{\pi}{2} < y \leq \pi$
Use the reciprocal identity: $\sec y = \dfrac{1}{\cos y}$.	$\dfrac{1}{\cos y} = x$
Simplify using algebraic techniques.	$\cos y = \dfrac{1}{x}$
Write the result in terms of the inverse cosine function.	$y = \cos^{-1}\left(\dfrac{1}{x}\right)$
Therefore, we have this relationship:	$\sec^{-1}x = \cos^{-1}\left(\dfrac{1}{x}\right)$ for $x \leq -1$ or $x \geq 1$

The other relationships will be found in the exercises. They are summarized below:

INVERSE SECANT, INVERSE COSECANT, AND INVERSE COTANGENT IDENTITIES

$$\sec^{-1}x = \cos^{-1}\left(\frac{1}{x}\right) \qquad \text{for} \quad x \leq -1 \quad \text{or} \quad x \geq 1$$

$$\csc^{-1}x = \sin^{-1}\left(\frac{1}{x}\right) \qquad \text{for} \quad x \leq -1 \quad \text{or} \quad x \geq 1$$

$$\cot^{-1}x = \begin{cases} \tan^{-1}\left(\dfrac{1}{x}\right) & \text{for} \quad x > 0 \\ \pi + \tan^{-1}\left(\dfrac{1}{x}\right) & \text{for} \quad x < 0 \end{cases}$$

EXAMPLE 8 **Using Inverse Identities**

a. Find the exact value of $\sec^{-1} 2$. Write your answer in radians.

b. Use a calculator to find the value of $\cot^{-1} 7$. Write your answer in degrees.

Solution (a):

Let $\theta = \sec^{-1} 2$. $\qquad\qquad\qquad \sec\theta = 2 \quad$ on $\left[0, \dfrac{\pi}{2}\right) \cup \left(\dfrac{\pi}{2}, \pi\right]$

Substitute the reciprocal identity. $\qquad\qquad \dfrac{1}{\cos\theta} = 2$

Solve for $\cos\theta$. $\qquad\qquad\qquad\qquad \cos\theta = \dfrac{1}{2}$

The restricted interval $\left[0, \dfrac{\pi}{2}\right) \cup \left(\dfrac{\pi}{2}, \pi\right]$ corresponds to quadrants I and II.

The cosine function is positive in quadrant I. $\qquad \theta = \dfrac{\pi}{3}$

$\cos\left(\dfrac{\pi}{3}\right) = \dfrac{1}{2}.$ $\qquad\qquad\qquad \boxed{\sec^{-1} 2 = \cos^{-1}\left(\dfrac{1}{2}\right) = \dfrac{\pi}{3}}$

Solution (b):

Since we do not know an exact value that would correspond to the cotangent function equal to 7 (or the tangent function equal to $\frac{1}{7}$), we proceed using identities and a calculator.

Select the correct identity given
that $x = 7 > 0$. $\qquad\qquad\qquad \cot^{-1} x = \tan^{-1}\left(\dfrac{1}{x}\right)$

Let $x = 7$. $\qquad\qquad\qquad\qquad\qquad \cot^{-1} 7 = \tan^{-1}\left(\dfrac{1}{7}\right)$

Use a calculator to evaluate the right side. $\quad \cot^{-1} 7 \approx \boxed{8.13°}$

6.1.5 Finding Exact Values for Expressions Involving Inverse Trigonometric Functions

Now we will find exact values of trigonometric expressions that involve both trigonometric functions and inverse trigonometric functions.

6.1.5 SKILL

Use identities to find the exact values of trigonometric expressions involving inverse trigonometric functions.

6.1.5 CONCEPTUAL

Visualize the quadrants in order to find exact values of trigonometric expressions involving inverse trigonometric functions.

EXAMPLE 9 **Finding Exact Values of Trigonometric Expressions Involving Inverse Trigonometric Functions**

Find the exact value of $\cos\left[\sin^{-1}\left(\frac{2}{3}\right)\right]$.

Solution: Let $\theta = \sin^{-1}\left(\frac{2}{3}\right)$ and find $\cos\theta$.

STEP 1 Let $\theta = \sin^{-1}\left(\frac{2}{3}\right)$. $\qquad\qquad \sin\theta = \dfrac{2}{3}$ when $-\dfrac{\pi}{2} \le \theta \le \dfrac{\pi}{2}$

The range $-\dfrac{\pi}{2} \le \theta \le \dfrac{\pi}{2}$ corresponds to quadrants I and IV.

The sine function is positive in quadrant I.

STEP 2 Draw angle θ in quadrant I.

Label the sides known from

$$\sin\theta = \frac{2}{3} = \frac{y}{r} = \frac{\text{opposite}}{\text{hypotenuse}}.$$

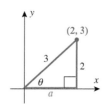

STEP 3 Find the unknown side length a.
Solve for a.
Since θ is in quadrant I, a is positive.

$$a^2 + 2^2 = 3^2$$
$$a = \pm\sqrt{5}$$
$$a = \sqrt{5}$$

STEP 4 Find $\cos\left[\sin^{-1}\left(\frac{2}{3}\right)\right]$.

Substitute $\theta = \sin^{-1}\left(\frac{2}{3}\right)$.

$$\cos\left[\sin^{-1}\left(\frac{2}{3}\right)\right] = \cos\theta$$

Find $\cos\theta$.

$$\cos\theta = \frac{\text{adjacent}}{\text{hypotenuse}} = \frac{x}{r} = \frac{\sqrt{5}}{3}$$

$$\boxed{\cos\left[\sin^{-1}\left(\frac{2}{3}\right)\right] = \frac{\sqrt{5}}{3}}$$

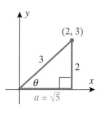

ANSWER
$\dfrac{2\sqrt{2}}{3}$

YOUR TURN Find the exact value of $\sin\left[\cos^{-1}\left(\frac{1}{3}\right)\right]$.

[CONCEPT CHECK]

TRUE OR FALSE $\sin(\arctan(A)) < 0$
if $A < 0$

ANSWER True

▶ **EXAMPLE 10** **Finding Exact Values of Trigonometric Expressions Involving Inverse Trigonometric Functions**

Find the exact value of $\tan\left[\cos^{-1}\left(-\frac{7}{12}\right)\right]$.

Solution: Let $\theta = \cos^{-1}\left(-\frac{7}{12}\right)$, then find $\tan\theta$.

STEP 1 Let $\theta = \cos^{-1}\left(-\frac{7}{12}\right)$.
$$\cos\theta = -\frac{7}{12} \quad \text{when } 0 \le \theta \le \pi$$

The range $0 \le \theta \le \pi$ corresponds to quadrants I and II.
The cosine function is negative in quadrant II.

STEP 2 Draw the reference triangle in quadrant II.

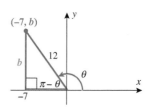

Label the sides known from the cosine value.

$$\cos\theta = -\frac{7}{12} = \frac{x}{r} = \frac{\text{adjacent}}{\text{hypotenuse}}$$

STEP 3 Find the unknown side length b. $b^2 + (-7)^2 = 12^2$

Solve for b. $b = \pm\sqrt{95}$

Since θ is in quadrant II, b is positive. $b = \sqrt{95}$

STEP 4 Find $\tan\left[\cos^{-1}\left(-\frac{7}{12}\right)\right]$.

Substitute $\theta = \cos^{-1}\left(-\frac{7}{12}\right)$. $\tan\left[\cos^{-1}\left(-\frac{7}{12}\right)\right] = \tan\theta$

Find $\tan\theta$. $\tan\theta = \dfrac{\text{opposite}}{\text{adjacent}} = \dfrac{y}{x} = \dfrac{\sqrt{95}}{-7}$

$$\boxed{\tan\left[\cos^{-1}\left(-\frac{7}{12}\right)\right] = -\frac{\sqrt{95}}{7}}$$

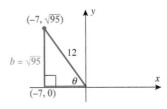

▼
YOUR TURN Find the exact value of $\tan\left[\sin^{-1}\left(-\frac{3}{7}\right)\right]$.

▼
ANSWER
$-\dfrac{3\sqrt{10}}{20}$

EXAMPLE 11   **Using Identities to Find Exact Values of Trigonometric Expressions Involving Inverse Trigonometric Functions**

Find the exact value of $\cos\left[\sin^{-1}\left(\frac{3}{5}\right) + \tan^{-1}1\right]$.

Solution:

Recall the cosine sum identity: $\cos(A + B) = \cos A \cos B - \sin A \sin B$

Let $A = \sin^{-1}\left(\frac{3}{5}\right)$ and $B = \tan^{-1}1$.

$$\cos\left[\sin^{-1}\left(\frac{3}{5}\right) + \tan^{-1}1\right] = \cos\left[\sin^{-1}\left(\frac{3}{5}\right)\right]\cos(\tan^{-1}1) - \sin\left[\sin^{-1}\left(\frac{3}{5}\right)\right]\sin(\tan^{-1}1)$$

From the figure,

$$A = \sin^{-1}\left(\tfrac{3}{5}\right) \Rightarrow \sin A = \tfrac{3}{5}$$

$$\cos\left[\sin^{-1}\left(\frac{3}{5}\right)\right] = \cos A = \frac{4}{5}$$

we see that

$$\sin\left[\sin^{-1}\left(\frac{3}{5}\right)\right] = \sin A = \frac{3}{5}$$

From the figure,

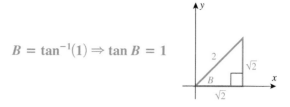

$$B = \tan^{-1}(1) \Rightarrow \tan B = 1$$

we see that

$$\cos(\tan^{-1}1) = \cos B = \frac{\sqrt{2}}{2}$$

$$\sin(\tan^{-1}1) = \sin B = \frac{\sqrt{2}}{2}$$

Substitute these values into the cosine sum identity:

$$\cos\left[\sin^{-1}\left(\frac{3}{5}\right) + \tan^{-1}1\right] = \left(\frac{4}{5}\right)\left(\frac{\sqrt{2}}{2}\right) - \left(\frac{3}{5}\right)\left(\frac{\sqrt{2}}{2}\right)$$

Simplify.

$$= \boxed{\frac{\sqrt{2}}{10}}$$

▼
ANSWER
$$\frac{7\sqrt{2}}{10}$$

▼
YOUR TURN Find the exact value of $\sin\left[\cos^{-1}\left(\frac{3}{5}\right) + \tan^{-1}1\right]$.

EXAMPLE 12 **Writing Trigonometric Expressions Involving Inverse Trigonometric Functions in Terms of a Single Variable**

Write the expression $\cos(\tan^{-1}u)$ as an equivalent expression in terms of only the variable u.

Solution: Let $\theta = \tan^{-1}u$; therefore, $\tan\theta = u = \dfrac{u}{1}$.

Realize that u can be positive or negative. Since the range

of the inverse tangent function is $\left(-\dfrac{\pi}{2}, \dfrac{\pi}{2}\right)$, sketch the angle

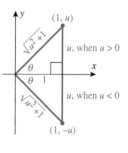

θ in both quadrants I and IV and draw the corresponding two right triangles. Recalling that the tangent ratio is opposite over adjacent, we label those corresponding sides with u and 1, respectively. Then solving for the hypotenuse using the Pythagorean theorem gives $\sqrt{u^2 + 1}$.

Substitute $\theta = \tan^{-1}u$ into $\cos(\tan^{-1}u)$.

$$\cos(\tan^{-1}u) = \cos\theta$$

Use the right triangle ratio for cosine: adjacent over hypotenuse.

$$= \frac{1}{\sqrt{u^2 + 1}}$$

Rationalize the denominator.

$$\cos(\tan^{-1}u) = \frac{1}{\sqrt{u^2 + 1}} \cdot \frac{\sqrt{u^2 + 1}}{\sqrt{u^2 + 1}}$$

$$= \boxed{\frac{\sqrt{u^2 + 1}}{u^2 + 1}}$$

▼
ANSWER
$$\frac{u\sqrt{u^2 + 1}}{u^2 + 1}$$

▼
YOUR TURN Write the expression $\sin(\tan^{-1}u)$ as an equivalent expression in terms of only the variable u.

▶[SECTION 6.1] SUMMARY

If a trigonometric function value of an angle is known, what is the measure of that angle? Inverse trigonometric functions determine the angle measure. To define the inverse trigonometric relations as functions, we first restrict the trigonometric functions to domains in which they are one-to-one functions. Exact values for the inverse trigonometric functions can be found when the function values are those of the special angles. Inverse trigonometric functions also provide a means for evaluating one trigonometric function when we are given the value of another. It is important to note that the -1 superscript indicates an inverse function, not a reciprocal.

INVERSE FUNCTION	$y = \sin^{-1}x$	$y = \cos^{-1}x$	$y = \tan^{-1}x$	$y = \cot^{-1}x$	$y = \sec^{-1}x$	$y = \csc^{-1}x$
DOMAIN	$[-1, 1]$	$[-1, 1]$	$(-\infty, \infty)$	$(-\infty, \infty)$	$(-\infty, -1] \cup [1, \infty)$	$(-\infty, -1] \cup [1, \infty)$
RANGE	$\left[-\dfrac{\pi}{2}, \dfrac{\pi}{2}\right]$	$[0, \pi]$	$\left(-\dfrac{\pi}{2}, \dfrac{\pi}{2}\right)$	$(0, \pi)$	$\left[0, \dfrac{\pi}{2}\right) \cup \left(\dfrac{\pi}{2}, \pi\right]$	$\left[-\dfrac{\pi}{2}, 0\right) \cup \left(0, \dfrac{\pi}{2}\right]$
GRAPHY						

[SECTION 6.1] EXERCISES

• SKILLS

In Exercises 1–12, find the exact value of each expression. Give the answer in radians.

1. $\arccos\left(\dfrac{\sqrt{2}}{2}\right)$ **2.** $\arccos\left(-\dfrac{\sqrt{2}}{2}\right)$ **3.** $\arcsin\left(-\dfrac{\sqrt{3}}{2}\right)$ **4.** $\arcsin\left(\dfrac{1}{2}\right)$

5. $\cot^{-1}(-1)$ **6.** $\tan^{-1}\left(\dfrac{\sqrt{3}}{3}\right)$ **7.** $\text{arcsec}\left(\dfrac{2\sqrt{3}}{3}\right)$ **8.** $\text{arccsc}(-1)$

9. $\csc^{-1}2$ **10.** $\sec^{-1}(-2)$ **11.** $\arctan(-\sqrt{3})$ **12.** $\text{arccot}(\sqrt{3})$

In Exercises 13–24, find the exact value of each expression. Give the answer in degrees.

13. $\cos^{-1}\left(\dfrac{1}{2}\right)$ **14.** $\cos^{-1}\left(-\dfrac{\sqrt{3}}{2}\right)$ **15.** $\sin^{-1}\left(\dfrac{\sqrt{2}}{2}\right)$ **16.** $\sin^{-1}0$ **17.** $\cot^{-1}\left(-\dfrac{\sqrt{3}}{3}\right)$ **18.** $\tan^{-1}(\sqrt{3})$

19. $\arctan\left(\dfrac{\sqrt{3}}{3}\right)$ **20.** $\text{arccot}\,1$ **21.** $\text{arccsc}(-2)$ **22.** $\csc^{-1}\left(-\dfrac{2\sqrt{3}}{3}\right)$ **23.** $\text{arcsec}(-\sqrt{2})$ **24.** $\text{arccsc}(-\sqrt{2})$

In Exercises 25–36, use a calculator to evaluate each expression. Give the answer in degrees and round to two decimal places.

25. $\cos^{-1}(0.5432)$ **26.** $\arccos(-0.3245)$ **27.** $\arcsin(0.1223)$ **28.** $\sin^{-1}(0.7821)$

29. $\tan^{-1}(1.895)$ **30.** $\tan^{-1}(3.2678)$ **31.** $\sec^{-1}(1.4973)$ **32.** $\sec^{-1}(2.7864)$

33. $\csc^{-1}(-3.7893)$ **34.** $\csc^{-1}(-6.1324)$ **35.** $\cot^{-1}(-4.2319)$ **36.** $\cot^{-1}(-0.8977)$

In Exercises 37–48, use a calculator to evaluate each expression. Give the answer in radians and round to two decimal places.

37. $\sin^{-1}(-0.5878)$ **38.** $\sin^{-1}(0.8660)$ **39.** $\cos^{-1}(0.1423)$ **40.** $\arccos(0.7469)$

41. $\tan^{-1}(1.3242)$ **42.** $\tan^{-1}(-0.9279)$ **43.** $\cot^{-1}(-0.5774)$ **44.** $\cot^{-1}(2.4142)$

45. $\text{arcsec}(3.462)$ **46.** $\sec^{-1}(-1.0422)$ **47.** $\csc^{-1}(3.2361)$ **48.** $\csc^{-1}(-2.9238)$

In Exercises 49–68, evaluate each expression exactly, if possible. If not possible, state why.

49. $\sin^{-1}\left[\sin\left(\dfrac{5\pi}{12}\right)\right]$ **50.** $\sin^{-1}\left[\sin\left(-\dfrac{5\pi}{12}\right)\right]$ **51.** $\sin[\sin^{-1}(1.03)]$ **52.** $\sin[\sin^{-1}(1.1)]$

53. $\sin^{-1}\left[\sin\left(-\dfrac{7\pi}{6}\right)\right]$ **54.** $\sin^{-1}\left[\sin\left(\dfrac{7\pi}{6}\right)\right]$ **55.** $\cos^{-1}\left[\cos\left(\dfrac{4\pi}{3}\right)\right]$ **56.** $\cos^{-1}\left[\cos\left(-\dfrac{5\pi}{3}\right)\right]$

57. $\arccos\left[\cos\left(\dfrac{11\pi}{6}\right)\right]$ **58.** $\arccos\left[\cos\left(\dfrac{4\pi}{3}\right)\right]$ **59.** $\cot[\cot^{-1}(\sqrt{3})]$ **60.** $\cot^{-1}\left[\cot\left(\dfrac{5\pi}{4}\right)\right]$

61. $\sec^{-1}\left[\sec\left(-\dfrac{\pi}{3}\right)\right]$ **62.** $\sec\left[\sec^{-1}\left(\dfrac{1}{2}\right)\right]$ **63.** $\csc\left[\csc^{-1}\left(\dfrac{1}{2}\right)\right]$ **64.** $\csc^{-1}\left[\csc\left(\dfrac{7\pi}{6}\right)\right]$

65. $\cot(\cot^{-1}0)$ **66.** $\cot^{-1}\left[\cot\left(-\dfrac{\pi}{4}\right)\right]$ **67.** $\tan^{-1}\left[\tan\left(-\dfrac{\pi}{4}\right)\right]$ **68.** $\tan^{-1}\left[\tan\left(\dfrac{\pi}{4}\right)\right]$

In Exercises 69–88, evaluate each expression exactly.

69. $\cos\left[\sin^{-1}\left(\dfrac{3}{4}\right)\right]$ **70.** $\sin\left[\cos^{-1}\left(\dfrac{2}{3}\right)\right]$ **71.** $\sin\left[\tan^{-1}\left(\dfrac{12}{5}\right)\right]$ **72.** $\cos\left[\tan^{-1}\left(\dfrac{7}{24}\right)\right]$

73. $\tan\left[\sin^{-1}\left(\dfrac{3}{5}\right)\right]$ **74.** $\tan\left[\cos^{-1}\left(\dfrac{2}{5}\right)\right]$ **75.** $\sec\left[\sin^{-1}\left(\dfrac{\sqrt{2}}{5}\right)\right]$ **76.** $\sec\left[\cos^{-1}\left(\dfrac{\sqrt{7}}{4}\right)\right]$

77. $\csc\left[\cos^{-1}\left(\dfrac{1}{4}\right)\right]$ **78.** $\csc\left[\sin^{-1}\left(\dfrac{1}{4}\right)\right]$ **79.** $\cot\left[\sin^{-1}\left(\dfrac{60}{61}\right)\right]$ **80.** $\cot\left[\sec^{-1}\left(\dfrac{41}{9}\right)\right]$

81. $\cos\left[\tan^{-1}\left(\dfrac{3}{4}\right) - \sin^{-1}\left(\dfrac{4}{5}\right)\right]$ **82.** $\cos\left[\tan^{-1}\left(\dfrac{12}{5}\right) + \sin^{-1}\left(\dfrac{3}{5}\right)\right]$

83. $\sin\left[\cos^{-1}\left(\dfrac{5}{13}\right) + \tan^{-1}\left(\dfrac{4}{3}\right)\right]$ **84.** $\sin\left[\cos^{-1}\left(\dfrac{3}{5}\right) - \tan^{-1}\left(\dfrac{5}{12}\right)\right]$

85. $\sin\left[2\cos^{-1}\left(\dfrac{3}{5}\right)\right]$ **86.** $\cos\left[2\sin^{-1}\left(\dfrac{3}{5}\right)\right]$ **87.** $\tan\left[2\sin^{-1}\left(\dfrac{5}{13}\right)\right]$ **88.** $\tan\left[2\cos^{-1}\left(\dfrac{5}{13}\right)\right]$

For each of the following expressions, write an equivalent expression in terms of only the variable u.

89. $\cos(\sin^{-1}u)$ **90.** $\sin(\cos^{-1}u)$ **91.** $\tan(\cos^{-1}u)$ **92.** $\tan(\sin^{-1}u)$

• **APPLICATIONS**

For Exercises 93 and 94, refer to the following:

Annual sales of a product are generally subject to seasonal fluctuations and are approximated by the function

$$s(t) = 4.3\cos\left(\frac{\pi}{6}t\right) + 56.2 \quad 0 \le t \le 11$$

where t represents time in months ($t = 0$ represents January) and $s(t)$ represents monthly sales of the product in thousands of dollars.

93. Business. Find the month(s) in which monthly sales are $56,200.

94. Business. Find the month(s) in which monthly sales are $51,900.

For Exercises 95 and 96, refer to the following:

Allergy sufferers' symptoms fluctuate with pollen levels. Pollen levels are often reported to the public on a scale of 0–12, which is meant to reflect the levels of pollen in the air. For example, a pollen level between 4.9 and 7.2 indicates that pollen levels will likely cause symptoms for many individuals allergic to the predominant pollen of the season (*Source:* http://www.pollen .com). The pollen levels at a single location were measured and averaged for each month. Over a period of 6 months, the levels fluctuated according to the model

$$p(t) = 5.5 + 1.5\sin\left(\frac{\pi}{6}t\right) \quad 0 \le t \le 6$$

where t is measured in months and $p(t)$ is the pollen level.

95. Biology/Health. In which month(s) was the monthly average pollen level 7.0?

96. Biology/Health. In which month(s) was the monthly average pollen level 6.25?

97. Alternating Current. Alternating electrical current in amperes (A) is modeled by the equation $i = I\sin(2\pi ft)$, where i is the induced current, I is the maximum current, t is time in seconds, and f is the frequency in hertz (Hz, the number of cycles per second). If the frequency is 5 hertz and the maximum current is 115 amperes, what time t corresponds to a current of 85 amperes? Find the smallest positive value of t.

98. Alternating Current. Refer to the formula in Exercise 97. If the frequency is 100 hertz and the maximum current is 240 amperes, what time t corresponds to a current of 100 amperes? Find the smallest positive value of t.

99. Hours of Daylight. The number of hours of daylight in San Diego, California, can be modeled with

$$H(t) = 12 + 2.4\sin(0.017t - 1.377)$$

where t is the day of the year (January 1, $t = 1$, etc.). For what value of t is the number of hours equal to 14.4? If May 31 is the 151st day of the year, what month and day correspond to that value of t?

100. Hours of Daylight. Refer to the formula in Exercise 99. For what value of t is the number of hours of daylight equal to 9.6? What month and day correspond to the value of t? (You may have to count backward.)

101. Debt. A young couple gets married and immediately starts saving money. They renovate a house and are left with less and less saved money. They have children after 10 years and are in debt until their children are in college. They then save until retirement. A formula that represents the percentage of their annual income that they either save (positive) or are in debt (negative) is given by $P(t) = 12.5\cos(0.157t) + 2.5$, where $t = 0$ corresponds to the year they get married. How many years into their marriage do they first accrue debt? (See graph below.)

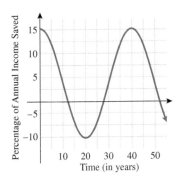

102. Savings. How many years into their marriage are the couple in Exercise 101 back to saving 15% of their annual income?

103. Viewing Angle of Painting. A museum patron whose eye level is 5 feet above the floor is studying a painting that is 8 feet in height and mounted on the wall 4 feet above the floor. If the patron is x feet from the wall, use $\tan(\alpha + \beta)$ to express $\tan\theta$, where θ is the angle that the patron's eye sweeps from the top to the bottom of the painting.

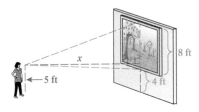

104. Viewing Angle of Painting. Using the expression for $\tan\theta$ in Exercise 103, solve for θ using the inverse tangent. Then find the measure of the angles θ for $x = 10$ and $x = 20$ (to the nearest degree).

105. Earthquake Movement. The horizontal movement M of a point that is k kilometers away from an earthquake's fault line can be estimated with

$$M = \frac{f}{2}\left[1 - \frac{2\tan^{-1}\left(\frac{k}{d}\right)}{\pi}\right]$$

where M is the movement of the point in meters, f is the total horizontal displacement occurring along the fault line, k is the distance of the point from the fault line, and d is the depth in kilometers of the focal point of the earthquake.

If an earthquake produces a displacement f of 2 meters and the depth of the focal point is 4 kilometers, then what are the movement M of a point that is 2 kilometers from the fault line? 10 kilometers from the fault line?

106. **Earthquake Movement.** Use the formula in Exercise 105. If an earthquake produces a displacement f of 3 meters and the depth of the focal point is 2.5 kilometers, then what is the movement M of a point that is 5 kilometers from the fault line? 10 kilometers from the fault line?

107. **Laser Communication.** A laser communication system depends on a narrow beam and a direct line of sight is necessary for communication links. If a transmitter/receiver for a laser system is placed between two buildings (see the figure) and the other end of the system is located on a low Earth orbit satellite, then the link is only operational when the satellite and the ground system have a line of sight (when the buildings are not in the way). Find the angle θ that corresponds to the system being operational (i.e., find the maximum value of θ that permits the system to be operational). Express θ in terms of inverse tangent functions and the distance from the shorter building.

108. **Laser Communication.** Repeat Exercise 107, assuming the ground system is on top of a 20-foot tower.

109. **Ski Slope.** Pete is at the bottom of a ski slope debating about whether he wants to try the run. The angle of elevation θ to the top of the slope is given by the equation $\theta = \tan^{-1}\left(\dfrac{h}{x}\right)$, where h is the vertical height of the slope in yards and x is the horizontal change also measured in yards. For this slope, $h = 750$ yards and $x = 550$ yards. Find the angle of elevation.

110. **Ski Slope.** MaryAnn is at the bottom of a ski slope debating about whether she wants to try the run. For this slope, $h = 1000$ yards and $x = 600$ yards. Find the angle of elevation.

111. **Vectors.** The angle between two vectors is found by taking the inverse cosine of the quotient of the dot product of the vectors and the product of the magnitudes of the vectors. Find the angle θ between two vectors if their dot product is 6 and the magnitudes of the vectors are $\sqrt{13}$ and $\sqrt{21}$.

112. **Vectors.** Find the angle θ between two vectors if their dot product is -11 and the magnitudes of the vectors are $\sqrt{10}$ and $\sqrt{13}$.

• **CATCH THE MISTAKE**

In Exercises 113–116, explain the mistake that is made.

113. Evaluate the expression exactly: $\sin^{-1}\left[\sin\left(\dfrac{3\pi}{5}\right)\right]$.

Solution:

Use the identity $\sin^{-1}(\sin x) = x$ on $0 \le x \le \pi$.

Since $\dfrac{3\pi}{5}$ is in the interval $[0, \pi]$, the identity can be used.

$\sin^{-1}\left[\sin\left(\dfrac{3\pi}{5}\right)\right] = \dfrac{3\pi}{5}$

This is incorrect. What mistake was made?

114. Evaluate the expression exactly: $\cos^{-1}\left[\cos\left(-\dfrac{\pi}{5}\right)\right]$.

Solution:

Use the identity $\cos^{-1}(\cos x) = x$ on $-\dfrac{\pi}{2} \le x \le \dfrac{\pi}{2}$.

Since $-\dfrac{\pi}{5}$ is in the interval $\left[-\dfrac{\pi}{2}, \dfrac{\pi}{2}\right]$, the identity can be used.

$\cos^{-1}\left[\cos\left(-\dfrac{\pi}{5}\right)\right] = -\dfrac{\pi}{5}$

This is incorrect. What mistake was made?

115. Evaluate the expression exactly: $\cot^{-1}(2.5)$.

Solution:

Use the reciprocal identity. $\quad \cot^{-1}(2.5) = \dfrac{1}{\tan^{-1}(2.5)}$

Evaluate $\tan^{-1}(2.5) = 1.19$. $\quad \cot^{-1}(2.5) = \dfrac{1}{1.19}$

Simplify. $\quad \cot^{-1}(2.5) = 0.8403$

This is incorrect. What mistake was made?

116. Evaluate the expression exactly: $\csc^{-1}\left(\frac{1}{4}\right)$.

Solution:

Use the reciprocal identity. $\quad \csc^{-1}\left(\dfrac{1}{4}\right) = \dfrac{1}{\sin^{-1}\left(\dfrac{1}{4}\right)}$

Evaluate $\sin^{-1}\left(\frac{1}{4}\right) = 14.478$. $\quad \csc^{-1}\left(\dfrac{1}{4}\right) = \dfrac{1}{14.478}$

Simplify. $\quad \csc^{-1}\left(\dfrac{1}{4}\right) = 0.0691$

This is incorrect. What mistake was made?

• **CONCEPTUAL**

In Exercises 117–120, determine whether each statement is true or false.

117. The inverse secant function is an even function.

118. The inverse cosecant function is an odd function.

119. $\arccos\left(-\frac{1}{2}\right) = -120°$

120. $\arcsin\left(-\frac{1}{2}\right) = 330°$

121. Explain why $\sec^{-1}\left(\frac{1}{2}\right)$ does not exist.

122. Explain why $\csc^{-1}\left(\frac{1}{2}\right)$ does not exist.

• **CHALLENGE**

123. Let $f(x) = 2 - 4\sin\left(x - \dfrac{\pi}{2}\right)$.

    a. State an accepted domain of $f(x)$ so that $f(x)$ is a one-to-one function.

    b. Find $f^{-1}(x)$ and state its domain.

124. Let $f(x) = 3 + \cos\left(x - \dfrac{\pi}{4}\right)$.

    a. State an accepted domain of $f(x)$ so that $f(x)$ is a one-to-one function.

    b. Find $f^{-1}(x)$ and state its domain.

125. Find the expression that corresponds to $\sin\left[\cos^{-1}\left(\dfrac{1}{x}\right)\right]$.

126. Find the expression that corresponds to $\cos\left[\sin^{-1}\left(\dfrac{1}{x}\right)\right]$.

• **TECHNOLOGY**

127. Use a graphing calculator to plot $Y_1 = \sin(\sin^{-1}x)$ and $Y_2 = x$ for the domain $-1 \le x \le 1$. If you then increase the domain to $-3 \le x \le 3$, you get a different result. Explain the result.

128. Use a graphing calculator to plot $Y_1 = \cos(\cos^{-1}x)$ and $Y_2 = x$ for the domain $-1 \le x \le 1$. If you then increase the domain to $-3 \le x \le 3$, you get a different result. Explain the result.

129. Use a graphing calculator to plot $Y_1 = \csc^{-1}(\csc x)$ and $Y_2 = x$. For what domain is the following statement true: $\csc^{-1}(\csc x) = x$? Give the domain in terms of π.

130. Use a graphing calculator to plot $Y_1 = \sec^{-1}(\sec x)$ and $Y_2 = x$. For what domain is the following statement true: $\sec^{-1}(\sec x) = x$? Give the domain in terms of π.

6.2 SOLVING TRIGONOMETRIC EQUATIONS THAT INVOLVE ONLY ONE TRIGONOMETRIC FUNCTION

SKILLS OBJECTIVES	CONCEPTUAL OBJECTIVES
▪ Solve trigonometric equations by inspection. ▪ Solve trigonometric equations using algebraic techniques. ▪ Solve trigonometric functions using inverse functions.	▪ When solving trigonometric equations by inspection remember to find all solutions, not just the ones in a single period. ▪ Utilize the concept of substitution in solving trigonometric equations when the results are simple linear or quadratic equations. ▪ Utilize a calculator to approximate solutions when solving trigonometric equations using inverse functions.

In this section, we will develop a strategy for solving trigonometric equations that involve only one trigonometric function. In the following section, we will solve more complicated equations that involve multiple trigonometric functions and require the use of trigonometric identities.

Recall in solving algebraic equations that the goal is to find the value for the variable that makes the equation true. For example, the linear equation $2x - 5 = 7$ has only one value, $x = 6$, which makes the statement true. A quadratic equation, however, can have two solutions. The equation $x^2 = 9$ has two values, $x = \pm 3$, which make the statement true. With trigonometric equations, the goal is the same: Find the value(s) that make the equation true.

We will start with simple trigonometric equations that can be solved by inspection. Then we will use inverse functions and algebraic techniques to solve trigonometric equations involving a single trigonometric function.

6.2.1 Solving Trigonometric Equations by Inspection

The goal in solving equations in one variable is to find the values for that variable that make the equation true. For example, $9x = 72$ can be solved by inspection by asking the question, "9 times what is 72?" The answer is $x = 8$. We approach simple trigonometric equations in the same way we can approach some simple algebraic equations: We inspect the equation and determine the solution.

EXAMPLE 1 **Solving a Trigonometric Equation by Inspection**

Solve each of the following equations over $[0, 2\pi)$:

a. $\sin x = \frac{1}{2}$ **b.** $\cos(2x) = \frac{1}{2}$

Solution (a):

Ask the question "sine of what angle yields $\frac{1}{2}$?" The sine function is positive only in quadrants I and II.

$$x = \frac{\pi}{6} \quad \text{or} \quad x = \frac{5\pi}{6}$$

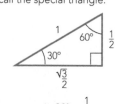

$$\sin 30° = \frac{1}{2}$$

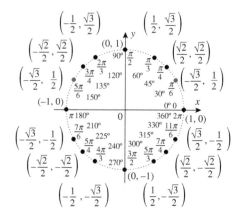

Solution (b):

Ask the question "cosine of what angle yields $\frac{1}{2}$?" The cosine function is positive only in quadrants I and IV. In this case, the angle is equal to $2x$.

$$2x = \frac{\pi}{3} \quad \text{or} \quad 2x = \frac{5\pi}{3}$$

$$\boxed{x = \frac{\pi}{6}} \quad \text{or} \quad \boxed{x = \frac{5\pi}{6}}$$

Since the angle is equal to $2x$, we need to consider two complete rotations.

$$2x = \frac{\pi}{3} + 2\pi \quad \text{or} \quad 2x = \frac{5\pi}{3} + 2\pi$$

$$2x = \frac{7\pi}{3} \quad \text{or} \quad 2x = \frac{11\pi}{3}$$

$$\boxed{x = \frac{7\pi}{6}} \quad \text{or} \quad \boxed{x = \frac{11\pi}{6}}$$

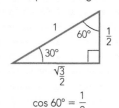

▼
YOUR TURN Solve each of the following equations over $[0, 2\pi)$:
a. $\cos x = \frac{1}{2}$ **b.** $\sin(2x) = \frac{1}{2}$

▼
ANSWER

a. $x = \frac{\pi}{3}, \frac{5\pi}{3}$

b. $x = \frac{\pi}{12}, \frac{5\pi}{12}, \frac{13\pi}{12}, \frac{17\pi}{12}$

EXAMPLE 2 **Solving a Trigonometric Equation by Inspection**

Solve the equation $\sin x = \frac{\sqrt{2}}{2}$.

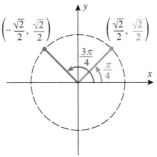

Solution:

STEP 1 Solve over one period, $[0, 2\pi)$.

Ask the question, "sine of what angle yields $\frac{\sqrt{2}}{2}$?"

The sine function is positive in quadrants I and II.

DEGREES	$x = 45°$ or $x = 135°$
RADIANS	$x = \frac{\pi}{4}$ or $x = \frac{3\pi}{4}$

STEP 2 Solve over all x.

Since the sine function has a period 360° or 2π, adding integer multiples of 360° or 2π will give the other (infinitely many) solutions.

DEGREES	$x = 45° + 360°n$ or $x = 135° + 360°n$
RADIANS	$x = \frac{\pi}{4} + 2n\pi$ or $x = \frac{3\pi}{4} + 2n\pi$, where n is any integer

▼
YOUR TURN Solve the equation $\cos x = \frac{1}{2}$.

[CONCEPT CHECK]

Solve $\sin(x) = 0$.

▼
ANSWER $x = n\pi$, where n is an integer

▼
ANSWER

DEGREES	$x = 60° + 360°n$ or $x = 300° + 360°n$
RADIANS	$x = \frac{\pi}{3} + 2n\pi$, or $x = \frac{5\pi}{3} + 2n\pi$, where n is any integer

Notice that the equations in Example 2 and the Your Turn have an infinite number of solutions. Unless the domain is restricted, you must find *all* solutions.

EXAMPLE 3 **Solving a Trigonometric Equation by Inspection**

Solve the equation $\tan(2x) = -\sqrt{3}$.

Solution:

STEP 1 Solve over one period, $[0, \pi)$.

Ask the question, "tangent of what angle yields $-\sqrt{3}$?" Note that the angle in this case is $2x$.

DEGREES	$2x = 120°$
RADIANS	$2x = \dfrac{2\pi}{3}$

The tangent function is negative in quadrants II and IV. Since $[0, \pi)$ includes quadrants I and II, we find only the angle in quadrant II. (The solution corresponding to quadrant IV will be found when we extend the solution over all real numbers.)

STEP 2 Solve over all x.

Since the tangent function has a period of 180° or π, adding integer multiples of 180° or π will give all of the other solutions.

Solve for x by dividing both sides by 2.

DEGREES	$2x = 120° + 180°n$
RADIANS	$2x = \dfrac{2\pi}{3} + n\pi,$ where n is any integer

DEGREES	$x = 60° + 90°n$
RADIANS	$x = \dfrac{\pi}{3} + \dfrac{n\pi}{2},$ where n is any integer

Note:

- There are infinitely many solutions. If we graph $y = \tan(2x)$ and $y = -\sqrt{3}$, we see that there are infinitely many points of intersection.

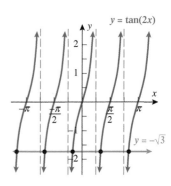

- Had we restricted the domain to $0 \le x < 2\pi$, the solutions (in radians) would be the values given to the right in the table.

Notice that only $n = 0, 1, 2, 3$ yield x-values in the domain $0 \le x < 2\pi$.

n	$x = \dfrac{\pi}{3} + \dfrac{n\pi}{2}$
0	$x = \dfrac{\pi}{3}$
1	$x = \dfrac{5\pi}{6}$
2	$x = \dfrac{4\pi}{3}$
3	$x = \dfrac{11\pi}{6}$

Notice in Step 2 of Example 2, $2n\pi$ was added to get all of the solutions, whereas in Step 2 of Example 3, we added $n\pi$ to the argument of the tangent function. We added $2n\pi$ in Example 2 and $n\pi$ in Example 3 because the sine function has period 2π, whereas the tangent function has period π.

6.2.2 Solving Trigonometric Equations Using Algebraic Techniques

6.2.2 **SKILL**

Solve trigonometric equations using algebraic techniques.

6.2.2 **CONCEPTUAL**

Utilize the concept of substitution in solving trigonometric equations when the results are simple linear or quadratic equations.

We now will use algebraic techniques to solve trigonometric equations. Let us first start with linear- and quadratic-type trigonometric equations. For linear equations, we solve for the variable by isolating it. For quadratic equations, we often employ factoring or the quadratic formula. If we can let x represent the trigonometric function and the resulting equation is either linear or quadratic, then we use techniques learned in solving algebraic equations.

TYPE	EQUATION	SUBSTITUTION	ALGEBRAIC EQUATION
Linear-type trigonometric equation	$4\sin\theta - 2 = -4$	$x = \sin\theta$	$4x - 2 = -4$
Quadratic-type trigonometric equation	$2\cos^2\theta + \cos\theta - 1 = 0$	$x = \cos\theta$	$2x^2 + x - 1 = 0$

It is not always necessary to make the substitution, though it is convenient. Often we can see how to factor a quadratic-type trigonometric equation without first converting it to an algebraic equation. In Example 4, we will not use the substitution. However, in Example 5, we will illustrate the use of a substitution.

EXAMPLE 4   **Solving a Linear-Type Trigonometric Equation**

Solve $4\sin\theta - 2 = -4$ on $0 \le \theta < 2\pi$.

Solution:

STEP 1 Solve for $\sin\theta$. $\qquad\qquad\qquad 4\sin\theta - 2 = -4$

　　　　Add 2 to both sides. $\qquad\qquad\qquad 4\sin\theta = -2$

　　　　Divide both sides by 4. $\qquad\qquad\qquad \sin\theta = -\dfrac{1}{2}$

STEP 2 Find the values of θ on $0 \le \theta < 2\pi$ that satisfy the equation $\sin\theta = -\frac{1}{2}$.

　　　　The sine function is negative in quadrants III and IV.

$$\sin\left(\frac{7\pi}{6}\right) = -\frac{1}{2} \quad\text{and}\quad \sin\left(\frac{11\pi}{6}\right) = -\frac{1}{2}. \qquad \boxed{\theta = \frac{7\pi}{6}} \text{ or } \boxed{\theta = \frac{11\pi}{6}}$$

▼ YOUR TURN Solve $2\cos\theta + 1 = 2$ on $0 \le \theta < 2\pi$.

▼
ANSWER
$\theta = \dfrac{\pi}{3}$ or $\dfrac{5\pi}{3}$

[CONCEPT CHECK]

Solve $\cos^2 x = 1$.

ANSWER $x = n\pi$, where n is an integer

▶ **EXAMPLE 5** **Solving a Quadratic-Type Trigonometric Equation**

Solve $2\cos^2\theta + \cos\theta - 1 = 0$ on $0 \le \theta < 2\pi$.

Solution:

STEP 1 Solve for $\cos\theta$. $2\cos^2\theta + \cos\theta - 1 = 0$

Let $x = \cos\theta$. $2x^2 + x - 1 = 0$

Factor the resulting quadratic expression. $(2x - 1)(x + 1) = 0$

Set each factor equal to 0. $2x - 1 = 0$ or $x + 1 = 0$

Solve each resulting equation for x. $x = \dfrac{1}{2}$ or $x = -1$

Substitute back to θ: $x = \cos\theta$. $\cos\theta = \dfrac{1}{2}$ or $\cos\theta = -1$

STEP 2 Find the values of θ on $0 \le \theta < 2\pi$ that satisfy the equation $\cos\theta = \frac{1}{2}$.

The cosine function is positive in quadrants I and IV.

$\cos\left(\dfrac{\pi}{3}\right) = \dfrac{1}{2}$ and $\cos\left(\dfrac{5\pi}{3}\right) = \dfrac{1}{2}$. $\boxed{\theta = \dfrac{\pi}{3}}$ or $\boxed{\theta = \dfrac{5\pi}{3}}$

STEP 3 Find the values of θ on $0 \le \theta < 2\pi$ that satisfy the equation $\cos\theta = -1$.

$\cos\pi = -1$ $\boxed{\theta = \pi}$

The solutions to $2\cos^2\theta + \cos\theta - 1 = 0$ on $0 \le \theta < 2\pi$ are $\theta = \dfrac{\pi}{3}, \theta = \dfrac{5\pi}{3}$, or $\theta = \pi$.

▼

YOUR TURN Solve $2\sin^2\theta - \sin\theta - 1 = 0$ on $0 \le \theta < 2\pi$.

▼
ANSWER
$\theta = \dfrac{\pi}{2}, \dfrac{7\pi}{6}$, or $\dfrac{11\pi}{6}$

6.2.3 **SKILL**

Solve trigonometric functions using inverse functions.

6.2.3 **CONCEPTUAL**

Utilize a calculator to approximate solutions when solving trigonometric equations using inverse functions.

6.2.3 Solving Trigonometric Equations That Require the Use of Inverse Functions

Thus far, we have been able to solve the trigonometric equations exactly. Now we turn our attention to those cases for which a calculator and inverse functions are required to approximate a solution to a trigonometric equation.

EXAMPLE 6   **Solving a Trigonometric Equation That Requires the Use of Inverse Functions**

Solve $\tan^2\theta - \tan\theta = 6$ on $0° \le \theta < 180°$.

Solution:

STEP 1   Solve for $\tan\theta$.

Noticing a quadratic type, get 0 on one side. $\tan^2\theta - \tan\theta - 6 = 0$

Factor the quadratic trigonometric expression on the left. $(\tan\theta - 3)(\tan\theta + 2) = 0$

Set each factor equal to 0. $\tan\theta - 3 = 0$ or $\tan\theta + 2 = 0$

Solve for $\tan\theta$. $\tan\theta = 3$ or $\tan\theta = -2$

STEP 2   Solve $\tan\theta = 3$ on $0° \le \theta < 180°$.

The tangent function is positive on $0° \le \theta < 180°$ only in quadrant I.

Write the equivalent inverse notation to $\tan\theta = 3$. $\theta = \tan^{-1}3$

Use a calculator to evaluate (approximate) θ. $\boxed{\theta \approx 71.6°}$

STEP 3   Solve $\tan\theta = -2$ on $0° \le \theta < 180°$.

The tangent function is negative on $0° \le \theta < 180°$ only in quadrant II.

A calculator gives values of the inverse tangent in quadrants I and IV.

We will call the reference angle in quadrant IV "α."

Write the equivalent inverse notation to $\tan\alpha = -2$. $\alpha = \tan^{-1}(-2)$

Use a calculator to evaluate (approximate) α. $\alpha \approx -63.4°$

To find the value of θ in quadrant II, add $180°$. $\theta = \alpha + 180°$

    $\boxed{\theta \approx 116.6°}$

The solutions to $\tan^2\theta - \tan\theta = 6$ on $0° \le \theta < 180°$ are $\theta = 71.6°$ or $\theta = 116.6°$.

▼

YOUR TURN   Solve $\tan^2\theta + \tan\theta = 6$ on $0° \le \theta < 180°$.

[CONCEPT CHECK]

Explain why $\cos^2 x + \cos x = 6$ does not have a solution.

▼ ....

ANSWER The range of cosine is $[-1, 1]$. If you rewrite the equation as $\cos^2 x + \cos x - 6 = 0$ and factor the left side as $(\cos x - 3)(\cos x + 2)$, then neither $\cos x = 2$ nor $\cos x = -3$ has a solution. So, there are no x-values that satisfy the original equation.

▼ ....
ANSWER

$\theta \approx 63.4°$ or $108.4°$

Recall in solving algebraic quadratic equations that one method (when factoring is not obvious or possible) is to use the Quadratic Formula.

$$ax^2 + bx + c = 0 \quad \text{has solutions} \quad x = \frac{-b \pm \sqrt{b^2 - 4ac}}{2a}$$

EXAMPLE 7 **Solving a Quadratic Trigonometric Equation That Requires the Use of the Quadratic Formula and Inverse Functions**

Solve $2\cos^2\theta + 5\cos\theta - 6 = 0$ on $0° \leq \theta < 360°$.

Solution:

STEP 1 Solve for $\cos\theta$. $\qquad\qquad\qquad\qquad\qquad 2\cos^2\theta + 5\cos\theta - 6 = 0$

Let $x = \cos\theta$. $\qquad\qquad\qquad\qquad\qquad\qquad\quad 2x^2 + 5x - 6 = 0$

Use the Quadratic Formula:
$a = 2, b = 5$, and $c = -6$. $\qquad\qquad x = \dfrac{-5 \pm \sqrt{5^2 - 4(2)(-6)}}{2(2)}$

Simplify. $\qquad\qquad\qquad\qquad\qquad\qquad x = \dfrac{-5 \pm \sqrt{73}}{4}$

Use a calculator to approximate
the solution. $\qquad\qquad\qquad\qquad\qquad x \approx -3.3860$ or 0.8860

Substitute back to θ: $x = \cos\theta$. $\qquad \cos\theta \approx -3.3860$ or 0.8860

STEP 2 Solve $\cos\theta = -3.3860$ on $0° \leq \theta < 360°$.

Recall that the range of the cosine function is $[-1, 1]$; therefore, the cosine function can never equal a number outside that range ($-3.3860 < -1$).

No solution from this equation.

STEP 3 Solve $\cos\theta \approx 0.8860$ on $0° \leq \theta < 360°$.

The cosine function is positive in quadrants I and IV. Since a calculator gives inverse cosine values only in quadrants I and II, we will have to use a reference angle to get the quadrant IV solution.

Write the equivalent inverse
notation for $\cos\theta = 0.8860$. $\qquad\qquad \theta = \cos^{-1}(0.8860)$

Use a calculator to approximate
the solution. $\qquad\qquad\qquad\qquad\qquad \boxed{\theta \approx 27.6°}$

To find the second solution
(in quadrant IV), subtract the
reference angle from $360°$. $\qquad\qquad \theta = 360° - 27.6°$

$\qquad\qquad\qquad\qquad\qquad\qquad\qquad \boxed{\theta \approx 332.4°}$

The solutions to $2\cos^2\theta + 5\cos\theta - 6 = 0$ on $0° \leq \theta < 360°$ are $\theta \approx 27.6°$ or $332.4°$.

▼
ANSWER

$\theta \approx 242.4°$ or $297.6°$

▼

YOUR TURN Solve $2\sin^2\theta - 5\sin\theta - 6 = 0$ on $0° \leq \theta < 360°$.

Applications

▶ **EXAMPLE 8** **Applications Involving Trigonometric Equations**

Light bends (refracts) according to Snell's law, which states:

$$n_i \sin(\theta_i) = n_r \sin(\theta_r)$$

- n_i is the refractive index of the medium the light is leaving.

- θ_i is the incident angle between the light ray and the normal (perpendicular) to the interface between mediums.

- n_r is the refractive index of the medium the light is entering.

- θ_r is the refractive angle between the light ray and the normal (perpendicular) to the interface between mediums.

Janis Christie /Getty Images, Inc.

Assume that light is going from air into a diamond. Calculate the refractive angle θ_r if the incidence angle is $\theta_i = 32°$ and the index of refraction values for air and diamond are $n_i = 1.00$ and $n_r = 2.417$, respectively.

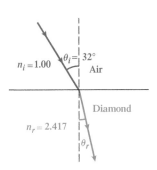

Solution:

Write Snell's law.

$$n_i \sin(\theta_i) = n_r \sin(\theta_r)$$

Substitute $\theta_i = 32°$, $n_i = 1.00$, and $n_r = 2.417$.

$$\sin 32° = 2.417 \sin(\theta_r)$$

Isolate $\sin \theta_r$ and simplify.

$$\sin(\theta_r) = \frac{\sin 32°}{2.417} \approx 0.21925$$

Solve for θ_r using the inverse sine function.

$$\theta_r \approx \sin^{-1}(0.21925) \approx 12.665°$$

Round to the nearest degree.

$$\boxed{\theta_r \approx 13°}$$

◉[SECTION 6.2] SUMMARY

In this section, we began by solving basic trigonometric equations that contained only one trigonometric function. Some such equations can be solved exactly by inspection; others can be solved exactly using algebraic techniques similar to those of linear and quadratic equations. Calculators and inverse functions are needed when exact values are not known. It is important to note that calculators give the inverse function only in one of the two relevant quadrants. The other quadrant solutions must be found using reference angles.

[SECTION 6.2] EXERCISES

• **SKILLS**

In Exercises 1–18, solve each of the trigonometric equations exactly over the indicated intervals.

1. $\cos\theta = -\dfrac{\sqrt{2}}{2}, 0 \le \theta < 2\pi$

2. $\sin\theta = -\dfrac{\sqrt{2}}{2}, 0 \le \theta < 2\pi$

3. $\sin\theta = \dfrac{\sqrt{3}}{2}, 0 \le \theta < 2\pi$

4. $\cos\theta = -\dfrac{\sqrt{3}}{2}, 0 \le \theta < 2\pi$

5. $\cos\theta = \dfrac{1}{2}, 0 \le \theta < 2\pi$

6. $\sin\theta = -\dfrac{1}{2}, 0 \le \theta < 2\pi$

7. $\csc\theta = -2, 0 \le \theta < 4\pi$

8. $\sec\theta = -2, 0 \le \theta < 4\pi$

9. $\tan\theta = 0$, all real numbers

10. $\cot\theta = 0$, all real numbers

11. $\sin(2\theta) = -\dfrac{1}{2}, 0 \le \theta < 2\pi$

12. $\cos(2\theta) = \dfrac{\sqrt{3}}{2}, 0 \le \theta < 2\pi$

13. $\sin\left(\dfrac{\theta}{2}\right) = -\dfrac{1}{2}$, all real numbers

14. $\cos\left(\dfrac{\theta}{2}\right) = -1$, all real numbers

15. $\tan(2\theta) = \sqrt{3}, -2\pi \le \theta < 2\pi$

16. $\tan(2\theta) = -\sqrt{3}$, all real numbers

17. $\sec\theta = -2, -2\pi \le \theta < 0$

18. $\csc\theta = \dfrac{2\sqrt{3}}{3}, -\pi \le \theta < \pi$

In Exercises 19–36, solve each of the trigonometric equations exactly on $0 \le \theta < 2\pi$.

19. $2\sin(2\theta) = \sqrt{3}$

20. $2\cos\left(\dfrac{\theta}{2}\right) = -\sqrt{2}$

21. $3\tan(2\theta) - \sqrt{3} = 0$

22. $4\tan\left(\dfrac{\theta}{2}\right) - 4 = 0$

23. $2\cos(2\theta) + 1 = 0$

24. $4\csc(2\theta) + 8 = 0$

25. $\sqrt{3}\cot\left(\dfrac{\theta}{2}\right) - 3 = 0$

26. $\sqrt{3}\sec(2\theta) + 2 = 0$

27. $\tan^2\theta - 1 = 0$

28. $\sin^2\theta + 2\sin\theta + 1 = 0$

29. $2\cos^2\theta = \cos\theta$

30. $\tan^2\theta = \sqrt{3}\tan\theta$

31. $\csc^2\theta + 3\csc\theta + 2 = 0$

32. $\cot^2\theta = 1$

33. $\sin^2\theta + 2\sin\theta = 3$

34. $2\sec^2\theta + \sec\theta = 1$

35. $4\cos^2\theta - 3 = 0$

36. $4\sin^2\theta = 3 + 4\sin\theta$

In Exercises 37–54, solve each of the trigonometric equations on $0° \le \theta < 360°$ and express answers in degrees to two decimal places.

37. $\sin(2\theta) = -0.7843$

38. $\cos(2\theta) = 0.5136$

39. $\tan\left(\dfrac{\theta}{2}\right) = -0.2343$

40. $\sec\left(\dfrac{\theta}{2}\right) = 1.4275$

41. $5\cot\theta - 9 = 0$

42. $5\sec\theta + 6 = 0$

43. $4\sin\theta + \sqrt{2} = 0$

44. $3\cos\theta - \sqrt{5} = 0$

45. $4\cos^2\theta + 5\cos\theta - 6 = 0$

46. $6\sin^2\theta - 13\sin\theta - 5 = 0$

47. $6\tan^2\theta = \tan\theta + 12$

48. $6\sec^2\theta = 7\sec\theta + 20$

49. $15\sin^2(2\theta) + \sin(2\theta) - 2 = 0$

50. $12\cos^2\left(\dfrac{\theta}{2}\right) - 13\cos\left(\dfrac{\theta}{2}\right) + 3 = 0$

51. $\cos^2\theta - 6\cos\theta + 1 = 0$

52. $\sin^2\theta + 3\sin\theta - 3 = 0$

53. $2\tan^2\theta - \tan\theta - 7 = 0$

54. $3\cot^2\theta + 2\cot\theta - 4 = 0$

• **APPLICATIONS**

For Exercises 55 and 56, refer to the following:

Computer sales are generally subject to seasonal fluctuations. The sales of QualComp computers during 2016–2018 is approximated by the function

$$s(t) = 0.120 \sin(0.790t - 2.380) + 0.387 \quad 1 \le t \le 12$$

where t represents time in quarters ($t = 1$ represents the end of the first quarter of 2016), and $s(t)$ represents computer sales (quarterly revenue) in millions of dollars.

55. **Business.** Find the quarter(s) in which the quarterly sales are $472,000.

56. **Business.** Find the quarter(s) in which the quarterly sales are $507,000.

For Exercises 57 and 58, refer to the following:

Allergy sufferers' symptoms fluctuate with the concentration of pollen in the air. At one location the pollen concentration, measured in grains per cubic meter, of grasses fluctuates throughout the day according to the function:

$$p(t) = 35 - 26 \cos\left(\frac{\pi}{12}t - \frac{7\pi}{6}\right), \quad 0 \le t \le 24$$

where t is measured in hours and $t = 0$ is 12:00 A.M.

57. **Biology/Health.** Find the time(s) of day when the grass pollen level is 41 grains per cubic meter. Round to the nearest hour.

58. **Biology/Health.** Find the time(s) of day when the grass pollen level is 17 grains per cubic meter. Round to the nearest hour.

59. **Sales.** Monthly sales of soccer balls are approximated by
$$S = 400 \sin\left(\frac{\pi}{6}x\right) + 2000, \text{ where } x \text{ is the number of the}$$
month (January is $x = 1$, etc.). During which month do sales reach 2400?

60. **Sales.** Use the formula given in Exercise 59. During which two months do sales reach 1800?

61. **Home Improvement.** A rain gutter is constructed from a single strip of sheet metal by bending as shown below, so that the base and sides are the same length. Express the area of the cross section of the rain gutter as a function of the angle θ (note that the expression will also involve x).

62. **Home Improvement.** A rain gutter is constructed from a single strip of sheet metal by bending as shown on the left below, so that the base and sides are the same length. When the area of the cross section of the rain gutter is expressed as a function of the angle θ, you can then use calculus to determine the value of θ that produces the cross section with the greatest possible area. The angle is found by solving the equation $\cos^2\theta - \sin^2\theta + \cos\theta = 0$. Which angle gives the maximum area? *Hint*: Use the Pythagorean identity to express the equation in terms of the cosine function.

63. **Deer Population.** The number of deer on an island is given
by $D = 200 + 100\sin\left(\frac{\pi}{2}x\right)$, where x is the number of years
since 2016. Which is the first year after 2016 that the number of deer reaches 300?

64. **Deer Population.** The number of deer on an island is given
by $D = 200 + 100\sin\left(\frac{\pi}{6}x\right)$, where x is the number of years
since 2016. Which is the first year after 2016 that the number of deer reaches 150?

65. **Optics.** Assume that light is going from air into a diamond. Calculate the refractive angle θ_r if the incidence angle is $\theta_i = 75°$, and the index of refraction values for air and diamond are $n_i = 1.00$ and $n_r = 2.417$, respectively. Round to the nearest degree. (See Example 8 for Snell's law.)

66. **Optics.** Assume that light is going from a diamond into air. Calculate the refractive angle θ_r if the incidence angle is $\theta_i = 15°$, and the index of refraction values for diamond and air are $n_i = 2.417$ and $n_r = 1.00$, respectively. Round to the nearest degree. (See Example 8 for Snell's law.)

67. **Calculus.** In calculus, the term "critical numbers" refers to values for x, where a graph may have a maximum or minimum. To find the critical numbers of the equation $y = \sin(2x) + 2$, solve the equation $2\cos(2x) = 0$ for all values of x.

68. **Calculus.** To find the critical numbers of the equation $y = \cos(2x) - 3$, solve the equation $-2\sin(2x) = 0$ for all values of x.

69. **Angle of Elevation.** If a 7-foot lamppost makes a 10-foot shadow on the sidewalk, find its angle of elevation to the sun.

70. **Angle of Elevation.** If a 6-foot lamppost makes an 8-foot shadow on the sidewalk, find its angle of elevation to the sun.

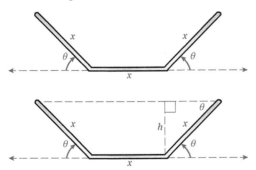

● CATCH THE MISTAKE

In Exercises 71–74, explain the mistake that is made.

71. Solve $\cos\theta = 0.5899$ on $0° \le \theta < 360°$.

Solution:

| Write in equivalent inverse notation. | $\theta = \cos^{-1}(0.5899)$ |
| Use a calculator to approximate inverse cosine. | $\theta = 53.85°$ |

This is incorrect. What mistake was made?

72. Find all solutions to the equation $\sin\theta = \frac{3}{5}$ on $0° \le \theta < 360°$.

Solution:

| Write in equivalent inverse notation. | $\theta = \sin^{-1}\left(\frac{3}{5}\right)$ |
| Use a calculator to approximate inverse sine. | $\theta = 36.87°$ |

This is incorrect. What mistake was made?

73. Solve $\sqrt{2 + \sin\theta} = \sin\theta$ on $0 \le \theta < 2\pi$.

Solution:

Square both sides.	$2 + \sin\theta = \sin^2\theta$
Gather all terms to one side.	$\sin^2\theta - \sin\theta - 2 = 0$
Factor.	$(\sin\theta - 2)(\sin\theta + 1) = 0$
Set each factor equal to zero.	$\sin\theta - 2 = 0$ or $\sin\theta + 1 = 0$
Solve for $\sin\theta$.	$\sin\theta = 2$ or $\sin\theta = -1$
Solve $\sin\theta = 2$ for θ.	no solution
Solve $\sin\theta = -1$ for θ.	$\theta = \frac{3\pi}{2}$

This is incorrect. What mistake was made?

74. Solve $\sqrt{3\sin\theta - 2} = -\sin\theta$ on $0 \le \theta < 2\pi$.

Solution:

Square both sides.	$3\sin\theta - 2 = \sin^2\theta$
Gather all terms to one side.	$\sin^2\theta - 3\sin\theta + 2 = 0$
Factor.	$(\sin\theta - 2)(\sin\theta - 1) = 0$
Set each factor equal to zero.	$\sin\theta - 2 = 0$ or $\sin\theta - 1 = 0$
Solve for $\sin\theta$.	$\sin\theta = 2$ or $\sin\theta = 1$
Solve $\sin\theta = 2$ for θ.	no solution
Solve $\sin\theta = 1$ for θ.	$\theta = \frac{\pi}{2}$

This is incorrect. What mistake was made?

● CONCEPTUAL

In Exercises 75–78, determine whether each statement is true or false.

75. Linear-type trigonometric equations always have one solution on $[0, 2\pi)$.

76. Quadratic-type trigonometric equations always have two solutions on $[0, 2\pi)$.

77. The solution set for the equation $\sin^2 x = 0.5\sin x$ for $0 \le x < 2\pi$ is $\left[\frac{\pi}{6}, \frac{5\pi}{6}\right]$.

78. The equation $\sin^4 x - 1 = 0$ has two solutions on the interval $[0, 2\pi)$.

79. How many solutions does the equation $\sin\left(kx + \frac{\pi}{2}\right) = 1$ have on the interval $[0, 2\pi)$ for integer k?

80. How many solutions does the equation $\cos\left(kx - \frac{\pi}{2}\right) = \frac{\sqrt{2}}{2}$ have on the interval $[0, 2\pi)$ for integer k?

● CHALLENGE

81. Solve $16\sin^4\theta - 8\sin^2\theta = -1$ over $0 \le \theta \le 2\pi$.

82. Solve $\left|\cos\left(\theta + \frac{\pi}{4}\right)\right| = \frac{\sqrt{3}}{2}$ over all real numbers.

83. Solve $|\sin x| < \frac{1}{2}$ over $0 \le x < 2\pi$.

84. Solve $|\cos x| \ge \frac{1}{2}$ over $0 \le x < 2\pi$.

85. Solve $2\cos^3 x - \cos^2 x - 2\cos x + 1 = 0$ over $0 \le x < 2\pi$.

86. Solve $\tan x = \sqrt{\tan x + 2}$ over $0 \le x < 2\pi$.

For Exercises 87–92, refer to the following:

Graphing calculators can be used to find approximate solutions to trigonometric equations. For the equation $f(x) = g(x)$, let $Y_1 = f(x)$ and $Y_2 = g(x)$. The x values that correspond to points of intersections represent solutions.

87. Use a graphing utility to solve the equation $\sin\theta = \cos(2\theta)$ on $0 \le \theta < \pi$.

88. Use a graphing utility to solve the equation $\csc\theta = \sec\theta$ on $0 \le \theta < \dfrac{\pi}{2}$.

89. Use a graphing utility to solve the equation $\sin\theta = \sec\theta$ on $0 \le \theta < \pi$.

90. Use a graphing utility to solve the equation $\cos\theta = \csc\theta$ on $0 \le \theta < \pi$.

91. Use a graphing utility to find all solutions to the equation $\sin\theta = e^\theta$ for $\theta \ge 0$.

92. Use a graphing utility to find all solutions to the equation $\cos\theta = e^\theta$ for $\theta \ge 0$.

6.3 SOLVING TRIGONOMETRIC EQUATIONS THAT INVOLVE MULTIPLE TRIGONOMETRIC FUNCTIONS

SKILLS OBJECTIVE	**CONCEPTUAL OBJECTIVE**
▪ Solve trigonometric equations (involving more than one trigonometric function) using trigonometric identities.	▪ Using trigonometric identities can often help convert a trigonometric equation involving more than one trigonometric function into trigonometric equations involving a single trigonometric function.

6.3.1 Solving Trigonometric Equations

We now consider trigonometric equations that involve more than one trigonometric function. Trigonometric identities are an important part of solving these types of equations. You will notice that we rely on the basic trigonometric identities and the identities discussed in Chapter 5 to transform equations involving different trigonometric functions into an equation involving only one trigonometric function. We also will use algebraic techniques to factor equations so that each factor contains only one trigonometric function.

6.3.1 SKILL

Solve trigonometric equations (involving more than one trigonometric function) using trigonometric identities.

6.3.1 CONCEPTUAL

Using trigonometric identities can often help convert a trigonometric equation involving more than one trigonometric function into a trigonometric equation involving a single trigonometric function.

EXAMPLE 1 **Using Trigonometric Identities in Solving Trigonometric Equations**

Solve $\sin x + \cos x = 1$ on $0 \le x < 2\pi$.

Solution:

Square both sides.	$\sin^2 x + 2\sin x\cos x + \cos^2 x = 1$
Recognize the Pythagorean identity.	$\underbrace{\sin^2 x + \cos^2 x}_{1} + 2\sin x\cos x = 1$
Subtract 1 from both sides.	$2\sin x\cos x = 0$
Use the zero product property.	$\sin x = 0 \quad$ or $\quad \cos x = 0$
Solve for x on $0 \le x < 2\pi$.	$x = 0$ or $x = \pi \quad$ or $\quad x = \dfrac{\pi}{2}$ or $x = \dfrac{3\pi}{2}$

Because we squared the equation, we have to check for extraneous solutions.

Check $x = 0$: $\qquad\qquad\qquad\qquad$ $\sin 0 + \cos 0 = 0 + 1 = 1$ ✓

Check $x = \pi$: $\qquad\qquad\qquad\qquad$ $\sin \pi + \cos \pi = 0 - 1 = -1$ X

Check $x = \dfrac{\pi}{2}$: $\qquad\qquad\qquad$ $\sin\left(\dfrac{\pi}{2}\right) + \cos\left(\dfrac{\pi}{2}\right) = 1 + 0 = 1$ ✓

Check $x = \dfrac{3\pi}{2}$: $\qquad\qquad\qquad$ $\sin\left(\dfrac{3\pi}{2}\right) + \cos\left(\dfrac{3\pi}{2}\right) = -1 + 0 = -1$ X

The solutions to $\sin x + \cos x = 1$ on $0 \le x < 2\pi$ are only $\boxed{x = 0}$ or $\boxed{x = \dfrac{\pi}{2}}$.

▼

ANSWER

$x = \dfrac{\pi}{2}$, or π

YOUR TURN Solve $\sin x - \cos x = 1$ on $0 \le x < 2\pi$.

[CONCEPT CHECK]

TRUE OR FALSE $\sin(2A) = \cos(A)$ has the same solution as $\sin(2A) = \cos(-A)$.

▼

ANSWER True

▼

CAUTION

Do not divide equations by trigonometric functions as they can sometimes equal zero.

▶ **EXAMPLE 2** **Using Trigonometric Identities in Solving Trigonometric Equations**

Solve $\sin(2x) = \sin x$ on $0 \le x < 2\pi$.

common mistake

Dividing by a trigonometric function (which could be equal to zero).

✓CORRECT	✗INCORRECT
$\sin(2x) = \sin x$	
Use the double-angle formula for sine.	
$2\sin x\cos x = \sin x$	$2\sin x\cos x = \sin x$
Subtract $\sin x$.	Divide by $\sin x$. **ERROR**
$2\sin x\cos x - \sin x = 0$	$2\cos x = 1$
Factor out the common $\sin x$.	**Incorrect Solution: Two additional solutions missing.**
$(\sin x)(2\cos x - 1) = 0$	

Use the zero product property.

$\sin x = 0$ or $2\cos x - 1 = 0$

$\sin x = 0$ or $\cos x = \dfrac{1}{2}$

Solve $\sin x = 0$ for x on $0 \le x < 2\pi$.

$\boxed{x = 0}$ or $\boxed{x = \pi}$

Solve $\cos x = \dfrac{1}{2}$ for x on $0 \le x < 2\pi$.

$\boxed{x = \dfrac{\pi}{3}}$ or $\boxed{x = \dfrac{5\pi}{3}}$

The solutions to $\sin(2x) = \sin x$ on $[0, 2\pi)$ are $x = 0, \dfrac{\pi}{3}, \pi$, and $\dfrac{5\pi}{3}$.

▼

ANSWER

$x = \dfrac{\pi}{2}, \dfrac{3\pi}{2}, \dfrac{\pi}{6}$, or $\dfrac{5\pi}{6}$

YOUR TURN Solve $\sin(2x) = \cos x$ on $0 \le x < 2\pi$.

EXAMPLE 3   **Using Trigonometric Identities to Solve Trigonometric Equations**

Solve $\sin x + \csc x = -2$.

Solution:

Apply the reciprocal identity
$\csc x = \dfrac{1}{\sin x}$.

$$\sin x + \underbrace{\csc x}_{\frac{1}{\sin x}} = -2$$

Get 0 on one side of the equation.

$$\sin x + 2 + \frac{1}{\sin x} = 0$$

Multiply by $\sin x$. *Note*: $\sin x \neq 0$.

$$\sin^2 x + 2\sin x + 1 = 0$$

Factor.

$$(\sin x + 1)^2 = 0$$

Since $\sin x + 1 = 0$, we can solve for $\sin x$.

$$\sin x = -1$$

Solve for x on one period of the sine function, $[0, 2\pi)$.

$$x = \frac{3\pi}{2}$$

Add integer multiples of 2π to obtain all solutions.

$$\boxed{x = \frac{3\pi}{2} + 2n\pi}$$

▶ **EXAMPLE 4**   **Using Trigonometric Identities and Inverse Functions to Solve Trigonometric Equations**

Solve $3\cos^2\theta + \sin\theta = 3$ on $0° \leq \theta < 360°$.

Solution:

Notice we can write the equation involving only $\sin x$ by using the Pythagorean identity.

$$3\underbrace{\cos^2\theta}_{1 - \sin^2\theta} + \sin\theta = 3$$

Since this is a quadratic type, get 0 on one side.

$$3(1 - \sin^2\theta) + \sin\theta - 3 = 0$$

Eliminate the parentheses.

$$3 - 3\sin^2\theta + \sin\theta - 3 = 0$$

Simplify.

$$-3\sin^2\theta + \sin\theta = 0$$

Factor the common $\sin\theta$.

$$(\sin\theta)(1 - 3\sin\theta) = 0$$

Set each factor equal to 0.

$$\sin\theta = 0 \quad \text{or} \quad 1 - 3\sin\theta = 0$$

Solve for $\sin\theta$.

$$\sin\theta = 0 \quad \text{or} \quad \sin\theta = \frac{1}{3}$$

Solve $\sin\theta = 0$ for θ on $0° \leq \theta < 360°$. $\boxed{\theta = 0°}$ or $\boxed{\theta = 180°}$

Solve $\sin\theta = \frac{1}{3}$ for θ on $0° \leq \theta < 360°$.

The sine function is positive in quadrants I and II. A calculator gives inverse values only in quadrant I.

Write the equivalent inverse notation for $\sin\theta = \frac{1}{3}$.

$$\theta = \sin^{-1}\left(\frac{1}{3}\right)$$

Use a calculator to approximate the quadrant I solution. $\boxed{\theta \approx 19.5°}$

19.5° is the reference angle. To find the quadrant II solution, subtract the 19.5° from 180°.

$$\theta = 180° - 19.5°$$

$$\boxed{\theta \approx 160.5°}$$

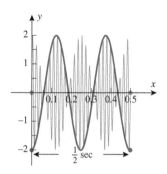

Applications

Musical tones can be represented mathematically with sinusoidal functions. A tone of 48 hertz (cycles per second) can be represented by the function $A\sin[2\pi(48)t]$, where A is the amplitude (loudness) and 48 hertz is the frequency. If we play two musical tones simultaneously, the combined tone can be found using the sum-to-product identities. For example, if a tone of 48 hertz is simultaneously played with a tone of 56 hertz, the result (assuming uniform amplitude, $A = 1$) is

$$\sin(96\pi t) + \sin(112\pi t) = 2\sin(104\pi t)\cos(8\pi t)$$

whose graph is given on the left.

The $\sin(104\pi t)$ term represents a sound of average frequency 52 hertz. The $2\cos(8\pi t)$ represents a time-varying amplitude and a "beat" frequency that corresponds to 8 beats per second. Notice in the graph that there are 4 beats in $\frac{1}{2}$ second.

EXAMPLE 5 Tuning Fork

A tuning fork is used to help musicians tune their instruments. They simultaneously listen to the vibrating fork and play a note and adjust the instrument until the beat frequency is eliminated.

If an E (659 hertz) tuning fork is used and the musician hears 4 beats per second, find the frequency of the note from the instrument if the loudest note is the third of each group. Assume each note has an amplitude of 1.

abzee/ E+/ Getty Images

Solution:

Express the E note mathematically.	$\sin[2\pi(659)t]$
Express the instrument's note mathematically.	$\sin[2\pi(f)t]$
Add the two notes together to get the combined sound.	$\sin[2\pi(659)t] + \sin(2\pi ft)$
Use the sum-to-product identity.	$= 2\sin\left[2\left(\dfrac{f + 659}{2}\right)\pi t\right]\cos\left[2\left(\dfrac{f - 659}{2}\right)\pi t\right]$
The beat frequency is 4 hertz.	$= 2\sin\left[2\left(\dfrac{f + 659}{2}\right)\pi t\right]\cos(4\pi t)$
Since the cosine function is an even function, $\cos(-x) = \cos x$, we equate the absolute value of the cosine arguments.	$\left\lvert 2\left(\dfrac{f - 659}{2}\right)\pi t\right\rvert = \lvert 4\pi t\rvert$
Solve for f.	$f - 659 = -4 \quad \text{or} \quad f - 659 = 4$
f is either 655 hertz or 663 hertz.	$f = 655 \quad \text{or} \quad f = 663$
The loudest note is the peak amplitude.	$\lvert A\rvert = \underbrace{\lvert 2\rvert}_{2}\,\underbrace{\lvert\sin(f + 659)\pi t\rvert}_{1}\,\underbrace{\lvert\cos(4\pi t)\rvert}_{1}$
The loudest sound, $\lvert A\rvert = 2$, is heard on the third beat of each four seconds, $t = \frac{3}{4}$.	$2 = 2\sin\left[(f + 659)\dfrac{3}{4}\pi\right]\cos\left(4\pi\dfrac{3}{4}\right)$
Simplify.	$\sin\left[(f + 659)\dfrac{3}{4}\pi\right] = -1$

The inverse does not exist unless we restrict the domain of the sine function. Therefore, we inspect the two choices, $\boxed{655 \text{ hertz}}$ or $\boxed{663 \text{ hertz}}$, and see that both satisfy the equation. The instrument can be playing either frequency based on the information given.

▶ [SECTION 6.3] SUMMARY

Trigonometric identities are useful for solving equations that involve more than one trigonometric function. With trigonometric identities, we can transform such equations into equations involving only one trigonometric function, and then we can apply algebraic techniques. A second strategy is to factor a trigonometric equation so that each factor contains a single trigonometric function.

[SECTION 6.3] EXERCISES

• SKILLS

In Exercises 1–36, solve each of the trigonometric equations exactly on the interval $0 \leq x < 2\pi$.

1. $\sin x = \cos x$

2. $\sin x = -\cos x$

3. $\sec x + \cos x = -2$

4. $\sin x + \csc x = 2$

5. $\sec x - \tan x = \dfrac{\sqrt{3}}{3}$

6. $\sec x + \tan x = 1$

7. $\csc x + \cot x = \sqrt{3}$

8. $\csc x - \cot x = \dfrac{\sqrt{3}}{3}$

9. $2\sin x - \csc x = 0$

10. $2\sin x + \csc x = 3$

11. $\sin(2x) = 4\cos x$

12. $\sin(2x) = \sqrt{3}\sin x$

13. $\sqrt{2}\sin x = \tan x$

14. $\cos(2x) = \sin x$

15. $\tan(2x) = \cot x$

16. $3\cot(2x) = \cot x$

17. $\sqrt{3}\sec x = 4\sin x$

18. $\sqrt{3}\tan x = 2\sin x$

19. $\sin^2 x - \cos(2x) = -\dfrac{1}{4}$

20. $\sin^2 x - 2\sin x = 0$

21. $\cos^2 x + 2\sin x + 2 = 0$

22. $2\cos^2 x = \sin x + 1$

23. $2\sin^2 x + 3\cos x = 0$

24. $4\cos^2 x - 4\sin x = 5$

25. $\cos(2x) + \cos x = 0$

26. $2\cot x = \csc x$

27. $2\sin^2 x + \sin(2x) + \cos(2x) = 0$

28. $\sin(2x) - \cos(2x) = 0$

29. $\tan(2x) + 1 = \sec(2x)$

30. $\tan(2x) - \tan x = 0$

31. $\dfrac{\cos(2x)}{\cos x + \sin x} = 0$

32. $\csc(2x) + \sec x = -1 - \dfrac{1}{2}\csc x$

33. $\cos(3x)\cos(2x) + \sin(3x)\sin(2x) = 1$

34. $\sin(3x)\cos(2x) - \cos(3x)\sin(2x) = 1$

35. $\sin(7x)\cos(5x) = \dfrac{1}{2} + \cos(7x)\sin(5x)$

36. $\sin(7x)\sin(5x) = -1 - \cos(7x)\cos(5x)$

In Exercises 37–48, solve each of the trigonometric equations on the interval $0° \leq \theta < 360°$. Give answers in degrees and round to two decimal places.

37. $\cos(2x) + \dfrac{1}{2}\sin x = 0$

38. $\sec^2 x = \tan x + 1$

39. $6\cos^2 x + \sin x = 5$

40. $\sec^2 x = 2\tan x + 4$

41. $\cot^2 x - 3\csc x - 3 = 0$

42. $\csc^2 x + \cot x = 7$

43. $2\sin^2 x + 2\cos x - 1 = 0$

44. $\sec^2 x + \tan x - 2 = 0$

45. $\cos(2x) + \sin x + 1 = 0$

46. $\sin^2(2x) + \sin(2x) - 2 = 0$

47. $\tan^2 x + \sec x = -1$

48. $\dfrac{\sin^2 x - 2\sin x}{\tan x} = 0$

• **APPLICATIONS**

For Exercises 49 and 50, refer to the following:

If a person breathes in and out every 3 seconds, the volume of air in his lungs can be modeled by $A = 2\sin\left(\dfrac{\pi}{3}x\right)\cos\left(\dfrac{\pi}{3}x\right) + 3$, where A is in liters of air and x is in seconds.

49. **Biology/Health.** How many seconds into the cycle is the volume of air equal to 4 liters?

50. **Biology/Health.** For the function given in Exercise 49, how many seconds into the cycle is the volume of air equal to 2 liters?

For Exercises 51 and 52, refer to the following:

The figure below shows the graph of $y = 2\cos x - \cos(2x)$ between -2π and 2π. The maximum and minimum values of the curve occur at the *turning points* and are found in the solutions of the equation $-2\sin x + 2\sin(2x) = 0$.

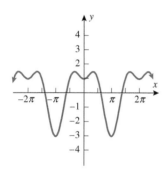

51. **Finding Turning Points.** Solve for the coordinates of the turning points of the curve between 0 and 2π (not including 0 or 2π).

52. **Finding Turning Points.** Solve for the coordinates of the turning points of the curve between -2π and 0 (not including -2π or 0).

53. **Treadmill.** Romona and Marci are running on treadmills beside each other. Both have their treadmills programmed to change speeds during the workout. Romona's speed S, in miles per hour, at time t minutes after starting can be modeled by the equation $S = 5 + 3\sin t$, whereas Marci's speed can be modeled by the equation $S = 5 - 2\cos t$. During the first 10 minutes of their workout, at what times are they both going the same speed?

54. **Treadmill.** Keith and Kevin are running on treadmills beside each other. Both have their treadmills programmed to change speeds during the workout. Keith's speed S, in miles per hour, at time t minutes after starting can be modeled by the equation $S = 5.5 + 4\sin t$, whereas Kevin's speed can be modeled by the equation $S = 5.5 - 3\cos t$. During the first 10 minutes of their workout, at what times are they both going the same speed?

55. **Trail System.** Alex is mapping out his city's walking trails. One small section of the trail system has two trails that can be modeled by the equations $y = \tan x + 1$ and $y = 2\sin x + 1$, where x and y represent the horizontal and vertical distances from a park in miles, where $0 < x \le 1.5$ and $0 < y \le 5$. Give the position of the point where the trails intersect as it relates to the park.

56. **Trail System.** Maria is mapping out her city's walking trails. One small section of the trail system has two trails that can be modeled by the equations $y = \sin x + 1$ and $y = 2\cot x + 1$, where x and y represent the horizontal and vertical distances from a park in miles, where $0 \le x \le 1.5$ and $0 \le y \le 5$. Give the position of the point where the trails intersect as it relates to the park.

57. **Business.** An analysis of a company's costs and revenue shows that annual costs of producing their product as well as annual revenues from the sale of a product are generally subject to seasonal fluctuations and are approximated by the function

$$C(t) = 2.3 + 0.25\sin\left(\dfrac{\pi}{6}t\right) \qquad 0 \le t \le 11$$

$$R(t) = 2.3 + 0.5\cos\left(\dfrac{\pi}{6}t\right) \qquad 0 \le t \le 11$$

where t represents time in months ($t = 0$ represents January), $C(t)$ represents the monthly costs of producing the product in millions of dollars, and $R(t)$ represents monthly revenue from sales of the product in millions of dollars. Find the month(s) in which the company breaks even. *Hint:* A company breaks even when its profit is zero.

58. **Business.** An analysis of a company's costs and revenue shows that the annual costs of producing its product as well as annual revenues from the sale of a product are generally subject to seasonal fluctuations and are approximated by the function

$$C(t) = 25.7 + 0.2\sin\left(\dfrac{\pi}{6}t\right) \qquad 0 \le t \le 11$$

$$R(t) = 25.7 + 9.6\cos\left(\dfrac{\pi}{6}t\right) \qquad 0 \le t \le 11$$

where t represents time in months ($t = 0$ represents January), $C(t)$ represents the monthly costs of producing the product in millions of dollars, and $R(t)$ represents monthly revenue from sales of the product in millions of dollars. Find the month(s) in which the company breaks even. *Hint:* A company breaks even when its profit is zero.

For Exercises 59 and 60, refer to the following:

By analyzing available empirical data, it has been determined that the body temperature of a species fluctuates according to the model

$$T(t) = 37.10 + 1.40\sin\left(\dfrac{\pi}{24}t\right)\cos\left(\dfrac{\pi}{24}t\right) \qquad 0 \le t \le 24$$

where T represents temperature in degrees Celsius and t represents time (in hours) measured from 12:00 A.M. (midnight).

59. **Biology/Health.** Find the time(s) of day the body temperature is 37.28 degrees Celsius. Round to the nearest hour.

60. **Biology/Health.** Find the time(s) of day the body temperature is 36.75 degrees Celsius. Round to the nearest hour.

• **CATCH THE MISTAKE**

In Exercises 61 and 62, explain the mistake that is made.

61. Solve $3\sin(2x) = 2\cos x$ on $0° \le \theta < 180°$.

Solution:

Use the double-angle identity for sine.	$3\underbrace{\sin(2x)}_{2\sin x\cos x} = 2\cos x$
Simplify.	$6\sin x\cos x = 2\cos x$
Divide by $2\cos x$.	$3\sin x = 1$
Divide by 3.	$\sin x = \dfrac{1}{3}$
Write equivalent inverse notation.	$x = \sin^{-1}\left(\dfrac{1}{3}\right)$
Use a calculator to approximate.	$x = 19.47°$, QI solution

The quadrant II solution is $x = 180° - 19.47° = 160.53°$.

This is incorrect. What mistake was made?

62. Solve $\sqrt{1 + \sin x} = \cos x$ on $0 \le x < 2\pi$.

Solution:

Square both sides.	$1 + \sin x = \cos^2 x$
Use the Pythagorean identity.	$1 + \sin x = \underbrace{\cos^2 x}_{1 - \sin^2 x}$
Simplify.	$\sin^2 x + \sin x = 0$
Factor.	$\sin x(\sin x + 1) = 0$
Set each factor equal to zero.	$\sin x = 0$ or $\sin x + 1 = 0$
Solve for $\sin x$.	$\sin x = 0$ or $\sin x = -1$
Solve for x.	$x = 0, \pi, \dfrac{3\pi}{2}$

This is incorrect. What mistake was made?

• **CONCEPTUAL**

In Exercises 63–66, determine whether each statement is true or false.

63. If a trigonometric equation has the set of all real numbers as its solution, then it is an identity.

64. If a trigonometric equation has an infinite number of solutions, then it is an identity.

65. The equations $\dfrac{\sin x}{\sec x} = 1$ and $\sin x = \sec x$ have the same solution set.

66. The equations $\tan x(\cos x + 1) = 0$ and $\dfrac{\cos x + 1}{\tan x} = 0$ have the same solution set.

67. Use the graphs of $y = \tan x$ and $y = \cos x + 1$ to determine the number of solutions to the equation $\tan x = \cos x + 1$ on the interval $[0, 2\pi)$.

68. Use the graphs of $y = \csc x$ and $y = \sin(2x)$ to determine the number of solutions to the equation $\csc x = \sin(2x)$ on the interval $[0, 2\pi)$.

• **CHALLENGE**

69. Solve for the smallest positive x that makes this statement true:

$$\sin\left(x + \frac{\pi}{4}\right) + \sin\left(x - \frac{\pi}{4}\right) = \frac{\sqrt{2}}{2}$$

70. Solve for the smallest positive x that makes this statement true:

$$\cos x\cos 15° + \sin x\sin 15° = 0.7$$

71. Find the exact value of the solutions to the equation $\cos\left(\frac{1}{2}x\right) = \cos x$ on the interval $[0, 2\pi)$.

72. Find the exact value of the solutions to the equation $\tan\left(\frac{1}{2}x\right) = \sin x$ on the interval $[0, 2\pi)$.

73. Solve the inequality exactly:

$$-\tfrac{1}{4} \le \sin x\cos x \le \tfrac{1}{4} \text{ on the interval } [0, \pi)$$

74. Solve the inequality exactly:

$$|\cos^2 x - \sin^2 x| \le \tfrac{1}{2} \text{ on the interval } [0, \pi)$$

• **TECHNOLOGY**

In Exercises 75–78, find the smallest positive value of x that makes the statement true. Give the answer in degrees and round to two decimal places.

75. $\sec(3x) + \csc(2x) = 5$ **76.** $\cot(5x) + \tan(2x) = -3$ **77.** $e^x - \tan x = 0$ **78.** $e^x + 2\sin x = 1$

▶[**CHAPTER 6 REVIEW**]

SECTION	CONCEPT	KEY IDEAS/FORMULAS
6.1	**Inverse trigonometric functions**	$\sin^{-1}x$ or $\arcsin x$ $\cot^{-1}x$ or $\text{arccot}\,x$ $\cos^{-1}x$ or $\arccos x$ $\sec^{-1}x$ or $\text{arcsec}\,x$ $\tan^{-1}x$ or $\arctan x$ $\csc^{-1}x$ or $\text{arccsc}\,x$
	Inverse sine function	**Definition:** $y = \sin^{-1}x$ means $x = \sin y$ $-1 \le x \le 1$ and $-\dfrac{\pi}{2} \le y \le \dfrac{\pi}{2}$ **Identities:** $\sin^{-1}(\sin x) = x$ for $-\dfrac{\pi}{2} \le x \le \dfrac{\pi}{2}$ $\sin(\sin^{-1}x) = x$ for $-1 \le x \le 1$
	Inverse cosine function	**Definition:** $y = \cos^{-1}x$ means $x = \cos y$ $-1 \le x \le 1$ and $0 \le y \le \pi$ **Identities:** $\cos^{-1}(\cos x) = x$ for $0 \le x \le \pi$ $\cos(\cos^{-1}x) = x$ for $-1 \le x \le 1$
	Inverse tangent function	**Definition:** $y = \tan^{-1}x$ means $x = \tan y$ $-\infty < x < \infty$ and $-\dfrac{\pi}{2} < y < \dfrac{\pi}{2}$ **Identities:** $\tan^{-1}(\tan x) = x$ for $-\dfrac{\pi}{2} < x < \dfrac{\pi}{2}$ $\tan(\tan^{-1}x) = x$ for $-\infty < x < \infty$

SECTION	CONCEPT	KEY IDEAS/FORMULAS
	Remaining inverse trigonometric functions	**Inverse Secant Function** **Definition:** $y = \sec^{-1} x$ means $x = \sec y$  $x \le -1$ or $x \ge 1$ and $0 \le y < \dfrac{\pi}{2}$ or $\dfrac{\pi}{2} < y \le \pi$ **Identity:** $\sec^{-1} x = \cos^{-1}\left(\dfrac{1}{x}\right)$ for $x \le -1$ or $x \ge 1$ **Inverse Cosecant Function** **Definition:** $y = \csc^{-1} x$ means $x = \csc y$ $x \le -1$ or $x \ge 1$ and $-\dfrac{\pi}{2} \le y < 0$ or $0 < y \le \dfrac{\pi}{2}$ **Identity:** $\csc^{-1} x = \sin^{-1}\left(\dfrac{1}{x}\right)$ for $x \le -1$ or $x \ge 1$ **Inverse Cotangent Function** **Definition:** $y = \cot^{-1} x$ means $x = \cot y$ $-\infty < x < \infty$ and $0 < y < \pi$ **Identity:** $\cot^{-1} x = \begin{cases} \tan^{-1}\left(\dfrac{1}{x}\right), x > 0 \\ \pi + \tan^{-1}\left(\dfrac{1}{x}\right), x < 0 \end{cases}$
	Finding exact values for expressions involving inverse trigonometric functions	Use sketches to identify angles, quadrants, and trigonometric values.

SECTION	CONCEPT	KEY IDEAS/FORMULAS
6.2	**Solving trigonometric equations that involve only one trigonometric function**	Goal: Find the value(s) of the variable that make the equation true.
	Solving trigonometric equations by inspection	Solve: $\sin\theta = \dfrac{\sqrt{2}}{2}$ on $0 \le \theta \le 2\pi$ Answer: $\theta = \dfrac{\pi}{4}$ or $\theta = \dfrac{3\pi}{4}$ Solve: $\sin\theta = \dfrac{\sqrt{2}}{2}$ on all real numbers Answer: $\theta = \begin{cases} \dfrac{\pi}{4} + 2n\pi \\[2mm] \dfrac{3\pi}{4} + 2n\pi \end{cases}$, where n is an integer
	Solving trigonometric equations using algebraic techniques	Trigonometric equations can sometimes be transformed into linear or quadratic algebraic equations by making a substitution such as $x = \sin\theta$. Algebraic methods are then used for solving linear and quadratic equations. *Note*: If an expression is squared, check for extraneous solutions.
	Solving trigonometric equations that require the use of inverse functions	When a trigonometric equation cannot be solved by inspection (i.e., not special angles), then inverse trigonometric functions are used.
6.3	**Solving trigonometric equations that involve multiple trigonometric functions**	
	Solving trigonometric equations	■ Use trigonometric identities to transform an equation with multiple trigonometric functions into an equation with only one trigonometric function. ■ Then use the methods outlined in Section 6.2.

[CHAPTER 6 REVIEW EXERCISES]

6.1 Inverse Trigonometric Functions

Find the exact value of each expression. Give the answer in radians.

1. $\arctan 1$
2. $\text{arccsc}(-2)$
3. $\cos^{-1} 0$
4. $\sin^{-1}(-1)$

Find the exact value of each expression. Give the answer in degrees.

5. $\csc^{-1}(-1)$
6. $\arctan(-1)$
7. $\text{arccot}\left(\dfrac{\sqrt{3}}{3}\right)$
8. $\cos^{-1}\left(\dfrac{\sqrt{2}}{2}\right)$

Use a calculator to evaluate each expression. Give the answer in degrees and round to two decimal places.

9. $\sin^{-1}(-0.6088)$
10. $\tan^{-1}(1.1918)$
11. $\sec^{-1}(1.0824)$
12. $\cot^{-1}(-3.7321)$

Use a calculator to evaluate each expression. Give the answer in radians and round to two decimal places.

13. $\cos^{-1}(-0.1736)$
14. $\tan^{-1}(0.1584)$
15. $\csc^{-1}(-10.0167)$
16. $\sec^{-1}(-1.1223)$

Evaluate each expression exactly, if possible. If not possible, state why.

17. $\sin^{-1}\left[\sin\left(-\dfrac{\pi}{4}\right)\right]$
18. $\cos\left[\cos^{-1}\left(-\dfrac{\sqrt{2}}{2}\right)\right]$
19. $\tan\left[\tan^{-1}(\sqrt{3})\right]$
20. $\cot^{-1}\left[\cot\left(\dfrac{11\pi}{6}\right)\right]$
21. $\csc^{-1}\left[\csc\left(\dfrac{2\pi}{3}\right)\right]$
22. $\sec\left[\sec^{-1}\left(-\dfrac{2\sqrt{3}}{3}\right)\right]$

Evaluate each expression exactly.

23. $\sin\left[\cos^{-1}\left(\dfrac{11}{61}\right)\right]$
24. $\cos\left[\tan^{-1}\left(\dfrac{40}{9}\right)\right]$
25. $\tan\left[\cot^{-1}\left(\dfrac{6}{7}\right)\right]$
26. $\cot\left[\sec^{-1}\left(\dfrac{25}{7}\right)\right]$
27. $\sec\left[\sin^{-1}\left(\dfrac{1}{6}\right)\right]$
28. $\csc\left[\cot^{-1}\left(\dfrac{5}{12}\right)\right]$

Applications

29. **Average Temperature.** If the average temperature in Chicago, Illinois, can be modeled with the formula $T(m) = 26\sin(0.48m - 1.84) + 47$, where m is the month of the year (January corresponds to $m = 1$, etc.) and T is in degrees Fahrenheit, then during which month is the average temperature 73 degrees?

30. **Average Temperature.** If the average temperature in Chicago, Illinois, can be modeled with the formula $T(m) = 26\sin(0.48m - 1.84) + 47$, where m is the month of the year (January corresponds to $m = 1$, etc.) and T is in degrees Fahrenheit, then during which month is the average temperature 21 degrees?

6.2 Solving Trigonometric Equations That Involve Only One Trigonometric Function

Solve each of the following trigonometric equations over the indicated interval.

31. $\sin(2\theta) = -\dfrac{\sqrt{3}}{2}, 0 \le \theta < 2\pi$

32. $\sec\left(\dfrac{\theta}{2}\right) = 2, -2\pi \le \theta < 2\pi$
33. $\sin\left(\dfrac{\theta}{2}\right) = -\dfrac{\sqrt{2}}{2}, -2\pi \le \theta < 2\pi$
34. $\csc(2\theta) = 2, 0 \le \theta < 2\pi$

Solve each of the following trigonometric equations exactly on $0 \le \theta < 2\pi$.

35. $4\cos(2\theta) + 2 = 0$
36. $\sqrt{3}\tan\left(\dfrac{\theta}{2}\right) - 1 = 0$
37. $2\tan(2\theta) + 2 = 0$
38. $2\sin^2\theta + \sin\theta - 1 = 0$
39. $\tan^2\theta + \tan\theta = 0$
40. $\sec^2\theta - 3\sec\theta + 2 = 0$

Solve the given trigonometric equation on $0° \le \theta < 360°$ and express the answer in degrees to two decimal places.

41. $\tan(2\theta) = -0.3459$
42. $6\sin\theta - 5 = 0$
43. $4\cos^2\theta + 3\cos\theta = 0$
44. $12\cos^2\theta - 7\cos\theta + 1 = 0$
45. $\csc^2\theta - 3\csc\theta - 1 = 0$
46. $2\cot^2\theta + 5\cot\theta - 4 = 0$

6.3 Solving Trigonometric Equations That Involve Multiple Trigonometric Functions

Solve the trigonometric equations exactly on the interval $0 \le \theta < 2\pi$.

47. $\sec x = 2\sin x$
48. $3\tan x + \cot x = 2\sqrt{3}$
49. $\sqrt{3}\tan x - \sec x = 1$
50. $2\sin(2x) = \cot x$
51. $\sqrt{3}\tan x = 2\sin x$
52. $2\sin x = 3\cot x$
53. $\cos^2 x + \sin x + 1 = 0$
54. $2\cos^2 x - \sqrt{3}\cos x = 0$
55. $\cos(2x) + 4\cos x + 3 = 0$
56. $\sin(2x) + \sin x = 0$

Solve the trigonometric equations on $0° \le \theta < 360°$. Give the answers in degrees and round to two decimal places.

57. $\csc^2 x + \cot x = 1$
58. $8\cos^2 x + 6\sin x = 9$
59. $\sin^2 x + 2 = 2\cos x$
60. $\cos(2x) = 3\sin x - 1$
61. $\cos x - 1 = \cos(2x)$
62. $12\cos^2 x + 4\sin x = 11$

Technology Exercises

Section 6.1

63. Given $\cos x = -\dfrac{1}{\sqrt{5}}$ and $\dfrac{\pi}{2} < x < \pi$:
   a. Find $\cos(2x)$ using the double-angle identity.
   b. Use the inverse of cosine to find x in quadrant II and to find $\cos(2x)$.
   c. Are the results in (a) and (b) the same?

64. Given $\cos x = \dfrac{5}{12}$ and $\dfrac{3\pi}{2} < x < 2\pi$:
   a. Find $\cos\left(\tfrac{1}{2}x\right)$ using the half-angle identity.
   b. Use the inverse of cosine to find x in quadrant IV and to find $\cos\left(\tfrac{1}{2}x\right)$. Round to five decimal places.
   c. Are the results in (a) and (b) the same?

Section 6.2

Find the smallest positive value of x that makes each statement true. Give the answer in radians and round to four decimal places.

65. $\ln x + \sin x = 0$
66. $\ln x + \cos x = 0$

[CHAPTER 6 PRACTICE TEST]

State the interval over x for which the indicated identity is valid.

1. $\sin(\sin^{-1}x) = x$

2. $\tan^{-1}(\tan x) = x$

3. $\cos^{-1}(\cos x) = x$

Evaluate *exactly*.

4. $\csc^{-1}\left(\sqrt{2}\right)$

5. $\cot^{-1}\left(-\sqrt{3}\right)$

6. $\sin^{-1}\left(-\dfrac{\sqrt{3}}{2}\right)$

7. $\cos^{-1}0$

Evaluate the expressions *exactly*.

8. $\cos^{-1}\left[\cos\left(\dfrac{4\pi}{3}\right)\right]$

9. $\tan^{-1}\left[\tan\left(\dfrac{5\pi}{6}\right)\right]$

10. $\sin^{-1}\left[\sin\left(\dfrac{3\pi}{4}\right)\right]$

11. $\cos(\tan^{-1}2)$

12. $\sec(\sec^{-1}2)$

Solve the trigonometric equations exactly, if possible; otherwise, use a calculator to approximate solution(s).

13. $2\sin\theta = -\sqrt{3}$ on all real numbers

14. $4 + 4\cos^2\theta = 6$ on $0 \le \theta < 2\pi$

15. $2 + 3\sec\theta = -10$ on $0 \le \theta < 2\pi$

16. $3\tan\theta = 1$ on $0 \le \theta < 180°$

17. $2\cos^2\theta + \cos\theta - 1 = 0$ on $0 \le \theta < 2\pi$

18. $\sin^2\theta + 3\sin\theta - 1 = 0$ on $0 \le \theta < 360°$

19. $\sin^2\theta + \cos\theta = -1$ on all real numbers

20. $6\tan^2\theta + \tan\theta - 2 = 0$ on $0 \le \theta < 2\pi$

21. $\sin(2\theta) = \frac{1}{2}\cos\theta$ over $0 \le \theta < 360°$

22. $\cos(2\theta) + \sin^2\theta = \cos\theta$ on $0 \le \theta < 2\pi$

23. $[\sin(2\theta)][(\sin(2\theta) + 1] + \cos^2(2\theta) + \cos(2\theta) = 1$ on $0 \le \theta < 2\pi$

24. $\sec^2(2\theta) - 5\sec(2\theta) - 6 = 0$ on $0° \le \theta < 360°$

25. $\cot^2(2\theta) + \csc^2(2\theta) = 3$ on $0 \le \theta < 2\pi$

26. $\sin^2\theta - 3\cos\theta - 3 = 0$ over $0 \le \theta < 360°$

27. $\sqrt{\sin x + \cos x} = -1$ over $0 \le \theta < 2\pi$

28. $(\sin\theta + \cos\theta)^2 + \cos(90° + \theta) = 1$ on $0° \le \theta < 360°$

29. $\sqrt{\tan\theta + 1} = \tan\theta$ on $0° \le \theta < 360°$

30. $\dfrac{\sin x + \cos x}{\cos(2x)} = 3$ on $0 \le \theta < 2\pi$

[CHAPTERS 1–6 CUMULATIVE TEST]

1. Use the triangle below to find $\cot\theta$.

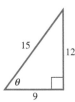

2. Use a calculator to approximate $\csc(64°39')$.
3. Given that $\cot\theta$ is positive and $\csc\theta$ is negative, find the quadrant in which the terminal side of θ must lie.
4. Give the exact value of $\sec(-225°)$.
5. Given $\cos\theta = 0.75$, use the reciprocal identity to find $\sec\theta$.
6. Simplify the expression $\cot\theta\,\sec^2\theta$. Leave your answer in terms of $\sin\theta$ and $\cos\theta$.
7. Find the measure (in radians) of a central angle θ that intercepts an arc on a circle of radius $r = \frac{3}{4}$ inch with arc length $s = \frac{5}{8}$ inch.
8. Find the area of a circular sector with radius 4.5 centimeters and central $\theta = 24°$. Round the answer to three significant digits.
9. A ladybug is clinging to the outer edge of a child's spinning disk. The disk is 5 inches in diameter and is spinning at 55 revolutions per minute. How fast is the ladybug traveling?
10. Find the exact value of $\csc 240°$.
11. Find all of the exact values of θ, when $\cot\theta$ is undefined and $0 \le \theta \le 4\pi$.

For Exercises 12–14, refer to the graph below of the cosine function:

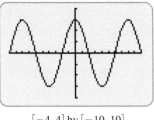

$[-4, 4]$ by $[-10, 10]$

12. Determine the period of the function.
13. Find the amplitude of the function.
14. Write an equation for the cosine function.

15. Sketch the graph of the function $y = \frac{1}{2} + \frac{1}{2}\sin\left(\frac{1}{2}x - \frac{\pi}{4}\right)$ over the interval $\left[-\frac{7\pi}{2}, \frac{9\pi}{2}\right]$.

16. Graph the function $y = 4\sec\left(\frac{x}{2}\right)$ over the interval $[-2\pi, 2\pi]$.

17. Verify the trigonometric identity $\cot x(\cos x - \sec x) = -\sin x$.

18. Find the exact value of $\cos(\alpha - \beta)$ if $\cos\alpha = \frac{1}{5}$ and $\cos\beta = -\frac{3}{5}$ and the terminal side of α lies in quadrant IV and the terminal side of β lies in quadrant II.

19. Simplify $\dfrac{2\tan\left(-\dfrac{\pi}{8}\right)}{1 - \tan^2\left(-\dfrac{\pi}{8}\right)}$ and evaluate exactly.

20. If $\tan x = \dfrac{7}{24}$ and $\pi < x < \dfrac{3\pi}{2}$, use half-angle identities to find $\cos\left(\dfrac{x}{2}\right)$.

21. Write the product $7\sin(-2x)\sin(5x)$ as a sum or difference of sines and/or cosines.

22. Simplify the trigonometric expression $\dfrac{\sin(7x) - \sin(3x)}{\cos(7x) + \cos(3x)}$.

23. Find the exact value of $\sec^{-1}\left(-\dfrac{2\sqrt{3}}{3}\right)$. Give the answer in degrees.

24. Evaluate exactly the expression $\tan\left[\sin^{-1}\left(\dfrac{5}{13}\right)\right]$.

25. Solve exactly the trigonometric equation $2\cos^2\theta - \cos\theta - 1 = 0$ over $0 \le \theta \le 2\pi$.

Applications of Trigonometry: Triangles and Vectors

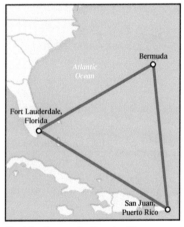

Florida, Bermuda, Puerto Rico

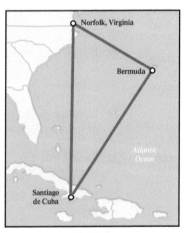

Virginia, Bermuda, Cuba

In recent decades, many people have come to believe that an imaginary area called the "Bermuda Triangle," located off the southeastern Atlantic coast of the United States, has been the site of a high incidence of losses of ships, small boats, and aircraft over the centuries. The U.S. Board of Geographic Names does not recognize the "Bermuda Triangle" as an official name and does not maintain an official file on the area.

Assume for the moment, without judging the merits of the hypothesis, that the "Bermuda Triangle" has vertices either in Miami (Florida), San Juan (Puerto Rico), and Bermuda, or in Norfolk (Virginia), Bermuda, and Santiago (Cuba). In this chapter, you will develop a formula that determines the area of a triangle from its perimeter and side lengths. Which "Bermuda Triangle" has a larger area: Miami-Bermuda-Puerto Rico or Norfolk-Bermuda-Cuba? You will calculate the answer in this chapter.*

*Section 7.3, Exercises 37 and 38.

LEARNING OBJECTIVES

- Solve oblique triangles using the Law of Sines.
- Solve oblique triangles using the Law of Cosines.
- Find areas of oblique triangles.
- Perform vector operations.
- Find the dot product of two vectors.

We discuss oblique (nonright) triangles. We use the Law of Sines and the Law of Cosines to solve oblique triangles. Then on the basis of the Law of Sines and the Law of Cosines and with trigonometric identities, we develop formulas for calculating the area of an oblique triangle. We also define vectors and use the Law of Cosines and the Law of Sines to determine resulting velocity and force vectors. Finally, we define dot products (product of two vectors), and see how they are applicable to physical problems such as calculating work.

APPLICATIONS OF TRIGONOMETRY: TRIANGLES AND VECTORS

7.1 OBLIQUE TRIANGLES AND THE LAW OF SINES	7.2 THE LAW OF COSINES	7.3 THE AREA OF A TRIANGLE	7.4 VECTORS	7.5 THE DOT PRODUCT
• Solving Oblique Triangles	• Solving Oblique Triangles	• The Area of a Triangle (SAS Case) • The Area of a Triangle (SSS Case)	• Vectors: Magnitude and Direction • Vector Operations • Horizontal and Vertical Components of a Vector • Unit Vectors • Resultant Vectors	• Multiplying Two Vectors: The Dot Product • Angle Between Two Vectors • Work

7.1 OBLIQUE TRIANGLES AND THE LAW OF SINES

SKILLS OBJECTIVE	CONCEPTUAL OBJECTIVE
■ Solve the AAS, ASA, and SSA right triangles using the Law of Sines.	■ Understand why an AAA case cannot be solved.

7.1.1 Solving Oblique Triangles

7.1.1 SKILL

Solve the AAS, ASA, and SSA right triangles using the Law of Sines.

7.1.1 CONCEPTUAL

Understand why an AAA case cannot be solved.

Thus far, we have discussed only *right* triangles. There are, however, two types of triangles, right and *oblique*. An **oblique triangle** is any triangle that does not have a right angle. An oblique triangle will be either an **acute triangle**, having three acute (less than 90°) angles; or an **obtuse triangle**, having one obtuse (between 90° and 180°) angle.

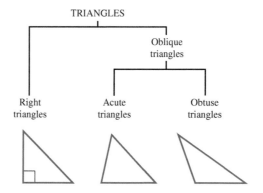

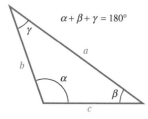

It is customary to label oblique triangles the following way:

■ angle α (alpha): opposite side a.

■ angle β (beta): opposite side b.

■ angle γ (gamma): opposite side c.

Remember that the sum of the three angles of any triangle must equal 180°. Recall that in Section 1.5 we solved right triangles. In this chapter, we solve oblique triangles, which means we find the lengths of all three sides and the measures of all three angles.

Four Cases

To solve an oblique triangle, *we need to know the length of one side* and one of the following three:

■ two angle measures

■ one angle measure and another side length

■ the other two side lengths

This requirement leads to four possible cases to consider:

REQUIRED INFORMATION TO SOLVE OBLIQUE TRIANGLES

CASE	WHAT'S GIVEN	EXAMPLES/NAMES
Case 1	One side and two angles	AAS: Angle-Angle-Side ASA: Angle-Side-Angle
Case 2	Two sides and the angle opposite one of them	SSA: Side-Side-Angle
Case 3	Two sides and the angle between them	SAS: Side-Angle-Side
Case 4	Three sides	SSS: Side-Side-Side

Notice that there is no AAA case. This is because two similar triangles can have the same angle measures but different side lengths, so at least one side must be known to determine a unique triangle.

In this section, we will derive the Law of Sines, which will enable us to solve Case 1 and Case 2 problems. In the next section, we will derive the Law of Cosines, which will enable us to solve Case 3 and Case 4 problems.

The Law of Sines

Let us start with two oblique triangles, an acute triangle and an obtuse triangle.

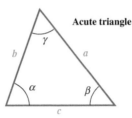 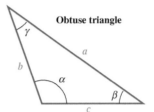

The following discussion applies to both triangles. First, construct an altitude h (perpendicular) from the vertex at angle γ to the side (or its extension) opposite γ.

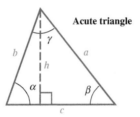 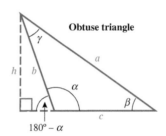

WORDS	MATH
Formulate sine ratios for the acute triangle.	$\sin\alpha = \dfrac{h}{b}$ and $\sin\beta = \dfrac{h}{a}$
Formulate sine ratios for the obtuse triangle.	$\sin(180° - \alpha) = \dfrac{h}{b}$ and $\sin\beta = \dfrac{h}{a}$
For the obtuse angle, apply the difference identity for the sine function.	$\begin{aligned}\sin(180° - \alpha) &= \sin 180° \cos\alpha - \cos 180° \sin\alpha \\ &= 0 \cdot \cos\alpha - (-1)\sin\alpha \\ &= \sin\alpha\end{aligned}$
Therefore, in both triangles we find the same two equations.	$\sin\alpha = \dfrac{h}{b}$ and $\sin\beta = \dfrac{h}{a}$
Solve for h in both equations.	$h = b\sin\alpha$ and $h = a\sin\beta$
Since h is equal to itself in each triangle, equate the expressions for h.	$b\sin\alpha = a\sin\beta$

Divide both sides by *ab*.

$$\frac{b\sin\alpha}{ab} = \frac{a\sin\beta}{ab}$$

Divide out common factors.

$$\frac{\sin\alpha}{a} = \frac{\sin\beta}{b}$$

In a similar manner, we can extend an altitude (perpendicular) from angle α, and we will find that $\dfrac{\sin\gamma}{c} = \dfrac{\sin\beta}{b}$. Equating these two expressions leads us to the third ratio of the *Law of Sines*: $\dfrac{\sin\alpha}{a} = \dfrac{\sin\gamma}{c}$.

THE LAW OF SINES

For a triangle with side lengths *a*, *b*, and *c* and opposite angle measures α, β, and γ, the following relationship is true:

$$\frac{\sin\alpha}{a} = \frac{\sin\beta}{b} = \frac{\sin\gamma}{c}$$

In other words, the ratio of the sine of an angle in a triangle to its opposite side is equal to the ratios of the sines of the other two angles to their opposite sides. That is, the ratio of the sine of an angle to its opposite side is constant in any triangle.

A few things to note before we begin solving oblique triangles:

- The angles and sides share the same progression of magnitude:
  - The longest side of a triangle is opposite the largest angle.
  - The shortest side of a triangle is opposite the smallest angle.
- Draw the triangle and label the angles and sides.
- If two angle measures are known, start by determining the third angle measure.
- Whenever possible, in successive steps, always return to given values rather than referring to calculated (approximate) values.

In this chapter, we will round answers so that the number of significant digits in the answer is equal to the number of significant digits of the given information: We use the *least* number of significant digits from the given information.

STUDY TIP

The longest side is opposite the largest angle; the shortest side is opposite the smallest angle.

STUDY TIP

When possible, use given values rather than calculated (approximated) values for better accuracy.

Case 1: Two Angles and One Side (AAS or ASA)

EXAMPLE 1 **Using the Law of Sines to Solve a Triangle (AAS)**

Solve the triangle.

Solution:

This is an AAS (angle-angle-side) case because two angle measures and a side length are given and the side is opposite one of the angles. Based on the given information, answers will be rounded to two significant digits.

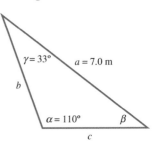

STEP 1 Find β.

The sum of the measures of the angles in a triangle is 180°.	$\alpha + \beta + \gamma = 180°$
Let $\alpha = 110°$ and $\gamma = 33°$.	$110° + \beta + 33° = 180°$
Solve for β.	$\boxed{\beta = 37°}$

STEP 2 Find b.

Use the Law of Sines with the known side a.

$$\frac{\sin\alpha}{a} = \frac{\sin\beta}{b}$$

Isolate b.

$$b = \frac{a\sin\beta}{\sin\alpha}$$

Let $\alpha = 110°$, $\beta = 37°$, and $a = 7$ meters.

$$b = \frac{7\sin 37°}{\sin 110°}$$

Use a calculator to approximate b.

$$b \approx 4.483067 \text{ m}$$

Round b to two significant digits.

$$\boxed{b \approx 4.5 \text{ m}}$$

STEP 3 Find c.

Use the Law of Sines with the known side a.

$$\frac{\sin\alpha}{a} = \frac{\sin\gamma}{c}$$

Isolate c.

$$c = \frac{a\sin\gamma}{\sin\alpha}$$

Let $\alpha = 110°$, $\gamma = 33°$, and $a = 7$ meters.

$$c = \frac{7\sin 33°}{\sin 110°}$$

Use a calculator to approximate c.

$$c \approx 4.057149 \text{ m}$$

Round c to two significant digits.

$$\boxed{c \approx 4.1 \text{ m}}$$

YOUR TURN Solve the triangle.

▼
ANSWER

$\gamma = 32°$, $a \approx 42$ ft, and $c \approx 23$ ft

STUDY TIP

Notice in Step 3 that we used a, which is given, as opposed to b, which has been calculated (approximated).

▶ **EXAMPLE 2**   **Using the Law of Sines to Solve a Triangle (ASA)**

Solve the triangle.

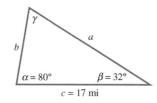

[CONCEPT CHECK]

TRUE OR FALSE If you know all three angles (AAA), then you can solve the triangle using the Law of Sines.

ANSWER False

Solution:

This is an ASA (angle-side-angle) case because the measure of two angles and a side length are given and the side is not opposite one of the angles. Based on the given information, answers will be rounded to two significant digits.

STEP 1 Find γ.

The sum of the measures of the angles in a triangle is 180°.

$$\alpha + \beta + \gamma = 180°$$

Let $\alpha = 80°$ and $\beta = 32°$.

$$80° + 32° + \gamma = 180°$$

Solve for γ.

$$\boxed{\gamma = 68°}$$

STEP 2 Find b.

Use the Law of Sines with the known side c.

$$\frac{\sin\beta}{b} = \frac{\sin\gamma}{c}$$

Isolate b.

$$b = \frac{c\sin\beta}{\sin\gamma}$$

Let $\beta = 32°$, $\gamma = 68°$, and $c = 17$ miles.

$$b = \frac{17\sin 32°}{\sin 68°}$$

Use a calculator to approximate b.

$$b \approx 9.716117$$

Round b to two significant digits.

$$\boxed{b \approx 9.7 \text{ mi}}$$

STEP 3 Find a.

Use the Law of Sines with the known side c.

$$\frac{\sin\alpha}{a} = \frac{\sin\gamma}{c}$$

Isolate a.

$$a = \frac{c\sin\alpha}{\sin\gamma}$$

Let $\alpha = 80°$, $\gamma = 68°$, and $c = 17$ miles.

$$a = \frac{17\sin 80°}{\sin 68°}$$

Use a calculator to approximate a.

$$a \approx 18.056539$$

Round a to two significant digits.

$$\boxed{a \approx 18 \text{ mi}}$$

[STUDY TIP]
Whenever possible, use given values as opposed to calculated (approximate) values.

▼

YOUR TURN   Solve the triangle.

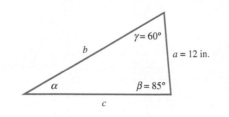

▼
ANSWER
  $\alpha = 35°$, $b \approx 21$ in., and $c \approx 18$ in.

Case 2 (Ambiguous Case): Two Sides and One Angle (SSA)

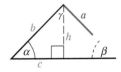

If we are given two side lengths and the measure of an angle opposite one of the sides, we call that Case 2, SSA (side-side-angle). This case is called the ambiguous case because the given information by itself can represent one triangle, two triangles, or no triangle at all. If the angle given is acute, then the possibilities are zero, one, or two triangles. If the angle given is obtuse, then the possibilities are zero or one triangle. The possibilities come from the fact that $\sin \alpha = k$, where $0 < k < 1$, has two solutions for α: one in quadrant I (acute angle) and one in quadrant II (obtuse angle). In the figure on the left, note that $h = b \sin \alpha$ by the definition of the sine ratio.

STUDY TIP

$h = b \sin \alpha$

When the Given Angle α Is Acute

CONDITION	PICTURE	NUMBER OF TRIANGLES
$0 < a < h$ $\sin \beta > 1$	**No Triangle**	0
$a = h$ $\sin \beta = 1$	**Right Triangle**	1
$h < a < b$ $0 < \sin \beta < 1$	**Acute Triangle** **Obtuse Triangle**	2
$a \geq b$ $0 < \sin \beta < 1$	**Acute Triangle**	1

Note: β is the unknown angle opposite the known side b.

STUDY TIP

Notice that if the angle given is obtuse, the side opposite that angle *must* be longer than the other given side (longest side opposite the largest angle).

When the Given Angle α Is Obtuse

CONDITION	PICTURE	NUMBER OF TRIANGLES
$a \leq b$ $\sin \beta \geq 1$		0
$a > b$ $0 < \sin \beta < 1$		1

▶ **EXAMPLE 3**   **Solving the Ambiguous Case (SSA)—One Triangle**

Solve the triangle $a = 23$ feet, $b = 11$ feet, and $\alpha = 122°$.

Solution:

This is an ambiguous case because two side lengths and the measure of an angle opposite one of those sides are given. Since the given angle α is obtuse, we know that the other angles in the triangle are acute. So we know that the Law of Sines will give the correct values for the other angles and only one triangle will exist since $23 > 11$. Based on the given information, answers will be rounded to two significant digits.

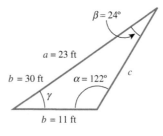

STEP 1 Find β.

Use the Law of Sines.

$$\frac{\sin\alpha}{a} = \frac{\sin\beta}{b}$$

Isolate $\sin\beta$.

$$\sin\beta = \frac{b\sin\alpha}{a}$$

Let $a = 23$ feet, $b = 11$ feet, and $\alpha = 122°$.

$$\sin\beta = \frac{11\sin 122°}{23}$$

Use a calculator to evaluate $\sin\beta$.

$$\sin\beta \approx 0.40558822$$

Solve for β using the inverse sine function.

$$\beta \approx \sin^{-1}(0.40558822)$$

Round the answer to two significant digits.

$$\boxed{\beta \approx 24°}$$

STEP 2 Find γ.

The measures of the angles in a triangle sum to 180°.

$$\alpha + \beta + \gamma = 180°$$

Let $\alpha = 122°$ and $\beta = 24°$.

$$122° + 24° + \gamma = 180°$$

Solve for γ.

$$\boxed{\gamma \approx 34°}$$

STEP 3 Find c.

Use the Law of Sines.

$$\frac{\sin\alpha}{a} = \frac{\sin\gamma}{c}$$

Isolate c.

$$c = \frac{a\sin\gamma}{\sin\alpha}$$

Let $a = 23$ feet, $\alpha = 122°$, and $\gamma = 34°$.

$$c = \frac{23\sin 34°}{\sin 122°}$$

Use a calculator to evaluate c and round to two significant digits.

$$\boxed{c \approx 15 \text{ ft}}$$

▼

YOUR TURN   Solve the triangle $\alpha = 133°$, $a = 48$ millimeters, and $c = 17$ millimeters.

▼ **ANSWER**
$\beta \approx 32°$, $\gamma \approx 15°$, and $b \approx 35$ mm

EXAMPLE 4   **Solving the Ambiguous Case (SSA)—Two Triangles**

Solve the triangle $a = 8.1$ meters, $b = 8.3$ meters, and $\alpha = 72°$.

Solution:

This is an ambiguous case because two side lengths and the measure of an angle opposite one of those sides are given. Since the given angle α is acute and $a < b$, we might expect two triangles. Based on the given information, answers will be rounded to two significant digits.

STEP 1   Find β.

Use the Law of Sines.

$$\dfrac{\sin\alpha}{a} = \dfrac{\sin\beta}{b}$$

Isolate $\sin\beta$.

$$\sin\beta = \dfrac{b\sin\alpha}{a}$$

Let $a = 8.1$ meters, $b = 8.3$ meters, and $\alpha = 72°$.

$$\sin\beta = \dfrac{8.3\sin 72°}{8.1}$$

Use a calculator to evaluate $\sin\beta$.

$$\sin\beta \approx 0.974539393$$

Solve for β using the inverse sine function. Note that β can be acute or obtuse.

$$\beta \approx \sin^{-1}(0.974539393)$$

This is the quadrant I (β is acute) solution.

$$\boxed{\beta_1 \approx 77°}$$

The quadrant II (β is obtuse) solution is $\beta_2 = 180° - \beta_1$.

$$\boxed{\beta_2 \approx 103°}$$

STEP 2   Find γ (two values).

The measures of the angles in a triangle sum to $180°$.

$$\alpha + \beta + \gamma = 180°$$

Let $\alpha = 72°$ and $\beta_1 \approx 77°$.

$$72° + 77° + \gamma_1 \approx 180°$$

Solve for γ_1.

$$\boxed{\gamma_1 \approx 31°}$$

Let $\alpha = 72°$ and $\beta_2 \approx 103°$.

$$72° + 103° + \gamma_2 \approx 180°$$

Solve for γ_2.

$$\boxed{\gamma_2 \approx 5°}$$

STEP 3   Find c.

Use the Law of Sines.

$$\dfrac{\sin\alpha}{a} = \dfrac{\sin\gamma}{c}$$

Isolate c.

$$c = \dfrac{a\sin\gamma}{\sin\alpha}$$

Let $a = 8.1$ meters, $\alpha = 72°$, and $\gamma_1 = 31°$.

$$c_1 = \dfrac{8.1\sin 31°}{\sin 72°}$$

Use a calculator to approximate c_1.

$$\boxed{c_1 \approx 4.4 \text{ m}}$$

Let $a = 8.1$ meters, $\alpha = 72°$, and $\gamma_2 = 5°$.

$$c_2 \approx \dfrac{8.1\sin 5°}{\sin 72°}$$

Use a calculator to approximate c_2.

$$\boxed{c_2 \approx 0.74 \text{ m}}$$

STEP 4 Draw and label the two triangles.

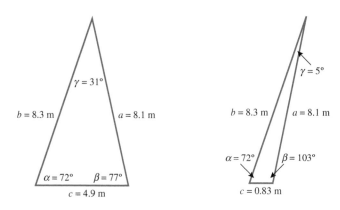

Notice that when there are two solutions for the SSA case, one of the triangles will be obtuse.

EXAMPLE 5   Solving the Ambiguous Case (SSA)—No Triangle

Solve the triangle $\alpha = 107°$, $a = 6$, and $b = 8$.

Solution:

This is an ambiguous case because two side lengths and the measure of an angle opposite one of those sides are given. Since the given angle α is obtuse and $a < b$, there is no triangle. (Notice $\alpha = 107°$ is the largest angle, which implies that side a is the longest side.) Notice what would happen in our calculations.

Use the Law of Sines.

$$\frac{\sin\alpha}{a} = \frac{\sin\beta}{b}$$

Isolate $\sin\beta$.

$$\sin\beta = \frac{b\sin\alpha}{a}$$

Let $\alpha = 107°$, $a = 6$, and $b = 8$.

$$\sin\beta = \frac{8\sin 107°}{6}$$

Use a calculator to approximate $\sin\beta$.

$$\sin\beta \approx 1.28 > 1$$

Since the range of the sine function is $[-1, 1]$, there is no angle β such that $\sin\beta \approx 1.28$.

Therefore, there is no triangle with the given measurements.

Applications

The solution of oblique triangles has applications in astronomy, surveying, aircraft design, piloting, and many other areas.

EXAMPLE 6 **How Far Over Is the Tower of Pisa Leaning?**

The Tower of Pisa was originally built 56 meters tall. Because of poor soil in the foundation, it started to lean. At a distance 44 meters from the base of the tower, the angle of elevation is 55°. How much is the Tower of Pisa leaning away from the vertical position?

Nathalie Michel/Photodisc/Getty Images, Inc.

Solution:

$\beta = 55°$, $c = 44$ meters, and $b = 56$ meters is the given information, so this is an SSA problem. There is only one triangle because $b > c$.

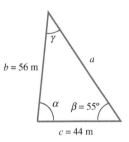

STEP 1 Find γ.

Use the Law of Sines.

$$\frac{\sin\beta}{b} = \frac{\sin\gamma}{c}$$

Isolate $\sin\gamma$.

$$\sin\gamma = \frac{c\sin\beta}{b}$$

Let $\beta = 55°$, $c = 44$ meters, and $b = 56$ meters.

$$\sin\gamma = \frac{44\sin55°}{56}$$

Evaluate the right side using a calculator.

$$\sin\gamma \approx 0.643619463$$

Solve for γ using the inverse sine function.

$$\gamma \approx \sin^{-1}(0.643619463)$$

Round to two significant digits.

$$\gamma \approx 40°$$

STEP 2 Find α.

The measures of angles in a triangle sum to 180°.

$$\alpha + \beta + \gamma = 180°$$

Let $\beta = 55°$ and $\gamma = 40°$.

$$\alpha + 55° + 40° \approx 180°$$

Solve for α.

$$\alpha \approx 85°$$

The Tower of Pisa makes an angle of 85° with the ground. It is leaning at an angle of $\boxed{5°}$.

▶[SECTION 7.1] SUMMARY

In this section, we solved oblique triangles. When given three measurements of a triangle (at least one side), we classify the triangle according to the data (sides and angles). Four cases arise:

- one side and two angles (AAS or ASA)
- two sides and the angle opposite one of the sides (SSA)
- two sides and the angle between sides (SAS)
- three sides (SSS)

The Law of Sines

$$\frac{\sin \alpha}{a} = \frac{\sin \beta}{b} = \frac{\sin \gamma}{c}$$

can be used to solve the first two cases (AAS or ASA and SSA). The SSA case is called the ambiguous case because any one of three results is possible: no triangle, one triangle, or two triangles.

[SECTION 7.1] EXERCISES

• SKILLS

In Exercises 1–6, classify each triangle as AAS, ASA, SAS, SSA, SAS, or SSS on the basis of the given information.

1. $c, a,$ and α
2. $c, a,$ and γ
3. $a, b,$ and c
4. $a, b,$ and γ
5. $\alpha, \beta,$ and c
6. $\beta, \gamma,$ and a

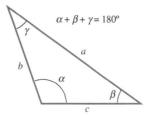

$\alpha + \beta + \gamma = 180°$

In Exercises 7–20, solve the given triangles.

7. $\alpha = 45°, \beta = 60°, a = 10$ m
8. $\beta = 75°, \gamma = 60°, b = 25$ in.
9. $\alpha = 46°, \gamma = 72°, b = 200$ cm
10. $\gamma = 100°, \beta = 40°, a = 16$ ft
11. $\alpha = 16.3°, \gamma = 47.6°, c = 211$ yd
12. $\beta = 104.2°, \gamma = 33.6°, a = 26$ in.
13. $\alpha = 30°, \beta = 30°, c = 12$ m
14. $\alpha = 45°, \gamma = 75°, c = 9$ in.
15. $\beta = 26°, \gamma = 57°, c = 100$ yd
16. $\alpha = 80°, \gamma = 30°, b = 3$ ft
17. $\alpha = 120°, \beta = 10°, c = 12$ cm
18. $\gamma = 46°, \beta = 37°, c = 10$ yd
19. $\alpha = 88.6°, \gamma = 13.4°, b = 57$ m
20. $\alpha = 5°, \beta = 15°, a = 2.3$ mm

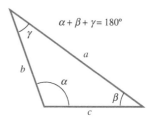

$\alpha + \beta + \gamma = 180°$

In Exercises 21–40, two sides and an angle are given. Determine whether a triangle (or two) exists, and if so, solve the triangle(s).

21. $a = 4, b = 5, \alpha = 16°$
22. $b = 30, c = 20, \beta = 70°$
23. $a = 12, c = 12, \gamma = 40°$
24. $b = 111, a = 80, \alpha = 25°$
25. $a = 21, b = 14, \beta = 100°$
26. $a = 13, b = 26, \alpha = 120°$
27. $b = 500, c = 330, \gamma = 40°$
28. $b = 16, a = 9, \beta = 137°$
29. $a = \sqrt{2}, b = \sqrt{7}, \beta = 106°$
30. $b = 15.3, c = 27.2, \gamma = 11.6°$
31. $\alpha = 27°, c = 12, a = 7$
32. $\gamma = 14.5°, b = 10, c = 4$
33. $\beta = 52°, b = 18, a = 20$
34. $\beta = 47°, b = 18, c = 22$
35. $\gamma = 20°, b = 6.2, c = 10$
36. $\alpha = 28°, a = 5, b = 3.8$
37. $a = 3, b = 5, \alpha = 40°$
38. $b = 13, c = 9, \gamma = 70°$
39. $a = 11, c = 1.5, \gamma = 10°$
40. $b = 8, c = 10, \beta = 80°$

• **APPLICATIONS**

For Exercises 41 and 42, refer to the following:

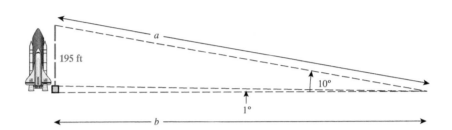

On the launch pad at Kennedy Space Center, when the Space Shuttle was in operation, there was an escape basket that could hold four astronauts. The basket slides down a wire that is attached 195 feet above the base of the launch pad. The angle of inclination measured from where the basket would touch the ground to the base of the launch pad is 1°, and the angle of inclination from that same point to where the wire is attached is 10°.

41. NASA. How long is the wire? Find length *a*.

42. NASA. How far from the launch pad does the basket touch the ground? Find length *b*.

43. Hot-Air Balloon. A hot-air balloon is sighted at the same time by two friends who are 1.0 mile apart on the same side of the balloon. The angles of elevation of the balloon from the two friends are 20.5° and 25.5°. How high is the balloon?

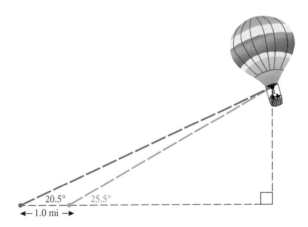

44. Hot-Air Balloon. A hot-air balloon is sighted at the same time by two friends who are 2 miles apart on the same side of the balloon. The angles of elevation of the balloon from the two friends are 10° and 15°. How high is the balloon?

45. Rocket Tracking. A tracking station has two telescopes that are 1.0 mile apart. Each telescope can lock onto a rocket after it is launched and record its angle of elevation to the rocket. If the angles of elevation from telescopes *A* and *B* are 30° and 80°, respectively, then how far is the rocket from telescope *A*?

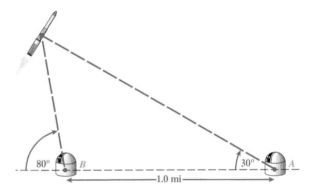

46. Rocket Tracking. Given the information in Exercise 45, how far is the rocket from telescope *B*?

47. Distance Across a River. An engineer wants to construct a bridge across a fast-moving river. Using a straight-line segment between two points that are 100 feet apart along his side of the river, he measures the angles formed when sighting the point *C* on the other side where he wants to have the bridge end. If the angles formed at points *A* and *B* are 65° and 15°, respectively, how far is it from point *A* to point *C* on the other side of the river? Round to the nearest foot.

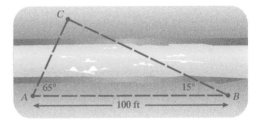

48. Distance Across a River. Given the data in Exercise 47, how far is it from point B to the point on the other side of the river?

49. Submarine. A submarine travels N45°E 15 miles. It then turns to a course of 120° measured clockwise off of due north and travels until it is due east of its original position. How far is it from where it started?

50. Submarine. A submarine travels N45°W 20 miles. It then turns to a course of 200° measured clockwise off of due north and travels until it is due west of its original position. How far is it from where it started?

51. Hiking. A photographer parks his car along a straight country road (that runs due north and south) to hike into the woods to take some pictures. Initially, he travels 850 yards on a course of 130° measured clockwise from due north. He changes direction only to arrive back at the road after walking 700 yards. How far from his starting point is he when he finds the road? *Hint:* There are two possibilities: one leading to an acute triangle and one leading to an obtuse triangle.

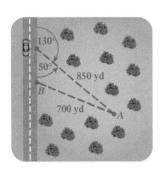

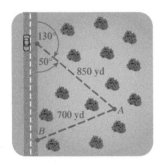

52. Hiking. Suppose the hiker in Exercise 51 initially travels 800 yards on a course of 135° measured clockwise from due north. He changes direction only to arrive back at the road after walking 700 yards. How far from his starting point is he when he finds the road? *Hint:* There are two possibilities: one leading to an acute triangle and one leading to an obtuse triangle.

53. Carpentry. A woodworker fashions a chair such that the legs come down at an angle to the floor as shown in the figure. If the legs are 28 inches long, how far apart are they along the floor?

54. Carpentry. A woodworker fashions a chair such that the legs come down at an angle to the floor as shown in the figure. If the legs are 30 inches long, how far apart are they along the floor?

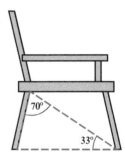

For Exercises 55 and 56, refer to the following:

To quantify the torque (rotational force) of the elbow joint of a human arm (see the figure below), it is necessary to identify angles A, B, and C as well as lengths a, b, and c. Measurements performed on an arm determine that the measure of angle C is 95°, the measure of angle A is 82°, and the length of the muscle a is 23 centimeters.

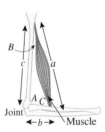

55. Health/Medicine. Find the length of the forearm from the elbow joint to the muscle attachment b.

56. Health/Medicine. Find the length of the upper arm from the muscle attachment to the elbow joint c.

● **CATCH THE MISTAKE**

In Exercises 57 and 58, explain the mistake that is made.

57. Solve the triangle $\alpha = 120°$, $a = 7$, and $b = 9$.

Solution:

Use the Law of Sines to find β. $\qquad \dfrac{\sin \alpha}{a} = \dfrac{\sin \beta}{b}$

Let $\alpha = 120°$, $a = 7$, and $b = 9$. $\qquad \dfrac{\sin 120°}{7} = \dfrac{\sin \beta}{9}$

Solve for $\sin \beta$. $\qquad \sin \beta \approx 1.113$

Solve for β. $\qquad \beta \approx 42°$

Sum the angle measures
to 180°. $\qquad 120° + 42° + \gamma = 180°$

Solve for γ. $\qquad \gamma \approx 18°$

Use the Law of Sines to find c. $\qquad \dfrac{\sin \alpha}{a} = \dfrac{\sin \gamma}{c}$

Let $\alpha = 120°$, $a = 7$,
and $\gamma = 18°$ $\qquad \dfrac{\sin 120°}{7} = \dfrac{\sin 18°}{c}$

Solve for c. $\qquad c \approx 2.5$

$\alpha = 120°$, $\beta = 42°$, $\gamma = 18°$, $a = 7$, $b = 9$, and $c = 2.5$.

This is incorrect. The longest side is not opposite the longest angle. There is no triangle that makes the original measurements work. What mistake was made?

58. Solve the triangle $\alpha = 40°$, $a = 7$, and $b = 9$.

Solution:

Use the Law of Sines to find β. $\qquad \dfrac{\sin \alpha}{a} = \dfrac{\sin \beta}{b}$

Let $\alpha = 40°$, $a = 7$, and $b = 9$. $\qquad \dfrac{\sin 40°}{7} = \dfrac{\sin \beta}{9}$

Solve for $\sin \beta$. $\qquad \sin \beta \approx 0.826441212$

Solve for β. $\qquad \beta \approx 56°$

Find γ. $\qquad 40° + 56° + \gamma = 180°$

$\qquad \gamma \approx 84°$

Use the Law of Sines to find c. $\qquad \dfrac{\sin \alpha}{a} = \dfrac{\sin \gamma}{c}$

Let $\alpha = 40°$, $a = 7$,
and $\gamma = 84°$. $\qquad \dfrac{\sin 40°}{7} \approx \dfrac{\sin 84°}{c}$

Solve for c. $\qquad c \approx 11$

$\alpha = 40°$, $\beta = 56°$, $\gamma = 84°$, $a = 7$, $b = 9$, and $c = 11$.

This is incorrect. What mistake was made?

● **CONCEPTUAL**

In Exercises 59–62, determine whether each statement is true or false.

59. The Law of Sines applies only to right triangles.

60. If you are given two sides and any angle, there is a unique solution for the triangle.

61. An acute triangle is an oblique triangle.

62. An obtuse triangle is an oblique triangle.

● **CHALLENGE**

63. Consider a triangle in which $\alpha = 30°$. If a and b are given, determine the values of a in relation to b whereby two triangles are possible.

64. Consider a triangle in which $\alpha = 60°$. If a and b are given, determine the values of a in relation to b whereby two triangles are possible.

65. Mollweide's Identity. For any triangle, the following identity is true. It is often used to check the solution of a triangle since all six pieces of information (three sides and three angles) are involved. Derive the identity using the Law of Sines.

$$(a + b)\sin\left(\frac{1}{2}\gamma\right) = c\cos\left[\frac{1}{2}(\alpha - \beta)\right]$$

66. The Law of Tangents. Use the Law of Sines and trigonometric identities to show that for any triangle, the following is true:

$$\frac{a - b}{a + b} = \frac{\tan\left[\dfrac{1}{2}(\alpha - \beta)\right]}{\tan\left[\dfrac{1}{2}(\alpha + \beta)\right]}$$

67. Let $\alpha = 30°$, $\beta = 45°$, and $a = \sqrt{2}$. Use the Law of Sines along with a sum and difference identity to find the exact value of c.

68. Let $\alpha = 120°$, $\beta = 45°$, and $b = \sqrt{6}$. Use the Law of Sines along with a sum and difference identity to find the exact value of c.

For Exercises 69–72, let *A*, *B*, and *C* be the lengths of the three sides with *X*, *Y*, and *Z* as the corresponding angles. Write a program using a calculator to solve the given triangle. Go to this book's website (www.wiley.com/college/young) for instructions or help on how to write a program in a graphing calculator.

69. $A = 10, Y = 45°,$ and $Z = 65°$
70. $B = 42.8, X = 31.6°,$ and $Y = 82.2°$
71. $A = 22, B = 17,$ and $X = 105°$
72. $B = 16.5, C = 9.8,$ and $Z = 79.2°$

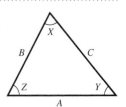

7.2 THE LAW OF COSINES

SKILLS OBJECTIVE	CONCEPTUAL OBJECTIVE
■ Solve SAS and the SSS triangles using the Law of Cosines.	■ Understand that the Pythagorean theorem is a special case of the Law of Cosines.

7.2.1 Solving Oblique Triangles

In the previous section (Section 7.1), we learned that to solve oblique triangles means to find all three side lengths and all three angle measures. At least one side length must be known. Two additional pieces of information are needed to solve a triangle (combinations of side lengths and/or angle measures). We found that there are four cases:

- Case 1: AAS or ASA (the measure of two angles and a side length are given)
- Case 2: SSA (two side lengths and the measure of an angle opposite one of the sides are given)
- Case 3: SAS (two side lengths and the measure of the angle between them are given)
- Case 4: SSS (three side lengths are given)

We used the Law of Sines to solve Case 1 and Case 2 triangles. Now, we use the *Law of Cosines* to solve Case 3 and Case 4 triangles.

WORDS	MATH
Start with an acute triangle. (The same can be shown for an obtuse triangle.)	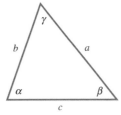
Drop a perpendicular line segment with height *h* from γ to the opposite side. The result is two triangles within the larger triangle.	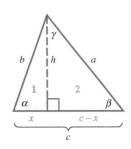

Use the Pythagorean theorem for both right triangles.	Triangle 1: $\qquad x^2 + h^2 = b^2$ Triangle 2: $(c - x)^2 + h^2 = a^2$
Solve for h^2.	Triangle 1: $h^2 = b^2 - x^2$ Triangle 2: $h^2 = a^2 - (c - x)^2$
Since the segment of length h is shared, set $h^2 = h^2$ for the two triangles.	$b^2 - x^2 = a^2 - (c - x)^2$
Multiply out the squared binomial on the right.	$b^2 - x^2 = a^2 - (c^2 - 2cx + x^2)$
Eliminate the parentheses.	$b^2 - x^2 = a^2 - c^2 + 2cx - x^2$
Add x^2 to both sides.	$b^2 = a^2 - c^2 + 2cx$
Isolate a^2.	$a^2 = b^2 + c^2 - 2cx$
Notice that $\cos\alpha = \dfrac{x}{b}$. Let $x = b\cos\alpha$.	$\boxed{a^2 = b^2 + c^2 - 2bc\cos\alpha}$

Note: If we instead drop the perpendicular line segment with length h from the angle α or the angle β to the opposite side, we can derive the other two forms of the Law of Cosines.

$$\boxed{b^2 = a^2 + c^2 - 2ac\cos\beta} \quad \text{and} \quad \boxed{c^2 = a^2 + b^2 - 2ab\cos\gamma}$$

THE LAW OF COSINES

For a triangle with sides a, b, and c, and opposite angles α, β, and γ, the following is true:

$$a^2 = b^2 + c^2 - 2bc\cos\alpha$$
$$b^2 = a^2 + c^2 - 2ac\cos\beta$$
$$c^2 = a^2 + b^2 - 2ab\cos\gamma$$

It is important to note that the Law of Cosines can be used to find unknown side lengths or angle measures. As long as three of the four variables in any of the equations are known, the fourth can be calculated.

Notice that in the special case of a right triangle (say, $\alpha = 90°$),

$$a^2 = b^2 + c^2 - 2\,bc\,\underbrace{\cos 90°}_{0}$$

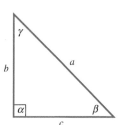

one of the components of the Law of Cosines reduces to the Pythagorean theorem:

$$\underbrace{a^2}_{\text{hyp}} = \underbrace{b^2}_{\text{leg}} + \underbrace{c^2}_{\text{leg}}$$

The Pythagorean theorem can thus be regarded as a special case of the Law of Cosines.

Case 3: Solving Oblique Triangles (SAS)

We can now solve SAS triangle problems, where the angle between two known sides is given. We start by using the Law of Cosines to solve for the length of the side opposite the given angle. We then can use either the Law of Sines or the Law of Cosines to find the measure of a second angle.

EXAMPLE 1   **Using the Law of Cosines to Solve a Triangle (SAS)**

Solve the triangle $a = 13$, $c = 6.0$, and $\beta = 20°$.

Solution:

Two side lengths and the measure of the angle between them are given (SAS).

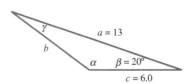

Notice that the Law of Sines cannot be used because it requires knowledge of at least one angle measure and the length of the side opposite that angle.

STEP 1 Find b.

Use the Law of Cosines that involves β.	$b^2 = a^2 + c^2 - 2ac\cos\beta$
Let $a = 13$, $c = 6.0$, and $\beta = 20°$.	$b^2 = 13^2 + 6^2 - 2(13)(6)\cos 20°$
Evaluate the right side using a calculator.	$b^2 \approx 58.40795$
Solve for b.	$b \approx \pm 7.6425$
Round to two significant digits; b can be only positive.	$\boxed{b \approx 7.6}$

STEP 2 Find the acute angle γ.

Use the Law of Sines to find the smaller angle γ.	$\dfrac{\sin\gamma}{c} = \dfrac{\sin\beta}{b}$
Isolate $\sin\gamma$.	$\sin\gamma = \dfrac{c\sin\beta}{b}$
Let $b = 7.6$, $c = 6.0$, and $\beta = 20°$.	$\sin\gamma \approx \dfrac{6\sin 20°}{7.6}$
Use the inverse sine function.	$\gamma \approx \sin^{-1}\left(\dfrac{6\sin 20°}{7.6}\right)$
Evaluate the right side with a calculator.	$\gamma \approx 15.66521°$
Round to the nearest degree.	$\boxed{\gamma \approx 16°}$

STEP 3 Find α.

The three angle measures must sum to 180°.	$\alpha + 20° + 16° \approx 180°$
Solve for α.	$\boxed{\alpha \approx 144°}$

YOUR TURN   Solve the triangle $b = 4.2$, $c = 1.8$, and $\alpha = 35°$.

▼
ANSWER
$a \approx 2.9$, $\gamma \approx 21°$, and
$\beta \approx 124°$

STUDY TIP

Although the Law of Sines can sometimes lead to the ambiguous case, the Law of Cosines never leads to the ambiguous case.

Notice the steps we took in solving a SAS triangle:

1. Find the length of the side opposite the given angle using the Law of Cosines.
2. Solve for the smaller angle (which has to be acute) using the Law of Sines.
3. Solve for the larger angle using properties of triangles.

You may be thinking, would it matter if we solved for α before solving for γ? Yes, it does matter—in this problem, you cannot solve for α by the Law of Sines before finding γ. The Law of Sines should be used only on the smaller angle (opposite the shortest side). If we had tried to use the Law of Sines with the obtuse angle α, the inverse sine would have resulted in $\alpha = 36°$. Since the sine function is positive in quadrants I and II, we would not know if that angle was $\alpha = 36°$ or its supplementary angle $\alpha = 144°$. Notice that $c < a$; therefore, the angles opposite those sides must have the same relationship $\gamma < \alpha$. We choose the smaller angle first to avoid the ambiguity with the Law of Sines. Alternatively, if we wanted to solve for the obtuse angle first, we could have used the Law of Cosines to solve for α.

EXAMPLE 2 **Using the Law of Cosines in an Application (SAS)**

In an AKC (American Kennel Club)-sanctioned field trial, a judge sets up a mark (bird) that requires the dog to swim across a body of water (the dogs are judged on how closely they adhere to the straight line to the bird, not the time it takes to retrieve the bird). The judge is trying to calculate how far the dog would have to swim to this mark, so she walks off the two legs across the land and measures the angle as shown in the figure. How far will the dog swim from the starting line to the bird?

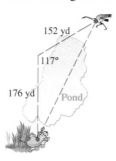

Solution:

Label the triangle.

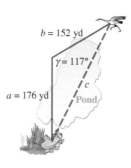

Use the Law of Cosines.	$c^2 = a^2 + b^2 - 2ab\cos\gamma$
Let $a = 176$, $b = 152$, and $\gamma = 117°$.	$c^2 = 176^2 + 152^2 - 2(176)(152)\cos 117°$
Use a calculator to approximate the right side.	$c^2 \approx 78370.3077$
Solve for c and round to the nearest yard (three significant digits).	$\boxed{c \approx 280 \text{ yd}}$

Case 4: Solving Oblique Triangles (SSS)

We now solve oblique triangles when all three side lengths are given (the SSS case). In this case, start by finding the measure of the largest angle (opposite the longest side) using the Law of Cosines. Then use the Law of Sines to find either of the remaining two angle measures. Finally, find the third angle measure using the fact that the three angles in a triangle always sum to 180°.

▶ **EXAMPLE 3**   **Using the Law of Cosines to Solve a Triangle (SSS)**

Solve the triangle $a = 8, b = 6$, and $c = 7$.

Solution:

STEP 1 Identify the largest angle, which is α.

Use the Law of Cosines that involves α. $a^2 = b^2 + c^2 - 2bc\cos\alpha$

Let $a = 8, b = 6$, and $c = 7$. $8^2 = 6^2 + 7^2 - 2(6)(7)\cos\alpha$

Simplify and isolate $\cos\alpha$. $\cos\alpha = \dfrac{6^2 + 7^2 - 8^2}{2(6)(7)} = 0.25$

Apply the inverse cosine function. $\alpha = \cos^{-1}(0.25)$

Approximate with a calculator.
Round to the nearest degree. $\boxed{\alpha \approx 76°}$

STEP 2 Find either of the remaining angle measures. To solve for acute angle β:

Use the Law of Sines. $\dfrac{\sin\alpha}{a} = \dfrac{\sin\beta}{b}$

Isolate $\sin\beta$. $\sin\beta = \dfrac{b\sin\alpha}{a}$

Let $a = 8, b = 6$, and $\alpha = 76°$. $\sin\beta = \dfrac{6\sin 76°}{8}$

Use the inverse sine function. $\beta = \sin^{-1}\left(\dfrac{6\sin 76°}{8}\right)$

Approximate with a calculator. $\boxed{\beta \approx 47°}$

STEP 3 Find the measure of the third angle γ.

The sum of the angle measures is 180°. $76° + 47° + \gamma = 180°$

Solve for γ. $\boxed{\gamma \approx 57°}$

▼

YOUR TURN   Solve the triangle $a = 5, b = 7$, and $c = 8$.

▼
ANSWER
$\alpha \approx 38°, \beta \approx 60°$,
and $\gamma \approx 82°$

In the next example, instead of immediately substituting values into the Law of Cosines equation, we will solve for the angle measure in general, and then substitute in values.

Virginia

α $c = 850$ nm

β Bermuda

$b = 894$ nm

$a = 810$ nm

γ

Cuba

EXAMPLE 4 **Using the Law of Cosines in an Application (SSS)**

In recent decades, many people have come to believe that an imaginary area called the Bermuda Triangle, located off the southeastern Atlantic coast of the United States, has been the site of a high incidence of losses of ships, small boats, and aircraft over the centuries. Assume for the moment, without judging the merits of the hypothesis, that the Bermuda Triangle has vertices in Norfolk (Virginia), Bermuda, and Santiago (Cuba). Find the angles of the Bermuda Triangle given the following distances:

LOCATION	LOCATION	DISTANCE (NAUTICAL MILES)
Norfolk	Bermuda	850
Bermuda	Santiago	810
Norfolk	Santiago	894

Ignore the curvature of the Earth in your calculations. Round answers to the nearest degree.

Solution:

STEP 1 Find β (the largest angle).

Use the Law of Cosines. $b^2 = a^2 + c^2 - 2ac\cos\beta$

Isolate $\cos\beta$. $\cos\beta = \dfrac{a^2 + c^2 - b^2}{2ac}$

Use the inverse cosine
function to solve for β. $\beta = \cos^{-1}\left(\dfrac{a^2 + c^2 - b^2}{2ac}\right)$

Let $a = 810$, $b = 894$, and $c = 850$. $\beta = \cos^{-1}\left(\dfrac{810^2 + 850^2 - 894^2}{2(810)(850)}\right)$

Use a calculator to approximate,
and round to the nearest degree. $\boxed{\beta \approx 65°}$

STEP 2 Find α.

Use the Law of Sines. $\dfrac{\sin\alpha}{a} = \dfrac{\sin\beta}{b}$

Isolate $\sin\alpha$. $\sin\alpha = \dfrac{a\sin\beta}{b}$

Use the inverse sine function
to solve for α. $\alpha = \sin^{-1}\left(\dfrac{a\sin\beta}{b}\right)$

Let $a = 810$, $b = 894$, and
$\beta \approx 65°$. $\alpha \approx \sin^{-1}\left(\dfrac{810\sin65°}{894}\right)$

Use a calculator to approximate,
and round to nearest degree. $\boxed{\alpha \approx 55°}$

STEP 3 Find γ.

The angle measures must sum
to 180°. $55° + 65° + \gamma \approx 180°$

Solve for γ. $\boxed{\gamma \approx 60°}$

▶[SECTION 7.2]   SUMMARY

We can solve any triangle given three pieces of information (measurements), as long as one of the pieces is a side length. Depending on the information given, we use either the **Law of Sines**

$$\frac{\sin\alpha}{a} = \frac{\sin\beta}{b} = \frac{\sin\gamma}{c}$$

or a combination of the **Law of Cosines**

$$a^2 = b^2 + c^2 - 2bc\cos\alpha$$
$$b^2 = a^2 + c^2 - 2ac\cos\beta$$
$$c^2 = a^2 + b^2 - 2ab\cos\gamma$$

and the Law of Sines. The table summarizes the strategies for solving oblique triangles.

OBLIQUE TRIANGLE	WHAT IS KNOWN	PROCEDURE FOR SOLVING
AAS or ASA	Two angles and a side	Step 1: Find the remaining angle measure using $\alpha + \beta + \gamma = 180°$. Step 2: Find the remaining sides using the Law of Sines.
SSA	Two sides and an angle opposite one of the sides	This is the ambiguous case, so there is either no triangle, one triangle, or two triangles. If the given angle is obtuse, then there is either one or no triangles. If the given angle is acute, then there is no triangle, one triangle, or two triangles. Step 1: Use the Law of Sines to find one of the angle measures. Step 2: Find the remaining angle measure using $\alpha + \beta + \gamma = 180°$. Step 3: Find the remaining side using the Law of Sines. If two triangles exist, then the angle measure found in Step 1 can be either acute or obtuse, and Steps 2 and 3 must be performed for each triangle.
SAS	Two sides and an angle between the sides	Step 1: Find the third side length using the Law of Cosines. Step 2: Find the smaller angle measure using the Law of Sines. Step 3: Find the remaining angle measure using $\alpha + \beta + \gamma = 180°$.
SSS	Three sides	Step 1: Find the largest angle measure using the Law of Cosines, which avoids the ambiguity with the Law of Sines. Step 2: Find either remaining angle measure using the Law of Sines. Step 3: Find the last remaining angle measure using $\alpha + \beta + \gamma = 180°$.

[SECTION 7.2]   EXERCISES

• SKILLS

In Exercises 1–8, for each of the given triangles, the angle sum identity, $\alpha + \beta + \gamma = 180°$, will be used in solving the triangle. Label the problem as S if only the Law of Sines is needed to solve the triangle. Label the problem as C if the Law of Cosines is needed to solve the triangle.

1. a, b, and c are given.

2. a, b, and γ are given.

3. α, β, and c are given.

4. a, b, and α are given.

5. a, β, and γ are given.

6. α, β, and a are given.

7. b, c, and α are given.

8. b, c, and β are given.

In Exercises 9–44, solve each triangle.

9. $a = 4, c = 3, \beta = 100°$

10. $a = 6, b = 10, \gamma = 80°$

11. $b = 7, c = 2, \alpha = 16°$

12. $b = 5, a = 6, \gamma = 170°$

13. $b = 5, c = 5, \alpha = 20°$

14. $a = 4.2, b = 7.3, \gamma = 25°$

15. $a = 9, c = 12, \beta = 23°$

16. $b = 6, c = 13, \alpha = 16°$

17. $a = 16, b = 12, \gamma = 12°$

18. $a = 3.6, c = 4.4, \beta = 50°$

19. $b = 70, c = 82, \alpha = 170°$

20. $a = 80, c = 20, \beta = 95°$

21. $a = 4, c = 8, \beta = 60°$

22. $b = 3, c = \sqrt{18}, \alpha = 45°$

23. $a = 8, b = 5, c = 6$

24. $a = 6, b = 9, c = 12$

25. $a = 4, b = 4, c = 5$

26. $a = 17, b = 20, c = 33$

27. $a = 14, b = 6, c = 12$

28. $a = 30, b = 19, c = 41$

29. $a = 3.8, b = 2.9, c = 1.2$

30. $a = 6, b = 8, c = 11$

31. $a = 8.2, b = 7.1, c = 6.3$

32. $a = 1492, b = 2001, c = 1776$

33. $a = 4, b = 5, c = 10$

34. $a = 1.3, b = 2.7, c = 4.2$

35. $a = 12, b = 5, c = 13$

36. $a = 4, b = 5, c = \sqrt{41}$

37. $\alpha = 40°, \beta = 35°, a = 6$

38. $b = 11.2, a = 19.0, \gamma = 13.3°$

39. $\alpha = 31°, b = 5, a = 12$

40. $a = 11, c = 12, \gamma = 60°$

41. $a = \sqrt{7}, b = \sqrt{8}, c = \sqrt{3}$

42. $\beta = 106°, \gamma = 43°, a = 1$

43. $b = 11, c = 2, \beta = 10°$

44. $\alpha = 25°, a = 6, c = 9$

• **APPLICATIONS**

45. **Aviation.** A plane flew due north at 500 mph for 3 hours. A second plane, starting at the same point and at the same time, flew southeast at an angle 150° clockwise from due north at 435 mph for 3 hours. At the end of the 3 hours, how far apart were the two planes? Round to the nearest mile.

46. **Aviation.** A plane flew due north at 400 mph for 4 hours. A second plane, starting at the same point and at the same time, flew southeast at an angle 120° clockwise from due north at 300 mph for 4 hours. At the end of the 4 hours, how far apart were the two planes? Round to the nearest mile.

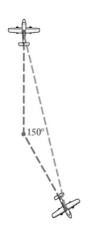

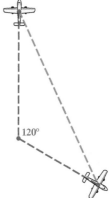

47. Baseball. A baseball diamond is actually a square that is 90 feet on each side. The pitcher's mound is located 60.5 feet from home plate. How far is it from third base to the pitcher's mound?

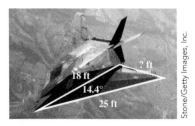

48. Aircraft Wing. Given the acute angle (14.4°) and two sides (18 feet and 25 feet) of the stealth bomber, what is the unknown length?

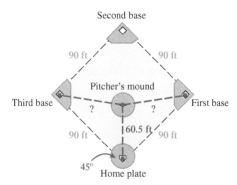

49. Sliding Board. A 40-foot slide leaning against the bottom of a building's window makes a 55° angle with the building. The angle formed with the building with the line of sight from the top of the window to the point on the ground where the slide ends is 40°. How tall is the window?

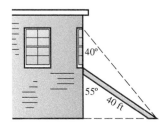

50. Airplane Slide. An airplane door is 6 feet high. If a slide attached to the bottom of the open door is at an angle of 40° with the ground, and the angle formed by the line of sight from where the slide touches the ground to the top of the door is 45°, then how long is the slide?

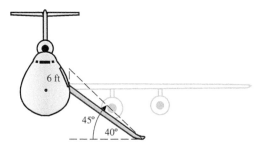

For Exercises 51 and 52, refer to the following:

To quantify the torque (rotational force) of the elbow joint of a human arm (see the figure to the right), it is necessary to identify angles *A*, *B*, and *C* as well as lengths *a*, *b*, and *c*. Measurements performed on an arm determine that the measure of angle *C* is 105°, the length of the muscle *a* is 25.5 centimeters, and the length of the forearm from the elbow joint to the muscle attachment *b* is 1.76 centimeters.

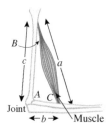

51. Health/Medicine. Find the length of the upper arm from the muscle attachment to the elbow joint *c*.

52. Health/Medicine. Find the measure of angle *B*.

53. Balloon. A hot-air balloon floating in the air is being tethered by two 50-foot ropes. If the ropes are staked to the ground 75 feet apart, what angle do the ropes make with each other at the balloon?

54. Balloon. A hot-air balloon floating in the air is being tethered by two 75-foot ropes. If the ropes are staked to the ground 100 feet apart, what angle do the ropes make with each other at the balloon?

55. Roof Construction. The roof of a house is longer on one side than on the other. If the length of one side of the roof is 27 feet and the length of the other side is 35 feet, find the distance between the ends of the roof if the angle at the top is 130°.

56. Roof Construction. The roof of a house is longer on one side than on the other. If the length of one side of the roof is 29 feet and the length of the other side is 36 feet, find the distance between the ends of the roof if the angle at the top is 135°.

For Exercises 57 and 58, refer to the following:

The term "triangulation" is often used to pinpoint a location. A person's cell phone always uses the closest tower. As the cell phone switches towers, authorities can locate the person's location based on known distances from each of the towers. Similarly, firefighters can determine the location of a fire from known distances.

57. Triangulation. Two lead firefighters (Beth and Tim) are 300 yards apart, and they estimate each of their distances to the fire as approximately 150 yards (Beth) and 200 yards (Tim). What angle with respect to their line of sight with each other should each lead their team in order to reach the fire?

58. Triangulation. Two cell phone towers are 100 meters apart. When the cell phone switches from tower *A* to tower *B*, it is estimated that the phone is 60 meters from tower *A* and 50 meters from tower *B*. Authorities looking to locate the owner of the cell phone will head 50 meters from tower *B* at what angle (with respect to the line of sight between towers *A* and *B*)?

• **CATCH THE MISTAKE**

In Exercises 59 and 60, explain the mistake that is made.

59. Solve the triangle $b = 3$, $c = 4$, and $\alpha = 30°$.

Solution:

STEP 1: Find a.

Use the Law of Cosines.

$$a^2 = b^2 + c^2 - 2bc\cos\alpha$$

Let $b = 3$, $c = 4$, and $\alpha = 30°$.

$$a^2 = 3^2 + 4^2 - 2(3)(4)\cos 30°$$

Solve for a.

$$a \approx 2.1$$

STEP 2: Find γ.

Use the Law of Sines.

$$\frac{\sin\alpha}{a} = \frac{\sin\gamma}{c}$$

Solve for $\sin\gamma$.

$$\sin\gamma = \frac{c\sin\alpha}{a}$$

Solve for γ.

$$\gamma = \sin^{-1}\left(\frac{c\sin\alpha}{a}\right)$$

Let $a = 2.1$, $c = 4$, and $\alpha = 30°$.

$$\gamma \approx 72°$$

STEP 3: Find β.

$$\alpha + \beta + \gamma = 180°. \qquad 30° + \beta + 72° = 180°$$

Solve for β.

$$\beta \approx 78°$$

$a \approx 2.1$, $b = 3$, $c = 4$, $\alpha = 30°$, $\beta \approx 78°$, and $\gamma \approx 72°$.

This is incorrect. The longest side is not opposite the largest angle. What mistake was made?

60. Solve the triangle $a = 6$, $b = 2$, and $c = 5$.

Solution:

STEP 1: Find β.

Use the Law of Cosines.

$$b^2 = a^2 + c^2 - 2ac\cos\beta$$

Solve for β.

$$\beta = \cos^{-1}\left(\frac{a^2 + c^2 - b^2}{2ac}\right)$$

Let $a = 6$, $b = 2$, and $c = 5$.

$$\beta \approx 18°$$

STEP 2: Find α.

Use the Law of Sines.

$$\frac{\sin\alpha}{a} = \frac{\sin\beta}{b}$$

Solve for α.

$$\alpha = \sin^{-1}\left(\frac{a\sin\beta}{b}\right)$$

Let $a = 6$, $b = 2$, and $\beta = 18°$.

$$\alpha \approx 68°$$

STEP 3: Find γ.

$$\alpha + \beta + \gamma = 180°. \qquad 68° + 18° + \gamma = 180°$$

Solve for γ.

$$\gamma \approx 94°$$

$a = 6$, $b = 2$, $c = 5$, $\alpha \approx 68°$, $\beta \approx 18°$, and $\gamma \approx 94°$.

This is incorrect. The longest side is not opposite the largest angle. What mistake was made?

• **CONCEPTUAL**

In Exercises 61–64, determine whether each statement is true or false.

61. Given three sides of a triangle, there is insufficient information to solve the triangle.

62. Given three angles of a triangle, there is insufficient information to solve the triangle.

63. The Pythagorean theorem is a special case of the Law of Cosines.

64. The Law of Cosines is a special case of the Pythagorean theorem.

65. In a triangle, the length of the shortest side of the triangle is $\frac{1}{2}$ the length of the longest side. The other side of the triangle is $\frac{3}{4}$ the length of the longest side. What is the size of the largest angle?

66. In a triangle, the length of one side is $\frac{1}{4}$ the length of an adjacent side. If the angle between the sides is $60°$, how does the length of the third side compare with that of the longer of the other two?

67. Show that $\dfrac{\cos\alpha}{a} + \dfrac{\cos\beta}{b} + \dfrac{\cos\gamma}{c} = \dfrac{a^2 + b^2 + c^2}{2abc}$.

Hint: Use the Law of Cosines.

68. Show that $a = c\cos\beta + b\cos\gamma$.

Hint: Use the Law of Cosines.

69. In an isosceles triangle, the longer side is 50% longer than the other two sides. What is the size of the vertex angle?

70. In an isosceles triangle, the longer side is 2 inches longer than the other two sides. If the vertex angle measures 80°, what are the lengths of the sides?

For Exercises 71–74, let A, B, and C be the lengths of the three sides with X, Y, and Z as the corresponding angle measures. Write a program using the TI calculator to solve the given triangle.

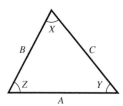

71. $B = 45$, $C = 57$, and $X = 43°$

72. $B = 24.5$, $C = 31.6$, and $X = 81.5°$

73. $A = 29.8$, $B = 37.6$, and $C = 53.2$

74. $A = 100$, $B = 170$, and $C = 250$

7.3 THE AREA OF A TRIANGLE

SKILLS OBJECTIVES	CONCEPTUAL OBJECTIVES
■ Find the area of a triangle in the SAS case. ■ Find the area of a triangle in the SSS case.	■ Understand how to derive a formula for the area of a triangle (SAS case) using the Law of Sines. ■ Understand how to derive a formula for the area of a triangle (SSS case) using the Law of Cosines.

In Sections 7.1 and 7.2, we used the Law of Sines and the Law of Cosines to solve oblique triangles, which means to find all of the side lengths and angle measures. Now, we use these laws to derive formulas for the area of a triangle (SAS and SSS cases).

Our starting point for both cases is the standard formula for the area of a triangle:

$$A = \frac{1}{2}bh$$

> **STUDY TIP**
>
> If the triangle given is not an SAS or SSS case, then the Law of Sines or Law of Cosines can be used to determine the side length and/or angle needed to use either of the two area formulas derived in this section.

7.3.1 The Area of a Triangle (SAS Case)

We now can use the general formula for the area of a triangle and the Law of Sines to develop a formula for the area of a triangle when the length of two sides and the measure of the angle between them are given.

WORDS	MATH
Start with an acute triangle (given b, c, and α).	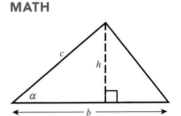
Write the sine ratio for angle α.	$\sin\alpha = \dfrac{h}{c}$
Solve for h.	$h = c\sin\alpha$
Write the formula for area of a triangle.	$A_{\text{triangle}} = \dfrac{1}{2}bh$
For the SAS case, substitute $h = c\sin\alpha$.	$\boxed{A_{\text{SAS}} = \dfrac{1}{2}bc\sin\alpha}$

Now, the area of this triangle can be calculated with the given information (two sides and the angle between them: b, c, and α). Similarly, it can be shown that the other formulas for SAS triangles are

$$\boxed{A_{\text{SAS}} = \dfrac{1}{2}ab\sin\gamma} \quad \text{and} \quad \boxed{A_{\text{SAS}} = \dfrac{1}{2}ac\sin\beta}$$

THE AREA OF A TRIANGLE (SAS)

For any triangle where two sides and the angle between them are known, the area for that triangle is given by one of the following formulas (depending on which angle and sides are given):

$$A_{\text{SAS}} = \tfrac{1}{2}bc\sin\alpha \quad \text{when } b, c, \text{ and } \alpha \text{ are known.}$$

$$A_{\text{SAS}} = \tfrac{1}{2}ab\sin\gamma \quad \text{when } a, b, \text{ and } \gamma \text{ are known.}$$

$$A_{\text{SAS}} = \tfrac{1}{2}ac\sin\beta \quad \text{when } a, c, \text{ and } \beta \text{ are known.}$$

In other words, the area of a triangle equals one-half the product of two of its side lengths and the sine of the angle between them.

▶ **EXAMPLE 1**   **Finding the Area of a Triangle (SAS Case)**

Find the area of the triangle $a = 7.0$ feet, $b = 9.3$ feet, and $\gamma = 86°$.

Solution:

Use the area formula where a, b, and γ are given.

$$A = \frac{1}{2} ab \sin \gamma$$

Substitute $a = 7.0$ feet, $b = 9.3$ feet, and $\gamma = 86°$.

$$A = \frac{1}{2}(7.0)(9.3)\sin 86°$$

Use a calculator to approximate.

$$A \approx 32.47071$$

Round to two significant digits.

$$\boxed{A \approx 32 \text{ ft}^2}$$

▼

YOUR TURN   Find the area of the triangle $a = 3.2$ meters, $c = 5.1$ meters, and $\beta = 49°$.

▼
ANSWER
6.2 m^2

7.3.2 The Area of a Triangle (SSS Case)

We used the Law of Sines to develop a formula for the area of an SAS triangle. That formula, several trigonometric identities, and the Law of Cosines can be used to develop a formula for the area of an SSS triangle, called **Heron's formula**.

7.3.2 SKILL

Find the area of a triangle in the SSS case.

WORDS	MATH
Start with any of the formulas for SAS triangles.	$A = \dfrac{1}{2} ab \sin \gamma$
Square both sides.	$A^2 = \dfrac{1}{4} a^2 b^2 \sin^2 \gamma$
Isolate $\sin^2 \gamma$.	$\dfrac{4A^2}{a^2 b^2} = \sin^2 \gamma$
Use the Pythagorean identity.	$\dfrac{4A^2}{a^2 b^2} = 1 - \cos^2 \gamma$
Factor the difference of the two squares on the right.	$\dfrac{4A^2}{a^2 b^2} = (1 - \cos \gamma)(1 + \cos \gamma)$
Solve the Law of Cosines, $c^2 = a^2 + b^2 - 2ab \cos \gamma$, for $\cos \gamma$.	$\cos \gamma = \dfrac{a^2 + b^2 - c^2}{2ab}$

7.3.2 CONCEPTUAL

Understand how to derive a formula for the area of a triangle (SSS case) using the Law of Cosines.

[CONCEPT CHECK]

TRUE OR FALSE Heron's formula can be used to find the area of a triangle where either all three sides of a triangle are known or all three angle measures are known.

▼

ANSWER False

Substitute $\cos\gamma = \dfrac{a^2 + b^2 - c^2}{2ab}$ into

$\dfrac{4A^2}{a^2 b^2} = (1 - \cos\gamma)(1 + \cos\gamma).$

$$\frac{4A^2}{a^2 b^2} = \left[1 - \frac{a^2 + b^2 - c^2}{2ab}\right]\left[1 + \frac{a^2 + b^2 - c^2}{2ab}\right]$$

Combine the expressions in brackets.

$$\frac{4A^2}{a^2 b^2} = \left[\frac{2ab - a^2 - b^2 + c^2}{2ab}\right]\left[\frac{2ab + a^2 + b^2 - c^2}{2ab}\right]$$

Group the terms in the numerators.

$$\frac{4A^2}{a^2 b^2} = \left[\frac{-(a^2 - 2ab + b^2) + c^2}{2ab}\right]\left[\frac{(a^2 + 2ab + b^2) - c^2}{2ab}\right]$$

Write the numerators as the difference of two squares.

$$\frac{4A^2}{a^2 b^2} = \left[\frac{c^2 - (a - b)^2}{2ab}\right]\left[\frac{(a + b)^2 - c^2}{2ab}\right]$$

Factor the numerators:
$x^2 - y^2 = (x - y)(x + y).$

$$\frac{4A^2}{a^2 b^2} = \left[\frac{[c - (a - b)][c + (a - b)]}{2ab}\right]\left[\frac{[(a + b) - c][(a + b) + c]}{2ab}\right]$$

Simplify.

$$\frac{4A^2}{a^2 b^2} = \left[\frac{(c - a + b)(c + a - b)}{2ab}\right]\left[\frac{(a + b - c)(a + b + c)}{2ab}\right]$$

$$\frac{4A^2}{a^2 b^2} = \frac{(c - a + b)(c + a - b)(a + b - c)(a + b + c)}{4a^2 b^2}$$

Multiply by $\dfrac{a^2 b^2}{4}.$

$$A^2 = \frac{1}{16}(c - a + b)(c + a - b)(a + b - c)(a + b + c)$$

The semiperimeter s is half the perimeter of the triangle.

$$s = \frac{1}{2}(a + b + c)$$

Manipulate each of the four factors.

$c - a + b = a + b + c - 2a = 2s - 2a = 2(s - a)$

$c + a - b = a + b + c - 2b = 2s - 2b = 2(s - b)$

$a + b - c = a + b + c - 2c = 2s - 2c = 2(s - c)$

$a + b + c = 2s$

Substitute these values for the four factors.

$$A^2 = \frac{1}{16} \cdot 2(s - a) \cdot 2(s - b) \cdot 2(s - c) \cdot 2s$$

Simplify.

$A^2 = s(s - a)(s - b)(s - c)$

Solve for A (area is always positive).

$$\boxed{A = \sqrt{s(s - a)(s - b)(s - c)}}$$

**THE AREA OF A TRIANGLE
(SSS CASE, HERON'S FORMULA)**

For any triangle where the lengths of the three sides are known, the area for that triangle is given by the following formula:

$$A_{SSS} = \sqrt{s(s - a)(s - b)(s - c)}$$

where a, b, and c are the lengths of the sides of the triangle and s is half the perimeter of the triangle, called the semiperimeter:

$$s = \frac{a + b + c}{2}$$

▶ **EXAMPLE 2**   **Finding the Area of a Triangle (SSS Case)**

Find the area of the triangle $a = 5$, $b = 6$, and $c = 9$.

Solution:

Find the semiperimeter s.	$s = \dfrac{a + b + c}{2}$
Substitute: $a = 5$, $b = 6$, and $c = 9$.	$s = \dfrac{5 + 6 + 9}{2}$
Simplify.	$s = 10$
Write the formula for the area of a triangle in the SSS case.	$A = \sqrt{s(s - a)(s - b)(s - c)}$
Substitute $a = 5$, $b = 6$, $c = 9$, and $s = 10$.	$A = \sqrt{10(10 - 5)(10 - 6)(10 - 9)}$
Simplify the radicand.	$A = \sqrt{10 \cdot 5 \cdot 4 \cdot 1}$
Evaluate the radical.	$\boxed{A = 10\sqrt{2} \approx 14 \text{ sq units}}$

▼

YOUR TURN   Find the area of the triangle $a = 3$, $b = 5$, and $c = 6$.

▼
ANSWER
$2\sqrt{14} \approx 7.5$ sq units

▶ **[SECTION 7.3]   SUMMARY**

In this section, we derived formulas for calculating the areas of triangles (SAS and SSS cases). The Law of Sines leads to three area formulas for the SAS case, depending on which angles and sides are given.

$$A_{SAS} = \tfrac{1}{2}bc\sin\alpha$$
$$A_{SAS} = \tfrac{1}{2}ab\sin\gamma$$
$$A_{SAS} = \tfrac{1}{2}ac\sin\beta$$

The Law of Cosines was instrumental in developing a formula for the area of a triangle (SSS case) when all three sides are given.

$$A_{SSS} = \sqrt{s(s - a)(s - b)(s - c)}$$

where $s = \dfrac{a + b + c}{2}$.

[SECTION 7.3]   EXERCISES

• **SKILLS**

In Exercises 1–36, find the area (in square units) of each triangle described.

1. $a = 8$, $c = 16$, $\beta = 60°$
2. $b = 6$, $c = 4\sqrt{3}$, $\alpha = 30°$
3. $a = 1$, $b = \sqrt{2}$, $\alpha = 45°$
4. $b = 2\sqrt{2}$, $c = 4$, $\beta = 45°$
5. $a = 6$, $b = 8$, $\gamma = 80°$
6. $b = 9$, $c = 10$, $\alpha = 100°$
7. $a = 12$, $c = 10$, $\alpha = 35°$
8. $a = 5$, $b = 6$, $\beta = 65°$
9. $a = 12$, $c = 6$, $\beta = 45°$
10. $b = 28.3$, $c = 1.9$, $\alpha = 75°$
11. $a = 4$, $c = 7$, $\beta = 27°$
12. $a = 6.3$, $b = 4.8$, $\gamma = 17°$
13. $b = 100$, $c = 150$, $\alpha = 36°$
14. $c = 0.3$, $a = 0.7$, $\beta = 145°$
15. $a = \sqrt{5}$, $b = 5\sqrt{5}$, $\gamma = 50°$
16. $b = \sqrt{11}$, $c = \sqrt{11}$, $\alpha = 21°$
17. $a = 15$, $b = 15$, $c = 15$
18. $a = 1$, $b = 1$, $c = 1$
19. $a = 6$, $b = 13$, $c = 10$
20. $a = 49$, $b = 33$, $c = 18$

21. $a = 1.3, b = 0.6, c = 1.5$

22. $a = 40, b = 50, c = 60$

23. $a = 7, b = \sqrt{51}, c = 10$

24. $a = 9, b = 40, c = 41$

25. $a = 6, b = 10, c = 9$

26. $a = 0.4, b = 0.5, c = 0.6$

27. $a = 14.3, b = 15.7, c = 20.1$

28. $a = 146.5, b = 146.5, c = 100$

29. $a = 14{,}000, b = 16{,}500, c = 18{,}700$

30. $a = \sqrt{2}, b = \sqrt{3}, c = \sqrt{5}$

31. $a = 80, b = 75, c = 160$

32. $a = 19, b = 23, c = 3$

33. $a = \frac{9}{4}, b = \frac{11}{4}, \beta = 27°$

34. $b = 25\frac{1}{2}, c = 13\frac{3}{4}, \beta = 57°$

35. $a = \frac{7}{3}, b = \frac{8}{3}, c = \frac{13}{3}$

36. $a = \frac{25}{6}, b = \frac{17}{6}, c = \frac{11}{6}$

• APPLICATIONS

37. Bermuda Triangle. Calculate the area (to the nearest square mile) of the so-called Bermuda Triangle, described in Example 4 of Section 7.2, if, as some people define it, its vertices are located in Norfolk, Bermuda, and Santiago.

LOCATION	LOCATION	DISTANCE (NAUTICAL MILES)
Norfolk	Bermuda	850
Bermuda	Santiago	810
Norfolk	Santiago	894

Again, ignore the curvature of the Earth in your calculations.

38. Bermuda Triangle. Calculate the area (to the nearest square mile) of the Bermuda Triangle if, as some people define it, its vertices are located in Miami, Bermuda, and San Juan.

LOCATION	LOCATION	DISTANCE (NAUTICAL MILES)
Miami	Bermuda	898
Bermuda	San Juan	831
Miami	San Juan	890

39. Triangular Tarp. A large triangular tarp is needed to cover a playground when it rains. If the sides of the tarp measure 160 feet, 140 feet, and 175 feet, then what is the area of the tarp (to the nearest square foot)?

40. Flower Seed. A triangular garden measures 41 feet by 16 feet by 28 feet. You are going to plant wildflower seed that costs $4 per bag. Each bag of flower seed covers an area of 20 square feet. How much will the seed cost? (Assume you have to buy a whole bag—you can't split one.)

41. Mural. Some students are painting a mural on the side of a building. They have enough paint for a 1000-square-foot area triangle. If two sides of the triangle measure 60 feet and 120 feet, then what angle (to the nearest degree) should the two sides form to create a triangle that uses up all the paint?

42. Mural. Some students are painting a mural on the side of a building. They have enough paint for a 500-square-foot area triangle. If two sides of the triangle measure 40 feet and 60 feet, then what angle (to the nearest degree) should the two sides form to create a triangle that uses up all the paint?

43. Insect Infestation. Some very destructive beetles have made their way into a forest preserve. The rangers are trying to keep track of their spread and how well preventive measures are working. In a triangular area that is 22.5 miles on one side, 28.1 miles on the second, and 38.6 miles on the third, what is the total area the rangers are covering?

44. Real Estate. A real estate agent needs to determine the area of a triangular lot. Two sides of the lot are 150 feet and 60 feet. The angle between the two measured sides is 43°. What is the area of the lot?

For Exercises 45 and 46, refer to the following:

A company is considering purchasing a triangular piece of property (see the figure below) for the construction of a new facility. The purchase is going to be "sight unseen," and the company is using old surveys to approximate area and costs.

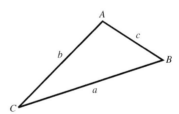

45. Business. If the survey indicates that one side b is 275 feet, a second side a is 310 feet, and the angle between the two sides is 79°,

a. Find the area of the property.

b. If the company wants to offer the seller $2.13 per square foot, what is the total cost of the property?

46. Business. If the survey indicates that one side b is 475 feet, a second side c is 310 feet, and the angle between the two sides is 118°,

a. Find the area of the property.

b. If the company wants to offer the seller $1.97 per square foot, what is the total cost of the property?

47. Parking Lot. A parking lot is to have the shape of a parallelogram that has adjacent sides measuring 200 feet and 260 feet. The angle between the two sides is 65°. What is the area of the parking lot? Round to the nearest square foot.

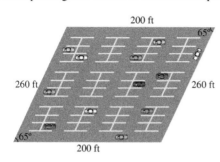

200 ft

65°

260 ft

260 ft

65°

200 ft

48. Parking Lot. A parking lot is to have the shape of a parallelogram that has adjacent sides measuring 250 feet and 300 feet. The angle between the two sides is 55°. What is the area of the parking lot? Round to the nearest square foot.

49. Regular Hexagon. A regular hexagon has sides measuring 3 feet. What is its area? Recall that the measure of an angle of a regular n-gon is given by the formula angle $= \dfrac{180(n-2)}{n}$.

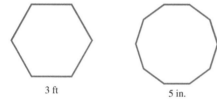

3 ft

5 in.

50. Regular Decagon. A regular decagon has sides measuring 5 inches. What is its area?

51. Pond Plot. A survey of a pond finds that it is roughly in the shape of a triangle that measures 250 feet by 275 feet by 295 feet. Find the area of the pond. Round to the nearest square foot.

52. Field Plot. A field is partly covered in marsh that makes it unusable for growing crops. The usable portion is roughly in the shape of a triangle that measures $\frac{1}{2}$ mile by $\frac{3}{8}$ mile by $\frac{2}{3}$ mile. What is the area of the usable portion of the field? Round to the nearest square mile.

53. Marine Biology. A marine biologist measures the dorsal fin on a shark (roughly in the shape of a triangle) and finds that two of its sides measure 12 inches and 15 inches. If the angle between the sides measures 42°, find the area of the shark's fin.

54. Marine Biology. A marine biologist measures the tail fin on a shark (roughly in the shape of a triangle) and finds that two of its sides measure 34 inches and 37 inches. If the angle between the sides measures 15°, find the area of the shark's fin.

● **CATCH THE MISTAKE**

In Exercises 55 and 56, explain the mistake that is made.

55. Calculate the area of the triangle $a = 2$, $b = 6$, and $c = 7$.

Solution:

Find the semiperimeter.	$s = a + b + c = 2 + 6 + 7 = 15$
Write the formula for the area of the triangle.	$A = \sqrt{s(s-a)(s-b)(s-c)}$
Let $a = 2$, $b = 6$, $c = 7$, and $s = 15$.	$A = \sqrt{15(15-2)(15-6)(15-7)}$
Simplify.	$A = \sqrt{14{,}040} \approx 118$

This is incorrect. What mistake was made?

56. Calculate the area of the triangle $a = 2$, $b = 6$, and $c = 5$.

Solution:

Find the semiperimeter.	$s = \dfrac{a+b+c}{2} = \dfrac{2+6+5}{2} = 6.5$
Write the formula for the area of the triangle.	$A = s(s-a)(s-b)(s-c)$
Let $a = 2$, $b = 6$, $c = 5$, and $s = 6.5$.	$A = 6.5(4.5)(0.5)(1.5)$
Simplify.	$A \approx 22$

This is incorrect. What mistake was made?

• CONCEPTUAL

In Exercises 57 and 58, determine whether each statement is true or false.

57. Heron's formula can be used to find the area of right triangles.

58. Heron's formula can be used to find the area of isosceles triangles.

59. Find the area of the triangle in terms of x.

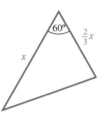

60. Find the area of the triangle in terms of x.

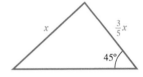

• CHALLENGE

61. Show that the area for an SAA triangle is given by

$$A = \frac{a^2 \sin\beta \sin\gamma}{2 \sin\alpha}$$

Assume that α, β, and a are given.

62. Show that the area of an isosceles triangle with equal sides of length s is given by

$$A_{\text{isosceles}} = \frac{1}{2} s^2 \sin\theta$$

where θ is the angle between the two equal sides.

63. The segment of a circle is the region bounded by a chord and the intersected arc. Find the area of the segment shown in the figure.

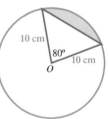

64. The segment of a circle is the region bounded by a chord and the intersected arc. Find the area of the segment shown in the figure.

• TECHNOLOGY

For Exercises 65–68, let A, B, and C be the lengths of the three sides with X, Y, and Z as the corresponding angle measures. Write a program using the TI calculator to find the area of the given triangle.

65. $A = 35$, $B = 47$, and $Z = 68°$

66. $A = 1241$, $B = 1472$, and $Z = 56°$

67. $A = 85$, $B = 92$, and $C = 123$

68. $A = \sqrt{167}$, $B = \sqrt{113}$, and $C = \sqrt{203}$

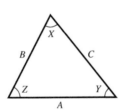

7.4 VECTORS

SKILLS OBJECTIVES	CONCEPTUAL OBJECTIVES
■ Find the magnitude and direction of a vector.	■ Understand the difference between scalars and vectors.
■ Add and subtract vectors and perform scalar multiplication of a vector.	■ Relate the geometric and algebraic representations of operations on vectors.
■ Express a vector in terms of its horizontal and vertical components.	■ Relate the right triangle trigonometric definitions of sine and cosine to the vertical and horizontal components of a vector.
■ Find unit vectors.	■ Understand that unit vectors have a magnitude (length) equal to one.
■ Find resultant vectors in application problems.	■ Use resultant vectors to find actual velocities or actual forces.

7.4.1 Vectors: Magnitude and Direction

What is the difference between velocity and speed? Speed has only *magnitude*, whereas velocity has *magnitude* and *direction*. We use **scalars**, which are real numbers, to denote magnitudes such as speed and weight. We use **vectors**, which have magnitude *and* direction, to denote quantities such as velocity (speed in a certain direction) and force (weight in a certain direction).

A vector quantity is geometrically denoted by a **directed line segment**, which is a line segment with an arrow representing direction. There are many ways to denote a vector. For example, the vector shown in the margin can be denoted as **u**, **AB**, $\vec{u}$, or $\overrightarrow{AB}$, where A is the **initial point** and B is the **terminal point**.

It is customary in books to use the bold letter to denote a vector and, when handwritten (as in your class notes and homework), to use the arrow on top to represent a vector.

In this section, we will limit our discussion to vectors in a plane (two-dimensional). It is important to note that geometric representation can be extended to three dimensions, and algebraic representation can be extended to any higher dimension, as you will see in the exercises and in later sections.

7.4.1 SKILL

Find the magnitude and direction of a vector.

7.4.1 CONCEPTUAL

Understand the difference between scalars and vectors.

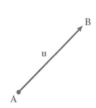

Geometric Interpretation of Vectors

The *magnitude* of a vector can be denoted one of two ways: $|\mathbf{u}|$ or $\|\mathbf{u}\|$. We will use the former notation.

STUDY TIP

The magnitude of a vector is the distance between the initial and terminal points of the vector.

MAGNITUDE: |U|

The **magnitude** of a vector $\mathbf{u}$, denoted $|\mathbf{u}|$, is the length of the directed line segment, which is the distance between the initial and terminal points of the vector.

Two vectors have the **same direction** if they are parallel and point in the same direction. Two vectors have the **opposite direction** if they are parallel and point in opposite directions.

[CONCEPT CHECK]

Match:
(1) speed
(2) velocity
(A) Vector
(B) Scalar

▼ ....

ANSWER 1. B   2. A

EQUAL VECTORS: U = V

Two vectors $\mathbf{u}$ and $\mathbf{v}$ are **equal** ($\mathbf{u} = \mathbf{v}$) if and only if they have the same magnitude ($|\mathbf{u}| = |\mathbf{v}|$) and the same direction.

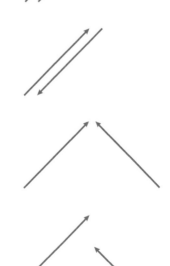

Equal Vectors
$\mathbf{u} = \mathbf{v}$

Same Magnitude but Opposite Direction
$\mathbf{u} = -\mathbf{v}$

Same Magnitude
$|\mathbf{u}| = |\mathbf{v}|$

Different Magnitude

It is important to note that vectors do not have to coincide to be equal. In fact, vectors are movable with magnitude and direction unchanged.

VECTOR ADDITION: U + V

Two vectors, **u** and **v**, can be added together using
either of the following approaches:

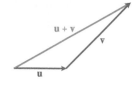

- The **tail-to-tip** (or head-to-tail) method: Sketch the
  initial point of one vector at the terminal point of
  the other vector. The **sum**, **u** + **v**, is the **resultant**
  vector from the tail end of **u** to the tip end of **v**.

  [or]

- The parallelogram method: Sketch the initial
  points of the vectors at the same point. The sum
  u + **v** is the diagonal of the parallelogram formed
  by **u** and **v**.

The difference, **u** − **v**, is the

- Resultant vector from the tip of **v** to the tip
  of **u**, when the tails of **v** and **u** coincide.

  [or]

- The other diagonal formed by the
  parallelogram method.

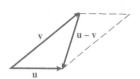

Algebraic Interpretation of Vectors

Since vectors that have the same direction and magnitude are equal, any vector can be
translated to an equal vector with its initial point located at the origin in the Cartesian
plane. Therefore, we will now consider vectors in a rectangular coordinate system.

A vector with its initial point at the origin is called a **position vector**, or a vector
in **standard position**. A position vector **u** with its terminal point at the point (a, b) is
denoted:

$$\mathbf{u} = \langle a, b \rangle$$

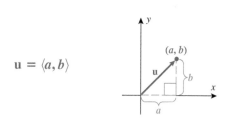

where the real numbers a and b are called the **components** of vector **u**.

Notice the subtle difference between coordinate notation and vector notation. The point is denoted with parentheses, (a, b), whereas the vector is denoted with angled brackets, $\langle a, b \rangle$. The notation $\langle a, b \rangle$ denotes a vector whose initial point is $(0, 0)$ and terminal point is (a, b).

Recall that the geometric definition of the *magnitude* of a vector is the length of the vector.

MAGNITUDE: $|\mathbf{U}|$

The **magnitude** (or norm) of a vector, $\mathbf{u} = \langle a, b \rangle$, is

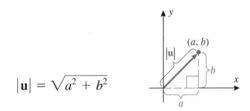

$$|\mathbf{u}| = \sqrt{a^2 + b^2}$$

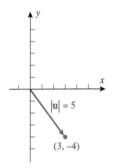

EXAMPLE 1 **Finding the Magnitude of a Vector**

Find the magnitude of the vector $\mathbf{u} = \langle 3, -4 \rangle$.

Solution:

Write the formula for magnitude of a vector. $|\mathbf{u}| = \sqrt{a^2 + b^2}$

Let $a = 3$ and $b = -4$. $|\mathbf{u}| = \sqrt{3^2 + (-4)^2}$

Simplify. $|\mathbf{u}| = \boxed{\sqrt{25} = 5}$

Note: If we graph the vector $\mathbf{u} = \langle 3, -4 \rangle$, we see that the distance from the origin to the point $(3, -4)$ is five units.

▼

ANSWER

$\sqrt{26}$

▼ **YOUR TURN** Find the magnitude of the vector $\mathbf{v} = \langle -1, 5 \rangle$.

DIRECTION ANGLE OF A VECTOR

The positive angle between the positive x-axis and a position vector is called the **direction angle**, denoted θ.

$$\tan \theta = \frac{b}{a} \quad \text{where } a \neq 0$$

or

$$\theta = \tan^{-1}\left(\frac{b}{a}\right) \quad \text{for} \quad \theta \text{ in quadrants I or IV}$$

$$\theta = \tan^{-1}\left(\frac{b}{a}\right) + 180° \quad \text{for} \quad \theta \text{ in quadrants II or III}$$

EXAMPLE 2 **Finding the Direction Angle of a Vector**

Find the direction angle of the vector $\mathbf{v} = \langle -1, 5 \rangle$.

Solution:

Start with $\tan \theta = \dfrac{b}{a}$ and let $a = -1$ and $b = 5$. $\tan \theta = \dfrac{5}{-1}$

Use a calculator to find $\tan^{-1}(-5)$. $\tan^{-1}(-5) = -78.7°$

The calculator gave a quadrant IV angle.

The point $(-1, 5)$ lies in quadrant II.

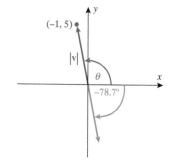

Add 180°. $\theta = -78.7° + 180° = 101.3°$

$$\boxed{\theta = 101.3°}$$

▼

YOUR TURN   Find the direction angle of the vector $\mathbf{u} = \langle 3, -4 \rangle$.

▼
ANSWER
306.9°

Recall that two vectors are equal if they have the same magnitude and direction. Algebraically, this corresponds to their components (a and b) being equal.

EQUAL VECTORS: U = V

The vectors $\mathbf{u} = \langle a, b \rangle$ and $\mathbf{v} = \langle c, d \rangle$ are **equal** ($\mathbf{u} = \mathbf{v}$) if and only if $a = c$ and $b = d$.

7.4.2 Vector Operations

Vector addition geometrically is done with the tail-to-tip rule. Algebraically, vector addition is performed component by component.

VECTOR ADDITION: U + V

If $\mathbf{u} = \langle a, b \rangle$ and $\mathbf{v} = \langle c, d \rangle$, then $\mathbf{u} + \mathbf{v} = \langle a + c, b + d \rangle$.

7.4.2 SKILL

Add and subtract vectors and perform scalar multiplication of a vector.

7.4.2 CONCEPTUAL

Relate the geometric and algebraic representations of operations on vectors.

EXAMPLE 3 **Adding Vectors**

Let $\mathbf{u} = \langle 2, -7 \rangle$ and $\mathbf{v} = \langle -3, 4 \rangle$. Find $\mathbf{u} + \mathbf{v}$.

Solution:

Let $\mathbf{u} = \langle 2, -7 \rangle$ and $\mathbf{v} = \langle -3, 4 \rangle$ in the addition formula.

$$\mathbf{u} + \mathbf{v} = \langle 2 + (-3), -7 + 4 \rangle$$

Simplify.

$$\mathbf{u} + \mathbf{v} = \boxed{\langle -1, -3 \rangle}$$

▼

ANSWER

$\mathbf{u} + \mathbf{v} = \langle -4, -2 \rangle$

▼

YOUR TURN Let $\mathbf{u} = \langle 1, 2 \rangle$ and $\mathbf{v} = \langle -5, -4 \rangle$. Find $\mathbf{u} + \mathbf{v}$.

We now summarize some vector operations. Addition and subtraction of vectors are performed algebraically component by component. Multiplication, however, is not as straightforward. To perform *scalar multiplication* of a vector (to multiply a vector by a real number), we multiply component by component. In Section 7.5, we will study a form of multiplication for two vectors that is defined as long as the vectors have the same number of components; it gives a result known as the *dot product* and is useful in solving common problems in physics.

SCALAR MULTIPLICATION: KU

If k is a scalar (real number) and $\mathbf{u} = \langle a, b \rangle$, then

$$k\mathbf{u} = k \langle a, b \rangle = \langle ka, kb \rangle$$

STUDY TIP

A scalar multiple of u, ku, is always parallel to u.

Scalar multiplication corresponds to:

- Increasing the length of the vector: $|k| > 1$
- Decreasing the length of the vector: $|k| < 1$
- Changing the direction of the vector: $k < 0$

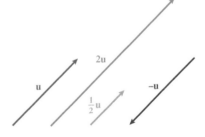

The following box is a summary of some basic vector operations:

CONCEPT CHECK

If a, b, c, and d are all positive with $a > c$ and $b < d$, then $u - v$ is a vector pointing into which quadrant?

▼

ANSWER IV

VECTOR OPERATIONS

If $\mathbf{u} = \langle a, b \rangle$ and $\mathbf{v} = \langle c, d \rangle$ and k is a scalar, then

$$\mathbf{u} + \mathbf{v} = \langle a + c, b + d \rangle$$
$$\mathbf{u} - \mathbf{v} = \langle a - c, b - d \rangle$$
$$k\mathbf{u} = k \langle a, b \rangle = \langle ka, kb \rangle$$

The zero vector $\mathbf{0} = \langle 0, 0 \rangle$ is a vector in any direction with a magnitude equal to zero. We now can state the algebraic properties of vectors:

ALGEBRAIC PROPERTIES OF VECTORS

$$\mathbf{u} + \mathbf{v} = \mathbf{v} + \mathbf{u}$$
$$(\mathbf{u} + \mathbf{v}) + \mathbf{w} = \mathbf{u} + (\mathbf{v} + \mathbf{w})$$
$$(k_1 k_2)\mathbf{u} = k_1(k_2\mathbf{u})$$
$$k(\mathbf{u} + \mathbf{v}) = k\mathbf{u} + k\mathbf{v}$$
$$(k_1 + k_2)\mathbf{u} = k_1\mathbf{u} + k_2\mathbf{u}$$
$$0\mathbf{u} = \mathbf{0} \quad 1\mathbf{u} = \mathbf{u} \quad -1\mathbf{u} = -\mathbf{u}$$
$$\mathbf{u} + (-\mathbf{u}) = \mathbf{0}$$

7.4.3 Horizontal and Vertical Components of a Vector

The **horizontal component** a and **vertical component** b of a position vector $\mathbf{u}$ are related to the magnitude of the vector $|\mathbf{u}|$ through the sine and cosine of the direction angle.

$$\cos\theta = \frac{a}{|\mathbf{u}|} \qquad \sin\theta = \frac{b}{|\mathbf{u}|}$$

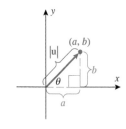

HORIZONTAL AND VERTICAL COMPONENTS OF A VECTOR

The horizontal and vertical components of a position vector $\mathbf{u}$, with magnitude $|\mathbf{u}|$ and direction angle θ, are given by

$$\text{horizontal component: } a = |\mathbf{u}|\cos\theta$$
$$\text{vertical component: } \quad b = |\mathbf{u}|\sin\theta$$

The vector $\mathbf{u}$ can then be written as $\mathbf{u} = \langle a, b \rangle = \langle |\mathbf{u}|\cos\theta, |\mathbf{u}|\sin\theta \rangle$.

EXAMPLE 4 **Finding the Horizontal and Vertical Components of a Vector**

Find the position vector that has a magnitude of 6 and a direction angle of 15°.

Solution:

Write the horizontal and vertical components of vector $\mathbf{u}$. $a = |\mathbf{u}|\cos\theta$ and $b = |\mathbf{u}|\sin\theta$

Let $|\mathbf{u}| = 6$ and $\theta = 15°$. $a = 6\cos 15°$ and $b = 6\sin 15°$

Evaluate the sine and cosine functions of 15°. $a \approx 5.8$ and $b \approx 1.6$

Let $\mathbf{u} = \langle a, b \rangle$. $\mathbf{u} = \boxed{\langle 5.8, 1.6 \rangle}$

▼

YOUR TURN Find the position vector that has a magnitude of 3 and a direction angle of 75°.

7.4.4 Unit Vectors

A **unit vector** is any vector with a magnitude equal to 1, or $|\mathbf{u}| = 1$. It is often useful to be able to find a unit vector in the same direction of some vector **v**. A unit vector can be formed from any nonzero vector as follows:

FINDING A UNIT VECTOR

If **v** is a nonzero vector, then

$$\mathbf{u} = \frac{\mathbf{v}}{|\mathbf{v}|} = \frac{1}{|\mathbf{v}|} \cdot \mathbf{v}$$

is the **unit vector** in the same direction as **v**.

In other words, multiplying any nonzero vector by the reciprocal of its magnitude results in a unit vector.

It is important to notice that since the magnitude is always a scalar, then the reciprocal of the magnitude is always a scalar. A scalar times a vector is a vector.

▶ **EXAMPLE 5** **Finding a Unit Vector**

Find the unit vector in the same direction as $\mathbf{v} = \langle -3, -4 \rangle$.

Solution:

Find the magnitude of the vector $\mathbf{v} = \langle -3, -4 \rangle$.

$$|\mathbf{v}| = \sqrt{(-3)^2 + (-4)^2}$$

Simplify.

$$|\mathbf{v}| = 5$$

Multiply **v** by the reciprocal of its magnitude.

$$\frac{1}{|\mathbf{v}|} \cdot \mathbf{v}$$

Let $|\mathbf{v}| = 5$ and $\mathbf{v} = \langle -3, -4 \rangle$.

$$\frac{1}{5} \langle -3, -4 \rangle$$

Simplify.

$$\boxed{\left\langle -\frac{3}{5}, -\frac{4}{5} \right\rangle}$$

Check: The unit vector $\langle -\frac{3}{5}, -\frac{4}{5} \rangle$ should have a magnitude of 1.

$$\sqrt{\left(-\frac{3}{5}\right)^2 + \left(-\frac{4}{5}\right)^2} = \sqrt{\frac{25}{25}} = 1$$

▼ YOUR TURN Find the unit vector in the same direction as $\mathbf{v} = \langle 5, -12 \rangle$.

Two important unit vectors are the horizontal and vertical unit vectors **i** and **j**. The unit vector **i** has an initial point at the origin and a terminal point at (1, 0). The unit vector **j** has an initial point at the origin and a terminal point at (0, 1). These unit vectors can be used to represent vectors algebraically: $\langle 3, -4 \rangle = 3\mathbf{i} - 4\mathbf{j}$.

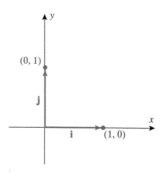

7.4.5 Resultant Vectors

Vectors arise in many applications. **Velocity vectors** and **force vectors** are two that we will discuss. For example, you might be at the beach and "think" that you are swimming straight out at a certain speed (magnitude and direction). This is your **apparent velocity** with respect to the water. But after a few minutes you turn around to look at the shore, and you are farther out than you thought and you also appear to have drifted down the beach. This is because of the current of the water. When the current velocity and the apparent velocity are added together, the result is the **actual** or **resultant velocity**.

7.4.5 SKILL

Find resultant vectors in application problems.

7.4.5 CONCEPTUAL

Use resultant vectors to find actual velocities or actual forces.

EXAMPLE 6   **Resultant Velocities**

A boat's speedometer reads 25 mph (which is relative to the water) and sets a course due east (90° from due north). If the river is moving 10 mph due north, what is the resultant (actual) velocity of the boat?

Solution:

Draw a picture.

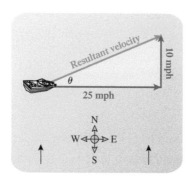

[CONCEPT CHECK]

The speedometer on a boat will correspond to the magnitude of which velocity, apparent or actual?

▼

ANSWER Apparent

Label the horizontal and vertical components of the resultant vector.

$\langle 25, 10 \rangle$

Determine the magnitude of the resultant vector.

$\sqrt{25^2 + 10^2} = 5\sqrt{29} \approx 27 \text{ mph}$

Determine the direction angle.

$\tan\theta = \dfrac{10}{25}$

Solve for θ.

$\theta \approx 22°$

The actual velocity of the boat has magnitude $\boxed{27 \text{ mph}}$, and the boat is headed $\boxed{22° \text{ north of east or } 68° \text{ east of north}}$.

In Example 6, the three vectors formed a right triangle. In Example 7, the three vectors form an oblique triangle.

▶ **EXAMPLE 7**   **Resultant Velocities**

A speedboat traveling 30 mph has a compass heading of 100° east of north. The current velocity has a magnitude of 15 mph, and its heading is 22° east of north. Find the resultant (actual) velocity of the boat.

Solution:

Draw a picture.

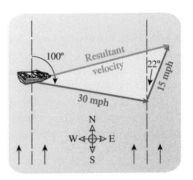

Label the supplementary angle to 100° and its equal alternate interior angle with parallel north/south lines.

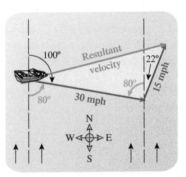

Draw and label the oblique triangle.

The magnitude of the actual (resultant) velocity is b.

The heading of the actual (resultant) velocity is $100° - \alpha$.

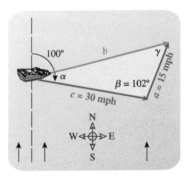

Use the Law of Sines and the Law of Cosines to solve for α and b.

Find b: Use the Law of Cosines. $b^2 = a^2 + c^2 - 2ac\cos\beta$

Let $a = 15$, $c = 30$,
and $\beta = 102°$. $b^2 = 15^2 + 30^2 - 2(15)(30)\cos 102°$

Solve for b. $\boxed{b \approx 36 \text{ mph}}$

Find α: Use the Law of Sines.
$$\frac{\sin\alpha}{a} = \frac{\sin\beta}{b}$$

    Isolate $\sin\alpha$.
$$\sin\alpha = \frac{a\sin\beta}{b}$$

    Let $a = 15$, $b = 36$, and $\beta = 102°$.
$$\sin\alpha = \frac{15\sin 102°}{36}$$

    Use the inverse sine function to solve for α.
$$\alpha = \sin^{-1}\left(\frac{15\sin 102°}{36}\right)$$

    Use a calculator to approximate α.
$$\alpha \approx 24°$$

Actual heading: $100° - \alpha = 100° - 24° = 76°$.

The actual velocity vector of the boat has magnitude $\boxed{36 \text{ mph}}$, and the boat is headed $\boxed{76° \text{ east of north}}$.

Two vectors, **u** and **v**, combine to yield a resultant vector, **u** + **v**. The vector **u** − **v** is called the **equilibrant**.

EXAMPLE 8 **Finding an Equilibrant**

A skier is being pulled up a handle lift. Let F_1 represent the vertical force due to gravity and F_2 represent the force of the skier pushing against the side of the mountain, which is at an angle of $35°$ to the horizontal. If the weight of the skier is 145 pounds, that is, $|F_1| = 145$, find the magnitude of the equilibrant force F_3 required to hold the skier in place (not let the skier slide down the mountain). Assume that the side of the mountain is a frictionless surface.

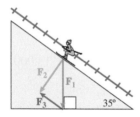

Solution:

The angle between vectors F_1 and F_2 is $35°$.

The magnitude of vector F_3 is the force required to hold the skier in place.

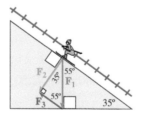

Relate the magnitudes (side lengths) to the given angle using the sine ratio.
$$\sin 35° = \frac{|\mathbf{F}_3|}{|\mathbf{F}_1|}$$

Solve for $|\mathbf{F}_3|$.
$$|\mathbf{F}_3| = |\mathbf{F}_1|\sin 35°$$

Let $|\mathbf{F}_1| = 145$.
$$|\mathbf{F}_3| = 145\sin 35°$$
$$|\mathbf{F}_3| \approx 83.16858$$

A force of approximately $\boxed{83 \text{ pounds}}$ is required to keep the skier from sliding down the hill.

EXAMPLE 9 **Resultant Forces**

A barge runs aground outside its channel. A single tugboat cannot generate enough force to pull the barge off the sandbar. A second tugboat comes to assist. The following diagram illustrates the force vectors, $\mathbf{F}_1$ and $\mathbf{F}_2$, from the tugboats. What is the resultant force vector of the two tugboats?

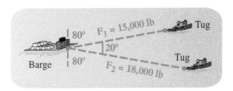

Solution:

Using the tail-to-tip rule, we can add these two vectors and form a triangle:

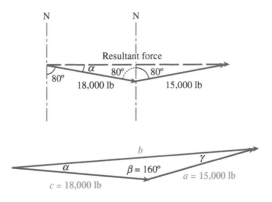

Solve for b:

Use the Law of Cosines. $b^2 = a^2 + c^2 - 2ac\cos\beta$

Let $a = 15{,}000$,
$c = 18{,}000$, and
$\beta = 160°$. $b^2 = 15{,}000^2 + 18{,}000^2 - 2(15{,}000)(18{,}000)\cos 160°$

Solve for b. $\boxed{b \approx 32{,}503 \text{ lb}}$

Solve for α:

Use the Law of Sines. $\dfrac{\sin\alpha}{a} = \dfrac{\sin\beta}{b}$

Isolate $\sin\alpha$. $\sin\alpha = \dfrac{a\sin\beta}{b}$

Let $a = 15{,}000$, $b = 32{,}503$, and $\beta = 160°$. $\sin\alpha = \dfrac{15{,}000 \cdot \sin 160°}{32{,}503}$

Use the inverse sine function to solve for α. $\alpha = \sin^{-1}\left(\dfrac{15{,}000 \cdot \sin 160°}{32{,}503}\right)$

Use a calculator to approximate α. $\boxed{\alpha \approx 9.08°}$

The resulting force is $\boxed{32{,}503 \text{ pounds}}$ at an angle of

$\boxed{9° \text{ from the tug pulling with a force of 18,000 pounds}}$, which is 91° east of north.

▶[SECTION 7.4] SUMMARY

In this section, we discussed scalars (real numbers) and vectors. Scalars have only magnitude, whereas vectors have both magnitude and direction.

$$\text{Vector:} \quad \mathbf{u} = \langle a, b \rangle$$

$$\text{Magnitude:} \quad |\mathbf{u}| = \sqrt{a^2 + b^2}$$

$$\text{Direction } (\theta): \quad \tan\theta = \frac{b}{a} \quad a \neq 0$$

We defined vectors both algebraically and geometrically and gave interpretations of magnitude and vector addition in both

ways. Vector addition is performed algebraically component by component.

$$\langle a, b \rangle + \langle c, d \rangle = \langle a + c, b + d \rangle$$

The trigonometric functions are used to express the horizontal and vertical components of a vector.

$$\text{Horizontal component: } a = |\mathbf{u}| \cos\theta$$
$$\text{Vertical component: } b = |\mathbf{u}| \sin\theta$$

Resultant velocity and force vectors can be found using the Law of Sines and the Law of Cosines.

[SECTION 7.4] EXERCISES

• **SKILLS**

In Exercises 1–6, find the magnitude of the vector $\overrightarrow{AB}$, given the points A and B.

1. $A = (2, 7)$ and $B = (5, 9)$

2. $A = (-2, 3)$ and $B = (3, -4)$

3. $A = (4, 1)$ and $B = (-3, 0)$

4. $A = (-1, -1)$ and $B = (2, -5)$

5. $A = (0, 7)$ and $B = (-24, 0)$

6. $A = (-2, 1)$ and $B = (4, 9)$

In Exercises 7–18, find the magnitude and direction angle of each vector.

7. $\mathbf{u} = \langle 3, 8 \rangle$

8. $\mathbf{u} = \langle 4, 7 \rangle$

9. $\mathbf{u} = \langle 5, -1 \rangle$

10. $\mathbf{u} = \langle -6, -2 \rangle$

11. $\mathbf{u} = \langle -4, 1 \rangle$

12. $\mathbf{u} = \langle -6, 3 \rangle$

13. $\mathbf{u} = \langle -8, 0 \rangle$

14. $\mathbf{u} = \langle 0, 7 \rangle$

15. $\mathbf{u} = \langle \sqrt{3}, 3 \rangle$

16. $\mathbf{u} = \langle -5, -5 \rangle$

17. $\mathbf{u} = \langle \frac{4}{5}, \frac{1}{3} \rangle$

18. $\mathbf{u} = \langle -\frac{3}{7}, \frac{1}{2} \rangle$

In Exercises 19–30, perform the indicated vector operation, given $\mathbf{u} = \langle -4, 3 \rangle$ and $\mathbf{v} = \langle 2, -5 \rangle$.

19. $\mathbf{u} + \mathbf{v}$

20. $\mathbf{u} - \mathbf{v}$

21. $3\mathbf{u}$

22. $-2\mathbf{u}$

23. $2\mathbf{u} + 4\mathbf{v}$

24. $5(\mathbf{u} + \mathbf{v})$

25. $6(\mathbf{u} - \mathbf{v})$

26. $-3(\mathbf{u} - \mathbf{v})$

27. $10\mathbf{v} - 2\mathbf{u} - 3\mathbf{v}$

28. $2\mathbf{u} - 3\mathbf{v} + 4\mathbf{u}$

29. $4(\mathbf{u} + 2\mathbf{v}) + 3(\mathbf{v} - 2\mathbf{u})$

30. $2(4\mathbf{u} - 2\mathbf{v}) - 5(3\mathbf{v} - \mathbf{u})$

In Exercises 31–42, find the position vector, given its magnitude and direction angle.

31. $|\mathbf{u}| = 7, \theta = 25°$

32. $|\mathbf{u}| = 5, \theta = 75°$

33. $|\mathbf{u}| = 16, \theta = 100°$

34. $|\mathbf{u}| = 8, \theta = 200°$

35. $|\mathbf{u}| = 4, \theta = 310°$

36. $|\mathbf{u}| = 8, \theta = 225°$

37. $|\mathbf{u}| = 9, \theta = 335°$

38. $|\mathbf{u}| = 3, \theta = 315°$

39. $|\mathbf{u}| = 2, \theta = 120°$

40. $|\mathbf{u}| = 6, \theta = 330°$

41. $|\mathbf{u}| = 3, \theta = 195°$

42. $|\mathbf{u}| = 12, \theta = 280°$

In Exercises 43–54, find the unit vector in the direction of the given vector.

43. $\mathbf{v} = \langle -5, -12 \rangle$ **44.** $\mathbf{v} = \langle 3, 4 \rangle$ **45.** $\mathbf{v} = \langle 60, 11 \rangle$ **46.** $\mathbf{v} = \langle -7, 24 \rangle$

47. $\mathbf{v} = \langle 24, -7 \rangle$ **48.** $\mathbf{v} = \langle -10, 24 \rangle$ **49.** $\mathbf{v} = \langle -9, -12 \rangle$ **50.** $\mathbf{v} = \langle 40, -9 \rangle$

51. $\mathbf{v} = \langle \sqrt{2}, 3\sqrt{2} \rangle$ **52.** $\mathbf{v} = \langle -4\sqrt{3}, -2\sqrt{3} \rangle$ **53.** $\mathbf{v} = \langle -2\sqrt{6}, 3\sqrt{6} \rangle$ **54.** $\mathbf{v} = \langle 6\sqrt{5}, -4 \rangle$

In Exercises 55–62, express the vector as a sum of unit vectors i and j.

55. $\langle 7, 3 \rangle$ **56.** $\langle -2, 4 \rangle$ **57.** $\langle 5, -3 \rangle$ **58.** $\langle -6, -2 \rangle$

59. $\langle -1, 0 \rangle$ **60.** $\langle 0, 2 \rangle$ **61.** $\langle 0.8, -3.6 \rangle$ **62.** $\langle -\frac{15}{4}, -\frac{10}{3} \rangle$

In Exercises 63–68, perform the indicated vector operation.

63. $(5\mathbf{i} - 2\mathbf{j}) + (-3\mathbf{i} + 2\mathbf{j})$ **64.** $(4\mathbf{i} - 2\mathbf{j}) + (3\mathbf{i} - 5\mathbf{j})$ **65.** $(-3\mathbf{i} + 3\mathbf{j}) - (2\mathbf{i} - 2\mathbf{j})$

66. $(\mathbf{i} - 3\mathbf{j}) - (-2\mathbf{i} + \mathbf{j})$ **67.** $(5\mathbf{i} + 3\mathbf{j}) + (2\mathbf{i} - 3\mathbf{j})$ **68.** $(-2\mathbf{i} + \mathbf{j}) + (2\mathbf{i} - 4\mathbf{j})$

• **A P P L I C A T I O N S**

69. Bullet Speed. A bullet is fired from ground level at a speed of 2200 feet per second at an angle of 30° from the horizontal. Find the magnitude of both the horizontal and vertical components of the velocity vector.

70. Weightlifting. A 50-pound weight lies on an inclined bench that makes an angle of 40° with the horizontal. Find the component of the weight directed perpendicular to the bench and also the component of the weight parallel to the inclined bench.

71. Weight of a Boat. A force of 630 pounds is needed to pull a speedboat and its trailer up a ramp that has an incline of 13°. What is the combined weight of the boat and its trailer to the nearest pound?

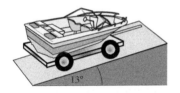

72. Weight of a Boat. A force of 500 pounds is needed to pull a speedboat and its trailer up a ramp that has an incline of 16°. What is the weight of the boat and its trailer to the nearest pound?

73. Speed and Direction of a Ship. A ship's captain sets a course due north at 10 mph. The water is moving at 6 mph due west. What is the actual velocity of the ship, and in what direction is it traveling?

74. Speed and Direction of a Ship. A ship's captain sets a course due west at 12 mph. The water is moving at 3 mph due north. What is the actual velocity of the ship, and in what direction is it traveling?

75. Heading and Airspeed. A plane has a compass heading of 60° east of due north and an airspeed of 300 mph. The wind is blowing at 40 mph with a heading of 30° west of due north. What are the plane's actual heading and airspeed?

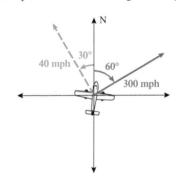

76. Heading and Airspeed. A plane has a compass heading of 30° east of due north and an airspeed of 400 mph. The wind is blowing at 30 mph with a heading of 60° west of due north. What are the plane's actual heading and airspeed?

77. Heading. An airplane takes off and flies at 175 mph for 1 hour on a compass heading of N 135° E. The pilot then turns and flies for 2 hours at 185 mph on a heading of N 80° E. How far is the plane from the airport, and what is its bearing from the airport?

78. Heading. An airplane takes off and flies at 205 mph for 1 hour on a compass heading of N 255° E. The pilot then turns and flies for 2 hours at 165 mph on a heading of N 170° E. How far is the plane from the airport, and what is its bearing from the airport?

79. **Wind Speed.** An airplane has an airspeed of 250 mph and a compass heading of 285°. With a 30 mph wind, its actual heading is 280°. When taking into effect the wind, what is the actual speed of the plane?

80. **Wind Speed.** An airplane has an airspeed of 220 mph and a compass heading of 165°. With a 40 mph wind, its actual heading is 173°. When taking into effect the wind, what is the actual speed of the plane?

81. **Sliding Box.** A box weighing 500 pounds is held in place on an inclined plane that has an angle of 30° with the ground. What force is required to hold it in place?

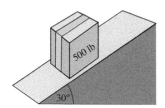

82. **Sliding Box.** A box weighing 500 pounds is held in place on an inclined plane that has an angle of 10° with the ground. What force is required to hold it in place?

83. **Baseball.** A baseball player throws a ball with an initial velocity of 80 feet per second at an angle of 40° with the horizontal. What are the vertical and horizontal components of the velocity?

84. **Baseball.** A baseball pitcher throws a ball with an initial velocity of 100 feet per second at an angle of 5° with the horizontal. What are the vertical and horizontal components of the velocity?

For Exercises 85 and 86, refer to the following:

A post pattern in football is when the receiver in motion runs past the quarterback parallel to the line of scrimmage (A), then runs 12 yards perpendicular to the line of scrimmage (B), and then cuts toward the goal post (C).

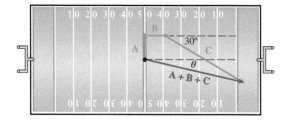

85. **Football.** A receiver runs the post pattern. If the magnitudes of the vectors are $|A| = 4$ yards, $|B| = 12$ yards, and $|C| = 20$ yards, find the magnitude of the resultant vector $\mathbf{A} + \mathbf{B} + \mathbf{C}$.

86. **Football.** A receiver runs the post pattern. If the magnitudes of the vectors are $|A| = 4$ yards, $|B| = 12$ yards, and $|C| = 20$ yards, find the direction angle θ.

87. **Resultant Force.** A force with a magnitude of 100 pounds and another with a magnitude of 400 pounds are acting on an object. The two forces have an angle of 60° between them. What is the direction angle of the resultant force with respect to the force pulling 400 pounds?

88. **Resultant Force.** A force with a magnitude of 100 pounds and another with a magnitude of 400 pounds are acting on an object. The two forces have an angle of 60° between them. What is the magnitude of the resultant force?

89. **Resultant Force.** A force of 1000 pounds is acting on an object at an angle of 45° from the horizontal. Another force of 500 pounds is acting at an angle of −40° from the horizontal. What is the magnitude of the resultant force?

90. **Resultant Force.** A force of 1000 pounds is acting on an object at an angle of 45° from the horizontal. Another force of 500 pounds is acting at an angle of −40° from the horizontal. What is the direction angle of the resultant force?

For Exercises 91 and 92, refer to the following:

Muscle A and muscle B are attached to a bone as indicated in the figure below. Muscle A exerts a force on the bone at angle α, while muscle B exerts a force on the bone at angle β.

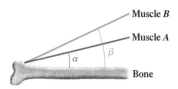

91. **Health/Medicine.** Assume muscle A exerts a force of 900 N on the bone at angle $\alpha = 8°$, while muscle B exerts a force of 750 N on the bone at angle $\beta = 33°$. Find the resultant force and the angle of the force due to muscle A and muscle B on the bone.

92. **Health/Medicine.** Assume muscle A exerts a force of 1000 N on the bone at angle $\alpha = 9°$, while muscle B exerts a force of 820 N on the bone at angle $\beta = 38°$. Find the resultant force and the angle of the force due to muscle A and muscle B on the bone.

• CATCH THE MISTAKE

In Exercises 93 and 94, explain the mistake that is made.

93. Find the magnitude of the vector $\langle -2, -8 \rangle$.

Solution:

Factor the -1. $\qquad -\langle 2, 8 \rangle$

Find the magnitude of $\langle 2, 8 \rangle$.
$$|\langle 2, 8 \rangle| = \sqrt{2^2 + 8^2}$$
$$= \sqrt{68} = 2\sqrt{17}$$

Write the magnitude of $\langle -2, -8 \rangle$.
$$|\langle -2, -8 \rangle| = -2\sqrt{17}$$

This is incorrect. What mistake was made?

94. Find the direction angle of the vector $\langle -2, -8 \rangle$.

Solution:

Write the formula for the direction angle of $\langle a, b \rangle$.
$$\tan\theta = \frac{b}{a}$$

Let $a = -2$ and $b = -8$.
$$\tan\theta = \frac{-8}{-2}$$

Use the inverse tangent function.
$$\theta = \tan^{-1}4$$

Use a calculator to evaluate. $\qquad \theta \approx 76°$

This is incorrect. What mistake was made?

• CONCEPTUAL

In Exercises 95–98, determine whether each statement is true or false.

95. The magnitude of the vector **i** is the imaginary number i.

96. Equal vectors must coincide.

97. The magnitude of a vector is always greater than or equal to the magnitude of its horizontal component.

98. The magnitude of a vector is always greater than or equal to the magnitude of its vertical component.

99. Would a scalar or a vector represent the following? A car is driving 72 mph due east ($90°$ with respect to north).

100. Would a scalar or a vector represent the following? The granite has a mass of 131 kilograms.

101. Find the magnitude of the vector $\langle -a, b \rangle$ if $a > 0$ and $b > 0$.

102. Find the direction angle of the vector $\langle -a, b \rangle$ if $a > 0$ and $b > 0$.

• CHALLENGE

103. Find the magnitude and direction of the vector $\langle 3a, -4a \rangle$. Assume $a > 0$.

104. Find the magnitude and direction of the vector $\langle \frac{3}{4}a, \frac{2}{3}a \rangle$. Assume $a > 0$.

105. Let $\mathbf{u} = \langle -2, 3 \rangle$ and $\mathbf{v} = \langle -4, -2 \rangle$. Find the magnitude and direction of $\mathbf{u} + \mathbf{v}$.

106. Let $\mathbf{u} = \langle -2, 3 \rangle$ and $\mathbf{v} = \langle -4, -2 \rangle$. Find the magnitude and direction of $\mathbf{u} - \mathbf{v}$.

107. Let $\mathbf{u} = \langle 2, 1 \rangle$ and $\mathbf{v} = \langle -2, 1 \rangle$. Find the angle θ between $\mathbf{u}$ and $\mathbf{v}$.

108. Let $\mathbf{u} = \langle 4, 1 \rangle$ and $\mathbf{v} = \langle -1, 5 \rangle$. Find the angle θ between $\mathbf{u}$ and $\mathbf{v}$.

• TECHNOLOGY

For Exercises 109–112, refer to the following:

Vectors can be represented as column matrices. For example, the vector $\mathbf{u} = \langle 3, -4 \rangle$ can be represented as a 2×1 column matrix $\begin{bmatrix} 3 \\ -4 \end{bmatrix}$. Using a TI-83, you can enter vectors as matrices in two ways, directly or via MATRIX.

Directly:

```
[[3][-4]]
        [[3 ]
         [-4]]
■
```

Matrix:

```
[A]
        [[3 ]
         [-4]]
```

Use a calculator to perform the vector operation given $\mathbf{u} = \langle 8, -5 \rangle$ and $\mathbf{v} = \langle -7, 11 \rangle$.

109. $\mathbf{u} + 3\mathbf{v}$

110. $-9(\mathbf{u} - 2\mathbf{v})$

Use a calculator to find the unit vector in the direction of the given vector.

111. $\mathbf{u} = \langle 10, -24 \rangle$

112. $\mathbf{u} = \langle -9, -40 \rangle$

7.5 THE DOT PRODUCT

SKILLS OBJECTIVES	CONCEPTUAL OBJECTIVES
▪ Find the dot product of two vectors. ▪ Use the dot product to find the angle between two vectors. ▪ Use the dot product to calculate the amount of work associated with a physical problem.	▪ Understand that the dot product of two vectors is a scalar. ▪ Understand why the dot product of two orthogonal (perpendicular) vectors is equal to zero. ▪ Understand that work is done when a force causes an object to move a certain distance.

7.5.1 Multiplying Two Vectors: The Dot Product

There are two types of multiplication defined for vectors: scalar multiplication and the dot product. Scalar multiplication (which we already demonstrated in Section 7.4) is multiplication of a vector by a scalar; the result is a parallel vector. Now, we discuss the *dot product* of two vectors. In this case, there are two important things to note: (1) The dot product of two vectors is defined only if the vectors have the same number of components, and (2) if the dot product does exist, the result is a scalar.

7.5.1 SKILL

Find the dot product of two vectors.

7.5.1 CONCEPTUAL

Understand that the dot product of two vectors is a scalar.

THE DOT PRODUCT

The **dot product** of two vectors $\mathbf{u} = \langle a, b \rangle$ and $\mathbf{v} = \langle c, d \rangle$ is given by

$$\mathbf{u} \cdot \mathbf{v} = ac + bd$$

$\mathbf{u} \cdot \mathbf{v}$ is pronounced "u dot v."

▶ **EXAMPLE 1**   **Finding the Dot Product of Two Vectors**

Find the dot product $\langle -7, 3 \rangle \cdot \langle 2, 5 \rangle$.

Solution:

Sum the products of the first components and the products of the second components.

$$\langle -7, 3 \rangle \cdot \langle 2, 5 \rangle = (-7)(2) + (3)(5)$$

Simplify.

$$= -14 + 15 = \boxed{1}$$

▼

YOUR TURN   Find the dot product $\langle 6, 1 \rangle \cdot \langle -2, 3 \rangle$.

ANSWER
-9

The following box summarizes the properties of the dot product:

PROPERTIES OF THE DOT PRODUCT

1. $\mathbf{u} \cdot \mathbf{v} = \mathbf{v} \cdot \mathbf{u}$
2. $\mathbf{u} \cdot \mathbf{u} = |\mathbf{u}|^2$
3. $0 \cdot \mathbf{u} = 0$
4. $k(\mathbf{u} \cdot \mathbf{v}) = (k\mathbf{u}) \cdot \mathbf{v} = \mathbf{u} \cdot (k\mathbf{v})$
5. $(\mathbf{u} + \mathbf{v}) \cdot \mathbf{w} = \mathbf{u} \cdot \mathbf{w} + \mathbf{v} \cdot \mathbf{w}$
6. $\mathbf{u} \cdot (\mathbf{v} + \mathbf{w}) = \mathbf{u} \cdot \mathbf{v} + \mathbf{u} \cdot \mathbf{w}$

[CONCEPT CHECK]

TRUE OR FALSE The dot product of two vectors is a scalar, and the product of two scalars is a vector.

▼

ANSWER False

7.5.2 Angle Between Two Vectors

We can use these properties to develop an equation that relates the angle between two vectors and the dot product of the vectors.

WORDS	MATH
Let **u** and **v** be two nonzero vectors with the same initial point and let θ, $0° \leq \theta \leq 180°$, be the angle between them.	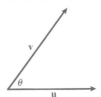
The vector **u** − **v** is opposite angle θ.	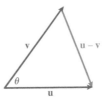
A triangle is formed with side lengths equal to the magnitudes of the three vectors.	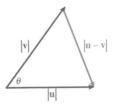

Use the Law of Cosines.

$$|\mathbf{u} - \mathbf{v}|^2 = |\mathbf{u}|^2 + |\mathbf{v}|^2 - 2|\mathbf{u}||\mathbf{v}|\cos\theta$$

Use properties of the dot product to rewrite the left side of equation:

Property (2):

$$|\mathbf{u} - \mathbf{v}|^2 = (\mathbf{u} - \mathbf{v}) \cdot (\mathbf{u} - \mathbf{v})$$

Property (6):

$$= \mathbf{u} \cdot (\mathbf{u} - \mathbf{v}) - \mathbf{v} \cdot (\mathbf{u} - \mathbf{v})$$

Property (6):

$$= \mathbf{u} \cdot \mathbf{u} - \mathbf{u} \cdot \mathbf{v} - \mathbf{v} \cdot \mathbf{u} + \mathbf{v} \cdot \mathbf{v}$$

Property (2):

$$= |\mathbf{u}|^2 - \mathbf{u} \cdot \mathbf{v} - \mathbf{v} \cdot \mathbf{u} + |\mathbf{v}|^2$$

Property (1):

$$= |\mathbf{u}|^2 - 2(\mathbf{u} \cdot \mathbf{v}) + |\mathbf{v}|^2$$

Substitute this last expression for the left side of the original Law of Cosines equation.

$$|\mathbf{u}|^2 - 2(\mathbf{u} \cdot \mathbf{v}) + |\mathbf{v}|^2 = |\mathbf{u}|^2 + |\mathbf{v}|^2 - 2|\mathbf{u}||\mathbf{v}|\cos\theta$$

Simplify.

$$-2(\mathbf{u} \cdot \mathbf{v}) = -2|\mathbf{u}||\mathbf{v}|\cos\theta$$

Isolate $\cos\theta$.

$$\cos\theta = \frac{\mathbf{u} \cdot \mathbf{v}}{|\mathbf{u}||\mathbf{v}|}$$

Notice that **u** and **v** have to be nonzero vectors since we divided by them in the last step.

ANGLE BETWEEN TWO VECTORS

If θ is the angle, $0° \le \theta \le 180°$, between two nonzero vectors **u** and **v**, then

$$\cos\theta = \frac{\mathbf{u} \cdot \mathbf{v}}{|\mathbf{u}||\mathbf{v}|}$$

In the Cartesian plane, there are two angles between two vectors: θ and $360° - \theta$. **We assume θ is the "smaller" angle, which will always be between 0° and 180°.**

▶ **EXAMPLE 2**   **Finding the Angle Between Two Vectors**

Find the "smaller" angle between $\langle 2, -3 \rangle$ and $\langle -4, 3 \rangle$.

Solution:

Let $\mathbf{u} = \langle 2, -3 \rangle$ and $\mathbf{v} = \langle -4, 3 \rangle$.

STEP 1 Find $\mathbf{u} \cdot \mathbf{v}$.

$$\begin{aligned}\mathbf{u} \cdot \mathbf{v} &= \langle 2, -3 \rangle \cdot \langle -4, 3 \rangle \\ &= (2)(-4) + (-3)(3) \\ &= -17\end{aligned}$$

STEP 2 Find $|\mathbf{u}|$.

$$|\mathbf{u}| = \sqrt{\mathbf{u} \cdot \mathbf{u}} = \sqrt{2^2 + (-3)^2} = \sqrt{13}$$

STEP 3 Find $|\mathbf{v}|$.

$$|\mathbf{v}| = \sqrt{\mathbf{v} \cdot \mathbf{v}} = \sqrt{(-4)^2 + 3^2} = \sqrt{25} = 5$$

STEP 4 Find θ.

$$\cos\theta = \frac{\mathbf{u} \cdot \mathbf{v}}{|\mathbf{u}||\mathbf{v}|} = \frac{-17}{5\sqrt{13}}$$

Use a calculator to approximate θ.

$$\theta = \cos^{-1}\left(-\frac{17}{5\sqrt{13}}\right) \approx 160.559965°$$

$$\boxed{\theta \approx 161°}$$

STEP 5 Draw a picture to confirm the answer.

Draw the vectors $\langle 2, -3 \rangle$ and $\langle -4, 3 \rangle$.

161° appears to be correct.

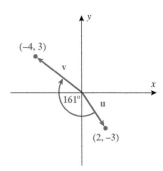

YOUR TURN Find the "smaller" angle between $\langle 1, 5 \rangle$ and $\langle -2, 4 \rangle$.

When two vectors are **parallel**, the angle between them is 0° or 180°.

When two vectors are **perpendicular (orthogonal)**, the angle between them is 90°.

Note: We did not include 270° because the angle θ between two vectors is taken to be the smaller angle (i.e., $0° \le \theta < 180°$).

STUDY TIP

The angle between two vectors $\theta = \cos^{-1}\left(\frac{\mathbf{u} \cdot \mathbf{v}}{|\mathbf{u}||\mathbf{v}|}\right)$ is an angle between 0° and 180° (the range of the inverse cosine function).

▼
ANSWER
$\theta \approx 38°$

WORDS	MATH
When two vectors **u** and **v** are perpendicular, $\theta = 90°$.	$\cos 90° = \dfrac{\mathbf{u} \cdot \mathbf{v}}{|\mathbf{u}||\mathbf{v}|}$
Substitute $\cos 90° = 0$.	$0 = \dfrac{\mathbf{u} \cdot \mathbf{v}}{|\mathbf{u}||\mathbf{v}|}$
Therefore, the dot product of **u** and **v** must be zero.	$\mathbf{u} \cdot \mathbf{v} = 0$

ORTHOGONAL VECTORS

Two vectors **u** and **v** are **orthogonal** (perpendicular) if and only if their dot product is zero.

$$\mathbf{u} \cdot \mathbf{v} = 0$$

EXAMPLE 3 **Determining Whether Vectors Are Orthogonal**

Determine whether each pair of vectors is an orthogonal pair.

a. $\mathbf{u} = \langle 2, -3 \rangle$ and $\mathbf{v} = \langle 3, 2 \rangle$ **b.** $\mathbf{u} = \langle -7, -3 \rangle$ and $\mathbf{v} = \langle 7, 3 \rangle$

Solution (a):

Find the dot product $\mathbf{u} \cdot \mathbf{v}$.	$\mathbf{u} \cdot \mathbf{v} = (2)(3) + (-3)(2)$
Simplify.	$\mathbf{u} \cdot \mathbf{v} = 0$

Vectors **u** and **v** are orthogonal since $\mathbf{u} \cdot \mathbf{v} = 0$.

Solution (b):

Find the dot product $\mathbf{u} \cdot \mathbf{v}$.	$\mathbf{u} \cdot \mathbf{v} = (-7)(7) + (-3)(3)$
Simplify.	$\mathbf{u} \cdot \mathbf{v} = -58$

Vectors **u** and **v** are not orthogonal since $\mathbf{u} \cdot \mathbf{v} \neq 0$.

7.5.3 Work

7.5.3 SKILL

Use the dot product to calculate the amount of work associated with a physical problem.

7.5.3 CONCEPTUAL

Understand that work is done when a force causes an object to move a certain distance.

If you had to carry either barbells with weights or pillows for 1 mile, which would you choose? You probably would pick the pillows over the barbell and weights because the pillows are lighter. It requires less work to carry the pillows than it does to carry the barbell with weights. If asked to carry either of them 1 mile or 10 miles, you would probably pick 1 mile, because it's a shorter distance and requires less work. **Work** *is done when a force causes an object to move a certain distance.*

The simplest case is when the force is in the same direction as the displacement—for example, a stagecoach (the horses pull with a force in the same direction). In this case, the work is defined as the magnitude of the force times the magnitude of the displacement, distance d.

$$W = |\mathbf{F}|d$$

Notice that the magnitude of the force is a scalar, the distance d is a scalar, and hence the product is a scalar.

If the horses pull with a force of 1000 pounds and they move the stagecoach 100 feet, then the work done by the force is

$$W = (1000 \text{ lb})(100 \text{ ft}) = 100{,}000 \text{ ft-lb}$$

In many physical applications, however, the force is not in the same direction as the displacement, and hence vectors (not just their magnitudes) are required.

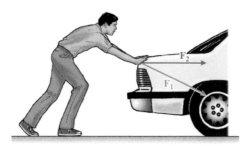

[CONCEPT CHECK]

TRUE OR FALSE Work increases only if either the magnitude of the force or the distance increases.

▼

ANSWER True

We often want to know how much of a force is applied in a certain direction. For example, when your car runs out of gasoline and you try to push it, some of the force vector F_1 you generate from pushing translates into the horizontal component F_2; hence, the car moves horizontally.

If we let θ be the angle between the vectors F_1 and F_2, then the horizontal component of F_1 is F_2 where $|F_2| = |F_1|\cos\theta$.

If the man in the picture pushes at an angle of 25° with a force of 150 pounds, then the horizontal component of the force vector F_1 is

$$(150\,\text{lb})(\cos 25°) \approx \boxed{136\,\text{lb}}$$

WORDS	MATH
To develop a generalized formula when the force exerted and the displacement are not in the same direction, we start with the formula for the angle between two vectors.	$\cos\theta = \dfrac{\mathbf{u}\cdot\mathbf{v}}{|\mathbf{u}||\mathbf{v}|}$
We then isolate the dot product $\mathbf{u}\cdot\mathbf{v}$.	$\mathbf{u}\cdot\mathbf{v} = |\mathbf{u}||\mathbf{v}|\cos\theta$
Let $\mathbf{u} = \mathbf{F}$ and $\mathbf{v} = \mathbf{d}$.	$W = \mathbf{F}\cdot\mathbf{d} = |\mathbf{F}||\mathbf{d}|\cos\theta = \underbrace{|\mathbf{F}|\cos\theta}_{\substack{\text{magnitude of force in} \\ \text{direction of displacement}}} \cdot \underbrace{|\mathbf{d}|}_{\text{distance}}$

WORK

If an object is moved from point A to point B by a constant force, then the work associated with this displacement is

$$W = \mathbf{F}\cdot\mathbf{d}$$

where $\mathbf{d}$ is the displacement vector and $\mathbf{F}$ is the force vector.

Work is typically expressed in one of two units:

SYSTEM	FORCE	DISTANCE	WORK
American	pound	foot	ft-lb
SI	newton	meter	N-m

EXAMPLE 4   **Calculating Work**

How much work is done when a force (in pounds) $\mathbf{F} = \langle 2, 4 \rangle$ moves an object from $(0, 0)$ to $(5, 9)$ (the distance is in feet)?

Solution:

Find the displacement vector $\mathbf{d}$.	$\mathbf{d} = \langle 5, 9 \rangle$
Use the work formula $W = \mathbf{F} \cdot \mathbf{d}$.	$W = \langle 2, 4 \rangle \cdot \langle 5, 9 \rangle$
Calculate the dot product.	$W = (2)(5) + (4)(9)$
Simplify.	$\boxed{W = 46 \text{ ft-lb}}$

▼
ANSWER

25 N-m

YOUR TURN   How much work is done when a force (in newtons) $\mathbf{F} = \langle 1, 3 \rangle$ moves an object from $(0, 0)$ to $(4, 7)$ (the distance is in meters)?

◗[SECTION 7.5]   SUMMARY

In this section, we defined the dot product as a form of multiplication of two vectors. A scalar times a vector results in a vector, whereas the dot product of two vectors is a scalar.

$$\langle a, b \rangle \cdot \langle c, d \rangle = ac + bd$$

We developed a formula that determines the angle θ between two vectors $\mathbf{u}$ and $\mathbf{v}$.

$$\cos\theta = \frac{\mathbf{u} \cdot \mathbf{v}}{|\mathbf{u}||\mathbf{v}|}$$

Orthogonal (perpendicular) vectors have an angle of 90° between them, and we showed that the dot product of two orthogonal vectors is equal to zero. Work is the result of a force displacing an object. When the force and displacement are in the same direction, the work is equal to the product of the magnitude of the force and the distance (magnitude of the displacement). When the force and displacement are not in the same direction, work is the dot product of the force vector and displacement vector, $W = \mathbf{F} \cdot \mathbf{d}$.

[SECTION 7.5]   EXERCISES

● **SKILLS**

In Exercises 1–16, find each of the following dot products.

1. $\langle 4, -2 \rangle \cdot \langle 3, 5 \rangle$ **2.** $\langle 7, 8 \rangle \cdot \langle 2, -1 \rangle$ **3.** $\langle -5, 6 \rangle \cdot \langle 3, 2 \rangle$ **4.** $\langle 6, -3 \rangle \cdot \langle 2, 1 \rangle$

5. $\langle -1, 3 \rangle \cdot \langle 4, 10 \rangle$ **6.** $\langle 2, -3 \rangle \cdot \langle 3, 5 \rangle$ **7.** $\langle 1, 8 \rangle \cdot \langle -2, 6 \rangle$ **8.** $\langle 10, 8 \rangle \cdot \langle 12, -6 \rangle$

9. $\langle -7, -4 \rangle \cdot \langle -2, -7 \rangle$ **10.** $\langle 5, -2 \rangle \cdot \langle -1, -1 \rangle$ **11.** $\langle \sqrt{3}, -2 \rangle \cdot \langle 3\sqrt{3}, -1 \rangle$ **12.** $\langle 4\sqrt{2}, \sqrt{7} \rangle \cdot \langle -\sqrt{2}, -\sqrt{7} \rangle$

13. $\langle 5, a \rangle \cdot \langle -3a, 2 \rangle$ **14.** $\langle 4x, 3y \rangle \cdot \langle 2y, -5x \rangle$ **15.** $\langle 0.8, -0.5 \rangle \cdot \langle 2, 6 \rangle$ **16.** $\langle -18, 3 \rangle \cdot \langle 10, -300 \rangle$

In Exercises 17–32, find the angle θ ($0° \leq \theta \leq 180°$; round to the nearest degree) between each pair of vectors.

17. $\langle -4, 3 \rangle$ and $\langle -5, -9 \rangle$ **18.** $\langle 2, -4 \rangle$ and $\langle 4, -1 \rangle$ **19.** $\langle -2, -3 \rangle$ and $\langle -3, 4 \rangle$ **20.** $\langle 6, 5 \rangle$ and $\langle 3, -2 \rangle$

21. $\langle -4, 6 \rangle$ and $\langle -6, 8 \rangle$ **22.** $\langle 1, 5 \rangle$ and $\langle -3, -2 \rangle$ **23.** $\langle -6, 2 \rangle$ and $\langle -3, 8 \rangle$ **24.** $\langle 7, -2 \rangle$ and $\langle 4, 1 \rangle$

25. $\langle -2, 2\sqrt{3} \rangle$ and $\langle -\sqrt{3}, 1 \rangle$

26. $\langle -3\sqrt{3}, -3 \rangle$ and $\langle -2\sqrt{3}, 2 \rangle$

27. $\langle -5\sqrt{3}, -5 \rangle$ and $\langle \sqrt{2}, -\sqrt{2} \rangle$

28. $\langle -5, -5\sqrt{3} \rangle$ and $\langle 2, -\sqrt{2} \rangle$

29. $\langle 4, 6 \rangle$ and $\langle -6, -9 \rangle$

30. $\langle 2, 8 \rangle$ and $\langle -12, 3 \rangle$

31. $\langle -1, 6 \rangle$ and $\langle 2, -4 \rangle$

32. $\langle -8, -2 \rangle$ and $\langle 10, -3 \rangle$

In Exercises 33–44, determine whether each pair of vectors is orthogonal.

33. $\langle -6, 8 \rangle$ and $\langle -8, 6 \rangle$

34. $\langle 5, -2 \rangle$ and $\langle -5, 2 \rangle$

35. $\langle 6, -4 \rangle$ and $\langle -6, -9 \rangle$

36. $\langle 8, 3 \rangle$ and $\langle -6, 16 \rangle$

37. $\langle 0.8, 4 \rangle$ and $\langle 3, -6 \rangle$

38. $\langle -7, 3 \rangle$ and $\langle \frac{1}{7}, \frac{1}{3} \rangle$

39. $\langle 5, -0.4 \rangle$ and $\langle 1.6, 20 \rangle$

40. $\langle 12, 9 \rangle$ and $\langle 3, -4 \rangle$

41. $\langle \sqrt{3}, \sqrt{6} \rangle$ and $\langle -\sqrt{2}, 1 \rangle$

42. $\langle \sqrt{7}, -\sqrt{3} \rangle$ and $\langle 3, 7 \rangle$

43. $\langle \frac{4}{3}, \frac{8}{15} \rangle$ and $\langle -\frac{1}{12}, \frac{5}{24} \rangle$

44. $\langle \frac{5}{6}, \frac{6}{7} \rangle$ and $\langle \frac{25}{36}, -\frac{49}{36} \rangle$

• **APPLICATIONS**

45. **Lifting Weights.** How much work does it take to lift 100 pounds vertically 4 feet?

46. **Lifting Weights.** How much work does it take to lift 150 pounds vertically 3.5 feet?

47. **Raising Wrecks.** How much work is done by a crane to lift a 2-ton car to a level of 20 feet?

48. **Raising Wrecks.** How much work is done by a crane to lift a 2.5-ton car to a level of 25 feet?

49. **Work.** To slide a crate across the floor, a force of 50 pounds at a 30° angle is needed. How much work is done if the crate is dragged 30 feet? Round to the nearest ft-lb.

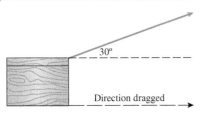

50. **Work.** To slide a crate across the floor, a force of 800 pounds at a 20° angle is needed. How much work is done if the crate is dragged 50 feet? Round to the nearest ft-lb.

51. **Close a Door.** A sliding door is closed by pulling a cord with a constant force of 35 pounds at a constant angle of 45°. The door is moved 6 feet to close it. How much work is done? Round to the nearest ft-lb.

52. **Close a Door.** A sliding door is closed by pulling a cord with a constant force of 45 pounds at a constant angle of 55°. The door is moved 6 feet to close it. How much work is done? Round to the nearest ft-lb.

53. **Braking Power.** A car that weighs 2500 pounds is parked on a hill in San Francisco with a slant of 40° from the horizontal. How much force will keep it from rolling down the hill? Round to the nearest pound.

54. **Braking Power.** A car that weighs 40,000 pounds is parked on a hill in San Francisco with a slant of 10° from the horizontal. How much force will keep it from rolling down the hill? Round to the nearest pound.

55. **Towing Power.** A semi-trailer truck that weighs 40,000 pounds is parked on a hill in San Francisco with a slant of 10° from the horizontal. A tow truck has to remove the truck from its parking spot and move it 100 feet up the hill. How much work is required? Round to the nearest ft-lb.

56. **Towing Power.** A car that weighs 2500 pounds is parked on a hill in San Francisco with a slant of 40° from the horizontal. A tow truck has to remove the car from its parking spot and move it 120 feet up the hill. How much work is required? Round to the nearest ft-lb.

57. **Travel Vector.** Two airplanes take off from the same airport and travel in different directions. One passes over town **A** known to be 30 miles north and 50 miles east of the airport. The other plane flies over town **B** known to be 10 miles south and 60 miles east of the airport. What is the angle between the direction of travel of the two planes?

58. **Travel Vector.** Two airplanes take off from the same airport and travel in different directions. One passes over town **A** known to be 60 miles north and 10 miles west of the airport. The other plane flies over town **B** known to be 40 miles south and 20 miles west of the airport. What is the angle between the direction of travel of the two planes?

59. **Push Lawn Mower.** How much work is required to push a lawn mower 100 feet if the force applied to the handle is 70 pounds and the handle makes an angle of 40° with the horizontal? Round to the nearest ft-lb.

60. **Push Lawn Mower.** How much work is required to push a lawn mower 75 feet if the force applied to the handle is 65 pounds and the handle makes an angle of 45° with the horizontal? Round to the nearest ft-lb.

• CATCH THE MISTAKE

In Exercises 61 and 62, explain the mistake that is made.

61. Find the dot product $\langle -3, 2 \rangle \cdot \langle 2, 5 \rangle$.

Solution:

Multiply component
by component. $\quad \langle -3, 2 \rangle \cdot \langle 2, 5 \rangle = \langle (-3)(2), (2)(5) \rangle$

Simplify. $\quad \langle -3, 2 \rangle \cdot \langle 2, 5 \rangle = \langle -6, 10 \rangle$

This is incorrect. What mistake was made?

62. Find the dot product $\langle 11, 12 \rangle \cdot \langle -2, 3 \rangle$.

Solution:

Multiply the outer and inner components.

$$\langle 11, 12 \rangle \cdot \langle -2, 3 \rangle = (11)(3) + (12)(-2)$$

Simplify. $\quad \langle 11, 12 \rangle \cdot \langle -2, 3 \rangle = 9$

This is incorrect. What mistake was made?

• CONCEPTUAL

In Exercises 63–66, determine whether each statement is true or false.

63. A dot product of two vectors is a vector.

64. A dot product of two vectors is a scalar.

65. Orthogonal vectors have a dot product equal to zero.

66. If the dot product of two nonzero vectors is equal to zero, then the vectors must be perpendicular.

For Exercises 67 and 68, refer to the following to find the dot product:

The dot product of vectors with n component is

$$\langle a_1, a_2, \ldots, a_n \rangle \cdot \langle b_1, b_2, \ldots, b_n \rangle = a_1 b_1 + a_2 b_2 + \cdots + a_n b_n$$

67. $\langle 3, 7, -5 \rangle \cdot \langle -2, 4, 1 \rangle$

68. $\langle 1, 0, -2, 3 \rangle \cdot \langle 5, 2, 3, 1 \rangle$

In Exercises 69–72, given $\mathbf{u} = \langle a, b \rangle$ and $\mathbf{v} = \langle c, d \rangle$, show that the following properties are true.

69. $\mathbf{u} \cdot \mathbf{v} = \mathbf{v} \cdot \mathbf{u}$

70. $\mathbf{u} \cdot \mathbf{u} = |\mathbf{u}|^2$

71. $\mathbf{0} \cdot \mathbf{u} = 0$

72. $k(\mathbf{u} \cdot \mathbf{v}) = (k\mathbf{u}) \cdot \mathbf{v} = \mathbf{u} \cdot (k\mathbf{v}), \quad k$ is a scalar

• CHALLENGE

73. Explain why $(\mathbf{u} \cdot \mathbf{v}) \cdot \mathbf{w}$ for vectors $\mathbf{u}, \mathbf{v}, \mathbf{w}$ does not exist.

74. Let $\mathbf{u} = \langle -2, 3 \rangle$, $\mathbf{v} = \langle 4, 1 \rangle$, and $\mathbf{w} = \langle -3, -5 \rangle$. Demonstrate that $\mathbf{u} \cdot (\mathbf{v} + \mathbf{w}) = \mathbf{u} \cdot \mathbf{v} + \mathbf{u} \cdot \mathbf{w}$.

75. Let $\mathbf{u} = \langle a, b \rangle$ and $\mathbf{v} = \langle b, a \rangle$. Find $|\mathbf{u}|^2 + (\mathbf{u} \cdot \mathbf{v})$.

76. Find the dot product of $\mathbf{u}$ and $\mathbf{v}$ if the angle between the vectors is $45°$ and $|\mathbf{u}| = 2\sqrt{2}$ and $|\mathbf{v}| = 4$.

77. Find the dot product of $\mathbf{u}$ and $\mathbf{v}$ if the angle between the vectors is $30°$ and $|\mathbf{u}| = \sqrt{10}$ and $|\mathbf{v}| = \sqrt{15}$.

78. Let $\mathbf{u} = \langle a, 0 \rangle$ and $\mathbf{v} = \langle a, \sqrt{3}a \rangle$. Find the angle θ between vectors $\mathbf{u}$ and $\mathbf{v}$.

• TECHNOLOGY

For Exercises 79 and 80, use a calculator to find the indicated dot product.

79. $\langle -11, 34 \rangle \cdot \langle 15, -27 \rangle$ **80.** $\langle 23, -350 \rangle \cdot \langle 45, 202 \rangle$

81. A rectangle has sides with lengths of 18 units and 11 units. Find the angle, to one decimal place, between the diagonal and the side with a length of 18 units. *Hint:* Set up a rectangular coordinate system and use vectors $\langle 18, 0 \rangle$ to represent the side with a length of 18 units and $\langle 18, 11 \rangle$ to represent the diagonal.

82. The definition of a dot product and the formula to find the angle between two vectors can be extended and applied to vectors with more than two components. A rectangular box has sides with lengths 12 feet, 7 feet, and 9 feet. Find the angle, to the nearest degree, between the diagonal and the side with length 7 feet.

▶[CHAPTER 7 REVIEW]

SECTION	CONCEPT	KEY IDEAS/FORMULAS
7.1	**Oblique triangles and the Law of Sines**	■ AAS (or ASA) triangles ■ SSA triangles (ambiguous case)
	Solving oblique triangles	Oblique (nonright) triangles The Law of Sines $$\frac{\sin\alpha}{a} = \frac{\sin\beta}{b} = \frac{\sin\gamma}{c}$$
7.2	**The Law of Cosines**	■ SAS triangles ■ SSS triangles
	Solving oblique triangles	The Law of Cosines $a^2 = b^2 + c^2 - 2bc\cos\alpha$ $b^2 = a^2 + c^2 - 2ac\cos\beta$ $c^2 = a^2 + b^2 - 2ab\cos\gamma$
7.3	**The area of a triangle**	
	The area of a triangle (SAS case)	Use the Law of Sines for the SAS case: $A_{SAS} = \frac{1}{2} bc \sin\alpha$ when b, c, and α are known. $A_{SAS} = \frac{1}{2} ab \sin\gamma$ when a, b, and γ are known. $A_{SAS} = \frac{1}{2} ac \sin\beta$ when a, c, and β are known.
	The area of a triangle (SSS case)	Use Heron's formula for the SSS case: $$A_{SSS} = \sqrt{s(s-a)(s-b)(s-c)}$$ where a, b, and c are the lengths of the sides of the triangle and s is half the perimeter of the triangle, called the semiperimeter: $$s = \frac{a+b+c}{2}$$

CHAPTER 7 REVIEW

SECTION	CONCEPT	KEY IDEAS/FORMULAS
7.4	**Vectors**	Vector **u** or $\overrightarrow{AB}$
	Vectors: Magnitude and direction	*Geometric interpretation* $\mathbf{u} = \langle a, b \rangle$ Magnitude: length of a vector $\mathbf{u} + \mathbf{v}$: tail-to-tip *Algebraic interpretation* $\tan\theta = \dfrac{b}{a}$ Magnitude: $\left|\mathbf{u}\right| = \sqrt{a^2 + b^2}$ $\mathbf{u} = \langle a, b \rangle$ and $\mathbf{v} = \langle c, d \rangle$ $\mathbf{u} + \mathbf{v} = \langle a + c, b + d \rangle$ $\mathbf{u} - \mathbf{v} = \langle a - c, b - d \rangle$
	Vector operations	Scalar multiplication: $k \langle a, b \rangle = \langle ka, kb \rangle$
	Horizontal and vertical components of a vector	Horizontal component: $a = \left|\mathbf{u}\right| \cos\theta$ Vertical component: $b = \left|\mathbf{u}\right| \sin\theta$
	Unit vectors	$\mathbf{u} = \dfrac{\mathbf{v}}{\left|\mathbf{v}\right|}$ $\qquad$ $\left|\mathbf{u}\right| = 1$
	Resultant vectors	Velocities and forces
7.5	**The dot product**	▪ The product of a scalar and vector is a vector. ▪ The dot product of two vectors is a scalar.
	Multiplying two vectors: The dot product	$\mathbf{u} = \langle a, b \rangle$ and $\mathbf{v} = \langle c, d \rangle$ $\mathbf{u} \cdot \mathbf{v} = ac + bd$
	Angle between two vectors	If θ is the smaller angle between two nonzero vectors **u** and **v** (i.e., $0 \leq \theta \leq 180°$), then $\qquad \cos\theta = \dfrac{\mathbf{u} \cdot \mathbf{v}}{\left|\mathbf{u}\right|\left|\mathbf{v}\right|}$ Orthogonal (perpendicular) vectors: $\mathbf{u} \cdot \mathbf{v} = 0$
	Work	When force and displacement are in the same direction: $W = \left|\mathbf{F}\right|\left|\mathbf{d}\right|$. When force and displacement are not in the same direction: $W = \mathbf{F} \cdot \mathbf{d}$.

[CHAPTER 7 REVIEW EXERCISES]

7.1 Oblique Triangles and the Law of Sines

Solve each triangle.

1. $\alpha = 10°, \beta = 20°, a = 4$
2. $\beta = 40°, \gamma = 60°, b = 10$
3. $\alpha = 5°, \beta = 45°, c = 10$
4. $\beta = 60°, \gamma = 70°, a = 20$
5. $\gamma = 11°, \alpha = 11°, c = 11$
6. $\beta = 20°, \gamma = 50°, b = 8$
7. $\alpha = 45°, \gamma = 45°, b = 2$
8. $\alpha = 60°, \beta = 20°, c = 17$
9. $\alpha = 12°, \gamma = 22°, a = 99$
10. $\beta = 102°, \gamma = 27°, a = 24$

Two side lengths and an angle measure are given. Determine whether a triangle (or two) exist and, if so, solve for all possible triangles.

11. $a = 7, b = 9, \alpha = 20°$
12. $b = 24, c = 30, \beta = 16°$
13. $a = 10, c = 12, \alpha = 24°$
14. $b = 100, c = 116, \beta = 12°$
15. $a = 40, b = 30, \beta = 150°$
16. $b = 2, c = 3, \gamma = 165°$
17. $a = 4, b = 6, \alpha = 10°$
18. $c = 25, a = 37, \gamma = 4°$

19. **Tightrope walker.** A man is walking on a tightrope between two platforms 50 feet apart. At point A, the tightrope creates an angle of depression 10° on one end and 15° on the other end as shown in the figure below. Find the length of the rope from each platform to A.

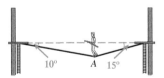

20. **Tightrope walker.** A man is walking on a tightrope between two platforms 40 feet apart. At point A, the tightrope creates an angle of depression 8° on one end and 12° on the other end as shown in the figure below. Find the length of the rope from each platform to A.

7.2 The Law of Cosines

Solve each triangle.

21. $a = 40, b = 60, \gamma = 50°$
22. $b = 15, c = 12, \alpha = 140°$
23. $a = 24, b = 25, c = 30$
24. $a = 6, b = 6, c = 8$
25. $a = \sqrt{11}, b = \sqrt{14}, c = 5$
26. $a = 22, b = 120, c = 122$
27. $b = 7, c = 10, \alpha = 14°$
28. $a = 6, b = 12, \gamma = 80°$
29. $b = 10, c = 4, \alpha = 90°$
30. $a = 4, b = 5, \gamma = 75°$
31. $a = 10, b = 11, c = 12$
32. $a = 22, b = 24, c = 25$
33. $b = 16, c = 18, \alpha = 100°$
34. $a = 25, c = 25, \beta = 9°$
35. $b = 12, c = 40, \alpha = 10°$
36. $a = 26, b = 20, c = 10$
37. $a = 26, b = 40, c = 13$
38. $a = 1, b = 2, c = 3$
39. $a = 6.3, b = 4.2, \alpha = 15°$
40. $b = 5, c = 6, \beta = 35°$

41. **How Far from Home?** Gary walked 8 miles due north, made a 130° turn to the southeast, and walked for another 6 miles. How far was Gary from home?

42. **How Far from Home?** Mary walked 10 miles due north, made a 55° turn to the northeast, and walked for another 3 miles. How far was Mary from home?

7.3 The Area of a Triangle

Find the area of each triangle in square units.

43. $b = 16, c = 18, \alpha = 100°$
44. $a = 25, c = 25, \beta = 9°$
45. $a = 10, b = 11, c = 12$
46. $a = 22, b = 24, c = 25$
47. $a = 26, b = 20, c = 10$
48. $a = 24, b = 32, c = 40$
49. $b = 12, c = 40, \alpha = 10°$
50. $a = 21, c = 75, \beta = 60°$

51. Area of Inscribed Triangle. The area of a triangle inscribed in a circle can be found if you know the lengths of the sides *a*, *b*, and *c* of the triangle and the radius of the circle. The formula is $A = \dfrac{abc}{4r}$. Find the radius of a circle that circumscribes a triangle if each side of the triangle measures 9 inches and the area of the triangle is 35 square inches.

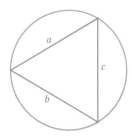

52. Area of Inscribed Triangle. The area of a triangle inscribed in a circle can be found if you know the lengths of the sides *a*, *b*, and *c* of the triangle and the radius of the circle. The formula is $A = \dfrac{abc}{4r}$. Find the radius of a circle that circumscribes a triangle if the sides of the triangle measure 9, 12, and 15 inches and the area of the triangle is 54 square inches.

7.4 Vectors

Find the magnitude of vector $\overrightarrow{AB}$.

53. $A = (4, -3)$ and $B = (-8, 2)$

54. $A = (-2, 11)$ and $B = (2, 8)$

55. $A = (0, -3)$ and $B = (5, 9)$

56. $A = (3, -11)$ and $B = (9, -3)$

Find the magnitude and direction angle of the given vector.

57. $\mathbf{u} = \langle -10, 24 \rangle$ **58.** $\mathbf{u} = \langle -5, -12 \rangle$

59. $\mathbf{u} = \langle 16, -12 \rangle$ **60.** $\mathbf{u} = \langle 0, 3 \rangle$

Perform the given vector operation, given that $\mathbf{u} = \langle 7, -2 \rangle$ and $\mathbf{v} = \langle -4, 5 \rangle$.

61. $2\mathbf{u} + 3\mathbf{v}$ **62.** $\mathbf{u} - \mathbf{v}$

63. $6\mathbf{u} + \mathbf{v}$ **64.** $-3(\mathbf{u} + 2\mathbf{v})$

Find the position vector, given its magnitude and direction angle.

65. $|\mathbf{u}| = 10, \theta = 75°$ **66.** $|\mathbf{u}| = 8, \theta = 225°$

67. $|\mathbf{u}| = 12, \theta = 105°$ **68.** $|\mathbf{u}| = 20, \theta = 15°$

Find the unit vector in the direction of the given vector.

69. $\mathbf{v} = \langle \sqrt{6}, -\sqrt{6} \rangle$ **70.** $\mathbf{v} = \langle -11, 60 \rangle$

Perform the indicated vector operation.

71. $(3\mathbf{i} - 4\mathbf{j}) + (2\mathbf{i} + 5\mathbf{j})$

72. $(-6\mathbf{i} + \mathbf{j}) - (9\mathbf{i} - \mathbf{j})$

73. Airspeed and Course. An airplane is traveling at 375 mph on a heading of 160° as measured clockwise from due north. If the wind is blowing 45 mph in the direction 25° as measured clockwise from due north, what is the actual speed and course of the airplane?

74. Airspeed and Course. An airplane is traveling at 425 mph on a heading of 110° as measured clockwise from due north. If the wind is blowing 25 mph in the direction 85° as measured clockwise from due north, what is the actual speed and course of the airplane?

7.5 The Dot Product

Find the indicated dot product.

75. $\langle 6, -3 \rangle \cdot \langle 1, 4 \rangle$

76. $\langle -6, 5 \rangle \cdot \langle -4, 2 \rangle$

77. $\langle 3, 3 \rangle \cdot \langle 3, -6 \rangle$

78. $\langle -2, -8 \rangle \cdot \langle -1, 1 \rangle$

79. $\langle 0, 8 \rangle \cdot \langle 1, 2 \rangle$

80. $\langle 4, -3 \rangle \cdot \langle -1, 0 \rangle$

Find the angle θ (round to the nearest degree) between each pair of vectors, $0° \leq \theta \leq 180°$.

81. $\langle 3, 4 \rangle$ and $\langle -5, 12 \rangle$

82. $\langle -4, 5 \rangle$ and $\langle 5, -4 \rangle$

83. $\langle 1, \sqrt{2} \rangle$ and $\langle -1, 3\sqrt{2} \rangle$

84. $\langle 7, -24 \rangle$ and $\langle -6, 8 \rangle$

85. $\langle 3, 5 \rangle$ and $\langle -4, -4 \rangle$

86. $\langle -1, 6 \rangle$ and $\langle 2, -2 \rangle$

Determine whether each pair of vectors is orthogonal.

87. $\langle 8, 3 \rangle$ and $\langle -3, 12 \rangle$

88. $\langle -6, 2 \rangle$ and $\langle 4, 12 \rangle$

89. $\langle 5, -6 \rangle$ and $\langle -12, -10 \rangle$

90. $\langle 1, 1 \rangle$ and $\langle -4, 4 \rangle$

91. $\langle 0, 4 \rangle$ and $\langle 0, -4 \rangle$

92. $\langle -7, 2 \rangle$ and $\langle \frac{1}{7}, -\frac{1}{2} \rangle$

93. $\langle 6z, a - b \rangle$ and $\langle a + b, -6z \rangle$

94. $\langle a - b, -1 \rangle$ and $\langle a + b, a^2 - b^2 \rangle$

95. Work. Tommy is pulling a wagon with a force of 25 pounds at an angle of 25° to the ground. How much work does he do in pulling the wagon 20 feet? Round to the nearest ft-lb.

96. Work. Annie is pulling a wagon with a force of 30 pounds at an angle of 22° to the ground. How much work does she do in pulling the wagon 25 feet? Round to the nearest ft-lb.

Technology Exercises

Section 7.1

Let A, B, and C be the lengths of the three sides of a triangle with X, Y, and Z as the corresponding angle measures. Write a program using a TI calculator to solve the given triangle.

97. $A = 31.6$, $C = 23.9$, $X = 42°$

98. $A = 137.2$, $B = 125.1$, $Y = 54°$

Section 7.2

Let A, B, and C be the lengths of the three sides of a triangle with X, Y, and Z as the corresponding angle measures. Write a program using a TI calculator to solve the given triangle.

99. $A = \sqrt{33}$, $B = \sqrt{29}$, $Z = 41.6°$

100. $A = 3412$, $B = 2178$, $C = 1576$

Section 7.3

Find the area of the given triangles. Round your answer to the nearest square unit.

101. $A = 312$, $B = 267$, $C = 189$

102. $A = 12.7$, $B = 29.9$, $Z = 104.8°$

Section 7.4

With the graphing calculator $\boxed{\text{SUM}}$ **command, find the magnitude of the given vector. Also, find the direction angle to the nearest degree.**

103. $\langle 25, -60 \rangle$

104. $\langle -70, 10\sqrt{15} \rangle$

Section 7.5

With the graphing calculator $\boxed{\text{SUM}}$ **command, find the angle (round to the nearest degree) between each pair of vectors.**

105. $\langle 14, 37 \rangle$, $\langle 9, -26 \rangle$

106. $\langle -23, -8 \rangle$, $\langle 18, -32 \rangle$

[CHAPTER 7 PRACTICE TEST]

Solve each of the following triangles (if possible).

1. $\alpha = 30°, \beta = 40°, b = 10$

2. $\alpha = 47°, \beta = 98°, \gamma = 35°$

3. $a = 7.0, b = 9.0, c = 12$

4. $\alpha = 45°, a = 8, b = 10$

5. $a = 10, b = 13, \gamma = 43°$

6. $a = 3, \beta = 26°, \gamma = 13°$

7. $a = 9, c = 10, \alpha = 130°$

8. $b = 20, c = 18, \beta = 12°$

In Exercises 9 and 10, find the area of each of the following triangles.

9. $\gamma = 72°, a = 10, b = 12$

10. $a = 7, b = 10, c = 13$

11. A triangular plot of land measures 250 yards by 375 yards by 305 yards. What is the area of the plot of land? Round to the nearest square yard.

12. A new pennant is being made for Big Town High School. If the pennant is to be in the shape of an isosceles triangle measuring 20 inches on a side with a vertex angle of 25°, what is the area of the pennant?

13. Find the magnitude and direction angle of the vector $\mathbf{u} = \langle -5, 12 \rangle$.

14. Find the magnitude and direction angle of the vector $\mathbf{u} = \langle -3, -6 \rangle$.

15. Find the unit vector pointing in the same direction as $\mathbf{v} = \langle -5, 2 \rangle$.

16. Find a unit vector pointing in the same direction as $\mathbf{v} = \langle -3, -4 \rangle$.

17. Perform the indicated operation:

    a. $2 \langle -1, 4 \rangle - 3 \langle 4, 1 \rangle$

    b. $\langle -7, -1 \rangle \cdot \langle 2, 2 \rangle$.

18. Are the vectors $\mathbf{u} = \langle 1, 5 \rangle$ and $\mathbf{v} = \langle -4, 1 \rangle$ perpendicular?

19. Find the smallest positive angle between the two vectors: $\mathbf{u} = \langle 3, 4 \rangle$ and $\mathbf{v} = \langle -5, 12 \rangle$.

20. Find the smallest positive angle between the two vectors: $\mathbf{u} = \langle -2, 4 \rangle$ and $\mathbf{v} = \langle 6, 1 \rangle$.

21. Find the smallest positive angle between the two vectors: $\mathbf{u} = \langle -4, -4 \rangle$ and $\mathbf{v} = \langle 1, 3 \rangle$.

22. Calculate the work involved in pushing a lawn mower 100 feet if the constant force you exert equals 30 pounds and makes an angle of 60° with the ground.

23. A boat is intending to cross a river on a heading of 40° measured clockwise off of due north. If the speed of the boat in still water is 18 mph and the speed of the current is 10 mph in the direction N 80° E, what is the actual speed and direction of the boat?

24. An airplane is flying at 300 mph on a compass heading of 250° measured clockwise off of due north. If the wind is blowing at 45 mph in the direction N10°E, what is the actual speed and direction of the plane?

25. Tom is pulling Janey on a sled with a force of 30 pounds at an angle of 35° to the ground. If he pulls her 50 feet, how much work has he done? Round to the nearest ft-lb.

26. A tractor is pulling a plow through a field with a force of 3500 pounds. If the tractor is hooked up such that it pulls at an angle of 40° to the horizontal, how much work does it do when pulling the plow 2500 feet? Round to the nearest ft-lb.

27. Explain why the magnitude of a vector can never be negative.

28. A post pattern in football is when the receiver in motion runs past the quarterback parallel to the line of scrimmage (A), runs 12 yards perpendicular to the line of scrimmage (B), and then cuts toward the goal post (C).

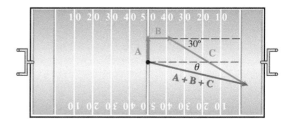

A receiver runs the post pattern. If the magnitudes of the vectors are $|A| = 3$ yards, $|B| = 12$ yards, and $|C| = 18$ yards, find the magnitude of the resultant vector, $A + B + C$ and the direction angle, θ.

Fill in the blanks so that each statement is true.

29. The dot product of two vectors is a _____ as long as the number of components of each vector is _____.

30. The product of a scalar and a vector is a _____.

[CHAPTERS 1–7 CUMULATIVE TEST]

1. For the angle $79°$, find
   a. the complement of this angle
   b. the supplement of this angle

2. Determine whether the statement $\sec\theta = \dfrac{\pi}{4}$ is possible.

3. Given $\sin\theta = \dfrac{\sqrt{5}}{3}$ and $\cos\theta = -\dfrac{2}{3}$, find $\cot\theta$.

4. Find the distance traveled (arc length) of a point that moves with a constant speed of $v = 55$ mph along a circle in 36 minutes.

5. Find all of the exact values of θ, when $\tan\theta = -\dfrac{\sqrt{3}}{3}$ and $0 \leq \theta \leq 2\pi$.

6. State the amplitude and period of the function $y = -8\sin(4\pi x)$.

7. Use addition of ordinates to graph the function
   $y = 3\cos\left(\dfrac{x}{2}\right) - \sin(2x)$, over the interval $0 \leq x \leq 4\pi$.

8. State the domain and range of the function $y = 3\sec(7x)$.

9. Simplify the trigonometric expression
   $\dfrac{\cos x + \sec x}{\cos x - \sec x} + 2\csc^2 x$.

10. Verify the trigonometric identity: $\dfrac{\csc x + \cot x}{\sec x + 1} = \cot x$.

11. Determine whether the equation $\tan(A + B) = \tan A + \tan B$ is conditional or an identity.

12. Graph $y = \sin x \cos\left(\dfrac{\pi}{4}\right) - \cos x \sin\left(\dfrac{\pi}{4}\right)$ by first rewriting as a sine or cosine of a difference or sum.

13. Verify the identity $\cos(3x) = \cos x(1 - 4\sin^2 x)$.

14. Simplify, using an identity, then graph $y = \dfrac{1 - \cos(2x)}{\sin(2x)}$.

15. Use half-angle identities to simplify $\sqrt{\dfrac{1 - \cos\left(\dfrac{3\pi}{4}\right)}{2}}$. Do not evaluate.

16. Write $\cos(0.2x) - \cos(0.8x)$ as a product of sines and/or cosines.

17. Find the exact value of $\tan^{-1}(-1)$. Give the answer in radians.

18. Use a calculator to evaluate $\csc^{-1}(1.0537)$. Give the answer in degrees and round to two decimal places.

19. Give the exact evaluation of the expression $\sec\left[\tan^{-1}\left(\frac{1}{4}\right)\right]$.

20. Solve the trigonometric equation $\cos(2\theta) = -\frac{1}{2}$ exactly, over the interval $0 \leq \theta \leq 2\pi$.

21. Solve the trigonometric equation $\sec^2(2\theta) - \sec(2\theta) - 12 = 0$ over the interval $0° \leq \theta < 360°$ and express the answer in degrees to two decimal places.

22. Solve the trigonometric equation $4\cos^2 x + 4\cos(2x) + 1 = 0$ exactly, over the interval $0 \leq \theta < 2\pi$.

23. Solve the triangle with $\beta = 106.3°$, $\gamma = 37.4°$, and $a = 76.1$ meters.

24. **Airplane Speed.** A plane flew due north at 450 mph for 2 hours. A second plane, starting at the same point and at the same time, flew southeast at an angle of $135°$ clockwise from due north at 375 mph for 2 hours. At the end of 2 hours, how far apart were the two planes? Round to the nearest mile.

25. Determine whether the vectors $\langle 0.5, 2 \rangle$ and $\langle -4, 1 \rangle$ are orthogonal.

Complex Numbers, Polar Coordinates, and Parametric Equations

David Cannon/Getty Images, Inc.

If a golfer tees off with an initial velocity of v_0 feet per second and an initial angle of trajectory θ, we can describe the position of the ball (x, y) with *parametric equations*. Parametric equations are a set of equations that express a set of quantities, such as x- and y-coordinates, as explicit functions of a number of independent variables, known as *parameters*. At some time t (seconds), the horizontal distance x (feet), from the golfer down the fairway, and the height above the ground y (feet) are given by the parametric equations:

$$x = (v_0\cos\theta)t \text{ and } y = (v_0\sin\theta)t - 16t^2$$

where we have neglected air resistance and t is the parameter. These *parametric equations* essentially map the path of the ball over time.

LEARNING OBJECTIVES

- Perform operations on complex numbers.
- Express complex numbers in polar form.

- Find products, quotients, powers, and roots of complex numbers using polar form.
- Convert between rectangular and polar coordinates.

- Use parametric equations to model paths: spirals and projectiles.

▶ **[IN THIS CHAPTER]**

We will review complex numbers. We will discuss the polar (trigonometric) form of complex numbers and operations on complex numbers. We will then introduce the polar coordinate system, which is often a preferred coordinate system over the rectangular system. We will graph polar equations in the polar coordinate system and finally discuss parametric equations and their graphs.

COMPLEX NUMBERS, POLAR COORDINATES, AND PARAMETRIC EQUATIONS

8.1 COMPLEX NUMBERS	8.2 POLAR (TRIGONOMETRIC) FORM OF COMPLEX NUMBERS	8.3 PRODUCTS, QUOTIENTS, POWERS, AND ROOTS OF COMPLEX NUMBERS; DE MOIVRE'S THEOREM	8.4 POLAR EQUATIONS AND GRAPHS	8.5 PARAMETRIC EQUATIONS AND GRAPHS
• The Imaginary Unit i • Adding and Subtracting Complex Numbers • Multiplying Complex Numbers • Dividing Complex Numbers • Raising Complex Numbers to Integer Powers	• Complex Numbers in Rectangular Form • Complex Numbers in Polar Form	• Products of Complex Numbers • Quotients of Complex Numbers • Powers of Complex Numbers • Roots of Complex Numbers	• Polar Coordinates • Converting Between Polar and Rectangular Coordinates • Graphs of Polar Equations	• Parametric Equations of a Curve

8.1 COMPLEX NUMBERS

SKILLS OBJECTIVES	CONCEPTUAL OBJECTIVES
▪ Write radicals with negative radicands as imaginary numbers. ▪ Add and subtract complex numbers. ▪ Multiply complex numbers. ▪ Divide complex numbers. ▪ Raise complex numbers to powers.	▪ Understand that real numbers and imaginary numbers are subsets of complex numbers. ▪ Recognize the real and imaginary parts of a complex number. ▪ Recognize that the square of i is -1. ▪ Understand how to eliminate imaginary numbers in denominators. ▪ Understand why i raised to a positive integer power can be reduced to 1, -1, i, or $-i$.

8.1.1 The Imaginary Unit i

8.1.1 SKILL

Write radicals with negative radicands as imaginary numbers.

8.1.1 CONCEPTUAL

Understand that real numbers and imaginary numbers are subsets of complex numbers.

[CONCEPT CHECK]

Which of the following are subsets of the set of complex numbers?
(A) real numbers
(B) purely imaginary numbers
(C) both A and B

▼
ANSWER C

For some equations like $x^2 = 1$, the solutions are always real numbers, $x = \pm 1$. However, some equations like $x^2 = -1$ do not have real solutions because the square of a real number cannot be negative. In order to solve such equations, mathematicians created a new set of numbers based on a number, called the *imaginary unit*, which when squared would give the negative quantity -1. This new set of numbers is called *imaginary numbers*.

DEFINITION The Imaginary Unit i

The **imaginary unit** is denoted by the letter i and is defined as

$$i = \sqrt{-1}$$

where $i^2 = -1$.

Recall that for positive real numbers a and b, we defined the principal square root as

$$b = \sqrt{a}, \quad \text{which means} \quad b^2 = a$$

Similarly, we define the *principal square root of a negative number* as $\sqrt{-a} = i\sqrt{a}$, since $(i\sqrt{a})^2 = i^2 a = -a$, for $a > 0$.

DEFINITION Principal Square Root

If $-a$ is a negative real number, then the principal square root of $-a$ is

$$\sqrt{-a} = i\sqrt{a}$$

where i is the imaginary unit and $i^2 = -1$.

It is customary to write $i\sqrt{a}$ instead of $\sqrt{a}i$ to avoid any confusion when defining a radical.

EXAMPLE 1 Using the Imaginary Unit i to Simplify Radicals

Simplify using imaginary numbers.
a. $\sqrt{-9}$　**b.** $\sqrt{-8}$

Solution:

a. $\sqrt{-9} = i\sqrt{9} = \boxed{3i}$ 　　　　　**b.** $\sqrt{-8} = i\sqrt{8} = i \cdot 2\sqrt{2} = \boxed{2i\sqrt{2}}$

▼
ANSWER

$12i$

▼
YOUR TURN Simplify $\sqrt{-144}$.

DEFINITION Complex Number

A **complex number** in standard form is defined as

$$a + bi$$

where a and b are real numbers and i is the imaginary unit. We denote a as the **real part** of the complex number and b as the **imaginary part** of the complex number.

A complex number written as $a + bi$ is said to be in **standard form**. If $a = 0$ and $b \neq 0$, then the resulting complex number bi is called a **pure imaginary number**. If $b = 0$, then $a + bi = a$ is a real number. The set of all real numbers and the set of all pure imaginary numbers are both subsets of the set of complex numbers.

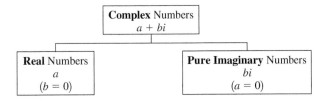

The following are examples of complex numbers:

$$17 \qquad 2 - 3i \qquad -5 + i \qquad 3 - i\sqrt{11} \qquad -9i$$

DEFINITION Equality of Complex Numbers

The complex numbers $a + bi$ and $c + di$ are **equal** if and only if $a = c$ and $b = d$. In other words, two complex numbers are equal if and only if both real parts are equal *and* both imaginary parts are equal.

8.1.2 Adding and Subtracting Complex Numbers

Complex numbers in the standard form $a + bi$ are treated in much the same way as binomials of the form $a + bx$. We can add, subtract, and multiply complex numbers the same way we performed these operations on binomials. When adding or subtracting complex numbers, combine real parts with real parts and combine imaginary parts with imaginary parts.

EXAMPLE 2 Adding and Subtracting Complex Numbers

Perform the indicated operation and simplify.
a. $(3 - 2i) + (-1 + i)$ **b.** $(2 - i) - (3 - 4i)$

Solution (a):

Eliminate the parentheses.	$= 3 - 2i - 1 + i$
Group real and imaginary numbers, respectively.	$= (3 - 1) + (-2i + i)$
Simplify.	$= \boxed{2 - i}$

Solution (b):

Eliminate the parentheses (distribute the negative).	$= 2 - i - 3 + 4i$
Group real and imaginary numbers, respectively.	$= (2 - 3) + (-i + 4i)$
Simplify.	$= \boxed{-1 + 3i}$

▼ **YOUR TURN** Perform the indicated operation and simplify: $(4 + i) - (3 - 5i)$.

8.1.2 SKILL

Add and subtract complex numbers.

8.1.2 CONCEPTUAL

Recognize the real and imaginary parts of a complex number.

[CONCEPT CHECK]

Add $(a + bi) + (c - di)$ and group the real parts and the imaginary parts of the resulting complex number.

▼ . . . .

ANSWER $(a + c) + (b - d)i$

▼ . . . .

ANSWER

$1 + 6i$

8.1.3 Multiplying Complex Numbers

When multiplying complex numbers, you apply all of the same methods as you did when multiplying binomials. It is important to remember that $i^2 = -1$.

WORDS	MATH
Multiply the complex numbers.	$(5 - i)(3 - 4i)$
Multiply using the distributive property.	$= 5(3) + 5(-4i) - i(3) - i(-4i)$
Eliminate the parentheses.	$= 15 - 20i - 3i + 4i^2$
Substitute $i^2 = -1$.	$= 15 - 20i - 3i + 4(-1)$
Simplify.	$= 15 - 20i - 3i - 4$
Combine real parts and imaginary parts, respectively.	$= 11 - 23i$

▶ **EXAMPLE 3** **Multiplying Complex Numbers**

Multiply the complex numbers and express the result in standard form: $a \pm bi$.

a. $(3 - i)(2 + i)$ **b.** $i(-3 + i)$

Solution (a):

Use the distributive property.	$(3 - i)(2 + i) = 3(2) + 3(i) - i(2) - i(i)$
Eliminate the parentheses.	$= 6 + 3i - 2i - i^2$
Substitute $i^2 = -1$.	$= 6 + 3i - 2i - (-1)$
Group like terms.	$= (6 + 1) + (3i - 2i)$
Simplify.	$= \boxed{7 + i}$

Solution (b):

Use the distributive property.	$i(-3 + i) = -3i + i^2$
Substitute $i^2 = -1$.	$= -3i - 1$
Write in standard form.	$= \boxed{-1 - 3i}$

▼ YOUR TURN Multiply the complex numbers and express the result in standard form, $a + bi$: $(4 - 3i)(-1 + 2i)$.

8.1.4 Dividing Complex Numbers

Recall the special product that produces a difference of two squares, $(a + b)(a - b) = a^2 - b^2$. This special product has only first and last terms because the products of the outer and inner terms subtract out and become zero. Similarly, if we multiply complex numbers in the same manner, the result is a real number because the imaginary terms cancel each other out.

COMPLEX CONJUGATE

The product of a complex number, $z = a + bi$, and its **complex conjugate**, $\bar{z} = a - bi$, is a real number.

$$z\bar{z} = (a + bi)(a - bi) = a^2 - b^2 i^2 = a^2 - b^2(-1) = a^2 + b^2$$

In order to write a quotient of complex numbers in standard form, $a + bi$, multiply the numerator and the denominator by the complex conjugate of the denominator. It is important to note that if i is present in the denominator, then the complex number is *not* in standard form, $a + bi$.

[CONCEPT CHECK]

Write the quotient in standard form: $\frac{1}{a + i}$.

ANSWER $\frac{a}{a^2 + 1} - \frac{1}{a^2 + 1}i$

EXAMPLE 4 **Dividing Complex Numbers**

Write the quotient in standard form: $\frac{2 - i}{1 + 3i}$.

Solution:

Multiply the numerator and the denominator by the complex conjugate of the denominator, $1 - 3i$. $\quad \left(\frac{2 - i}{1 + 3i}\right)\left(\frac{1 - 3i}{1 - 3i}\right)$

Multiply the numerators and denominators, respectively. $\quad = \frac{(2 - i)(1 - 3i)}{(1 + 3i)(1 - 3i)}$

Use the FOIL method (or distributive property). $\quad = \frac{2 - 6i - i + 3i^2}{1 - 3i + 3i - 9i^2}$

Combine imaginary parts. $\quad = \frac{2 - 7i + 3i^2}{1 - 9i^2}$

Substitute $i^2 = -1$. $\quad = \frac{2 - 7i - 3}{1 - 9(-1)}$

Simplify the numerator and denominator. $\quad = \frac{-1 - 7i}{10}$

Write in standard form. Recall that $\frac{a + b}{c} = \frac{a}{c} + \frac{b}{c}$. $\quad = \boxed{-\frac{1}{10} - \frac{7}{10}i}$

YOUR TURN Write the quotient in standard form: $\frac{3 + 2i}{4 - i}$.

ANSWER

$\frac{10}{17} + \frac{11}{17}i$

[STUDY TIP

When denominators are multiplied by their complex conjugate, the result is a real number.

$$(a + bi)(a - bi) = a^2 - b^2i^2 = a^2 + b^2$$
]

8.1.5 Raising Complex Numbers to Integer Powers

8.1.5 SKILL

Raise complex numbers to powers.

8.1.5 CONCEPTUAL

Understand why i raised to a positive integer power can be reduced to 1, -1, i, or $-i$.

Note that i raised to the fourth power is 1. In simplifying integer powers of the imaginary unit, i, we factor out i raised to the largest multiple of 4.

$$i = \sqrt{-1}$$
$$i^2 = -1$$
$$i^3 = i^2 \cdot i = (-1)i = -i$$
$$i^4 = i^2 \cdot i^2 = (-1)(-1) = 1$$
$$i^5 = i^4 \cdot i = (1)(i) = i$$
$$i^6 = i^4 \cdot i^2 = (1)(-1) = -1$$
$$i^7 = i^4 \cdot i^3 = (1)(-i) = -i$$
$$i^8 = (i^4)^2 = 1$$

EXAMPLE 5 **Raising the Imaginary Unit to Integer Powers**

Simplify:

a. i^{11} **b.** i^{13} **c.** i^{100}

Solution:

a. $i^{11} = i^8 \cdot i^3 = (i^4)^2 \cdot i^3 = 1^2(-i) = \boxed{-i}$

b. $i^{13} = i^{12} \cdot i = (i^4)^3 \cdot i = 1^3 \cdot i = \boxed{i}$

c. $i^{100} = (i^4)^{25} = 1^{25} = \boxed{1}$

▼

YOUR TURN Simplify i^{27}.

EXAMPLE 6 **Raising a Complex Number to an Integer Power**

Write $(2 - i)^3$ in standard form.

Solution:

Recall the formula for cubing a binomial.	$(a - b)^3 = a^3 - 3a^2b + 3ab^2 - b^3$
Let $a = 2$ and $b = i$.	$(2 - i)^3 = 2^3 - 3(2)^2(i) + 3(2)(i)^2 - i^3$
Substitute $i^2 = -1$ and $i^3 = -i$.	$= 2^3 - 3(2)^2(i) + (3)(2)(-1) - (-i)$
Eliminate parentheses and rearrange terms.	$= 8 - 6 - 12i + i$
Combine the real parts and imaginary parts, respectively.	$= \boxed{2 - 11i}$

▼

YOUR TURN Write $(2 + i)^3$ in standard form.

▶[SECTION 8.1] SUMMARY

The Imaginary Unit i

- $i = \sqrt{-1}$
- $i^2 = -1$

Complex Numbers

- *Standard Form*: $a + bi$, where a is the real part and b is the imaginary part.
- The set of real numbers and the set of pure imaginary numbers are both subsets of the set of complex numbers.

Adding and Subtracting Complex Numbers

- $(a + bi) + (c + di) = (a + c) + (b + d)i$
- $(a + bi) - (c + di) = (a - c) + (b - d)i$
- To add or subtract complex numbers, add or subtract the real parts and the imaginary parts, respectively.

Multiplying Complex Numbers

- $(a + bi)(c + di) = (ac - bd) + (ad + bc)i$
- Apply the same methods for multiplying binomials (FOIL). It is important to remember that $i^2 = -1$.

Dividing Complex Numbers

- The complex conjugate of $a + bi$ is $a - bi$.
- In order to write a quotient of complex numbers in standard form, multiply the numerator and the denominator by the complex conjugate of the denominator:

$$\frac{(a + bi)}{(c + di)} \cdot \frac{(c - di)}{(c - di)}$$

[SECTION 8.1] EXERCISES

• **SKILLS**

In Exercises 1–12, write each expression as a complex number in standard form. If an expression simplifies to either a real number or a pure imaginary number, leave in that form.

1. $\sqrt{-16}$

2. $\sqrt{-100}$

3. $\sqrt{-20}$

4. $\sqrt{-24}$

5. $\sqrt[3]{-64}$

6. $\sqrt[3]{-27}$

7. $\sqrt{-64}$

8. $\sqrt{-27}$

9. $3 - \sqrt{-100}$

10. $4 - \sqrt{-121}$

11. $-10 - \sqrt{-144}$

12. $7 - \sqrt[3]{-125}$

In Exercises 13–40, perform the indicated operation, simplify, and express in standard form.

13. $(3 - 7i) + (-1 - 2i)$

14. $(1 + i) + (9 - 3i)$

15. $(3 - 4i) + (7 - 10i)$

16. $(5 + 7i) + (-10 - 2i)$

17. $(4 - 5i) - (2 - 3i)$

18. $(-2 + i) - (1 - i)$

19. $(-3 + i) - (-2 - i)$

20. $(4 + 7i) - (5 + 3i)$

21. $3(4 - 2i)$

22. $4(7 - 6i)$

23. $12(8 - 5i)$

24. $-3(16 + 4i)$

25. $-3(16 - 9i)$

26. $5(-6i + 3)$

27. $-6(17 - 5i)$

28. $-12(8 + 3i)$

29. $(1 - i)(3 + 2i)$

30. $(-3 + 2i)(1 - 3i)$

31. $(5 - 7i)(-3 + 4i)$

32. $(16 - 5i)(-2 - i)$

33. $(7 - 5i)(6 + 9i)$

34. $(-3 - 2i)(7 - 4i)$

35. $(12 - 18i)(-2 + i)$

36. $(-4 + 3i)(-4 - 3i)$

37. $\left(\frac{1}{2} + 2i\right)\left(\frac{4}{9} - 3i\right)$

38. $\left(-\frac{3}{4} + \frac{9}{16}i\right)\left(\frac{2}{3} + \frac{4}{9}i\right)$

39. $(-i + 17)(2 + 3i)$

40. $(-3i - 2)(-2 - 3i)$

For Exercises 41–48, for each complex number z, write the complex conjugate $\bar{z}$, and find $z\bar{z}$.

41. $z = 4 + 7i$

42. $z = 2 + 5i$

43. $z = 2 - 3i$

44. $z = 5 - 3i$

45. $z = 6 + 4i$

46. $z = -2 + 7i$

47. $z = -2 - 6i$

48. $z = -3 - 9i$

For Exercises 49–64, write each quotient in standard form.

49. $\dfrac{2}{i}$

50. $\dfrac{3}{i}$

51. $\dfrac{1}{3 - i}$

52. $\dfrac{2}{7 - i}$

53. $\dfrac{1}{3 + 2i}$

54. $\dfrac{1}{4 - 3i}$

55. $\dfrac{2}{7 + 2i}$

56. $\dfrac{8}{1 + 6i}$

57. $\dfrac{1 - i}{1 + i}$

58. $\dfrac{3 - i}{3 + i}$

59. $\dfrac{2 + 3i}{3 - 5i}$

60. $\dfrac{2 + i}{3 - i}$

61. $\dfrac{4 - 5i}{7 + 2i}$

62. $\dfrac{7 + 4i}{9 - 3i}$

63. $\dfrac{8 + 3i}{9 - 2i}$

64. $\dfrac{10 - i}{12 + 5i}$

For Exercises 65–76, simplify and express in standard form.

65. i^{15}

66. i^{99}

67. i^{40}

68. i^{18}

69. $(5 - 2i)^2$

70. $(3 - 5i)^2$

71. $(2 + 3i)^2$

72. $(4 - 9i)^2$

73. $(3 + i)^3$

74. $(2 + i)^3$

75. $(1 - i)^3$

76. $(4 - 3i)^3$

• **APPLICATIONS**

In Exercises 77 and 78, refer to the following:

Electrical impedance is the ratio of voltage to current in AC circuits. Let Z represent the total impedance of an electrical circuit. If there are two resistors in a circuit, let $Z_1 = 3 - 6i$ ohms and $Z_2 = 5 + 4i$ ohms.

77. Electrical Circuits in Series. When the resistors in the circuit are placed in series, the total impedance is the sum of the two impedances $Z = Z_1 + Z_2$. Find the total impedance of the electrical circuit in series.

78. Electrical Circuits in Parallel. When the resistors in the circuit are placed in parallel, the total impedance is given by $\dfrac{1}{Z} = \dfrac{1}{Z_1} + \dfrac{1}{Z_2}$. Find the total impedance of the electrical circuit in parallel.

• CATCH THE MISTAKE

In Exercises 79 and 80, explain the mistake that is made.

79. Write the quotient in standard form: $\dfrac{2}{4-i}$.

Solution:

Multiply the numerator and the denominator by $4 - i$.	$\dfrac{2}{(4-i)} \cdot \dfrac{(4-i)}{(4-i)}$
Multiply the numerator using the distributive property and the denominator using the FOIL method.	$\dfrac{8-2i}{16-1}$
Simplify.	$\dfrac{8-2i}{15}$
Write in standard form.	$\dfrac{8}{15} - \dfrac{2}{15}i$

This is incorrect. What mistake was made?

80. Write the product in standard form: $(2 - 3i)(5 + 4i)$.

Solution:

Use the FOIL method to multiply the complex numbers.	$10 - 7i - 12i^2$
Simplify.	$-2 - 7i$

This is incorrect. What mistake was made?

• CONCEPTUAL

In Exercises 81–84, determine whether each statement is true or false.

81. The product $(a + bi)(a - bi)$ is a real number.

82. The set of pure imaginary numbers is a subset of the set of complex numbers.

83. The set of real numbers is a subset of the set of complex numbers.

84. There is no complex number that is equal to its conjugate.

• CHALLENGE

85. Factor completely over the complex numbers: $x^4 + 2x^2 + 1$.

86. Factor completely over the complex numbers: $x^4 + 18x^2 + 81$.

• TECHNOLOGY

In Exercises 87–90, apply a graphing utility to simplify the expression. Write your answer in standard form.

87. $(1 + 2i)^5$

88. $(3 - i)^6$

89. $\dfrac{1}{(2 - i)^3}$

90. $\dfrac{1}{(4 + 3i)^2}$

8.2 POLAR (TRIGONOMETRIC) FORM OF COMPLEX NUMBERS

SKILLS OBJECTIVES	CONCEPTUAL OBJECTIVES
▪ Calculate the modulus of a complex number. ▪ Convert complex numbers from rectangular form to polar form and vice versa.	▪ Understand that the modulus, or magnitude, of a complex number is the distance from the origin to the point in the complex plane. ▪ Understand that imaginary numbers lie along the vertical axis and real numbers lie along the horizontal axis of the complex plane.

8.2.1 Complex Numbers in Rectangular Form

We are already familiar with the **rectangular coordinate system**, where the horizontal axis is called the *x*-axis and the vertical axis is called the *y*-axis. In our study of complex numbers, we refer to the **standard (rectangular) form** as $a + bi$, where *a* represents the real part and *b* represents the imaginary part. If we let the horizontal axis be the **real axis** and the vertical axis be the **imaginary axis**, the result is the **complex plane**. The number $a + bi$ is located in the complex plane by finding the point with coordinates (a, b).

8.2.1 SKILL

Calculate the modulus of a complex number.

8.2.1 CONCEPTUAL

Understand that the modulus, or magnitude, of a complex number is the distance from the origin to the point in the complex plane.

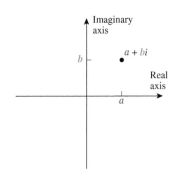

When $b = 0$, the result is a real number, and therefore all numbers represented by a point along the horizontal axis are real numbers. When $a = 0$, the result is an imaginary number, so all numbers represented by a point along the vertical axis are imaginary numbers.

The variable *z* is often used to represent a complex number: $z = x + iy$. Complex numbers are analogous to vectors. Suppose we have a vector $\mathbf{z} = \langle x, y \rangle$, whose initial point is the origin and terminal point is (x, y); then the magnitude of that vector is $|\mathbf{z}| = \sqrt{x^2 + y^2}$. Similarly, the magnitude, or *modulus*, of a complex number is defined like the magnitude of a position vector in the *xy*-plane: as the distance from the origin $(0, 0)$ to the point (x, y) in the complex plane.

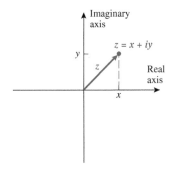

> **DEFINITION** Modulus of a Complex Number
>
> The **modulus**, or magnitude, of a complex number $z = x + iy$ is the distance from the origin to the point (x, y) in the complex plane given by
>
> $$|z| = \sqrt{x^2 + y^2}$$

Recall from Section 8.1 that a complex number $z = x + iy$ has a complex conjugate $\bar{z} = x - iy$. The bar above a complex number denotes its conjugate. Notice that

$$z\bar{z} = (x + iy)(x - iy) = x^2 - i^2y^2 = x^2 + y^2$$

and therefore the modulus can also be written as

$$\boxed{|z| = \sqrt{z\bar{z}}}$$

EXAMPLE 1 **Finding the Modulus of a Complex Number**

Find the modulus of $z = -3 + 2i$.

> **common mistake**
>
> Including the i in the imaginary part.
>
> **✓CORRECT**
>
> Let $x = -3$ and $y = 2$ in $|z| = \sqrt{x^2 + y^2}$.
>
> $$|-3 + 2i| = \sqrt{(-3)^2 + 2^2}$$
>
> Eliminate the parentheses.
>
> $$|-3 + 2i| = \sqrt{9 + 4}$$
>
> Simplify.
>
> $$\boxed{|z| = |-3 + 2i| = \sqrt{13}}$$
>
> **✗INCORRECT**
>
> $$|-3 + 2i| = \sqrt{(-3)^2 + (2i)^2}$$
>
> **ERROR**
>
> The i is not included in the formula; only the imaginary part (coefficient of i) is used.

▼

YOUR TURN Find the modulus of $z = 2 - 5i$.

Impedance is a term used in circuit theory that describes a measure of opposition to a sinusoidal alternating current (AC). The complex number Z denotes impedance and is given by

$$Z = R + iX$$

where R is the resistance and X is the reactance.

EXAMPLE 2   **Calculating the Magnitude of Impedance**

Calculate the magnitude of impedance in terms of the resistive and reactive parts.

Solution:

Write the impedance.

$$Z = R + iX$$

The magnitude of a complex number is
the square root of the sum of the squares
of the real and imaginary parts.

$$\boxed{|Z| = \sqrt{R^2 + X^2}}$$

8.2.2 Complex Numbers in Polar Form

We say that a complex number $z = x + iy$ is in *rectangular* form because it is located
at the point (x, y), which is expressed in rectangular coordinates in the complex plane.
Another convenient way of expressing complex numbers is in *polar* form (sometimes
called trigonometric form). Recall in our study of vectors (Section 7.4) that vectors
have both magnitude and a direction angle. The same is true of points in the complex
plane. Let r represent the magnitude, or distance from the origin to the point (x, y), and
θ represent the direction angle; then we have the following relationships:

$$r = \sqrt{x^2 + y^2}$$

$$\sin\theta = \frac{y}{r} \quad \cos\theta = \frac{x}{r} \quad \tan\theta = \frac{y}{x} \quad (x \neq 0)$$

Note: If $x = 0$, then the result is a pure imaginary number that corresponds to a point
on the y-axis. Therefore, in that case, $\theta = 90°$ or $270°$ $\left(\dfrac{\pi}{2} \text{ or } \dfrac{3\pi}{2}, \text{ respectively} \right)$.

Isolating x and y in the equations above, we find:

$$x = r\cos\theta \quad y = r\sin\theta$$

Using these expressions for x and y, a complex number can be written in *polar* form:

$$z = x + yi = (r\cos\theta) + (r\sin\theta)i = r(\cos\theta + i\sin\theta)$$

**POLAR (TRIGONOMETRIC) FORM
OF COMPLEX NUMBERS**

The following expression is the **polar form** of a complex number:

$$z = r(\cos\theta + i\sin\theta)$$

where r represents the **modulus** (magnitude) of the complex number and θ is
called the **argument** of z.

The following is standard notation for modulus and argument:

$$r = \operatorname{mod} z = |z| \quad \text{and} \quad \theta = \arg z \quad 0 \le \theta < 2\pi \text{ or } 0° \le \theta < 360°$$

Converting Complex Numbers Between Rectangular and Polar Forms

We can convert back and forth between rectangular and polar (trigonometric) forms of
complex numbers using the modulus and trigonometric ratios:

$$r = \sqrt{x^2 + y^2} \quad \sin\theta = \frac{y}{r} \quad \cos\theta = \frac{x}{r} \quad \tan\theta = \frac{y}{x} \quad (x \neq 0)$$

8.2.2 SKILL

Convert complex numbers from
rectangular form to polar form
and vice verse.

8.2.2 CONCEPTUAL

Understand that imaginary
numbers lie along the vertical
axis and real numbers lie along
the horizontal axis of the
complex plane.

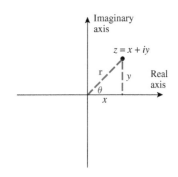

[CONCEPT CHECK]

TRUE OR FALSE In the complex
plane, a real number or purely
imaginary number lies along one
of the axes, and all the other
complex numbers lie in one of
the four quadrants.

▼

ANSWER True

CONVERTING COMPLEX NUMBERS FROM RECTANGULAR FORM TO POLAR FORM

Step 1: Plot the point $z = x + yi$ in the complex plane (note the quadrant).

Step 2: Find r. Use $r = \sqrt{x^2 + y^2}$.

Step 3: Find θ. Use $\theta = \tan^{-1}\left(\dfrac{y}{x}\right)$ or $\tan\theta = \dfrac{y}{x}, x \neq 0$, where θ is in the quadrant found in Step 1 and $0° \leq \theta < 360°$ or $0 \leq \theta < 2\pi$.

Step 4: Write the complex number in polar form: $z = r(\cos\theta + i\sin\theta)$.

EXAMPLE 3 Converting from Rectangular to Polar Form

Express the complex number $z = \sqrt{3} - i$ in polar form.

Solution:

STEP 1 Plot the point.

The point lies in **quadrant IV**.

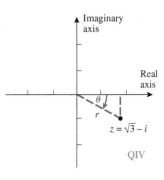

STEP 2 Find r.

Let $x = \sqrt{3}$ and $y = -1$

in $r = \sqrt{x^2 + y^2}$. $r = \sqrt{(\sqrt{3})^2 + (-1)^2}$

Eliminate the parentheses. $r = \sqrt{3 + 1}$

Simplify. $\boxed{r = 2}$

STEP 3 Find θ.

Let $x = \sqrt{3}$ and $y = -1$ in $\tan\theta = \dfrac{y}{x}$. $\tan\theta = -\dfrac{1}{\sqrt{3}}$

Solve for θ. $\theta = \tan^{-1}\left(-\dfrac{1}{\sqrt{3}}\right) = -\dfrac{\pi}{6}$

Find the reference angle.   reference angle $= \dfrac{\pi}{6}$

The complex number lies in quadrant IV. $\boxed{\theta = \dfrac{11\pi}{6}}$

STEP 4 Write the complex number in polar form, $z = r(\cos\theta + i\sin\theta)$. $\boxed{z = 2\left[\cos\left(\dfrac{11\pi}{6}\right) + i\sin\left(\dfrac{11\pi}{6}\right)\right]}$

Note: An alternative form is in degrees: $z = 2(\cos 330° + i\sin 330°)$.

▼

ANSWER

$z = 2\left[\cos\left(\dfrac{5\pi}{3}\right) + i\sin\left(\dfrac{5\pi}{3}\right)\right]$ or

$2(\cos 300° + i\sin 300°)$

▼ YOUR TURN Express the complex number $z = 1 - i\sqrt{3}$ in polar form.

You must be very careful in converting from rectangular to polar form. Remember that the inverse tangent function is a one-to-one function and will yield values in quadrants I and IV. If the point lies in quadrant II or III, add 180° to the angle in degrees found through the inverse tangent function (for θ in radians, add π).

▶ **EXAMPLE 4**   **Converting from Rectangular to Polar Form**

Express the complex number $z = -2 + i$ in polar form.

common mistake

Forgetting to confirm the quadrant in which the point lies.

✓CORRECT   ✗INCORRECT

STEP 1 Plot the point.
The point lies in **quadrant II**.

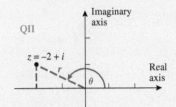

STEP 2 Find r.
Let $x = -2$ and $y = 1$ in

$$r = \sqrt{x^2 + y^2}.$$
$$r = \sqrt{(-2)^2 + 1^2}$$

Simplify.

$$\boxed{r = \sqrt{5}}$$

STEP 3 Find θ.
Let $x = -2$ and $y = 1$ in

$$\tan\theta = \frac{y}{x}.$$

$$\tan\theta = -\frac{1}{2}$$

$$\theta = \tan^{-1}\left(-\frac{1}{2}\right)$$
$$\approx -26.565°$$

The complex number lies in quadrant II.

$$\boxed{\theta \approx -26.6° + 180° = 153.4°}$$

STEP 4 Write the complex number in polar form, $z = r(\cos\theta + i\sin\theta)$.

$$\boxed{z \approx \sqrt{5}[\cos(153.4°) + i\sin(153.4°)]}$$

$$\boxed{\theta = \tan^{-1}\left(-\frac{1}{2}\right) \approx -26.565°}$$

Write the complex number in polar form, $z = r(\cos\theta + i\sin\theta)$.

$$z = \sqrt{5}[\cos(-26.6°) + i\sin(-26.6°)]$$

Note: $\theta = -26.565°$ lies in quadrant IV, whereas the original point lies in quadrant II. Therefore, we should have added 180° to θ to arrive at a point in quadrant II.

▼

YOUR TURN   Express the complex number $z = -1 + 2i$ in polar form.

▼
ANSWER

$$z \approx \sqrt{5}[\cos(116.6°) + i\sin(116.6°)]$$

To convert from polar to rectangular form, simply evaluate the trigonometric functions.

EXAMPLE 5 **Converting from Polar to Rectangular Form**

Express $z = 4(\cos 120° + i \sin 120°)$ in rectangular form.

Solution:

Evaluate the trigonometric functions exactly.

$$z = 4(\underbrace{\cos 120°}_{-\frac{1}{2}} + i \underbrace{\sin 120°}_{\frac{\sqrt{3}}{2}})$$

Distribute the 4.

$$z = 4\left(-\frac{1}{2}\right) + 4\left(\frac{\sqrt{3}}{2}\right)i$$

Simplify.

$$\boxed{z = -2 + 2\sqrt{3}i}$$

▼
ANSWER

$z = -\sqrt{3} - i$

YOUR TURN Express $z = 2(\cos 210° + i \sin 210°)$ in rectangular form.

EXAMPLE 6 **Using a Calculator to Convert from Polar to Rectangular Form**

Express $z = 3(\cos 109° + i \sin 109°)$ in rectangular form. Round values to four decimal places.

Solution:

Use a calculator to evaluate the trigonometric functions.

$$z = 3(\underbrace{\cos 109°}_{-0.325568} + i \underbrace{\sin 109°}_{0.945519})$$

Simplify.

$$\boxed{z \approx -0.9767 + 2.8366i}$$

▼
ANSWER

$z \approx -5.5904 - 4.2127i$

YOUR TURN Express $z = 7(\cos 217° + i \sin 217°)$ in rectangular form. Round to four decimal places.

▶[SECTION 8.2] SUMMARY

In the complex plane, the horizontal axis is the real axis and the vertical axis is the imaginary axis. Complex numbers can be expressed in either rectangular form, $z = x + iy$, or polar form, $z = r(\cos\theta + i\sin\theta)$. The modulus of a complex number $z = x + iy$ is given by $|z| = \sqrt{x^2 + y^2}$. To convert from rectangular to polar form, we use the relationships $r = \sqrt{x^2 + y^2}$

and $\tan\theta = \dfrac{y}{x}, x \neq 0$, where $0 \leq \theta < 2\pi$ or $0° \leq \theta < 360°$. It is important to note in which quadrant the point lies. To convert from polar to rectangular form, simply evaluate the trigonometric expressions for x and y.

$$x = r\cos\theta \quad \text{and} \quad y = r\sin\theta$$

[SECTION 8.2] EXERCISES

• **SKILLS**

In Exercises 1–12, graph each complex number in the complex plane.

1. $7 + 8i$

2. $3 + 5i$

3. $-2 - 4i$

4. $-3 - 2i$

5. 2

6. 7

7. $-3i$

8. $-5i$

9. $\dfrac{2}{3} + \dfrac{11}{4}i$

10. $-\dfrac{7}{2} + \dfrac{15}{2}i$

11. $4 - \dfrac{5}{2}i$

12. $-\dfrac{47}{10} - \dfrac{19}{10}i$

In Exercises 13–28, express each complex number in polar form.

13. $1 - i$

14. $2 + 2i$

15. $1 + \sqrt{3}i$

16. $-3 - \sqrt{3}i$

17. $-4 + 4i$

18. $\sqrt{5} - \sqrt{5}i$

19. $\sqrt{3} - 3i$

20. $-\sqrt{3} + i$

21. $3 + 0i$

22. $-2 + 0i$

23. $2\sqrt{3} - 2i$

24. $-8 - 8\sqrt{3}i$

25. $-\dfrac{1}{2} + \dfrac{\sqrt{3}}{2}i$

26. $-\dfrac{5}{3} - \dfrac{5}{3}i$

27. $\dfrac{\sqrt{3}}{8} - \dfrac{1}{8}i$

28. $\dfrac{7}{16} + \dfrac{7}{16}i$

In Exercises 29–44, use a calculator to express each complex number in polar form. Express Exercises 29–36 in degrees and Exercises 37–44 in radians.

29. $3 - 7i$

30. $2 + 3i$

31. $-6 + 5i$

32. $-4 - 3i$

33. $-5 + 12i$

34. $24 + 7i$

35. $8 - 6i$

36. $-3 + 4i$

37. $-6 + 2i$

38. $-5 - 8i$

39. $35 - 10i$

40. $20 + i$

41. $\dfrac{3}{4} + \dfrac{5}{3}i$

42. $-\dfrac{12}{6} + \dfrac{5}{6}i$

43. $-\dfrac{25}{2} - \dfrac{13}{4}i$

44. $\dfrac{27}{13} - \dfrac{3}{11}i$

In Exercises 45–60, express each complex number in exact rectangular form.

45. $5(\cos 180° + i\sin 180°)$

46. $2(\cos 135° + i\sin 135°)$

47. $2(\cos 315° + i\sin 315°)$

48. $3(\cos 270° + i\sin 270°)$

49. $-4(\cos 60° + i\sin 60°)$

50. $-4(\cos 210° + i\sin 210°)$

51. $\sqrt{3}(\cos 150° + i\sin 150°)$

52. $\sqrt{3}(\cos 330° + i\sin 330°)$

53. $\sqrt{2}\left[\cos\left(\dfrac{\pi}{4}\right) + i\sin\left(\dfrac{\pi}{4}\right)\right]$

54. $2\left[\cos\left(\dfrac{5\pi}{6}\right) + i\sin\left(\dfrac{5\pi}{6}\right)\right]$

55. $5\left[\cos\left(\dfrac{\pi}{3}\right) + i\sin\left(\dfrac{\pi}{3}\right)\right]$

56. $8\left[\cos\left(\dfrac{7\pi}{4}\right) + i\sin\left(\dfrac{7\pi}{4}\right)\right]$

57. $\dfrac{3}{2}\left[\cos\left(\dfrac{7\pi}{6}\right) + i\sin\left(\dfrac{7\pi}{6}\right)\right]$

58. $\dfrac{9}{5}\left[\cos\left(\dfrac{3\pi}{2}\right) + i\sin\left(\dfrac{3\pi}{2}\right)\right]$

59. $10\left[\cos\left(\dfrac{5\pi}{3}\right) + i\sin\left(\dfrac{5\pi}{3}\right)\right]$

60. $15(\cos \pi + i\sin \pi)$

In Exercises 61–72, use a calculator to express each complex number in rectangular form.

61. $5(\cos 295° + i\sin 295°)$

62. $4(\cos 35° + i\sin 35°)$

63. $3(\cos 100° + i\sin 100°)$

64. $6(\cos 250° + i\sin 250°)$

65. $-7(\cos 140° + i\sin 140°)$

66. $-5(\cos 320° + i\sin 320°)$

67. $3\left[\cos\left(\dfrac{11\pi}{12}\right) + i\sin\left(\dfrac{11\pi}{12}\right)\right]$

68. $2\left[\cos\left(\dfrac{4\pi}{7}\right) + i\sin\left(\dfrac{4\pi}{7}\right)\right]$

69. $-2\left[\cos\left(\dfrac{3\pi}{5}\right) + i\sin\left(\dfrac{3\pi}{5}\right)\right]$

70. $-4\left[\cos\left(\dfrac{15\pi}{11}\right) + i\sin\left(\dfrac{15\pi}{11}\right)\right]$

71. $6\left[\cos\left(\dfrac{\pi}{8}\right) + i\sin\left(\dfrac{\pi}{8}\right)\right]$

72. $14\left[\cos\left(\dfrac{5\pi}{12}\right) + i\sin\left(\dfrac{5\pi}{12}\right)\right]$

● **APPLICATIONS**

73. Resultant Force. Force A, at 100 pounds, and force B, at 120 pounds, make an angle of 30° with each other. Represent their respective vectors as complex numbers written in polar form, and determine the resultant force.

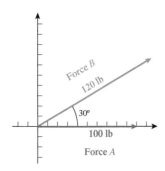

74. Resultant Force. Force A, at 40 pounds, and force B, at 50 pounds, make an angle of 45° with each other. Represent their respective vectors as complex numbers written in polar form, and determine the resultant force.

75. Resultant Force. Force A, at 80 pounds, and force B, at 150 pounds, make an angle of 30° with each other. Represent their respective vectors as complex numbers written in polar form, and determine the resultant angle.

76. Resultant Force. Force A, at 20 pounds, and force B, at 60 pounds, make an angle of 60° with each other. Represent their respective vectors as complex numbers written in polar form, and determine the resultant angle.

77. Actual Speed and True Course. An airplane is flying on a course of 285° as measured from due north at 300 mph. The wind is blowing due south at 30 mph. Represent their respective vectors as complex numbers written in polar form, and determine the resultant speed and direction vector.

78. Actual Speed and True Course. An airplane is flying on a course of 80° as measured from due north at 150 mph. The wind is blowing due north at 20 mph. Represent their respective vectors as complex numbers written in polar form, and determine the resultant speed and direction vector.

79. Boating. A boat is moving across a river at 15 mph on a bearing of N 50° W. The current is running from east to west at 5 mph. Represent their vectors as complex numbers written in polar form, and determine the resultant speed and direction vector.

80. Boating. A boat is moving across a river at 22 mph on a bearing of S 50° E. The current is running from north to south at 9 mph. Represent their vectors as complex numbers written in polar form, and determine the resultant speed and direction vector.

● **CATCH THE MISTAKE**

In Exercises 81 and 82, explain the mistake that is made.

81. Express $z = -3 - 8i$ in polar form.

Solution:

Find r. $\qquad r = \sqrt{x^2 + y^2} = \sqrt{9 + 64} = \sqrt{73}$

Find θ. $\qquad \tan\theta = \dfrac{8}{3}$

$$\theta = \tan^{-1}\left(\dfrac{8}{3}\right) \approx 69.44°$$

Write the complex number in polar form.

$$z \approx \sqrt{73}\left[\cos(69.44°) + i\sin(69.44°)\right]$$

This is incorrect. What mistake was made?

82. Express $z = -3 + 8i$ in polar form.

Solution:

Find r. $\qquad r = \sqrt{x^2 + y^2} = \sqrt{9 + 64} = \sqrt{73}$

Find θ. $\qquad \tan\theta = -\dfrac{8}{3}$

$$\theta = \tan^{-1}\left(-\dfrac{8}{3}\right) \approx -69.44°$$

Write the complex number in polar form.

$$z \approx \sqrt{73}\left[\cos(-69.44°) + i\sin(-69.44°)\right]$$

This is incorrect. What mistake was made?

● **CONCEPTUAL**

In Exercises 83–86, determine whether each statement is true or false.

83. In the complex plane, any point that lies along the horizontal axis represents a real number.

84. In the complex plane, any point that lies along the vertical axis represents an imaginary number.

85. The modulus of z and the modulus of $\bar{z}$ are equal.

86. The argument of z and the argument of $\bar{z}$ are equal.

87. Find the argument of $z = a$, where a is a positive real number.

88. Find the argument of $z = bi$, where b is a positive real number.

89. Find the modulus of $z = bi$, where b is a negative real number.

90. Find the modulus of $z = a$, where a is a negative real number.

In Exercises 91 and 92, express the complex number in polar form.

91. $a - 2ai$, where $a > 0$

92. $-3a - 4ai$, where $a > 0$

● **CHALLENGE**

93. Use identities to express the complex number
$$4\left[\cos\left(\frac{\pi}{12}\right) + i\sin\left(\frac{\pi}{12}\right)\right]$$ exactly in rectangular form.

94. Use identities to express the complex number
$$4\left[\cos\left(\frac{5\pi}{8}\right) + i\sin\left(\frac{5\pi}{8}\right)\right]$$ exactly in rectangular form.

95. Perform the given operations and then convert to polar form: $3i(2 + 4i)(3 - 2i)$.

96. Perform the given operations and then convert to polar form: $-2i^3(1 + 4i)(2 - 5i)$.

97. Let $z = 1 + 2i$. Find and graph $z^0, z^1, z^2, z^3, z^4, z^5$ on the same coordinate plane.

98. Let $z = -1 + i$. Find and graph $z^0, z^1, z^2, z^3, z^4, z^5$ on the same coordinate plane.

● **TECHNOLOGY**

For Exercises 99 and 100, use a graphing calculator to convert complex numbers from rectangular to polar form. Use the ⟨Abs⟩ command to find the modulus and the ⟨Angle⟩ command to find the angle.

99. Find abs$(1 + i)$. Find angle$(1 + i)$. Write $1 + i$ in polar form.

100. Find abs$(1 - i)$. Find angle$(1 - i)$. Write $1 - i$ in polar form.

For Exercises 101 and 102, use a graphing calculator to convert between rectangular and polar coordinates with the ⟨Pol⟩ and ⟨Rec⟩ commands.

101. Find Pol$(2, 1)$. Write $2 + i$ in polar form.

102. Find Rec$(3, 45°)$. Write $3(\cos 45° + i\sin 45°)$ in rectangular form.

8.3 PRODUCTS, QUOTIENTS, POWERS, AND ROOTS OF COMPLEX NUMBERS; DE MOIVRE'S THEOREM

SKILLS OBJECTIVES	CONCEPTUAL OBJECTIVES
■ Find the product of two complex numbers. ■ Find the quotient of two complex numbers. ■ Raise a complex number to an integer power. ■ Find the nth root of a complex number.	■ Understand that when two complex numbers are multiplied, the magnitudes are multiplied and the arguments are added. ■ Understand that when two complex numbers are divided, the magnitudes are divided and the arguments are subtracted. ■ Understand that when a complex number is raised to an integer power n, the magnitude is raised to the power n and the argument is multiplied by n. ■ We can often solve a polynomial equation by finding complex roots.

8.3.1 SKILL

Find the product of two complex numbers.

8.3.1 CONCEPTUAL

Understand that when two complex numbers are multiplied, the magnitudes are multiplied and the arguments are added.

In this section, we will multiply complex numbers, divide complex numbers, raise complex numbers to powers, and find roots of complex numbers.

8.3.1 Products of Complex Numbers

We will first derive a formula for the product of two complex numbers that are given in polar form.

WORDS	MATH
Start with two complex numbers z_1 and z_2 in polar form.	$z_1 = r_1(\cos\theta_1 + i\sin\theta_1)$ and $z_2 = r_2(\cos\theta_2 + i\sin\theta_2)$
Multiply z_1 and z_2.	$z_1 z_2 = r_1 r_2(\cos\theta_1 + i\sin\theta_1)(\cos\theta_2 + i\sin\theta_2)$
Use the FOIL method to multiply the expressions in parentheses.	$z_1 z_2 = r_1 r_2(\cos\theta_1\cos\theta_2 + i\cos\theta_1\sin\theta_2 + i\sin\theta_1\cos\theta_2 + \underset{-1}{i^2}\sin\theta_1\sin\theta_2)$
Group the real parts and the imaginary parts.	$z_1 z_2 = r_1 r_2[(\cos\theta_1\cos\theta_2 - \sin\theta_1\sin\theta_2) + i(\cos\theta_1\sin\theta_2 + \sin\theta_1\cos\theta_2)]$
Use the cosine and sine sum identities (Section 5.2).	$z_1 z_2 = r_1 r_2\left[\underset{\cos(\theta_1 + \theta_2)}{\underbrace{(\cos\theta_1\cos\theta_2 - \sin\theta_1\sin\theta_2)}} + i\underset{\sin(\theta_1 + \theta_2)}{\underbrace{(\cos\theta_1\sin\theta_2 + \sin\theta_1\cos\theta_2)}}\right]$
Simplify.	$z_1 z_2 = r_1 r_2[\cos(\theta_1 + \theta_2) + i\sin(\theta_1 + \theta_2)]$

PRODUCT OF TWO COMPLEX NUMBERS

Let $z_1 = r_1(\cos\theta_1 + i\sin\theta_1)$ and $z_2 = r_2(\cos\theta_2 + i\sin\theta_2)$ be two complex numbers. The **complex product** $z_1 z_2$ is given by

$$z_1 z_2 = r_1 r_2[\cos(\theta_1 + \theta_2) + i\sin(\theta_1 + \theta_2)]$$

In other words, *when multiplying two complex numbers, the magnitudes are multiplied and the arguments are added.*

EXAMPLE 1   **Multiplying Complex Numbers**

Find the product of $z_1 = 3(\cos 35° + i \sin 35°)$ and $z_2 = 2(\cos 10° + i \sin 10°)$.

Solution:

Set up the product.

$$z_1 z_2 = 3(\cos 35° + i \sin 35°) \cdot 2(\cos 10° + i \sin 10°)$$

Multiply the magnitudes and add the arguments.

$$z_1 z_2 = 3 \cdot 2[\cos(35° + 10°) + i \sin(35° + 10°)]$$

Simplify.

$$z_1 z_2 = 6(\cos 45° + i \sin 45°) = 6\left[\cos\left(\frac{\pi}{4}\right) + i \sin\left(\frac{\pi}{4}\right)\right]$$

The product is in polar form. To express the product in rectangular form, evaluate the trigonometric functions.

$$z_1 z_2 = 6\left[\frac{\sqrt{2}}{2} + i\frac{\sqrt{2}}{2}\right] = \boxed{3\sqrt{2} + 3i\sqrt{2}}$$

▼

YOUR TURN   Find the product of $z_1 = 2(\cos 55° + i \sin 55°)$ and $z_2 = 5(\cos 65° + i \sin 65°)$. Express the answer in both polar and rectangular form.

▼
ANSWER

$z_1 z_2 = 10(\cos 120° + i \sin 120°)$
or $z_1 z_2 = -5 + 5i\sqrt{3}$

EXAMPLE 2   **Multiplying Complex Numbers**

Find the product of $z_1 = 5\left[\cos\left(\frac{\pi}{6}\right) + i \sin\left(\frac{\pi}{6}\right)\right]$ and $z_2 = 2\left[\cos\left(\frac{\pi}{3}\right) + i \sin\left(\frac{\pi}{3}\right)\right]$.

Solution:

Set up the product.

$$z_1 z_2 = 5\left[\cos\left(\frac{\pi}{6}\right) + i \sin\left(\frac{\pi}{6}\right)\right] \cdot 2\left[\cos\left(\frac{\pi}{3}\right) + i \sin\left(\frac{\pi}{3}\right)\right]$$

Multiply the magnitudes and add the arguments.

$$z_1 z_2 = 5 \cdot 2\left[\cos\left(\frac{\pi}{6} + \frac{\pi}{3}\right) + i \sin\left(\frac{\pi}{6} + \frac{\pi}{3}\right)\right]$$

Simplify. The result is the product in polar form.

$$\boxed{z_1 z_2 = 10\left[\cos\left(\frac{\pi}{2}\right) + i \sin\left(\frac{\pi}{2}\right)\right]}$$

Evaluate the trigonometric functions.

$$z_1 z_2 = 10[0 + i(1)]$$

The result is the product in rectangular form.

$$\boxed{z_1 z_2 = 10i}$$

▼

YOUR TURN   Find the product of $z_1 = 3\left[\cos\left(\frac{\pi}{4}\right) + i \sin\left(\frac{\pi}{4}\right)\right]$ and $z_2 = 2\left[\cos\left(\frac{\pi}{2}\right) + i \sin\left(\frac{\pi}{2}\right)\right]$. Express the answer in both polar and rectangular forms.

[CONCEPT CHECK]

TRUE OR FALSE When two complex numbers are multiplied, their magnitudes and arguments are multiplied, respectively.

▼
ANSWER False

▼
ANSWER

Polar form:

$$z_1 z_2 = 6\left[\cos\left(\frac{3\pi}{4}\right) + i \sin\left(\frac{3\pi}{4}\right)\right]$$

Rectangular form:

$$z_1 z_2 = -3\sqrt{2} + 3i\sqrt{2}$$

8.3.2 Quotients of Complex Numbers

We now derive a formula for the quotient of two complex numbers.

WORDS	MATH
Start with two complex numbers z_1 and z_2, in polar form.	$z_1 = r_1(\cos\theta_1 + i\sin\theta_1)$ and $z_2 = r_2(\cos\theta_2 + i\sin\theta_2)$
Divide z_1 by z_2.	$\dfrac{z_1}{z_2} = \dfrac{r_1(\cos\theta_1 + i\sin\theta_1)}{r_2(\cos\theta_2 + i\sin\theta_2)} = \left(\dfrac{r_1}{r_2}\right)\left(\dfrac{\cos\theta_1 + i\sin\theta_1}{\cos\theta_2 + i\sin\theta_2}\right)$
Multiply the second expression in parentheses by the conjugate of the denominator, $\cos\theta_2 - i\sin\theta_2$.	$\dfrac{z_1}{z_2} = \left(\dfrac{r_1}{r_2}\right)\left(\dfrac{\cos\theta_1 + i\sin\theta_1}{\cos\theta_2 + i\sin\theta_2}\right)\left(\dfrac{\cos\theta_2 - i\sin\theta_2}{\cos\theta_2 - i\sin\theta_2}\right)$
Use the FOIL method to multiply the expressions in parentheses in the last two expressions.	$\dfrac{z_1}{z_2} = \left(\dfrac{r_1}{r_2}\right)\left(\dfrac{\cos\theta_1\cos\theta_2 - i^2\sin\theta_1\sin\theta_2 + i\sin\theta_1\cos\theta_2 - i\sin\theta_2\cos\theta_1}{\cos^2\theta_2 - i^2\sin^2\theta_2}\right)$
Substitute $i^2 = -1$ and group the real parts and the imaginary parts. Apply the Pythagorean identity to the denominator inside the brackets.	$\dfrac{z_1}{z_2} = \left(\dfrac{r_1}{r_2}\right)\left[\dfrac{(\cos\theta_1\cos\theta_2 + \sin\theta_1\sin\theta_2) + i(\sin\theta_1\cos\theta_2 - \sin\theta_2\cos\theta_1)}{\underbrace{\cos^2\theta_2 + \sin^2\theta_2}_{1}}\right]$
Simplify.	$\dfrac{z_1}{z_2} = \left(\dfrac{r_1}{r_2}\right)[(\cos\theta_1\cos\theta_2 + \sin\theta_1\sin\theta_2) + i(\sin\theta_1\cos\theta_2 - \sin\theta_2\cos\theta_1)]$
Use the cosine and sine difference identities (Section 5.2).	$\dfrac{z_1}{z_2} = \left(\dfrac{r_1}{r_2}\right)[\underbrace{(\cos\theta_1\cos\theta_2 + \sin\theta_1\sin\theta_2)}_{\cos(\theta_1 - \theta_2)} + i\underbrace{(\sin\theta_1\cos\theta_2 - \sin\theta_2\cos\theta_1)}_{\sin(\theta_1 - \theta_2)}]$
Simplify.	$z_1 z_2 = \dfrac{r_1}{r_2}[\cos(\theta_1 - \theta_2) + i\sin(\theta_1 - \theta_2)]$

8.3.2 SKILL

Find the quotient of two complex numbers.

It is important to notice that the argument of the quotient is the argument of the numerator minus the argument of the denominator.

8.3.2 CONCEPTUAL

Understand that when two complex numbers are divided, the magnitudes are divided and the arguments are subtracted.

QUOTIENT OF TWO COMPLEX NUMBERS

Let $z_1 = r_1(\cos\theta_1 + i\sin\theta_1)$ and $z_2 = r_2(\cos\theta_2 + i\sin\theta_2)$ be two complex numbers. The **complex quotient** $\dfrac{z_1}{z_2}$ is given by

$$\frac{z_1}{z_2} = \frac{r_1}{r_2}[\cos(\theta_1 - \theta_2) + i\sin(\theta_1 - \theta_2)]$$

In other words, *when dividing two complex numbers, the magnitudes are divided and the arguments are subtracted. The argument of the quotient is the argument of the complex number in the numerator minus the argument of the complex number in the denominator.*

EXAMPLE 3 **Dividing Complex Numbers**

Let $z_1 = 6(\cos 125° + i\sin 125°)$ and $z_2 = 3(\cos 65° + i\sin 65°)$. Find $\dfrac{z_1}{z_2}$.

Solution:

Set up the quotient.

$$\frac{z_1}{z_2} = \frac{6(\cos 125° + i\sin 125°)}{3(\cos 65° + i\sin 65°)}$$

Divide the magnitudes and subtract the arguments.

$$\frac{z_1}{z_2} = \frac{6}{3}[\cos(125° - 65°) + i\sin(125° - 65°)]$$

Simplify.

$$\frac{z_1}{z_2} = 2(\cos 60° + i\sin 60°)$$

The quotient is in polar form. To express the product in rectangular form, evaluate the trigonometric functions.

$$\frac{z_1}{z_2} = 2\left(\frac{1}{2} + i\frac{\sqrt{3}}{2}\right) = 1 + i\sqrt{3}$$

Polar form:

$$\boxed{\frac{z_1}{z_2} = 2(\cos 60° + i\sin 60°)}$$

Rectangular form:

$$\boxed{\frac{z_1}{z_2} = 1 + i\sqrt{3}}$$

▼

YOUR TURN Let $z_1 = 10(\cos 275° + i\sin 275°)$ and $z_2 = 5(\cos 65° + i\sin 65°)$.

Find $\dfrac{z_1}{z_2}$. Express the answer in both polar and rectangular forms.

▼
ANSWER

$\dfrac{z_1}{z_2} = 2(\cos 210° + i\sin 210°)$

or $\dfrac{z_1}{z_2} = -\sqrt{3} - i$

When multiplying or dividing complex numbers, we have considered only those values of θ such that $0° \le \theta < 360°$ or $0 \le \theta < 2\pi$. When the value of θ is negative or greater than $360°$ or 2π, find the coterminal angle in the interval $[0°, 360°)$ or $[0, 2\pi)$.

8.3.3 Powers of Complex Numbers

Raising a number to a positive integer power is the same as multiplying that number by itself repeated times.

$$x^3 = x \cdot x \cdot x \qquad (a + b)^2 = (a + b)(a + b)$$

Therefore, raising a complex number to a power that is a positive integer is the same as multiplying the complex number by itself multiple times. Let us illustrate this with the complex number $z = r(\cos\theta + i\sin\theta)$, which we will raise to positive integer powers (n).

8.3.3 SKILL

Raise a complex number to an integer power.

8.3.3 CONCEPTUAL

Understand that when a complex number is raised to an integer power n, the magnitude is raised to the power n and the argument is multiplied by n.

WORDS	**MATH**
Take the case $n = 2$.	$z^2 = [r(\cos\theta + i\sin\theta)][r(\cos\theta + i\sin\theta)]$
Apply the complex product rule (multiply the magnitudes and add the arguments).	$z^2 = r^2[\cos(2\theta) + i\sin(2\theta)]$
Take the case $n = 3$.	$z^3 = z^2 z = \{r^2[\cos(2\theta) + i\sin(2\theta)]\}[r(\cos\theta + i\sin\theta)]$
Apply the complex product rule (multiply the magnitudes and add the arguments).	$z^3 = r^3[\cos(3\theta) + i\sin(3\theta)]$
Take the case $n = 4$.	$z^4 = z^3 z = \{r^3[\cos(3\theta) + i\sin(3\theta)]\}[r(\cos\theta + i\sin\theta)]$
Apply the complex product rule (multiply the magnitudes and add the arguments).	$z^4 = r^4[\cos(4\theta) + i\sin(4\theta)]$
The pattern observed for any positive integer n is:	$z^n = r^n[\cos(n\theta) + i\sin(n\theta)]$

Although we will not prove this generalized representation of a complex number raised to a power, it was proved by Abraham De Moivre (1667–1754) and hence its name.

> **DE MOIVRE'S THEOREM**
>
> If $z = r(\cos\theta + i\sin\theta)$ is a complex number, then
>
> $$z^n = r^n[\cos(n\theta) + i\sin(n\theta)]$$
>
> where n is a positive integer.
>
> In other words, when raising a complex number to a positive integer power n, raise the magnitude to the same power n and multiply the argument by n.

[CONCEPT CHECK]

TRUE OR FALSE When a complex number is raised to a positive integer power n, then the magnitude is raised to the nth power and the argument is multiplied by n.

▼

ANSWER True

Although De Moivre's theorem has been proven for all real numbers n, we will use it only for positive integer values of n and their reciprocals (nth roots). This is a very powerful theorem. For example, if asked to find $(\sqrt{3} + i)^{10}$, you have two choices: (1) Multiply out the expression algebraically, which we will call the "long way," or (2) convert to polar coordinates and use De Moivre's theorem, which we will call the "short way." We will use De Moivre's theorem.

▶ **EXAMPLE 4** **Finding a Power of a Complex Number**

Find $(\sqrt{3} + i)^{10}$ and express the answer in rectangular form.

Solution:

Convert to polar form.	$(\sqrt{3} + i)^{10} = [2(\cos 30° + i\sin 30°)]^{10}$
Apply De Moivre's theorem with $n = 10$.	$(\sqrt{3} + i)^{10} = 2^{10}[\cos(10 \cdot 30°) + i\sin(10 \cdot 30°)]$
Simplify.	$(\sqrt{3} + i)^{10} = 2^{10}(\cos 300° + i\sin 300°)$
Evaluate 2^{10} and the sine and cosine functions.	$= 1024\left(\dfrac{1}{2} - i\dfrac{\sqrt{3}}{2}\right)$
	$\boxed{= 512 - 512i\sqrt{3}}$

[STUDY TIP]
$\sqrt{3} + i$ in polar form:
$$x = \sqrt{3} \quad y = 1$$
$$r = \sqrt{3 + 1} = 2$$
$$\tan\theta = \frac{1}{\sqrt{3}} \quad \text{or} \quad \theta = 30°.$$

▼

ANSWER

$-512 - 512i\sqrt{3}$

YOUR TURN Find $(1 + i\sqrt{3})^{10}$ and express the answer in rectangular form.

8.3.4 Roots of Complex Numbers

De Moivre's theorem is the basis for the *nth root theorem*. Before we proceed, let us motivate it with a problem: Solve $x^3 - 1 = 0$. Recall that a polynomial of degree n has n solutions (roots in the complex number system). So the polynomial $P(x) = x^3 - 1$ is of degree 3 and has three solutions (roots). We can solve it algebraically.

8.3.4 CONCEPTUAL

We can often solve a polynomial equation by finding complex roots.

WORDS	MATH
List the potential rational roots of the polynomial $P(x) = x^3 - 1$.	$x = \pm 1$
Use synthetic division to test $x = 1$.	$\begin{array}{c|cccc} 1 & 1 & 0 & 0 & -1 \\ & & 1 & 1 & 1 \\ \hline & 1 & 1 & 1 & \boxed{0} \end{array}$ $\underbrace{\qquad\qquad}_{x^2+x+1}$
Since $x = 1$ is a zero, then the polynomial can be written as a product of the linear factor $(x - 1)$ and a quadratic factor.	$P(x) = (x - 1)(x^2 + x + 1)$
Use the quadratic formula on $x^2 + x + 1 = 0$ to solve for x.	$x = \dfrac{-1 \pm \sqrt{1 - 4}}{2} = \dfrac{-1 \pm \sqrt{-3}}{2} = -\dfrac{1}{2} \pm \dfrac{i\sqrt{3}}{2}$
So the three solutions to the equation $x^3 - 1 = 0$ are	$\boxed{x = 1,\ x = -\dfrac{1}{2} + \dfrac{i\sqrt{3}}{2},\ \text{and } x = -\dfrac{1}{2} - \dfrac{i\sqrt{3}}{2}.}$

An alternative approach to solving $x^3 - 1 = 0$ is to use the *nth root theorem* to find the additional complex cube roots of 1.

Derivation of the *n*th Root Theorem

WORDS	MATH
Let z and w be complex numbers such that w is the nth root of z.	$w = z^{1/n}$ or $w = \sqrt[n]{z}$, where n is a positive integer
Raise both sides of the equation to the nth power.	$w^n = z$
Let $z = r(\cos\theta + i\sin\theta)$ and $w = s(\cos\alpha + i\sin\alpha)$.	$[s(\cos\alpha + i\sin\alpha)]^n = r(\cos\theta + i\sin\theta)$
Apply De Moivre's theorem to the left side of the equation.	$s^n[\cos(n\alpha) + i\sin(n\alpha)] = r(\cos\theta + i\sin\theta)$
For these two expressions to be equal, their magnitudes must be equal and their angles must be coterminal.	$s^n = r$ and $n\alpha = \theta + 2k\pi$, where k is any integer
Solve for s and α.	$s = r^{1/n}$ and $\alpha = \dfrac{\theta + 2k\pi}{n}$
Substitute $s = r^{1/n}$ and $\alpha = \dfrac{\theta + 2k\pi}{n}$ into $w = z^{1/n}$.	$z^{1/n} = r^{1/n}\left[\cos\left(\dfrac{\theta + 2k\pi}{n}\right) + i\sin\left(\dfrac{\theta + 2k\pi}{n}\right)\right]$

Notice that when $k = n$, the arguments $\dfrac{\theta}{n} + 2\pi$ and $\dfrac{\theta}{n}$ are coterminal. Therefore, to get distinct roots, let $k = 0, 1, \ldots, n - 1$. If we let z be a given complex number and w be any complex number that satisfies the relationship $z^{1/n} = w$ or $z = w^n$, where $n \geq 2$, then we say that w is a **complex nth root** of z.

nTH ROOT THEOREM

The **nth roots** of the complex number $z = r(\cos\theta + i\sin\theta)$ are given by

$$w_k = r^{1/n}\left[\cos\left(\frac{\theta}{n} + \frac{2k\pi}{n}\right) + i\sin\left(\frac{\theta}{n} + \frac{2k\pi}{n}\right)\right] \quad \theta \text{ in radians}$$

or

$$w_k = r^{1/n}\left[\cos\left(\frac{\theta}{n} + \frac{k \cdot 360°}{n}\right) + i\sin\left(\frac{\theta}{n} + \frac{k \cdot 360°}{n}\right)\right] \quad \theta \text{ in degrees}$$

where $k = 0, 1, 2, \ldots, n - 1$.

EXAMPLE 5 **Finding Roots of Complex Numbers**

Find the three distinct cube roots of $-4 - 4i\sqrt{3}$ and plot the roots in the complex plane.

Solution:

STEP 1 Write $-4 - 4i\sqrt{3}$ in polar form. $\qquad\qquad 8(\cos 240° + i\sin 240°)$

STEP 2 Find the three cube roots.

$$w_k = r^{1/n}\left[\cos\left(\frac{\theta}{n} + \frac{k \cdot 360°}{n}\right) + i\sin\left(\frac{\theta}{n} + \frac{k \cdot 360°}{n}\right)\right]$$

$$\theta = 240°, r = 8, n = 3, k = 0, 1, 2$$

For $k = 0$: $w_0 = 8^{1/3}\left[\cos\left(\dfrac{240°}{3} + \dfrac{0 \cdot 360°}{3}\right) + i\sin\left(\dfrac{240°}{3} + \dfrac{0 \cdot 360°}{3}\right)\right]$

Simplify. $\boxed{w_0 = 2(\cos 80° + i\sin 80°)}$

For $k = 1$: $w_1 = 8^{1/3}\left[\cos\left(\dfrac{240°}{3} + \dfrac{1 \cdot 360°}{3}\right) + i\sin\left(\dfrac{240°}{3} + \dfrac{1 \cdot 360°}{3}\right)\right]$

Simplify. $\boxed{w_1 = 2(\cos 200° + i\sin 200°)}$

For $k = 2$: $w_2 = 8^{1/3}\left[\cos\left(\dfrac{240°}{3} + \dfrac{2 \cdot 360°}{3}\right) + i\sin\left(\dfrac{240°}{3} + \dfrac{2 \cdot 360°}{3}\right)\right]$

Simplify. $\boxed{w_2 = 2(\cos 320° + i\sin 320°)}$

STEP 3 Plot the three complex cube roots in the complex plane.

Notice the following:

- The roots all have a magnitude of 2, and hence all lie on a circle of radius 2.

- The roots are equally spaced around the circle (120° apart).

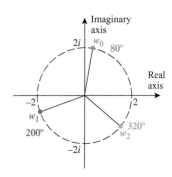

> **STUDY TIP**
>
> $-4 - 4i\sqrt{3}$ in polar form:
> $\quad x = -4 \qquad y = -4\sqrt{3}$
> $\quad r = \sqrt{16 + 48} = \sqrt{64} = 8$
> $\tan\theta = \dfrac{y}{x} = \dfrac{-4\sqrt{3}}{-4} = \sqrt{3}$
> $\theta = \tan^{-1}(\sqrt{3}) = 60°$, but the point is in quadrant III; therefore, $\theta = 60° + 180° = 240°$.
>
> *Note:*
> - The modulus of the *n*th root will always be the *n*th root of *r*.
> - The first angle will always be $\dfrac{\theta}{n}$.
> - The angles always increase by a factor of $\dfrac{2\pi}{n}$ or $\dfrac{360°}{n}$.

▼

ANSWER

$w_0 = 2(\cos 100° + i\sin 100°)$
$w_1 = 2(\cos 220° + i\sin 220°)$
$w_2 = 2(\cos 340° + i\sin 340°)$

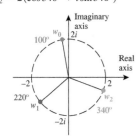

▼

YOUR TURN Find the three distinct complex cube roots of $4 - 4i\sqrt{3}$ and plot the roots in the complex plane.

Solving Equations Using Roots of Complex Numbers

Let us return to solving the equation $x^3 - 1 = 0$. We have solved this equation using known algebraic techniques; now let us solve it using the nth root theorem.

▶ **EXAMPLE 6** **Solving Equations Using Complex Roots**

Find all complex solutions to $x^3 - 1 = 0$.

Solution: $x^3 = 1$

STEP 1 Write 1 in polar form. $\qquad 1 = 1 + 0i = \cos 0° + i \sin 0°$

STEP 2 Find the three cube roots of 1.

$$w_k = r^{1/n}\left[\cos\left(\frac{\theta}{n} + \frac{k \cdot 360°}{n}\right) + i \sin\left(\frac{\theta}{n} + \frac{k \cdot 360°}{n}\right)\right]$$

$$r = 1, \theta = 0°, n = 3, k = 0, 1, 2$$

For $k = 0$: $\quad w_0 = 1^{1/3}\left[\cos\left(\frac{0°}{3} + \frac{0 \cdot 360°}{3}\right) + i \sin\left(\frac{0°}{3} + \frac{0 \cdot 360°}{3}\right)\right]$

Simplify. $\quad w_0 = \cos 0° + i \sin 0°$

For $k = 1$: $\quad w_1 = 1^{1/3}\left[\cos\left(\frac{0°}{3} + \frac{1 \cdot 360°}{3}\right) + i \sin\left(\frac{0°}{3} + \frac{1 \cdot 360°}{3}\right)\right]$

Simplify. $\quad w_1 = \cos 120° + i \sin 120°$

For $k = 2$: $\quad w_2 = 1^{1/3}\left[\cos\left(\frac{0°}{3} + \frac{2 \cdot 360°}{3}\right) + i \sin\left(\frac{0°}{3} + \frac{2 \cdot 360°}{3}\right)\right]$

Simplify. $\quad w_2 = \cos 240° + i \sin 240°$

STEP 3 Write the roots in rectangular form.

For w_0: $\qquad w_0 = \underset{1}{\underline{\cos 0°}} + \underset{0}{\underline{i \sin 0°}} = 1$

For w_1: $\qquad w_1 = \underset{-\frac{1}{2}}{\underline{\cos 120°}} + \underset{\frac{\sqrt{3}}{2}}{\underline{i \sin 120°}} = -\frac{1}{2} + i\frac{\sqrt{3}}{2}$

For w_2: $\qquad w_2 = \underset{-\frac{1}{2}}{\underline{\cos 240°}} + \underset{-\frac{\sqrt{3}}{2}}{\underline{i \sin 240°}} = -\frac{1}{2} - i\frac{\sqrt{3}}{2}$

STEP 4 Write the solutions to the equation $x^3 - 1 = 0$.

$$\boxed{x = 1} \quad \boxed{x = -\frac{1}{2} + i\frac{\sqrt{3}}{2}} \quad \boxed{x = -\frac{1}{2} - i\frac{\sqrt{3}}{2}}$$

Notice that there is one real solution and there are two (nonreal) complex solutions and that the two (nonreal) complex solutions are complex conjugates.

It is always a good idea to check that the solutions indeed satisfy the equation. The equation $x^3 - 1 = 0$ can also be written as $x^3 = 1$, so the check in this case is to cube the three solutions and confirm that the result is 1.

$$x = 1: \ 1^3 = 1 \ \checkmark$$

$$x = -\frac{1}{2} + i\frac{\sqrt{3}}{2}: \ \left(-\frac{1}{2} + i\frac{\sqrt{3}}{2}\right)^3 = \left(-\frac{1}{2} + i\frac{\sqrt{3}}{2}\right)^2 \left(-\frac{1}{2} + i\frac{\sqrt{3}}{2}\right)$$

$$= \left(-\frac{1}{2} - i\frac{\sqrt{3}}{2}\right)\left(-\frac{1}{2} + i\frac{\sqrt{3}}{2}\right)$$

$$= \frac{1}{4} + \frac{3}{4}$$

$$= 1 \ \checkmark$$

$$x = -\frac{1}{2} - i\frac{\sqrt{3}}{2}: \ \left(-\frac{1}{2} - i\frac{\sqrt{3}}{2}\right)^3 = \left(-\frac{1}{2} - i\frac{\sqrt{3}}{2}\right)^2 \left(-\frac{1}{2} - i\frac{\sqrt{3}}{2}\right)$$

$$= \left(-\frac{1}{2} + i\frac{\sqrt{3}}{2}\right)\left(-\frac{1}{2} - i\frac{\sqrt{3}}{2}\right)$$

$$= \frac{1}{4} + \frac{3}{4}$$

$$= 1 \ \checkmark$$

> **STUDY TIP**
>
> *Note:* You could use the polar form in Step 3 and De Moivre's theorem for the integer power $n = 3$ to check the nonreal complex solutions.

▶[SECTION 8.3] SUMMARY

In this section, we multiplied and divided complex numbers given in polar form and, using De Moivre's theorem, raised complex numbers to integer powers and found the nth roots of complex numbers, as follows.

Let $z_1 = r_1(\cos\theta_1 + i\sin\theta_1)$ and $z_2 = r_2(\cos\theta_2 + i\sin\theta_2)$ be two complex numbers.

The **product** $z_1 z_2$ is given by

$$z_1 z_2 = r_1 r_2[\cos(\theta_1 + \theta_2) + i\sin(\theta_1 + \theta_2)]$$

The **quotient** $\dfrac{z_1}{z_2}$ is given by

$$\frac{z_1}{z_2} = \frac{r_1}{r_2}[\cos(\theta_1 - \theta_2) + i\sin(\theta_1 - \theta_2)]$$

Let $z = r(\cos\theta + i\sin\theta)$ be a complex number. Then for a positive integer n: z raised to a **power** n is given by

$$z^n = r^n[\cos(n\theta) + i\sin(n\theta)]$$

The n **nth roots** of z are given by

$$w_k = r^{1/n}\left[\cos\left(\frac{\theta}{n} + \frac{k \cdot 360°}{n}\right) + i\sin\left(\frac{\theta}{n} + \frac{k \cdot 360°}{n}\right)\right]$$

where θ is in degrees or

$$w_k = r^{1/n}\left[\cos\left(\frac{\theta}{n} + \frac{k \cdot 2\pi}{n}\right) + i\sin\left(\frac{\theta}{n} + \frac{k \cdot 2\pi}{n}\right)\right]$$

where θ is in radians and $k = 0, 1, 2, \ldots, n - 1$.

[SECTION 8.3] EXERCISES

• **SKILLS**

In Exercises 1–20, find the product $z_1 z_2$ and express it in rectangular form.

1. $z_1 = 4(\cos 40° + i \sin 40°)$ and $z_2 = 3(\cos 80° + i \sin 80°)$

2. $z_1 = 2(\cos 100° + i \sin 100°)$ and $z_2 = 5(\cos 50° + i \sin 50°)$

3. $z_1 = 4(\cos 80° + i \sin 80°)$ and $z_2 = 2(\cos 145° + i \sin 145°)$

4. $z_1 = 3(\cos 130° + i \sin 130°)$ and $z_2 = 4(\cos 170° + i \sin 170°)$

5. $z_1 = 2(\cos 10° + i \sin 10°)$ and $z_2 = 4(\cos 80° + i \sin 80°)$

6. $z_1 = 3(\cos 190° + i \sin 190°)$ and $z_2 = 5(\cos 80° + i \sin 80°)$

7. $z_1 = 6(\cos 20° + i \sin 20°)$ and $z_2 = 8(\cos 10° + i \sin 10°)$

8. $z_1 = 5(\cos 200° + i \sin 200°)$ and $z_2 = 2(\cos 40° + i \sin 40°)$

9. $z_1 = \frac{1}{2}(\cos 280° + i \sin 280°)$ and $z_2 = \frac{2}{9}(\cos 50° + i \sin 50°)$

10. $z_1 = \frac{5}{6}(\cos 15° + i \sin 15°)$ and $z_2 = \frac{12}{5}(\cos 195° + i \sin 195°)$

11. $z_1 = \sqrt{3}\left[\cos\left(\frac{\pi}{12}\right) + i \sin\left(\frac{\pi}{12}\right)\right]$ and $z_2 = \sqrt{27}\left[\cos\left(\frac{\pi}{6}\right) + i \sin\left(\frac{\pi}{6}\right)\right]$

12. $z_1 = \sqrt{5}\left[\cos\left(\frac{\pi}{15}\right) + i \sin\left(\frac{\pi}{15}\right)\right]$ and $z_2 = \sqrt{5}\left[\cos\left(\frac{4\pi}{15}\right) + i \sin\left(\frac{4\pi}{15}\right)\right]$

13. $z_1 = 4\left[\cos\left(\frac{3\pi}{8}\right) + i \sin\left(\frac{3\pi}{8}\right)\right]$ and $z_2 = 3\left[\cos\left(\frac{\pi}{8}\right) + i \sin\left(\frac{\pi}{8}\right)\right]$

14. $z_1 = 6\left[\cos\left(\frac{2\pi}{9}\right) + i \sin\left(\frac{2\pi}{9}\right)\right]$ and $z_2 = 5\left[\cos\left(\frac{2\pi}{9}\right) + i \sin\left(\frac{2\pi}{9}\right)\right]$

15. $z_1 = 9\left[\cos\left(\frac{4\pi}{3}\right) + i \sin\left(\frac{4\pi}{3}\right)\right]$ and $z_2 = 1\left[\cos\left(\frac{\pi}{3}\right) + i \sin\left(\frac{\pi}{3}\right)\right]$

16. $z_1 = 13\left[\cos\left(\frac{3\pi}{10}\right) + i \sin\left(\frac{3\pi}{10}\right)\right]$ and $z_2 = 4\left[\cos\left(\frac{7\pi}{10}\right) + i \sin\left(\frac{7\pi}{10}\right)\right]$

17. $z_1 = 3\left[\cos\left(\frac{7\pi}{12}\right) + i \sin\left(\frac{7\pi}{12}\right)\right]$ and $z_2 = 5\left[\cos\left(\frac{\pi}{4}\right) + i \sin\left(\frac{\pi}{4}\right)\right]$

18. $z_1 = 18\left[\cos\left(\frac{\pi}{3}\right) + i \sin\left(\frac{\pi}{3}\right)\right]$ and $z_2 = 2\left[\cos\left(\frac{3\pi}{2}\right) + i \sin\left(\frac{3\pi}{2}\right)\right]$

19. $z_1 = \frac{\sqrt{7}}{2}\left[\cos\left(\frac{\pi}{2}\right) + i \sin\left(\frac{\pi}{2}\right)\right]$ and $z_2 = \frac{\sqrt{7}}{4}\left[\cos\left(\frac{\pi}{4}\right) + i \sin\left(\frac{\pi}{4}\right)\right]$

20. $z_1 = \frac{\sqrt{3}}{3}\left[\cos\left(\frac{8\pi}{7}\right) + i \sin\left(\frac{8\pi}{7}\right)\right]$ and $z_2 = \frac{\sqrt{2}}{5}\left[\cos\left(\frac{6\pi}{7}\right) + i \sin\left(\frac{6\pi}{7}\right)\right]$

In Exercises 21–40, find the quotient $\dfrac{z_1}{z_2}$ and express it in rectangular form.

21. $z_1 = 6(\cos 100° + i \sin 100°)$ and $z_2 = 2(\cos 40° + i \sin 40°)$

22. $z_1 = 8(\cos 80° + i \sin 80°)$ and $z_2 = 2(\cos 35° + i \sin 35°)$

23. $z_1 = 10(\cos 200° + i \sin 200°)$ and $z_2 = 5(\cos 65° + i \sin 65°)$

24. $z_1 = 4(\cos 280° + i \sin 280°)$ and $z_2 = 4(\cos 55° + i \sin 55°)$

25. $z_1 = \sqrt{12}(\cos 350° + i \sin 350°)$ and $z_2 = \sqrt{3}(\cos 80° + i \sin 80°)$

26. $z_1 = \sqrt{40}(\cos 110° + i\sin 110°)$ and $z_2 = \sqrt{10}(\cos 20° + i\sin 20°)$

27. $z_1 = 2(\cos 213° + i\sin 213°)$ and $z_2 = 4(\cos 33° + i\sin 33°)$

28. $z_1 = 12(\cos 315° + i\sin 315°)$ and $z_2 = 3(\cos 15° + i\sin 15°)$

29. $z_1 = \frac{3}{5}(\cos 295° + i\sin 295°)$ and $z_2 = \frac{4}{10}(\cos 55° + i\sin 55°)$

30. $z_1 = \frac{2}{3}(\cos 355° + i\sin 355°)$ and $z_2 = \frac{8}{9}(\cos 235° + i\sin 235°)$

31. $z_1 = 9\left[\cos\left(\frac{5\pi}{12}\right) + i\sin\left(\frac{5\pi}{12}\right)\right]$ and $z_2 = 3\left[\cos\left(\frac{\pi}{12}\right) + i\sin\left(\frac{\pi}{12}\right)\right]$

32. $z_1 = 8\left[\cos\left(\frac{5\pi}{8}\right) + i\sin\left(\frac{5\pi}{8}\right)\right]$ and $z_2 = 4\left[\cos\left(\frac{3\pi}{8}\right) + i\sin\left(\frac{3\pi}{8}\right)\right]$

33. $z_1 = 45\left[\cos\left(\frac{22\pi}{15}\right) + i\sin\left(\frac{22\pi}{15}\right)\right]$ and $z_2 = 9\left[\cos\left(\frac{2\pi}{15}\right) + i\sin\left(\frac{2\pi}{15}\right)\right]$

34. $z_1 = 22\left[\cos\left(\frac{11\pi}{18}\right) + i\sin\left(\frac{11\pi}{18}\right)\right]$ and $z_2 = 11\left[\cos\left(\frac{5\pi}{18}\right) + i\sin\left(\frac{5\pi}{18}\right)\right]$

35. $z_1 = 25\left[\cos\left(\frac{13\pi}{9}\right) + i\sin\left(\frac{13\pi}{9}\right)\right]$ and $z_2 = 5\left[\cos\left(\frac{7\pi}{9}\right) + i\sin\left(\frac{7\pi}{9}\right)\right]$

36. $z_1 = 30\left[\cos\left(\frac{3\pi}{2}\right) + i\sin\left(\frac{3\pi}{2}\right)\right]$ and $z_2 = 15\left[\cos\left(\frac{\pi}{6}\right) + i\sin\left(\frac{\pi}{6}\right)\right]$

37. $z_1 = \frac{1}{2}\left[\cos\left(\frac{17\pi}{12}\right) + i\sin\left(\frac{17\pi}{12}\right)\right]$ and $z_2 = \frac{3}{8}\left[\cos\left(\frac{\pi}{4}\right) + i\sin\left(\frac{\pi}{4}\right)\right]$

38. $z_1 = \frac{5}{12}\left[\cos\left(\frac{61\pi}{36}\right) + i\sin\left(\frac{61\pi}{36}\right)\right]$ and $z_2 = \frac{1}{8}\left[\cos\left(\frac{31\pi}{36}\right) + i\sin\left(\frac{31\pi}{36}\right)\right]$

39. $z_1 = \frac{15}{2}\left[\cos\left(\frac{35\pi}{18}\right) + i\sin\left(\frac{35\pi}{18}\right)\right]$ and $z_2 = \frac{25}{4}\left[\cos\left(\frac{5\pi}{18}\right) + i\sin\left(\frac{5\pi}{18}\right)\right]$

40. $z_1 = \frac{7}{12}\left[\cos\left(\frac{13\pi}{12}\right) + i\sin\left(\frac{13\pi}{12}\right)\right]$ and $z_2 = \frac{14}{3}\left[\cos\left(\frac{\pi}{12}\right) + i\sin\left(\frac{\pi}{12}\right)\right]$

In Exercises 41–50, evaluate each expression using De Moivre's theorem. Write the answer in rectangular form.

41. $(-1 + i)^5$ **42.** $(1 - i)^4$ **43.** $(-\sqrt{3} + i)^6$ **44.** $(\sqrt{3} - i)^8$ **45.** $(1 - \sqrt{3}i)^4$ **46.** $(-1 + \sqrt{3}i)^5$

47. $(4 - 4i)^8$ **48.** $(-3 + 3i)^{10}$ **49.** $(4\sqrt{3} + 4i)^7$ **50.** $(-5 + 5\sqrt{3}i)^7$

In Exercises 51–62, find all nth roots of z. Write the answers in polar form, and plot the roots in the complex plane.

51. $2 - 2i\sqrt{3}, n = 2$ **52.** $2 + 2i\sqrt{3}, n = 2$ **53.** $3\sqrt{2} - 3i\sqrt{2}, n = 2$ **54.** $-\sqrt{2} + i\sqrt{2}, n = 2$

55. $4 + 4i\sqrt{3}, n = 3$ **56.** $-\frac{27}{2} + \frac{27\sqrt{3}}{2}i, n = 3$ **57.** $\sqrt{3} - i, n = 3$ **58.** $4\sqrt{2} + 4i\sqrt{2}, n = 3$

59. $8\sqrt{2} - 8i\sqrt{2}, n = 4$ **60.** $-8\sqrt{2} + 8i\sqrt{2}, n = 4$ **61.** $10\sqrt{3} - 10i, n = 4$ **62.** $-5 - 5i\sqrt{3}, n = 4$

In Exercises 63–74, find all complex solutions to the given equations.

63. $x^4 - 16 = 0$ **64.** $x^3 - 8 = 0$ **65.** $x^3 + 8 = 0$ **66.** $x^3 + 1 = 0$

67. $x^4 + 16 = 0$ **68.** $x^6 + 1 = 0$ **69.** $x^6 - 1 = 0$ **70.** $4x^2 + 1 = 0$

71. $x^2 + i = 0$ **72.** $x^2 - i = 0$ **73.** $x^3 + 8i = 0$ **74.** $x^3 - 8i = 0$

• **APPLICATIONS**

75. Complex Pentagon. When you graph the five fifth roots of $-\dfrac{\sqrt{2}}{2} - \dfrac{\sqrt{2}}{2}i$ and connect the points around the circle of radius r, you form a pentagon. Find the roots and draw the pentagon.

76. Complex Square. When you graph the four fourth roots of $16i$ and connect the points around the circle of radius r, you form a square. Find the roots and draw the square.

77. Hexagon. Compute the six sixth roots of $\dfrac{1}{2} - \dfrac{\sqrt{3}}{2}i$, and form a hexagon by connecting successive roots.

78. Octagon. Compute the eight eighth roots of $2i$, and form an octagon by connecting successive roots.

• **CATCH THE MISTAKE**

In Exercises 79–82, explain the mistake that is made.

79. Let $z_1 = 6(\cos 65° + i \sin 65°)$ and $z_2 = 3(\cos 125° + i \sin 125°)$. Find $\dfrac{z_1}{z_2}$.

Solution:

Use the quotient formula.
$$\frac{z_1}{z_2} = \frac{r_1}{r_2}[\cos(\theta_1 - \theta_2) + i\sin(\theta_1 - \theta_2)]$$

Substitute values.
$$\frac{z_1}{z_2} = \frac{6}{3}[\cos(125° - 65°) + i\sin(125° - 65°)]$$

Simplify. $\quad \dfrac{z_1}{z_2} = 2(\cos 60° + i\sin 60°)$

Evaluate the trigonometric functions.
$$\frac{z_1}{z_2} = 2\left(\frac{1}{2} + i\frac{\sqrt{3}}{2}\right) = 1 + i\sqrt{3}$$

This is incorrect. What mistake was made?

80. Let $z_1 = 6(\cos 65° + i \sin 65°)$ and $z_2 = 3(\cos 125° + i \sin 125°)$. Find $z_1 z_2$.

Solution:

Write the product.
$$z_1 z_2 = 6(\cos 65° + i\sin 65°) \cdot 3(\cos 125° + i\sin 125°)$$

Multiply the magnitudes.
$$z_1 z_2 = 18(\cos 65° + i\sin 65°)(\cos 125° + i\sin 125°)$$

Multiply the cosine terms and sine terms (add arguments).
$$z_1 z_2 = 18[\cos(65° + 125°) + i^2\sin(65° + 125°)]$$

Simplify $(i^2 = -1)$.
$$z_1 z_2 = 18(\cos 190° - \sin 190°)$$

This is incorrect. What mistake was made?

81. Find $\left(\sqrt{2} + i\sqrt{2}\right)^6$.

Solution:

Raise each term to the sixth power. $\quad (\sqrt{2})^6 + i^6(\sqrt{2})^6$

Simplify. $\quad 8 + 8i^6$

Let $i^6 = i^4 \cdot i^2 = -1$. $\quad 8 - 8 = 0$

This is incorrect. What mistake was made?

82. Find all complex solutions to $x^5 - 1 = 0$.

Solution:

Add 1 to both sides. $\quad x^5 = 1$

Raise both sides to the fifth power. $\quad x = 1^{1/5}$

Simplify. $\quad x = 1$

This is incorrect. What mistake was made?

• **CONCEPTUAL**

In Exercises 83 and 84, determine whether each statement is true or false.

83. The product of two complex numbers is a complex number.

84. The quotient of two complex numbers is a complex number.

85. Find the square roots of the complex number in standard form for $n + ni\sqrt{3}$, where n is a positive integer.

86. Find the square roots of the complex number in standard form for $n - ni\sqrt{3}$, where n is a positive integer.

• **CHALLENGE**

In Exercises 87–90, use the following identity:

There is an identity you will see in calculus called Euler's formula, or identity $e^{i\theta} = \cos\theta + i\sin\theta$. Notice that when $\theta = \pi$, the identity can be written as $e^{i\pi} + 1 = 0$, which is a beautiful identity in that it relates the fundamental numbers $(e, \pi, 1, \text{ and } 0)$ and fundamental operations (multiplication, addition, exponents, and equality) in mathematics.

87. Let $z_1 = r_1(\cos\theta_1 + i\sin\theta_1) = r_1e^{i\theta_1}$ and $z_2 = r_2(\cos\theta_2 + i\sin\theta_2) = r_2e^{i\theta_2}$ be two complex numbers, and use the properties of exponents to show that $z_1z_2 = r_1r_2[\cos(\theta_1 + \theta_2) + i\sin(\theta_1 + \theta_2)]$.

88. Let $z_1 = r_1(\cos\theta_1 + i\sin\theta_1) = r_1e^{i\theta_1}$ and $z_2 = r_2(\cos\theta_2 + i\sin\theta_2) = r_2e^{i\theta_2}$ be two complex numbers, and use the properties of exponents to show that $\dfrac{z_1}{z_2} = \dfrac{r_1}{r_2}[\cos(\theta_1 - \theta_2) + i\sin(\theta_1 - \theta_2)]$.

89. Let $z = r(\cos\theta + i\sin\theta) = re^{i\theta}$, and use properties of exponents to show that $z^n = r^n[\cos(n\theta) + i\sin(n\theta)]$.

90. Let $z = r(\cos\theta + i\sin\theta) = re^{i\theta}$, and use properties of exponents to show that
$$w_k = r^{1/n}\left[\cos\left(\frac{\theta}{n} + \frac{2k\pi}{n}\right) + i\sin\left(\frac{\theta}{n} + \frac{2k\pi}{n}\right)\right].$$

• **TECHNOLOGY**

91. Find the five fifth roots of $\dfrac{\sqrt{3}}{2} - \dfrac{1}{2}i$ and use a graphing utility to plot the roots.

92. Find the four fourth roots of $-\dfrac{\sqrt{2}}{2} + \dfrac{\sqrt{2}}{2}i$ and use a graphing utility to plot the roots.

93. Complex hexagon. Find the six sixth roots of $-\dfrac{1}{2} - \dfrac{\sqrt{3}}{2}i$ and use a graphing utility to draw the hexagon by connecting the points around the circle of radius r.

94. Complex pentagon. Find the five fifth roots of $-4 + 4i$ and use a graphing utility to draw the pentagon by connecting the points around the circle of radius r.

8.4 POLAR EQUATIONS AND GRAPHS

SKILLS OBJECTIVES	CONCEPTUAL OBJECTIVES
■ Plot points in the polar coordinate system. ■ Convert between rectangular and polar coordinates. ■ Graph polar equations.	■ Understand that the name of a point (r, θ) in the polar coordinate system is not unique. ■ Relate the rectangular coordinate system to the polar coordinate system. ■ Classify common shapes that arise from plotting certain types of polar equations.

We have discussed the rectangular and polar (trigonometric) forms of complex numbers in the complex plane. We now turn our attention back to the familiar Cartesian plane, where the horizontal axis represents the x-variable and the vertical axis represents the y-variable and points in this plane represent pairs of real numbers. It is often convenient to instead represent real-number plots in the *polar coordinate system*.

8.4.1 Polar Coordinates

The **polar coordinate system** is anchored by a point, called the **pole** (taken to be the **origin**), and a ray with its endpoint at the pole, called the **polar axis**. The polar axis is normally shown where we expect to find the positive x-axis in the Cartesian plane.

If you align the pole with the origin on the rectangular graph and the polar axis with the positive x-axis, you can label a point either with rectangular coordinates (x, y) or with an ordered pair (r, θ) in **polar coordinates**.

Typically, polar graph paper is used that gives the angles and radii. The graph shown below in the margin gives the angles in radians (the angles also can be given in degrees) and shows the radii from 0 through 5.

When plotting points in the polar coordinate system, $|r|$ represents the distance from the origin to the point. The following procedure guides us in plotting points in the polar coordinate system.

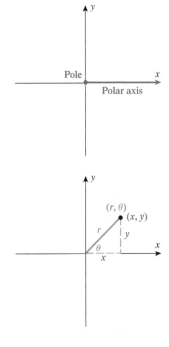

POINT-PLOTTING POLAR COORDINATES

To plot a point (r, θ):

1. Start on the polar axis and rotate the terminal side of an angle to the value θ.
2. If $r > 0$, the point is r units from the origin in the *same direction* of the terminal side of θ.
3. If $r < 0$, the point is $|r|$ units from the origin in the *opposite direction* of the terminal side of θ.

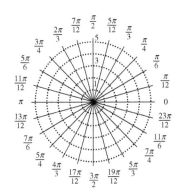

In polar form, it is important to note that (r, θ), the name of the point, is not unique, whereas in rectangular form, (x, y), it is unique. For example, $(2, 30°) = (-2, 210°)$.

8.4.1 SKILL

Plot points in the polar coordinate system.

8.4.1 CONCEPTUAL

Understand that the name of a point (r, θ) in the polar coordinate system is not unique.

[CONCEPT CHECK]

Match the following:
(1) (x, y)
(2) (r, θ)
(A) not unique
(B) unique

▼ ....

ANSWER 1. B   2. A

▼ ....

ANSWER

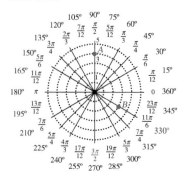

EXAMPLE 1   Plotting Points in the Polar Coordinate System

Plot the following points in the polar coordinate system.

a. $\left(3, \dfrac{3\pi}{4}\right)$ **b.** $(-2, 60°)$

Solution (a):

Start by placing a pencil along the polar axis (positive x-axis).

Rotate the pencil to the angle $\dfrac{3\pi}{4}$.

Go out (in the direction of the pencil) three units.

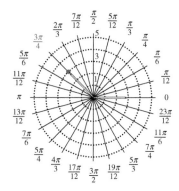

Solution (b):

Start by placing a pencil along the polar axis.

Rotate the pencil to the angle $60°$.

Go out (opposite the direction of the pencil) two units.

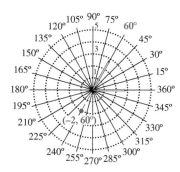

▼ ....

YOUR TURN Plot the following points in the polar coordinate system.

a. $\left(-4, \dfrac{3\pi}{2}\right)$ **b.** $(3, 330°)$

8.4.2 Converting Between Polar and Rectangular Coordinates

8.4.2 SKILL

Convert between rectangular and polar coordinates.

8.4.2 CONCEPTUAL

Relate the rectangular coordinate system to the polar coordinate system.

The relationships between polar and rectangular coordinates are the familiar relationships:

$$\sin\theta = \frac{y}{r} \quad (\text{or } y = r\sin\theta)$$

$$r^2 = x^2 + y^2$$

$$\cos\theta = \frac{x}{r} \quad (\text{or } x = r\cos\theta)$$

$$\tan\theta = \frac{y}{x} \quad (x \neq 0)$$

If $x = 0$, $\theta = \dfrac{\pi}{2}$ or $\dfrac{3\pi}{2}$ according to the sign of y.

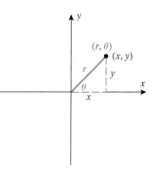

CONVERTING BETWEEN POLAR AND RECTANGULAR COORDINATES

FROM	TO	IDENTITIES
Polar (r, θ)	Rectangular (x, y)	$x = r\cos\theta \qquad y = r\sin\theta$
Rectangular (x, y)	Polar (r, θ)	$r = \sqrt{x^2 + y^2} \qquad \tan\theta = \dfrac{y}{x}, x \neq 0$ Make sure that θ is in the correct quadrant.

EXAMPLE 2 **Converting Between Polar and Rectangular Coordinates**

a. Convert the rectangular coordinate $(-1, \sqrt{3})$ to polar coordinates.
b. Convert the polar coordinate $(6\sqrt{2}, 135°)$ to rectangular coordinates.

Solution (a): $(-1, \sqrt{3})$ lies in quadrant II.

Identify x and y. $\qquad\qquad x = -1 \qquad y = \sqrt{3}$

Find r. $\qquad\qquad r = \sqrt{x^2 + y^2} = \sqrt{(-1)^2 + (\sqrt{3})^2} = \sqrt{4} = 2$

Find θ. $\qquad\qquad \tan\theta = \dfrac{\sqrt{3}}{-1} \qquad \theta$ lies in quadrant II

Identify θ from the unit circle. $\qquad \theta = \dfrac{2\pi}{3}$

Write the point in polar coordinates. $\qquad \boxed{\left(2, \dfrac{2\pi}{3}\right)}$

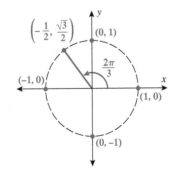

Note: Other polar coordinates like $\left(2, -\dfrac{4\pi}{3}\right)$ and $\left(-2, \dfrac{5\pi}{3}\right)$ also correspond to the point $(-1, \sqrt{3})$.

Solution (b): $(6\sqrt{2}, 135°)$ lies in quadrant II.

Identify r and θ. $\qquad\qquad r = 6\sqrt{2} \qquad \theta = 135°$

Find x. $\qquad\qquad x = r\cos\theta = 6\sqrt{2}\cos 135° = 6\sqrt{2}\left(-\dfrac{\sqrt{2}}{2}\right) = -6$

Find y. $\qquad\qquad y = r\sin\theta = 6\sqrt{2}\sin 135° = 6\sqrt{2}\left(\dfrac{\sqrt{2}}{2}\right) = 6$

Write the point in rectangular coordinates. $\qquad \boxed{(-6, 6)}$

8.4.3 Graphs of Polar Equations

8.4.3 **SKILL**

Graph polar equations.

8.4.3 **CONCEPTUAL**

Classify common shapes that arise from plotting certain types of polar coordinates.

We are familiar with equations in rectangular form such as

$$y = 3x + 5 \qquad y = x^2 + 2 \qquad x^2 + y^2 = 9$$
$$\text{(line)} \qquad\qquad \text{(parabola)} \qquad\qquad \text{(circle)}$$

We now discuss equations in polar form (known as **polar equations**) such as

$$r = 5\theta \qquad r = 2\cos\theta \qquad r = \sin(5\theta)$$

which you will learn to recognize in this section as typical polar equations whose plots are examples of some general shapes.

Our first example deals with two of the simplest forms of polar equations: when r or θ is constant. The results are a circle centered at the origin and a line that passes through the origin, respectively.

[CONCEPT CHECK]

Match the polar equations with their graphs:
(1) r = constant
(2) θ = constant
(3) r = constant $\cdot\ \theta$
(A) Spiral
(B) Circle
(C) Line

▼ ·····························

ANSWER 1. B 2. C 3. A

EXAMPLE 3 **Graphing a Polar Equation of the Form**
r = Constant or θ = Constant

Graph the polar equations.

a. $r = 3$ **b.** $\theta = \dfrac{\pi}{4}$

Solution (a): Constant value of r

Approach 1: (polar coordinates) $r = 3$ (θ can take on any value).

Plot points for arbitrary θ and $r = 3$.

Connect the points; they make a circle with radius 3.

Approach 2: (rectangular coordinates) $r = 3$

Square both sides. $r^2 = 9$

Remember that in rectangular coordinates $r^2 = x^2 + y^2$. $x^2 + y^2 = 3^2$

This is a circle, centered at the origin, with radius 3.

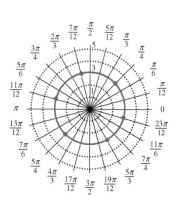

Solution (b): Constant value of θ

Approach 1: $\theta = \dfrac{\pi}{4}$ (r can take on any value, positive or negative).

Plot points for $\theta = \dfrac{\pi}{4}$ at several arbitrary values of r.

Connect the points. The result is a line passing through the origin with slope $= 1\left[m = \tan\left(\dfrac{\pi}{4}\right)\right]$.

Approach 2: $\theta = \dfrac{\pi}{4}$

Take the tangent of both sides. $\tan\theta = \underset{1}{\underline{\tan\left(\dfrac{\pi}{4}\right)}}$

Use the identity $\tan\theta = \dfrac{y}{x}$. $\dfrac{y}{x} = 1$

Multiply by x. $y = x$

The result is a line passing through the origin with slope $= 1$.

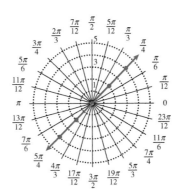

Rectangular equations that depend on varying (not constant) values of x or y can be graphed by point-plotting (making a table and plotting the points). We will use this same procedure for graphing polar equations that depend on varying (not constant) values of r and θ.

 **EXAMPLE 4** **Graphing a Polar Equation of the Form**
$r = C \cdot \cos\theta$ or $r = C \cdot \sin\theta$

Graph $r = 4\cos\theta$.

Solution:

STEP 1 Make a table and find several key values.

θ	$r = 4\cos\theta$	(r, θ)
0	$4(1) = 4$	$(4, 0)$
$\dfrac{\pi}{4}$	$4\left(\dfrac{\sqrt{2}}{2}\right) \approx 2.8$	$\left(2.8, \dfrac{\pi}{4}\right)$
$\dfrac{\pi}{2}$	$4(0) = 0$	$\left(0, \dfrac{\pi}{2}\right)$
$\dfrac{3\pi}{4}$	$4\left(-\dfrac{\sqrt{2}}{2}\right) \approx -2.8$	$\left(-2.8, \dfrac{3\pi}{4}\right)$
π	$4(-1) = -4$	$(-4, \pi)$
$\dfrac{5\pi}{4}$	$4\left(-\dfrac{\sqrt{2}}{2}\right) \approx -2.8$	$\left(-2.8, \dfrac{5\pi}{4}\right)$
$\dfrac{3\pi}{2}$	$4(0) = 0$	$\left(0, \dfrac{3\pi}{2}\right)$
$\dfrac{7\pi}{4}$	$4\left(\dfrac{\sqrt{2}}{2}\right) \approx 2.8$	$\left(2.8, \dfrac{7\pi}{4}\right)$
2π	$4(1) = 4$	$(4, 2\pi)$

STEP 2 Plot the points in polar coordinates.

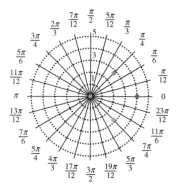

> **STUDY TIP**
>
> Graphs of $r = a\sin\theta$ and $r = a\cos\theta$ are circles.

STEP 3 Connect the points with a smooth curve.

Notice that $(4, 0)$ and $(-4, \pi)$ correspond to the same point. There is no need to continue with angles beyond π, as the result would be to go around the same circle again.

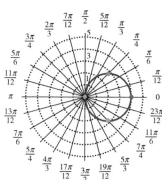

▼
ANSWER

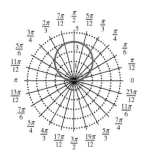

▼

YOUR TURN Graph $r = 4\sin\theta$.

Compare the result of Example 4, the graph of $r = 4\cos\theta$, with the result of Your Turn, the graph of $r = 4\sin\theta$. Notice that they are 90° out of phase (we simply rotate one graph 90° about the pole to get the other graph).

In general, graphs of polar equations of the form $r = a\sin\theta$ and $r = a\cos\theta$ are circles.

WORDS	MATH	
Polar equation	$r = a\sin\theta$	$r = a\cos\theta$
Use trigonometric ratios: $\sin\theta = \dfrac{y}{r}$ and $\cos\theta = \dfrac{x}{r}$.	$r = a\dfrac{y}{r}$	$r = a\dfrac{x}{r}$
Multiply equations by r.	$r^2 = ay$	$r^2 = ax$
Let $r^2 = x^2 + y^2$.	$x^2 + y^2 = ay$	$x^2 + y^2 = ax$
Group x terms together and y terms together.	$x^2 + (y^2 - ay) = 0$	$(x^2 - ax) + y^2 = 0$
Complete the square on the expressions in parentheses.	$x^2 + \left(y^2 - ay + \left(\dfrac{a}{2}\right)^2\right) = \left(\dfrac{a}{2}\right)^2$	$\left(x^2 - ax + \left(\dfrac{a}{2}\right)^2\right) + y^2 = \left(\dfrac{a}{2}\right)^2$
	$x^2 + \left(y - \dfrac{a}{2}\right)^2 = \left(\dfrac{a}{2}\right)^2$	$\left(x - \dfrac{a}{2}\right)^2 + y^2 = \left(\dfrac{a}{2}\right)^2$
The result is a graph of a circle.	Center: $\left(0, \dfrac{a}{2}\right)$ Radius: $\dfrac{a}{2}$	Center: $\left(\dfrac{a}{2}, 0\right)$ Radius: $\dfrac{a}{2}$

EXAMPLE 5   Graphing a Polar Equation of the Form $r = C \cdot \sin(2\theta)$ or $r = C \cdot \cos(2\theta)$

Graph $r = 5\sin(2\theta)$.

Solution:

STEP 1 Make a table and find key values. Since the argument of the sine function is doubled, the period is halved. Therefore, instead of steps of $\dfrac{\pi}{4}$, take steps of $\dfrac{\pi}{8}$.

θ	$r = 5\sin(2\theta)$	(r, θ)
0	$5(0) = 0$	$(0, 0)$
$\dfrac{\pi}{8}$	$5\left(\dfrac{\sqrt{2}}{2}\right) \approx 3.5$	$\left(3.5, \dfrac{\pi}{8}\right)$
$\dfrac{\pi}{4}$	$5(1) = 5$	$\left(5, \dfrac{\pi}{4}\right)$
$\dfrac{3\pi}{8}$	$5\left(\dfrac{\sqrt{2}}{2}\right) \approx 3.5$	$\left(3.5, \dfrac{3\pi}{8}\right)$
$\dfrac{\pi}{2}$	$5(0) = 0$	$\left(0, \dfrac{\pi}{2}\right)$

STEP 2 Label the polar coordinates.

These values in the table represent what happens in quadrant I. The same pattern repeats in the other three quadrants. The result is a **four-leaf rose**.

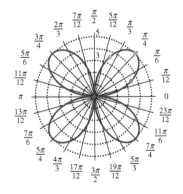

▼
ANSWER

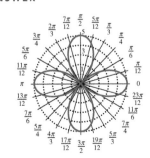

STEP 3 Connect the points in each quadrant with a smooth curve.

▼
YOUR TURN Graph $r = 5\cos(2\theta)$.

Compare the result of Example 5, the graph of $r = 5\sin(2\theta)$, with the result of Your Turn, the graph of $r = 5\cos(2\theta)$. Notice that they are 45° out of phase (we rotate one graph 45° about the pole to get the other graph).

In general, for $r = a\sin(n\theta)$ or $r = a\cos(n\theta)$, the graph is a rose with n leaves (petals) if n is odd and $2n$ leaves if n is even. As r increases, the leaves (petals) get longer.

The next class of graphs, so-called **limaçons**, have equations of the form $r = a \pm b\cos\theta$ or $r = a \pm b\sin\theta$. When $a = b$, the result is a **cardioid** (heart shape).

EXAMPLE 6 **The Cardioid as a Polar Equation**

Graph $r = 2 + 2\cos\theta$.

Solution:

STEP 1 Make a table and find key values.

This behavior repeats in quadrant III and quadrant IV because the cosine function has corresponding values in quadrant I and quadrant IV and in quadrant II and quadrant III.

θ	$r = 2 + 2\cos\theta$	(r, θ)
0	$2 + 2(1) = 4$	$(4, 0)$
$\dfrac{\pi}{4}$	$2 + 2\left(\dfrac{\sqrt{2}}{2}\right) \approx 3.4$	$\left(3.4, \dfrac{\pi}{4}\right)$
$\dfrac{\pi}{2}$	$2 + 2(0) = 2$	$\left(2, \dfrac{\pi}{2}\right)$
$\dfrac{3\pi}{4}$	$2 + 2\left(-\dfrac{\sqrt{2}}{2}\right) \approx 0.6$	$\left(0.6, \dfrac{3\pi}{4}\right)$
π	$2 + 2(-1) = 0$	$(0, \pi)$

STEP 2 Plot the points in polar coordinates.

STEP 3 Connect the points with a smooth curve. The curve is a *cardioid*, a term formed from Greek roots meaning "heart-shaped."

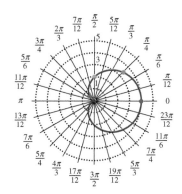

EXAMPLE 7 **Graphing a Polar Equation of the Form $r = C \cdot \theta$**

Graph $r = 0.5\theta$

Solution:

STEP 1 Make a table and find key values.

θ	$r = 0.50\theta$	(r, θ)
0	$0.5(0) = 0$	$(0, 0)$
$\dfrac{\pi}{2}$	$0.5\left(\dfrac{\pi}{2}\right) \approx 0.8$	$\left(0.8, \dfrac{\pi}{2}\right)$
π	$0.5(\pi) \approx 1.6$	$(1.6, \pi)$
$\dfrac{3\pi}{2}$	$0.5\left(\dfrac{3\pi}{2}\right) \approx 2.4$	$\left(2.4, \dfrac{3\pi}{2}\right)$
π	$0.5(2\pi) \approx 3.1$	$(3.1, 2\pi)$

STEP 2 Plot the points in polar coordinates.

STEP 3 Connect the points with a smooth curve starting at $\theta = 0$.

The curve is a *spiral*.

Notice that the larger θ gets, the larger r gets, creating a spiral about the origin.

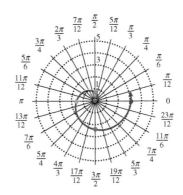

EXAMPLE 8 Graphing a Polar Equation of the Form
$r^2 = C \cdot \cos(2\theta)$ or $r^2 = C \cdot \sin(2\theta)$

Graph $r^2 = 4\cos(2\theta)$.

Solution:

STEP 1 Make a table and find key values.

Solving for r yields $r = \pm 2\sqrt{\cos(2\theta)}$. All coordinates $(-r, \theta)$ can be expressed as $(r, \theta + \pi)$. The following table does not have values for $\dfrac{\pi}{4} < \theta < \dfrac{3\pi}{4}$ because the corresponding values of $\cos(2\theta)$ are negative, and hence r is an imaginary number. The table also does not have values for $\theta > \pi$ because $2\theta > 2\pi$, and the corresponding points are repeated.

θ	$\cos(2\theta)$	$r = \pm 2\sqrt{\cos(2\theta)}$	(r, θ)
0	1	$r = \pm 2$	$(2, 0)$ and $(-2, 0) = (2, \pi)$
$\dfrac{\pi}{6}$	0.5	$r = \pm 1.4$	$\left(1.4, \dfrac{\pi}{6}\right)$ and $\left(-1.4, \dfrac{\pi}{6}\right) = \left(1.4, \dfrac{7\pi}{6}\right)$
$\dfrac{\pi}{4}$	0	$r = 0$	$\left(0, \dfrac{\pi}{4}\right)$
$\dfrac{3\pi}{4}$	0	$r = 0$	$\left(0, \dfrac{3\pi}{4}\right)$
$\dfrac{5\pi}{6}$	0.5	$r = \pm 1.4$	$\left(1.4, \dfrac{5\pi}{6}\right)$ and $\left(-1.4, \dfrac{5\pi}{6}\right) = \left(1.4, \dfrac{11\pi}{6}\right)$
π	1	$r = \pm 2$	$(2, \pi)$ and $(-2, \pi) = (2, 2\pi)$

STEP 2 Plot the points in polar coordinates.

STEP 3 Connect the points with a smooth curve. The resulting curve is known as a *lemniscate*.

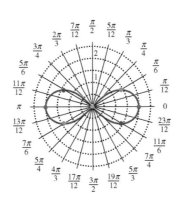

Converting Equations Between Polar and Rectangular Form

It is not always advantageous to plot an equation in the form in which it is given. It is sometimes easier to first convert to rectangular form and then plot. For example, to plot $r = \dfrac{2}{\cos\theta + \sin\theta}$, we could make a table with values. However, as you will see in Example 9, it is much easier to convert this equation to rectangular coordinates.

▶ **EXAMPLE 9** **Converting an Equation from Polar to Rectangular Form**

Graph $r = \dfrac{2}{\cos\theta + \sin\theta}$.

Solution:

Multiply the equation by $\cos\theta + \sin\theta$.

$$r(\cos\theta + \sin\theta) = 2$$

Eliminate parentheses.

$$r\cos\theta + r\sin\theta = 2$$

Convert the result to rectangular form.

$$\underset{x}{\underline{r\cos\theta}} + \underset{y}{\underline{r\sin\theta}} = 2$$

Simplify. The result is a straight line.

$$\boxed{y = -x + 2}$$

Graph the line.

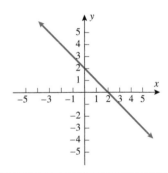

▼

ANSWER

$y = x - 2$

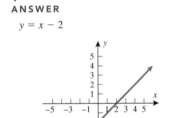

▼

YOUR TURN Graph $r = \dfrac{2}{\cos\theta - \sin\theta}$.

▶[SECTION 8.4] SUMMARY

Polar coordinates (r, θ) are graphed in the polar coordinate system by first rotating a ray from the positive x-axis position (polar axis) to get the terminal side of the angle. Then, if r is positive, go out r units from the origin in the direction of the resulting ray, or, if r is negative, go out $|r|$ units in the opposite direction of the angle. Conversions between polar and rectangular forms are given by

FROM	TO	IDENTITIES
Polar (r, θ)	Rectangular (x, y)	$x = r\cos\theta \qquad y = r\sin\theta$
Rectangular (x, y)	Polar (r, θ)	$r = \sqrt{x^2 + y^2} \qquad \tan\theta = \dfrac{y}{x}, x \neq 0$
		Be careful to note the proper quadrant for θ.

Polar equations can be graphed by point-plotting. Common shapes that arise are given in the following table. Similar polar equations that only differ by $\sin\theta$ or $\cos\theta$ have the same shapes (just rotated). If more than one equation is given, then the top equation corresponds to the actual graph. In this table, a and b are assumed to be positive.

CLASSIFICATION	SPECIAL NAME	POLAR EQUATIONS	GRAPH
Line	Radial line	$\theta = a$	
Circle	Circle centered at the origin	$r = a$	
Circle	Circle that touches the pole and whose center is on the polar axis	$r = a\cos\theta$	
Circle	Circle that touches the pole and whose center is on the line $\theta = \dfrac{\pi}{2}$	$r = a\sin\theta$	
Limaçon	Cardioid	$r = a + a\cos\theta$ $r = a + a\sin\theta$	
Limaçon	Without inner loop $a > b$	$r = a + b\cos\theta$ $r = a + b\sin\theta$	
Limaçon	With inner loop $a < b$	$r = a + b\sin\theta$ $r = a + b\cos\theta$	
Lemniscate		$r^2 = a^2\cos(2\theta)$ $r^2 = a^2\sin(2\theta)$	
Rose	Three* rose petals	$r = a\sin(3\theta)$ $r = a\cos(3\theta)$	
Rose	Four* rose petals	$r = a\sin(2\theta)$ $r = a\cos(2\theta)$	
Spiral		$r = a\theta$	

*In the argument $n\theta$, if n is odd, then there are n petals (leaves), and if n is even, then there are $2n$ petals (leaves).

[SECTION 8.4] EXERCISES

• SKILLS

In Exercises 1–10, plot each indicated polar point in a polar coordinate system.

1. $\left(3, \dfrac{5\pi}{6}\right)$ **2.** $\left(2, \dfrac{5\pi}{4}\right)$ **3.** $\left(4, \dfrac{11\pi}{6}\right)$ **4.** $\left(1, \dfrac{2\pi}{3}\right)$ **5.** $\left(-2, \dfrac{\pi}{6}\right)$ **6.** $\left(-4, \dfrac{7\pi}{4}\right)$

7. $(-4, 270°)$ **8.** $(3, 135°)$ **9.** $(4, 225°)$ **10.** $(-2, 60°)$

In Exercises 11–20, convert each point given in rectangular coordinates to exact polar coordinates. Assume $0 \le \theta < 2\pi$.

11. $(2, 2\sqrt{3})$ **12.** $(3, -3)$ **13.** $(-1, -\sqrt{3})$ **14.** $(6, 6\sqrt{3})$ **15.** $(-4, 4)$ **16.** $(0, \sqrt{2})$

17. $(3, 0)$ **18.** $(-7, -7)$ **19.** $(-\sqrt{3}, -1)$ **20.** $(2\sqrt{3}, -2)$

In Exercises 21–40, convert each point given in polar coordinates to exact rectangular coordinates.

21. $\left(4, \dfrac{5\pi}{3}\right)$ **22.** $\left(2, \dfrac{3\pi}{4}\right)$ **23.** $\left(-1, \dfrac{5\pi}{6}\right)$ **24.** $\left(-2, \dfrac{7\pi}{4}\right)$ **25.** $\left(0, \dfrac{11\pi}{6}\right)$ **26.** $(6, 0)$

27. $\left(-6, \dfrac{\pi}{2}\right)$ **28.** $\left(-4, \dfrac{3\pi}{2}\right)$ **29.** $\left(8, \dfrac{\pi}{3}\right)$ **30.** $\left(10, \dfrac{4\pi}{3}\right)$ **31.** $(2, 240°)$ **32.** $(-3, 150°)$

33. $(-1, 135°)$ **34.** $(5, 315°)$ **35.** $\left(\dfrac{5}{2}, 330°\right)$ **36.** $\left(\dfrac{3}{4}, 225°\right)$ **37.** $(3, 180°)$ **38.** $(5, 270°)$

39. $(7, 315°)$ **40.** $(2, 120°)$

In Exercises 41–44, match the polar graphs with their corresponding equations.

41. $r = 4\cos\theta$ **42.** $r = 2\theta$ **43.** $r = 3 + 3\sin\theta$ **44.** $r = 3\sin(2\theta)$

a.

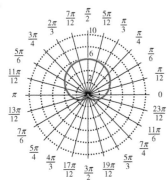

b.

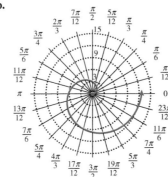

c.

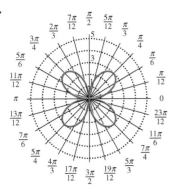

d.

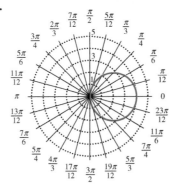

In Exercises 45–62, graph each equation. In Exercises 63–68, convert the equation from polar to rectangular form first and identify the resulting equation as a line, parabola, or circle.

45. $r = 5$

46. $r = -3$

47. $\theta = \dfrac{4\pi}{3}$

48. $\theta = -\dfrac{\pi}{3}$

49. $r = 2\cos\theta$

50. $r = 3\sin\theta$

51. $r = 4\sin(2\theta)$

52. $r = 5\cos(2\theta)$

53. $r = 3\sin(3\theta)$

54. $r = 4\cos(3\theta)$

55. $r^2 = 9\cos(2\theta)$

56. $r^2 = 16\sin(2\theta)$

57. $r = -2\cos\theta$

58. $r = -3\sin(3\theta)$

59. $r = 4\theta$

60. $r = -2\theta$

61. $r = -3 + 2\cos\theta$

62. $r = 2 + 3\sin\theta$

63. $r(\sin\theta + 2\cos\theta) = 1$

64. $r(\sin\theta - 3\cos\theta) = 2$

65. $r^2\cos^2\theta - 2r\cos\theta + r^2\sin^2\theta = 8$

66. $r^2\cos^2\theta - r\sin\theta = -2$

67. $r^2\sin^2\theta + 2r\cos\theta = 3$

68. $r^2\cos^2\theta + r\sin\theta = 3$

• APPLICATIONS

69. Halley's Comet. Halley's comet travels an elliptical path that can be modeled with the polar equation $r = \dfrac{0.587(1 + 0.967)}{1 - 0.967\cos\theta}$. Sketch the graph of the path of Halley's comet.

70. Dwarf Planet Pluto. The dwarf planet Pluto travels in an elliptical orbit that can be modeled with the polar equation $r = \dfrac{29.62(1 + 0.249)}{1 - 0.249\cos\theta}$. Sketch the graph of Pluto's orbit.

For Exercises 71 and 72, refer to the following:

Spirals are seen in nature—for example, in the swirl of a pine cone. They are also used in machinery to convert motions. An Archimedes spiral has the general equation $r = a\theta$. A more general form for the equation of a spiral is $r = a\theta^{1/n}$, where n is a constant that determines how tightly the spiral is wrapped.

71. Archimedes Spiral. Compare the Archimedes spiral $r = \theta$ with the spiral $r = \theta^{1/2}$ by graphing both on the same polar graph.

72. Archimedes Spiral. Compare the Archimedes spiral $r = \theta$ with the spiral $r = \theta^{4/3}$ by graphing both on the same polar graph.

For Exercises 73 and 74, refer to the following:

The *lemniscate motion* occurs naturally in the flapping of birds' wings. The bird's vertical lift and wing sweep create the distinctive figure-eight pattern. The patterns vary with different wing profiles.

73. Flapping Wings of Birds. Compare the two following possible lemniscate patterns by graphing them on the same polar graph: $r^2 = 4\cos(2\theta)$ and $r^2 = \frac{1}{4}\cos(2\theta)$.

74. Flapping Wings of Birds. Compare the two following possible lemniscate patterns by graphing them on the same polar graph: $r^2 = 4\cos(2\theta)$ and $r^2 = 4\cos(2\theta + 2)$.

For Exercises 75 and 76, refer to the following:

Many microphone manufacturers advertise their exceptional pickup capabilities that isolate the sound source and minimize background noise. The name of these microphones comes from the pattern formed by the range of the pickup.

75. Cardioid Pickup Pattern. Graph the cardioid curve to see what the range of a microphone might look like: $r = 2 + 2\sin\theta$.

76. Cardioid Pickup Pattern. Graph the cardioid curve to see what the range of a microphone might look like: $r = -4 - 4\sin\theta$.

For Exercises 77 and 78, refer to the following:

The sword artistry of the samurai is legendary in Japanese folklore and myth. The elegance with which a samurai could wield a sword rivals the grace exhibited by modern figure skaters. In more modern times, such legends have been rendered digitally in many different video games (e.g., *Ominusha*). In order to make the characters realistically move across the screen—and in particular, wield various sword motions true to the legends—trigonometric functions are extensively used in constructing the underlying graphics module. One famous movement is a figure eight, swept out with two hands on the sword. The actual path of the tip of the blade as the movement progresses in this figure-eight motion depends essentially on the length L of the sword and the speed with which the figure is swept out. Such a path is modeled using a polar equation of the form $r^2(\theta) = L\cos(A\theta)$ or $r^2(\theta) = L\sin(A\theta)$, $\theta_1 \le \theta \le \theta_2$.

77. Video Games. Graph the following equations:

a. $r^2(\theta) = 5\cos\theta, 0 \le \theta \le 2\pi$

b. $r^2(\theta) = 5\cos(2\theta), 0 \le \theta \le \pi$

c. $r^2(\theta) = 5\cos(4\theta), 0 \le \theta \le \dfrac{\pi}{2}$

What do you notice about all of these graphs? Suppose that the movement of the tip of the sword in a game is governed by these graphs. Describe what happens if you change the domain in (b) and (c) to $0 \le \theta \le 2\pi$.

78. Video Games. Write a polar equation that would describe the motion of a sword 12 units long that makes 8 complete motions in $[0, 2\pi]$.

• **CATCH THE MISTAKE**

In Exercises 79 and 80, explain the mistake that is made.

79. Convert the rectangular coordinate $(-2, -2)$ to polar coordinates.

Solution:

Label x and y. $x = -2, y = -2$

Find r. $r = \sqrt{x^2 + y^2} = \sqrt{4 + 4} = \sqrt{8} = 2\sqrt{2}$

Find θ. $\tan\theta = \dfrac{-2}{-2} = 1$

$$\theta = \tan^{-1}(1) = \frac{\pi}{4}$$

Write the point in polar coordinates. $\left(2\sqrt{2}, \dfrac{\pi}{4}\right)$

This is incorrect. What mistake was made?

80. Convert the rectangular coordinate $(-\sqrt{3}, 1)$ to polar coordinates.

Solution:

Label x and y. $x = -\sqrt{3}, y = 1$

Find r. $r = \sqrt{x^2 + y^2} = \sqrt{3 + 1} = \sqrt{4} = 2$

Find θ. $\tan\theta = \dfrac{1}{-\sqrt{3}} = -\dfrac{1}{\sqrt{3}}$

$$\theta = \tan^{-1}\left(-\frac{1}{\sqrt{3}}\right) = -\frac{\pi}{4}$$

Write the point in polar coordinates. $\left(2, -\dfrac{\pi}{4}\right)$

This is incorrect. What mistake was made?

• **CONCEPTUAL**

In Exercises 81 and 82, determine whether each statement is true or false.

81. All cardioids are limaçons, but not all limaçons are cardioids.

82. All limaçons are cardioids, but not all cardioids are limaçons.

83. Find the polar equation that is equivalent to a vertical line, $x = a$.

84. Find the polar equation that is equivalent to a horizontal line, $y = b$.

85. Give another pair of polar coordinates for the point (a, θ).

86. Convert $(-a, b)$ to polar coordinates. Assume $a > 0$ and $b > 0$.

• **CHALLENGE**

87. Algebraically, find the polar coordinates (r, θ) where $0 \le \theta < 2\pi$ that the graphs $r_1 = 1 - 2\sin(2\theta)$ and $r_2 = 1 - 2\cos(2\theta)$ have in common.

88. Algebraically, find the polar coordinates (r, θ) where $0 \le \theta < 2\pi$ that the graphs $r_1 = 1 - 2\sin^2\theta$ and $r_2 = 1 - 6\cos^2\theta$ have in common.

89. Determine an algebraic method for testing a polar equation for symmetry to the x-axis, the y-axis, and the origin. Apply the test to determine what symmetry the graph with equation $r = \sin(3\theta)$ has.

90. Determine an algebraic method for testing a polar equation for symmetry to the x-axis, the y-axis, and the origin. Apply the test to determine what symmetry the graph with equation $r^2 = \cos(4\theta)$ has.

• **TECHNOLOGY**

91. Given $r = \cos\left(\dfrac{\theta}{2}\right)$, find the θ-intervals for the inner loop above the x-axis.

92. Given $r = 2\cos\left(\dfrac{3\theta}{2}\right)$, find the θ-intervals for the petal in the first quadrant.

93. Given $r = 1 + 3\cos\theta$, find the θ-intervals for the inner loop.

94. Given $r = 1 + \sin(2\theta)$ and $r = 1 - \cos(2\theta)$, find all points of intersection.

8.5 PARAMETRIC EQUATIONS AND GRAPHS

SKILLS OBJECTIVE	CONCEPTUAL OBJECTIVE
▪ Graph parametric equations.	▪ Understand that the results of increasing the value of the parameter reveal the orientation of a curve or the direction of motion along it.

8.5.1 Parametric Equations of a Curve

Thus far we have talked about graphs in planes. For example, the equation $x^2 + y^2 = 1$ when graphed in the Cartesian plane is the unit circle. Similarly, the function $f(x) = \sin x$ when graphed in the Cartesian plane is a sinusoidal curve. Now, we consider the **path (orientation) along a curve**. For example, if a car is being driven on a circular racetrack, we want to see the movement along the circle. We can determine where (the position) along the circle the car is at some time t using *parametric equations*. Before we define *parametric equations* in general, let us start with a simple example.

Let $x = \cos t$ and $y = \sin t$ and $t \geq 0$. We then can make a table of some corresponding values.

8.5.1 SKILL

Graph parametric equations.

8.5.1 CONCEPTUAL

Understand that the results of increasing the value of the parameter reveal the orientation of a curve or the direction of motion along it.

[CONCEPT CHECK]

Describe the curve defined by the parametric equations: $x = \sin t$, $y = \cos t$, where t goes from 0 to $2 \cdot \pi$.

▼

ANSWER Circle whose path is traced clockwise.

t SECONDS	$x = \cos t$	$y = \sin t$	(x, y)
0	$x = \cos 0 = 1$	$y = \sin 0 = 0$	$(1, 0)$
$\dfrac{\pi}{2}$	$x = \cos\left(\dfrac{\pi}{2}\right) = 0$	$y = \sin\left(\dfrac{\pi}{2}\right) = 1$	$(0, 1)$
π	$x = \cos \pi = -1$	$y = \sin \pi = 0$	$(-1, 0)$
$\dfrac{3\pi}{2}$	$x = \cos\left(\dfrac{3\pi}{2}\right) = 0$	$y = \sin\left(\dfrac{3\pi}{2}\right) = -1$	$(0, -1)$
2π	$x = \cos(2\pi) = 1$	$y = \sin(2\pi) = 0$	$(1, 0)$

If we plot these points in the Cartesian plane and note the correspondence to time (by converting all numbers to decimals), we will be tracing a path counterclockwise along the unit circle, $x^2 + y^2 = 1$.

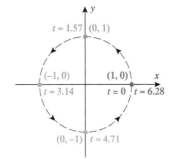

TIME (SECONDS)	$t = 0$	$t \approx 1.57$	$t \approx 3.14$	$t \approx 4.71$
POSITION	$(1, 0)$	$(0, 1)$	$(-1, 0)$	$(0, -1)$

Notice that at time $t \approx 6.28$ seconds, we are back to the point $(1, 0)$.

We can see that the path is along the unit circle, since $x^2 + y^2 = \cos^2 t + \sin^2 t = 1$.

DEFINITION Parametric Equations

Let $x = f(t)$ and $y = g(t)$ be functions defined for t on some interval. The set of points $(x, y) = (f(t), g(t))$ represents a **plane curve**. The equations

$$x = f(t) \qquad \text{and} \qquad y = g(t)$$

are called **parametric equations** of the curve. The variable t is called the **parameter**.

Parametric equations are useful for showing movement along a curve. We insert arrows in the graph to show **direction**, or **orientation**, along the curve as t increases.

EXAMPLE 1 **Graphing a Curve Defined by Parametric Equations**

Graph the curve defined by the parametric equations:

$$x = t^2 \qquad y = (t - 1) \qquad t \text{ in } [-2, 2]$$

Indicate the orientation with arrows.

Solution:

STEP 1 Make a table and find values for t, x, and y.

t	$x = t^2$	$y = (t - 1)$	(x, y)
$t = -2$	$x = (-2)^2 = 4$	$y = (-2 - 1) = -3$	$(4, -3)$
$t = -1$	$x = (-1)^2 = 1$	$y = (-1 - 1) = -2$	$(1, -2)$
$t = 0$	$x = 0^2 = 0$	$y = (0 - 1) = -1$	$(0, -1)$
$t = 1$	$x = 1^2 = 1$	$y = (1 - 1) = 0$	$(1, 0)$
$t = 2$	$x = 2^2 = 4$	$y = (2 - 1) = 1$	$(4, 1)$

STEP 2 Plot the points in the xy-plane.

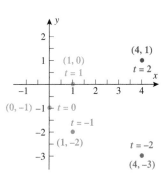

STEP 3 Connect the points with a smooth curve and use arrows to indicate direction.

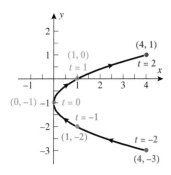

The shape of the curve appears to be part of a parabola. The parametric equations are $x = t^2$ and $y = (t - 1)$. If we solve the second equation for t, getting $t = y + 1$, and then substitute this expression for t into $x = t^2$, the result is $x = (y + 1)^2$. The graph of $x = (y + 1)^2$ is a parabola with its vertex at the point $(0, -1)$ and opening to the right. Notice, however, that the limited domain of the parameter, t, only gives a path along part of the parabola.

▼
ANSWER

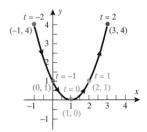

YOUR TURN Graph the curve defined by the parametric equations:

$$x = t + 1 \qquad y = t^2 \qquad t \text{ in } [-2, 2]$$

Indicate the orientation with arrows.

Sometimes it is easier to give the rectangular equivalent of the curve by eliminating the parameter. However, direction along the curve must be noted.

▶ **EXAMPLE 2** **Graphing a Curve Defined by Parametric Equations by First Finding an Equivalent Rectangular Equation**

Graph the curve defined by the parametric equations:

$$x = 4\cos t \qquad y = 3\sin t \qquad t \text{ is any real number}$$

Indicate the orientation with arrows.

Solution:

One approach is to point-plot as in Example 1. A second approach is to find the equivalent rectangular equation that represents the curve.

Use the Pythagorean identity. $\sin^2 t + \cos^2 t = 1$

Find $\sin^2 t$ from the parametric
equation for y. $y = 3\sin t$

   Square both sides. $y^2 = 9\sin^2 t$

   Divide by 9. $\sin^2 t = \dfrac{y^2}{9}$

Similarly find $\cos^2 t$. $x = 4\cos t$

   Square both sides. $x^2 = 16\cos^2 t$

   Divide by 16. $\cos^2 t = \dfrac{x^2}{16}$

Substitute $\sin^2 t = \dfrac{y^2}{9}$ and $\cos^2 t = \dfrac{x^2}{16}$ $\dfrac{y^2}{9} + \dfrac{x^2}{16} = 1$

into $\sin^2 t + \cos^2 t = 1$.

The curve is an ellipse centered at the origin
and elongated horizontally.

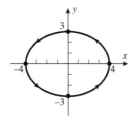

The orientation is *counterclockwise*. For example, when $t = 0$, the position is $(4, 0)$; when $t = \dfrac{\pi}{2}$, the position is $(0, 3)$, and when $t = \pi$, the position is $(-4, 0)$.

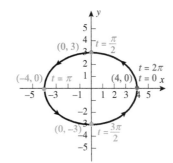

STUDY TIP

For open curves, the orientation can be determined from two values of *t*. However, for closed curves three points should be chosen to ensure clockwise or counterclockwise orientation.

It is important to point out the difference between the graph of the ellipse $\dfrac{y^2}{9} + \dfrac{x^2}{16} = 1$ and the curve defined by $x = 4\cos t$ and $y = 3\sin t$, where t is any real number. The graph is the ellipse. The curve is infinitely many counterclockwise rotations around the ellipse, since t is any real number.

Applications of Parametric Equations

Parametric equations can be used to describe motion in many applications. Two that we will discuss are the *cycloid* and a *projectile*. Suppose you paint a red **X** on a bicycle tire. As the bicycle moves in a straight line, if you watch the motion of the red **X**, it follows the path of a **cycloid**.

The parametric equations that define a cycloid are

$$\boxed{x = a(t - \sin t)} \quad \text{and} \quad \boxed{y = a(1 - \cos t)}$$

where t is any real number.

EXAMPLE 3 **Graphing a Cycloid**

Graph the cycloid given by $x = 2(t - \sin t)$ and $y = 2(1 - \cos t)$ for t in $[0, 4\pi]$.

Solution:

STEP 1 Make a table and find key values for t, x, and y.

t	$x = 2(t - \text{SIN } t)$	$y = 2(1 - \text{COS } t)$	(x, y)
$t = 0$	$x = 2(0 - 0) = 0$	$y = 2(1 - 1) = 0$	$(0, 0)$
$t = \pi$	$x = 2(\pi - 0) = 2\pi$	$y = 2[1 - (-1)] = 4$	$(2\pi, 4)$
$t = 2\pi$	$x = 2(2\pi - 0) = 4\pi$	$y = 2(1 - 1) = 0$	$(4\pi, 0)$
$t = 3\pi$	$x = 2(3\pi - 0) = 6\pi$	$y = 2[1 - (-1)] = 4$	$(6\pi, 4)$
$t = 4\pi$	$x = 2(4\pi - 0) = 8\pi$	$y = 2(1 - 1) = 0$	$(8\pi, 0)$

STEP 2 Plot points in the Cartesian plane and connect them with a smooth curve.

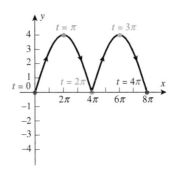

Another example of parametric equations describing real-world phenomena is *projectile motion*. The accompanying photo of a golfer hitting a golf ball illustrates an example of a projectile.

Joshua Dalsimer/©Corbis

Let v_0 be the initial velocity of an object, θ be the initial angle of inclination with the horizontal, and h be the initial height above the ground. Then the parametric equations describing the **projectile motion** (which will be developed in calculus) are

$$x = (v_0 \cos \theta)t \qquad \text{and} \qquad y = -\tfrac{1}{2}gt^2 + (v_0 \sin \theta)t + h$$

where t is the time and g is the constant acceleration due to gravity (9.8 meters per square second or 32 feet per square second).

EXAMPLE 4 Graphing Projectile Motion

Suppose a golfer hits his golf ball with an initial velocity of 160 feet per second at an angle of 30° with the ground. How far is his drive, assuming the length of the drive is from the tee to where the ball first hits the ground? Graph the curve representing the path of the golf ball. Assume that he hits the ball straight off the tee and down the fairway.

Solution:

STEP 1 Find the parametric equations that describe the golf ball that the golfer drove.

First, write the parametric equations for projectile motion.

$$x = (v_0 \cos\theta)t \quad \text{and} \quad y = -\tfrac{1}{2}gt^2 + (v_0\sin\theta)t + h$$

Let $g = 32$ ft/sec^2, $v_0 = 160$ ft/sec, $h = 0$, and $\theta = 30°$.

$$x = (160 \cdot \cos 30°)t \quad \text{and} \quad y = -16t^2 + (160 \cdot \sin 30°)t$$

Evaluate the sine and cosine functions and simplify.

$$x = 80\sqrt{3}\,t \quad \text{and} \quad y = -16t^2 + 80t$$

STEP 2 Graph the projectile motion.

t	$x = 80\sqrt{3}\,t$	$y = -16t^2 + 80t$	(x, y)
$t = 0$	$x = 80\sqrt{3}(0) = 0$	$y = -16(0)^2 + 80(0) = 0$	$(0, 0)$
$t = 1$	$x = 80\sqrt{3}(1) \approx 139$	$y = -16(1)^2 + 80(1) = 64$	$(139, 64)$
$t = 2$	$x = 80\sqrt{3}(2) \approx 277$	$y = -16(2)^2 + 80(2) = 96$	$(277, 96)$
$t = 3$	$x = 80\sqrt{3}(3) \approx 416$	$y = -16(3)^2 + 80(3) = 96$	$(416, 96)$
$t = 4$	$x = 80\sqrt{3}(4) \approx 554$	$y = -16(4)^2 + 80(4) = 64$	$(554, 64)$
$t = 5$	$x = 80\sqrt{3}(5) \approx 693$	$y = -16(5)^2 + 80(5) = 0$	$(693, 0)$

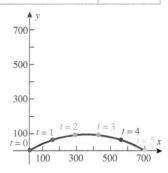

We can see that we selected our time increments well (the last point, $(693, 0)$, corresponds to the ball hitting the ground 693 feet from the tee).

STEP 3 Identify the horizontal distance from the tee to where the ball first hits the ground.

Algebraically, we can determine the distance of the tee shot by setting the height y equal to zero.

$$y = -16t^2 + 80t = 0$$

Factor (divide) the common, $-16t$.

$$-16t(t - 5) = 0$$

Solve for t.

$$t = 0 \text{ or } t = 5$$

The ball hits the ground after 5 seconds.

Let $t = 5$ in the horizontal distance, $x = 80\sqrt{3}\,t$.

$$x = 80\sqrt{3}(5) \approx 693$$

> The ball hits the ground 693 feet from the tee.

With parametric equations, we can also determine when the ball lands (5 seconds).

▶[SECTION 8.5] SUMMARY

Parametric equations provide a way of describing the path an object takes along a curve in the xy-plane. Often, t is a parameter used, where $x = f(t)$ and $y = g(t)$ describe the coordinates (x, y) that lie along the curve. Parametric equations have equivalent rectangular equations. Typically, the method of graphing a set of parametric equations is to eliminate t and graph the corresponding rectangular equation. Once the curve is found, orientation along the curve can be determined by finding points for different t-values. Two important applications are cycloids and projectiles, whose paths can be traced using parametric equations.

[SECTION 8.5] EXERCISES

• SKILLS

In Exercises 1–20, graph the curve defined by the following sets of parametric equations. Be sure to indicate the direction of movement along the curve.

1. $x = t + 1, y = \sqrt{t}, t \geq 0$

2. $x = 3t, y = t^2 - 1, t \text{ in } [0, 4]$

3. $x = -3t, y = t^2 + 1, t \text{ in } [0, 4]$

4. $x = t^2 - 1, y = t^2 + 1, t \text{ in } [-3, 3]$

5. $x = t^2, y = t^3, t \text{ in } [-2, 2]$

6. $x = t^3 + 1, y = t^3 - 1, t \text{ in } [-2, 2]$

7. $x = \sqrt{t}, y = t, t \text{ in } [0, 10]$

8. $x = t, y = \sqrt{t^2 + 1}, t \text{ in } [0, 10]$

9. $x = (t + 1)^2, y = (t + 2)^3, t \text{ in } [0, 1]$

10. $x = (t - 1)^3, y = (t - 2)^2, t \text{ in } [0, 4]$

11. $x = 3 \sin t, y = 2 \cos t, t \text{ in } [0, 2\pi]$

12. $x = \cos(2t), y = \sin t, t \text{ in } [0, 2\pi]$

13. $x = \sin t + 1, y = \cos t - 2, t \text{ in } [0, 2\pi]$

14. $x = \tan t, y = 1, t \text{ in } \left[-\dfrac{\pi}{4}, \dfrac{\pi}{4}\right]$

15. $x = 1, y = \sin t, t \text{ in } [-2\pi, 2\pi]$

16. $x = \sin t, y = 2, t \text{ in } [0, 2\pi]$

17. $x = \sin^2 t, y = \cos^2 t, t \text{ in } [0, 2\pi]$

18. $x = 2 \sin^2 t, y = 2 \cos^2 t, t \text{ in } [0, 2\pi]$

19. $x = 2 \sin(3t), y = 3 \cos(2t), t \text{ in } [0, 2\pi]$

20. $x = 4 \cos(2t), y = t, t \text{ in } [0, 2\pi]$

In Exercises 21–36, each set of parametric equations defines a plane curve. Find an equation in rectangular form that also corresponds to the plane curve.

21. $x = \dfrac{1}{t}, y = t^2$

22. $x = t^2 - 1, y = t^2 + 1$

23. $x = t^3 + 1, y = t^3 - 1$

24. $x = 3t, y = t^2 - 1$

25. $x = t, y = \sqrt{t^2 + 1}$

26. $x = \sin^2 t, y = \cos^2 t$

27. $x = 2 \sin^2 t, y = 2 \cos^2 t$

28. $x = \sec^2 t, y = \tan^2 t$

29. $x = 4(t^2 + 1), y = 1 - t^2$

30. $x = \sqrt{t - 1}, y = \sqrt{t}$

31. $x = 2t, y = 2 \sin t \cos t$

32. $x = \dfrac{t}{2}, y = 2 \tan t$

33. $x = \sqrt{t}, y = t + 2$

34. $x = \sqrt{t + 1}, y = \dfrac{t}{4} - 1$

35. $x = \sin t, y = \sin t$

36. $x = \cos t, y = -\cos^2 t$

• **APPLICATIONS**

For Exercises 37–46, recall that the flight of a projectile can be modeled with the parametric equations

$$x = (v_0 \cos \theta)t \qquad y = -16t^2 + (v_0 \sin \theta)t + h$$

where t is in seconds, v_0 is the initial velocity in feet per second, θ is the initial angle with the horizontal, and h is the initial height above ground, where x and y are in feet.

37. **Flight of a Projectile.** A projectile is launched from the ground at a speed of 400 feet per second at an angle of 45° with the horizontal. After how many seconds does the projectile hit the ground?

38. **Flight of a Projectile.** A projectile is launched from the ground at a speed of 400 feet per second at an angle of 45° with the horizontal. How far does the projectile travel (what is the horizontal distance), and what is its maximum altitude? (Note the symmetry of the projectile path.)

39. **Flight of a Baseball.** A baseball is hit at an initial speed of 105 mph and an angle of 20° at a height of 3 feet above the ground. If home plate is 420 feet from the back fence, which is 15 feet tall, will the baseball clear the back fence for a home run?

40. **Flight of a Baseball.** A baseball is hit at an initial speed of 105 mph and an angle of 20° at a height of 3 feet above the ground. If there is no back fence or other obstruction, how far does the baseball travel (horizontal distance), and what is its maximum height? (Note the symmetry of the projectile path.)

41. **Bullet Fired.** A gun is fired from the ground at an angle of 60°, and the bullet has an initial speed of 700 feet per second. How high does the bullet go? What is the horizontal (ground) distance between where the gun was fired and where the bullet hit the ground?

42. **Bullet Fired.** A gun is fired from the ground at an angle of 60°, and the bullet has an initial speed of 2000 feet per second. How high does the bullet go? What is the horizontal (ground) distance between where the gun was fired and where the bullet hit the ground?

43. **Missile Fired.** A missile is fired from a ship at an angle of 30°, an initial height of 20 feet above the water's surface, and at a speed of 4000 feet per second. How long will it be before the missile hits the water?

44. **Missile Fired.** A missile is fired from a ship at an angle of 40°, an initial height of 20 feet above the water's surface, and at a speed of 5000 feet per second. Will the missile be able to hit a target that is 2 miles away?

45. **Path of a Projectile.** A projectile is launched at a speed of 100 feet per second at an angle of 35° with the horizontal. Plot the path of the projectile on a graph. Assume $h = 0$.

46. **Path of a Projectile.** A projectile is launched at a speed of 150 feet per second at an angle of 55° with the horizontal. Plot the path of the projectile on a graph. Assume $h = 0$.

For Exercises 47 and 48, refer to the following:

Modern amusement park rides are often designed to push the envelope in terms of speed, angle, and ultimately *g*-force, and usually take the form of gargantuan roller coasters or skyscraping towers. However, even just a couple of decades ago, such creations were depicted only in fantasy-type drawings, with their creators never truly believing their construction

would become a reality. Nevertheless, thrill rides still capable of nauseating any would-be rider were still able to be constructed; one example is the *Calypso*. This ride is a not-too-distant cousin of the better-known *Scrambler*. It consists of four rotating arms (instead of three like the Scrambler), and on each of these arms, four cars (equally spaced around the circumference of a circular frame) are attached. Once in motion, the main piston to which the four arms are connected rotates clockwise, while each of the four arms themselves rotates counterclockwise. The combined motion appears as a blur to any onlooker from the crowd, but the motion of a single rider is much less chaotic. In fact, a single rider's path can be modeled by the following graph:

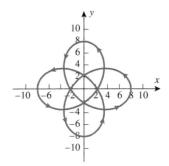

The equation of this graph is defined parametrically by

$$x(t) = A\cos t + B\cos(-3t)$$
$$y(t) = A\sin t + B\sin(-3t), 0 \le t \le 2\pi$$

47. **Amusement Rides.** What is the location of the rider at $t = 0, t = \dfrac{\pi}{2}, t = \pi, t = \dfrac{3\pi}{2}$, and $t = 2\pi$?

48. **Amusement Rides.** Suppose the ride conductor was rather sinister and speeded up the ride to twice the speed. How would you modify the parametric equations to model such a change? Now vary the values of A and B. What do you conjecture these parameters are modeling in this problem?

49. **Fan Blade.** The position on the tip of a ceiling fan is given by the parametric equations $x = \sin(10t)$ and $y = \cos(10t)$, where x and y are the vertical and lateral positions relative to the center of the fan respectively, and t is the time in seconds. How long does it take for the fan blade to make one complete revolution?

50. **Fan Blade.** If the fan is reversed, how can the equations be altered to represent the motion of the fan in the opposite direction?

51. **Bicycle Racing.** A boy on a bicycle racing around an oval track has a position given by the equations

$x = -100\sin\left(\dfrac{t}{4}\right)$ and $y = 75\cos\left(\dfrac{t}{4}\right)$, where x and y are

the horizontal and vertical positions in feet relative to the center of the track t seconds after the start of the race. Find the boy's position at $t = 10, 20$, and 30.

52. **Bicycle Racing.** A boy on a bicycle racing around an oval track has a position given by the equations

$x = -100\sin\left(\dfrac{t}{4}\right)$ and $y = 75\cos\left(\dfrac{t}{4}\right)$, where x and y are

the horizontal and vertical positions in feet relative to the center of the track t seconds after the start of the race. Find out how long it takes the boy to complete one lap.

53. **Bicycle Racing.** A boy on a bicycle racing around an oval track has a position given by the equations

$$x = -100\sin\left(\frac{t}{4}\right) \text{ and } y = 75\cos\left(\frac{t}{4}\right), \text{ where } x \text{ and } y \text{ are}$$

the horizontal and vertical positions in feet relative to the center of the track t seconds after the start of the race. Another racer has a position given by the equations

$$x = -100\sin\left(\frac{t}{3}\right) \text{ and } y = 75\cos\left(\frac{t}{3}\right). \text{ Which racer is going}$$

faster?

54. **Bicycle Racing.** For the two boys racing in Exercise 53, if they were to continue according to the same formula, how long would it take one to lap the other?

• **CATCH THE MISTAKE**

In Exercises 55 and 56, explain the mistake that is made.

55. Find the rectangular equation that corresponds to the plane curve defined by the parametric equations $x = t + 1$ and $y = \sqrt{t}$. Describe the plane curve.

Solution:

   Square $y = \sqrt{t}$. $y^2 = t$

   Substitute $t = y^2$ into $x = t + 1$. $x = y^2 + 1$

   The graph of $x = y^2 + 1$ is a parabola opening to the right with vertex at $(1, 0)$.

   This is incorrect. What mistake was made?

56. Find the rectangular equations that correspond to the plane curve defined by the parametric equations $x = \sqrt{t}$ and $y = t - 1$. Describe the plane curve.

Solution:

   Square $x = \sqrt{t}$. $x^2 = t$

   Substitute $t = x^2$ into $y = t - 1$. $y = x^2 - 1$

   The graph of $y = x^2 - 1$ is a parabola opening up with vertex at $(0, -1)$.

   This is incorrect. What mistake was made?

• **CONCEPTUAL**

In Exercises 57 and 58, determine whether each statement is true or false.

57. Curves given by equations in rectangular form have orientation.

58. Curves given by parametric equations have orientation.

59. Describe the graph given by the parametric equations $x = nt$ and $y = \sin t$ for positive integer n.

60. Describe the graph given by the parametric equations

$$x = \frac{t}{n} \text{ and } y = \cot t \text{ for positive integer } n.$$

• **CHALLENGE**

61. Determine which type of curve the parametric equations $x = \sqrt{t}$ and $y = \sqrt{1 - t}$ define.

62. Determine which type of curve the parametric equations $x = \ln t$ and $y = t$ define.

63. Determine which type of curve the parametric equations $x = \tan t$ and $y = \sec t$ define.

64. Determine which type of curve the parametric equations $x = e^t$ and $y = -t$ define.

• **TECHNOLOGY**

65. Consider the parametric equations $x = a\sin t - \sin(at)$ and $y = a\cos t + \cos(at)$. Use a graphing utility to explore the graphs for $a = 2, 3,$ and 4.

66. Consider the parametric equations $x = a\cos t - b\cos(at)$ and $y = a\sin t + \sin(at)$. Use a graphing utility to explore the graphs for $a = 3$ and $b = 1$, $a = 4$ and $b = 2$, and $a = 6$ and $b = 2$. Find the t-interval that gives one cycle of the curve.

67. Consider the parametric equations $x = \cos(at)$ and $y = \sin(bt)$. Use a graphing utility to explore the graphs for $a = 2$ and $b = 4$, $a = 4$ and $b = 2$, $a = 1$ and $b = 3$, and $a = 3$ and $b = 1$. Find the t-interval that gives one cycle of the curve.

68. Consider the parametric equations $x = a\sin(at) - \sin t$ and $y = a\cos(at) - \cos t$. Use a graphing utility to explore the graphs for $a = 2$ and 3. If $y = a\cos(at) + \cos t$, explore the graphs for $a = 2$ and 3. Describe the t-interval for a complete cycle for each case.

[CHAPTER 8 REVIEW]

SECTION	CONCEPT	KEY IDEAS/FORMULAS
8.1	**Complex numbers**	
	The imaginary unit i	$i = \sqrt{-1}$, where $i^2 = -1$
	Adding and subtracting complex numbers	Add corresponding real parts and corresponding imaginary parts.
	Multiplying complex numbers	Remember that $i^2 = -1$.
	Dividing complex numbers	Complex number: $a + bi$ Complex conjugate: $a - bi$ $(a + bi)(a - bi) = a^2 + b^2$
	Raising complex numbers to integer powers	$i^2 = -1, i^3 = -i, i^4 = 1$
8.2	**Polar (trigonometric) form of complex numbers**	
	Complex numbers in rectangular form	 **Modulus**, or magnitude, of a complex number $z = x + iy$ is the distance from the origin to the point (x, y) in the complex plane given by $$\lvert z \rvert = \sqrt{x^2 + y^2}$$
	Complex numbers in polar form	**Polar form** of a complex number is $$z = r(\cos\theta + i\sin\theta)$$ where r represents the **modulus** (magnitude) of the complex number and θ is called the **argument** of z. *Converting complex numbers between rectangular and polar forms:* Step 1: Plot the point $z = x + yi$ in the complex plane (note the quadrant). Step 2: Find r. Use $r = \sqrt{x^2 + y^2}$. Step 3: Find θ. Use $\tan\theta = \dfrac{y}{x}$, $x \neq 0$, where θ is in the quadrant found in Step 1 and $0 \leq \theta < 2\pi$.
8.3	**Products, quotients, powers, and roots of complex numbers; De Moivre's theorem**	
	Products of complex numbers	Let $z_1 = r_1(\cos\theta_1 + i\sin\theta_1)$ and $z_2 = r_2(\cos\theta_2 + i\sin\theta_2)$ be two complex numbers. The product z_1z_2 is given by $z_1z_2 = r_1r_2[\cos(\theta_1 + \theta_2) + i\sin(\theta_1 + \theta_2)]$. *Multiply the magnitudes and add the arguments.*

SECTION	CONCEPT	KEY IDEAS/FORMULAS
	Quotients of complex numbers	Let $z_1 = r_1(\cos\theta_1 + i\sin\theta_1)$ and $z_2 = r_2(\cos\theta_2 + i\sin\theta_2)$ be two complex numbers. The quotient $\dfrac{z_1}{z_2}$ is given by $\dfrac{z_1}{z_2} = \dfrac{r_1}{r_2}[\cos(\theta_1 - \theta_2) + i\sin(\theta_1 - \theta_2)]$. *Divide the magnitudes and subtract the arguments.*
	Powers of complex numbers	**De Moivre's theorem** If $z = r(\cos\theta + i\sin\theta)$ is a complex number, then $z^n = r^n[\cos(n\theta) + i\sin(n\theta)]$, where n is a positive integer.
	Roots of complex numbers	The n **nth roots** of the complex number $z = r(\cos\theta + i\sin\theta)$ are given by $w_k = r^{1/n}\left[\cos\left(\dfrac{\theta}{n} + \dfrac{k \cdot 360°}{n}\right) + i\sin\left(\dfrac{\theta}{n} + \dfrac{k \cdot 360°}{n}\right)\right]$ with θ in degrees, and where $k = 0, 1, 2, \ldots, n - 1$.
8.4	**Polar equations and graphs**	
	Polar coordinates	 To plot a point (r, θ): ■ Start on the polar axis and rotate a ray to form the terminal side of an angle θ. ■ If $r > 0$, the point is r units from the origin in the *same direction* of the terminal side of θ. ■ If $r < 0$, the point is $\lvert r \rvert$ units from the origin in the *opposite direction* of the terminal side of θ.
	Converting between polar and rectangular coordinates	From polar (r, θ) to rectangular (x, y): $\quad x = r\cos\theta \qquad y = r\sin\theta$ From rectangular (x, y) to polar (r, θ): $\quad r = \sqrt{x^2 + y^2} \qquad \tan\theta = \dfrac{y}{x}, \qquad x \neq 0$
	Graphs of polar equations	Radial line, circle, spiral, rose petals, lemniscate, and limaçon
8.5	**Parametric equations and graphs**	
	Parametric equations of a curve	Parametric equations: $x = f(t)$ and $y = g(t)$ Plane curve: $(x, y) = \big(f(t), g(t)\big)$

[CHAPTER 8 REVIEW EXERCISES]

8.1 Complex Numbers

Simplify each complex number.

1. i^{39}

2. i^{100}

3. $\sqrt{-49}$

4. $\sqrt{-50}$

Perform the indicated operation, simplify, and express in standard complex form, $a + bi$.

5. $(2 - 3i) + (4 + 6i)$

6. $(-3 - i) - (5 - 2i)$

7. $(1 + 4i)(-6 - 2i)$

8. $(2 + 5i)(-3 + 2i)$

9. $\dfrac{3 - 6i}{2 + i}$

10. $\dfrac{4 - i}{2 + 3i}$

11. $\dfrac{8 - 12i}{4i}$

12. $\dfrac{1 - 2i}{1 + 2i}$

13. $i(2 + 3i) + 4(-3 + 2i)$

14. $-3(2 - 3i) + i(-6 + i)$

Solve the quadratic equations. For all complex solutions, write the answers in standard complex form, $a + bi$.

15. $x^2 - 4x + 9 = 0$

16. $2x^2 + 2x + 11 = 0$

17. $5x^2 + 6x + 5 = 0$

18. $x^2 + 4x + 5 = 0$

Write each expression in terms of imaginary numbers, and then perform the indicated operation and simplify.

19. $\sqrt{-25}\sqrt{-9}$

20. $(2 + \sqrt{-100})(-3 + \sqrt{-4})$

8.2 Polar (Trigonometric) Form of Complex Numbers

Graph each complex number in the complex plane.

21. $-3 - 4i$

22. -8

23. $-6 + 2i$

24. $5i$

Express each complex number in exact polar form.

25. $\sqrt{2} - \sqrt{2}i$

26. $\sqrt{3} + i$

27. $-8i$

28. $-8 - 8i$

Use a calculator to express each complex number in polar form.

29. $-60 + 11i$

30. $9 - 40i$

31. $15 + 8i$

32. $-10 - 24i$

Express each complex number in exact rectangular form.

33. $6(\cos 300° + i \sin 300°)$

34. $4(\cos 210° + i \sin 210°)$

35. $\sqrt{2}(\cos 135° + i \sin 135°)$

36. $4(\cos 150° + i \sin 150°)$

Use a calculator to express each complex number in rectangular form.

37. $4(\cos 200° + i \sin 200°)$

38. $3(\cos 350° + i \sin 350°)$

8.3 Products, Quotients, Powers, and Roots of Complex Numbers; De Moivre's Theorem

Find the product $z_1 z_2$.

39. $z_1 = 3(\cos 200° + i \sin 200°)$ and $z_2 = 4(\cos 70° + i \sin 70°)$

40. $z_1 = 3(\cos 20° + i \sin 20°)$ and $z_2 = 4(\cos 220° + i \sin 220°)$

41. $z_1 = 7(\cos 100° + i \sin 100°)$ and $z_2 = 3(\cos 140° + i \sin 140°)$

42. $z_1 = (\cos 290° + i \sin 290°)$ and $z_2 = 4(\cos 40° + i \sin 40°)$

Find the quotient $\dfrac{z_1}{z_2}$.

43. $z_1 = \sqrt{6}(\cos 200° + i \sin 200°)$ and $z_2 = \sqrt{6}(\cos 50° + i \sin 50°)$

44. $z_1 = 18(\cos 190° + i \sin 190°)$ and $z_2 = 2(\cos 100° + i \sin 100°)$

45. $z_1 = 24(\cos 290° + i \sin 290°)$ and $z_2 = 4(\cos 110° + i \sin 110°)$

46. $z_1 = \sqrt{200}(\cos 93° + i \sin 93°)$ and $z_2 = \sqrt{2}(\cos 48° + i \sin 48°)$

Evaluate each expression using De Moivre's theorem. Write the answer in rectangular form.

47. $(3 + 3i)^4$

48. $(3 + \sqrt{3}i)^4$

49. $(1 + \sqrt{3}i)^5$

50. $(-2 - 2i)^7$

Find all n nth roots of z. Write the roots in polar form, and plot the roots in the complex plane.

51. $z = 2 + 2\sqrt{3}i, n = 2$

52. $z = -8 + 8\sqrt{3}i, n = 4$

53. $z = -256, n = 4$

54. $z = -18i, n = 2$

Find all complex solutions to the given equations.

55. $x^3 + 216 = 0$

56. $x^4 - 1 = 0$

57. $x^4 + 1 = 0$

58. $x^3 - 125 = 0$

8.4 Polar Equations and Graphs

Convert each rectangular point to exact polar coordinates (assuming that $0 \leq \theta < 2\pi$), and then graph the point in the polar coordinate system.

59. $(-2, 2)$ **60.** $\left(4, -4\sqrt{3}\right)$ **61.** $\left(-5\sqrt{3}, -5\right)$

62. $\left(\sqrt{3}, \sqrt{3}\right)$ **63.** $(0, -2)$ **64.** $(11, 0)$

Convert each polar point to exact rectangular coordinates.

65. $\left(-3, \dfrac{5\pi}{3}\right)$ **66.** $\left(4, \dfrac{5\pi}{4}\right)$ **67.** $\left(2, \dfrac{\pi}{3}\right)$

68. $\left(6, \dfrac{7\pi}{6}\right)$ **69.** $\left(1, \dfrac{4\pi}{3}\right)$ **70.** $\left(-3, \dfrac{7\pi}{4}\right)$

Graph each equation in the polar coordinate plane.

71. $r = 4\cos(2\theta)$ **72.** $r = \sin(3\theta)$

73. $r = -\theta$ **74.** $r = 4 - 3\sin\theta$

8.5 Parametric Equations and Graphs

Graph the curve defined by each set of parametric equations.

75. $x = \sin t, y = 4\cos t$ for t in $[-\pi, \pi]$

76. $x = 5\sin^2 t, y = 2\cos^2 t$ for t in $[-\pi, \pi]$

77. $x = 4 - t^2, y = t^2$ for t in $[-3, 3]$

78. $x = t + 3, y = 4$ for t in $[-4, 4]$

Each given set of parametric equations defines a plane curve. Find an equation in rectangular form that also corresponds to the plane curve. Let t be any real number.

79. $x = 4 - t^2, y = t$ **80.** $x = 5\sin^2 t, y = 2\cos^2 t$

81. $x = 2\tan^2 t, y = 4\sec^2 t$ **82.** $x = 3t^2 + 4, y = 3t^2 - 5$

Technology Exercises

Section 8.1

In Exercises 83 and 84, apply a graphing utility to simplify the expression. Write your answer in standard form.

83. $(3 + 5i)^5$ **84.** $\dfrac{1}{(1 + 3i)^4}$

Section 8.2

Another way of using a graphing calculator to represent complex numbers in rectangular form is to enter the real and imaginary parts as a list of two numbers and use the $\boxed{\text{SUM}}$ command to find the modulus.

85. Write $-\sqrt{23} - 11i$ in polar form using the $\boxed{\text{SUM}}$ command to find its modulus, and round the angle to the nearest degree.

86. Write $11 + \sqrt{23}i$ in polar form using the $\boxed{\text{SUM}}$ command to find its modulus, and round the angle to the nearest degree.

Section 8.3

87. Find the fourth roots of $-8 + 8\sqrt{3}i$, and draw the complex rectangle with the calculator.

88. Find the fourth roots of $8\sqrt{3} + 8i$, and draw the complex rectangle with the calculator.

Section 8.4

89. Given $r = 1 - 2\sin(3\theta)$, find the angles of all points of intersection (where $r = 0$), such that $0° \leq \theta < 360°$.

90. Given $r = 1 + 2\cos(3\theta)$, find the angles of all points of intersection (where $r = 0$), such that $0° \leq \theta < 360°$.

Section 8.5

91. Consider the parametric equations
$x = a\cos(at) + b\sin(bt)$ and $y = a\sin(at) + b\cos(bt)$.
Use a graphing utility to explore the graphs for
$(a, b) = (2, 3)$ and $(a, b) = (3, 2)$. Describe the
t-interval for each case.

92. Consider the parametric equations
$x = a\sin(at) - b\cos(bt)$ and $y = a\cos(at) - b\sin(bt)$.
Use a graphing utility to explore the graphs for
$(a, b) = (1, 2)$ and $(a, b) = (2, 1)$. Describe the
t-interval for each case.

REVIEW EXERCISES

[CHAPTER 8 PRACTICE TEST]

1. Simplify $i^{20} - i^{13} + i^{17} - i^{1000}$.

2. Plot the complex number $-3 - 5i$ in the complex plane.

3. Simplify $\dfrac{2 - 4i}{3i}$ and write in standard form.

4. Simplify $\dfrac{8 - i}{3 + i}$ and write in standard form.

5. Convert the complex number $3\sqrt{2}(1 - i)$ to polar form.

6. Convert the point $2\sqrt{6}(3\sqrt{2} - \sqrt{6}i)$ to polar form.

7. Find the modulus and argument of the complex number $z = -3 + 3i\sqrt{3}$.

8. Find the modulus and argument of the complex number $z = -5 - 5i$.

Given the complex numbers $z_1 = 4(\cos 75° + i \sin 75°)$ and $z_2 = 2(\cos 15° + i \sin 15°)$, find:

9. $z_1 \cdot z_2$

10. $\dfrac{z_1}{z_2}$

In Exercises 11 and 12, for the complex number $z = 16(\cos 120° + i \sin 120°)$, find:

11. z^4

12. four complex distinct fourth roots of z

13. Give another pair of polar coordinates for the point (a, θ).

14. Convert the point $(3, 210°)$ to rectangular coordinates.

15. Use a calculator to convert the rectangular point $(-3, -4)$ to polar coordinates.

16. Graph $r = 6 \sin(2\theta)$.

17. Graph $r = 1 + \cos(3\theta)$.

18. Graph $r = 4\theta$.

19. Graph $r^2 = 9 \cos(2\theta)$.

20. Describe (classify) the plane curve defined by the parametric equations $x = \sqrt{1 - t}$ and $y = \sqrt{t}$ for t in $[0, 1]$.

21. A golf ball is hit with an initial speed of 120 feet per second at an angle of 45° with the ground. How long will the ball stay in the air? How far will the ball travel (horizontal distance) before it hits the ground? (See Section 8.5, Example 4, for equations governing projectile motion.)

22. Force A, at 150 pounds, and force B, at 250 pounds, make an angle of 45° with each other, with force A operating due east and force B above it. Represent their respective vectors as complex numbers written in trigonometric form, and solve for the resultant force.

23. An airplane is flying on a course of 245° as measured from due north at 400 miles per hour. The wind is blowing due south at 24 miles per hour. Represent their respective vectors as complex numbers written in trigonometric form, and solve for the resultant actual speed and direction vector.

24. An arrow is shot from the ground at an angle of 60°. If the arrow has an initial speed of 50 feet per second, how far (horizontal ground distance) will the arrow travel?

25. A woman on a motorcycle racing around an oval track has a position given by the equations $x = -100 \sin\left(\dfrac{t}{2}\right)$ and $y = 75 \cos\left(\dfrac{t}{2}\right)$, where x and y are the horizontal and vertical positions, respectively, in feet relative to the center of the track t seconds after the start of the race. Find the woman's position relative to the center of the track 10 seconds after the start of the race.

26. When you connect the three cube roots of -8 in the complex plane, you form an equilateral triangle. Find the roots and draw the triangle.

[CHAPTERS 1–8 CUMULATIVE TEST]

1. In the right triangle with legs of length a and b and hypotenuse length c, if $b = 5$ and $c = 13$, find a.

2. **Mountain Grade.** From the top of a mountain road looking down, the angle of depression is given by the angle θ. If the grade of the road is given by tangent of θ, written as a percentage, find the grade of a road given that
$$\sin\theta = \frac{\sqrt{401}}{401} \text{ and } \cos\theta = \frac{20\sqrt{401}}{401}.$$

3. If $\sin\theta = \frac{8}{11}$ and the terminal side of θ lies in quadrant II, find the exact value of $\sec\theta$.

4. Convert $140°$ to radians. Leave the answer in terms of π.

5. Find the exact length of the arc with central angle $\theta = 15°$ and radius $r = 900$ kilometers.

6. A point travels along a circle over the time $t = 5$ minutes, with angular speed $\omega = \frac{\pi \text{ rad}}{10 \text{ sec}}$, and radius of the circle $r = 4.2$ inches. Find the distance s. Round to three significant digits.

7. **Sound Waves.** If a sound wave is represented by $y = 0.009 \sin(800\pi t)$ centimeters, what are the amplitude and frequency of the sound wave?

8. Sketch the graph of the function $y = -2 + 3\sin[\pi(x - 1)]$ over the interval $[0, 4]$.

9. Graph the function $y = \cot(\frac{1}{4}x)$ over the interval $0 \le x \le 4\pi$.

10. Determine whether the equation $\cot^2 x - \csc^2 x = 1$ is conditional or an identity.

11. Write $\dfrac{\tan\left(\frac{7\pi}{15}\right) - \tan\left(\frac{2\pi}{15}\right)}{1 + \tan\left(\frac{7\pi}{15}\right)\tan\left(\frac{2\pi}{15}\right)}$ as a single trigonometric function.

12. If $\sin x = \frac{4}{5}$ and $\tan x < 0$, find $\sec(2x)$.

13. Verify the identity $\sec^2\left(\frac{x}{2}\right) - \csc^2\left(\frac{x}{2}\right) = -4\cot x \csc x$.

14. Verify the identity $\dfrac{\cos A + \cos B}{\sin A + \sin B} = \cot\left(\dfrac{A + B}{2}\right)$.

15. Use a calculator to evaluate $\tan^{-1}(-92.6235)$. Give the answer in radians and round to two decimal places.

16. Give the exact evaluation of the expression
$$\tan\left[\sec^{-1}\left(\frac{\sqrt{29}}{2}\right)\right].$$

17. Solve the trigonometric equation $5\tan\theta + 7 = 0$ over the interval $0° \le \theta < 360°$, and express the answer in degrees to two decimal places.

18. Solve the trigonometric equation $\sin(2\theta) + \frac{1}{3}\sin\theta = 0$ over the interval $0° \le \theta < 360°$, and express the answer in degrees to two decimal places.

19. Given two lengths, $a = 211$ and $b = 143$, and angle $\beta = 104°$, determine whether a triangle (or two) exists, and if so, solve the triangle(s).

20. Solve the triangle with two side lengths, $b = 8$ and $a = 13$, and angle $\gamma = 19°$.

21. Find the area of the triangle with $a = \sqrt{13}$, $b = \sqrt{11}$, and $\gamma = 31°$.

22. Find the vector with magnitude $|\mathbf{u}| = 15$ and direction angle $\theta = 110°$.

23. Perform the operation, simplify, and express in standard form: $\dfrac{4 + i}{2 - 3i}$.

24. Find all complex solutions to $x^3 - 27 = 0$.

25. Graph $\theta = -\dfrac{\pi}{4}$.

Algebraic Prerequisites and Review

A.1	A.2	A.3	A.4	A.5	A.6	A.7	A.8
FACTORING POLYNOMIALS	BASIC TOOLS: CARTESIAN PLANE, DISTANCE, AND MIDPOINT	GRAPHING EQUATIONS: POINT-PLOTTING, INTERCEPTS, AND SYMMETRY	FUNCTIONS	GRAPHS OF FUNCTIONS; PIECEWISE-DEFINED FUNCTIONS; INCREASING AND DECREASING FUNCTIONS; AVERAGE RATE OF CHANGE	GRAPHING TECHNIQUES: TRANSFOR-MATIONS	OPERATIONS ON FUNCTIONS AND COMPOSITION OF FUNCTIONS	ONE-TO-ONE FUNCTIONS AND INVERSE FUNCTIONS
• Greatest Common Factor • Factoring Formulas: Special Polynomial Forms • Factoring a Trinomial as a Product of Two Binomials • Factoring by Grouping	• Cartesian Plane • Distance Between Two Points • Midpoint of a Line Segment Joining Two Points	• Point-Plotting • Intercepts • Symmetry • Using Intercepts and Symmetry as Graphing Aids	• Relations and Functions • Functions Defined by Equations • Function Notation • Domain of a Function	• Graphs of Functions • Average Rate of Change • Piecewise-Defined Functions	• Horizontal and Vertical Shifts • Reflection about the Axes • Stretching and Compressing	• Adding, Subtracting, Multiplying, and Dividing Functions • Composition of Functions	• Determine Whether a Function Is One-to-One • Inverse Functions • Graphical Interpretation of Inverse Functions • Finding the Inverse Function

- Factor binomials and trinomials.
- Calculate the distance between two points and the midpoint of a line segment joining two points.

- Sketch the graph of an equation by point-plotting using intercepts and symmetry as graphing aids.
- Find the domain and range of a function.
- Find the average rate of change of a function.

- Sketch graphs of general functions using translations of common functions.
- Perform composition of functions.
- Find the inverse of a function.

A.1 FACTORING POLYNOMIALS

SKILLS OBJECTIVES

- Factor out the greatest common factor.
- Factor polynomials using special forms.
- Factor a trinomial as a product of binomials.
- Factor polynomials by grouping.
- Follow a strategy to factor polynomials.

CONCEPTUAL OBJECTIVES

- Understand that factoring has its basis in the distributive property.
- Learn formulas for factoring special forms.
- Know the general strategy for factoring a trinomial as a product of binomials.
- Be able to identify polynomials that may be factored by grouping.
- Develop a general strategy for factoring polynomials.

You already know how to multiply polynomials. In this section, we examine the reverse of that process, which is called *factoring*. Consider the following product:

$$(x + 3)(x + 1) = x^2 + 4x + 3$$

To *factor* the resulting polynomial, you reverse the process to undo the multiplication:

$$x^2 + 4x + 3 = (x + 3)(x + 1)$$

The polynomials $(x + 3)$ and $(x + 1)$ are called **factors** of the polynomial $x^2 + 4x + 3$. The process of writing a polynomial as a product is called **factoring**.

We will restrict our discussion to factoring polynomials with integer coefficients, which is called **factoring over the integers**. If a polynomial cannot be factored using integer coefficients, then it is **prime** or irreducible over the integers. When a polynomial is written as a product of prime polynomials, then the polynomial is said to be **factored completely**.

A.1.1 Greatest Common Factor

A.1.1 SKILL

Factor out the greatest common factor.

A.1.1 CONCEPTUAL

Understand that factoring has its basis in the distributive property.

The simplest type of factoring of polynomials occurs when there is a factor common to every term of the polynomial. This **common factor** is a monomial that can be "factored out" by applying the distributive property in reverse:

$$ab + ac = a(b + c)$$

For example, $4x^2 - 6x$ can be written as $2x(2x) - 2x(3)$. Notice that $2x$ is a common factor to both terms, so the distributive property tells us we can factor this polynomial to yield $2x(2x - 3)$. Although 2 is a common factor and x is a common factor, the monomial $2x$ is called the *greatest common factor*.

GREATEST COMMON FACTOR

The monomial ax^k is called the **greatest common factor (GCF)** of a polynomial in x with integer coefficients if *both* of the following are true:

- a is the *greatest* integer factor common to all of the polynomial coefficients.
- k is the *smallest* exponent on x found in all of the terms of the polynomial.

POLYNOMIAL	GCF	WRITE EACH TERM AS A PRODUCT OF GCF AND REMAINING FACTOR	FACTORED FORM
$7x + 21$	7	$7(x) + 7(3)$	$7(x + 3)$
$3x^2 + 12x$	$3x$	$3x(x) + 3x(4)$	$3x(x + 4)$
$4x^3 + 2x + 6$	2	$2(2x^3) + 2x + 2(3)$	$2(2x^3 + x + 3)$
$6x^4 - 9x^3 + 12x^2$	$3x^2$	$3x^2(2x^2) - 3x^2(3x) + 3x^2(4)$	$3x^2(2x^2 - 3x + 4)$
$-5x^4 + 25x^3 - 20x^2$	$-5x^2$	$-5x^2 x^2 - 5x^2(-5x) - 5x^2(4)$	$-5x^2(x^2 - 5x + 4)$

[CONCEPT CHECK]

Find the greatest common factor of the polynomial $a^3x^6 + a^4x^4$

▼ ....

ANSWER a^3x^4

EXAMPLE 1 **Factoring Polynomials by Extracting the Greatest Common Factor**

Factor:

a. $6x^5 - 18x^4$ **b.** $6x^5 - 10x^4 - 8x^3 + 12x^2$

Solution (a):

Identify the greatest common factor. $6x^4$

Write each term as a product with the GCF as a factor. $6x^5 - 18x^4 = 6x^4(x) - 6x^4(3)$

Factor out the GCF. $= \boxed{6x^4(x - 3)}$

Solution (b):

Identify the greatest common factor. $2x^2$

Write each term as a product with the GCF as a factor.

$$6x^5 - 10x^4 - 8x^3 + 12x^2 = 2x^2(3x^3) - 2x^2(5x^2) - 2x^2(4x) + 2x^2(6)$$

Factor out the GCF. $= \boxed{2x^2(3x^3 - 5x^2 - 4x + 6)}$

▼
ANSWER

a. $4x(3x^2 - 1)$
b. $3x^2(x^3 - 3x^2 + 4x - 2)$

YOUR TURN Factor:

a. $12x^3 - 4x$ **b.** $3x^5 - 9x^4 + 12x^3 - 6x^2$

A.1.2 Factoring Formulas: Special Polynomial Forms

The first step in factoring polynomials is to look for a common factor. If there is no common factor, then we look for special polynomial forms.

A.1.2 SKILL

Factor polynomials using special forms.

Difference of two squares	$a^2 - b^2 = (a + b)(a - b)$
Perfect squares	$a^2 + 2ab + b^2 = (a + b)^2$
	$a^2 - 2ab + b^2 = (a - b)^2$
Sum of two cubes	$a^3 + b^3 = (a + b)(a^2 - ab + b^2)$
Difference of two cubes	$a^3 - b^3 = (a - b)(a^2 + ab + b^2)$

A.1.2 CONCEPTUAL

Learn formulas for factoring special forms.

EXAMPLE 2 **Factoring the Difference of Two Squares**

Factor:

a. $x^2 - 9$ **b.** $4x^2 - 25$ **c.** $x^4 - 16$

Solution (a):

Rewrite as the difference of two squares. $x^2 - 9 = x^2 - 3^2$

Let $a = x$ and $b = 3$ in $a^2 - b^2 = (a + b)(a - b)$. $= \boxed{(x + 3)(x - 3)}$

Solution (b):

Rewrite as the difference of two squares. $4x^2 - 25 = (2x)^2 - 5^2$

Let $a = 2x$ and $b = 5$ in $a^2 - b^2 = (a + b)(a - b)$. $= \boxed{(2x + 5)(2x - 5)}$

Solution (c):

Rewrite as the difference of two squares. $x^4 - 16 = (x^2)^2 - 4^2$

Let $a = x^2$ and $b = 4$ in $a^2 - b^2 = (a + b)(a - b)$. $= (x^2 + 4)(x^2 - 4)$

Note that $x^2 - 4$ is also a difference of two squares (part a below). $= \boxed{(x + 2)(x - 2)(x^2 + 4)}$

$\begin{bmatrix} \text{CONCEPT CHECK} \end{bmatrix}$

Factor: $a^2x^4 - b^2y^2$

▼ ....

ANSWER $(ax^2 + by)(ax^2 - by)$

▼

YOUR TURN Factor:

a. $x^2 - 4$ **b.** $9x^2 - 16$ **c.** $x^4 - 81$

▼

ANSWER

a. $(x + 2)(x - 2)$
b. $(3x + 4)(3x - 4)$
c. $(x^2 + 9)(x - 3)(x + 3)$

A trinomial is a perfect square if it has the form $a^2 \pm 2ab + b^2$. Notice that:

- The first term and third term are perfect squares.
- The middle term is twice the product of the bases of these two perfect squares.
- The sign of the middle term determines the sign of the factored form:

$$a^2 \pm 2ab + b^2 = (a \pm b)^2$$

EXAMPLE 3 **Factoring Trinomials That Are Perfect Squares**

Factor:

a. $x^2 + 6x + 9$ **b.** $x^2 - 10x + 25$ **c.** $9x^2 - 12x + 4$

Solution (a):

Rewrite the trinomial so that
the first and third terms
are perfect squares. $x^2 + 6x + 9 = x^2 + 6x + 3^2$

Notice that if we let $a = x$ and $b = 3$ in
$a^2 + 2ab + b^2 = (a + b)^2$, then the
middle term $6x$ is $2ab$. $x^2 + 6x + 9 = x^2 + 2(3x) + 3^2 = \boxed{(x + 3)^2}$

Solution (b):

Rewrite the trinomial so that
the first and third terms
are perfect squares. $x^2 - 10x + 25 = x^2 - 10x + 5^2$

Notice that if we let $a = x$ and $b = 5$
in $a^2 - 2ab + b^2 = (a - b)^2$,
then the middle
term $-10x$ is $-2ab$. $x^2 - 10x + 25 = x^2 - 2(5x) + 5^2 = \boxed{(x - 5)^2}$

Solution (c):

Rewrite the trinomial so that
the first and third terms
are perfect squares. $9x^2 - 12x + 4 = (3x)^2 - 2(3x)(2) + 2^2$

Notice that if we let $a = 3x$ and $b = 2$
in $a^2 - 2ab + b^2 = (a - b)^2$,
then the middle
term $-12x$ is $-2ab$. $9x^2 - 12x + 4 = (3x)^2 - 2(3x)(2) + 2^2 = \boxed{(3x - 2)^2}$

▼

ANSWER

a. $(x + 4)^2$

b. $(x - 2)^2$

c. $(5x - 2)^2$

▼

YOUR TURN Factor:

a. $x^2 + 8x + 16$ **b.** $x^2 - 4x + 4$ **c.** $25x^2 - 20x + 4$

EXAMPLE 4 **Factoring the Sum of Two Cubes**

Factor $x^3 + 27$.

Solution:

Rewrite as the sum of two cubes. $x^3 + 27 = x^3 + 3^3$

Write the sum of two cubes formula. $a^3 + b^3 = (a + b)(a^2 - ab + b^2)$

Let $a = x$ and $b = 3$. $x^3 + 27 = x^3 + 3^3 = \boxed{(x + 3)(x^2 - 3x + 9)}$

EXAMPLE 5   **Factoring the Difference of Two Cubes**

Factor $x^3 - 125$.

Solution:

Rewrite as the difference of two cubes.

$$x^3 - 125 = x^3 - 5^3$$

Write the difference of two cubes formula.

$$a^3 - b^3 = (a - b)(a^2 + ab + b^2)$$

Let $a = x$ and $b = 5$.

$$x^3 - 125 = x^3 - 5^3 = (x - 5)(x^2 + 5x + 25)$$

▼

YOUR TURN Factor:

a. $x^3 + 8$ **b.** $x^3 - 64$

▼
ANSWER
a. $(x + 2)(x^2 - 2x + 4)$
b. $(x - 4)(x^2 + 4x + 16)$

A.1.3 Factoring a Trinomial as a Product of Two Binomials

The first step in factoring is to look for a common factor. If there is no common factor, look to see whether the polynomial is a special form for which we know the factoring formula. If it is not of such a special form, then if it is a trinomial, we proceed with a general factoring strategy.

A.1.3 SKILL

Factor a trinomial as a product of binomials.

We know that $(x + 3)(x + 2) = x^2 + 5x + 6$, so we say the **factors** of $x^2 + 5x + 6$ are $(x + 3)$ and $(x + 2)$. In factored form we have $x^2 + 5x + 6 = (x + 3)(x + 2)$. Recall the FOIL method: The product of the last terms (3 and 2) is 6, and the sum of the products of the inner terms (3x) and the outer terms (2x) is 5x. Let's pretend for a minute that we didn't know this factored form but had to work with the general form:

A.1.3 CONCEPTUAL

Know the general strategy for factoring a trinomial as a product of binomials.

$$x^2 + 5x + 6 = (x + a)(x + b)$$

The goal is to find a and b. We start by multiplying the two binomials on the right.

$$x^2 + 5x + 6 = (x + a)(x + b) = x^2 + ax + bx + ab = x^2 + (a + b)x + ab$$

Compare the expression we started with on the left with the expression on the far right: $x^2 + 5x + 6 = x^2 + (a + b)x + ab$. We see that $ab = 6$ and $(a + b) = 5$. *Start* with the possible combinations of a and b whose product is 6, and *then* look among those for the combination whose sum is 5.

$ab = 6$	a, b:	1, 6	−1, −6	2, 3	−2, −3
$a + b$		7	−7	**5**	−5

All of the possible a, b combinations in the first row have a product equal to 6, but only one of those has a sum equal to 5. Therefore, the factored form is

$$x^2 + 5x + 6 = (x + a)(x + b) = (x + 2)(x + 3)$$

▶ **EXAMPLE 6** **Factoring a Trinomial Whose Leading Coefficient Is 1**

Factor $x^2 + 10x + 9$.

Solution:

Write the trinomial as a product of two binomials in general form.

$$x^2 + 10x + 9 = (x + \square)(x + \square)$$

Write all of the integers whose product is 9.

Integers whose product is 9	1, 9	−1, −9	3, 3	−3, −3

Determine the sum of the integers.

Integers whose product is 9	**1, 9**	−1, −9	3, 3	−3,−3
Sum	**10**	−10	6	−6

Select 1, 9 because the product is 9 (last term of the trinomial) and the sum is 10 (middle-term coefficient of the trinomial).

$$\boxed{x^2 + 10x + 9 = (x + 9)(x + 1)}$$

Check: $(x + 9)(x + 1) = x^2 + 9x + 1x + 9 = x^2 + 10x + 9$ ✓

▼
ANSWER

$(x + 4)(x + 5)$

▼
YOUR TURN Factor $x^2 + 9x + 20$.

In Example 6, all terms in the trinomial are positive. When the constant term is negative, then (regardless of whether the middle term is positive or negative) the factors will be opposite in sign, as illustrated in Example 7.

EXAMPLE 7 **Factoring a Trinomial Whose Leading Coefficient Is 1**

Factor $x^2 - 3x - 28$.

Solution:

Write the trinomial as a product of two binomials in general form.

$$x^2 - 3x + -28 = (x + \square)(x - \square)$$

Write all of the integers whose product is −28.

Integers whose product is −28	1, −28	−1, 28	2, −14	−2, 14	4, −7	−4, 7

Determine the sum of the integers.

Integers whose product is −28	1, −28	−1, 28	2, −14	−2, 14	**4, −7**	−4, 7
Sum	−27	27	−12	12	**−3**	3

Select 4, −7 because the product is −28 (last term of the trinomial) and the sum is −3 (middle-term coefficient of the trinomial).

$$\boxed{x^2 - 3x - 28 = (x + 4)(x - 7)}$$

Check: $(x + 4)(x - 7) = x^2 - 7x + 4x + -28 = x^2 - 3x + -28$ ✓

▼
ANSWER

$(x + 6)(x - 3)$

▼
YOUR TURN Factor $x^2 + 3x - 18$.

When the leading coefficient of the trinomial is not equal to 1, then we consider all possible factors using the following procedure, which is based on the FOIL method in reverse.

**FACTORING A TRINOMIAL WHOSE
LEADING COEFFICIENT IS NOT 1**

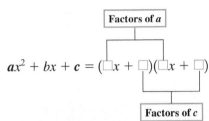

Step 1: Find two **First** terms whose product is the first term of the trinomial.
Step 2: Find two **Last** terms whose product is the last term of the trinomial.
Step 3: Consider all possible combinations found in Steps 1 and 2 until the sum of the **Outer** and **Inner** products is equal to the middle term of the trinomial.

▶ **EXAMPLE 8**   **Factoring a Trinomial Whose Leading Coefficient Is Not 1**

Factor $5x^2 + 9x - 2$.

Solution:

STEP 1 Start with the <u>first</u> term. Note that $5x \cdot x = 5x^2$. $(5x \pm \square)(x \pm \square)$

STEP 2 The product of the <u>last</u> terms should yield -2. $-1, 2$ or $1, -2$

STEP 3 Consider all possible factors based on Steps 1 and 2.
$$(5x - 1)(x + 2)$$
$$(5x + 1)(x - 2)$$
$$(5x + 2)(x - 1)$$
$$(5x - 2)(x + 1)$$

Since the <u>outer</u> and <u>inner</u> products must sum to $9x$, the factored form must be: $\boxed{5x^2 + 9x - 2 = (5x - 1)(x + 2)}$

Check: $(5x - 1)(x + 2) = 5x^2 + 10x - 1x - 2 = 5x^2 + 9x - 2$ ✓

▼
YOUR TURN   Factor $2t^2 + t - 3$.

▼
ANSWER
$(2t + 3)(t - 1)$

STUDY TIP

In Example 9, Step 3, we can eliminate any factors that have a common factor since there is no common factor to the terms in the trinomial.

EXAMPLE 9 **Factoring a Trinomial Whose Leading Coefficient Is Not 1**

Factor $15x^2 - x - 6$.

Solution:

STEP 1 Start with the <u>first</u> term. $(5x \pm \square)(3x \pm \square)$ or $(15x \pm \square)(x \pm \square)$

STEP 2 The product of the <u>last</u> terms
should yield -6. $-1, 6$ or $1, -6$ or $2, -3$ or $-2, 3$

STEP 3 Consider all possible factors
based on Steps 1 and 2.

$(5x - 1)(3x + 6)$ $(15x - 1)(x + 6)$
$(5x + 6)(3x - 1)$ $(15x + 6)(x - 1)$
$(5x + 1)(3x - 6)$ $(15x + 1)(x - 6)$
$(5x - 6)(3x + 1)$ $(15x - 6)(x + 1)$
$(5x + 2)(3x - 3)$ $(15x + 2)(x - 3)$
$(5x - 3)(3x + 2)$ $(15x - 3)(x + 2)$
$(5x - 2)(3x + 3)$ $(15x - 2)(x + 3)$
$(5x + 3)(3x - 2)$ $(15x + 3)(x - 2)$

Since the <u>outer</u> and <u>inner</u> products must
sum to $-x$, the factored form must be:

$$15x^2 - x - 6 = (5x + 3)(3x - 2)$$

Check: $(5x + 3)(3x - 2) = 15x^2 - 10x + 9x - 6 = 15x^2 - x - 6$ ✓

▼
ANSWER
$(3x - 4)(2x + 3)$

▼
YOUR TURN   Factor $6x^2 + x - 12$.

[**CONCEPT CHECK**]

TRUE OR FALSE A trinomial that is degree 2 and has integer coefficients with the leading coefficient equal to 1 is always factorable.

▼
ANSWER False

EXAMPLE 10 **Identifying Prime (Irreducible) Polynomials**

Factor $x^2 + x - 8$.

Solution:

Write the trinomial as a product of
two binomials in general form. $x^2 + x - 8 = (x + \square)(x - \square)$

Write all of
the integers whose
product is -8.

Integers whose product is -8	$1, -8$	$-1, 8$	$4, -2$	$-4, 2$

Determine the sum
of the integers.

Integers whose product is -8	$1, -8$	$-1, 8$	$4, -2$	$-4, 2$
Sum	-7	7	2	-2

The middle term of the trinomial is $-x$, so we look for the sum of the integers
that equals -1. Since no sum exists for the given combinations, we say that this
polynomial is *prime* (irreducible) over the integers.

A.1.4 Factoring by Grouping

Much of our attention in this section has been on factoring trinomials. For polynomials with more than three terms, we first look for a common factor to all terms. If there is no common factor to all terms of the polynomial, we look for a group of terms that have a common factor. This strategy is called **factoring by grouping**.

A.1.4 SKILL

Factor polynomials by grouping.

A.1.4 CONCEPTUAL

Be able to identify polynomials that may be factored by grouping.

EXAMPLE 11   **Factoring a Polynomial by Grouping**

Factor $x^3 - x^2 + 2x - 2$.

Solution:

Group the terms that have a common factor.	$= (x^3 - x^2) + (2x - 2)$
Factor out the common factor in each pair of parentheses.	$= x^2(x - 1) + 2(x - 1)$
Use the distributive property.	$= \boxed{(x^2 + 2)(x - 1)}$

$\begin{bmatrix}\text{CONCEPT CHECK}\end{bmatrix}$

TRUE OR FALSE Trinomials cannot be factored by grouping.

ANSWER True

▶ **EXAMPLE 12**   **Factoring a Polynomial by Grouping**

Factor $2x^2 + 2x - x - 1$.

Solution:

Group the terms that have a common factor.	$= (2x^2 + 2x) + (-x - 1)$
Factor out the common factor in each pair of parentheses.	$= 2x(x + 1) - 1(x + 1)$
Use the distributive property.	$= \boxed{(2x - 1)(x + 1)}$

▼

YOUR TURN Factor $x^3 + x^2 - 3x - 3$.

▼

ANSWER

$(x + 1)(x^2 - 3)$

A.1.5 A Strategy for Factoring Polynomials

The first step in factoring a polynomial is to look for the greatest common factor. When specifically factoring trinomials, look for special known forms like a perfect square. A general approach to factoring a trinomial uses the FOIL method in reverse. Finally, we look for factoring by grouping. The following strategy for factoring polynomials is based on the techniques discussed in this section.

A.1.5 SKILL

Follow a strategy to factor polynomials.

A.1.5 CONCEPTUAL

Develop a general strategy for factoring polynomials.

STRATEGY FOR FACTORING POLYNOMIALS

1. Factor out the greatest common factor (monomial).
2. Identify any special polynomial forms and apply factoring formulas.
3. Factor a trinomial into a product of two binomials: $(ax + b)(cx + d)$.
4. Factor by grouping.

$\begin{bmatrix}\text{STUDY TIP}\end{bmatrix}$

When factoring, always start by factoring out the GCF.

EXAMPLE 13 **Factoring Polynomials**

Factor:

a. $3x^2 - 6x + 3$ **b.** $-4x^3 + 2x^2 + 6x$ **c.** $15x^2 + 7x - 2$ **d.** $x^3 - x + 2x^2 - 2$

Solution (a):

Factor out the greatest common factor. $3x^2 - 6x + 3 = 3(x^2 - 2x + 1)$

The trinomial is a perfect square. $= \boxed{3(x - 1)^2}$

Solution (b):

Factor out the greatest common factor. $-4x^3 + 2x^2 + 6x = -2x(2x^2 - x - 3)$

Use the FOIL method in reverse to factor the trinomial. $= \boxed{-2x(2x - 3)(x + 1)}$

Solution (c):

There is no common factor. $15x^2 + 7x - 2$

Use the FOIL method in reverse to factor the trinomial. $= \boxed{(3x + 2)(5x - 1)}$

Solution (d):

Factor by grouping.

$$(x^3 - x) + (2x^2 - 2)$$
$$= x(x^2 - 1) + 2(x^2 - 1)$$
$$= (x + 2)(x^2 - 1)$$
$$= \boxed{(x + 2)(x - 1)(x + 1)}$$

▶[SECTION A.1] SUMMARY

In this section, we discussed factoring polynomials, which is the reverse process of multiplying polynomials. Four main techniques were discussed.

GREATEST COMMON FACTOR: ax^k

- a is the greatest common factor for all coefficients of the polynomial.
- k is the smallest exponent found on all of the terms in the polynomial.

FACTORING FORMULAS: SPECIAL POLYNOMIAL FORMS

Difference of two squares: $a^2 - b^2 = (a + b)(a - b)$

Perfect squares: $a^2 + 2ab + b^2 = (a + b)^2$
$a^2 - 2ab + b^2 = (a - b)^2$

Sum of two cubes: $a^3 + b^3 = (a + b)(a^2 - ab + b^2)$

Difference of two cubes: $a^3 - b^3 = (a - b)(a^2 + ab + b^2)$

FACTORING A TRINOMIAL AS A PRODUCT OF TWO BINOMIALS

- $x^2 + bx + c = (x + \square)(x + \square)$

  1. Find all possible combinations of factors whose product is c.
  2. Of the combinations in Step 1, look for the sum of factors that equals b.

- $ax^2 + bx + c = (\square x + \square)(\square x + \square)$

  1. Find all possible combinations of the first terms whose product is ax^2.
  2. Find all possible combinations of the last terms whose product is c.
  3. Consider all possible factors based on Steps 1 and 2.

FACTORING BY GROUPING

- Group terms that have a common factor.
- Use the distributive property.

[SECTION A.1] EXERCISES

• **SKILLS**

In Exercises 1–12, factor each expression. Start by finding the greatest common factor (GCF).

1. $5x + 25$
2. $x^2 + 2x$
3. $4t^2 - 2$
4. $16z^2 - 20z$

5. $2x^3 - 50x$
6. $4x^2y - 8xy^2 + 16x^2y^2$
7. $3x^3 - 9x^2 + 12x$
8. $14x^4 - 7x^2 + 21x$

9. $x^3 - 3x^2 - 40x$
10. $-9y^2 + 45y$
11. $4x^2y^3 + 6xy$
12. $3z^3 - 6z^2 + 18z$

In Exercises 13–20, factor the difference of two squares.

13. $x^2 - 9$
14. $x^2 - 25$
15. $4x^2 - 9$
16. $1 - x^4$

17. $2x^2 - 98$
18. $144 - 81y^2$
19. $225x^2 - 169y^2$
20. $121y^2 - 49x^2$

In Exercises 21–32, factor the perfect squares.

21. $x^2 + 8x + 16$
22. $y^2 - 10y + 25$
23. $x^4 - 4x^2 + 4$
24. $1 - 6y + 9y^2$

25. $4x^2 + 12xy + 9y^2$
26. $x^2 - 6xy + 9y^2$
27. $9 - 6x + x^2$
28. $25x^2 - 20xy + 4y^2$

29. $x^4 + 2x^2 + 1$
30. $x^6 - 6x^3 + 9$
31. $p^2 + 2pq + q^2$
32. $p^2 - 2pq + q^2$

In Exercises 33–42, factor the sum or difference of two cubes.

33. $t^3 + 27$
34. $z^3 + 64$
35. $y^3 - 64$
36. $x^3 - 1$

37. $8 - x^3$
38. $27 - y^3$
39. $y^3 + 125$
40. $64x - x^4$

41. $27 + x^3$
42. $216x^3 - y^3$

In Exercises 43–52, factor each trinomial into a product of two binomials.

43. $x^2 - 6x + 5$
44. $t^2 - 5t - 6$
45. $y^2 - 2y - 3$
46. $y^2 - 3y - 10$

47. $2y^2 - 5y - 3$
48. $2z^2 - 4z - 6$
49. $3t^2 + 7t + 2$
50. $4x^2 - 2x - 12$

51. $-6t^2 + t + 2$
52. $-6x^2 - 17x + 10$

In Exercises 53–60, factor by grouping.

53. $x^3 - 3x^2 + 2x - 6$
54. $x^5 + 5x^3 - 3x^2 - 15$
55. $a^4 + 2a^3 - 8a - 16$
56. $x^4 - 3x^3 - x + 3$

57. $3xy - 5rx - 10rs + 6sy$
58. $6x^2 - 10x + 3x - 5$
59. $20x^2 + 8xy - 5xy - 2y^2$
60. $9x^5 - a^2x^3 - 9x^2 + a^2$

In Exercises 61–92, factor each of the polynomials completely, if possible. If the polynomial cannot be factored, state that it is prime.

61. $x^2 - 4y^2$
62. $a^2 + 5a + 6$
63. $3a^2 + a - 14$
64. $ax + b + bx + a$

65. $x^2 + 16$
66. $x^2 + 49$
67. $4z^2 + 25$
68. $\frac{1}{16} - b^4$

69. $6x^2 + 10x + 4$
70. $x^2 + 7x + 5$
71. $6x^2 + 13xy - 5y^2$
72. $15x + 15xy$

73. $36s^2 - 9t^2$
74. $3x^3 - 108x$
75. $a^2b^2 - 25c^2$
76. $2x^3 + 54$

77. $4x^2 - 3x - 10$
78. $10x - 25 - x^2$
79. $3x^3 - 5x^2 - 2x$
80. $2y^3 + 3y^2 - 2y$

81. $x^3 - 9x$
82. $w^3 - 25w$
83. $xy - x - y + 1$
84. $a + b + ab + b^2$

85. $x^4 + 5x^2 + 6$
86. $x^6 - 7x^3 - 8$
87. $x^2 - 2x - 24$
88. $25x^2 + 30x + 9$

89. $x^4 + 125x$
90. $x^4 - 1$
91. $x^4 - 81$
92. $10x^2 - 31x + 15$

• APPLICATIONS

93. Geometry. The perimeter p of a rectangle is given by $p = 2l + 2w$, where l is length, $l = 2x + 4$, and w is width, $w = x$. Express the perimeter of the rectangle as a factored polynomial in x.

94. Geometry. The volume v of a box is given by the expression $v = l \cdot w \cdot h = x^3 + 7x^2 + 12x$. Express the volume as a factored polynomial in the variable x. Give an expression for the length, width, and height in terms of x.

95. Business. The profit P of a business is given by the expression $P = 2x^2 - 15x + 4x - 30$, where x is the number of units produced. Express the profit as a factored polynomial in the variable x.

96. Business. The break-even point for a company is given by solving the equation $3x^2 + 9x - 4x - 12 = 0$, where x is the number of units produced. Factor the polynomial on the left side of the equation.

97. Engineering. The height h of a projectile is given by the equation $h = -16t^2 - 78t + 10$, where t is the time in seconds after the projectile is launched. Factor the expression on the right side of the equal sign.

98. Engineering. The electrical field at a point P between two charges is given by $k = \dfrac{10x - x^2}{100}$, where x is the distance between the two charges. Factor the numerator of this expression.

• CATCH THE MISTAKE

In Exercises 99 and 100, explain the mistake that is made.

99. Factor $x^3 - x^2 - 9x + 9$.

Solution:

Group terms with common factors.	$(x^3 - x^2) + (-9x + 9)$
Factor out common factors.	$x^2(x - 1) - 9(x - 1)$
Distributive property.	$(x - 1)(x^2 - 9)$
Factor $x^2 - 9$.	$(x - 1)(x - 3)^2$

This is incorrect. What mistake was made?

100. Factor $4x^2 + 12x - 40$.

Solution:

Factor the trinomial into a product of binomials.	$(2x - 4)(2x + 10)$
Factor out a 2.	$= 2(x - 2)(x + 5)$

This is incorrect. What mistake was made?

• CONCEPTUAL

In Exercises 101–104, determine whether each of the following statements is true or false.

101. All trinomials can be factored into a product of two binomials.

102. All polynomials can be factored into prime factors with respect to the integers.

103. $x^2 - y^2 = (x - y)(x + y)$

104. $x^2 + y^2 = (x + y)^2$

• CHALLENGE

105. Factor $a^{2n} - b^{2n}$ completely, assuming a, b, and n are positive integers.

106. Find all the values of c such that the trinomial $x^2 + cx - 14$ can be factored.

• TECHNOLOGY

107. Use a graphing utility to plot the graphs of the three expressions $8x^3 + 1$, $(2x + 1)(4x^2 - 2x + 1)$, and $(2x - 1)(4x^2 + 2x + 1)$. Which two graphs agree with each other?

108. Use a graphing utility to plot the graphs of the three expressions $27x^3 - 1$, $(3x - 1)^3$, and $(3x - 1)(9x^2 + 3x + 1)$. Which two graphs agree with each other?

A.2 BASIC TOOLS: CARTESIAN PLANE, DISTANCE, AND MIDPOINT

SKILLS OBJECTIVES

- Plot points on the Cartesian plane.
- Calculate the distance between two points in the Cartesian plane.
- Find the midpoint of a line segment joining two points in the Cartesian plane.

CONCEPTUAL OBJECTIVES

- Expand the concept of a one-dimensional number line to a two-dimensional plane.
- Derive the distance formula using the Pythagorean theorem.
- Conceptualize the midpoint as the average of the x- and y-coordinates.

A.2.1 Cartesian Plane

HIV infection rates, stock prices, and temperature conversions are all examples of relationships between two quantities that can be expressed in a two-dimensional graph. Because it is two-dimensional, such a graph lies in a **plane**.

Two perpendicular real number lines, known as the **axes** in the plane, intersect at a point we call the **origin**. Typically, the horizontal axis is called the **x-axis**, and the vertical axis is denoted as the **y-axis**. The axes divide the plane into four **quadrants**, numbered by Roman numerals and ordered counterclockwise.

Points in the plane are represented by **ordered pairs**, denoted (x, y). The first number of the ordered pair indicates the position in the horizontal direction and is often called the *x-coordinate* or **abscissa**. The second number indicates the position in the vertical direction and is often called the *y-coordinate* or **ordinate**. The origin is denoted $(0, 0)$.

Examples of other coordinates are given on the graph to the right.

The point $(2, 4)$ lies in quadrant I. To **plot** this point, start at the origin $(0, 0)$ and move to the right two units and up four units.

All points in quadrant I have positive coordinates, and all points in quadrant III have negative coordinates. Quadrant II has negative x-coordinates and positive y-coordinates; quadrant IV has positive x-coordinates and negative y-coordinates.

This representation is called the **rectangular coordinate system** or **Cartesian coordinate system**, named after the French mathematician René Descartes (1596–1650).

EXAMPLE 1 Plotting Points in a Cartesian Plane

a. Plot and label the points $(-1, -4)$, $(2, 2)$, $(-2, 3)$, $(2, -3)$, $(0, 5)$, and $(-3, 0)$ in the Cartesian plane.

b. List the points and corresponding quadrant or axis in a table.

Solution:

a.

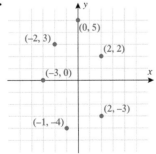

b.

POINT	QUADRANT
$(2, 2)$	I
$(-2, 3)$	II
$(-1, -4)$	III
$(2, -3)$	IV
$(0, 5)$	y-axis
$(-3, 0)$	x-axis

A.2.1 SKILL

Plot points on the Cartesian plane.

A.2.1 CONCEPTUAL

Expand the concept of a one-dimensional number line to a two-dimensional plane.

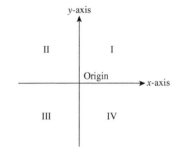

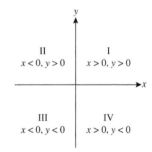

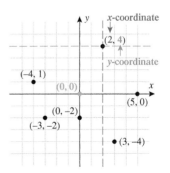

A.2.2 Distance Between Two Points

Suppose you want to find the distance between any two points in the plane. In the previous graph, to find the distance between the points $(2, -3)$ and $(2, 2)$, count the units between the two points. The distance is 5. What if the two points do not lie along a horizontal or vertical line? Example 2 uses the Pythagorean theorem to help find the distance between any two points.

EXAMPLE 2 **Finding the Distance Between Two Points**

Find the distance between the points $(-2, -1)$ and $(1, 3)$.

Solution:

STEP 1 Plot and label the two points in the Cartesian plane and draw a segment indicating the distance d between the two points.

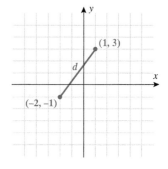

STEP 2 Form a right triangle by connecting the points to a third point, $(1, -1)$.

STEP 3 Calculate the length of the horizontal segment. $3 = |1 - (-2)|$

Calculate the length of the vertical segment. $4 = |3 - (-1)|$

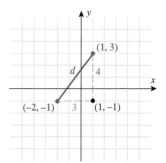

STEP 4 Use the Pythagorean theorem to calculate the distance d.

$$d^2 = 3^2 + 4^2$$
$$d^2 = 9 + 16 = 25$$
$$\boxed{d = 5}$$

WORDS

For any two points, (x_1, y_1) and (x_2, y_2):

MATH

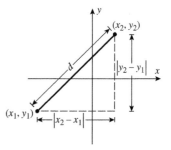

The distance along the horizontal segment is the absolute value of the difference between the x-values.

$$|x_2 - x_1|$$

The distance along the vertical segment is the absolute value of the difference between the y-values.

$$|y_2 - y_1|$$

Use the Pythagorean theorem to calculate the distance d.

$$d^2 = |x_2 - x_1|^2 + |y_2 - y_1|^2$$

$|a|^2 = a^2$ for all real numbers a.

$$d^2 = (x_2 - x_1)^2 + (y_2 - y_1)^2$$

Use the square-root property.

$$d = \pm\sqrt{(x_2 - x_1)^2 + (y_2 - y_1)^2}$$

Distance can be only positive.

$$d = \sqrt{(x_2 - x_1)^2 + (y_2 - y_1)^2}$$

DEFINITION Distance Formula

The **distance** d between two points $P_1 = (x_1, y_1)$ and $P_2 = (x_2, y_2)$ is given by

$$d = \sqrt{(x_2 - x_1)^2 + (y_2 - y_1)^2}$$

The distance between two points is the square root of the sum of the square of the distance between the x-coordinates and the square of the distance between the y-coordinates.

STUDY TIP

It does not matter which point is taken to be the first point or the second point.

You will prove in the exercises that it does not matter which point you take to be the first point when applying the distance formula.

▶ **EXAMPLE 3** **Using the Distance Formula to Find the Distance Between Two Points**

Find the distance between $(-3, 7)$ and $(5, -2)$.

Solution:

Write the distance formula.

$$d = \sqrt{(x_2 - x_1)^2 + (y_2 - y_1)^2}$$

Substitute $(x_1, y_1) = (-3, 7)$ and $(x_2, y_2) = (5, -2)$.

$$d = \sqrt{[5 - (-3)]^2 + (-2 - 7)^2}$$

Simplify.

$$d = \sqrt{(5 + 3)^2 + (-2 - 7)^2}$$

$$d = \sqrt{8^2 + (-9)^2} = \sqrt{64 + 81} = \sqrt{145}$$

Solve for d.

$$\boxed{d = \sqrt{145}}$$

▼

YOUR TURN Find the distance between $(4, -5)$ and $(-3, -2)$.

ANSWER
$d = \sqrt{58}$

A.2.3 Midpoint of a Line Segment Joining Two Points

The **midpoint** (x_m, y_m) of a line segment joining two points (x_1, y_1) and (x_2, y_2) is defined as the point that lies on the segment which has the same distance d from both points. In other words, the midpoint of a segment lies halfway between the given endpoints. The coordinates of the midpoint are found by averaging the *x*-coordinates and averaging the *y*-coordinates.

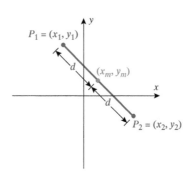

DEFINITION **Midpoint Formula**

The **midpoint** (x_m, y_m) of the line segment with endpoints (x_1, y_1) and (x_2, y_2) is given by

$$(x_m, y_m) = \left(\frac{x_1 + x_2}{2}, \frac{y_1 + y_2}{2} \right)$$

The midpoint can be found by averaging the x-coordinates and averaging the y-coordinates.

▶ **EXAMPLE 4** **Finding the Midpoint of a Line Segment**

Find the midpoint of the line segment joining the points $(2, 6)$ and $(-4, -2)$.

Solution:

Write the midpoint formula. $\qquad (x_m, y_m) = \left(\dfrac{x_1 + x_2}{2}, \dfrac{y_1 + y_2}{2} \right)$

Substitute $(x_1, y_1) = (2, 6)$
and $(x_2, y_2) = (-4, -2)$. $\qquad (x_m, y_m) = \left(\dfrac{2 + (-4)}{2}, \dfrac{6 + (-2)}{2} \right)$

Simplify. $\qquad \boxed{(x_m, y_m) = (-1, 2)}$

One way to verify your answer is to plot the given points and the midpoint to make sure your answer looks reasonable.

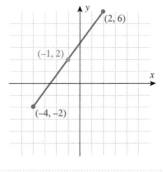

▼
YOUR TURN Find the midpoint of the line segment joining the points $(3, -4)$ and $(5, 8)$.

▶[SECTION A.2] SUMMARY

CARTESIAN PLANE

- Plotting coordinates: (x, y)
- Quadrants: I, II, III, and IV
- Origin: $(0, 0)$

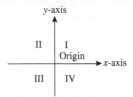

DISTANCE BETWEEN TWO POINTS

$$d = \sqrt{(x_2 - x_1)^2 + (y_2 - y_1)^2}$$

MIDPOINT OF SEGMENT JOINING TWO POINTS

$$\text{Midpoint} = (x_m, y_m) = \left(\frac{x_1 + x_2}{2}, \frac{y_1 + y_2}{2} \right)$$

[SECTION A.2] EXERCISES

- **SKILLS**

In Exercises 1–6, give the coordinates for each point labeled.

1. Point A
2. Point B
3. Point C
4. Point D
5. Point E
6. Point F

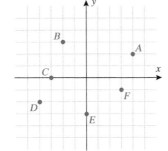

In Exercises 7 and 8, plot each point in the Cartesian plane and indicate in which quadrant or on which axis the point lies.

7. $A: (-2, 3)$ $B: (1, 4)$ $C: (-3, -3)$ $D: (5, -1)$ $E: (0, -2)$ $F: (4, 0)$
8. $A: (-1, 2)$ $B: (1, 3)$ $C: (-4, -1)$ $D: (3, -2)$ $E: (0, 5)$ $F: (-3, 0)$
9. Plot the points $(-3, 1), (-3, 4), (-3, -2), (-3, 0), (-3, -4)$. Describe the line containing points of the form $(-3, y)$.
10. Plot the points $(-1, 2), (-3, 2), (0, 2), (3, 2), (5, 2)$. Describe the line containing points of the form $(x, 2)$.

In Exercises 11–32, calculate the distance between the given points, and find the midpoint of the segment joining them.

11. $(1, 3)$ and $(5, 3)$
12. $(-2, 4)$ and $(-2, -4)$
13. $(-1, 4)$ and $(3, 0)$
14. $(-3, -1)$ and $(1, 3)$
15. $(-10, 8)$ and $(-7, -1)$
16. $(-2, 12)$ and $(7, 15)$
17. $(-3, -1)$ and $(-7, 2)$
18. $(-4, 5)$ and $(-9, -7)$
19. $(-6, -4)$ and $(-2, -8)$
20. $(0, -7)$ and $(-4, -5)$
21. $\left(-\frac{1}{2}, \frac{1}{3}\right)$ and $\left(\frac{7}{2}, \frac{10}{3}\right)$
22. $\left(\frac{1}{5}, \frac{7}{3}\right)$ and $\left(\frac{9}{5}, -\frac{2}{3}\right)$
23. $\left(-\frac{2}{3}, -\frac{1}{5}\right)$ and $\left(\frac{1}{4}, \frac{1}{3}\right)$
24. $\left(\frac{7}{5}, \frac{1}{9}\right)$ and $\left(\frac{1}{2}, -\frac{7}{3}\right)$
25. $(-1.5, 3.2)$ and $(2.1, 4.7)$
26. $(-1.2, -2.5)$ and $(3.7, 4.6)$
27. $(-14.2, 15.1)$ and $(16.3, -17.5)$
28. $(1.1, 2.2)$ and $(3.3, 4.4)$
29. $\left(\sqrt{3}, 5\sqrt{2}\right)$ and $\left(\sqrt{3}, \sqrt{2}\right)$
30. $\left(3\sqrt{5}, -3\sqrt{3}\right)$ and $\left(-\sqrt{5}, -\sqrt{3}\right)$
31. $\left(1, \sqrt{3}\right)$ and $\left(-\sqrt{2}, -2\right)$
32. $\left(2\sqrt{5}, 4\right)$ and $\left(1, 2\sqrt{3}\right)$

In Exercises 33 and 34, calculate (to two decimal places) the perimeter of the triangle with the following vertices.

33. Points A, B, and C
34. Points C, D, and E

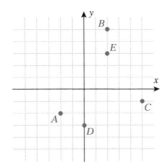

In Exercises 35–38, determine whether the triangle with the given vertices is a right triangle, an isosceles triangle, neither, or both. (Recall that a right triangle satisfies the Pythagorean theorem and an isosceles triangle has at least two sides of equal length.)

35. $(0, -3), (3, -3),$ and $(3, 5)$

36. $(0, 2), (-2, -2),$ and $(2, -2)$

37. $(1, 1), (3, -1),$ and $(-2, -4)$

38. $(-3, 3), (3, 3),$ and $(-3, -3)$

• APPLICATIONS

39. Cell Phones. A cellular phone company currently has three towers: one in Tampa, one in Orlando, and one in Gainesville to serve the central Florida region. If Orlando is 80 miles east of Tampa and Gainesville is 100 miles north of Tampa, what is the distance from Orlando to Gainesville?

40. Cell Phones. The same cellular phone company in Exercise 39 has decided to add additional towers at each "halfway" between cities. How many miles from Tampa is each "halfway" tower?

41. Travel. A retired couple who live in Columbia, South Carolina, decide to take their motor home and visit two children who live in Atlanta and in Savannah, Georgia. Savannah is 160 miles south of Columbia, and Atlanta is 215 miles west of Columbia. How far apart do the children live from each other?

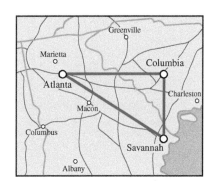

42. Sports. In the 1984 Orange Bowl, Doug Flutie, the 5-foot 9-inch quarterback for Boston College, shocked the world as he threw a "Hail Mary" pass that was caught in the end zone with no time left on the clock, defeating the Miami Hurricanes 47–45. Although the record books have it listed as a 48-yard pass, what was the actual distance the ball was thrown? The following illustration depicts the path of the ball.

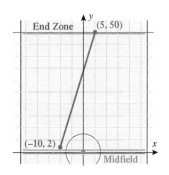

43. NASCAR Revenue. Action Performance Inc., the leading seller of NASCAR merchandise, recorded revenues of $260 million in 2002 and $400 million in 2004. Calculate the midpoint to estimate the revenue Action Performance Inc. recorded in 2003. Assume the horizontal axis represents the year and the vertical axis represents the revenue in millions.

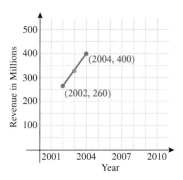

44. Ticket Price. In 1993, the average Miami Dolphins ticket price was $28, and in 2001 the average price was $56. Find the midpoint of the segment joining these two points to estimate the ticket price in 1997.

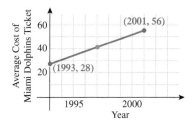

● **CATCH THE MISTAKE**

In Exercises 45–48, explain the mistake that is made.

45. Calculate the distance between $(2, 7)$ and $(9, 10)$.

Solution:

Write the distance formula. $d = \sqrt{(x_2 - x_1)^2 + (y_2 - y_1)^2}$

Substitute $(2, 7)$ and $(9, 10)$. $d = \sqrt{(7 - 2)^2 + (10 - 9)^2}$

Simplify. $d = \sqrt{(5)^2 + (1)^2} = \sqrt{26}$

This is incorrect. What mistake was made?

46. Calculate the distance between $(-2, 1)$ and $(3, -7)$.

Solution:

Write the distance formula. $d = \sqrt{(x_2 - x_1)^2 + (y_2 - y_1)^2}$

Substitute $(-2, 1)$ and $(3, -7)$. $d = \sqrt{(3 - 2)^2 + (-7 - 1)^2}$

Simplify. $d = \sqrt{(1)^2 + (-8)^2} = \sqrt{65}$

This is incorrect. What mistake was made?

47. Compute the midpoint of the segment with endpoints $(-3, 4)$ and $(7, 9)$.

Solution:

Write the midpoint formula. $(x_m, y_m) = \left(\dfrac{x_1 + x_2}{2}, \dfrac{y_1 + y_2}{2} \right)$

Substitute $(-3, 4)$ and $(7, 9)$. $(x_m, y_m) = \left(\dfrac{-3 + 4}{2}, \dfrac{7 + 9}{2} \right)$

Simplify. $(x_m, y_m) = \left(\dfrac{1}{2}, \dfrac{16}{2} \right) = \left(\dfrac{1}{2}, 4 \right)$

This is incorrect. What mistake was made?

48. Compute the midpoint of the segment with endpoints $(-1, -2)$ and $(-3, -4)$.

Solution:

Write the midpoint formula. $(x_m, y_m) = \left(\dfrac{x_1 - x_2}{2}, \dfrac{y_1 - y_2}{2} \right)$

Substitute $(-1, -2)$ and $(-3, -4)$. $(x_m, y_m) = \left(\dfrac{-1 - (-3)}{2}, \dfrac{-2 - (-4)}{2} \right)$

Simplify. $(x_m, y_m) = (1, 1)$

This is incorrect. What mistake was made?

● **CONCEPTUAL**

In Exercises 49–52, determine whether each statement is true or false.

49. The distance from the origin to the point (a, b) is $d = \sqrt{a^2 + b^2}$.

50. The midpoint of the line segment joining the origin and the point (a, a) is $\left(\dfrac{a}{2}, \dfrac{a}{2} \right)$.

51. The midpoint of any segment joining two points in quadrant I also lies in quadrant I.

52. The midpoint of any segment joining a point in quadrant I to a point in quadrant III also lies in either quadrant I or III.

53. Calculate the length and the midpoint of the line segment joining the points (a, b) and (b, a).

54. Calculate the length and the midpoint of the line segment joining the points (a, b) and $(-a, -b)$.

• CHALLENGE

55. Assume that two points (x_1, y_1) and (x_2, y_2) are connected by a segment. Prove that the distance from the midpoint of the segment to either of the two points is the same.

56. Prove that the diagonals of a parallelogram in the figure intersect at their midpoints.

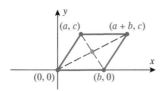

57. Assume that two points (a, b) and (c, d) are the endpoints of a line segment. Calculate the distance between the two points. Prove that it does not matter which point is labeled as the "first" point in the distance formula.

58. Show that the points $(-1, -1)$, $(0, 0)$, and $(2, 2)$ are collinear (lie on the same line) by showing that the sum of the distance from $(-1, -1)$ to $(0, 0)$ and the distance from $(0, 0)$ to $(2, 2)$ is equal to the distance from $(-1, -1)$ to $(2, 2)$.

• TECHNOLOGY

In Exercises 59–62, calculate the distance between the two points. Use a graphing utility to graph the segment joining the two points and find the midpoint of the segment.

59. $(-2.3, 4.1)$ and $(3.7, 6.2)$

60. $(-4.9, -3.2)$ and $(5.2, 3.4)$

61. $(1.1, 2.2)$ and $(3.3, 4.4)$

62. $(-1.3, 7.2)$ and $(2.3, -4.5)$

A.3 GRAPHING EQUATIONS: POINT-PLOTTING, INTERCEPTS, AND SYMMETRY

SKILLS OBJECTIVES	CONCEPTUAL OBJECTIVES
▪ Sketch graphs of equations by plotting points. ▪ Find intercepts for graphs of equations. ▪ Determine if the graph of an equation is symmetric about the x-axis, y-axis, or origin. ▪ Use intercepts and symmetry as graphing aids.	▪ Understand that if a point (a, b) satisfies the equation, then that point lies on its graph. ▪ Understand that intercepts are points that lie on the graph and either the x-axis or y-axis. ▪ Relate symmetry graphically and algebraically. ▪ Understand that intercepts are good starting points, but not the only points, and that symmetry eliminates the need to find points in other quadrants.

A.3.1 Point-Plotting

A.3.1 SKILL

Sketch graphs of equations by plotting points.

A.3.1 CONCEPTUAL

Understand that if a point (a, b) satisfies the equation, then that point lies on its graph.

Most equations in two variables, such as $y = x^2$, have an infinite number of ordered pairs as solutions. For example, $(0, 0)$ is a solution to $y = x^2$ because when $x = 0$ and $y = 0$, the equation is true. Two other solutions are $(-1, 1)$ and $(1, 1)$.

The **graph of an equation** in two variables, x and y, consists of all the points in the xy-plane whose coordinates (x, y) satisfy the equation. A procedure for plotting the graphs of equations is outlined below and is illustrated with the example $y = x^2$.

WORDS

Step 1: In a table, list several pairs of coordinates that make the equation true.

MATH

x	y = x²	(x, y)
0	0	(0, 0)
−1	1	(−1, 1)
1	1	(1, 1)
−2	4	(−2, 4)
2	4	(2, 4)

Step 2: Plot these points on a graph and connect the points with a smooth curve. Use arrows to indicate that the graph continues.

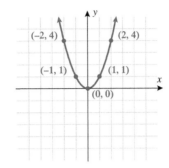

In graphing an equation, first select arbitrary values for x and then use the equation to find the corresponding value of y, or vice versa.

EXAMPLE 1   **Graphing an Equation of a Line by Plotting Points**

Graph the equation $y = 2x - 1$.

Solution:

STEP 1 In a table, list several pairs of coordinates that make the equation true.

x	y = 2x − 1	(x, y)
0	−1	(0, −1)
−1	−3	(−1, −3)
1	1	(1, 1)
−2	−5	(−2, −5)
2	3	(2, 3)

STEP 2 Plot these points on a graph and connect the points, resulting in a line.

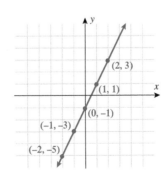

▼ **ANSWER**

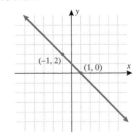

▼

YOUR TURN   The graph of the equation $y = -x + 1$ is a line. Graph the line.

EXAMPLE 2 **Graphing an Equation by Plotting Points**

Graph the equation $y = x^2 - 5$.

Solution:

STEP 1 In a table, list several pairs of coordinates that make the equation true.

x	y = x² − 5	(x, y)
0	−5	(0, −5)
−1	−4	(−1, −4)
1	−4	(1, −4)
−2	−1	(−2, −1)
2	−1	(2, −1)
−3	4	(−3, 4)
3	4	(3, 4)

STEP 2 Plot these points on a graph and connect the points with a smooth curve, indicating with arrows that the curve continues.

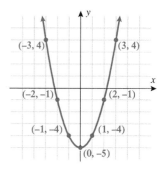

Recall from algebra that this graph is called a *parabola*.

YOUR TURN Graph the equation $y = x^2 - 1$.

▼
ANSWER

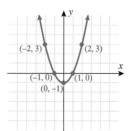

EXAMPLE 3 **Graphing an Equation by Plotting Points**

Graph the equation $y = x^3$.

Solution:

STEP 1 In a table, list several pairs of coordinates that satisfy the equation.

x	y = x³	(x, y)
0	0	(0, 0)
−1	−1	(−1, −1)
1	1	(1, 1)
−2	−8	(−2, −8)
2	8	(2, 8)

STEP 2 Plot these points on a graph and connect the points with a smooth curve, indicating with arrows that the curve continues in both the positive and negative directions.

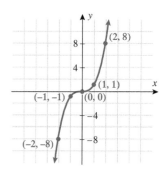

[CONCEPT CHECK]

Show the points (1, 0) and (2, 3) satisfy the equation $y = x^2 - 1$.

▼
ANSWER (1, 0): $0 = 1 - 1$ and (2, 3): $3 = 2^2 - 1$

A.3.2 Intercepts

When point-plotting graphs of equations, which points should be selected? Points where a graph crosses (or touches) either the x-axis or y-axis are called *intercepts*, and identifying these points helps define the graph unmistakably.

An **x-intercept** of a graph is a point where the graph intersects the x-axis. Specifically, an x-intercept is the x-coordinate of such a point. For example, if a graph intersects the x-axis at the point $(3, 0)$, then we say that 3 is the x-intercept. Since the value for y along the x-axis is zero, all points corresponding to x-intercepts have the form $(a, 0)$.

A **y-intercept** of a graph is a point where the graph intersects the y-axis. Specifically, a y-intercept is the y-coordinate of such a point. For example, if a graph intersects the y-axis at the point $(0, 2)$, then we say that 2 is the y-intercept. Since the value for x along the y-axis is zero, all points corresponding to y-intercepts have the form $(0, b)$.

It is important to note that graphs of equations do not have to have intercepts, and if they do have intercepts, they can have one or more of each type.

A.3.2 **SKILL**

Find intercepts for graphs of equations.

A.3.2 **CONCEPTUAL**

Understand that intercepts are points that lie on the graph and either the x-axis or y-axis.

> **STUDY TIP**
> Identifying the intercepts help define the graph unmistakably.

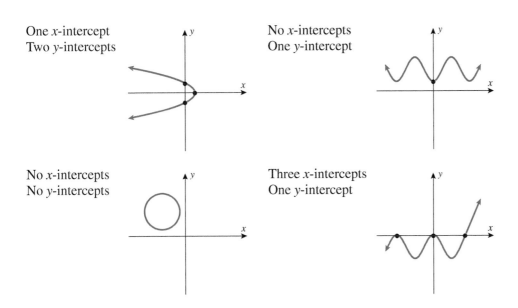

One x-intercept
Two y-intercepts

No x-intercepts
One y-intercept

No x-intercepts
No y-intercepts

Three x-intercepts
One y-intercept

Note: The origin $(0, 0)$ corresponds to both an x-intercept and a y-intercept.

The graph given in the margin has two y-intercepts and one x-intercept.

- The x-intercept is -1, which corresponds to the point $(-1, 0)$.
- The y-intercepts are -1 and 1, which correspond to the points $(0, -1)$ and $(0, 1)$, respectively.

Algebraically, how do we find intercepts from an equation? The graph in the margin corresponds to the equation $x = y^2 - 1$. The x-intercepts are located on the x-axis, which corresponds to $y = 0$. If we let $y = 0$ in the equation $x = y^2 - 1$ and solve for x, the result is $x = -1$. This corresponds to the x-intercept we identified above. Similarly, the y-intercepts are located on the y-axis, which corresponds to $x = 0$. If we let $x = 0$ in the equation $x = y^2 - 1$ and solve for y, the result is $y = \pm 1$. These correspond to the y-intercepts we identified above.

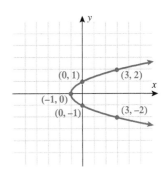

EXAMPLE 4 Finding Intercepts from an Equation

Given the equation $y = x^2 + 1$, find the indicated intercepts of its graph, if any.

a. x-intercept(s) **b.** y-intercept(s)

Solution (a):

Let $y = 0$.

$$0 = x^2 + 1$$

Solve for x.

$$x^2 = -1 \quad \text{no real solution}$$

There are no x-intercepts.

Solution (b):

Let $x = 0$.

$$y = 0^2 + 1$$

Solve for y.

$$y = 1$$

The y-intercept is located at the point $(0, 1)$.

▼ YOUR TURN For the equation $y = x^2 - 4$, find

a. x-intercept(s), if any. **b.** y-intercept(s), if any.

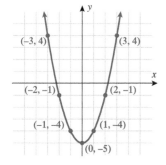

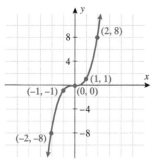

A.3.3 Symmetry

The word **symmetry** conveys balance. Suppose you have two pictures to hang on a wall. If you space them equally apart on the wall, then you prefer a symmetric décor. This is an example of symmetry about a line. The word (water) written below is identical if you rotate the word 180 degrees (or turn the page upside down). This is an example of symmetry about a point. Symmetric graphs have the characteristic that their mirror image can be obtained about a reference, typically a line or a point.

In Example 2, the points $(-2, -1)$ and $(2, -1)$ both lie on the graph, as do the points $(-1, -4)$ and $(1, -4)$. Notice that the graph on the right side of the y-axis is a mirror image of the part of the graph to the left of the y-axis. This graph illustrates *symmetry* with respect to the *y-axis* (the line $x = 0$).

In Example 3, the points $(-1, -1)$ and $(1, 1)$ both lie on the graph. Notice that rotating this graph 180 degrees (or turning your page upside down) results in an identical graph. This is an example of *symmetry* with respect to the *origin* $(0, 0)$.

Symmetry aids in graphing by giving information "for free." For example, if a graph is symmetric about the y-axis, then once the graph to the right of the y-axis is found, the left side of the graph is the mirror image of that. If a graph is symmetric about the origin, then once the graph is known in quadrant I, the graph in quadrant III is found by rotating the known graph 180 degrees.

It would be beneficial to know whether a graph of an equation is symmetric about a line or point before the graph of the equation is sketched. Although a graph can be symmetric about any line or point, we will discuss only symmetry about the x-axis, y-axis, and origin. These types of symmetry and the algebraic procedures for testing for symmetry are outlined below.

Types and Tests for Symmetry

TYPE OF SYMMETRY	GRAPH	IF THE POINT (a, b) IS ON THE GRAPH, THEN THE POINT . . .	ALGEBRAIC TEST FOR SYMMETRY
Symmetric with respect to the x-axis	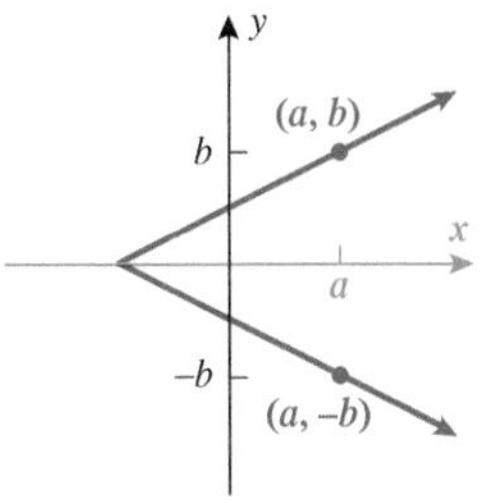	$(a, -b)$ is on the graph.	Replacing y with $-y$ leaves the equation unchanged.
Symmetric with respect to the y-axis	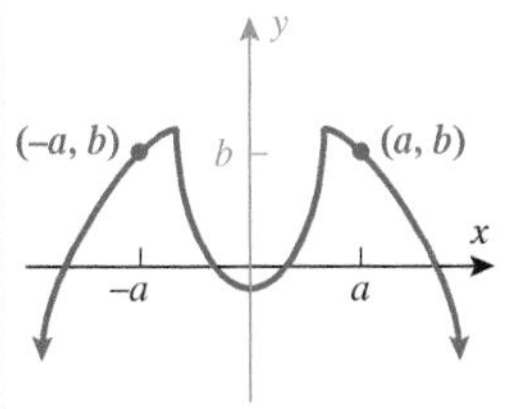	$(-a, b)$ is on the graph.	Replacing x with $-x$ leaves the equation unchanged.
Symmetric with respect to the origin	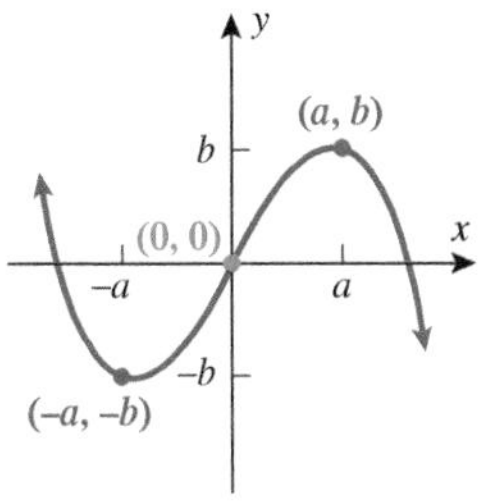	$(-a, -b)$ is on the graph.	Replacing x with $-x$ and y with $-y$ leaves the equation unchanged.

> **STUDY TIP**
>
> Symmetry gives us information about the graph "for free."

EXAMPLE 5   **Testing for Symmetry with Respect to the Axes**

Test the equation $y^2 = x^3$ for symmetry with respect to the axes.

Solution:

Test for symmetry with respect to the x-axis.

Replace y with $-y$. $(-y)^2 = x^3$

Simplify. $y^2 = x^3$

The resulting equation is the same as the original equation, $y^2 = x^3$.

> Therefore, $y^2 = x^3$ is **symmetric with respect to the x-axis**.

Test for symmetry with respect to the y-axis.

Replace x with $-x$. $y^2 = (-x)^3$

Simplify. $y^2 = -x^3$

The resulting equation, $y^2 = -x^3$, is not the same as the original equation, $y^2 = x^3$.

> Therefore, $y^2 = x^3$ is **not** symmetric with respect to the y-axis.

When testing for symmetry about the *x*-axis, *y*-axis, and origin, there are *five* possibilities:

- No symmetry
- Symmetry with respect to the *x*-axis
- Symmetry with respect to the *y*-axis
- Symmetry with respect to the origin
- Symmetry with respect to the *x*-axis, *y*-axis, and origin

[CONCEPT CHECK]

If the point (a, b) and the point (a, $-b$) is on the graph, then the graph is symmetric about:

(A) the *y*-axis,

(B) the *x*-axis. or

(C) the origin.

▼

ANSWER (B) The *x*-axis

▶ **EXAMPLE 6**   **Testing for Symmetry**

Determine what type of symmetry (if any) the graphs of the equations exhibit.

a. $y = x^2 + 1$ **b.** $y = x^3 + 1$

Solution (a):

Replace *x* with $-x$. $y = (-x)^2 + 1$

Simplify. $y = x^2 + 1$

The resulting equation is equivalent to the original equation, so the graph of the equation $y = x^2 + 1$ is **symmetric with respect to the *y*-axis**.

Replace *y* with $-y$. $(-y) = x^2 + 1$

Simplify. $y = -x^2 - 1$

The resulting equation, $y = -x^2 - 1$, is not equivalent to the original equation, $y = x^2 + 1$, so the graph of the equation $y = x^2 + 1$ is **not symmetric with respect to the *x*-axis**.

Replace *x* with $-x$ and *y* with $-y$. $(-y) = (-x)^2 + 1$

Simplify. $-y = x^2 + 1$

                                  $y = -x^2 - 1$

The resulting equation, $y = -x^2 - 1$, is not equivalent to the original equation, $y = x^2 + 1$, so the graph of the equation $y = x^2 + 1$ is **not symmetric with respect to the origin**.

> The graph of the equation $y = x^2 + 1$ is **symmetric with respect to the *y*-axis**.

Solution (b):

Replace *x* with $-x$. $y = (-x)^3 + 1$

Simplify. $y = -x^3 + 1$

The resulting equation, $y = -x^3 + 1$, is not equivalent to the original equation, $y = x^3 + 1$. Therefore, the graph of the equation $y = x^3 + 1$ is **not symmetric with respect to the *y*-axis**.

Replace *y* with $-y$. $(-y) = x^3 + 1$

Simplify. $y = -x^3 - 1$

The resulting equation, $y = -x^3 - 1$, is not equivalent to the original equation, $y = x^3 + 1$. Therefore, the graph of the equation $y = x^3 + 1$ is **not symmetric with respect to the *x*-axis**.

Replace *x* with $-x$ and *y* with $-y$. $(-y) = (-x)^3 + 1$

Simplify. $-y = -x^3 + 1$

                                  $y = x^3 - 1$

The resulting equation, $y = x^3 - 1$, is not equivalent to the original equation, $y = x^3 + 1$. Therefore, the graph of the equation $y = x^3 + 1$ is **not symmetric with respect to the origin**.

> The graph of the equation $y = x^3 + 1$ exhibits **no symmetry**.

▼

ANSWER

The graph of the equation is symmetric with respect to the *x*-axis.

▼

YOUR TURN   Determine the symmetry (if any) for $x = y^2 - 1$.

A.3.4 Using Intercepts and Symmetry as Graphing Aids

How can we use intercepts and symmetry to assist us in graphing? Intercepts are a good starting point—though not the only one. For symmetry, look back at Example 2, $y = x^2 - 5$. We selected seven x-coordinates and solved the equation to find the corresponding y-coordinates. If we had known that this graph was symmetric with respect to the y-axis, then we would have had to find the solutions to only the positive x-coordinates, since we get the negative x-coordinates for free. For example, we found the point $(1, -4)$ to be a solution to the equation. The rules of symmetry tell us that $(-1, -4)$ is also on the graph.

A.3.4 SKILL

Use intercepts and symmetry as graphing aids.

A.3.4 CONCEPTUAL

Understand that intercepts are good starting points, but not the only points, and that symmetry eliminates the need to find points in other quadrants.

EXAMPLE 7 **Using Intercepts and Symmetry as Graphing Aids**

For the equation $x^2 + y^2 = 25$, use intercepts and symmetry to help you graph the equation using the point-plotting technique.

Solution:

STEP 1 **Find the intercepts.**

For the x-intercepts, let $y = 0$. $\qquad\qquad x^2 + 0^2 = 25$

Solve for x. $\qquad\qquad\qquad\qquad\qquad\qquad x = \pm 5$

The two x-intercepts correspond to the points $(-5, 0)$ and $(5, 0)$.

For the y-intercepts, let $x = 0$. $\qquad\qquad 0^2 + y^2 = 25$

Solve for y. $\qquad\qquad\qquad\qquad\qquad\qquad y = \pm 5$

The two y-intercepts correspond to the points $(0, -5)$ and $(0, 5)$.

STEP 2 **Identify the points on the graph corresponding to the intercepts.**

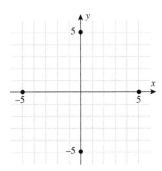

STEP 3 **Test for symmetry with respect to the y-axis, x-axis, and origin.**

Test for symmetry with respect to the y-axis.

Replace x with $-x$. $\qquad\qquad\qquad (-x)^2 + y^2 = 25$

Simplify. $\qquad\qquad\qquad\qquad\qquad\qquad x^2 + y^2 = 25$

The resulting equation is equivalent to the original, so the graph of $x^2 + y^2 = 25$ is **symmetric with respect to the y-axis.**

Test for symmetry with respect to the x-axis.

Replace y with $-y$. $\qquad\qquad\qquad x^2 + (-y)^2 = 25$

Simplify. $\qquad\qquad\qquad\qquad\qquad\qquad x^2 + y^2 = 25$

The resulting equation is equivalent to the original, so the graph of $x^2 + y^2 = 25$ is **symmetric with respect to the x-axis.**

$\Big[$CONCEPT CHECK$\Big]$

If the graph is symmetric about the x-axis, y-axis, and the origin, then once you find a point (a, b) in quadrant I that lies on the graph, what points do you get for free from the symmetry?
(A) $(-a, b)$,
(B) $(a, -b)$,
(C) $(-a, -b)$, or
(D) all of the above.

▼ ....

ANSWER (D) All of the above

Test for symmetry with respect to the origin.

Replace x with $-x$ and y with $-y$.	$(-x)^2 + (-y)^2 = 25$
Simplify.	$x^2 + y^2 = 25$

The resulting equation is equivalent to the original, so the graph of $x^2 + y^2 = 25$ is symmetric with respect to the origin.

Since the graph is symmetric with respect to the *y*-axis, *x*-axis, and origin, we need to determine solutions to the equation on only the positive *x*- and *y*-axes and in quadrant I because of the following symmetries:

- **Symmetry with respect to the *y*-axis gives the solutions in quadrant II.**
- Symmetry with respect to the origin gives the solutions in quadrant III.
- **Symmetry with respect to the *x*-axis yields solutions in quadrant IV.**

Solutions to $x^2 + y^2 = 25$.

Quadrant I: $(3, 4), (4, 3)$

Additional points due to symmetry:

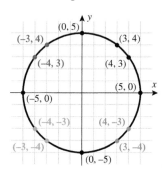

Quadrant II: $(-3, 4), (-4, 3)$

Quadrant III: $(-3, -4), (-4, -3)$

Quadrant IV: $(3, -4), (4, -3)$

Connecting the points with a smooth curve yields a **circle**.

▶[SECTION A.3] SUMMARY

Sketching graphs of equations can be accomplished using a point-plotting technique. Intercepts are defined as points where a graph intersects an axis or the origin.

Symmetry about the *x*-axis, *y*-axis, and origin is defined both algebraically and graphically. Intercepts and symmetry provide much of the information useful for sketching graphs of equations.

INTERCEPTS	POINT WHERE THE GRAPH INTERSECTS THE . . .	HOW TO FIND INTERCEPTS	POINT ON GRAPH
x-intercept	*x*-axis	Let $y = 0$ and solve for x.	$(a, 0)$
y-intercept	*y*-axis	Let $x = 0$ and solve for y.	$(0, b)$

[SECTION A.3] EXERCISES

• SKILLS

In Exercises 1–8, determine whether each point lies on the graph of the equation.

1. $y = 3x - 5$ **a.** $(1, 2)$ **b.** $(-2, -11)$ **2.** $y = -2x + 7$ **a.** $(-1, 9)$ **b.** $(2, -4)$

3. $y = \frac{2}{5}x - 4$ **a.** $(5, -2)$ **b.** $(-5, 6)$ **4.** $y = -\frac{3}{4}x + 1$ **a.** $(8, 5)$ **b.** $(-4, 4)$

5. $y = x^2 - 2x + 1$ **a.** $(-1, 4)$ **b.** $(0, -1)$ **6.** $y = x^3 - 1$ **a.** $(-1, 0)$ **b.** $(-2, -9)$

7. $y = \sqrt{x} + 2$ **a.** $(7, 3)$ **b.** $(-6, 4)$ **8.** $y = 2 + |3 - x|$ **a.** $(9, -4)$ **b.** $(-2, 7)$

In Exercises 9–14, complete the table and use the table to sketch a graph of the equation.

9.

x	y = 2 + x	(x, y)
−2		
0		
1		

10.

x	y = 3x − 1	(x, y)
−1		
0		
2		

11.

x	y = x² − x	(x, y)
−1		
0		
$\frac{1}{2}$		
1		
2		

12.

x	y = 1 − 2x − x²	(x, y)
−3		
−2		
−1		
0		
1		

13.

x	y = √x − 1	(x, y)
1		
2		
5		
10		

14.

x	y = − √x + 2	(x, y)
−2		
−1		
5		
7		

In Exercises 15–22, graph the equation by plotting points.

15. $y = -3x + 2$ **16.** $y = 4 - x$ **17.** $y = x^2 - x - 2$ **18.** $y = x^2 - 2x + 1$

19. $x = y^2 - 1$ **20.** $x = |y + 1| + 2$ **21.** $y = \frac{1}{2}x - \frac{3}{2}$ **22.** $y = 0.5|x - 1|$

In Exercises 23–32, find the x-intercept(s) and y-intercepts(s) (if any) of the graphs of the given equations.

23. $2x - y = 6$ **24.** $4x + 2y = 10$ **25.** $y = x^2 - 9$ **26.** $y = 4x^2 - 1$

27. $y = \sqrt{x - 4}$ **28.** $y = \sqrt[3]{x - 8}$ **29.** $y = \dfrac{1}{x^2 + 4}$ **30.** $y = \dfrac{x^2 - x - 12}{x}$

31. $4x^2 + y^2 = 16$ **32.** $x^2 - y^2 = 9$

In Exercises 33–38, match the graph with the corresponding symmetry.

a. No symmetry   **b.** Symmetry with respect to the x-axis   **c.** Symmetry with respect to the y-axis

d. Symmetry with respect to the origin   **e.** Symmetry with respect to the x-axis, y-axis, and origin

33.

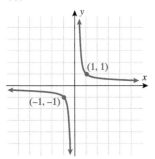

34.

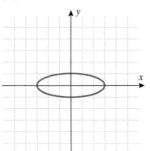

35.

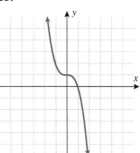

36.

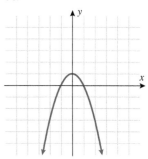

37.

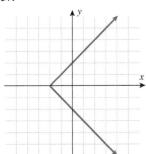

38.

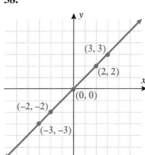

In Exercises 39–44, a point that lies on a graph is given along with that graph's symmetry. State the other known points that must also lie on the graph.

Point on a Graph	The Graph Is Symmetric about the	Point on a Graph	The Graph Is Symmetric about the
39. $(-1, 3)$	x-axis	**40.** $(-2, 4)$	y-axis
41. $(7, -10)$	origin	**42.** $(-1, -1)$	origin
43. $(3, -2)$	x-axis, y-axis, and origin	**44.** $(-1, 7)$	x-axis, y-axis, and origin

In Exercises 45–58, test algebraically to determine whether the equation's graph is symmetric with respect to the x-axis, y-axis, or origin.

45. $x = y^2 + 4$ **46.** $x = 2y^2 + 3$ **47.** $y = x^3 + x$ **48.** $y = x^5 + 1$ **49.** $x = |y|$ **50.** $x = |y| - 2$

51. $x^2 - y^2 = 100$ **52.** $x^2 + 2y^2 = 30$ **53.** $y = x^{2/3}$ **54.** $x = y^{2/3}$ **55.** $x^2 + y^3 = 1$ **56.** $y = \sqrt{1 + x^2}$

57. $y = \dfrac{2}{x}$ **58.** $xy = 1$

In Exercises 59–72, plot the graph of the given equation.

59. $y = x$ **60.** $y = -\frac{1}{2}x + 3$ **61.** $y = x^2 - 1$ **62.** $y = 9 - 4x^2$

63. $y = \dfrac{x^3}{2}$ **64.** $x = y^2 + 1$ **65.** $y = \dfrac{1}{x}$ **66.** $xy = -1$

67. $y = |x|$ **68.** $|x| = |y|$ **69.** $x^2 + y^2 = 16$ **70.** $\dfrac{x^2}{4} + \dfrac{y^2}{9} = 1$

71. $x^2 - y^2 = 16$ **72.** $x^2 - \dfrac{y^2}{25} = 1$

• **APPLICATIONS**

73. Sprinkler. A sprinkler will water a grassy area in the shape of $x^2 + y^2 = 9$. Apply symmetry to draw the watered area, assuming the sprinkler is located at the origin.

74. Sprinkler. A sprinkler will water a grassy area in the shape of $x^2 + \dfrac{y^2}{9} = 1$. Apply symmetry to draw the watered area, assuming the sprinkler is located at the origin.

75. Electronic Signals: Radio Waves. The received power of an electromagnetic signal is a fraction of the power transmitted. The relationship is given by

$$P_{received} = P_{transmitted} \cdot \frac{1}{R^2}$$

where R is the distance that the signal has traveled in meters. Plot the percentage of transmitted power that is received for $R = 100$ meters, 1 kilometer, and 10,000 kilometers.

76. Electronic Signals: Laser Beams. The wavelength λ and the frequency f of a signal are related by the equation

$$f = \frac{c}{\lambda}$$

where c is the speed of light in a vacuum, $c = 3.0 \times 10^8$ meters per second. For the values, $\lambda = 0.001$, $\lambda = 1$, and $\lambda = 100$ millimeters, plot the points corresponding to frequency f. What do you notice about the relationship between frequency and wavelength? Note that the frequency will have units Hz = 1/second.

77. Profit. The profit associated with making a particular product is given by the equation

$$y = -x^2 + 6x - 8$$

where y represents the profit in millions of dollars and x represents the number of thousands of units sold. ($x = 1$ corresponds to 1000 units and $y = 1$ corresponds to $1 million.) Graph this equation and determine how many units must be sold to break even (profit $= 0$). Determine the range of units sold that correspond to making a profit.

78. Profit. The profit associated with making a particular product is given by the equation

$$y = -x^2 + 4x - 3$$

where y represents the profit in millions of dollars and x represents the number of thousands of units sold. ($x = 1$ corresponds to 1000 units and $y = 1$ corresponds to $1 million.) Graph this equation and determine how many units must be sold to break even (profit $= 0$). Determine the range of units sold that correspond to making a profit.

• CATCH THE MISTAKE

In Exercises 79–82, explain the mistake that is made.

79. Graph the equation $y = x^2 + 1$.

Solution:

x	$y = x^2 + 1$	(x, y)
0	1	(0, 1)
1	2	(1, 2)

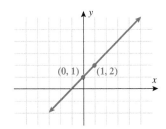

This is incorrect. What mistake was made?

80. Test $y = -x^2$ for symmetry with respect to the y-axis.

Solution:

Replace x with $-x$. $y = -(-x)^2$

Simplify. $y = x^2$

The resulting equation is not equivalent to the original equation; $y = -x^2$ is not symmetric with respect to the y-axis.

This is incorrect. What mistake was made?

81. Test $x = |y|$ for symmetry with respect to the y-axis.

Solution:

Replace y with $-y$. $x = |-y|$

Simplify. $x = |y|$

The resulting equation is equivalent to the original equation; $x = |y|$ is symmetric with respect to the y-axis.

This is incorrect. What mistake was made?

82. Use symmetry to help you graph $x^2 = y - 1$.

Solution:

Replace x with $-x$. $(-x)^2 = y - 1$

Simplify. $x^2 = y - 1$

$x^2 = y - 1$ is symmetric with respect to the x-axis.

Determine the points that lie on the graph in quadrant I.

y	$x^2 = y - 1$	(x, y)
1	0	(0, 1)
2	1	(1, 2)
5	2	(2, 5)

Symmetry with respect to the x-axis implies that $(0, -1)$, $(1, -2)$, and $(2, -5)$ are also points that lie on the graph.

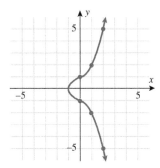

This is incorrect. What mistake was made?

• CONCEPTUAL

In Exercises 83–86, determine whether each statement is true or false.

83. If the point (a, b) lies on a graph that is symmetric about the x-axis, then the point $(-a, b)$ also must lie on the graph.

84. If the point (a, b) lies on a graph that is symmetric about the y-axis, then the point $(-a, b)$ also must lie on the graph.

85. If the point $(a, -b)$ lies on a graph that is symmetric about the x-axis, y-axis, and origin, then the points (a, b), $(-a, -b)$, and $(-a, b)$ must also lie on the graph.

86. Two points are all that is needed to plot the graph of an equation.

• CHALLENGE

87. Determine whether the graph of $y = \dfrac{ax^2 + b}{cx^3}$ has any symmetry, where a, b, and c are real numbers.

88. Find the intercepts of $y = (x - a)^2 - b^2$, where a and b are real numbers.

• TECHNOLOGY

In Exercises 89–94, graph the equation using a graphing utility and state whether there is any symmetry.

89. $y = 16.7x^4 - 3.3x^2 + 7.1$

90. $y = 0.4x^5 + 8.2x^3 - 1.3x$

91. $2.3x^2 = 5.5|y|$

92. $3.2x^2 - 5.1y^2 = 1.3$

93. $1.2x^2 + 4.7y^2 = 19.4$

94. $2.1y^2 = 0.8|x + 1|$

A.4 FUNCTIONS

SKILLS OBJECTIVES	CONCEPTUAL OBJECTIVES
■ Determine whether a relation is a function. ■ Determine whether an equation represents a function. ■ Use function notation to evaluate functions for particular arguments. ■ Determine the domain and range of a function.	■ Understand that all functions are relations but not all relations are functions. ■ Understand why the vertical line test determines if a relation is a function. ■ Think of function notation as a placeholder or mapping. ■ Understand the difference between implicit domain and explicit domain.

A.4.1 Relations and Functions

A.4.1 SKILL

Determine whether a relation is a function.

A.4.1 CONCEPTUAL

Understand that all functions are relations but not all relations are functions.

What do the following pairs have in common?

■ Every person has a blood type.

■ Temperature at a particular time of day.

■ Every working household phone in the United States has a 10-digit phone number.

■ First-class postage rates correspond to the weight of a letter.

■ Certain times of the day are start times of sporting events at a university.

They all describe a particular correspondence between two groups. **A relation** is a correspondence between two sets. The first set is called the **domain**, and the corresponding second set is called the **range**. Members of these sets are called **elements**.

A **relation** is a correspondence between two sets where each element in the first set, called the **domain**, corresponds to *at least* one element in the second set, called the **range**.

A relation is a set of ordered pairs. The domain is the set of all the first components of the ordered pairs, and the range is the set of all the second components of the ordered pairs.

PERSON	BLOOD TYPE	ORDERED PAIR
Michael	A	(Michael, A)
Tania	A	(Tania, A)
Dylan	AB	(Dylan, AB)
Trevor	O	(Trevor, O)
Megan	O	(Megan, O)

WORDS

The domain is the set of all the first components.

The range is the set of all the second components.

MATH

{Michael, Tania, Dylan, Trevor, Megan}

{A, AB, O}

A relation in which each element in the domain corresponds to exactly one element in the range is a **function**.

A **function** is a correspondence between two sets where each element in the first set, called the **domain**, corresponds to *exactly* one element in the second set, called the **range**.

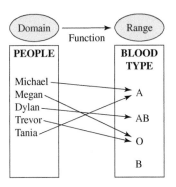

Note that the definition of a function is more restrictive than the definition of a relation. For a relation, each input corresponds to *at least* one output, whereas, for a function, each input corresponds to *exactly* one output. The blood-type example given is both a relation and a function.

Also note that the range (set of values to which the elements of the domain correspond) is a subset of the set of all blood types. However, although all functions are relations, not all relations are functions.

For example, at a university, four primary sports typically overlap in the late fall: football, volleyball, soccer, and basketball. On a given Saturday, the table on the right indicates the start times for the competitions.

TIME OF DAY	COMPETITION
1:00 P.M.	Football
2:00 P.M.	Volleyball
7:00 P.M.	Soccer
7:00 P.M.	Basketball

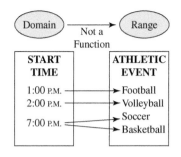

WORDS		MATH
The 1:00 start time corresponds to exactly one event, Football.		(1:00 P.M., Football)
The 2:00 start time corresponds to exactly one event, Volleyball.		2:00 P.M., Volleyball)
The 7:00 start time corresponds to two events, Soccer and Basketball.		(7:00 P.M., Soccer) (7:00 P.M., Basketball)

STUDY TIP

All functions are relations, but not all relations are functions.

Because an element in the domain, 7:00 P.M., corresponds to more than one element in the range, Soccer and Basketball, this is not a function. It is, however, a relation.

EXAMPLE 1 Determining Whether a Relation Is a Function

Determine whether the following relations are functions:

a. $\{(-3, 4), (2, 4), (3, 5), (6, 4)\}$

b. $\{(-3, 4), (2, 4), (3, 5), (2, 2)\}$

c. Domain = Set of all items for sale in a grocery store; Range = Price

Solution:

a. No x-value is repeated. Therefore, each x-value corresponds to exactly one y-value. This relation is a function.

b. The value $x = 2$ corresponds to *both* $y = 2$ and $y = 4$. This relation is not a function.

c. Each item in the grocery store corresponds to exactly one price. This relation is a function.

CONCEPT CHECK

If the domain consists of all physical (home) addresses in a particular county and the range is the persons living in that county, does this describe a relation? And if so, is that relation a function?

▼
ANSWER This is a relation but not a function.

▼
ANSWER

 a. function

 b. not a function

 c. function

▼
YOUR TURN Determine whether the following relations are functions:

a. $\{(1, 2), (3, 2), (5, 6), (7, 6)\}$

b. $\{(1, 2), (1, 3), (5, 6), (7, 8)\}$

c. $\{(11:00\ \text{A.M.}, 83°\text{F}), (2:00\ \text{P.M.}, 89°\text{F}), (6:00\ \text{P.M.}, 85°\text{F})\}$

All of the examples we have discussed thus far are **discrete** sets in that they represent a countable set of distinct pairs of (x, y). A function can also be defined algebraically by an equation.

A.4.2 Functions Defined by Equations

Let's start with the equation $y = x^2 - 3x$, where x can be any real number. This equation assigns to each x-value exactly one corresponding y-value. For example,

A.4.2 **SKILL**

Determine whether an equation represents a function.

A.4.2 **CONCEPTUAL**

Understand why the vertical line test determines if a relation is a function.

x	$y = x^2 - 3x$	y
1	$y = (1)^2 - 3(1)$	-2
5	$y = (5)^2 - 3(5)$	10
$-\frac{2}{3}$	$y = \left(-\frac{2}{3}\right)^2 - 3\left(-\frac{2}{3}\right)$	$\frac{22}{9}$
1.2	$y = (1.2)^2 - 3(1.2)$	-2.16

Since the variable y *depends* on what value of x is selected, we denote y as the **dependent variable**. The variable x can be any number in the domain; therefore, we denote x as the **independent variable**.

Although functions are defined by equations, it is important to recognize that *not all equations define functions*. The requirement for an equation to define a function is that each element in the domain corresponds to exactly one element in the range. Throughout the ensuing discussion, we assume x to be the independent variable and y to be the dependent variable.

Equations that represent functions of x: $y = x^2$ $y = |x|$ $y = x^3$

Equations that do not represent functions of x: $x = y^2$ $x^2 + y^2 = 1$ $x = |y|$

In the "equations that represent functions of x," every x-value corresponds to exactly one y-value. Some ordered pairs that correspond to these functions are

$$y = x^2: \qquad (-1, 1)\,(0, 0)\,(1, 1)$$
$$y = |x|: \qquad (-1, 1)\,(0, 0)\,(1, 1)$$
$$y = x^3: \qquad (-1, -1)\,(0, 0)\,(1, 1)$$

The fact that $x = -1$ and $x = 1$ both correspond to $y = 1$ in the first two examples does not violate the definition of a function.

In the "equations that do not represent functions of x," some x-values correspond to *more than one* y-value. Some ordered pairs that correspond to these equations are

$x = y^2$: $(1, -1)\,(0, 0)\,(1, 1)$ $x = 1$ maps to **both** $y = -1$ and $y = 1$

$x^2 + y^2 = 1$: $(0, -1)\,(0, 1)\,(-1, 0)\,(1, 0)$ $x = 0$ maps to **both** $y = -1$ and $y = 1$

$x = |y|$: $(1, -1)\,(0, 0)\,(1, 1)$ $x = 1$ maps to **both** $y = -1$ and $y = 1$

Let's look at the graphs of the three **functions of x**:

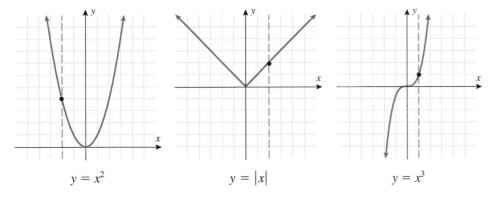

$$y = x^2 \qquad\qquad y = |x| \qquad\qquad y = x^3$$

Let's take any value for x, say, $x = a$. The graph of $x = a$ corresponds to a vertical line. A function of x maps each x-value to exactly one y-value; therefore, there should be at most one point of intersection with any vertical line. We see in the three graphs of the functions above that if a vertical line is drawn at any value of x on any of the three graphs, the vertical line only intersects the graph in one place. Look at the graphs of the three equations that do **not** represent **functions of x**.

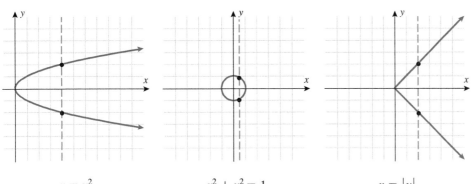

$$x = y^2 \qquad\qquad x^2 + y^2 = 1 \qquad\qquad x = |y|$$

A vertical line can be drawn on any of the three graphs such that the vertical line will intersect each of these graphs at two points. Thus, there is more than one *y*-value that corresponds to some *x*-value in the domain, which is why these equations do not define functions of *x*.

> **DEFINITION** Vertical Line Test
>
> Given the graph of an equation, if any vertical line that can be drawn intersects the graph at no more than one point, the equation defines a function of *x*. This test is called the **vertical line test**.

EXAMPLE 2 Using the Vertical Line Test

Use the vertical line test to determine whether the graphs of equations define functions of *x*.

a.

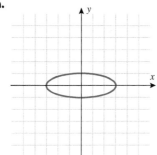

b.

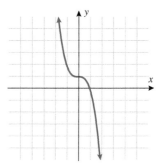

Solution:

Apply the vertical line test.

a.

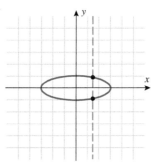

b.

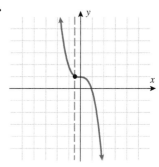

a. Because the vertical line intersects the graph of the equation at two points, this equation does not represent a function .

b. Because any vertical line will intersect the graph of this equation at no more than one point, this equation represents a function .

YOUR TURN Determine whether the equation $(x - 3)^2 + (y + 2)^2 = 16$ is a function of *x*.

▼
ANSWER

The graph of the equation is a circle, which does not pass the vertical line test. Therefore, the equation does not define a function.

To recap, a function can be expressed in one of four ways: verbally, numerically, algebraically, and graphically. This is sometimes called the Rule of 4.

Expressing a Function

VERBALLY	NUMERICALLY	ALGEBRAICALLY	GRAPHICALLY
Every real number has a corresponding absolute value.	$\{(-3, 3), (-1, 1), (0, 0), (1, 1), (5, 5)\}$	$y = |x|$	

A.4.3 Function Notation

We know that the equation $y = 2x + 5$ is a function because its graph is a nonvertical line and thus passes the vertical line test. We can select x-values (input) and determine unique corresponding y-values (output). The output is found by taking 2 times the input and then adding 5. If we give the function a name, say, "f," then we can use **function notation**:

$$f(x) = 2x + 5$$

The symbol $f(x)$ is read "f evaluated at x" or "f of x" and represents the y-value that corresponds to a particular x-value. In other words, $y = f(x)$.

A.4.3 SKILL

Use function notation to evaluate functions for particular arguments.

A.4.3 CONCEPTUAL

Think of function notation as a placeholder or mapping.

INPUT	FUNCTION	OUTPUT	EQUATION
x	f	$f(x)$	$f(x) = 2x + 5$
Independent variable	Mapping	Dependent variable	Mathematical rule

It is important to note that f is the function name, whereas $f(x)$ is the value of the function. In other words, the function f maps some value x in the domain to some value $f(x)$ in the range.

x	$f(x) = 2x + 5$	$f(x)$
0	$2(0) + 5$	$f(0) = 5$
1	$2(1) + 5$	$f(1) = 7$
2	$2(2) + 5$	$f(2) = 9$

The independent variable is also referred to as the **argument** of a function. To evaluate functions, it is often useful to think of the independent variable or argument as a placeholder. For example, $f(x) = x^2 - 3x$ can be thought of as

$$f(\square) = (\square)^2 - 3(\square)$$

In other words, "f of the argument is equal to the argument squared minus 3 times the argument." Any expression can be substituted for the argument:

$$f(1) = (1)^2 - 3(1)$$
$$f(x + 1) = (x + 1)^2 - 3(x + 1)$$
$$f(-x) = (-x)^2 - 3(-x)$$

It is important to note:

- $f(x)$ does *not* mean f times x.
- The most common function names are f and F since the word function begins with an "f." Other common function names are g and G, but any letter can be used.

STUDY TIP

It is important to note that $f(x)$ does not mean f times x.

■ The letter most commonly used for the independent variable is x. The letter t is also common because in real-world applications it represents time, but any letter can be used.

■ Although we can think of y and $f(x)$ as interchangeable, the function notation is useful when we want to consider two or more functions of the same independent variable or when we want to evaluate a function at more than one argument.

EXAMPLE 3 Evaluating Functions by Substitution

Given the function $f(x) = 2x^3 - 3x^2 + 6$, find $f(-1)$.

Solution:

Consider the independent variable x to be a placeholder.

$$f(\square) = 2(\square)^3 - 3(\square)^2 + 6$$

To find $f(-1)$, substitute $x = -1$ into the function.

$$f(-1) = 2(-1)^3 - 3(-1)^2 + 6$$

Evaluate the right side.

$$f(-1) = -2 - 3 + 6$$

Simplify.

$$\boxed{f(-1) = 1}$$

EXAMPLE 4 Finding Function Values from the Graph of a Function

The graph of f is given on the right.

a. Find $f(0)$.

b. Find $f(1)$.

c. Find $f(2)$.

d. Find $4f(3)$.

e. Find x such that $f(x) = 10$.

f. Find x such that $f(x) = 2$.

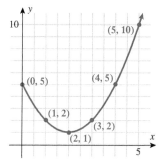

Solution (a): The value $x = 0$ corresponds to the value $y = 5$. $\boxed{f(0) = 5}$

Solution (b): The value $x = 1$ corresponds to the value $y = 2$. $\boxed{f(1) = 2}$

Solution (c): The value $x = 2$ corresponds to the value $y = 1$. $\boxed{f(2) = 1}$

Solution (d): The value $x = 3$ corresponds to the value $y = 2$. $4f(3) = 4 \cdot 2 = \boxed{8}$

Solution (e): The value $y = 10$ corresponds to the value $\boxed{x = 5}$.

Solution (f): The value $y = 2$ corresponds to the values $\boxed{x = 1}$ and $\boxed{x = 3}$.

▼

ANSWER

a. $f(-1) = 2$

b. $f(0) = 1$

c. $3f(2) = -21$

d. $x = 1$

YOUR TURN For the following graph of a function, find

a. $f(-1)$ **b.** $f(0)$ **c.** $3f(2)$

d. the value of x that corresponds to $f(x) = 0$

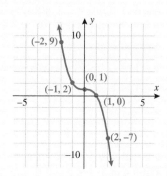

▶ **EXAMPLE 5** **Evaluating Functions with Variable Arguments (Inputs)**

For the given function $f(x) = x^2 - 3x$, evaluate $f(x + 1)$.

common mistake

A common misunderstanding is to interpret the notation $f(x + 1)$ as a sum: $f(x + 1) \neq f(x) + f(1)$.

✓CORRECT

Write the original function.

$$f(x) = x^2 - 3x$$

Replace the argument x with a placeholder.

$$f(\square) = (\square)^2 - 3(\square)$$

Substitute $x + 1$ for the argument.

$$f(x + 1) = (x + 1)^2 - 3(x + 1)$$

Eliminate the parentheses.

$$f(x + 1) = x^2 + 2x + 1 - 3x - 3$$

Combine like terms.

$$\boxed{f(x + 1) = x^2 - x - 2}$$

✗INCORRECT

The **ERROR** is in interpreting the notation as a sum.

$$f(x + 1) \neq f(x) + f(1)$$
$$\neq x^2 - 3x - 2$$

▼
CAUTION

$f(x + 1) \neq f(x) + f(1)$

▼
YOUR TURN For the given function $g(x) = x^2 - 2x + 3$, evaluate $g(x - 1)$.

▼
ANSWER

$g(x - 1) = x^2 - 4x + 6$

EXAMPLE 6 **Evaluating Functions: Sums**

For the given function $H(x) = x^2 + 2x$, evaluate

a. $H(x + 1)$ **b.** $H(x) + H(1)$

Solution (a):

Write the function H in placeholder notation. $H(\square) = (\square)^2 + 2(\square)$

Substitute $x + 1$ for the argument of H. $H(x + 1) = (x + 1)^2 + 2(x + 1)$

Eliminate the parentheses on the right side. $H(x + 1) = x^2 + 2x + 1 + 2x + 2$

Combine like terms on the right side. $\boxed{H(x + 1) = x^2 + 4x + 3}$

Solution (b):

Write $H(x)$. $H(x) = x^2 + 2x$

Evaluate H at $x = 1$. $H(1) = (1)^2 + 2(1) = 3$

Evaluate the sum $H(x) + H(1)$. $H(x) + H(1) = x^2 + 2x + 3$

$\boxed{H(x) + H(1) = x^2 + 2x + 3}$

Note: Comparing the results of parts (a) and (b), we see that $H(x + 1) \neq H(x) + H(1)$.

EXAMPLE 7 **Evaluating Functions: Negatives**

For the given function $G(t) = t^2 - t$, evaluate

a. $G(-t)$ **b.** $-G(t)$

Solution (a):

Write the function G in placeholder notation.	$G(\square) = (\square)^2 - (\square)$
Substitute $-t$ for the argument of G.	$G(-t) = (-t)^2 - (-t)$
Eliminate the parentheses on the right side.	$\boxed{G(-t) = t^2 + t}$

Solution (b):

Write $G(t)$.	$G(t) = t^2 - t$
Multiply by -1.	$-G(t) = -(t^2 - t)$
Eliminate the parentheses on the right side.	$\boxed{-G(t) = -t^2 + t}$

Note: Comparing the results of parts (a) and (b), we see that $G(-t) \neq -G(t)$.

EXAMPLE 8 **Evaluating Functions: Quotients**

For the given function $F(x) = 3x + 5$, evaluate

a. $F\left(\dfrac{1}{2}\right)$ **b.** $\dfrac{F(1)}{F(2)}$

Solution (a):

Write F in placeholder notation.	$F(\square) = 3(\square) + 5$
Replace the argument with $\frac{1}{2}$.	$F\left(\dfrac{1}{2}\right) = 3\left(\dfrac{1}{2}\right) + 5$
Simplify the right side.	$\boxed{F\left(\dfrac{1}{2}\right) = \dfrac{13}{2}}$

Solution (b):

Evaluate $F(1)$.	$F(1) = 3(1) + 5 = 8$
Evaluate $F(2)$.	$F(2) = 3(2) + 5 = 11$
Divide $F(1)$ by $F(2)$.	$\boxed{\dfrac{F(1)}{F(2)} = \dfrac{8}{11}}$

▼ **CAUTION**

$f\left(\dfrac{a}{b}\right) \neq \dfrac{f(a)}{f(b)}$

Note: Comparing the results of parts (a) and (b), we see that $F\left(\dfrac{1}{2}\right) \neq \dfrac{F(1)}{F(2)}$.

▼ **ANSWER**

a. $G(t - 2) = 3t - 10$

b. $G(t) - G(2) = 3t - 6$

c. $\dfrac{G(1)}{G(3)} = -\dfrac{1}{5}$

d. $G\left(\dfrac{1}{3}\right) = -3$

▼ **YOUR TURN** Given the function $G(t) = 3t - 4$, evaluate

a. $G(t - 2)$ **b.** $G(t) - G(2)$ **c.** $\dfrac{G(1)}{G(3)}$ **d.** $G\left(\dfrac{1}{3}\right)$

Examples 6, 7, and 8 illustrate the following, in general:

$$f(a + b) \neq f(a) + f(b) \qquad f(-t) \neq -f(t) \qquad f\left(\frac{a}{b}\right) \neq \frac{f(a)}{f(b)}$$

We now turn our attention to one of the fundamental expressions in calculus: the **difference quotient**.

$$\frac{f(x + h) - f(x)}{h} \qquad h \neq 0$$

Example 9 illustrates the difference quotient, which will be discussed in detail in Section A.5. For now, we will concentrate on the algebra involved when finding the difference quotient. In Section A.5, the application of the difference quotient will be the emphasis.

EXAMPLE 9 **Evaluating the Difference Quotient**

For the function $f(x) = x^2 - x$, find $\dfrac{f(x + h) - f(x)}{h}$.

Solution:

Use placeholder notation for the function $f(x) = x^2 - x$. $f(\square) = (\square)^2 - (\square)$

Calculate $f(x + h)$. $f(x + h) = (x + h)^2 - (x + h)$

Write the difference quotient. $\dfrac{f(x + h) - f(x)}{h}$

Let $f(x + h) = (x + h)^2 - (x + h)$ and $f(x) = x^2 - x$.

$$\frac{f(x + h) - f(x)}{h} = \frac{\overbrace{[(x + h)^2 - (x + h)]}^{f(x + h)} - \overbrace{(x^2 - x)}^{f(x)}}{h} \qquad h \neq 0$$

Eliminate the parentheses inside the first set of brackets.

$$= \frac{(x^2 + 2xh + h^2 - x - h) - (x^2 - x)}{h}$$

Eliminate the brackets in the numerator.

$$= \frac{x^2 + 2xh + h^2 - x - h - x^2 + x}{h}$$

Combine like terms.

$$= \frac{2xh + h^2 - h}{h}$$

Factor the numerator.

$$= \frac{h(2x + h - 1)}{h}$$

Divide out the common factor h.

$$= \boxed{2x + h - 1} \qquad h \neq 0$$

▼ YOUR TURN Evaluate the difference quotient for $f(x) = x^2 - 1$.

▼ ANSWER
$2x + h$

A.4.4 Domain of a Function

Sometimes the domain of a function is stated *explicitly*. For example,

$$f(x) = |x| \qquad \underbrace{x < 0}_{\text{domain}}$$

Here, the **explicit domain** is the set of all negative real numbers, $(-\infty, 0)$. Every negative real number in the domain is mapped to a positive real number in the range through the absolute value function.

A.4.4 SKILL

Determine the domain and range of a function.

A.4.4 CONCEPTUAL

Understand the difference between implicit domain and explicit domain.

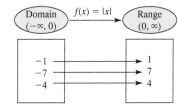

Find the implicit domain for
$f(x) = 1\sqrt{(x - a)}$.

ANSWER (a, ∞)

If the expression that defines the function is given but the domain is not stated explicitly, then the domain is implied. The **implicit domain** is the largest set of real numbers for which the function is defined and the output value $f(x)$ is a real number. For example,

$$f(x) = \sqrt{x}$$

does not have the domain explicitly stated. There is, however, an implicit domain. Note that if the argument is negative, that is, if $x < 0$, then the result is an imaginary number. In order for the output of the function $f(x)$ to be a real number, we must restrict the domain to nonnegative numbers, that is, if $x \geq 0$.

FUNCTION	IMPLICIT DOMAIN
$f(x) = \sqrt{x}$	$[0, \infty)$

In general, we ask the question, "what can x be?" The implicit domain of a function excludes values that cause a function to be undefined or have outputs that are not real numbers.

EXPRESSION THAT DEFINES THE FUNCTION	EXCLUDED x-VALUES	EXAMPLE	IMPLICIT DOMAIN
Polynomial	None	$f(x) = x^3 - 4x^2$	All real numbers
Rational	x-values that make the denominator equal to 0	$g(x) = \dfrac{2}{x^2 - 9}$	$x \neq \pm 3$ or $(-\infty, -3) \cup (-3, 3) \cup (3, \infty)$
Radical	x-values that result in a square (even) root of a negative number	$h(x) = \sqrt{x - 5}$	$x \geq 5$ or $[5, \infty)$

▶ **EXAMPLE 10** **Determining the Domain of a Function**

State the domain of the given functions.

a. $F(x) = \dfrac{3}{x^2 - 25}$ **b.** $H(x) = \sqrt[4]{9 - 2x}$ **c.** $G(x) = \sqrt[3]{x - 1}$

Solution (a):

Write the original equation.	$F(x) = \dfrac{3}{x^2 - 25}$
Determine any restrictions on the values of x.	$x^2 - 25 \neq 0$
Solve the restriction equation.	$x^2 \neq 25$ or $x \neq \pm\sqrt{25} = \pm 5$
State the domain restrictions.	$x \neq \pm 5$
Write the domain in interval notation.	$\boxed{(-\infty, -5) \cup (-5, 5) \cup (5, \infty)}$

Solution (b):

Write the original equation.	$H(x) = \sqrt[4]{9 - 2x}$
Determine any restrictions on the values of x.	$9 - 2x \geq 0$
Solve the restriction equation.	$9 \geq 2x$
State the domain restrictions.	$x \leq \dfrac{9}{2}$
Write the domain in interval notation.	$\boxed{\left(-\infty, \dfrac{9}{2}\right]}$

Solution (c):

Write the original equation. $G(x) = \sqrt[3]{x - 1}$

Determine any restrictions on the values of x. no restrictions

State the domain. $\mathbb{R}$

Write the domain in interval notation. $\boxed{(-\infty, \infty)}$

▼

YOUR TURN State the domain of the given functions.

a. $f(x) = \sqrt{x - 3}$ **b.** $g(x) = \dfrac{1}{x^2 - 4}$

Applications

Functions that are used in applications often have restrictions on the domains due to physical constraints. For example, the volume of a cube is given by the function $V(x) = x^3$, where x is the length of a side. The function $f(x) = x^3$ has no restrictions on x, and therefore the domain is the set of all real numbers. However, the volume of any cube has the restriction that the length of a side can never be negative or zero.

EXAMPLE 11 **Price of Gasoline**

Following the capture of Saddam Hussein in Iraq in 2003, gas prices in the United States escalated and then finally returned to their precapture prices. Over a 6-month period, the average price of a gallon of 87 octane gasoline was given by the function $C(x) = -0.05x^2 + 0.3x + 1.7$, where C is the cost function and x represents the number of months after the capture.

a. Determine the domain of the cost function.

b. What was the average price of gas per gallon 3 months after the capture?

Solution (a):

Since the cost function $C(x) = -0.05x^2 + 0.3x + 1.7$ modeled the price of gas only for 6 months after the capture, the domain is $0 \leq x \leq 6$ or $\boxed{[0, 6]}$.

Solution (b):

Write the cost function. $C(x) = -0.05x^2 + 0.3x + 1.7 \quad 0 \leq x \leq 6$

Find the value of the function
when $x = 3$. $C(3) = -0.05(3)^2 + 0.3(3) + 1.7$

Simplify. $C(3) = 2.15$

The average price per gallon 3 months after the capture was $\boxed{\$2.15}$.

EXAMPLE 12 **The Dimensions of a Pool**

Express the volume of a 30 foot $\times$ 10 foot rectangular swimming pool as a function of its depth.

Solution:

The volume of any rectangular box is $V = lwh$, where V is the volume, l is the length, w is the width, and h is the height. In this example, the length is 30 feet, the width is 10 feet, and the height represents the depth d of the pool.

Write the volume as a function of depth d. $V(d) = (30)(10)d$

Simplify. $\boxed{V(d) = 300d}$

Determine any restrictions on the domain. $d > 0$

▶[SECTION A.4] SUMMARY

Relations and Functions (Let _x_ represent the independent variable and _y_ the dependent variable.)

TYPE	MAPPING/CORRESPONDENCE	EQUATION	GRAPH
Relation	Every _x_-value in the domain maps to **at least one** _y_-value in the range.	$x = y^2$	
Function	Every _x_-value in the domain maps to **exactly one** _y_-value in the range.	$y = x^2$	Passes vertical line test

All functions are relations, but not all relations are functions. Functions can be represented by equations. In the following table, each column illustrates an alternative notation.

INPUT	CORRESPONDENCE	OUTPUT	EQUATION
x	Function	y	$y = 2x + 5$
Independent variable	Mapping	Dependent variable	Mathematical rule
Argument	f	$f(x)$	$f(x) = 2x + 5$

The **domain** is the set of all inputs (_x_-values), and the **range** is the set of all corresponding outputs (_y_-values). Placeholder notation is useful when evaluating functions.

$$f(x) = 3x^2 + 2x$$
$$f(\square) = 3(\square)^2 + 2(\square)$$

An explicit domain is stated, whereas an **implicit domain** is found by _excluding_ _x_-values that

- make the function undefined (denominator = 0).
- result in a nonreal output (even roots of negative real numbers).

[SECTION A.4] EXERCISES

• **SKILLS**

In Exercises 1–24, determine whether each relation is a function. Assume that the coordinate pair (x, y) represents the independent variable _x_ and the dependent variable _y_.

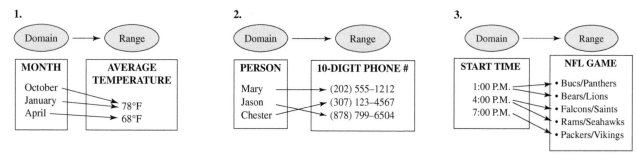

1.

Domain → Range

MONTH	AVERAGE TEMPERATURE
October	
January	78°F
April	68°F

2.

Domain → Range

PERSON	10-DIGIT PHONE #
Mary	(202) 555–1212
Jason	(307) 123–4567
Chester	(878) 799–6504

3.

Domain → Range

START TIME	NFL GAME
1:00 P.M.	• Bucs/Panthers
4:00 P.M.	• Bears/Lions
7:00 P.M.	• Falcons/Saints
	• Rams/Seahawks
	• Packers/Vikings

4.

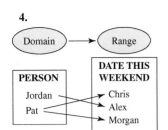

5.

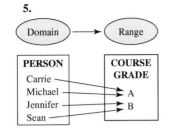

6.

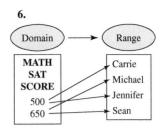

7. $\{(0, -3), (0, 3), (-3, 0), (3, 0)\}$

8. $\{(2, -2), (2, 2), (5, -5), (5, 5)\}$

9. $\{(0, 0), (9, -3), (4, -2), (4, 2), (9, 3)\}$

10. $\{(0, 0), (-1, -1), (-2, -8), (1, 1), (2, 8)\}$

11. $\{(0, 1), (1, 0), (2, 1), (-2, 1), (5, 4), (-3, 4)\}$

12. $\{(0, 1), (1, 1), (2, 1), (3, 1)\}$

13. $x^2 + y^2 = 9$ **14.** $x = |y|$ **15.** $x = y^2$ **16.** $y = x^3$ **17.** $y = |x - 1|$ **18.** $y = 3$

19.

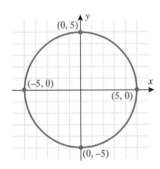

20.

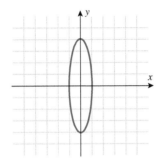

21.

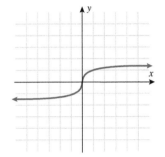

22.

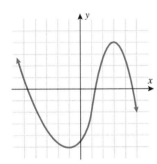

23.

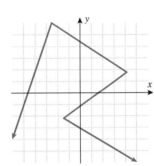

24.

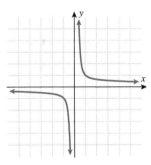

In Exercises 25–32, use the given graphs to evaluate the functions.

25. $y = f(x)$

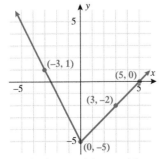

a. $f(2)$ **b.** $f(0)$ **c.** $f(-2)$

26. $y = g(x)$

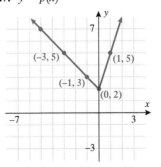

a. $g(-3)$ **b.** $g(0)$ **c.** $g(5)$

27. $y = p(x)$

a. $p(-1)$ **b.** $p(0)$ **c.** $p(1)$

28. $y = r(x)$

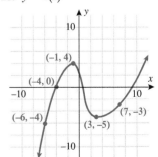

a. $r(-4)$ **b.** $r(-1)$ **c.** $r(3)$

29. $y = C(x)$

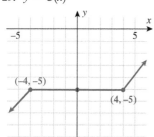

a. $C(2)$ **b.** $C(0)$ **c.** $C(-2)$

30. $y = q(x)$

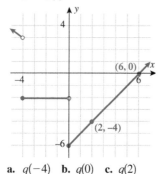

a. $q(-4)$ **b.** $q(0)$ **c.** $q(2)$

31. $y = S(x)$

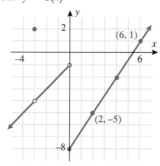

a. $S(-3)$ **b.** $S(0)$ **c.** $S(2)$

32. $y = T(x)$

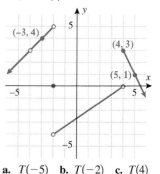

a. $T(-5)$ **b.** $T(-2)$ **c.** $T(4)$

33. Find x if $f(x) = 3$ in Exercise 25.

34. Find x if $g(x) = -2$ in Exercise 26.

35. Find x if $p(x) = 5$ in Exercise 27.

36. Find x if $C(x) = -7$ in Exercise 29.

37. Find x if $C(x) = -5$ in Exercise 29.

38. Find x if $q(x) = -2$ in Exercise 30.

39. Find x if $S(x) = 1$ in Exercise 31.

40. Find x if $T(x) = 4$ in Exercise 32.

In Exercises 41–56, evaluate the given quantities applying the following four functions:

$$f(x) = 2x - 3 \qquad F(t) = 4 - t^2 \qquad g(t) = 5 + t \qquad G(x) = x^2 + 2x - 7$$

41. $f(-2)$

42. $G(-3)$

43. $g(1)$

44. $F(-1)$

45. $f(-2) + g(1)$

46. $G(-3) - F(-1)$

47. $3f(-2) - 2g(1)$

48. $2F(-1) - 2G(-3)$

49. $\dfrac{f(-2)}{g(1)}$

50. $\dfrac{G(-3)}{F(-1)}$

51. $\dfrac{f(0) - f(-2)}{g(1)}$

52. $\dfrac{G(0) - G(-3)}{F(-1)}$

53. $f(x + 1) - f(x - 1)$

54. $F(t + 1) - F(t - 1)$

55. $g(x + a) - f(x + a)$

56. $G(x + b) + F(b)$

In Exercises 57–64, evaluate the difference quotients using the same f, F, G, and g given for Exercises 41–56.

57. $\dfrac{f(x + h) - f(x)}{h}$

58. $\dfrac{F(t + h) - F(t)}{h}$

59. $\dfrac{g(t + h) - g(t)}{h}$

60. $\dfrac{G(x + h) - G(x)}{h}$

61. $\dfrac{f(-2 + h) - f(-2)}{h}$

62. $\dfrac{F(-1 + h) - F(-1)}{h}$

63. $\dfrac{g(1 + h) - g(1)}{h}$

64. $\dfrac{G(-3 + h) - G(-3)}{h}$

In Exercises 65–96, find the domain of the given function. Express the domain in interval notation.

65. $f(x) = 2x - 5$

66. $f(x) = -2x - 5$

67. $g(t) = t^2 + 3t$

68. $h(x) = 3x^4 - 1$

69. $P(x) = \dfrac{x + 5}{x - 5}$

70. $Q(t) = \dfrac{2 - t^2}{t + 3}$

71. $T(x) = \dfrac{2}{x^2 - 4}$

72. $R(x) = \dfrac{1}{x^2 - 1}$

73. $F(x) = \dfrac{1}{x^2 + 1}$

74. $G(t) = \dfrac{2}{t^2 + 4}$

75. $q(x) = \sqrt{7 - x}$

76. $k(t) = \sqrt{t - 7}$

77. $f(x) = \sqrt{2x + 5}$

78. $g(x) = \sqrt{5 - 2x}$

79. $G(t) = \sqrt{t^2 - 4}$

80. $F(x) = \sqrt{x^2 - 25}$

81. $F(x) = \dfrac{1}{\sqrt{x - 3}}$

82. $G(x) = \dfrac{2}{\sqrt{5 - x}}$

83. $f(x) = \sqrt[3]{1 - 2x}$

84. $g(x) = \sqrt[5]{7 - 5x}$

85. $P(x) = \dfrac{1}{\sqrt[5]{x + 4}}$

86. $Q(x) = \dfrac{x}{\sqrt[3]{x^2 - 9}}$

87. $R(x) = \dfrac{x + 1}{\sqrt[4]{3 - 2x}}$

88. $p(x) = \dfrac{x^2}{\sqrt{25 - x^2}}$

89. $H(t) = \dfrac{t}{\sqrt{t^2 - t - 6}}$

90. $f(t) = \dfrac{t - 3}{\sqrt[4]{t^2 + 9}}$

91. $f(x) = (x^2 - 16)^{1/2}$

92. $g(x) = (2x - 5)^{1/3}$

93. $r(x) = x^2(3 - 2x)^{-1/2}$

94. $p(x) = (x - 1)^2(x^2 - 9)^{-3/5}$

95. $f(x) = \frac{2}{5}x - \frac{2}{4}$

96. $g(x) = \frac{2}{3}x^2 - \frac{1}{6}x - \frac{3}{4}$

97. Let $g(x) = x^2 - 2x - 5$ and find the values of x that correspond to $g(x) = 3$.

98. Let $g(x) = \frac{5}{6}x - \frac{3}{4}$ and find the value of x that corresponds to $g(x) = \frac{2}{3}$.

99. Let $f(x) = 2x(x - 5)^3 - 12(x - 5)^2$ and find the values of x that correspond to $f(x) = 0$.

100. Let $f(x) = 3x(x + 3)^2 - 6(x + 3)^3$ and find the values of x that correspond to $f(x) = 0$.

• **APPLICATIONS**

101. **Budget: Event Planning.** The cost associated with a catered wedding reception is $45 per person for a reception for more than 75 people. Write the cost of the reception in terms of the number of guests and state any domain restrictions.

102. **Budget: Long-Distance Calling.** The cost of a local home phone plan is $35 for basic service and $.10 per minute for any domestic long-distance calls. Write the cost of monthly phone service in terms of the number of monthly long-distance minutes and state any domain restrictions.

103. **Temperature.** The average temperature in Tampa, Florida, in the springtime is given by the function $T(x) = -0.7x^2 + 16.8x - 10.8$, where T is the temperature in degrees Fahrenheit and x is the time of day in military time and is restricted to $6 \le x \le 18$ (sunrise to sunset). What is the temperature at 6 A.M.? What is the temperature at noon?

104. **Falling Objects: Firecrackers.** A firecracker is launched straight up, and its height is a function of time, $h(t) = -16t^2 + 128t$, where h is the height in feet and t is the time in seconds with $t = 0$ corresponding to the instant it launches. What is the height 4 seconds after launch? What is the domain of this function?

105. **Collectibles.** The price of a signed Alex Rodriguez baseball card is a function of how many are for sale. When Rodriguez was traded from the Texas Rangers to the New York Yankees in 2004, the going rate for a signed baseball card on eBay was $P(x) = 10 + \sqrt{400,000 - 100x}$, where x represents the number of signed cards for sale. What was the value of the card when there were 10 signed cards for sale? What was the value of the card when there were 100 signed cards for sale?

106. **Collectibles.** In Exercise 105, what was the lowest price on eBay, and how many cards were available then? What was the highest price on eBay, and how many cards were available then?

107. **Volume.** An open box is constructed from a square 10-inch piece of cardboard by cutting squares of length x inches out of each corner and folding the sides up. Express the volume of the box as a function of x, and state the domain.

108. **Volume.** A cylindrical water basin will be built to harvest rainwater. The basin is limited in that the largest radius it can have is 10 feet. Write a function representing the volume of water V as a function of height h. How many additional gallons of water will be collected if you increase the height by 2 feet? *Hint:* 1 cubic foot = 7.48 gallons.

For Exercises 109 and 110, refer to the table below. It illustrates the average federal funds rate for the month of January (2008–2016).

109. **Finance.** Is the relation whose domain is the year and whose range is the average federal funds rate for the month of January a function? Explain.

110. **Finance.** Write five ordered pairs whose domain is the set of even years from 2008 to 2016 and whose range is the set of corresponding average federal funds rate for the month of January.

YEAR	FED. RATE
2008	5.45
2009	5.98
2010	1.73
2011	1.24
2012	1.00
2013	2.25
2014	4.50
2015	5.25
2016	3.50

For Exercises 111 and 112, use the following figure:

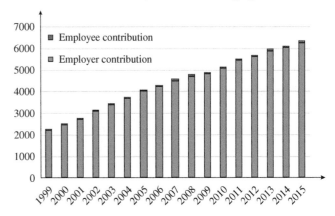

111. **Health-Care Costs:** Fill in the following table. Round dollars to the nearest $1000.

YEAR	TOTAL HEALTH-CARE COST FOR FAMILY PLANS
2000	
2005	
2010	
2015	

Write the five ordered pairs resulting from the table.

112. **Health-Care Costs**. Using the table found in Exercise 111, let the years correspond to the domain and the total costs correspond to the range. Is this relation a function? Explain.

For Exercises 113 and 114, use the following information:

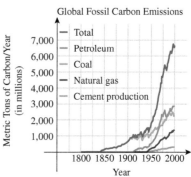

Source: http:/www.naftc.wvu.edu.

Let the functions f, F, g, G, and H represent the number of tons of carbon emitted per year as a function of year corresponding to cement production, natural gas, coal, petroleum, and the total amount, respectively. Let t represent the year, with $t = 0$ corresponding to 1900.

113. **Environment: Global Climate Change.** Estimate (to the nearest thousand) the value of

    a. $F(50)$ **b.** $g(50)$ **c.** $H(50)$

114. **Environment: Global Climate Change.** Explain what the sum $F(100) + g(100) + G(100)$ represents.

• CATCH THE MISTAKE

In Exercises 115–120, explain the mistake that is made.

115. Determine whether the relationship is a function.

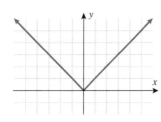

Solution:

Apply the horizontal line test.

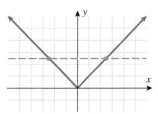

Because the horizontal line intersects the graph in two places, this is not a function.

This is incorrect. What mistake was made?

116. Given the function $H(x) = 3x - 2$, evaluate the quantity $H(3) - H(-1)$.

Solution: $H(3) - H(-1) = H(3) + H(1) = 7 + 1 = 8$

This is incorrect. What mistake was made?

117. Given the function $f(x) = x^2 - x$, evaluate the quantity $f(x + 1)$.

Solution: $f(x + 1) = f(x) + f(1) = x^2 - x + 0$
        $f(x + 1) = x^2 - x$

This is incorrect. What mistake was made?

118. Determine the domain of the function $g(t) = \sqrt{3 - t}$ and express it in interval notation.

Solution:

What can t be? Any nonnegative real number.

$3 - t > 0$
$3 > t$ or $t < 3$
Domain: $(-\infty, 3)$

This is incorrect. What mistake was made?

119. Given the function $G(x) = x^2$, evaluate
$$\frac{G(-1 + h) - G(-1)}{h}.$$

Solution:

$$\frac{G(-1 + h) - G(-1)}{h} = \frac{G(-1) + G(h) - G(-1)}{h}$$

$$= \frac{G(h)}{h} = \frac{h^2}{h} = h$$

This is incorrect. What mistake was made?

120. Given the functions $f(x) = |x - A| - 1$ and $f(1) = -1$, find A.

Solution:

Since $f(1) = -1$, the point $(-1, 1)$ must satisfy the function. $-1 = |-1 - A| - 1$

Add 1 to both sides of the equation. $|-1 - A| = 0$

The absolute value of zero is zero, so there is no need for the absolute value signs: $-1 - A = 0 \Rightarrow A = -1$.

This is incorrect. What mistake was made?

• **CONCEPTUAL**

In Exercises 121–124, determine whether each statement is true or false.

121. If a vertical line does not intersect the graph of an equation, then that equation does not represent a function.

122. If a horizontal line intersects a graph of an equation more than once, the equation does not represent a function.

123. If $f(-a) = f(a)$, then f does not represent a function.

124. If $f(-a) = f(a)$, then f may or may not represent a function.

125. If $f(x) = Ax^2 - 3x$ and $f(1) = -1$, find A.

126. If $g(x) = \dfrac{1}{b - x}$ and $g(3)$ is undefined, find b.

• **CHALLENGE**

127. If $F(x) = \dfrac{C - x}{D - x}$, $F(-2)$ is undefined, and $F(-1) = 4$, find C and D.

128. Construct a function that is undefined at $x = 5$ and whose graph passes through the point $(1, -1)$.

In Exercises 129 and 130, find the domain of each function, where a is any positive real number.

129. $f(x) = \dfrac{-100}{x^2 - a^2}$

130. $f(x) = -5\sqrt{x^2 - a^2}$

• **TECHNOLOGY**

131. Using a graphing utility, graph the temperature function in Exercise 103. What time of day is it the warmest? What is the temperature? Looking at this function, explain why this model for Tampa, Florida, is valid only from sunrise to sunset (6 to 18).

132. Using a graphing utility, graph the height of the firecracker in Exercise 104. How long after liftoff is the firecracker airborne? What is the maximum height that the firecracker attains? Explain why this height model is valid only for the first 8 seconds.

133. Using a graphing utility, graph the price function in Exercise 105. What are the lowest and highest prices of the cards? Does this agree with what you found in Exercise 106?

134. The makers of malted milk balls are considering increasing the size of the spherical treats. The thin chocolate coating on a malted milk ball can be approximated by the surface area, $S(r) = 4\pi r^2$. If the radius is increased 3 millimeters, what is the resulting increase in required chocolate for the thin outer coating?

135. Let $f(x) = x^2 + 1$. Graph $y_1 = f(x)$ and $y_2 = f(x - 2)$ in the same viewing window. Describe how the graph of y_2 can be obtained from the graph of y_1.

136. Let $f(x) = 4 - x^2$. Graph $y_1 = f(x)$ and $y_2 = f(x + 2)$ in the same viewing window. Describe how the graph of y_2 can be obtained from the graph of y_1.

A.5 GRAPHS OF FUNCTIONS; PIECEWISE-DEFINED FUNCTIONS; INCREASING AND DECREASING FUNCTIONS; AVERAGE RATE OF CHANGE

SKILLS OBJECTIVES	CONCEPTUAL OBJECTIVES
▪ Classify functions as even, odd, or neither. ▪ Determine whether functions are increasing, decreasing, or constant. ▪ Calculate the average rate of change of a function. ▪ Graph piecewise-defined functions.	▪ Identify common functions and understand that even functions have graphs that are symmetric about the *y*-axis and odd functions have graphs that are symmetric about the origin. ▪ Understand that intervals of increasing, decreasing, and constant correspond to the *x*-coordinates. ▪ Understand that the difference quotient is just another form of the average rate of change. ▪ Understand points of discontinuity and domain and range of piecewise-defined functions.

A.5.1 Graphs of Functions

Common Functions

A.5.1 SKILL

Classify functions as even, odd, or neither.

A.5.1 CONCEPTUAL

Identify common functions and understand that even functions have graphs that are symmetric about the *y*-axis and odd functions have graphs that are symmetric about the origin.

Point-plotting techniques were introduced in Section A.3, and we noted there that we would explore some more efficient ways of graphing functions. The nine main functions you will read about in this section will constitute a "library" of functions that you should commit to memory. We will draw on this library of functions in the next section when graphing transformations are discussed. Several of these functions have been shown previously in this appendix, but now we will classify them specifically by name and identify properties that each function exhibits.

In Section A.3, we discussed graphs of equations and lines. All lines (with the exception of vertical lines) pass the vertical line test and hence are classified as functions. Instead of the traditional notation of a line, $y = mx + b$, we use function notation and classify a function whose graph is a *line* as a *linear* function.

LINEAR FUNCTION

$$f(x) = mx + b \qquad m \text{ and } b \text{ are real numbers.}$$

The domain of a linear function $f(x) = mx + b$ is the set of all real numbers $\mathbb{R}$. The graph of this function has slope m and *y*-intercept b.

LINEAR FUNCTION: *f(x) = mx + b*	SLOPE: *m*	*y*-INTERCEPT: *b*
$f(x) = 2x - 7$	$m = 2$	$b = -7$
$f(x) = -x + 3$	$m = -1$	$b = 3$
$f(x) = x$	$m = 1$	$b = 0$
$f(x) = 5$	$m = 0$	$b = 5$

One special case of the linear function is the *constant function* ($m = 0$).

CONSTANT FUNCTION

$$f(x) = b \qquad b \text{ is any real number.}$$

The graph of a constant function $f(x) = b$ is a horizontal line. The y-intercept corresponds to the point $(0, b)$. The domain of a constant function is the set of all real numbers $\mathbb{R}$. The range, however, is a single value b. In other words, all x-values correspond to a single y-value.

Points that lie on the graph of a constant function $f(x) = b$ are

$(-5, b)$
$(-1, b)$
$(0, b)$
$(2, b)$
$(4, b)$
$\ldots$
(x, b)

Domain: $(-\infty, \infty)$ Range: $[b, b]$ or $\{b\}$

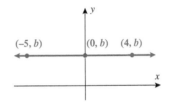

Another specific example of a linear function is the function having a slope of 1 ($m = 1$) and a y-intercept of 0 ($b = 0$). This special case is called the *identity function*.

IDENTITY FUNCTION

$$f(x) = x$$

The graph of the identity function has the following properties: It passes through the origin, and every point that lies on the line has equal x- and y-coordinates. Both the domain and the range of the identity function are the set of all real numbers $\mathbb{R}$.

A function that squares the input is called the *square function*.

Identity Function
Domain: $(-\infty, \infty)$ Range: $(-\infty, \infty)$

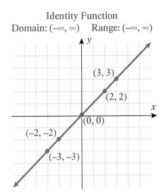

SQUARE FUNCTION

$$f(x) = x^2$$

The graph of the square function is called a parabola. The domain of the square function is the set of all real numbers $\mathbb{R}$. Because squaring a real number always yields a positive number or zero, the range of the square function is the set of all nonnegative numbers. Note that the only intercept is the origin and the square function is symmetric about the y-axis. This graph is contained in quadrants I and II.

Square Function
Domain: $(-\infty, \infty)$ Range: $[0, \infty)$

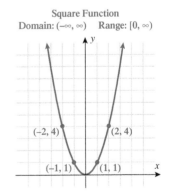

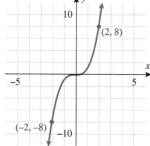

Cube Function
Domain: $(-\infty, \infty)$ Range: $(-\infty, \infty)$

A function that cubes the input is called the *cube function*.

CUBE FUNCTION

$$f(x) = x^3$$

The domain of the cube function is the set of all real numbers $\mathbb{R}$. Because cubing a negative number yields a negative number, cubing a positive number yields a positive number, and cubing 0 yields 0, the range of the cube function is also the set of all real numbers $\mathbb{R}$. Note that the only intercept is the origin and the cube function is symmetric about the origin. This graph extends only into quadrants I and III.

The next two functions are counterparts of the previous two functions: square root and cube root. When a function takes the square root of the input or the cube root of the input, the function is called the *square-root function* or the *cube-root function*, respectively.

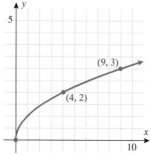

Square-Root Function
Domain: $[0, \infty)$ Range: $[0, \infty)$

SQUARE-ROOT FUNCTION

$$f(x) = \sqrt{x} \quad \text{or} \quad f(x) = x^{1/2}$$

In Section A.4, we found the domain to be $[0, \infty)$. The output of the function will be all real numbers greater than or equal to zero. Therefore, the range of the square-root function is $[0, \infty)$. The graph of this function will be contained in quadrant I.

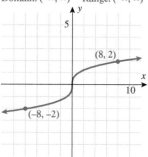

Cube-Root Function
Domain: $(-\infty, \infty)$ Range: $(-\infty, \infty)$

CUBE-ROOT FUNCTION

$$f(x) = \sqrt[3]{x} \quad \text{or} \quad f(x) = x^{1/3}$$

In Section A.4, we stated the domain of the cube-root function to be $(-\infty, \infty)$. We see by the graph that the range is also $(-\infty, \infty)$. This graph is contained in quadrants I and III and passes through the origin. This function is symmetric about the origin.

In algebra, you learned about absolute value. Now we shift our focus to the graph of the *absolute value function*.

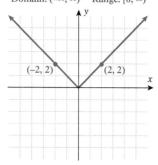

Absolute Value Function
Domain: $(-\infty, \infty)$ Range: $[0, \infty)$

ABSOLUTE VALUE FUNCTION

$$f(x) = |x|$$

Some points that are on the graph of the absolute value function are $(-1, 1)$, $(0, 0)$, and $(1, 1)$. The domain of the absolute value function is the set of all real numbers $\mathbb{R}$, yet the range is the set of nonnegative real numbers. The graph of this function is symmetric with respect to the y-axis and is contained in quadrants I and II.

A function whose output is the reciprocal of its input is called the *reciprocal function.*

Reciprocal Function
Domain: $(-\infty, 0) \cup (0, \infty)$
Range: $(-\infty, 0) \cup (0, \infty)$

RECIPROCAL FUNCTION

$$f(x) = \frac{1}{x} \qquad x \neq 0$$

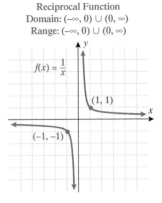

The only restriction on the domain of the reciprocal function is that $x \neq 0$. Therefore, we say the domain is the set of all real numbers excluding zero. The graph of the reciprocal function illustrates that its range is also the set of all real numbers except zero. Note that the reciprocal function is symmetric with respect to the origin and is contained in quadrants I and III.

Even and Odd Functions

Of the nine functions discussed above, several have similar properties of symmetry. The constant function, square function, and absolute value function are all symmetric with respect to the y-axis. The identity function, cube function, cube-root function, and reciprocal function are all symmetric with respect to the origin. The term **even** is used to describe functions that are symmetric with respect to the y-axis, or vertical axis, and the term **odd** is used to describe functions that are symmetric with respect to the origin. Recall from Section A.3 that symmetry can be determined both graphically and algebraically. The box below summarizes the graphic and algebraic characteristics of even and odd functions.

[CONCEPT CHECK]

Classify the functions $f(x) = x^{2n}$ and $g(x) = x^{(2n+1)}$ where n is a positive integer (1, 2, 3, …) as even, odd, or neither.

▼

ANSWER $f(x)$ is even; $g(x)$ is odd.

EVEN AND ODD FUNCTIONS

Function	Symmetric with Respect to	On Replacing x with $-x$
Even	y-axis, or vertical axis	$f(-x) = f(x)$
Odd	origin	$f(-x) = -f(x)$

The algebraic method for determining symmetry with respect to the y-axis, or vertical axis, is to substitute in $-x$ for x. If the result is an equivalent equation, the function is symmetric with respect to the y-axis. Some examples of even functions are $f(x) = b$, $f(x) = x^2$, $f(x) = x^4$, and $f(x) = |x|$. In any of these equations, if $-x$ is substituted for x, the result is the same, that is, $f(-x) = f(x)$. Also note that, with the exception of the absolute value function, these examples are all even-degree polynomial equations. All constant functions are degree zero and are even functions.

The algebraic method for determining symmetry with respect to the origin is to substitute $-x$ for x. If the result is the negative of the original function, that is, if $f(-x) = -f(x)$, then the function is symmetric with respect to the origin and, hence, classified as an odd function. Examples of odd functions are $f(x) = x$, $f(x) = x^3$, $f(x) = x^5$, and $f(x) = x^{1/3}$. In any of these functions, if $-x$ is substituted in for x, the result is the negative of the original function. Note that with the exception of the cube-root function, these equations are odd-degree polynomials.

Be careful, though, because functions that are combinations of even- and odd-degree polynomials can turn out to be neither even nor odd, as we will see in Example 1.

EXAMPLE 1 **Determining Whether a Function Is Even, Odd, or Neither**

Determine whether the functions are even, odd, or neither.

a. $f(x) = x^2 - 3$ **b.** $g(x) = x^5 + x^3$ **c.** $h(x) = x^2 - x$

Solution (a):

Original function.	$f(x) = x^2 - 3$
Replace x with $-x$.	$f(-x) = (-x)^2 - 3$
Simplify.	$f(-x) = x^2 - 3 = f(x)$

Because $f(-x) = f(x)$, we say that $\boxed{f(x) \text{ is an } even \text{ function}}$.

Solution (b):

Original function.	$g(x) = x^5 + x^3$
Replace x with $-x$.	$g(-x) = (-x)^5 + (-x)^3$
Simplify.	$g(-x) = -x^5 - x^3 = -(x^5 + x^3) = -g(x)$

Because $g(-x) = -g(x)$, we say that $\boxed{g(x) \text{ is an } odd \text{ function}}$.

Solution (c):

Original function.	$h(x) = x^2 - x$
Replace x with $-x$.	$h(-x) = (-x)^2 - (-x)$
Simplify.	$h(-x) = x^2 + x$

$h(-x)$ is neither $-h(x)$ nor $h(x)$; therefore the function $h(x)$ is $\boxed{\text{neither even nor odd}}$.

In parts (a), (b), and (c), we classified these functions as either even, odd, or neither, using the algebraic test. Look back at them now and reflect on whether these classifications agree with your intuition. In part (a), we combined two functions: the square function and the constant function. Both of these functions are even, and adding even functions yields another even function. In part (b), we combined two odd functions: the fifth-power function and the cube function. Both of these functions are odd, and adding two odd functions yields another odd function. In part (c), we combined two functions: the square function and the identity function. The square function is even, and the identity function is odd. In this part, combining an even function with an odd function yields a function that is neither even nor odd and, hence, has no symmetry with respect to the vertical axis or the origin.

▼

YOUR TURN Classify the functions as even, odd, or neither.

a. $f(x) = |x| + 4$ **b.** $f(x) = x^3 - 1$

▼
ANSWER

a. even b. neither

A.5.2 Increasing and Decreasing Functions

Look at the figure on the right. Graphs are read from **left to right**. If we start at the left side of the graph and trace the red curve with our pen, we see that the function values (values in the vertical direction) are decreasing until arriving at the point $(-2, -2)$. Then, the function values increase until arriving at the point $(-1, 1)$. The values then remain constant ($y = 1$) between the points $(-1, 1)$ and $(0, 1)$. Proceeding beyond the point $(0, 1)$, the function values decrease again until the point $(2, -2)$. Beyond the point $(2, -2)$, the function values increase again until the point $(6, 4)$. Finally, the function values decrease and continue to do so.

When specifying a function as increasing, decreasing, or constant, the **intervals are classified according to the x-coordinate**. For instance, in this graph, we say the function is increasing when x is between $x = -2$ and $x = -1$ and again when x is between $x = 2$ and $x = 6$. The graph is classified as decreasing when x is less than -2, again when x is between 0 and 2, and again when x is greater than 6. The graph is classified as constant when x is between -1 and 0. In interval notation, this is summarized as

Decreasing	**Increasing**	**Constant**
$(-\infty, -2) \cup (0, 2) \cup (6, \infty)$	$(-2, -1) \cup (2, 6)$	$(-1, 0)$

An algebraic test for determining whether a function is increasing, decreasing, or constant is to compare the value $f(x)$ of the function for particular points in the intervals.

A.5.2 SKILL

Determine whether functions are increasing, decreasing, or constant.

A.5.2 CONCEPTUAL

Understand that intervals of increasing, decreasing, and constant correspond to the x-coordinates.

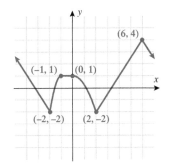

INCREASING, DECREASING, AND CONSTANT FUNCTIONS

1. A function f is **increasing** on an open interval I if for any x_1 and x_2 in I, where $x_1 < x_2$, then $f(x_1) < f(x_2)$.
2. A function f is **decreasing** on an open interval I if for any x_1 and x_2 in I, where $x_1 < x_2$, then $f(x_1) > f(x_2)$.
3. A function f is **constant** on an open interval I if for any x_1 and x_2 in I, then $f(x_1) = f(x_2)$.

STUDY TIP

• Graphs are read from left to right.
• Intervals correspond to the x-coordinates.

In addition to classifying a function as increasing, decreasing, or constant, we can also determine the domain and range of a function by inspecting its graph from left to right:

- The domain is the set of all x-values where the function is defined.
- The range is the set of all y-values that the graph of the function corresponds to.
- A solid dot on the left or right end of a graph indicates that the graph terminates there and the point is included in the graph.
- An open dot indicates that the graph terminates there and the point is not included in the graph.
- Unless a dot is present, it is assumed that a graph continues indefinitely in the same direction. (An arrow is used in some books to indicate direction.)

STUDY TIP

Increasing: Graph of function rises from left to right.
Decreasing: Graph of function falls from left to right.
Constant: Graph of function does not change height from left to right.

EXAMPLE 2 **Finding Intervals When a Function Is Increasing or Decreasing**

Given the graph of a function:

a. State the domain and range of the function.

b. Find the intervals when the function is increasing, decreasing, or constant.

Solution

Domain: $[-5, \infty)$

Range: $[0, \infty)$

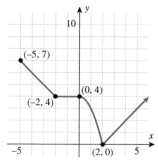

Reading the graph from **left to right**, we see that the graph

- decreases from the point $(-5, 7)$ to the point $(-2, 4)$.
- is constant from the point $(-2, 4)$ to the point $(0, 4)$.
- decreases from the point $(0, 4)$ to the point $(2, 0)$.
- increases from the point $(2, 0)$ on.

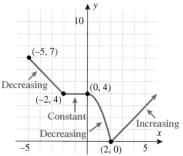

The intervals of increasing and decreasing correspond to the *x*-coordinates.

We say that this function is

- increasing on the interval $(2, \infty)$.
- decreasing on the interval $(-5, -2) \cup (0, 2)$.
- constant on the interval $(-2, 0)$.

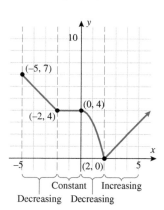

Note: The intervals of increasing or decreasing are defined on *open* intervals. This should not be confused with the domain. For example, the point $x = -5$ is included in the domain of the function but not in the interval where the function is classified as decreasing.

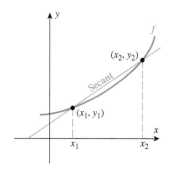

A.5.3 Average Rate of Change

How do we know *how much* a function is increasing or decreasing? For example, is the price of a stock slightly increasing or is it doubling every week? One way we determine how much a function is increasing or decreasing is by calculating its *average rate of change.*

Let (x_1, y_1) and (x_2, y_2) be two points that lie on the graph of a function f. Draw the line that passes through these two points (x_1, y_1) and (x_2, y_2). This line is called a secant line.

Note that the slope of the secant line is given by $m = \dfrac{y_2 - y_1}{x_2 - x_1}$, and recall that the slope of a line is the rate of change of that line. The **slope of the secant line** is used to represent the *average rate of change* of the function.

AVERAGE RATE OF CHANGE

Let $(x_1, f(x_1))$ and $(x_2, f(x_2))$ be two distinct points, $(x_1 \neq x_2)$, on the graph of the function f. The **average rate of change** of f between x_1 and x_2 is given by

Average rate of change $= \dfrac{f(x_2) - f(x_1)}{x_2 - x_1}$

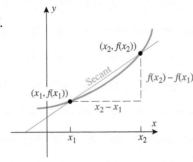

EXAMPLE 3 **Average Rate of Change**

Find the average rate of change of $f(x) = x^4$ from

a. $x = -1$ to $x = 0$ **b.** $x = 0$ to $x = 1$ **c.** $x = 1$ to $x = 2$

Solution (a):

Write the average rate of change formula.

$$\dfrac{f(x_2) - f(x_1)}{x_2 - x_1}$$

Let $x_1 = -1$ and $x_2 = 0$.

$$= \dfrac{f(0) - f(-1)}{0 - (-1)}$$

Substitute $f(-1) = (-1)^4 = 1$ and $f(0) = 0^4 = 0$.

$$= \dfrac{0 - 1}{0 - (-1)}$$

Simplify.

$$= \boxed{-1}$$

Solution (b):

Write the average rate of change formula.

$$\dfrac{f(x_2) - f(x_1)}{x_2 - x_1}$$

Let $x_1 = 0$ and $x_2 = 1$.

$$= \dfrac{f(1) - f(0)}{1 - 0}$$

Substitute $f(0) = 0^4 = 0$ and $f(1) = (1)^4 = 1$.

$$= \dfrac{1 - 0}{1 - 0}$$

Simplify.

$$= \boxed{1}$$

Solution (c):

Write the average rate of change formula.

$$\dfrac{f(x_2) - f(x_1)}{x_2 - x_1}$$

Let $x_1 = 1$ and $x_2 = 2$.

$$= \dfrac{f(2) - f(1)}{2 - 1}$$

Substitute $f(1) = 1^4 = 1$ and $f(2) = (2)^4 = 16$.

$$= \dfrac{16 - 1}{2 - 1}$$

Simplify.

$$= \boxed{15}$$

STUDY TIP

The average rate of change of a function equals the slope of the secant line.

Graphical Interpretation: Slope of the Secant Line

a. Between $(-1, 1)$ and $(0, 0)$, this function is decreasing at an average rate of 1.

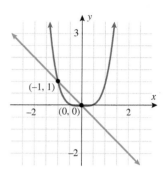

b. Between $(0, 0)$ and $(1, 1)$, this function is increasing at an average rate of 1.

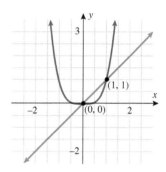

c. Between $(1, 1)$ and $(2, 16)$, this function is increasing at an average rate of 15.

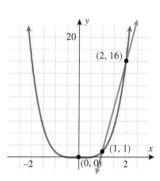

▼ ANSWER

a. -2 **b.** 2

YOUR TURN Find the average rate of change of $f(x) = x^2$ from

a. $x = -2$ to $x = 0$ **b.** $x = 0$ to $x = 2$

The average rate of change can also be written in terms of the difference quotient.

WORDS	MATH
Let the difference between x_1 and x_2 be h.	$x_2 - x_1 = h$
Solve for x_2.	$x_2 = x_1 + h$
Substitute $x_2 - x_1 = h$ into the denominator and $x_2 = x_1 + h$ into the numerator of the average rate of change.	Average rate of change $= \dfrac{f(x_2) - f(x_1)}{x_2 - x_1}$ $= \dfrac{f(x_1 + h) - f(x_1)}{h}$
Let $x_1 = x$.	$= \dfrac{f(x + h) - f(x)}{h}$

When written in this form, the average rate of change is called the **difference quotient**.

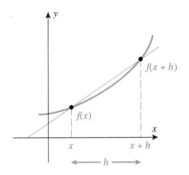

DEFINITION Difference Quotient

The expression $\dfrac{f(x + h) - f(x)}{h}$, where $h \neq 0$, is called the **difference quotient**.

The difference quotient is more meaningful when h is small. In calculus the difference quotient is used to define a derivative.

EXAMPLE 4 **Calculating the Difference Quotient**

Calculate the difference quotient for the function $f(x) = 2x^2 + 1$.

Solution:

Find $f(x + h)$.

$$f(x + h) = 2(x + h)^2 + 1$$
$$= 2(x^2 + 2xh + h^2) + 1$$
$$= 2x^2 + 4xh + 2h^2 + 1$$

Find the difference quotient.

$$\frac{f(x + h) - f(x)}{h} = \frac{\overbrace{2x^2 + 4xh + 2h^2 + 1}^{f(x+h)} - \overbrace{(2x^2 + 1)}^{f(x)}}{h}$$

Simplify.

$$\frac{f(x + h) - f(x)}{h} = \frac{2x^2 + 4xh + 2h^2 + 1 - 2x^2 - 1}{h}$$

$$\frac{f(x + h) - f(x)}{h} = \frac{4xh + 2h^2}{h}$$

Factor the numerator.

$$\frac{f(x + h) - f(x)}{h} = \frac{h(4x + 2h)}{h}$$

Cancel (divide out) the common h.

$$\frac{f(x + h) - f(x)}{h} = \boxed{4x + 2h} \quad h \neq 0$$

YOUR TURN Calculate the difference quotient for the function $f(x) = -x^2 + 2$.

A.5.4 Piecewise-Defined Functions

Most of the functions that we have seen in this text are functions defined by polynomials. Sometimes the need arises to define functions in terms of *pieces*. For example, most plumbers charge a flat fee for a house call and then an additional hourly rate for the job. For instance, if a particular plumber charges $100 to drive out to your house and work for 1 hour and then an additional $25 an hour for every additional hour he or she works on your job, we would define this function in pieces. If we let h be the number of hours worked, then the charge is defined as

$$\text{Plumbing charge} = \begin{cases} 100 & h \leq 1 \\ 100 + 25(h - 1) & h > 1 \end{cases}$$

If we were to graph this function, we would see that there is 1 hour that is constant and after that the function continually increases.

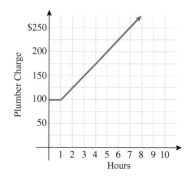

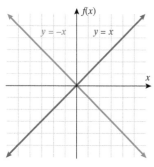

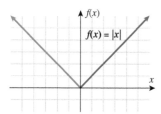

Another piecewise-defined function is the absolute value function. The absolute value function can be thought of as two pieces: the line $y = -x$ (when x is negative) and the line $y = x$ (when x is nonnegative). We start by graphing these two lines on the same graph.

The absolute value function behaves like the line $y = -x$ when x is negative (erase the blue graph in quadrant IV) and like the line $y = x$ when x is positive (erase the red graph in quadrant III).

Absolute value piecewise-defined function

$$f(x) = |x| = \begin{cases} -x & x < 0 \\ x & x \geq 0 \end{cases}$$

The next example is a piecewise-defined function given in terms of functions in our "library of functions." Because the function is defined in terms of pieces of other functions, we draw the graph of each individual function and, then, for each function darken the piece corresponding to its part of the domain. This is like the procedure above for the absolute value function.

EXAMPLE 5 **Graphing Piecewise-Defined Functions**

Graph the piecewise-defined function, and state the domain, range, and intervals when the function is increasing, decreasing, or constant.

$$G(x) = \begin{cases} x^2 & x < -1 \\ 1 & -1 \leq x \leq 1 \\ x & x > 1 \end{cases}$$

Solution:

Graph each of the functions on the same plane.

Square function: $f(x) = x^2$

Constant function: $f(x) = 1$

Identity function: $f(x) = x$

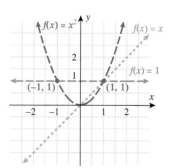

The points to focus on in particular are the x-values where the pieces change over—that is, $x = -1$ and $x = 1$.

Let's now investigate each piece. When $x < -1$, this function is defined by the square function, $f(x) = x^2$, so darken that particular function to the left of $x = -1$. When $-1 \leq x \leq 1$, the function is defined by the constant function, $f(x) = 1$, so darken that particular function between the x values of -1 and 1. When $x > 1$, the function is defined by the identity function, $f(x) = x$, so darken that function to the right of $x = 1$. Erase everything that is not darkened, and the resulting graph of the piecewise-defined function is given on the right.

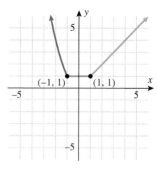

This function is defined for all real values of x, so the domain of this function is the set of all real numbers. The values that this function yields in the vertical direction are all real numbers greater than or equal to 1. Hence, the range of this function is $[1, \infty)$. The intervals of increasing, decreasing, and constant are as follows:

Decreasing: $(-\infty, -1)$

Constant: $(-1, 1)$

Increasing: $(1, \infty)$

The term **continuous** implies that there are no holes or jumps and that the graph can be drawn without picking up your pencil. A function that does have holes or jumps and cannot be drawn in one motion without picking up your pencil is classified as **discontinuous**, and the points where the holes or jumps occur are called *points of discontinuity*.

The previous example illustrates a *continuous* piecewise-defined function. At the $x = -1$ junction, the square function and constant function both pass through the point $(-1, 1)$. At the $x = 1$ junction, the constant function and the identity function both pass through the point $(1, 1)$. Since the graph of this piecewise-defined function has no holes or jumps, we classify it as a continuous function.

The next example illustrates a *discontinuous* piecewise-defined function.

EXAMPLE 6   **Graphing a Discontinuous Piecewise-Defined Function**

Graph the piecewise-defined function, and state the intervals where the function is increasing, decreasing, or constant, along with the domain and range.

$$f(x) = \begin{cases} 1 - x & x < 0 \\ x & 0 \le x < 2 \\ -1 & x > 2 \end{cases}$$

Solution:

Graph these functions on the same plane.

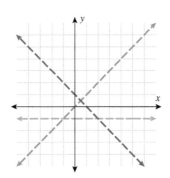

Linear function: $f(x) = 1 - x$

Identity function: $f(x) = x$

Constant function: $f(x) = -1$

Darken the piecewise-defined function on the graph. For all values less than zero $(x < 0)$, the function is defined by the **linear function**. Note the use of an open circle, indicating up to but not including $x = 0$. For values $0 \le x < 2$, the function is defined by the **identity function**.

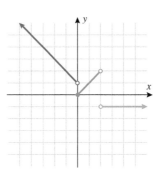

The circle is filled in at the left endpoint, $x = 0$. An open circle is used at $x = 2$. For all values greater than 2, $x > 2$, the function is defined by the **constant function**. Because this interval does not include the point $x = 2$, an open circle is used.

At what intervals is the function increasing, decreasing, or constant? Remember that the intervals correspond to the *x*-values.

Decreasing: $(-\infty, 0)$ Increasing: $(0, 2)$ Constant: $(2, \infty)$

The function is defined for all values of x except $x = 2$.

Domain: $(-\infty, 2) \cup (2, \infty)$

The output of this function (vertical direction) takes on the *y*-values $y \ge 0$ and the additional single value $y = -1$.

Range: $[-1, -1] \cup [0, \infty)$ or $\{-1\} \cup [0, \infty)$

We mentioned earlier that a discontinuous function has a graph that exhibits holes or jumps. In Example 6, the point $x = 0$ corresponds to a jump because you would have to pick up your pencil to continue drawing the graph. The point $x = 2$ corresponds to both a hole and a jump. The hole indicates that the function is not defined at that point, and there is still a jump because the identity function and the constant function do not meet at the same y-value at $x = 2$.

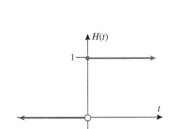
▼
YOUR TURN Graph the piecewise-defined function, and state the intervals where the function is increasing, decreasing, or constant, along with the domain and range.

$$f(x) = \begin{cases} -x & x \le -1 \\ 2 & -1 < x < 1 \\ x & x > 1 \end{cases}$$

Piecewise-defined functions whose "pieces" are constants are called **step functions**. The reason for this name is that the graph of a step function looks like steps of a staircase. A common step function used in engineering is the **Heaviside step function** (also called the **unit step function**):

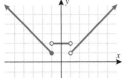

$$H(t) = \begin{cases} 0 & t < 0 \\ 1 & t \ge 0 \end{cases}$$

This function is used in signal processing to represent a signal that turns on at some time and stays on indefinitely.

A common step function used in business applications is the *greatest integer function*.

GREATEST INTEGER FUNCTION

$f(x) = [[x]]$ = greatest integer less than or equal to x.

x	1.0	1.3	1.5	1.7	1.9	2.0
$f(x) = [[x]]$	1	1	1	1	1	2

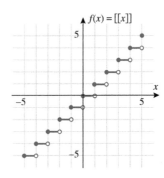

▶[SECTION A.5] SUMMARY

NAME	FUNCTION	DOMAIN	RANGE	GRAPH	EVEN/ODD
Linear	$f(x) = mx + b, m \ne 0$	$(-\infty, \infty)$	$(-\infty, \infty)$		Neither (unless $y = x$)
Constant	$f(x) = c$	$(-\infty, \infty)$	$[c, c]$ or $\{c\}$		Even

NAME	FUNCTION	DOMAIN	RANGE	GRAPH	EVEN/ODD
Identity	$f(x) = x$	$(-\infty, \infty)$	$(-\infty, \infty)$		Odd
Square	$f(x) = x^2$	$(-\infty, \infty)$	$[0, \infty)$		Even
Cube	$f(x) = x^3$	$(-\infty, \infty)$	$(-\infty, \infty)$		Odd
Square Root	$f(x) = \sqrt{x}$	$[0, \infty)$	$[0, \infty)$		Neither
Cube Root	$f(x) = \sqrt[3]{x}$	$(-\infty, \infty)$	$(-\infty, \infty)$		Odd
Absolute Value	$f(x) = |x|$	$(-\infty, \infty)$	$[0, \infty)$		Even
Reciprocal	$f(x) = \dfrac{1}{x}$	$(-\infty, 0) \cup (0, \infty)$	$(-\infty, 0) \cup (0, \infty)$		Odd

Domain and Range of a Function

- Implied domain: Exclude any values that lead to the function being undefined (dividing by zero) or imaginary outputs (square root of a negative real number).
- Inspect the graph to determine the set of all inputs (domain) and the set of all outputs (range).

Finding Intervals Where a Function Is Increasing, Decreasing, or Constant

- Increasing: Graph of function rises from left to right.
- Decreasing: Graph of function falls from left to right.
- Constant: Graph of function does not change height from left to right.

Average Rate of Change $\dfrac{f(x_2) - f(x_1)}{x_2 - x_1}$ $x_1 \neq x_2$

Difference Quotient $\dfrac{f(x + h) - f(x)}{h}$ $h \neq 0$

Piecewise-Defined Functions

- Continuous: You can draw the graph of a function without picking up the pencil.
- Discontinuous: Graph has holes and/or jumps.

[SECTION A.5] EXERCISES

• **SKILLS**

In Exercises 1–24, determine whether the function is even, odd, or neither.

1. $G(x) = x + 4$

2. $h(x) = 3 - x$

3. $f(x) = 3x^2 + 1$

4. $F(x) = x^4 + 2x^2$

5. $g(t) = 5t^3 - 3t$

6. $f(x) = 3x^5 + 4x^3$

7. $h(x) = x^2 + 2x$

8. $G(x) = 2x^4 + 3x^3$

9. $h(x) = x^{1/3} - x$

10. $g(x) = x^{-1} + x$

11. $f(x) = |x| + 5$

12. $f(x) = |x| + x^2$

13. $f(x) = |x|$

14. $f(x) = |x^3|$

15. $G(t) = |t - 3|$

16. $g(t) = |t + 2|$

17. $G(t) = \sqrt{t - 3}$

18. $f(x) = \sqrt{2 - x}$

19. $g(x) = \sqrt{x^2 + x}$

20. $f(x) = \sqrt{x^2 + 2}$

21. $h(x) = \dfrac{1}{x} + 3$

22. $h(x) = \dfrac{1}{x} - 2x$

23.

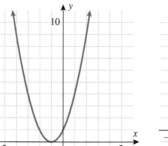

24.

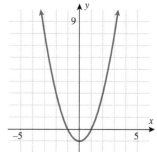

In Exercises 25–36, state the (a) domain, (b) range, and (c) x-interval(s) where the function is increasing, decreasing, or constant. Find the values of (d) $f(0)$, (e) $f(-2)$, and (f) $f(2)$.

25.

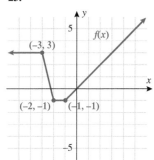

26.

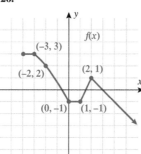

27.

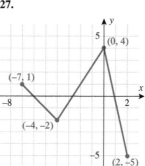

28.

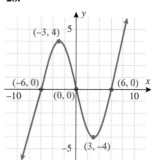

29.

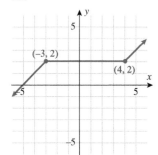

30.

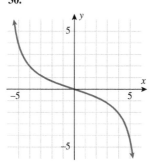

31.

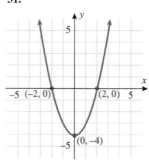

32.

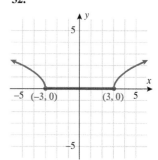

33.

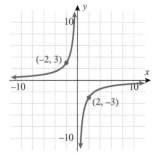

34.

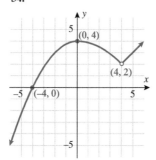

35.

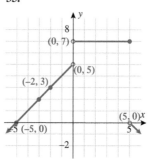

36.

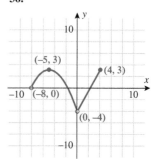

In Exercises 37–44, find the difference quotient $\dfrac{f(x + h) - f(x)}{h}$ for each function.

37. $f(x) = x^2 - x$
38. $f(x) = x^2 + 2x$
39. $f(x) = 3x + x^2$
40. $f(x) = 5x - x^2$
41. $f(x) = x^2 - 3x + 2$
42. $f(x) = x^2 - 2x + 5$
43. $f(x) = -3x^2 + 5x - 4$
44. $f(x) = -4x^2 + 2x - 3$

In Exercises 45–52, find the average rate of change of the function from $x = 1$ to $x = 3$.

45. $f(x) = x^3$
46. $f(x) = \dfrac{1}{x}$
47. $f(x) = |x|$
48. $f(x) = 2x$

49. $f(x) = 1 - 2x$
50. $f(x) = 9 - x^2$
51. $f(x) = |5 - 2x|$
52. $f(x) = \sqrt{x^2 - 1}$

In Exercises 53–76, graph the piecewise-defined functions. State the domain and range in interval notation. Determine the intervals where the function is increasing, decreasing, or constant.

53. $f(x) = \begin{cases} x & x < 2 \\ 2 & x \geq 2 \end{cases}$
54. $f(x) = \begin{cases} -x & x < -1 \\ -1 & x \geq -1 \end{cases}$
55. $f(x) = \begin{cases} 1 & x < -1 \\ x^2 & x \geq -1 \end{cases}$
56. $f(x) = \begin{cases} x^2 & x < 2 \\ 4 & x \geq 2 \end{cases}$

57. $f(x) = \begin{cases} x & x < 0 \\ x^2 & x \geq 0 \end{cases}$
58. $f(x) = \begin{cases} -x & x \leq 0 \\ x^2 & x > 0 \end{cases}$
59. $f(x) = \begin{cases} -x + 2 & x < 1 \\ x^2 & x \geq 1 \end{cases}$
60. $f(x) = \begin{cases} 2 + x & x \leq -1 \\ x^2 & x > -1 \end{cases}$

61. $G(x) = \begin{cases} -1 & x < -1 \\ x & -1 \leq x \leq 3 \\ 3 & x > 3 \end{cases}$
62. $G(x) = \begin{cases} -1 & x < -1 \\ x & -1 < x < 3 \\ 3 & x > 3 \end{cases}$

63. $G(t) = \begin{cases} 1 & t < 1 \\ t^2 & 1 \leq t \leq 2 \\ 4 & t > 2 \end{cases}$
64. $G(t) = \begin{cases} 1 & t < 1 \\ t^2 & 1 < t < 2 \\ 4 & t > 2 \end{cases}$

65. $f(x) = \begin{cases} -x - 1 & x < -2 \\ x + 1 & -2 < x < 1 \\ -x + 1 & x \geq 1 \end{cases}$
66. $f(x) = \begin{cases} -x - 1 & x \leq -2 \\ x + 1 & -2 < x < 1 \\ -x + 1 & x > 1 \end{cases}$

67. $G(x) = \begin{cases} 0 & x < 0 \\ \sqrt{x} & x \geq 0 \end{cases}$
68. $G(x) = \begin{cases} 1 & x < 1 \\ \sqrt[3]{x} & x > 1 \end{cases}$

69. $G(x) = \begin{cases} 0 & x = 0 \\ \dfrac{1}{x} & x \neq 0 \end{cases}$
70. $G(x) = \begin{cases} 0 & x = 0 \\ -\dfrac{1}{x} & x \neq 0 \end{cases}$

71. $G(x) = \begin{cases} -\sqrt[3]{x} & x \leq -1 \\ x & -1 < x < 1 \\ -\sqrt{x} & x > 1 \end{cases}$
72. $G(x) = \begin{cases} -\sqrt[3]{x} & x < -1 \\ x & -1 \leq x < 1 \\ \sqrt{x} & x > 1 \end{cases}$

73. $f(x) = \begin{cases} x + 3 & x \leq -2 \\ |x| & -2 < x < 2 \\ x^2 & x \geq 2 \end{cases}$
74. $f(x) = \begin{cases} |x| & x < -1 \\ 1 & -1 < x < 1 \\ |x| & x > 1 \end{cases}$

75. $f(x) = \begin{cases} x & x \leq -1 \\ x^3 & -1 < x < 1 \\ x^2 & x > 1 \end{cases}$
76. $f(x) = \begin{cases} x^2 & x \leq -1 \\ x^3 & -1 < x < 1 \\ x & x \geq 1 \end{cases}$

• **APPLICATIONS**

77. Budget: Costs. The Kappa Kappa Gamma sorority decides to order custom-made T-shirts for its Kappa Krush mixer with the Sigma Alpha Epsilon fraternity. If the sorority orders 50 or fewer T-shirts, the cost is $10 per shirt. If it orders more than 50 but fewer than 100, the cost is $9 per shirt. If it orders more than 100, the cost is $8 per shirt. Find the cost function $C(x)$ as a function of the number of T-shirts x ordered.

78. Budget: Costs. The marching band at a university is ordering some additional uniforms to replace existing uniforms that are worn out. If the band orders 50 or fewer, the cost is $176.12 per uniform. If it orders more than 50 but fewer than 100, the cost is $159.73 per uniform. Find the cost function $C(x)$ as a function of the number of new uniforms x ordered.

79. Budget: Costs. The Richmond rowing club is planning to enter the Head of the Charles race in Boston and is trying to figure out how much money to raise. The entry fee is $250 per boat for the first 10 boats and $175 for each additional boat. Find the cost function $C(x)$ as a function of the number of boats x the club enters.

80. Phone Cost: Long-Distance Calling. A phone company charges $0.39 per minute for the first 10 minutes of an international long-distance phone call and $0.12 per minute every minute after that. Find the cost function $C(x)$ as a function of the length of the phone call x in minutes.

81. Event Planning. A young couple are planning their wedding reception at a yacht club. The yacht club charges a flat rate of $1000 to reserve the dining room for a private party. The cost of food is $35 per person for the first 100 people and $25 per person for every additional person beyond the first 100. Write the cost function $C(x)$ as a function of the number of people x attending the reception.

82. Home Improvement. An irrigation company gives you an estimate for an eight-zone sprinkler system. The parts are $1400, and the labor is $25 per hour. Write a function $C(x)$ that determines the cost of a new sprinkler system if you choose this irrigation company.

83. Sales. A famous author negotiates with her publisher the monies she will receive for her next suspense novel. She will receive $50,000 up front and a 15% royalty rate on the first 100,000 books sold, and 20% on any books sold beyond that. If the book sells for $20 and royalties are based on the selling price, write a royalties function $R(x)$ as a function of total number x of books sold.

84. Sales. Rework Exercise 83 if the author receives $35,000 up front, 15% for the first 100,000 books sold, and 25% on any books sold beyond that.

85. Profit. A group of artists are trying to decide whether they will make a profit if they set up a Web-based business to market and sell stained glass that they make. The costs associated with this business are $100 per month for the website and $700 per month for the studio they rent. The materials cost $35 for each work in stained glass, and the artists charge $100 for each unit they sell. Write the monthly profit as a function of the number of stained-glass units they sell.

86. Profit. Philip decides to host a shrimp boil at his house as a fundraiser for his daughter's AAU basketball team. He orders gulf shrimp to be flown in from New Orleans. The shrimp costs $5 per pound. The shipping costs $30. If he charges $10 per person, write a function $F(x)$ that represents either his loss or profit as a function of the number of people x that attend. Assume that each person will eat 1 pound of shrimp.

87. Postage Rates. The following table corresponds to first-class postage rates for the U.S. Postal Service. Write a piecewise-defined function in terms of the greatest integer function that models this cost of mailing flat envelopes first class.

WEIGHT LESS THAN (OUNCES)	FIRST-CLASS RATE (FLAT ENVELOPES)
1	$0.98
2	1.20
3	1.42
4	1.64
5	1.86
6	2.08
7	2.30
8	2.52
9	2.74
10	2.96
11	3.18
12	3.40
13	3.62

88. Postage Rates. The following table corresponds to first-class postage rates for the U.S. Postal Service. Write a piecewise-defined function in terms of the greatest integer function that models this cost of mailing parcels first class.

WEIGHT LESS THAN (OUNCES)	PARCELS
1	$2.14
2	2.34
3	2.54
4	2.74
5	2.94
6	3.14
7	3.34
8	3.54
9	3.74
10	3.94
11	4.14
12	4.34
13	4.54

For Exercises 89 and 90, refer to the following:

A square wave is a waveform used in electronic circuit testing and signal processing. A square wave alternates regularly and instantaneously between two levels.

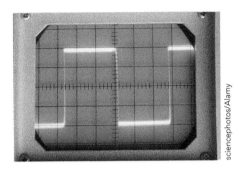

sciencephotos/Alamy

89. **Electronics: Signals.** Write a step function $f(t)$ that represents the following square wave.

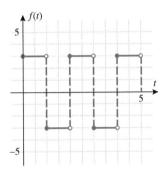

90. **Electronics: Signals.** Write a step function $f(x)$ that represents the following square wave, where x represents frequency in Hertz.

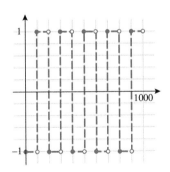

For Exercises 91 and 92, refer to the following table:

Global Carbon Emissions from Fossil Fuel Burning

YEAR	MILLIONS OF TONS OF CARBON
1900	500
1925	1000
1950	1500
1975	5000
2000	7000

91. **Climate Change: Global Warming.** What is the average rate of change in global carbon emissions from fossil fuel burning from
    a. 1900 to 1950   **b.** 1950 to 2000?

92. **Climate Change: Global Warming.** What is the average rate of change in global carbon emissions from fossil fuel burning from
    a. 1950 to 1975   **b.** 1975 to 2000?

For Exercises 93 and 94, use the following information:

The height (in feet) of a falling object with an initial velocity of 48 feet per second launched straight upward from the ground is given by $h(t) = -16t^2 + 48t$, where t is time (in seconds).

93. **Falling Objects.** What is the average rate of change of the height as a function of time from $t = 1$ to $t = 2$?

94. **Falling Objects.** What is the average rate of change of the height as a function of time from $t = 1$ to $t = 3$?

● **CATCH THE MISTAKE**

In Exercises 95–98, explain the mistake that is made.

95. Graph the piecewise-defined function. State the domain and range.

$$f(x) = \begin{cases} -x & x < 0 \\ x & x > 0 \end{cases}$$

Solution:

Draw the graphs of $f(x) = -x$ and $f(x) = x$.

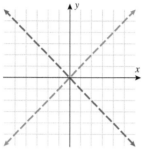

Darken the graph of $f(x) = -x$ when $x < 0$ and the graph $f(x) = x$ when $x > 0$. This gives us the familiar absolute value graph.

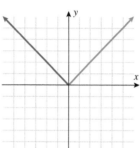

Domain: $(-\infty, \infty)$ or $\mathbb{R}$
Range: $[0, \infty)$

This is incorrect. What mistake was made?

96. Graph the piecewise-defined function. State the domain and range.

$$f(x) = \begin{cases} -x & x \leq 1 \\ x & x > 1 \end{cases}$$

Solution:

Draw the graphs of $f(x) = -x$ and $f(x) = x$.

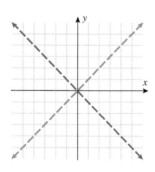

Darken the graph of $f(x) = -x$ when $x < 1$ and the graph of $f(x) = x$ when $x > 1$.

The resulting graph is as shown.

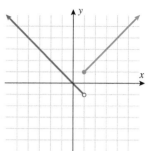

Domain: $(-\infty, \infty)$ or $\mathbb{R}$
Range: $(-1, \infty)$

This is incorrect. What mistake was made?

97. The cost of airport Internet access is $15 for the first 30 minutes and $1 per minute for each additional minute. Write a function describing the cost of the service as a function of minutes used online.

Solution: $C(x) = \begin{cases} 15 & x \leq 30 \\ 15 + x & x > 30 \end{cases}$

This is incorrect. What mistake was made?

98. Most money market accounts pay a higher interest with a higher principal. If the credit union is offering 2% on accounts with less than or equal to $10,000 and 4% on the additional money over $10,000, write the interest function $I(x)$ that represents the interest earned on an account as a function of dollars in the account.

Solution: $I(x) = \begin{cases} 0.02x & x \leq 10{,}000 \\ 0.02(10{,}000) + 0.04x & x > 10{,}000 \end{cases}$

This is incorrect. What mistake was made?

- CONCEPTUAL

In Exercises 99–102, determine whether each statement is true or false.

99. The identity function is a special case of the linear function.

100. The constant function is a special case of the linear function.

101. If an odd function has an interval where the function is increasing, then it also has to have an interval where the function is decreasing.

102. If an even function has an interval where the function is increasing, then it also has to have an interval where the function is decreasing.

- CHALLENGE

In Exercises 103 and 104, for a and b real numbers, can the function given ever be a continuous function? If so, specify the value for a and b that would make it so.

103. $f(x) = \begin{cases} ax & x \le 2 \\ bx^2 & x > 2 \end{cases}$

104. $f(x) = \begin{cases} -\dfrac{1}{x} & x < a \\ \dfrac{1}{x} & x \ge a \end{cases}$

- TECHNOLOGY

105. In trigonometry you will learn about the sine function, sin x. Plot the function $f(x) = \sin x$, using a graphing utility. It should look like the graph on the right. Is the sine function even, odd, or neither?

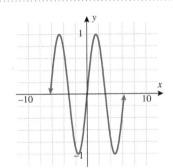

106. In trigonometry you will learn about the cosine function, cos x. Plot the function $f(x) = \cos x$, using a graphing utility. It should look like the graph on the right. Is the cosine function even, odd, or neither?

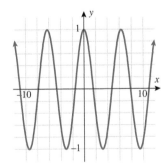

107. In trigonometry you will learn about the tangent function, tan x. Plot the function $f(x) = \tan x$, using a graphing utility. If you restrict the values of x so that $-\dfrac{\pi}{2} < x < \dfrac{\pi}{2}$, the graph should resemble the graph below. Is the tangent function even, odd, or neither?

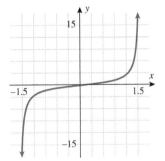

108. Plot the function $f(x) = \dfrac{\sin x}{\cos x}$. What function is this?

109. Graph the function $f(x) = [[3x]]$ using a graphing utility. State the domain and range.

110. Graph the function $f(x) = \left[\left[\frac{1}{3}x\right]\right]$ using a graphing utility. State the domain and range.

A.6 GRAPHING TECHNIQUES: TRANSFORMATIONS

SKILLS OBJECTIVES	CONCEPTUAL OBJECTIVES
▪ Sketch the graph of a function using horizontal and vertical shifting of common functions.	▪ Understand why a shift in the argument inside the function corresponds to a horizontal shift and why a shift outside the function corresponds to a vertical shift.
▪ Sketch the graph of a function by reflecting a common function about the *x*-axis or *y*-axis.	▪ Understand why a "negative" argument inside the function corresponds to a reflection about the *y*-axis and why a negative outside the function corresponds to a reflection about the *x*-axis.
▪ Sketch the graph of a function by stretching or compressing a common function.	▪ Understand the difference between rigid and nonrigid transformations.

The focus of the previous section was to learn the graphs that correspond to particular functions such as identity, square, cube, square root, cube root, absolute value, and reciprocal. Therefore, at this point, you should be able to recognize and generate the graphs of $y = x$, $y = x^2$, $y = x^3$, $y = \sqrt{x}$, $y = \sqrt[3]{x}$, $y = |x|$, and $y = \dfrac{1}{x}$. In this section, we will discuss how to sketch the graphs of functions that are very simple modifications of these functions. For instance, a common function may be shifted (horizontally or vertically), reflected, or stretched (or compressed). Collectively, these techniques are called **transformations**.

A.6.1 Horizontal and Vertical Shifts

A.6.1 SKILL

Sketch the graph of a function using horizontal and vertical shifting of common functions.

A.6.1 CONCEPTUAL

Understand why a shift in the argument inside the function corresponds to a horizontal shift and why a shift outside the function corresponds to a vertical shift.

Let's take the absolute value function as an example. The graph of $f(x) = |x|$ was given in the last section. Now look at two examples that are much like this function: $g(x) = |x| + 2$ and $h(x) = |x - 1|$. Graphing these functions by point-plotting yields

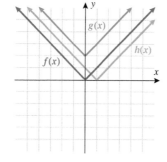

x	f(x)
−2	2
−1	1
0	0
1	1
2	2

x	g(x)
−2	4
−1	3
0	2
1	3
2	4

x	h(x)
−2	3
−1	2
0	1
1	0
2	1

Instead of point-plotting the function $g(x) = |x| + 2$, we could have started with the function $f(x) = |x|$ and shifted the entire graph *up* two units. Similarly, we could have generated the graph of the function $h(x) = |x - 1|$ by shifting the function $f(x) = |x|$ to the *right* one unit. In both cases, the base or starting function is $f(x) = |x|$. Why did we go up for $g(x)$ and to the right for $h(x)$?

Note that we could rewrite the functions $g(x)$ and $h(x)$ in terms of $f(x)$:

$$g(x) = |x| + 2 = f(x) + 2$$

$$h(x) = |x - 1| = f(x - 1)$$

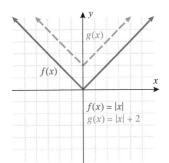

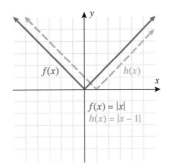

In the case of $g(x)$, the shift $(+2)$ occurs "outside" the function—that is, outside the parentheses showing the argument. Therefore, the output for $g(x)$ is 2 more than the typical output for $f(x)$. Because the output corresponds to the vertical axis, this results in a shift *upward* of two units. In general, shifts that occur *outside* the function correspond to a *vertical* shift corresponding to the sign of the shift. For instance, had the function been $G(x) = |x| - 2$, this graph would have started with the graph of the function $f(x)$ and shifted down two units.

In the case of $h(x)$, the shift occurs "inside" the function—that is, inside the parentheses showing the argument. Note that the point $(0, 0)$ that lies on the graph of $f(x)$ was shifted to the point $(1, 0)$ on the graph of the function $h(x)$. The y-value remained the same, but the x-value shifted to the right one unit. Similarly, the points $(-1, 1)$ and $(1, 1)$ were shifted to the points $(0, 1)$ and $(2, 1)$, respectively. In general, shifts that occur *inside* the function correspond to a *horizontal* shift opposite the sign. In this case, the graph of the function $h(x) = |x - 1|$ shifted the graph of the function $f(x)$ to the right one unit. If, instead, we had the function $H(x) = |x + 1|$, this graph would have started with the graph of the function $f(x)$ and shifted to the left one unit.

VERTICAL SHIFTS

Assuming that c is a positive constant,

To Graph	Shift the Graph of $f(x)$
$f(x) + c$	c units upward
$f(x) - c$	c units downward

Adding or subtracting a constant **outside** the function corresponds to a **vertical** shift that goes **with the sign**.

HORIZONTAL SHIFTS

Assuming that c is a positive constant,

To Graph	Shift the Graph of $f(x)$
$f(x + c)$	c units to the left
$f(x - c)$	c units to the right

Adding or subtracting a constant **inside** the function corresponds to a **horizontal** shift that goes **opposite the sign**.

EXAMPLE 1   **Horizontal and Vertical Shifts**

Sketch the graphs of the given functions using horizontal and vertical shifts.

a. $g(x) = x^2 - 1$

b. $H(x) = (x + 1)^2$

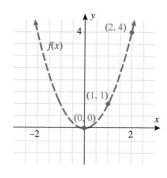

Solution:

In both cases, the function to start with is $f(x) = x^2$.

a. $g(x) = x^2 - 1$ can be rewritten as
$g(x) = f(x) - 1$.

   1. The shift (one unit) occurs *outside* of the function. Therefore, we expect a vertical shift that goes with the sign.

   2. Since the sign is *negative*, this corresponds to a *downward* shift.

   3. Shifting the graph of the function $f(x) = x^2$ down one unit yields the graph of
$g(x) = x^2 - 1$.

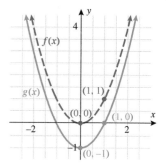

b. $H(x) = (x + 1)^2$ can be rewritten as $H(x) = f(x + 1)$.

   1. The shift (one unit) occurs *inside* of the function. Therefore, we expect a horizontal shift that goes *opposite* the sign.

   2. Since the sign is *positive*, this corresponds to a shift to the *left*.

   3. Shifting the graph of the function $f(x) = x^2$ to the left one unit yields the graph of
$H(x) = (x + 1)^2$.

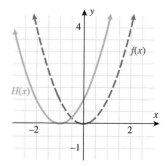

▼

ANSWER

a.

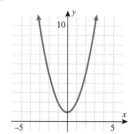

b.

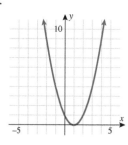

▼

YOUR TURN   Sketch the graphs of the given functions using horizontal and vertical shifts.

a. $g(x) = x^2 + 1$ **b.** $H(x) = (x - 1)^2$

It is important to note that the domain and range of the resulting function can be thought of as also being shifted. Shifts in the domain correspond to horizontal shifts, and shifts in the range correspond to vertical shifts.

EXAMPLE 2 **Horizontal and Vertical Shifts and Changes in the Domain and Range**

Graph the functions using translations and state the domain and range of each function.

a. $g(x) = \sqrt{x + 1}$

b. $G(x) = \sqrt{x} - 2$

Solution:

In both cases, the function to start with is $f(x) = \sqrt{x}$.

$$\text{Domain: } [0, \infty)$$

$$\text{Range: } [0, \infty)$$

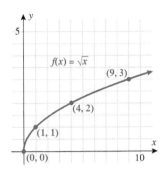

a. $g(x) = \sqrt{x + 1}$ can be rewritten as

$g(x) = f(x + 1)$.

1. The shift (one unit) is *inside* the function, which corresponds to a *horizontal* shift *opposite the sign*.

2. Shifting the graph of $f(x) = \sqrt{x}$ to the *left* one unit yields the graph of $g(x) = \sqrt{x + 1}$. Notice that the point $(0, 0)$, which lies on the graph of $f(x)$, gets shifted to the point $(-1, 0)$ on the graph of $g(x)$.

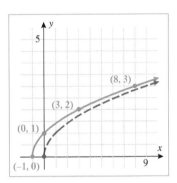

Although the original function $f(x) = \sqrt{x}$ had an implicit restriction on the domain $[0, \infty)$, the function $g(x) = \sqrt{x + 1}$ has the implicit restriction that $x \geq -1$. We see that the output or range of $g(x)$ is the same as the output of the original function $f(x)$.

> **Domain:** $[-1, \infty)$ **Range:** $[0, \infty)$

b. $G(x) = \sqrt{x} - 2$ can be rewritten as

$G(x) = f(x) - 2$.

1. The shift (two units) is *outside* the function, which corresponds to a *vertical* shift *with the sign*.

2. The graph of $G(x) = \sqrt{x} - 2$ is found by shifting $f(x) = \sqrt{x}$ down two units. Note that the point $(0, 0)$, which lies on the graph of $f(x)$, gets shifted to the point $(0, -2)$ on the graph of $G(x)$.

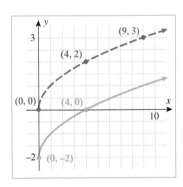

The original function $f(x) = \sqrt{x}$ has an implicit restriction on the domain: $[0, \infty)$. The function $G(x) = \sqrt{x} - 2$ also has the implicit restriction that $x \geq 0$. The output or range of $G(x)$ is always two units less than the output of the original function $f(x)$.

> **Domain:** $[0, \infty)$ **Range:** $[-2, \infty)$

▼

YOUR TURN Sketch the graph of the functions using shifts and state the domain and range.

a. $G(x) = \sqrt{x} - 2$ **b.** $h(x) = |x| + 1$

[**CONCEPT CHECK**]

For the functions $F(x) = \sqrt{(x - a)}$ and $G(x) = \sqrt{x} - a$ where $a > 0$, explain the shifts on $y = \sqrt{x}$.

▼ ·················

ANSWER $F(x)$ is shifted a units to the right. $G(x)$ is shifted a units down.

▼ ·················

ANSWER

a. $G(x) = \sqrt{x} - 2$

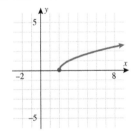

Domain: $[2, \infty)$, Range: $[0, \infty)$

b. $h(x) = |x| + 1$

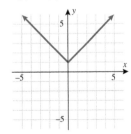

Domain: $(-\infty, \infty)$, Range: $[1, \infty)$

The previous examples have involved graphing functions by shifting a known function either in the horizontal or vertical direction. Let us now look at combinations of horizontal and vertical shifts.

> **EXAMPLE 3** **Combining Horizontal and Vertical Shifts**

Sketch the graph of the function $F(x) = (x + 1)^2 - 2$. State the domain and range of F.

Solution:

The base function is $y = x^2$.

1. The shift (one unit) is *inside* the function, so it represents a *horizontal* shift *opposite the sign.*

2. The -2 shift is *outside* the function, which represents a *vertical* shift *with the sign.*

3. Therefore, we shift the graph of $y = x^2$ to the left one unit and down two units. For instance, the point $(0, 0)$ on the graph of $y = x^2$ shifts to the point $(-1, -2)$ on the graph of $F(x) = (x + 1)^2 - 2$.

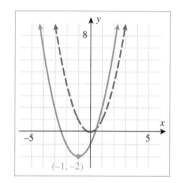

| Domain: $(-\infty, \infty)$ | Range: $[-2, \infty)$ |

▼

YOUR TURN Sketch the graph of the function $f(x) = |x - 2| + 1$. State the domain and range of f.

ANSWER

Domain: $(-\infty, \infty)$

Range: $[1, \infty)$

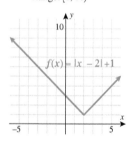

All of the previous transformation examples involve starting with a common function and shifting the function in either the horizontal or vertical direction (or a combination of both). Now, let's investigate *reflections* of functions about the x-axis or y-axis.

A.6.2 Reflection about the Axes

A.6.2 SKILL

Sketch the graph of a function by reflecting a common function about the x-axis or y-axis.

A.6.2 CONCEPTUAL

Understand why a "negative" argument inside the function corresponds to a reflection about the y-axis and why a negative outside the function corresponds to a reflection about the x-axis.

To sketch the graphs of $f(x) = x^2$ and $g(x) = -x^2$, start by listing points that are on each of the graphs and then connecting the points with smooth curves.

x	$f(x)$
-2	4
-1	1
0	0
1	1
2	4

x	$g(x)$
-2	-4
-1	-1
0	0
1	-1
2	-4

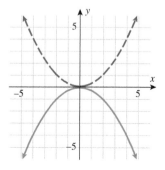

Note that if the graph of $f(x) = x^2$ is reflected about the x-axis, the result is the graph of $g(x) = -x^2$. Also note that the function $g(x)$ can be written as the negative of the function $f(x)$; that is, $g(x) = -f(x)$. In general, **reflection about the x-axis** is produced by multiplying a function by -1.

Let's now investigate reflection about the y-axis. To sketch the graphs of $f(x) = \sqrt{x}$ and $g(x) = \sqrt{-x}$, start by listing points that are on each of the graphs and then connecting the points with smooth curves.

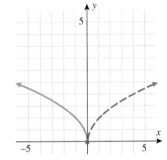

x	$f(x)$
0	0
1	1
4	2
9	3

x	$g(x)$
−9	3
−4	2
−1	1
0	0

Note that if the graph of $f(x) = \sqrt{x}$ is reflected about the y-axis, the result is the graph of $g(x) = \sqrt{-x}$. Also note that the function $g(x)$ can be written as $g(x) = f(-x)$. In general, **reflection about the y-axis** is produced by replacing x with $-x$ in the function. Notice that the domain of f is $[0, \infty)$, whereas the domain of g is $(-\infty, 0]$.

REFLECTION ABOUT THE AXES

The graph of $-f(x)$ is obtained by reflecting the graph of $f(x)$ about the x-axis.

The graph of $f(-x)$ is obtained by reflecting the graph of $f(x)$ about the y-axis.

EXAMPLE 4 **Sketching the Graph of a Function Using Both Shifts and Reflections**

Sketch the graph of the function $G(x) = -\sqrt{x + 1}$.

Solution:

Start with the square-root function.

$$f(x) = \sqrt{x}$$

Shift the graph of $f(x)$ to the left one unit to arrive at the graph of $f(x + 1)$.

$$f(x + 1) = \sqrt{x + 1}$$

Reflect the graph of $f(x + 1)$ about the x-axis to arrive at the graph of $-f(x + 1)$.

$$-f(x + 1) = -\sqrt{x + 1}$$

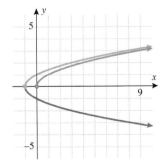

[CONCEPT CHECK]

For any even function $f(x)$ describe the graph of $f(-x)$.

▼

ANSWER The graph of $f(-x)$ is the same as the graph of $f(x)$ because for even functions $f(-x) = f(x)$.

▶ **EXAMPLE 5** **Sketching the Graph of a Function Using Both Shifts and Reflections**

Sketch the graph of the function $f(x) = \sqrt{2 - x} + 1$.

Solution:

Start with the square-root function.	$g(x) = \sqrt{x}$
Shift the graph of $g(x)$ to the left two units to arrive at the graph of $g(x + 2)$.	$g(x + 2) = \sqrt{x + 2}$
Reflect the graph of $g(x + 2)$ about the y-axis to arrive at the graph of $g(-x + 2)$.	$g(-x + 2) = \sqrt{-x + 2}$
Shift the graph $g(-x + 2)$ up one unit to arrive at the graph of $g(-x + 2) + 1$.	$g(-x + 2) + 1 = \sqrt{2 - x} + 1$

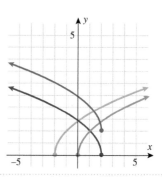

▼

ANSWER

Domain: $[1, \infty)$
Range: $(-\infty, 2]$

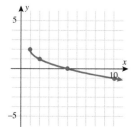

▼ **YOUR TURN** Use shifts and reflections to sketch the graph of the function $f(x) = -\sqrt{x - 1} + 2$. State the domain and range of $f(x)$.

Look back at the order in which transformations were performed in Example 5: horizontal shift, reflection, and then vertical shift. Let us consider an alternate order of transformations.

WORDS	**MATH**
Start with the square-root function.	$g(x) = \sqrt{x}$
Shift the graph of $g(x)$ up one unit to arrive at the graph of $g(x) + 1$.	$g(x) + 1 = \sqrt{x} + 1$
Reflect the graph of $g(x) + 1$ about the y-axis to arrive at the graph of $g(-x) + 1$.	$g(-x) + 1 = \sqrt{-x} + 1$
Replace x with $x - 2$, which corresponds to a shift of the graph of $g(-x) + 1$ to the right two units to arrive at the graph of $g[-(x - 2)] + 1$.	$g(-x + 2) + 1 = \sqrt{2 - x} + 1$

In the last step, we replaced x with $x - 2$, which required us to think ahead knowing the desired result was $2 - x$ inside the radical. To avoid any possible confusion, follow this order of transformations:

1. Horizontal shifts: $f(x \pm c)$
2. Reflection: $f(-x)$ and/or $-f(x)$
3. Vertical shifts: $f(x) \pm c$

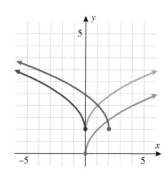

A.6.3 Stretching and Compressing

Horizontal shifts, vertical shifts, and reflections change only the position of the graph in the Cartesian plane, leaving the basic shape of the graph unchanged. These transformations (shifts and reflections) are called **rigid transformations** because they alter only the *position*. **Nonrigid transformations**, on the other hand, distort the *shape* of the original graph. We now consider *stretching* and *compressing* of graphs in both the vertical and horizontal direction.

A vertical stretch or compression of a graph occurs when the function is multiplied by a positive constant. For example, the graphs of the functions $f(x) = x^2$, $g(x) = 2f(x) = 2x^2$, and $h(x) = \frac{1}{2}f(x) = \frac{1}{2}x^2$ are illustrated below. Depending on if the constant is larger than 1 or smaller than 1 will determine whether it corresponds to a stretch (expansion) or compression (contraction) in the vertical direction.

A.6.3 SKILL

Sketch the graph of a function by stretching or compressing a common function.

A.6.3 CONCEPTUAL

Understand the difference between rigid and nonrigid transformations.

x	f(x)
−2	4
−1	1
0	0
1	1
2	4

x	g(x)
−2	8
−1	2
0	0
1	2
2	8

x	h(x)
−2	2
−1	$\frac{1}{2}$
0	0
1	$\frac{1}{2}$
2	2

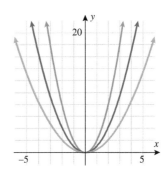

Note that when the function $f(x) = x^2$ is multiplied by 2, so that $g(x) = 2f(x) = 2x^2$, the result is a graph stretched in the vertical direction. When the function $f(x) = x^2$ is multiplied by $\frac{1}{2}$, so that $h(x) = \frac{1}{2}f(x) = \frac{1}{2}x^2$, the result is a graph that is compressed in the vertical direction.

VERTICAL STRETCHING AND VERTICAL COMPRESSING OF GRAPHS

The graph of $cf(x)$ is found by:

- **Vertically stretching** the graph of $f(x)$ if $c > 1$
- **Vertically compressing** the graph of $f(x)$ if $0 < c < 1$

Note: c is any positive real number.

EXAMPLE 6 **Vertically Stretching and Compressing Graphs**

Graph the function $h(x) = \frac{1}{4}x^3$.

Solution:

1. Start with the cube function. $f(x) = x^3$

2. Vertical compression is expected because $\frac{1}{4}$ is less than 1. $h(x) = \frac{1}{4}x^3$

3. Determine a few points that lie on the graph of h.

$$(0, 0) \quad (2, 2) \quad (-2, -2)$$

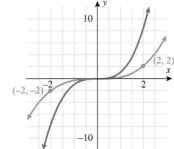

Conversely, if the argument x of a function f is multiplied by a positive real number c, then the result is a *horizontal* stretch of the graph of f if $0 < c < 1$. If $c > 1$, then the result is a *horizontal* compression of the graph of f.

HORIZONTAL STRETCHING AND HORIZONTAL COMPRESSING OF GRAPHS

The graph of $f(cx)$ is found by:

- **Horizontally stretching** the graph of $f(x)$ if $0 < c < 1$
- **Horizontally compressing** the graph of $f(x)$ if $c > 1$

Note: c is any positive real number.

EXAMPLE 7   **Vertically Stretching and Horizontally Compressing Graphs**

Given the graph of $f(x)$, graph

a. $2f(x)$

b. $f(2x)$

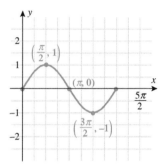

Solution (a):

Since the function is multiplied (on the outside) by 2, the result is that each **y-value** of $f(x)$ is ***multiplied* by 2**, which corresponds to vertical stretching.

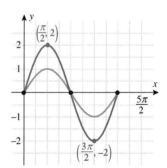

Solution (b):

Since the argument of the function is multiplied (on the inside) by 2, the result is that each **x-value** of $f(x)$ is **divided by 2**, which corresponds to horizontal compression.

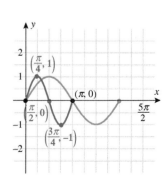

▼
ANSWER

Expansion of the graph $f(x) = x^3$.

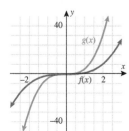

▼ YOUR TURN   Graph the function $g(x) = 4x^3$.

EXAMPLE 8   **Sketching the Graph of a Function Using Multiple Transformations**

Sketch the graph of the function $H(x) = -2(x - 3)^2$.

Solution:

Start with the square function.

$$f(x) = x^2$$

Shift the graph of $f(x)$ to the right three units to arrive at the graph of $f(x - 3)$.

$$f(x - 3) = (x - 3)^2$$

Vertically stretch the graph of $f(x - 3)$ by a factor of 2 to arrive at the graph of $2f(x - 3)$.

$$2f(x - 3) = 2(x - 3)^2$$

Reflect the graph $2f(x - 3)$ about the x-axis to arrive at the graph of $-2f(x - 3)$.

$$-2f(x - 3) = -2(x - 3)^2$$

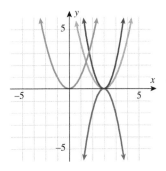

In Example 8, we followed the same "inside out" approach with the functions to determine the order for the transformations: horizontal shift, vertical stretch, and reflection.

▶[SECTION A.6]   SUMMARY

TRANSFORMATION	TO GRAPH THE FUNCTION . . .	DRAW THE GRAPH OF *f* AND THEN . . .	DESCRIPTION
Horizontal shifts	$f(x + c)$ $f(x - c)$	Shift the graph of f to the left c units. Shift the graph of f to the right c units.	Replace x by $x + c$. Replace x by $x - c$.
Vertical shifts	$f(x) + c$ $f(x) - c$	Shift the graph of f up c units. Shift the graph of f down c units.	Add c to $f(x)$. Subtract c from $f(x)$.
Reflection about the x-axis	$-f(x)$	Reflect the graph of f about the x-axis.	Multiply $f(x)$ by -1.
Reflection about the y-axis	$f(-x)$	Reflect the graph of f about the y-axis.	Replace x by $-x$.
Vertical stretch	$cf(x)$, where $c > 1$	Vertically stretch the graph of f.	Multiply $f(x)$ by c.
Vertical compression	$cf(x)$, where $0 < c < 1$	Vertically compress the graph of f.	Multiply $f(x)$ by c.
Horizontal stretch	$f(cx)$, where $0 < c < 1$	Horizontally stretch the graph of f.	Replace x by cx.
Horizontal compression	$f(cx)$, where $c > 1$	Horizontally compress the graph of f.	Replace x by cx.

[SECTION A.6] **EXERCISES**

• **SKILLS**

In Exercises 1–12, match the function to the graph (a)–(l).

1. $f(x) = x^2 + 1$

2. $f(x) = (x - 1)^2$

3. $f(x) = -(1 - x)^2$

4. $f(x) = -x^2 - 1$

5. $f(x) = -(x + 1)^2$

6. $f(x) = -(1 - x)^2 + 1$

7. $f(x) = \sqrt{x - 1} + 1$

8. $f(x) = -\sqrt{x} - 1$

9. $f(x) = \sqrt{1 - x} - 1$

10. $f(x) = \sqrt{-x} + 1$

11. $f(x) = -\sqrt{-x} + 1$

12. $f(x) = -\sqrt{1 - x} - 1$

a.

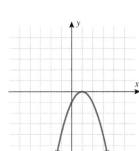

b.

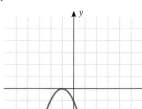

c.

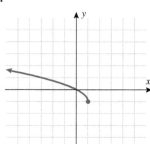

d.

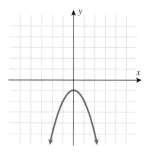

e.

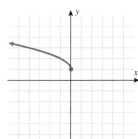

f.

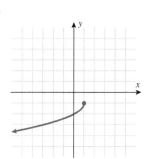

g.

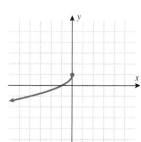

h.

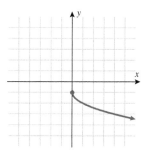

i.

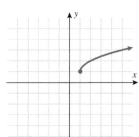

j.

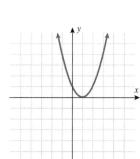

k.

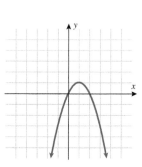

l.

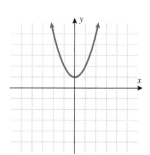

In Exercises 13–18, write the function whose graph is the graph of $y = |x|$ but is transformed accordingly.

13. Shifted up three units

14. Shifted to the left four units

15. Reflected about the y-axis

16. Reflected about the x-axis

17. Vertically stretched by a factor of 3

18. Vertically compressed by a factor of 3

In Exercises 19–24, write the function whose graph is the graph of $y = x^3$ but is transformed accordingly.

19. Shifted down four units

20. Shifted to the right three units

21. Shifted up three units and to the left one unit

22. Reflected about the x-axis

23. Reflected about the y-axis

24. Reflected about both the x-axis and the y-axis

In Exercises 25–48, use the given graph to sketch the graph of the indicated functions.

25.

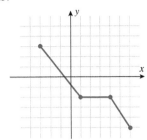

 a. $y = f(x - 2)$
 b. $y = f(x) - 2$

26.

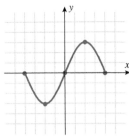

 a. $y = f(x + 2)$
 b. $y = f(x) + 2$

27.

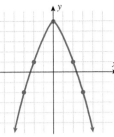

 a. $y = f(x) - 3$
 b. $y = f(x - 3)$

28.

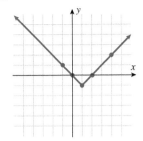

 a. $y = f(x) + 3$
 b. $y = f(x + 3)$

29.

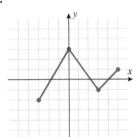

 a. $y = -f(x)$
 b. $y = f(-x)$

30.

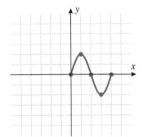

 a. $y = -f(x)$
 b. $y = f(-x)$

31.

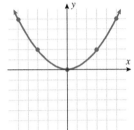

 a. $y = 2f(x)$
 b. $y = f(2x)$

32.

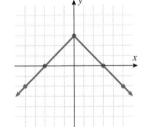

 a. $y = 2f(x)$
 b. $y = f(2x)$

33. $y = f(x - 2) - 3$

34. $y = f(x + 1) - 2$

35. $y = -f(x - 1) + 2$

36. $y = -2f(x) + 1$

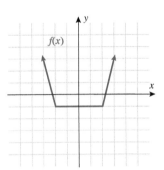

37. $y = -\frac{1}{2}g(x)$

38. $y = \frac{1}{4}g(-x)$

39. $y = -g(2x)$

40. $y = g\left(\frac{1}{2}x\right)$

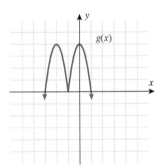

41. $y = \frac{1}{2}F(x - 1) + 2$

42. $y = \frac{1}{2}F(-x)$

43. $y = -F(1 - x)$

44. $y = -F(x - 2) - 1$

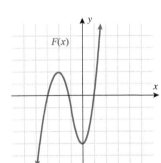

45. $y = 2G(x + 1) - 4$

46. $y = 2G(-x) + 1$

47. $y = -2G(x - 1) + 3$

48. $y = -G(x - 2) - 1$

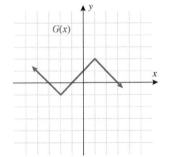

In Exercises 49–74, graph the function using transformations.

49. $y = x^2 - 2$

50. $y = x^2 + 3$

51. $y = (x + 1)^2$

52. $y = (x - 2)^2$

53. $y = (x - 3)^2 + 2$

54. $y = (x + 2)^2 + 1$

55. $y = -(1 - x)^2$

56. $y = -(x + 2)^2$

57. $y = |-x|$

58. $y = -|x|$

59. $y = -|x + 2| - 1$

60. $y = |1 - x| + 2$

61. $y = 2x^2 + 1$

62. $y = 2|x| + 1$

63. $y = -\sqrt{x - 2}$

64. $y = \sqrt{2 - x}$

65. $y = -\sqrt{2 + x} - 1$

66. $y = \sqrt{2 - x} + 3$

67. $y = \sqrt[3]{x - 1} + 2$

68. $y = \sqrt[3]{x + 2} - 1$

69. $y = \dfrac{1}{x + 3} + 2$

70. $y = \dfrac{1}{3 - x}$

71. $y = 2 - \dfrac{1}{x + 2}$

72. $y = 2 - \dfrac{1}{1 - x}$

73. $y = 5\sqrt{-x}$

74. $y = -\frac{1}{5}\sqrt{x}$

In Exercises 75–80, transform the function into the form $f(x) = c(x - h)^2 + k$, where c, k, and h are constants, by completing the square. Use graph-shifting techniques to graph the function.

75. $y = x^2 - 6x + 11$

76. $f(x) = x^2 + 2x - 2$

77. $f(x) = -x^2 - 2x$

78. $f(x) = -x^2 + 6x - 7$

79. $f(x) = 2x^2 - 8x + 3$

80. $f(x) = 3x^2 - 6x + 5$

• APPLICATIONS

81. **Salary.** A manager hires an employee at a rate of $10 per hour. Write the function that describes the current salary of the employee as a function of the number of hours worked per week, x. After a year, the manager decides to award the employee a raise equivalent to paying him for an additional 5 hours per week. Write a function that describes the salary of the employee after the raise.

82. **Profit.** The profit associated with St. Augustine sod in Florida is typically $P(x) = -x^2 + 14{,}000x - 48{,}700{,}000$, where x is the number of pallets sold per year in a normal year. In rainy years Sod King gives away 10 free pallets per year. Write the function that describes the profit of x pallets of sod in rainy years.

83. **Taxes.** Every year in the United States each working American typically pays in taxes a percentage of his or her earnings

(minus the standard deduction). Karen's 2005 taxes were calculated based on the formula $T(x) = 0.22(x - 6500)$. That year the standard deduction was $6500 and her tax bracket paid 22% in taxes. Write the function that will determine her 2006 taxes, assuming she receives the raise that places her in the 33% bracket.

84. **Medication.** The amount of medication that an infant requires is typically a function of the baby's weight. The number of milliliters of an antiseizure medication A is given by $A(x) = \sqrt{x} + 2$, where x is the weight of the infant in ounces. In emergencies there is often not enough time to weigh the infant, so nurses have to estimate the baby's weight. What is the function that represents the actual amount of medication the infant is given if his weight is overestimated by 3 ounces?

• CATCH THE MISTAKE

In Exercises 85–88, explain the mistake that is made.

85. Describe a procedure for graphing the function $f(x) = \sqrt{x - 3} + 2$.

Solution:

   a. Start with the function $f(x) = \sqrt{x}$.

   b. Shift the function to the left three units.

   c. Shift the function up two units.

This is incorrect. What mistake was made?

86. Describe a procedure for graphing the function $f(x) = -\sqrt{x + 2} - 3$.

Solution:

   a. Start with the function $f(x) = \sqrt{x}$.

   b. Shift the function to the left two units.

   c. Reflect the function about the y-axis.

   d. Shift the function down three units.

This is incorrect. What mistake was made?

87. Describe a procedure for graphing the function $f(x) = |3 - x| + 1$.

Solution:

a. Start with the function $f(x) = |x|$.

b. Reflect the function about the y-axis.

c. Shift the function to the left three units.

d. Shift the function up one unit.

This is incorrect. What mistake was made?

88. Describe a procedure for graphing the function $f(x) = -2x^2 + 1$.

Solution:

a. Start with the function $f(x) = x^2$.

b. Reflect the function about the y-axis.

c. Shift the function up one unit.

d. Expand in the vertical direction by a factor of 2.

This is incorrect. What mistake was made?

• **CONCEPTUAL**

In Exercises 89–92, determine whether each statement is true or false.

89. The graph of $y = |-x|$ is the same as the graph of $y = |x|$.

90. The graph of $y = \sqrt{-x}$ is the same as the graph of $y = \sqrt{x}$.

91. If the graph of an odd function is reflected about the x-axis and then the y-axis, the result is the graph of the original odd function.

92. If the graph of $y = \frac{1}{x}$ is reflected about the x-axis, it produces the same graph as if it had been reflected about the y-axis.

• **CHALLENGE**

93. The point (a, b) lies on the graph of the function $y = f(x)$. What point is guaranteed to lie on the graph of $f(x - 3) + 2$?

94. The point (a, b) lies on the graph of the function $y = f(x)$. What point is guaranteed to lie on the graph of $-f(-x) + 1$?

• **TECHNOLOGY**

95. Use a graphing utility to graph

a. $y = x^2 - 2$ and $y = |x^2 - 2|$

b. $y = x^3 - 1$ and $y = |x^3 + 1|$

What is the relationship between $f(x)$ and $|f(x)|$?

96. Use a graphing utility to graph

a. $y = x^2 - 2$ and $y = |x|^2 - 2$

b. $y = x^3 + 1$ and $y = |x|^3 + 1$

What is the relationship between $f(x)$ and $f(|x|)$?

97. Use a graphing utility to graph

a. $y = \sqrt{x}$ and $y = \sqrt{0.1x}$

b. $y = \sqrt{x}$ and $y = \sqrt{10x}$

What is the relationship between $f(x)$ and $f(ax)$, assuming that a is positive?

98. Use a graphing utility to graph

a. $y = \sqrt{x}$ and $y = 0.1\sqrt{x}$

b. $y = \sqrt{x}$ and $y = 10\sqrt{x}$

What is the relationship between $f(x)$ and $af(x)$, assuming that a is positive?

99. Use a graphing utility to graph $y = f(x) = [[0.5x]] + 1$. Use transforms to describe the relationship between $f(x)$ and $y = [[x]]$.

100. Use a graphing utility to graph $y = g(x) = 0.5[[x]] + 1$. Use transforms to describe the relationship between $g(x)$ and $y = [[x]]$.

A.7 OPERATIONS ON FUNCTIONS AND COMPOSITION OF FUNCTIONS

SKILLS OBJECTIVES

- Perform arithmetic operations on functions.
- Evaluate composite functions and determine the corresponding domains.

CONCEPTUAL OBJECTIVES

- Understand domain restrictions when dividing functions.
- Realize that the domain of a composition of functions excludes values that are not in the domain of the inside function.

Two different functions can be combined using mathematical operations such as addition, subtraction, multiplication, and division. Also, there is an operation on functions called composition, which can be thought of as a function of a function. When we combine functions, we do so algebraically. Special attention must be paid to the domain and range of the combined functions.

A.7.1 Adding, Subtracting, Multiplying, and Dividing Functions

A.7.1 SKILL

Perform arithmetic operations on functions.

A.7.1 CONCEPTUAL

Understand domain restrictions when dividing functions.

Consider the two functions $f(x) = x^2 + 2x - 3$ and $g(x) = x + 1$. The domain of both of these functions is the set of all real numbers. Therefore, we can add, subtract, or multiply these functions for any real number x.

$$\text{Addition: } f(x) + g(x) = x^2 + 2x - 3 + x + 1 = x^2 + 3x - 2$$

The result is in fact a new function, which we denote:

$$(f + g)(x) = x^2 + 3x - 2 \qquad \text{This is the } \textbf{sum function.}$$

$$\text{Subtraction: } f(x) - g(x) = x^2 + 2x - 3 - (x + 1) = x^2 + x - 4$$

The result is in fact a new function, which we denote:

$$(f - g)(x) = x^2 + x - 4 \qquad \text{This is the } \textbf{difference function.}$$

$$\text{Multiplication: } f(x) \cdot g(x) = (x^2 + 2x - 3)(x + 1) = x^3 + 3x^2 - x - 3$$

The result is in fact a new function, which we denote:

$$(f \cdot g)(x) = x^3 + 3x^2 - x - 3 \qquad \text{This is the } \textbf{product function.}$$

Although both f and g are defined for all real numbers x, we must restrict x so that $x \neq -1$ to form the quotient $\dfrac{f}{g}$.

$$\text{Division: } \frac{f(x)}{g(x)} = \frac{x^2 + 2x - 3}{x + 1}, \quad x \neq -1$$

The result is in fact a new function, which we denote:

$$\left(\frac{f}{g}\right)(x) = \frac{x^2 + 2x - 3}{x + 1}, \quad x \neq -1 \qquad \text{This is called the } \textbf{quotient function.}$$

Two functions can be added, subtracted, and multiplied. The resulting function domain is therefore the intersection of the domains of the two functions. However, for division, any value of x (input) that makes the denominator equal to zero must be eliminated from the domain.

The previous examples involved polynomials. The domain of any polynomial is the set of all real numbers. Adding, subtracting, and multiplying polynomials result in other polynomials, which have domains of all real numbers. Let's now investigate operations applied to functions that have a restricted domain.

The domain of the sum function, difference function, or product function is the *intersection* of the individual domains of the two functions. The quotient function has a similar domain in that it is the intersection of the two domains. However, any values that make the denominator zero must also be eliminated.

FUNCTION	NOTATION	DOMAIN
Sum	$(f + g)(x) = f(x) + g(x)$	{domain of f} $\cap$ {domain of g}
Difference	$(f - g)(x) = f(x) - g(x)$	{domain of f} $\cap$ {domain of g}
Product	$(f \cdot g)(x) = f(x) \cdot g(x)$	{domain of f} $\cap$ {domain of g}
Quotient	$\left(\dfrac{f}{g}\right)(x) = \dfrac{f(x)}{g(x)}$	{domain of f} $\cap$ {domain of g} $\cap$ $\{g(x) \neq 0\}$

We can think of this in the following way: Any number that is in the domain of *both* the functions is in the domain of the combined function. The exception to this thinking is the quotient function, which also eliminates values that make the denominator equal to zero.

EXAMPLE 1 **Operations on Functions: Determining Domains of New Functions**

For the functions $f(x) = \sqrt{x - 1}$ and $g(x) = \sqrt{4 - x}$, determine the sum function, difference function, product function, and quotient function. State the domain of these four new functions.

Solution:

Sum function:
$$f(x) + g(x) = \sqrt{x - 1} + \sqrt{4 - x}$$

Difference function:
$$f(x) - g(x) = \sqrt{x - 1} - \sqrt{4 - x}$$

Product function:
$$f(x) \cdot g(x) = \sqrt{x - 1} \cdot \sqrt{4 - x}$$
$$= \sqrt{(x - 1)(4 - x)} = \sqrt{-x^2 + 5x - 4}$$

Quotient function:
$$\frac{f(x)}{g(x)} = \frac{\sqrt{x - 1}}{\sqrt{4 - x}} = \sqrt{\frac{x - 1}{4 - x}}$$

The domain of the square-root function is determined by setting the argument under the radical greater than or equal to zero.

Domain of $f(x)$: $[1, \infty)$

Domain of $g(x)$: $(-\infty, 4]$

The domain of the sum, difference, and product functions is

$$[1, \infty) \cap (-\infty, 4] = [1, 4]$$

The quotient function has the additional constraint that the denominator cannot be zero. This implies that $x \neq 4$, so the domain of the quotient function is $[1, 4)$.

Let $f(x) = \sqrt{x + 1}$ and $g(x) = \dfrac{1}{x}$;

find the domain of $\dfrac{g(x)}{f(x)}$.

▼

ANSWER $(-1, 0) \cup (0, \infty)$

▼
YOUR TURN Given the function $f(x) = \sqrt{x + 3}$ and $g(x) = \sqrt{1 - x}$, find $(f + g)(x)$ and state its domain.

▼
ANSWER

$(f + g)(x) = \sqrt{x + 3} + \sqrt{1 - x}$

Domain: $[-3, 1]$

EXAMPLE 2 **Quotient Function and Domain Restrictions**

Given the functions $F(x) = \sqrt{x}$ and $G(x) = |x - 3|$, find the quotient function, $\left(\dfrac{F}{G}\right)(x)$, and state its domain.

Solution:

The quotient function is written as

$$\left(\frac{F}{G}\right)(x) = \frac{F(x)}{G(x)} = \frac{\sqrt{x}}{|x - 3|}$$

Domain of $F(x)$: $[0, \infty)$ Domain of $G(x)$: $(-\infty, \infty)$

The real numbers that are in both the domain for $F(x)$ and the domain for $G(x)$ are represented by the intersection $[0, \infty) \cap (-\infty, \infty) = [0, \infty)$. Also, the denominator of the quotient function is equal to zero when $x = 3$, so we must eliminate this value from the domain.

> Domain of $\left(\dfrac{F}{G}\right)(x)$: $[0, 3) \cup (3, \infty)$

▼
ANSWER

$\left(\dfrac{G}{F}\right)(x) = \dfrac{G(x)}{F(x)} = \dfrac{|x - 3|}{\sqrt{x}}$

Domain: $(0, \infty)$

▼
YOUR TURN For the functions given in Example 2, determine the quotient function $\left(\dfrac{G}{F}\right)(x)$, and state its domain.

A.7.2 Composition of Functions

A.7.2 SKILL

Evaluate composite functions and determine the corresponding domains.

A.7.2 CONCEPTUAL

Realize that the domain of a composition of functions excludes values that are not in the domain of the inside function.

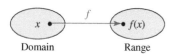

Domain   Range

Recall that a function maps every element in the domain to exactly one corresponding element in the range, as shown in the figure in the margin.

Suppose there is a sales rack of clothes in a department store. Let x correspond to the original price of each item on the rack. These clothes have recently been marked down 20%. Therefore, the function $g(x) = 0.80x$ represents the current sale price of each item. You have been invited to a special sale that lets you take 10% off the current sale price and an additional \$5 off every item at checkout. The function $f(g(x)) = 0.90g(x) - 5$ determines the checkout price. Note that the output of the function g is the input of the function f, as shown in the figure below.

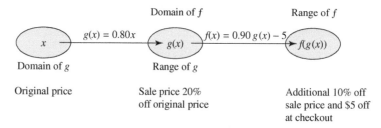

This is an example of a **composition of functions**, when the output of one function is the input of another function. It is commonly referred to as a function of a function.

An algebraic example of this is the function $y = \sqrt{x^2 - 2}$. Suppose we let $g(x) = x^2 - 2$ and $f(x) = \sqrt{x}$. Recall that the independent variable in function notation is a placeholder. Since $f(\square) = \sqrt{(\square)}$, then $f(g(x)) = \sqrt{(g(x))}$. Substituting the expression for $g(x)$, we find $f(g(x)) = \sqrt{x^2 - 2}$. The function $y = \sqrt{x^2 - 2}$ is said to be a composite function, $y = f(g(x))$.

Note that the domain of $g(x)$ is the set of all real numbers, and the domain of $f(x)$ is the set of all nonnegative numbers. The domain of a composite function is the set of all x such that $g(x)$ is in the domain of f. For instance, in the composite function $y = f(g(x))$, we know that the allowable inputs into f are all numbers greater than or equal to zero. Therefore, we restrict the outputs of $g(x) \geq 0$ and find the corresponding x-values. Those x-values are the only allowable inputs and constitute the domain of the composite function $y = f(g(x))$.

The symbol that represents composition of functions is a small open circle; thus, $(f \circ g)(x) = f(g(x))$ and is read aloud as "f of g." It is important not to confuse this with the multiplication sign: $(f \cdot g)(x) = f(x)g(x)$.

CAUTION

$f \circ g \neq f \cdot g$

COMPOSITION OF FUNCTIONS

Given two functions f and g, there are two **composite functions** that can be formed.

NOTATION	WORDS	DEFINITION	DOMAIN
$f \circ g$	f composed with g	$f(g(x))$	The set of all real numbers x in the domain of g such that $g(x)$ is also in the domain of f
$g \circ f$	g composed with f	$g(f(x))$	The set of all real numbers x in the domain of f such that $f(x)$ is also in the domain of g

It is important to realize that there are two "filters" that allow certain values of x into the domain. The first filter is $g(x)$. If x is not in the domain of $g(x)$, it cannot be in the domain of $(f \circ g)(x) = f(g(x))$. Of those values for x that are in the domain of $g(x)$, only some pass through because we restrict the output of $g(x)$ to values that are allowable as input into f. This adds an additional filter.

The domain of $f \circ g$ is always a subset of the domain of g, and the range of $f \circ g$ is always a subset of the range of f.

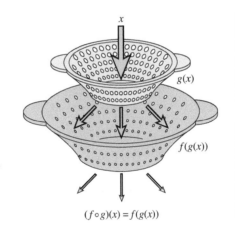

$(f \circ g)(x) = f(g(x))$

STUDY TIP

Order is important:
$(f \circ g)(x) = f(g(x))$
$(g \circ f)(x) = g(f(x))$

STUDY TIP

The domain of $f \circ g$ is always a subset of the domain of g, and the range of $f \circ g$ is always a subset of the range of f.

EXAMPLE 3 Finding a Composite Function

Given the functions $f(x) = x^2 + 1$ and $g(x) = x - 3$, find $(f \circ g)(x)$.

Solution:

Write $f(x)$ using placeholder notation.	$f(\square) = (\square)^2 + 1$
Express the composite function $f \circ g$.	$f(g(x)) = (g(x))^2 + 1$
Substitute $g(x) = x - 3$ into f.	$f(g(x)) = (x - 3)^2 + 1$
Eliminate the parentheses on the right side.	$f(g(x)) = x^2 - 6x + 10$

$$\boxed{(f \circ g)(x) = f(g(x)) = x^2 - 6x + 10}$$

YOUR TURN Given the functions in Example 3, find $(g \circ f)(x)$.

ANSWER

$(g \circ f)(x) = g(f(x)) = x^2 - 2$

▶ **EXAMPLE 4** **Determining the Domain of a Composite Function**

Given the functions $f(x) = \dfrac{1}{x-1}$ and $g(x) = \dfrac{1}{x}$, determine $f \circ g$, and state its domain.

Solution:

Write $f(x)$ using placeholder notation.

$$f(\square) = \frac{1}{(\square) - 1}$$

Express the composite function $f \circ g$.

$$f(g(x)) = \frac{1}{g(x) - 1}$$

Substitute $g(x) = \dfrac{1}{x}$ into f.

$$f(g(x)) = \frac{1}{\dfrac{1}{x} - 1}$$

Multiply the right side by $\dfrac{x}{x}$.

$$f(g(x)) = \frac{1}{\dfrac{1}{x} - 1} \cdot \frac{x}{x} = \frac{x}{1 - x}$$

$$\boxed{f \circ g = f(g(x)) = \frac{x}{1-x}}$$

What is the domain of $(f \circ g)(x) = f(g(x))$? By inspecting the final result of $f(g(x))$, we see that the denominator is zero when $x = 1$. Therefore, $x \neq 1$. Are there any other values for x that are not allowed? The function $g(x)$ has the domain $x \neq 0$; therefore, we must also exclude zero.

The domain of $(f \circ g)(x) = f(g(x))$ excludes $x = 0$ and $x = 1$ or, in interval notation,

$$\boxed{(-\infty, 0) \cup (0, 1) \cup (1, \infty)}$$

▼
ANSWER

$g(f(x)) = x - 1$. Domain of $g \circ f$ is $x \neq 1$, or in interval notation, $(-\infty, 1) \cup (1, \infty)$.

▼
CAUTION

The domain of the composite function cannot always be determined by examining the final form of $f \circ g$.

[CONCEPT CHECK]

For the functions $f(x) = x - a$ and $g(x) = \dfrac{1}{x + a}$, find $g(f(x))$ and state its domain.

▼
ANSWER $g(f(x)) = \dfrac{1}{x}$ and domain is all real numbers except $x = 0$.

▼ **YOUR TURN** For the functions f and g given in Example 4, determine the composite function $g \circ f$ and state its domain.

The domain of the composite function cannot always be determined by examining the final form of $f \circ g$.

EXAMPLE 5 **Determining the Domain of a Composite Function (Without Finding the Composite Function)**

Let $f(x) = \dfrac{1}{x - 2}$ and $g(x) = \sqrt{x + 3}$. Find the domain of $f(g(x))$. Do not find the composite function.

Solution:

Find the domain of g. $[-3, \infty)$

Find the range of g. $[0, \infty)$

In $f(g(x))$, the output of g becomes the input for f. Since the domain of f is the set of all real numbers except 2, we eliminate any values of x in the domain of g that correspond to $g(x) = 2$.

Let $g(x) = 2$. $\sqrt{x + 3} = 2$

Square both sides. $x + 3 = 4$

Solve for x. $x = 1$

Eliminate $x = 1$ from the domain of g, $[-3, \infty)$.

State the domain of $f(g(x))$. $\boxed{[-3, 1) \cup (1, \infty)}$

EXAMPLE 6 **Evaluating a Composite Function**

Given the functions $f(x) = x^2 - 7$ and $g(x) = 5 - x^2$, evaluate

a. $f(g(1))$ **b.** $f(g(-2))$ **c.** $g(f(3))$ **d.** $g(f(-4))$

Solution:

One way of evaluating these composite functions is to calculate the two individual composites in terms of x: $f(g(x))$ and $g(f(x))$. Once those functions are known, the values can be substituted for x and evaluated.

Another way of proceeding is as follows:

a. Write the desired quantity. $f(g(1))$

  Find the value of the inner function g. $g(1) = 5 - 1^2 = 4$

  Substitute $g(1) = 4$ into f. $f(g(1)) = f(4)$

  Evaluate $f(4)$. $f(4) = 4^2 - 7 = 9$

$$\boxed{f(g(1)) = 9}$$

b. Write the desired quantity. $f(g(-2))$

  Find the value of the inner function g. $g(-2) = 5 - (-2)^2 = 1$

  Substitute $g(-2) = 1$ into f. $f(g(-2)) = f(1)$

  Evaluate $f(1)$. $f(1) = 1^2 - 7 = -6$

$$\boxed{f(g(-2)) = -6}$$

c. Write the desired quantity. $g(f(3))$

  Find the value of the inner function f. $f(3) = 3^2 - 7 = 2$

  Substitute $f(3) = 2$ into g. $g(f(3)) = g(2)$

  Evaluate $g(2)$. $g(2) = 5 - 2^2 = 1$

$$\boxed{g(f(3)) = 1}$$

d. Write the desired quantity. $g(f(-4))$

  Find the value of the inner function f. $f(-4) = (-4)^2 - 7 = 9$

  Substitute $f(-4) = 9$ into g. $g(f(-4)) = g(9)$

  Evaluate $g(9)$. $g(9) = 5 - 9^2 = -76$

$$\boxed{g(f(-4)) = -76}$$

▼ YOUR TURN Given the functions $f(x) = x^3 - 3$ and $g(x) = 1 + x^3$, evaluate $f(g(1))$ and $g(f(1))$.

▼ ANSWER

$f(g(1)) = 5$ and $g(f(1)) = -7$

Application Problems

Recall the example at the beginning of this chapter regarding the clothes that are on sale. Often, real-world applications are modeled with composite functions. In the clothes example, x is the original price of each item. The first function maps its input (original price) to an output (sale price). The second function maps its input (sale price) to an output (checkout price). Example 7 is another real-world application of composite functions.

Three temperature scales are commonly used:

- The degree (°Celsius) scale
  - This scale was devised by dividing the range between the freezing (0°C) and boiling (100°C) points of pure water at sea level into 100 equal parts. This scale is used in science and is one of the standards of the "metric" (SI) system of measurements.

- The Kelvin (K) temperature scale
  - This scale shifts the Celsius scale down so that the zero point is equal to absolute zero (about $-273.15°C$), a hypothetical temperature at which there is a complete absence of heat energy.
    - Temperatures on this scale are called **kelvins**, *not* degrees kelvin, and kelvin is not capitalized. The symbol for the kelvin is K.
- The degree Fahrenheit (°F) scale
  - This scale evolved over time and is still widely used mainly in the United States, although Celsius is the preferred "metric" scale.
    - With respect to pure water at sea level, the **degrees Fahrenheit** are gauged by the spread from 32°F (freezing) to 212°F (boiling).

The equations that relate these temperature scales are

$$F = \frac{9}{5}C + 32 \qquad C = K - 273.15$$

EXAMPLE 7 **Applications Involving Composite Functions**

Determine degrees Fahrenheit as a function of kelvins.

Solution:

Degrees Fahrenheit is a function of degrees Celsius.	$F = \dfrac{9}{5}C + 32$
Now substitute $C = K - 273.15$ into the equation for F.	$F = \dfrac{9}{5}(K - 273.15) + 32$
Simplify.	$F = \dfrac{9}{5}K - 491.67 + 32$

$$\boxed{F = \frac{9}{5}K - 459.67}$$

▶[SECTION A.7] SUMMARY

Operations on Functions

Function	Notation
Sum	$(f + g)(x) = f(x) + g(x)$
Difference	$(f - g)(x) = f(x) - g(x)$
Product	$(f \cdot g)(x) = f(x) \cdot g(x)$
Quotient	$\left(\dfrac{f}{g}\right)(x) = \dfrac{f(x)}{g(x)} \qquad g(x) \neq 0$

The domain of the sum, difference, and product functions is the intersection of the domains, or common domain shared by both f and g. The domain of the quotient function is also the intersection of the domain shared by both f and g with an additional restriction that $g(x) \neq 0$.

Composition of Functions

$$(f \circ g)(x) = f(g(x))$$

The domain restrictions cannot always be determined simply by inspecting the final form of $f(g(x))$. Rather, the domain of the composite function is a subset of the domain of $g(x)$. Values of x must be eliminated if their corresponding values of $g(x)$ are not in the domain of f.

[SECTION A.7] **EXERCISES**

• **SKILLS**

In Exercises 1–10, given the functions f and g, find $f + g$, $f - g$, $f \cdot g$, and $\dfrac{f}{g}$, and state the domain of each.

1. $f(x) = 2x + 1$
    $g(x) = 1 - x$

2. $f(x) = 3x + 2$
    $g(x) = 2x - 4$

3. $f(x) = 2x^2 - x$
    $g(x) = x^2 - 4$

4. $f(x) = 3x + 2$
    $g(x) = x^2 - 25$

5. $f(x) = \dfrac{1}{x}$
    $g(x) = x$

6. $f(x) = \dfrac{2x + 3}{x - 4}$
    $g(x) = \dfrac{x - 4}{3x + 2}$

7. $f(x) = \sqrt{x}$
    $g(x) = 2\sqrt{x}$

8. $f(x) = \sqrt{x - 1}$
    $g(x) = 2x^2$

9. $f(x) = \sqrt{4 - x}$
    $g(x) = \sqrt{x + 3}$

10. $f(x) = \sqrt{1 - 2x}$
    $g(x) = \dfrac{1}{x}$

In Exercises 11–20, for the given functions f and g, find the composite functions $f \circ g$ and $g \circ f$, and state their domains.

11. $f(x) = 2x + 1$
     $g(x) = x^2 - 3$

12. $f(x) = x^2 - 1$
     $g(x) = 2 - x$

13. $f(x) = \dfrac{1}{x - 1}$
     $g(x) = x + 2$

14. $f(x) = \dfrac{2}{x - 3}$
     $g(x) = 2 + x$

15. $g(x) = |x|$
     $g(x) = \dfrac{1}{x - 1}$

16. $f(x) = |x - 1|$
     $g(x) = \dfrac{1}{x}$

17. $f(x) = \sqrt{x - 1}$
     $g(x) = x + 5$

18. $f(x) = \sqrt{2 - x}$
     $g(x) = x^2 + 2$

19. $f(x) = x^3 + 4$
     $g(x) = (x - 4)^{1/3}$

20. $f(x) = \sqrt[3]{x^2 - 1}$
     $g(x) = x^{2/3} + 1$

In Exercises 21–38, evaluate the functions for the specified values, if possible.

$$f(x) = x^2 + 10 \qquad g(x) = \sqrt{x - 1}$$

21. $(f + g)(2)$

22. $(f + g)(10)$

23. $(f - g)(2)$

24. $(f - g)(5)$

25. $(f \cdot g)(4)$

26. $(f \cdot g)(5)$

27. $\left(\dfrac{f}{g}\right)(10)$

28. $\left(\dfrac{f}{g}\right)(2)$

29. $f(g(2))$

30. $f(g(1))$

31. $g(f(-3))$

32. $g(f(4))$

33. $f(g(0))$

34. $g(f(0))$

35. $f(g(-3))$

36. $g(f(\sqrt{7}))$

37. $(f \circ g)(4)$

38. $(g \circ f)(-3)$

In Exercises 39–50, evaluate $f(g(1))$ and $g(f(2))$, if possible.

39. $f(x) = \dfrac{1}{x}$, $g(x) = 2x + 1$

40. $f(x) = x^2 + 1$, $g(x) = \dfrac{1}{2 - x}$

41. $f(x) = \sqrt{1 - x}$, $g(x) = x^2 + 2$

42. $f(x) = \sqrt{3 - x}$, $g(x) = x^2 + 1$

43. $f(x) = \dfrac{1}{|x - 1|}$, $g(x) = x + 3$

44. $f(x) = \dfrac{1}{x}$, $g(x) = |2x - 3|$

45. $f(x) = \sqrt{x - 1}$, $g(x) = x^2 + 5$

46. $f(x) = \sqrt[3]{x - 3}$, $g(x) = \dfrac{1}{x - 3}$

47. $f(x) = \dfrac{1}{x^2 - 3}$, $g(x) = \sqrt{x - 3}$

48. $f(x) = \dfrac{x}{2 - x}$, $g(x) = 4 - x^2$

49. $f(x) = (x - 1)^{1/3}$, $g(x) = x^2 + 2x + 1$

50. $f(x) = (1 - x^2)^{1/2}$, $g(x) = (x - 3)^{1/3}$

In Exercises 51–60, show that $f(g(x)) = x$ and $g(f(x)) = x$.

51. $f(x) = 2x + 1$, $g(x) = \dfrac{x-1}{2}$

52. $f(x) = \dfrac{x-2}{3}$, $g(x) = 3x + 2$

53. $f(x) = \sqrt{x-1}$, $g(x) = x^2 + 1$ for $x \geq 1$

54. $f(x) = 2 - x^2$, $g(x) = \sqrt{2-x}$ for $x \leq 2$

55. $f(x) = \dfrac{1}{x}$, $g(x) = \dfrac{1}{x}$ for $x \neq 0$

56. $f(x) = (5 - x)^{1/3}$, $g(x) = 5 - x^3$

57. $f(x) = 4x^2 - 9$, $g(x) = \dfrac{\sqrt{x+9}}{2}$ for $x \geq 0$

58. $f(x) = \sqrt[3]{8x - 1}$, $g(x) = \dfrac{x^3 + 1}{8}$

59. $f(x) = \dfrac{1}{x-1}$, $g(x) = \dfrac{x+1}{x}$ for $x \neq 0$, $x \neq 1$

60. $f(x) = \sqrt{25 - x^2}$, $g(x) = \sqrt{25 - x^2}$ for $0 \leq x \leq 5$

In Exercises 61–66, write the function as a composite of two functions f and g. (More than one answer is correct.)

61. $f(g(x)) = 2(3x - 1)^2 + 5(3x - 1)$

62. $f(g(x)) = \dfrac{1}{1 + x^2}$

63. $f(g(x)) = \dfrac{2}{|x - 3|}$

64. $f(g(x)) = \sqrt{1 - x^2}$

65. $f(g(x)) = \dfrac{3}{\sqrt{x+1} - 2}$

66. $f(g(x)) = \dfrac{\sqrt{x}}{3\sqrt{x} + 2}$

• **APPLICATIONS**

Exercises 67 and 68 depend on the relationship between degrees Fahrenheit, degrees Celsius, and kelvins:

$$F = \frac{9}{5}C + 32 \qquad C = K - 273.15$$

67. Temperature. Write a composite function that converts kelvins into degrees Fahrenheit.

68. Temperature. Convert the following degrees Fahrenheit to kelvins: 32°F and 212°F.

69. Dog Run. Suppose that you want to build a *square* fenced-in area for your dog. Fencing is purchased in linear feet.

 a. Write a composite function that determines the area of your dog pen as a function of how many linear feet are purchased.

 b. If you purchase 100 linear feet, what is the area of your dog pen?

 c. If you purchase 200 linear feet, what is the area of your dog pen?

70. Dog Run. Suppose that you want to build a *circular* fenced-in area for your dog. Fencing is purchased in linear feet.

 a. Write a composite function that determines the area of your dog pen as a function of how many linear feet are purchased.

 b. If you purchase 100 linear feet, what is the area of your dog pen?

 c. If you purchase 200 linear feet, what is the area of your dog pen?

71. Market Price. Typical supply and demand relationships state that as the number of units for sale increases, the market price decreases. Assume that the market price p and the number of units for sale x are related by the demand equation:

$$p = 3000 - \frac{1}{2}x$$

Assume that the cost $C(x)$ of producing x items is governed by the equation

$$C(x) = 2000 + 10x$$

and the revenue $R(x)$ generated by selling x units is governed by

$$R(x) = 100x$$

 a. Write the cost as a function of price p.

 b. Write the revenue as a function of price p.

 c. Write the profit as a function of price p.

72. Market Price. Typical supply and demand relationships state that as the number of units for sale increases, the market price decreases. Assume that the market price p and the number of units for sale x are related by the demand equation:

$$p = 10{,}000 - \frac{1}{4}x$$

Assume that the cost $C(x)$ of producing x items is governed by the equation

$$C(x) = 30{,}000 + 5x$$

and the revenue $R(x)$ generated by selling x units is governed by

$$R(x) = 1000x$$

 a. Write the cost as a function of price p.

 b. Write the revenue as a function of price p.

 c. Write the profit as a function of price p.

73. Environment: Oil Spill. An oil spill makes a circular pattern around a ship such that the radius in feet grows as a function of time in hours $r(t) = 150\sqrt{t}$. Find the area of the spill as a function of time.

74. Pool Volume. A 20 foot $\times$ 10 foot rectangular pool has been built. If 50 cubic feet of water is pumped into the pool per hour, write the water-level height (feet) as a function of time (hours).

75. Fireworks. A family is watching a fireworks display. If the family is 2 miles from where the fireworks are being launched and the fireworks travel vertically, what is the distance between the family and the fireworks as a function of height above ground?

76. Real Estate. A couple are about to put their house up for sale. They bought the house for $172,000 a few years ago, and when they list it with a real estate agent they will pay a 6% commission. Write a function that represents the amount of money they will make on their home as a function of the asking price p.

• **CATCH THE MISTAKE**

In Exercises 77–81, for the functions $f(x) = x + 2$ and $g(x) = x^2 - 4$, find the indicated function and state its domain. Explain the mistake that is made in each problem.

77. $\dfrac{g}{f}$

Solution:
$$\frac{g(x)}{f(x)} = \frac{x^2 - 4}{x + 2}$$
$$= \frac{(x - 2)(x + 2)}{x + 2}$$
$$= x - 2$$

Domain: $(-\infty, \infty)$

This is incorrect. What mistake was made?

78. $\dfrac{f}{g}$

Solution:
$$\frac{f(x)}{g(x)} = \frac{x + 2}{x^2 - 4}$$
$$= \frac{x + 2}{(x - 2)(x + 2)} = \frac{1}{x - 2}$$
$$= \frac{1}{x - 2}$$

Domain: $(-\infty, 2) \cup (2, \infty)$

This is incorrect. What mistake was made?

79. $f \circ g$

Solution:
$$f \circ g = f(x)g(x)$$
$$= (x + 2)(x^2 - 4)$$
$$= x^3 + 2x^2 - 4x - 8$$

Domain: $(-\infty, \infty)$

This is incorrect. What mistake was made?

80. $(f + g)(2) = (x + 2 + x^2 - 4)(2)$
$$= (x^2 + x - 2)(2)$$
$$= 2x^2 + 2x - 4$$

Domain: $(-\infty, \infty)$

This is incorrect. What mistake was made?

81. $f(x) - g(x) = x + 2 - x^2 - 4$
$$= -x^2 + x - 2$$

Domain: $(-\infty, \infty)$

This is incorrect. What mistake was made?

82. Given the function $f(x) = x^2 + 7$ and $g(x) = \sqrt{x - 3}$, find $f \circ g$, and state the domain.

Solution:
$$f \circ g = f(g(x)) = \left(\sqrt{x - 3}\right)^2 + 7$$
$$= f(g(x)) = x - 3 + 7$$
$$= x - 4$$

Domain: $(-\infty, \infty)$

This is incorrect. What mistake was made?

• **CONCEPTUAL**

In Exercises 83–86, determine whether each statement is true or false.

83. When adding, subtracting, multiplying, or dividing two functions, the domain of the resulting function is the union of the domains of the individual functions.

84. For any functions f and g, $f(g(x)) = g(f(x))$ for all values of x that are in the domain of both f and g.

85. For any functions f and g, $(f \circ g)(x)$ exists for all values of x that are in the domain of $g(x)$, provided the range of g is a subset of the domain of f.

86. The domain of a composite function can be found by inspection, without knowledge of the domain of the individual functions.

• CHALLENGE

87. For the functions $f(x) = x + a$ and $g(x) = \dfrac{1}{x - a}$, find $g \circ f$ and state its domain.

88. For the functions $f(x) = ax^2 + bx + c$ and $g(x) = \dfrac{1}{x - c}$, find $g \circ f$ and state its domain.

89. For the functions $f(x) = \sqrt{x + a}$ and $g(x) = x^2 - a$, find $g \circ f$ and state its domain.

90. For the functions $f(x) = \dfrac{1}{x^a}$ and $g(x) = \dfrac{1}{x^b}$, find $g \circ f$ and state its domain. Assume $a > 1$ and $b > 1$.

• TECHNOLOGY

91. Using a graphing utility, plot $y_1 = \sqrt{x + 7}$, $y_2 = \sqrt{9 - x}$, and $y_3 = y_1 + y_2$. What is the domain of y_3?

92. Using a graphing utility, plot $y_1 = \sqrt[3]{x + 5}$, $y_2 = \dfrac{1}{\sqrt{3 - x}}$, and $y_3 = \dfrac{y_1}{y_2}$. What is the domain of y_3?

93. Using a graphing utility, plot $y_1 = \sqrt{x^2 - 3x - 4}$,

$y_2 = \dfrac{1}{x^2 - 14}$, and $y_3 = \dfrac{1}{y_1^2 - 14}$. If y_1 represents a function f and y_2 represents a function g, then y_3 represents the composite function $g \circ f$. The graph of y_3 is only defined for the domain of $g \circ f$. State the domain of $g \circ f$.

94. Using a graphing utility, plot $y_1 = \sqrt{1 - x}$, $y_2 = x^2 + 2$, and $y_3 = y_1^2 + 2$. If y_1 represents a function f and y_2 represents a function g, then y_3 represents the composite function $g \circ f$. The graph of y_3 is only defined for the domain of $g \circ f$. State the domain of $g \circ f$.

A.8 ONE-TO-ONE FUNCTIONS AND INVERSE FUNCTIONS

SKILLS OBJECTIVES	CONCEPTUAL OBJECTIVES
▪ Determine whether a function is a one-to-one function. ▪ Verify that two functions are inverses of one another. ▪ Graph the inverse function given the graph of the function. ▪ Find the inverse of a function.	▪ Understand why a function that passes the horizontal line test is one-to-one. ▪ Visualize the relationships between the domain and range of a function and the domain and range of its inverse. ▪ Understand why functions and their inverses are symmetric about $y = x$. ▪ Understand why a function has to be one-to-one in order for its inverse to exist.

Every human being has a blood type, and every human being has a DNA sequence. These are examples of functions, where a person is the input and the output is blood type or DNA sequence. These relationships are classified as functions because each person can have one and only one blood type or DNA strand. The difference between these functions is that many people have the same blood type, but DNA is unique to each individual. Can we map backwards? For instance, if you know the blood type, do you know specifically which person it came from? No, but, if you know the DNA sequence, you know that the sequence belongs to only one person. When a function has a one-to-one correspondence, like the DNA example, then mapping backwards is possible. The map back is called the *inverse function*.

A.8.1 Determine Whether a Function Is One-to-One

In Section A.4, we defined a function as a relationship that maps an input (contained in the domain) to exactly one output (found in the range). Algebraically, each value for x can correspond to only a single value for y. Recall the square, identity, absolute value, and reciprocal functions from our library of functions in Section A.5.

All of the graphs of these functions satisfy the vertical line test. Although the square function and the absolute value function map each value of x to exactly one value for y, these two functions map two values of x to the same value for y. For example, $(-1, 1)$ and $(1, 1)$ lie on both graphs. The identity and reciprocal functions, on the other hand, map each x to a single value for y, and no two x-values map to the same y-value. These two functions are examples of *one-to-one functions*.

DEFINITION One-to-One Function

A function $f(x)$ is **one-to-one** if no two elements in the domain correspond to the same element in the range; that is,

$$\text{if } x_1 \neq x_2, \text{ then } f(x_1) \neq f(x_2).$$

In other words, it is one-to-one if no two inputs map to the same output.

EXAMPLE 1 Determining Whether a Function Defined as a Set of Points Is a One-to-One Function

For each of the three relations, determine whether the relation is a function. If it is a function, determine whether it is a one-to-one function.

$$f = \{(0, 0), (1, 1), (1, -1)\}$$
$$g = \{(-1, 1), (0, 0), (1, 1)\}$$
$$h = \{(-1, -1), (0, 0), (1, 1)\}$$

Solution:

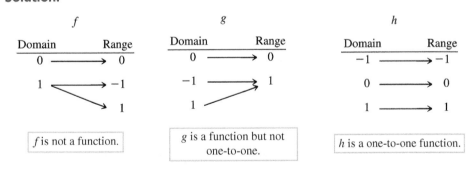

Just as there is a graphical test for functions, the vertical line test, there is a graphical test for one-to-one functions, the *horizontal line test*. Note that a horizontal line can be drawn on the square and absolute value functions so that it intersects the graph of each function at two points. The identity and reciprocal functions, however, will intersect a horizontal line in at most only one point. This leads us to the horizontal line test for one-to-one functions.

DEFINITION Horizontal Line Test

If every horizontal line intersects the graph of a function in at most one point, then the function is classified as a one-to-one function.

Draw a horizontal line at $y = 4$ on the graph of $y = x^2$. What are the two points of intersection? Explain why the equation $y = x^2$ cannot be a one-to-one function.

▼ ·

ANSWER $(-2, 4)$ and $(2, 4)$: Two different x values map to the same y value, so this graph fails the horizontal line test.

EXAMPLE 2 **Using the Horizontal Line Test to Determine Whether a Function Is One-to-One**

For each of the three relations, determine whether the relation is a function. If it is a function, determine whether it is a one-to-one function. Assume that x is the independent variable and y is the dependent variable.

$$x = y^2 \qquad y = x^2 \qquad y = x^3$$

Solution:

$$x = y^2 \qquad\qquad\qquad y = x^2 \qquad\qquad\qquad y = x^3$$

Not a function	Function but not one-to-one	One-to-one function
(Fails vertical line test)	(Passes vertical line test, but fails horizontal line test)	(Passes both horizontal and vertical line tests)

▼

YOUR TURN Determine whether each of the functions is a one-to-one function.

a. $f(x) = x + 2$ **b.** $f(x) = x^2 + 1$

▼
ANSWER

a. yes

b. no

Another way of writing the definition of a one-to-one function is:

$$\text{If } f(x_1) = f(x_2), \text{ then } x_1 = x_2.$$

In the Your Turn following Example 2, we found (using the horizontal line test) that $f(x) = x + 2$ is a one-to-one function, but that $f(x) = x^2 + 1$ is not a one-to-one function. We can also use this alternative definition to determine algebraically whether a function is one-to-one.

WORDS	MATH
State the function.	$f(x) = x + 2$
Let there be two real numbers, x_1 and x_2, such that $f(x_1) = f(x_2)$.	$x_1 + 2 = x_2 + 2$
Subtract 2 from both sides of the equation.	$x_1 = x_2$

$f(x) = x + 2$ is a one-to-one function.

WORDS	MATH
State the function.	$f(x) = x^2 + 1$
Let there be two real numbers, x_1 and x_2, such that $f(x_1) = f(x_2)$.	$x_1^2 + 1 = x_2^2 + 1$
Subtract 1 from both sides of the equation.	$x_1^2 = x_2^2$
Solve for x_1.	$x_1 = \pm x_2$

$f(x) = x^2 + 2$ is *not* a one-to-one function.

EXAMPLE 3 **Determining Algebraically Whether a Function Is One-to-One**

Determine algebraically whether these functions are one-to-one.

a. $f(x) = 5x^3 - 2$ **b.** $f(x) = |x + 1|$

Solution (a):

Find $f(x_1)$ and $f(x_2)$.	$f(x_1) = 5x_1^3 - 2$ and $f(x_2) = 5x_2^3 - 2$
Let $f(x_1) = f(x_2)$.	$5x_1^3 - 2 = 5x_2^3 - 2$
Add 2 to both sides of the equation.	$5x_1^3 = 5x_2^3$
Divide both sides of the equation by 5.	$x_1^3 = x_2^3$
Take the cube root of both sides of the equation.	$\left(x_1^3\right)^{1/3} = \left(x_2^3\right)^{1/3}$
Simplify.	$x_1 = x_2$

$$\boxed{f(x) = 5x^3 - 2 \text{ is a one-to-one function.}}$$

Solution (b):

Find $f(x_1)$ and $f(x_2)$.	$f(x_1) = |x_1 + 1|$ and $f(x_2) = |x_2 + 1|$
Let $f(x_1) = f(x_2)$.	$|x_1 + 1| = |x_2 + 1|$
Solve the absolute value equation.	$(x_1 + 1) = (x_2 + 1)$ or $(x_1 + 1) = -(x_2 + 1)$
	$x_1 = x_2$ or $x_1 = -x_2 - 2$

$$\boxed{f(x) = |x + 1| \text{ is } \textbf{not} \text{ a one-to-one function.}}$$

A.8.2 Inverse Functions

If a function is one-to-one, then the function maps each x to exactly one y, and no two x-values map to the same y-value. This implies that there is a one-to-one correspondence between the inputs (domain) and outputs (range) of a one-to-one function $f(x)$. In the special case of a one-to-one function, it would be possible to map from the output (range of f) back to the input (domain of f), and this mapping would also be a function. The function that maps the output back to the input of a function f is called the **inverse function** and is denoted $f^{-1}(x)$.

A one-to-one function f maps every x in the domain to a unique and distinct corresponding y in the range. Therefore, the inverse function f^{-1} maps every y back to a unique and distinct x.

The function notations $f(x) = y$ and $f^{-1}(y) = x$ indicate that if the point (x, y) satisfies the function, then the point (y, x) satisfies the inverse function.

For example, let the function $h(x) = \{(-1, 0), (1, 2), (3, 4)\}$.

$$h = \{(-1, 0), (1, 2), (3, 4)\}$$

Domain   Range
$-1 \rightleftarrows 0$

$1 \rightleftarrows 2$ h is a one-to-one function

$3 \rightleftarrows 4$
Range   Domain

$$h^{-1} = \{(0, -1), (2, 1), (4, 3)\}$$

A.8.2 SKILL

Verify that two functions are inverses of one another.

A.8.2 CONCEPTUAL

Visualize the relationships between the domain and range of a function and the domain and range of its inverse.

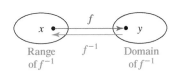

Range   Domain
of f^{-1} of f^{-1}

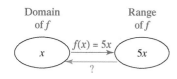

Domain of f Range of f

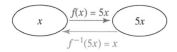

The inverse function undoes whatever the function does. For example, if $f(x) = 5x$, then the function f maps any value x in the domain to a value $5x$ in the range. If we want to map backwards or undo the $5x$, we develop a function called the inverse function that takes $5x$ as input and maps back to x as output. The inverse function is $f^{-1}(x) = \frac{1}{5}x$. Note that if we input $5x$ into the inverse function, the output is x: $f^{-1}(5x) = \frac{1}{5}(5x) = x$.

DEFINITION | **Inverse Function**

If f and g denote two one-to-one functions such that

$$f(g(x)) = x \text{ for every } x \text{ in the domain of } g$$

and

$$g(f(x)) = x \text{ for every } x \text{ in the domain of } f,$$

then g is the **inverse** of the function f. The function g is denoted by f^{-1} (read "f-inverse").

▼
CAUTION

$f^{-1} \neq \dfrac{1}{f}$

Note: f^{-1} is used to denote the inverse of f. The -1 is not used as an exponent and, therefore, does not represent the reciprocal of f: $\dfrac{1}{f}$.

Two properties hold true relating one-to-one functions to their inverses: (1) The range of the function is the domain of the inverse, and the range of the inverse is the domain of the function, and (2) the composite function that results with a function and its inverse (and vice versa) is the identity function x.

$$\text{Domain of } f = \text{range of } f^{-1} \text{ and range of } f = \text{domain of } f^{-1}$$

$$f^{-1}(f(x)) = x \quad \text{and} \quad f(f^{-1}(x)) = x$$

EXAMPLE 4 | **Verifying Inverse Functions**

Verify that $f^{-1}(x) = \frac{1}{2}x - 2$ is the inverse of $f(x) = 2x + 4$.

Solution:

Show that $f^{-1}(f(x)) = x$ and $f(f^{-1}(x)) = x$.

Write f^{-1} using placeholder notation.
$$f^{-1}(\square) = \frac{1}{2}(\square) - 2$$

Substitute $f(x) = 2x + 4$ into f^{-1}.
$$f^{-1}(f(x)) = \frac{1}{2}(2x + 4) - 2$$

Simplify.
$$f^{-1}(f(x)) = x + 2 - 2 = x$$
$$f^{-1}(f(x)) = x$$

Write f using placeholder notation.
$$f(\square) = 2(\square) + 4$$

Substitute $f^{-1}(x) = \frac{1}{2}x - 2$ into f.
$$f(f^{-1}(x)) = 2\left(\frac{1}{2}x - 2\right) + 4$$

Simplify.
$$f(f^{-1}(x)) = x - 4 + 4 = x$$
$$f(f^{-1}(x)) = x$$

Note the relationship between the domain and range of f and f^{-1}.

	DOMAIN	RANGE
$f(x) = 2x + 4$	$(-\infty, \infty)$	$(-\infty, \infty)$
$f^{-1}(x) = \frac{1}{2}x - 2$	$(-\infty, \infty)$	$(-\infty, \infty)$

EXAMPLE 5   **Verifying Inverse Functions with Domain Restrictions**

Verify that $f^{-1}(x) = x^2$, for $x \geq 0$, is the inverse of $f(x) = \sqrt{x}$.

Solution:

Show that $f^{-1}(f(x)) = x$ and $f(f^{-1}(x)) = x$.

Write f^{-1} using placeholder notation.	$f^{-1}(\square) = (\square)^2$
Substitute $f(x) = \sqrt{x}$ into f^{-1}.	$f^{-1}(f(x)) = (\sqrt{x})^2 = x$
	$f^{-1}(f(x)) = x \quad$ for $\quad x \geq 0$
Write f using placeholder notation.	$f(\square) = \sqrt{(\square)}$
Substitute $f^{-1}(x) = x^2$, $x \geq 0$ into f.	$f(f^{-1}(x)) = \sqrt{x^2} = x, x \geq 0$
	$f(f^{-1}(x)) = x \quad$ for $\quad x \geq 0$

	DOMAIN	RANGE
$f(x) = \sqrt{x}$	$[0, \infty)$	$[0, \infty)$
$f^{-1}(x) = x^2, x \geq 0$	$[0, \infty)$	$[0, \infty)$

[CONCEPT CHECK]

If a one-to-one function f has domain $[a, \infty)$ and range $[b, \infty)$ then what are the domain and range of it's inverse, f^{-1}?

ANSWER The inverse function, f^{-1} has domain $[b, \infty)$ and range $[a, \infty)$.

A.8.3 Graphical Interpretation of Inverse Functions

In Example 4, we showed that $f^{-1}(x) = \frac{1}{2}x - 2$ is the inverse of $f(x) = 2x + 4$. Let's now investigate the graphs that correspond to the function f and its inverse f^{-1}.

A.8.3 SKILL

Graph the inverse function given the graph of the function.

A.8.3 CONCEPTUAL

Understand why functions and their inverses are symmetric about $y = x$.

$f(x)$	
x	y
-3	-2
-2	0
-1	2
0	4

$f^{-1}(x)$	
x	y
-2	-3
0	-2
2	-1
4	0

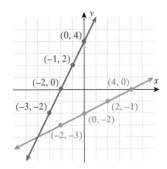

Note that the point $(-3, -2)$ lies on the function and the point $(-2, -3)$ lies on the inverse. In fact, every point (a, b) that lies on the function corresponds to a point (b, a) that lies on the inverse.

Draw the line $y = x$ on the graph. In general, the point (b, a) on the inverse $f^{-1}(x)$ is the reflection (about $y = x$) of the point (a, b) on the function $f(x)$.

In general, if the point (a, b) is on the graph of a function, then the point (b, a) is on the graph of its inverse.

[STUDY TIP]

If the point (a, b) is on the function, then the point (b, a) is on the inverse. Notice the interchanging of the x- and y-coordinates.

▶ **EXAMPLE 6** **Graphing the Inverse Function**

Given the graph of the function $f(x)$, plot the graph of its inverse $f^{-1}(x)$.

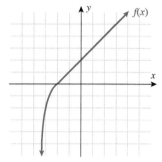

Solution:
Because the points $(-3, -2), (-2, 0),$ $(0, 2),$ and $(2, 4)$ lie on the graph of f, then the points $(-2, -3), (0, -2), (2, 0),$ and $(4, 2)$ lie on the graph of f^{-1}.

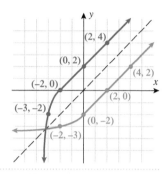

ANSWER

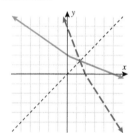

YOUR TURN Given the graph of a function f, plot the inverse function.

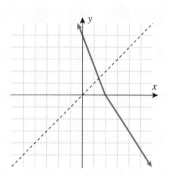

We have developed the definition of an inverse function and properties of inverses. At this point, you should be able to determine whether two functions are inverses of one another. Let's turn our attention to another problem: How do you find the inverse of a function?

A.8.4 SKILL

Find the inverse of a function.

A.8.4 CONCEPTUAL

Understand why a function has to be one-to-one in order for its inverse to exist.

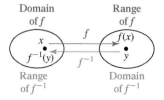

A.8.4 Finding the Inverse Function

If the point (a, b) lies on the graph of a function, then the point (b, a) lies on the graph of the inverse function. The symmetry about the line $y = x$ tells us that the roles of x and y interchange. Therefore, if we start with every point (x, y) that lies on the graph of a function, then every point (y, x) lies on the graph of its inverse. Algebraically, this corresponds to interchanging x and y. Finding the inverse of a finite set of ordered pairs is easy: Simply interchange the x- and y-coordinates. Earlier, we found that if $h(x) = \{(-1, 0), (1, 2), (3, 4)\}$, then $h^{-1}(x) = \{(0, -1), (2, 1), (4, 3)\}$. But how do we find the inverse of a function defined by an equation?

Recall the mapping relationship if f is a one-to-one function. This relationship implies that $f(x) = y$ and $f^{-1}(y) = x$. Let's use these two identities to find the inverse. Now consider the function defined by $f(x) = 3x - 1$. To find f^{-1}, we let $f(x) = y$, which yields $y = 3x - 1$. Solve for the variable x: $x = \frac{1}{3}y + \frac{1}{3}$.

Recall that $f^{-1}(y) = x$, so we have found the inverse to be $f^{-1}(y) = \frac{1}{3}y + \frac{1}{3}$. It is customary to write the independent variable as x, so we write the inverse as $f^{-1}(x) = \frac{1}{3}x + \frac{1}{3}$. Now that we have found the inverse, let's confirm that the properties $f^{-1}(f(x)) = x$ and $f(f^{-1}(x)) = x$ hold.

$$f(f^{-1}(x)) = 3\left(\frac{1}{3}x + \frac{1}{3}\right) - 1 = x + 1 - 1 = x$$

$$f^{-1}(f(x)) = \frac{1}{3}(3x - 1) + \frac{1}{3} = x - \frac{1}{3} + \frac{1}{3} = x$$

FINDING THE INVERSE OF A FUNCTION

Let f be a one-to-one function; then the following procedure can be used to find the inverse function f^{-1} if the inverse exists.

STEP	PROCEDURE	EXAMPLE
1	Let $y = f(x)$.	$f(x) = -3x + 5$ $y = -3x + 5$
2	Solve the resulting equation for x in terms of y (if possible).	$3x = -y + 5$ $x = -\frac{1}{3}y + \frac{5}{3}$
3	Let $x = f^{-1}(y)$.	$f^{-1}(y) = -\frac{1}{3}y + \frac{5}{3}$
4	Let $y = x$ (interchange x and y).	$f^{-1}(x) = -\frac{1}{3}x + \frac{5}{3}$

The same result is found if we first interchange x and y and then solve for y in terms of x.

STEP	PROCEDURE	EXAMPLE
1	Let $y = f(x)$.	$f(x) = -3x + 5$ $y = -3x + 5$
2	Interchange x and y	$x = -3y + 5$
3	Solve for y in terms of x.	$3y = -x + 5$ $y = -\frac{1}{3}x + \frac{5}{3}$
4	Let $y = f^{-1}(x)$.	$f^{-1}(x) = -\frac{1}{3}x + \frac{5}{3}$

Note the following:

- Verify first that a function is one-to-one prior to finding an inverse (if it is not one-to-one, then the inverse does not exist).
- State the domain restrictions on the inverse function. The domain of f is the range of f^{-1} and vice versa.
- To verify that you have found the inverse, show that $f(f^{-1}(x)) = x$ for all x in the domain of f^{-1} and $f^{-1}(f(x)) = x$ for all x in the domain of f.

EXAMPLE 7 **The Inverse of a Square-Root Function**

Find the inverse of the function $f(x) = \sqrt{x + 2}$ and state the domain and range of both f and f^{-1}.

Solution:

$f(x)$ is a one-to-one function because it passes the horizontal line test.

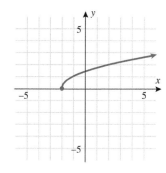

STEP 1 Let $y = f(x)$. $y = \sqrt{x + 2}$

STEP 2 Interchange x and y. $x = \sqrt{y + 2}$

STEP 3 Solve for y.

Square both sides of the equation. $x^2 = y + 2$

Subtract 2 from both sides. $x^2 - 2 = y$ or $y = x^2 - 2$

STEP 4 Let $y = f^{-1}(x)$. $f^{-1}(x) = x^2 - 2$

Note any domain restrictions. (State the domain and range of both f and f^{-1}.)

$$f: \quad \text{Domain: } [-2, \infty) \qquad \text{Range: } [0, \infty)$$
$$f^{-1}: \quad \text{Domain: } [0, \infty) \qquad \text{Range: } [-2, \infty)$$

The inverse of $f(x) = \sqrt{x + 2}$ is $\boxed{f^{-1}(x) = x^2 - 2 \text{ for } x \geq 0}$.

Check.

$f^{-1}(f(x)) = x$ for all x in the domain of f.

$f^{-1}(f(x)) = \left(\sqrt{x + 2}\right)^2 - 2$
$\qquad = x + 2 - 2 \quad \text{for } x \geq -2$
$\qquad = x$

$f(f^{-1}(x)) = x$ for all x in the domain of f^{-1}.

$f(f^{-1}(x)) = \sqrt{(x^2 - 2) + 2}$
$\qquad = \sqrt{x^2} \quad \text{for } x \geq 0$
$\qquad = x$

Note that the function $f(x) = \sqrt{x + 2}$ and its inverse $f^{-1}(x) = x^2 - 2$ for $x \geq 0$ are symmetric about the line $y = x$.

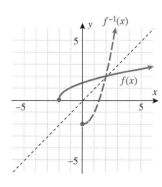

YOUR TURN Find the inverse of the given function and state the domain and range of the inverse function.

a. $f(x) = 7x - 3$ b. $g(x) = \sqrt{x - 1}$

EXAMPLE 8 **A Function That Does Not Have an Inverse Function**

Find the inverse of the function $f(x) = |x|$ if it exists.

Solution:

The function $f(x) = |x|$ fails the horizontal line test and therefore is not a one-to-one function. Because f is not a one-to-one function, its inverse function does not exist.

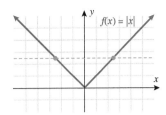

▶ EXAMPLE 9 **Finding the Inverse Function**

The function $f(x) = \dfrac{2}{x + 3}, x \neq -3$, is a one-to-one function. Find its inverse.

STUDY TIP

The range of a function is equal to the domain of its inverse function.

Solution:

STEP 1 Let $y = f(x)$.

$$y = \frac{2}{x + 3}$$

STEP 2 Interchange x and y.

$$x = \frac{2}{y + 3}$$

STEP 3 Solve for y.

Multiply the equation by $(y + 3)$.

$$x(y + 3) = 2$$

Eliminate the parentheses.

$$xy + 3x = 2$$

Subtract $3x$ from both sides.

$$xy = -3x + 2$$

Divide the equation by x.

$$y = \frac{-3x + 2}{x} = -3 + \frac{2}{x}$$

STEP 4 Let $y = f^{-1}(x)$.

$$f^{-1}(x) = -3 + \frac{2}{x}$$

Note any domain restrictions on $f^{-1}(x)$.

$$x \neq 0$$

The inverse of the function $f(x) = \dfrac{2}{x + 3}, x \neq -3$, is $\boxed{f^{-1}(x) = -3 + \dfrac{2}{x}, x \neq 0}$.

Check:

$$f^{-1}(f(x)) = -3 + \frac{2}{\left(\dfrac{2}{x + 3}\right)} = -3 + (x + 3) = x, x \neq -3$$

$$f(f^{-1}(x)) = \frac{2}{\left(-3 + \dfrac{2}{x}\right) + 3} = \frac{2}{\left(\dfrac{2}{x}\right)} = x, x \neq 0$$

▼ YOUR TURN The function $f(x) = \dfrac{4}{x - 1}, x \neq 1$, is a one-to-one function. Find its inverse.

▼ **ANSWER**

$$f^{-1}(x) = 1 + \frac{4}{x}, x \neq 0$$

Note in Example 9 that the domain of f is $(-\infty, -3) \cup (-3, \infty)$ and the domain of f^{-1} is $(-\infty, 0) \cup (0, \infty)$. Therefore, we know that the range of f is $(-\infty, 0) \cup (0, \infty)$, and the range of f^{-1} is $(-\infty, -3) \cup (-3, \infty)$.

▶[SECTION A.8] **SUMMARY**

One-to-One Functions

Each input in the domain corresponds to exactly one output in the range, and no two inputs map to the same output. There are three ways to test a function to determine whether it is a one-to-one function.

1. Discrete points: For the set of all points (a, b) verify that no y-values are repeated.
2. Algebraic equations: Let $f(x_1) = f(x_2)$; if it can be shown that $x_1 = x_2$, then the function is one-to-one.
3. Graphs: Use the horizontal line test; if any horizontal line intersects the graph of the function in more than one point, then the function is not one-to-one.

Properties of Inverse Functions

1. If f is a one-to-one function, then f^{-1} exists.
2. Domain and range
   - Domain of f = range of f^{-1}
   - Domain of f^{-1} = range of f
3. Composition of inverse functions
   - $f^{-1}(f(x)) = x$ for all x in the domain of f.
   - $f(f^{-1}(x)) = x$ for all x in the domain of f^{-1}.
4. The graphs of f and f^{-1} are symmetric with respect to the line $y = x$.

Procedure for Finding the Inverse of a Function

1. Let $y = f(x)$.
2. Interchange x and y.
3. Solve for y.
4. Let $y = f^{-1}(x)$.

[SECTION A.8] **EXERCISES**

● **SKILLS**

In Exercises 1–16, determine whether the given relation is a function. If it is a function, determine whether it is a one-to-one function.

1.
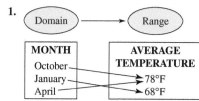

2.
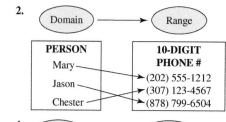

3.
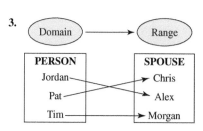

4.
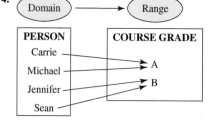

5. $\{(0, 1), (1, 2), (2, 3), (3, 4)\}$

6. $\{(0, -2), (2, 0), (5, 3), (-5, -7)\}$

7. $\{(0, 0), (9, -3), (4, -2), (4, 2), (9, 3)\}$

8. $\{(0, 1), (1, 1), (2, 1), (3, 1)\}$

9. $\{(0, 1), (1, 0), (2, 1), (-2, 1), (5, 4), (-3, 4)\}$

10. $\{(0, 0), (-1, -1), (-2, -8), (1, 1), (2, 8)\}$

11.
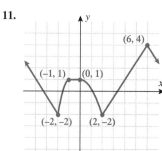

12.
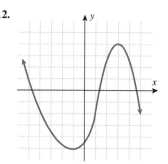

13.
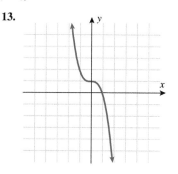

14.

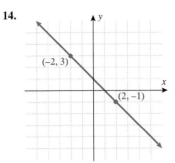

15.

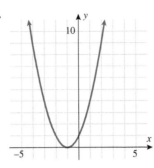

16.

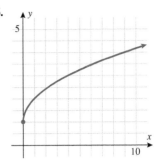

In 17–24, determine algebraically and graphically whether the function is one-to-one.

17. $f(x) = |x - 3|$

18. $f(x) = (x - 2)^2 + 1$

19. $f(x) = \dfrac{1}{x - 1}$

20. $f(x) = \sqrt[3]{x}$

21. $f(x) = x^2 - 4$

22. $f(x) = \sqrt{x + 1}$

23. $f(x) = x^3 - 1$

24. $f(x) = \dfrac{1}{x + 2}$

In Exercises 25–34, verify that the function $f^{-1}(x)$ is the inverse of $f(x)$ by showing that $f(f^{-1}(x)) = x$ and $f^{-1}(f(x)) = x$. Graph $f(x)$ and $f^{-1}(x)$ on the same axes to show the symmetry about the line $y = x$.

25. $f(x) = 2x + 1;\ f^{-1}(x) = \dfrac{x - 1}{2}$

26. $f(x) = \dfrac{x - 2}{3};\ f^{-1}(x) = 3x + 2$

27. $f(x) = \sqrt{x - 1},\ x \geq 1;\ f^{-1}(x) = x^2 + 1,\ x \geq 0$

28. $f(x) = 2 - x^2,\ x \geq 0;\ f^{-1}(x) = \sqrt{2 - x},\ x \leq 2$

29. $f(x) = \dfrac{1}{x};\ f^{-1}(x) = \dfrac{1}{x},\ x \neq 0$

30. $f(x) = (5 - x)^{1/3};\ f^{-1}(x) = 5 - x^3$

31. $f(x) = \dfrac{1}{2x + 6},\ x \neq -3;\ f^{-1}(x) = \dfrac{1}{2x} - 3,\ x \neq 0$

32. $f(x) = \dfrac{3}{4 - x},\ x \neq 4;\ f^{-1}(x) = 4 - \dfrac{3}{x},\ x \neq 0$

33. $f(x) = \dfrac{x + 3}{x + 4},\ x \neq -4;\ f^{-1}(x) = \dfrac{3 - 4x}{x - 1},\ x \neq 1$

34. $f(x) = \dfrac{x - 5}{3 - x},\ x \neq 3;\ f^{-1}(x) = \dfrac{3x + 5}{x + 1},\ x \neq -1$

In Exercises 35–42, graph the inverse of the one-to-one function that is given.

35.

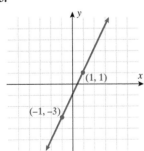

36.

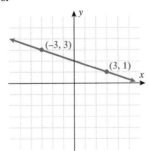

37.

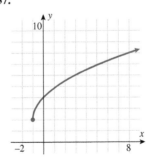

38.

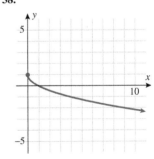

39.

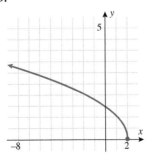

40.

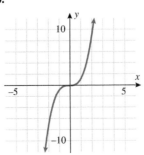

41.

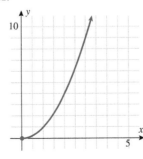

42.

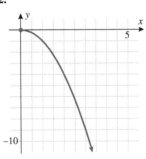

In Exercises 43–60, the function f is one-to-one. Find its inverse, and check your answer. State the domain and range of both f and f^{-1}.

43. $f(x) = x - 1$ **44.** $f(x) = 7x$ **45.** $f(x) = -3x + 2$ **46.** $f(x) = 2x + 3$

47. $f(x) = x^3 + 1$ **48.** $f(x) = x^3 - 1$ **49.** $f(x) = \sqrt{x - 3}$ **50.** $f(x) = \sqrt{3 - x}$

51. $f(x) = x^2 - 1, x \geq 0$ **52.** $f(x) = 2x^2 + 1, x \geq 0$ **53.** $f(x) = (x + 2)^2 - 3, x \geq -2$ **54.** $f(x) = (x - 3)^2 - 2, x \geq 3$

55. $f(x) = \dfrac{2}{x}$ **56.** $f(x) = -\dfrac{3}{x}$ **57.** $f(x) = \dfrac{2}{3 - x}$ **58.** $f(x) = \dfrac{7}{x + 2}$

59. $f(x) = \dfrac{7x + 1}{5 - x}$ **60.** $f(x) = \dfrac{2x + 5}{7 + x}$

In Exercises 61–64, graph the piecewise-defined function to determine whether it is a one-to-one function. If it is a one-to-one function, find its inverse.

61. $G(x) = \begin{cases} 0 & x < 0 \\ \sqrt{x} & x \geq 0 \end{cases}$ **62.** $G(x) = \begin{cases} \dfrac{1}{x} & x < 0 \\ x \\ \sqrt{x} & x \geq 0 \end{cases}$ **63.** $f(x) = \begin{cases} x & x \leq -1 \\ x^3 & -1 < x < 1 \\ x & x \geq 1 \end{cases}$ **64.** $f(x) = \begin{cases} x + 3 & x \leq -2 \\ |x| & -2 < x < 2 \\ x^2 & x \geq 2 \end{cases}$

• APPLICATIONS

65. Temperature. The equation used to convert from degrees Celsius to degrees Fahrenheit is $f(x) = \frac{9}{5}x + 32$. Determine the inverse function $f^{-1}(x)$. What does the inverse function represent?

66. Temperature. The equation used to convert from degrees Fahrenheit to degrees Celsius is $C(x) = \frac{5}{9}(x - 32)$. Determine the inverse function $C^{-1}(x)$. What does the inverse function represent?

67. Budget. The Richmond rowing club is planning to enter the Head of the Charles race in Boston and is trying to figure out how much money to raise. The entry fee is $250 per boat for the first 10 boats and $175 for each additional boat. Find the cost function $C(x)$ as a function of the number of boats the club enters x. Find the inverse function that will yield how many boats the club can enter as a function of how much money it will raise.

68. Long-Distance Calling Plans. A phone company charges $0.39 per minute for the first 10 minutes of a long-distance phone call and $0.12 per minute every minute after that. Find the cost function $C(x)$ as a function of the length of the phone call in minutes x. Suppose you buy a "prepaid" phone card that is planned for a single call. Find the inverse function that determines how many minutes you can talk as a function of how much you prepaid.

69. Salary. A student works at Target making $7 per hour and the weekly number of hours worked per week x varies. If Target withholds 25% of his earnings for taxes and Social Security, write a function $E(x)$ that expresses the student's take-home pay each week. Find the inverse function $E^{-1}(x)$. What does the inverse function tell you?

70. Salary. A grocery store pays you $8 per hour for the first 40 hours per week and time and a half for overtime. Write a piecewise-defined function that represents your weekly earnings $E(x)$ as a function of the number of hours worked x. Find the inverse function $E^{-1}(x)$. What does the inverse function tell you?

• CATCH THE MISTAKE

In Exercises 71–74, explain the mistake that is made.

71. Is $x = y^2$ a one-to-one function?

Solution:

Yes, this graph represents a one-to-one function because it passes the horizontal line test.

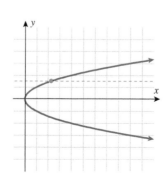

This is incorrect. What mistake was made?

72. A linear one-to-one function is graphed below. Draw its inverse.

Solution:

Note that the points $(3, 3)$ and $(0, -4)$ lie on the graph of the function.

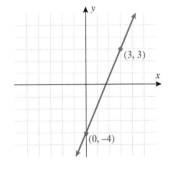

By symmetry, the points $(-3, -3)$ and $(0, 4)$ lie on the graph of the inverse.

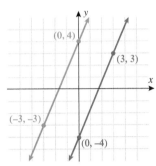

This is incorrect. What mistake was made?

73. Given the function $f(x) = x^2$, find the inverse function $f^{-1}(x)$.

Solution:

Step 1: Let $y = f(x)$. $\qquad y = x^2$

Step 2: Solve for x. $\qquad x = \sqrt{y}$

Step 3: Interchange x and y. $\qquad y = \sqrt{x}$

Step 4: Let $y = f^{-1}(x)$. $\qquad f^{-1}(x) = \sqrt{x}$

Check: $f(f^{-1}(x)) = (\sqrt{x})^2 = x$ and $f^{-1}(f(x)) = \sqrt{x^2} = x$.

The inverse of $f(x) = x^2$ is $f^{-1}(x) = \sqrt{x}$.

This is incorrect. What mistake was made?

74. Given the function $f(x) = \sqrt{x - 2}$, find the inverse function $f^{-1}(x)$, and state the domain restrictions on $f^{-1}(x)$.

Solution:

Step 1: Let $y = f(x)$. $\qquad y = \sqrt{x - 2}$

Step 2: Interchange x and y. $\qquad x = \sqrt{y - 2}$

Step 3: Solve for y. $\qquad y = x^2 + 2$

Step 4: Let $f^{-1}(x) = y$. $\qquad f^{-1}(x) = x^2 + 2$

Step 5: Domain restrictions $f(x) = \sqrt{x - 2}$ has the domain restriction that $x \geq 2$.

The inverse of $f(x) = \sqrt{x - 2}$ is $f^{-1}(x) = x^2 + 2$.

The domain of $f^{-1}(x)$ is $x \geq 2$.

This is incorrect. What mistake was made?

• **CONCEPTUAL**

In Exercises 75–78, determine whether each statement is true or false.

75. Every even function is a one-to-one function.

76. Every odd function is a one-to-one function.

77. It is not possible that $f = f^{-1}$.

78. A function f has an inverse. If the function lies in quadrant II, then its inverse lies in quadrant IV.

79. If $(0, b)$ is the y-intercept of a one-to-one function f, what is the x-intercept of the inverse f^{-1}?

80. If $(a, 0)$ is the x-intercept of a one-to-one function f, what is the y-intercept of the inverse f^{-1}?

• **CHALLENGE**

81. The unit circle is not a function. If we restrict ourselves to the semicircle that lies in quadrants I and II, the graph represents a function, but it is not a one-to-one function. If we further restrict ourselves to the quarter circle lying in quadrant I, the graph does represent a one-to-one function. Determine the equations of both the one-to-one function and its inverse. State the domain and range of both.

82. Find the inverse of $f(x) = \dfrac{c}{x}, c \neq 0$.

83. Under what conditions is the linear function $f(x) = mx + b$ a one-to-one function?

84. Assuming that the conditions found in Exercise 83 are met, determine the inverse of the linear function.

• **TECHNOLOGY**

In Exercises 85–88, graph the following functions and determine whether they are one-to-one.

85. $f(x) = |4 - x^2|$

86. $f(x) = \dfrac{3}{x^3 + 2}$

87. $f(x) = x^{1/3} - x^5$

88. $f(x) = \dfrac{1}{x^{1/2}}$

In Exercises 89–92, graph the functions f and g and the line $y = x$ in the same screen. Do the two functions appear to be inverses of each other?

89. $f(x) = \sqrt{3x - 5}$; $g(x) = \dfrac{x^2}{3} + \dfrac{5}{3}$

90. $f(x) = \sqrt{4 - 3x}$; $g(x) = \dfrac{4}{3} - \dfrac{x^2}{3}, x \geq 0$

91. $f(x) = (x - 7)^{1/3} + 2$; $g(x) = x^3 - 6x^2 + 12x - 1$

92. $f(x) = \sqrt[3]{x + 3} - 2$; $g(x) = x^3 + 6x^2 + 12x + 6$

APPENDIX A REVIEW ▶

SECTION	CONCEPT	KEY IDEAS/FORMULAS
A.1	**Factoring polynomials**	
	Greatest common factor	Factor out using distributive property: ax^k
	Factoring formulas: Special polynomial forms	Difference of Two Squares $$a^2 - b^2 = (a + b)(a - b)$$ Perfect Squares $$a^2 + 2ab + b^2 = (a + b)^2$$ $$a^2 - 2ab + b^2 = (a - b)^2$$ Sum of Two Cubes $$a^3 + b^3 = (a + b)(a^2 - ab + b^2)$$ Difference of Two Cubes $$a^3 - b^3 = (a - b)(a^2 + ab + b^2)$$
	Factoring a trinomial as a product of two binomials	▪ $x^2 + bx + c = (x + ?)(x + ?)$ ▪ $ax^2 + bx + c = (?x + ?)(?x + ?)$
	Factoring by grouping	Group terms with common factors
A.2	**Basic tools: Cartesian plane, distance, and midpoint**	Two points in the xy-plane: (x_1, y_1) and (x_2, y_2)
	Cartesian plane	x-axis, y-axis, origin, and quadrants
	Distance between two points	$d = \sqrt{(x_2 - x_1)^2 + (y_2 - y_1)^2}$
	Midpoint of a line segment joining two points	$(x_m, y_m) = \left(\dfrac{x_1 + x_2}{2}, \dfrac{y_1 + y_2}{2} \right)$
A.3	**Graphing equations: Point-plotting, intercepts, and symmetry**	
	Point-plotting	List a table with several coordinates that are solutions to the equation; plot and connect.
	Intercepts	x-intercept: let $y = 0$ y-intercept: let $x = 0$
	Symmetry	The graph of an equation can be symmetric about the x-axis, y-axis, or origin.
	Using intercepts and symmetry as graphing aids	If (a, b) is on the graph of the equation, then $(-a, b)$ is on the graph if symmetric about the y-axis, $(a, -b)$ is on the graph if symmetric about the x-axis, and $(-a, -b)$ is on the graph if symmetric about the origin.
A.4	**Functions**	
	Relations and functions	
	Functions defined by equations	A vertical line can intersect a function in at most one point.
	Function notation	Placeholder notation: $f(x) = x^2 + 3$ $f(\square) = (\square)^2 + 3$ Difference quotient: $\dfrac{f(x + h) - f(x)}{h}$ $h \neq 0$
	Domain of a function	Are there any restrictions on x?

SECTION	CONCEPT	KEY IDEAS/FORMULAS
A.5	Graphs of functions; piecewise-defined functions; increasing and decreasing functions; average rate of change	
	Graphs of functions	**Common functions** $f(x) = mx + b$, $f(x) = x$, $f(x) = x^2$, $f(x) = x^3$, $f(x) = \sqrt{x}$, $f(x) = \sqrt[3]{x}$, $f(x) = \lvert x \rvert$, $f(x) = \dfrac{1}{x}$ **Even and odd functions** Even (symmetry about y-axis): $f(-x) = f(x)$ Odd (symmetry about origin): $f(-x) = -f(x)$
	Average rate of change	$\dfrac{f(x_2) - f(x_1)}{x_2 - x_1} \quad x_1 \neq x_2$
	Piecewise-defined functions	Points of discontinuity
A.6	Graphing techniques: Transformations	Shift the graph of $f(x)$.
	Horizontal and vertical shifts	$f(x + c)$ c units to the left $f(x - c)$ c units to the right $f(x) + c$ c units upward $\left.\vphantom{\begin{matrix}a\\a\\a\\a\end{matrix}}\right\} c > 0$ $f(x) - c$ c units downward
	Reflection about the axes	$-f(x)$ Reflection about the x-axis $f(-x)$ Reflection about the y-axis
	Stretching and compressing	$cf(x)$ if $c > 1$; Stretch vertically $cf(x)$ if $0 < c < 1$; Compress vertically $f(cx)$ if $c > 1$; Compress horizontally $f(cx)$ if $0 < c < 1$; Stretch horizontally
A.7	Operations on functions and composition of functions	
	Adding, subtracting, multiplying, and dividing functions	$(f + g)(x) = f(x) + g(x)$ $(f - g)(x) = f(x) - g(x)$ $(f \cdot g)(x) = f(x) \cdot g(x)$ The domain of the resulting function is the intersection of the individual domains. $\left(\dfrac{f}{g}\right)(x) = \dfrac{f(x)}{g(x)}$, $g(x) \neq 0$ The domain of the quotient is the intersection of the domains of f and g, and any points when $g(x) = 0$ must be eliminated.
	Composition of functions	$(f \circ g)(x) = f(g(x))$ The domain of the composite function is a subset of the domain of $g(x)$. Values for x must be eliminated if their corresponding values $g(x)$ are not in the domain of f.

SECTION	CONCEPT	KEY IDEAS/FORMULAS
A.8	**One-to-one functions and inverse functions**	
	Determine whether a function is one-to-one	▪ No two x-values map to the same y-value. If $f(x_1) = f(x_2)$, then $x_1 = x_2$. ▪ A horizontal line may intersect a one-to-one function in at most one point.
	Inverse functions	▪ Only one-to-one functions have inverses. ▪ $f^{-1}(f(x)) = x$ and $f(f^{-1}(x)) = x$. ▪ Domain of f = range of f^{-1}. Range of f = domain of f^{-1}.
	Graphical interpretation of inverse functions	▪ The graph of a function and its inverse are symmetric about the line $y = x$. ▪ If the point (a, b) lies on the graph of a function, then the point (b, a) lies on the graph of its inverse.
	Finding the inverse function	1. Let $y = f(x)$. 2. Interchange x and y. 3. Solve for y. 4. Let $y = f^{-1}(x)$.

[APPENDIX A REVIEW EXERCISES]

A.1 Factoring Polynomials

Factor out the common factor.

1. $14x^2y^2 - 10xy^3$

2. $30x^4 - 20x^3 + 10x^2$

Factor the trinomial into a product of two binomials.

3. $2x^2 + 9x - 5$

4. $6x^2 - 19x - 7$

5. $16x^2 - 25$

6. $9x^2 - 30x + 25$

Factor the sum or difference of two cubes.

7. $x^3 + 125$

8. $1 - 8x^3$

Factor into a product of three polynomials.

9. $2x^3 + 4x^2 - 30x$

10. $6x^3 - 5x^2 + x$

Factor into a product of two binomials by grouping.

11. $x^3 + x^2 - 2x - 2$

12. $2x^3 - x^2 + 6x - 3$

A.2 Basic Tools: Cartesian Plane, Distance, and Midpoint

Plot each point and indicate which quadrant the point lies in.

13. $(-4, 2)$

14. $(4, 7)$

15. $(-1, -6)$

16. $(2, -1)$

Calculate the distance between the two points.

17. $(-2, 0)$ and $(4, 3)$

18. $(1, 4)$ and $(4, 4)$

19. $(-4, -6)$ and $(2, 7)$

20. $\left(\frac{1}{4}, \frac{1}{12}\right)$ and $\left(\frac{1}{3}, -\frac{7}{3}\right)$

Calculate the midpoint of the segment joining the two points.

21. $(2, 4)$ and $(3, 8)$

22. $(-2, 6)$ and $(5, 7)$

23. $(2.3, 3.4)$ and $(5.4, 7.2)$

24. $(-a, 2)$ and $(a, 4)$

Applications

25. **Sports.** A quarterback drops back to pass. At the point $(-5, -20)$, he throws the ball to his wide receiver located at $(10, 30)$. Find the distance the ball has traveled. Assume the width of the football field is $[-15, 15]$ and the length is $[-50, 50]$.

26. **Sports.** Suppose that in the above exercise a defender was midway between the quarterback and the receiver. At what point was the defender located when the ball was thrown over his head?

A.3 Graphing Equations: Point-Plotting, Intercepts, and Symmetry

Find the x-intercept(s) and y-intercept(s) if any.

27. $x^2 + 4y^2 = 4$

28. $y = x^2 - x + 2$

29. $y = \sqrt{x^2 - 9}$

30. $y = \dfrac{x^2 - x - 12}{x - 12}$

Use algebraic tests to determine symmetry with respect to the x-axis, y-axis, or origin.

31. $x^2 + y^3 = 4$

32. $y = x^2 - 2$

33. $xy = 4$

34. $y^2 = 5 + x$

Use symmetry as a graphing aid and point-plot the given equations.

35. $y = x^2 - 3$

36. $y = |x| - 4$

37. $y = \sqrt[3]{x}$

38. $x = y^2 - 2$

39. $y = x\sqrt{9 - x^2}$

40. $x^2 + y^2 = 36$

Applications

41. **Sports.** A track around a high school football field is in the shape of the graph $8x^2 + y^2 = 8$. Graph using symmetry and by plotting points.

42. **Transportation.** A "bypass" around a town follows the graph $y = x^3 + 2$, where the origin is the center of town. Graph the equation.

A.4 Functions

Determine whether each relation is a function.

43.

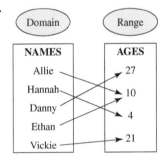

44. $\{(1, 2), (3, 4), (2, 4), (3, 7)\}$

45. $\{(-2, 3), (1, -3), (0, 4), (2, 6)\}$

46. $\{(4, 7), (2, 6), (3, 8), (1, 7)\}$

47. $x^2 + y^2 = 36$

48. $x = 4$

49. $y = |x + 2|$

50. $y = \sqrt{x}$

51.

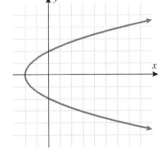

52.

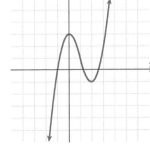

Use the graphs of the functions to find:

53.

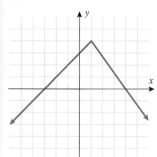

54.

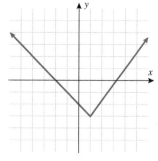

a. $f(-1)$ **b.** $f(1)$

c. x, where $f(x) = 0$

a. $f(-4)$ **b.** $f(0)$

c. x, where $f(x) = 0$

55.

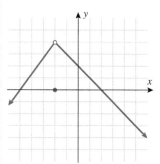

56.

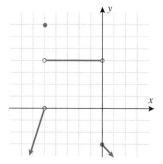

a. $f(-2)$ **b.** $f(4)$

c. x, where $f(x) = 0$

a. $f(-5)$ **b.** $f(0)$

c. x, where $f(x) = 0$

Evaluate the given quantities using the following three functions.

$$f(x) = 4x - 7 \qquad F(t) = t^2 + 4t - 3 \qquad g(x) = |x^2 + 2x + 4|$$

57. $f(3)$ **58.** $F(4)$ **59.** $f(-7) \cdot g(3)$ **60.** $\dfrac{F(0)}{g(0)}$

61. $\dfrac{f(2) - F(2)}{g(0)}$ **62.** $f(3 + h)$

63. $\dfrac{f(3 + h) - f(3)}{h}$ **64.** $\dfrac{F(t + h) - F(t)}{h}$

Find the domain of the given function. Express the domain in interval notation.

65. $f(x) = -3x - 4$ **66.** $g(x) = x^2 - 2x + 6$

67. $h(x) = \dfrac{1}{x + 4}$ **68.** $F(x) = \dfrac{7}{x^2 + 3}$

69. $G(x) = \sqrt{x - 4}$ **70.** $H(x) = \dfrac{1}{\sqrt{2x - 6}}$

Challenge

71. If $f(x) = \dfrac{D}{x^2 - 16}$, $f(4)$ and $f(-4)$ are undefined, and

$f(5) = 2$, find D.

72. Construct a function that is undefined at $x = -3$ and $x = 2$ such that the point $(0, -4)$ lies on the graph of the function.

A.5 Graphs of Functions; Piecewise-Defined Functions; Increasing and Decreasing Functions; Average Rate of Change

Determine whether the function is even, odd, or neither.

73. $f(x) = 2x - 7$ **74.** $g(x) = 7x^5 + 4x^3 - 2x$

75. $h(x) = x^3 - 7x$ **76.** $f(x) = x^4 + 3x^2$

77. $f(x) = x^{1/4} + x$ **78.** $f(x) = \sqrt{x + 4}$

79. $f(x) = \dfrac{1}{x^3} + 3x$ **80.** $f(x) = \dfrac{1}{x^2} + 3x^4 + |x|$

For Exercises 81 and 82, use the graph of the functions to find:

a. Domain

b. Range

c. Intervals on which the function is increasing, decreasing, or constant.

81.

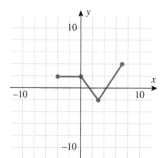

82.

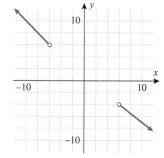

83. Find the average rate of change of $f(x) = 4 - x^2$ from $x = 0$ to $x = 2$.

84. Find the average rate of change of $f(x) = |2x - 1|$ from $x = 1$ to $x = 5$.

Graph the piecewise-defined function. State the domain and range in interval notation.

85. $F(x) = \begin{cases} x^2 & x < 0 \\ 2 & x \geq 0 \end{cases}$

86. $f(x) = \begin{cases} -2x - 3 & x \leq 0 \\ 4 & 0 < x \leq 1 \\ x^2 + 4 & x > 1 \end{cases}$

87. $f(x) = \begin{cases} x^2 & x \leq 0 \\ -\sqrt{x} & 0 < x \leq 1 \\ |x + 2| & x > 1 \end{cases}$

88. $F(x) = \begin{cases} x^2 & x < 0 \\ x^3 & 0 < x < 1 \\ -|x| - 1 & x \geq 1 \end{cases}$

Applications

89. Tutoring Costs. A tutoring company charges $25 for the first hour of tutoring and $10.50 for every 30-minute period after that. Find the cost function $C(x)$ as a function of the length of the tutoring session. Let x = number of 30-minute periods.

90. Salary. An employee who makes $30 per hour also earns time and a half for overtime (any hours worked above the normal 40-hour work week). Write a function $E(x)$ that describes her weekly earnings as a function of the number of hours worked x.

A.6 Graphing Techniques: Transformations

Graph the following functions using graphing aids.

91. $y = -(x - 2)^2 + 4$ **92.** $y = |-x + 5| - 7$

93. $y = \sqrt[3]{x - 3} + 2$ **94.** $y = \dfrac{1}{x - 2} - 4$

95. $y = -\frac{1}{2}x^3$ **96.** $y = 2x^2 + 3$

Use the given graph to graph the following:

97. **98.**

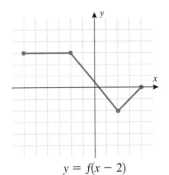

$y = f(x - 2)$

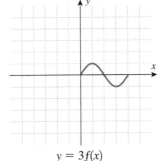

$y = 3f(x)$

99. **100.**

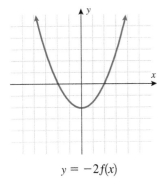

$y = -2f(x)$

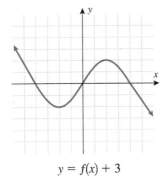

$y = f(x) + 3$

Write the function whose graph is the graph of $y = \sqrt{x}$, but is transformed accordingly, and state the domain of the resulting function.

101. Shifted to the left three units

102. Shifted down four units

103. Shifted to the right two units and up three units

104. Reflected about the y-axis

105. Stretched by a factor of 5 and shifted down six units

106. Compressed by a factor of 2 and shifted up three units

Transform the function into the form $f(x) = c(x - h)^2 + k$ by completing the square and graph the resulting function using transformations.

107. $y = x^2 + 4x - 8$ **108.** $y = 2x^2 + 6x - 5$

A.7 Operations on Functions and Composition of Functions

Given the functions g and h, find $g + h$, $g - h$, $g \cdot h$, and $\dfrac{g}{h}$, and state the domain.

109. $g(x) = -3x - 4$ **110.** $g(x) = 2x + 3$
   $h(x) = x - 3$ $h(x) = x^2 + 6$

111. $g(x) = \dfrac{1}{x^2}$ **112.** $g(x) = \dfrac{x + 3}{2x - 4}$
   $h(x) = \sqrt{x}$ $h(x) = \dfrac{3x - 1}{x - 2}$

113. $g(x) = \sqrt{x - 4}$ **114.** $g(x) = x^2 - 4$
   $h(x) = \sqrt{2x + 1}$ $h(x) = x + 2$

For the given functions f and g, find the composite functions $f \circ g$ and $g \circ f$, and state the domains.

115. $f(x) = 3x - 4$ **116.** $f(x) = x^3 + 2x - 1$
   $g(x) = 2x + 1$ $g(x) = x + 3$

117. $f(x) = \dfrac{2}{x + 3}$ **118.** $f(x) = \sqrt{2x^2 - 5}$
   $g(x) = \dfrac{1}{4 - x}$ $g(x) = \sqrt{x + 6}$

119. $f(x) = \sqrt{x - 5}$ **120.** $f(x) = \dfrac{1}{\sqrt{x}}$
   $g(x) = x^2 - 4$
       $g(x) = \dfrac{1}{x^2 - 4}$

Evaluate $f(g(3))$ and $g(f(-1))$, if possible.

121. $f(x) = 4x^2 - 3x + 2$ **122.** $f(x) = \sqrt{4 - x}$
   $g(x) = 6x - 3$ $g(x) = x^2 + 5$

123. $f(x) = \dfrac{x}{|2x - 3|}$ **124.** $f(x) = \dfrac{1}{x - 1}$
   $g(x) = |5x + 2|$ $g(x) = x^2 - 1$

125. $f(x) = x^2 - x + 10$ **126.** $f(x) = \dfrac{4}{x^2 - 2}$
   $g(x) = \sqrt[3]{x - 4}$
       $g(x) = \dfrac{1}{x^2 - 9}$

Write the function as a composite $f(g(x))$ of two functions f and g.

127. $h(x) = 3(x - 2)^2 + 4(x - 2) + 7$

128. $h(x) = \dfrac{\sqrt[3]{x}}{1 - \sqrt[3]{x}}$

129. $h(x) = \dfrac{1}{\sqrt{x^2 + 7}}$

130. $h(x) = \sqrt{|3x + 4|}$

Applications

131. Rain. A rain drop hitting a lake makes a circular ripple. If the radius, in inches, grows as a function of time, in minutes, $r(t) = 25\sqrt{t + 2}$, find the area of the ripple as a function of time.

132. Geometry. Let the area of a rectangle be given by $42 = l \cdot w$, and let the perimeter be $36 = 2 \cdot l + 2 \cdot w$. Express the perimeter in terms of w.

A.8 One-to-One Functions and Inverse Functions

Determine whether the given function is a one-to-one function.

133.

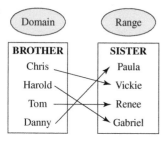

134.

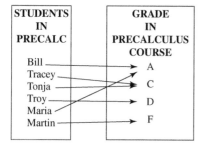

135. $\{(2, 3), (-1, 2), (3, 3), (-3, -4), (-2, 1)\}$

136. $\{(-3, 9), (5, 25), (2, 4), (3, 9)\}$

137. $\{(-2, 0), (4, 5), (3, 7)\}$

138. $\{(-8, -6), (-4, 2), (0, 3), (2, -8), (7, 4)\}$

139. $y = \sqrt{x}$ **140.** $y = x^2$

141. $f(x) = x^3$ **142.** $f(x) = \dfrac{1}{x^2}$

Verify that the function $f^{-1}(x)$ is the inverse of $f(x)$ by showing that $f(f^{-1}(x)) = x$. Graph $f(x)$ and $f^{-1}(x)$ on the same graph and show the symmetry about the line $y = x$.

143. $f(x) = 3x + 4;\ f^{-1}(x) = \dfrac{x - 4}{3}$

144. $f(x) = \dfrac{1}{4x - 7};\ f^{-1}(x) = \dfrac{1 + 7x}{4x}$

145. $f(x) = \sqrt{x + 4};\ f^{-1}(x) = x^2 - 4\quad x \geq 0$

146. $f(x) = \dfrac{x + 2}{x - 7};\ f^{-1}(x) = \dfrac{7x + 2}{x - 1}$

The function f is one-to-one. Find its inverse and check your answer. State the domain and range of both f and f^{-1}.

147. $f(x) = 2x + 1$ **148.** $f(x) = x^5 + 2$

149. $f(x) = \sqrt{x + 4}$ **150.** $f(x) = (x + 4)^2 + 3\quad x \geq -4$

151. $f(x) = \dfrac{x + 6}{x + 3}$ **152.** $f(x) = 2\sqrt[3]{x - 5} - 8$

Applications

153. Salary. A pharmaceutical salesperson makes $22,000 base salary a year plus 8% of the total products sold. Write a function $S(x)$ that represents her yearly salary as a function of the total dollars worth of products sold x. Find $S^{-1}(x)$. What does this inverse function tell you?

154. Volume. Express the volume V of a rectangular box that has a square base of length s and is 3 feet high as a function of the square length. Find V^{-1}. If a certain volume is desired, what does the inverse tell you?

[APPENDIX A PRACTICE TEST]

Factor.

1. $x^2 - 16$

2. $3x^2 + 15x + 18$

3. $4x^2 + 12xy + 9y^2$

4. $x^4 - 2x^2 + 1$

5. $2x^2 - x - 1$

6. $6y^2 - y - 1$

7. $2t^3 - t^2 - 3t$

8. $2x^3 - 5x^2 - 3x$

9. $x^2 + 3yx - 4yx - 12y^2$

10. $x^4 + 5x^2 - 3x^2 - 15$

11. $81 + 3x^3$

12. $27x - x^4$

13. Find the distance between the points $(-7, -3)$ and $(2, -2)$.

14. Find the midpoint between $(-3, 5)$ and $(5, -1)$.

15. Determine the length and the midpoint of a segment that joins the points $(-2, 4)$ and $(3, 6)$.

16. **Research Triangle.** The Research Triangle in North Carolina was established as a collaborative research center among Duke University (Durham), North Carolina State University (Raleigh), and the University of North Carolina (Chapel Hill).

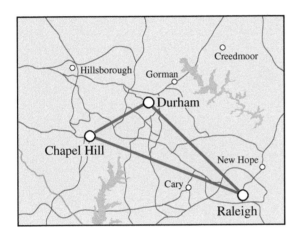

Durham is 10 miles north and 8 miles east of Chapel Hill, and Raleigh is 28 miles east and 15 miles south of Chapel Hill. What is the perimeter of the research triangle? Round your answer to the nearest mile.

17. Determine the two values for y so that the point $(3, y)$ is five units away from the point $(6, 5)$.

18. If the point $(3, -4)$ is on a graph that is symmetric with respect to the y-axis, what point must also be on the graph?

19. Determine whether the graph of the equation $x - y^2 = 5$ has any symmetry (x-axis, y-axis, and origin).

20. Find the x-intercept(s) and the y-intercept(s), if any: $4x^2 - 9y^2 = 36$.

Graph the following equations.

21. $2x^2 + y^2 = 8$

22. $y = \dfrac{4}{x^2 + 1}$

Assuming that x represents the independent variable and y represents the dependent variable, classify the relationships as:

a. not a function

b. a function, but not one-to-one

c. a one-to-one function

23. $f(x) = |2x + 3|$

24. $x = y^2 + 2$

25. $y = \sqrt[3]{x + 1}$

Use $f(x) = \sqrt{x - 2}$ and $g(x) = x^2 + 11$, and determine the desired quantity or expression. In the case of an expression, state the domain.

26. $f(11) - 2g(-1)$

27. $\left(\dfrac{f}{g}\right)(x)$

28. $\left(\dfrac{g}{f}\right)(x)$

29. $g(f(x))$

30. $(f + g)(6)$

31. $f(g(\sqrt{7}))$

Determine whether the function is odd, even, or neither.

32. $f(x) = |x| - x^2$

33. $f(x) = 9x^3 + 5x - 3$

34. $f(x) = \dfrac{2}{x}$

Graph the functions. State the domain and range of each function.

35. $f(x) = -\sqrt{x - 3} + 2$

36. $f(x) = -2(x - 1)^2$

37. $f(x) = \begin{cases} -x & x < -1 \\ 1 & -1 < x < 2 \\ x^2 & x \geq 2 \end{cases}$

Use the graphs of the function to find:

38.

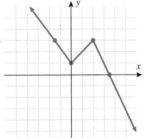

$y = f(x)$

a. $f(3)$ b. $f(0)$ c. $f(-4)$

d. x, where $f(x) = 3$ e. x, where $f(x) = 0$

39.

$y = g(x)$

a. $g(3)$ b. $g(0)$ c. $g(-4)$

d. x, where $g(x) = 0$

40.

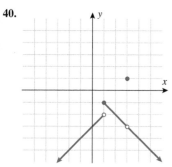

$y = f(x)$

a. $p(0)$ b. x, where $p(x) = 0$

c. $p(1)$ d. $p(3)$

Find $\dfrac{f(x + h) - f(x)}{h}$ for:

41. $f(x) = 3x^2 - 4x + 1$ **42.** $f(x) = 5 - 7x$

Given the function f, find the inverse if it exists. State the domain and range of both f and f^{-1}.

43. $f(x) = \sqrt{x - 5}$ **44.** $f(x) = x^2 + 5$

45. $f(x) = \dfrac{2x + 1}{5 - x}$ **46.** $f(x) = \begin{cases} -x & x \le 0 \\ -x^2 & x > 0 \end{cases}$

47. What domain restriction can be made so that $f(x) = x^2$ has an inverse?

48. If the point $(-2, 5)$ lies on the graph of a function, what point lies on the graph of its inverse function?

49. **Discount.** Suppose a suit has been marked down 40% off the original price. An advertisement in the newspaper has an "additional 30% off the sale price" coupon. Write a function that determines the "checkout" price of the suit.

50. **Temperature.** Degrees Fahrenheit (°F), degrees Celsius (°C), and kelvins (K) are related by the two equations: $F = \frac{9}{5}C + 32$ and $K = C - 273.15$. Write a function whose input is kelvins and output is degrees Fahrenheit.

51. Determine whether the triangle with the given vertices is a right triangle, isosceles triangle, neither, or both.

$$(-8.4, 16.8), (0, 37.8), (12.6, 8.4)$$

52. Graph the given equation using a graphing utility and state whether there is any symmetry.

$$0.25y^2 + 0.04x^2 = 1$$

53. Use a graphing utility to graph the function and determine whether it is one-to-one.

$$y = x^3 - 12x^2 + 48x - 65$$

APPENDIX

[B] Conic Sections

B.1 CONIC BASICS	B.2 THE PARABOLA	B.3 THE ELLIPSE	B.4 THE HYPERBOLA	B.5 ROTATION OF AXES	B.6 POLAR EQUATIONS OF CONICS
• Classifying Conic Sections: Parabola, Ellipse, and Hyperbola	• Parabola with Its Vertex at the Origin • Parabola with Vertex (h, k)	• Ellipse with Its Center at the Origin • Ellipse with Center (h, k)	• Hyperbola, with Its Center at the Origin • Hyperbola with Center (h, k)	• Transforming Second-Degree Equations Using Rotation of Axes • Determine the Angle of Rotation Necessary to Transform a General Second-Degree Equation into an Equation of a Conic • Graphing a Rotated Conic	• Eccentricity • Equations of Conics in Polar Coordinates • Graphing a Conic Given in Polar Form

LEARNING OBJECTIVES

- Visualize conics in terms of a plane intersecting a right double cone.
- Develop the equation for a parabola and graph parabolas in rectangular coordinates.

- Develop the equation for an ellipse and graph ellipses in rectangular coordinates.
- Develop the equation for a hyperbola and graph hyperbolas in rectangular coordinates.

- Graph general second-degree polynomial functions by rotation of axes.
- Graph parabolas, ellipses, and hyperbolas in polar coordinates.

B.1 CONIC BASICS

SKILLS OBJECTIVE	CONCEPTUAL OBJECTIVE
▪ Classify a conic as an ellipse, a parabola, or a hyperbola.	▪ Understand that conics are graphs that correspond to a second-degree equation in two variables.

B.1.1 Classifying Conic Sections: Parabola, Ellipse, and Hyperbola

Names of Conics

The word *conic* is derived from the word *cone*. Let's start with a (right circular) **double cone** (see the figure on the left).

    Conic sections are curves that result from the intersection of a plane and a double cone. The four conic sections are a **circle**, an **ellipse**, a **parabola**, and a **hyperbola**. **Conics** is an abbreviation for conic sections.

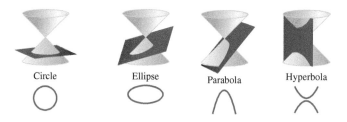

Circle   Ellipse   Parabola   Hyperbola

    There are two ways in which we usually describe conics: graphically and algebraically. An entire section will be devoted to each of the three conics, but in the present section we will summarize the definitions of a parabola, ellipse, and hyperbola and show how to identify the equations of these conics.

> **STUDY TIP**
>
> A circle is a special type of ellipse. All circles are ellipses, but not all ellipses are circles.

Definitions

You already know that a circle consists of all points equidistant (at a distance equal to the radius) from a point (the center). Ellipses, parabolas, and hyperbolas have similar definitions in that they all have a constant distance (or a sum or difference of distances) to some reference point(s).

    A **parabola** is the set of all points that are **equidistant from both a line and a point**. An **ellipse** is the set of all points, the **sum of whose distances to two fixed points is constant**. A **hyperbola** is the set of all points, the **difference of whose distances to two fixed points is a constant**.

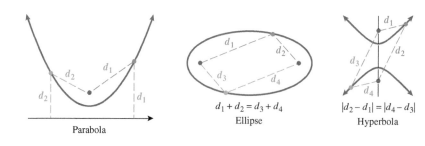

Parabola $d_1 + d_2 = d_3 + d_4$ $|d_2 - d_1| = |d_4 - d_3|$
              Ellipse   Hyperbola

The **general form of a second-degree equation in two variables**, x and y, is given by

$$Ax^2 + Bxy + Cy^2 + Dx + Ey + F = 0$$

If we let $A = 1, B = 0, C = 1, D = 0, E = 0$, and $F = -r^2$, this general equation reduces to the equation of a circle centered at the origin: $x^2 + y^2 = r^2$. In fact, all three conics (parabolas, ellipses, and hyperbolas) are special cases of the general second-degree equation.

Recall from algebra that when solving quadratic equations, the discriminant, $b^2 - 4ac$, determines what types of solutions result from solving a second-degree equation in one variable. If the discriminant is positive, the solutions are two distinct real roots. If the discriminant is zero, the solution is a real repeated root. If the discriminant is negative, the solutions are two complex conjugate roots.

The concept of discriminant has been generalized to conic sections. The discriminant determines the *shape* of the conic section.

CONIC	DISCRIMINANT
Ellipse	$B^2 - 4AC < 0$
Parabola	$B^2 - 4AC = 0$
Hyperbola	$B^2 - 4AC > 0$

STUDY TIP

All circles are ellipses since $B^2 - 4AC < 0$.

Using the discriminant to identify the shape of the conic will not work for degenerate cases (when the polynomial factors). For example,

$$2x^2 - xy - y^2 = 0$$

At first glance, one may think this is a hyperbola because $B^2 - 4AC > 0$, but this is a degenerative case.

$$(2x + y)(x - y) = 0$$

$$2x + y = 0 \quad \text{or} \quad x - y = 0$$

$$y = -2x \quad \text{or} \quad y = x$$

The graph is two intersecting lines.

We now identify conics from the general form of a second-degree equation in two variables.

TRUE OR FALSE The graph of a third-degree equation in two variables is also a conic section.

▼ . . . .

ANSWER False

EXAMPLE 1 **Determining the Type of Conic**

Determine what type of conic corresponds to each of the following equations:

a. $\dfrac{x^2}{a^2} + \dfrac{y^2}{b^2} = 1$ **b.** $y = x^2$ **c.** $\dfrac{x^2}{a^2} - \dfrac{y^2}{b^2} = 1$

Solution:

Write the general form of the second-degree equation:

$$Ax^2 + Bxy + Cy^2 + Dx + Ey + F = 0$$

a. Identify A, B, C, D, E, and F. $A = \dfrac{1}{a^2}, B = 0, C = \dfrac{1}{b^2}, D = 0, E = 0, F = -1$

Calculate the discriminant. $B^2 - 4AC = -\dfrac{4}{a^2b^2} < 0$

Since the discriminant is negative, the equation $\dfrac{x^2}{a^2} + \dfrac{y^2}{b^2} = 1$ is that of an **ellipse**.

Notice that if $a = b = r$, then this equation of an ellipse reduces to the general equation of a circle, $x^2 + y^2 = r^2$, centered at the origin, with radius r.

b. Identify A, B, C, D, E, and F. $A = 1, B = 0, C = 0, D = 0, E = -1, F = 0$

Calculate the discriminant. $B^2 - 4AC = 0$

Since the discriminant is zero, the equation $y = x^2$ is a **parabola**.

c. Identify A, B, C, D, E, and F. $A = \dfrac{1}{a^2}, B = 0, C = -\dfrac{1}{b^2}, D = 0, E = 0, F = -1$

Calculate the discriminant. $B^2 - 4AC = \dfrac{4}{a^2b^2} > 0$

Since the discriminant is positive, the equation $\dfrac{x^2}{a^2} - \dfrac{y^2}{b^2} = 1$ is a **hyperbola**.

▼ . . . .

ANSWER

a. ellipse

b. hyperbola

c. parabola

▼

YOUR TURN Determine what type of conic corresponds to each of the following equations:

a. $2x^2 + y^2 = 4$ **b.** $2x^2 = y^2 + 4$ **c.** $2y^2 = x$

In the next three sections, we will discuss the standard forms of equations and the graphs of parabolas, ellipses, and hyperbolas.

▶[SECTION B.1] SUMMARY

In this section, we defined the three conic sections and determined their general equations with respect to the general form of a second-degree equation in two variables:

$$Ax^2 + Bxy + Cy^2 + Dx + Ey + F = 0$$

The following table summarizes the three conics: ellipse, parabola, and hyperbola.

CONIC	GEOMETRIC DEFINITION: THE SET OF ALL POINTS	DISCRIMINANT
Ellipse	The sum of whose distances to two fixed points is constant	Negative: $B^2 - 4AC < 0$
Parabola	Equidistant to both a line and a point	Zero: $B^2 - 4AC = 0$
Hyperbola	The difference of whose distances to two fixed points is a constant	Positive: $B^2 - 4AC > 0$

It is important to note that a circle is a special type of ellipse.

[SECTION B.1] EXERCISES

• SKILLS

In Exercises 1–12, identify the conic section as a parabola, ellipse, circle, or hyperbola.

1. $x^2 + xy - y^2 + 2x = -3$
2. $x^2 + xy + y^2 + 2x = -3$
3. $2x^2 + 2y^2 = 10$
4. $x^2 - 4x + y^2 + 2y = 4$
5. $2x^2 - y^2 = 4$
6. $2y^2 - x^2 = 16$
7. $5x^2 + 20y^2 = 25$
8. $4x^2 + 8y^2 = 30$
9. $x^2 - y = 1$
10. $y^2 - x = 2$
11. $x^2 + y^2 = 10$
12. $x^2 + y^2 = 100$

B.2 THE PARABOLA

SKILLS OBJECTIVES

- Find the equation of a parabola whose vertex is at the origin.
- Find the equation of a parabola whose vertex is at the point (h, k).
- Solve applied problems that involve parabolas.

CONCEPTUAL OBJECTIVES

- Understand that a parabola is the set of all points that are equidistant from a fixed line (directrix) and a fixed point not on that line (focus).
- Employ completing the square to transform an equation into the standard form of a parabola.
- Understand that it is the focus that is the key to applications.

B.2.1 Parabola with Its Vertex at the Origin

Definition of a Parabola

Recall that the graphs of quadratic functions such as

$$f(x) = a(x - h)^2 + k \quad \text{or} \quad y = ax^2 + bx + c$$

are *parabolas* that open either upward or downward. We now expand our discussion to *parabolas* that open to the **right** or **left**. We did not discuss these types of parabolas before because they are not functions (they fail the vertical line test).

B.2.1 SKILL

Find the equation of a parabola whose vertex is at the origin.

B.2.1 CONCEPTUAL

Understand that a parabola is the set of all points that are equidistant from a fixed line (directrix) and a fixed point not on that line (focus).

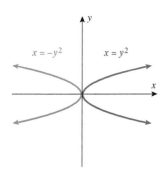

$x = -y^2$ $x = y^2$

> **DEFINITION** **Parabola**
>
> A **parabola** is the set of all points in a plane that are equidistant from a fixed line, the **directrix**, and a fixed point not on the line, the **focus**. The line through the focus and perpendicular to the directrix is the **axis of symmetry**. The **vertex** of the parabola is located at the midpoint between the directrix and the focus along the axis of symmetry.

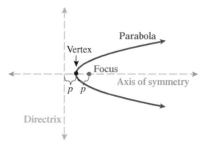

> Here, p is the distance along the axis of symmetry from the directrix to the vertex and from the vertex to the focus.

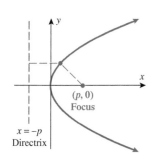

$(p, 0)$
Focus

$x = -p$
Directrix

Let's consider a parabola with the vertex at the origin and the focus on the positive x-axis. Let the distance from the vertex to the focus be p. Therefore, the focus is located at the point $(p, 0)$. Since the distance from the vertex to the focus is p, the distance from the vertex to the directrix must also be p. Since the axis of symmetry is the positive axis, the directrix must be perpendicular to the positive axis. Therefore, the directrix is given by $x = -p$. Any point (x, y) must have the same distance to the focus $(p, 0)$ as does the directrix $(-p, y)$.

Derivation of the Equation of a Parabola

WORDS	MATH
Calculate the distance from (x, y) to $(p, 0)$ with the distance formula.	$\sqrt{(x - p)^2 + y^2}$
Calculate the distance from (x, y) to $(-p, y)$ with the distance formula.	$\sqrt{[x - (-p)]^2 + 0^2}$
Set the two distances equal to one another.	$\sqrt{(x - p)^2 + y^2} = \sqrt{(x + p)^2}$
Recall that $\sqrt{x^2} = |x|$.	$\sqrt{(x - p)^2 + y^2} = |x + p|$
Square both sides of the equation.	$(x - p)^2 + y^2 = (x + p)^2$
Square the binomials inside the parentheses.	$x^2 - 2px + p^2 + y^2 = x^2 + 2px + p^2$
Simplify.	$\boxed{y^2 = 4px}$

The equation $y^2 = 4px$ represents a parabola opening right with the vertex at the origin. The following box summarizes parabolas that have a vertex at the origin and a focus along either the x-axis or the y-axis:

EQUATION OF A PARABOLA WITH A VERTEX AT THE ORIGIN

The standard (conic) form of the equation of a **parabola** with a vertex at the origin is given by

EQUATION	$y^2 = 4px$	$x^2 = 4py$
VERTEX	$(0, 0)$	$(0, 0)$
FOCUS	$(p, 0)$	$(0, p)$
DIRECTRIX	$x = -p$	$y = -p$
AXIS OF SYMMETRY	x-axis	y-axis
$p > 0$	opens to the right	opens upward
$p < 0$	opens to the left	opens downward
GRAPH ($p > 0$)		

▶ **EXAMPLE 1 Finding the Focus and Directrix of a Parabola Whose Vertex Is Located at the Origin**

Find the focus and directrix of a parabola whose equation is $y^2 = 8x$.

Solution:

Compare this parabola with the general equation of a parabola.

$$y^2 = 4px$$
$$y^2 = 8x$$

Let $y^2 = 8x$.

$$4px = 8x$$

Solve for p.

$$4p = 8$$
$$p = 2$$

The focus of a parabola of the form $y^2 = 4px$ is $(p, 0)$.

The directrix of a parabola of the form $y^2 = 4px$ is $x = -p$.

The focus is $\boxed{(2, 0)}$ and the directrix is $\boxed{x = -2}$.

▼

YOUR TURN Find the focus and directrix of a parabola whose equation is $y^2 = 16x$.

▼ ANSWER

The focus is $(4, 0)$ and the directrix is $x = -4$.

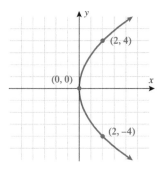

Graphing a Parabola with a Vertex at the Origin

When a seamstress starts with a pattern for a custom-made suit, the pattern is used as a guide. The pattern is not sewn into the suit, but rather removed once it is used to determine the exact shape and size of the fabric to be sewn together. The focus and directrix of a parabola are similar to the pattern used by a seamstress. Although the focus and directrix define a parabola, they do not appear on the graph of a parabola.

We can draw an approximate sketch of a parabola whose vertex is at the origin with three pieces of information. We know that the vertex is located at $(0, 0)$. Additional information that we seek is the direction in which the parabola opens and approximately how wide or narrow to draw the parabolic curve. The direction toward which the parabola opens is found from the equation. An equation of the form $y^2 = 4px$ opens either left or right. It opens right if $p > 0$ and opens left if $p < 0$. An equation of the form $x^2 = 4py$ opens either up or down. It opens up if $p > 0$ and opens down if $p < 0$. How narrow or wide should we draw the parabolic curve? If we select a few points that satisfy the equation, we can use those as graphing aids.

In Example 1, we found that the focus of that parabola is located at $(2, 0)$. If we select the x-coordinate of the focus $x = 2$ and substitute that value into the equation of the parabola $y^2 = 8x$, we find the corresponding y values to be $y = -4$ and $y = 4$. If we plot the three points $(0, 0)$, $(2, -4)$, and $(2, 4)$ and then connect the points with a parabolic curve, we get the graph on the right.

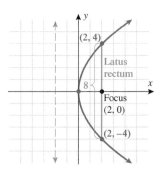

The line segment that passes through the focus, $(2, 0)$, is parallel to the directrix $x = -2$, and whose endpoints are on the parabola is called the **latus rectum**. The latus rectum in this case has length 8. The latus rectum is a graphing aid that helps us find the width of a parabola.

In general, the points on a parabola of the form $y^2 = 4px$ that lie above and below the focus, $(p, 0)$, satisfy the equation $y^2 = 4p^2$ and are located at $(p, -2p)$ and $(p, 2p)$. The latus rectum will have length $4|p|$. Similarly, a parabola of the form $x^2 = 4py$ will have a horizontal latus rectum of length $4|p|$. We will use the latus rectum as a graphing aid to determine the parabola's width.

EXAMPLE 2 **Graphing a Parabola Whose Vertex Is at the Origin, Using the Focus, Directrix, and Latus Rectum as Graphing Aids**

Determine the focus, directrix, and length of the latus rectum of the parabola $x^2 = -12y$. Employ these to assist in graphing the parabola.

Solution:

Compare this parabola with the general equation of a parabola.

$$x^2 = 4py \quad x^2 = -12y$$

Solve for p.

$$4p = -12$$
$$p = -3$$

A parabola of the form $x^2 = 4py$ has focus $(0, p)$, directrix $y = -p$, and a latus rectum of length $4|p|$. For this parabola, $p = -3$; therefore, the focus is $\boxed{(0, -3)}$, the directrix is $\boxed{y = 3}$, and the length of the latus rectum is $\boxed{12}$.

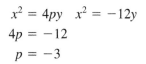

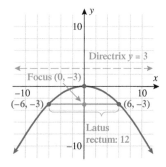

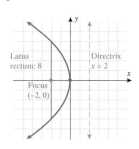

▼

YOUR TURN   Find the focus, directrix, and length of the latus rectum of the parabola $y^2 = -8x$, and use these to graph the parabola.

Finding the Equation of a Parabola with a Vertex at the Origin

Thus far, we have started with the equation of a parabola and then determined its focus and directrix. Let's now reverse the process. For example, if we know the focus and directrix of a parabola, how do we find the equation of the parabola? If we are given the focus and directrix, then we can find the vertex, which is the midpoint between the focus and the directrix. If the vertex is at the origin, then we know the general equation of the parabola that corresponds with the focus.

EXAMPLE 3 **Finding the Equation of a Parabola Given the Focus and Directrix When the Vertex Is at the Origin**

Find the equation of a parabola whose focus is at the point $\left(0, \frac{1}{2}\right)$ and whose directrix is $y = -\frac{1}{2}$. Graph the equation.

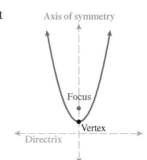

Solution:

The midpoint of the segment joining the focus and the directrix along the axis of symmetry is the vertex.

Calculate the midpoint between $\left(0, \frac{1}{2}\right)$ and $\left(0, -\frac{1}{2}\right)$.

$$\text{Vertex} = \left(\frac{0 + 0}{2}, \frac{\frac{1}{2} - \frac{1}{2}}{0}\right) = (0, 0).$$

A parabola with its vertex at $(0, 0)$, focus at $(0, p)$, and directrix $y = -p$ corresponds to the equation $x^2 = 4py$.

Identify p given that the focus is $(0, p) = \left(0, \frac{1}{2}\right)$.

$$p = \frac{1}{2}$$

Substitute $p = \frac{1}{2}$ into the standard equation of a parabola with vertex at the origin $x^2 = 4py$.

$$x^2 = 2y$$

Now that the equation is known, a few points can be selected, and the parabola can be point-plotted. Alternatively, the length of the latus rectum can be calculated to sketch the approximate width of the parabola.

To graph $x^2 = 2y$, first calculate the latus rectum.

$$4|p| = 4\left(\frac{1}{2}\right) = 2$$

Label the focus, directrix, and latus rectum, and draw a parabolic curve whose vertex is at the origin that intersects with the latus rectum's endpoints.

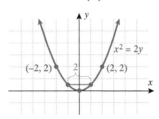

▼

YOUR TURN Find the equation of a parabola whose focus is at the point $(-5, 0)$ and whose directrix is $x = 5$.

▼
ANSWER

$y^2 = -20x$

Before we proceed to parabolas with general vertices, let's first make a few observations: The larger the latus rectum, the wider the parabola. An alternative approach for graphing the parabola is to plot a few points that satisfy the equation of the parabola, which is the approach in most textbooks.

B.2.2 Parabola with Vertex (h, k)

Graphing a Parabola with a Vertex at (h, k)

Recall that the graph of $x^2 + y^2 = r^2$ is a circle with radius r centered at the origin, whereas the graph of $(x - h)^2 + (y - k)^2 = r^2$ is a circle with radius r centered at the point (h, k). In other words, the center is shifted from the origin to the point (h, k). This same translation (shift) can be used to describe parabolas whose vertex is at the point (h, k).

EQUATION OF A PARABOLA WITH VERTEX AT THE POINT (h, k)

The standard (conic) form of the equation of a parabola with its vertex at the point (h, k) is given by

EQUATION	$(y - k)^2 = 4p(x - h)$	$(x - h)^2 = 4p(y - k)$
VERTEX	(h, k)	(h, k)
FOCUS	$(p + h, k)$	$(h, p + k)$
DIRECTRIX	$x = -p + h$	$y = -p + k$
AXIS OF SYMMETRY	$y = k$	$x = h$
$p > 0$	opens to the right	opens upward
$p < 0$	opens to the left	opens downward

In order to find the vertex of a parabola given a general second-degree equation, first complete the square in order to identify (h, k). Then determine whether the parabola opens up, down, left, or right. Identify points that lie on the graph of the parabola. It is important to note that intercepts are often the easiest points to find, since one of the variables is set equal to zero.

▶ **EXAMPLE 4** **Graphing a Parabola with Vertex (h, k)**

Graph the parabola given by the equation $y^2 - 6y - 2x + 8 = 0$.

Solution:

Transform this equation into the form $(y - k)^2 = 4p(x - h)$, since this equation is of degree two in y and degree one in x. We know this parabola opens either to the left or right.

Complete the square on y: $\qquad\qquad\qquad\qquad y^2 - 6y - 2x + 8 = 0$

Isolate the y terms. $\qquad\qquad\qquad\qquad\qquad\quad y^2 - 6y = 2x - 8$

Add 9 to both sides to complete the square. $\quad y^2 - 6y + 9 = 2x - 8 + 9$

Write the left side as a perfect square. $\qquad\quad (y - 3)^2 = 2x + 1$

Factor out a 2 on the right side. $\qquad\qquad\qquad (y - 3)^2 = 2\left(x + \dfrac{1}{2}\right)$

Compare with $(y - k)^2 = 4p(x - h)$ and identify (h, k) and p. $\qquad\qquad (h, k) = \left(-\dfrac{1}{2}, 3\right)$

$$4p = 2 \Rightarrow p = \dfrac{1}{2}$$

The vertex is at the point $\left(-\dfrac{1}{2}, 3\right)$, and since $p = \dfrac{1}{2}$ is positive, the parabola opens to the right. Since the parabola's vertex lies in quadrant II and it opens to the right, we know there are two y-intercepts and one x-intercept. Apply the general equation $y^2 - 6y - 2x + 8 = 0$ to find the intercepts.

Find the y-intercepts (set $x = 0$). $\qquad$ $y^2 - 6y + 8 = 0$

   Factor. $\qquad$ $(y - 2)(y - 4) = 0$

   Solve for y. $\qquad$ $y = 2$ or $y = 4$

Find the x-intercept (set $y = 0$). $\qquad$ $-2x + 8 = 0$

   Solve for x. $\qquad$ $x = 4$

Label the following points and connect them
with a smooth curve:

Vertex:	$\left(-\frac{1}{2}, 3\right)$
y-intercepts:	$(0, 2)$ and $(0, 4)$
x-intercept:	$(4, 0)$

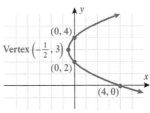

ANSWER

Vertex: $\left(-\frac{9}{5}, -2\right)$

x-intercept: $x = -1$

y-intercepts: $y = -5$ and $y = 1$

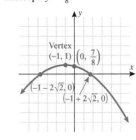

▼

YOUR TURN   For the equation $y^2 + 4y - 5x - 5 = 0$, identify the vertex and
the intercepts, and graph.

EXAMPLE 5 **Graphing a Parabola with Vertex (h, k)**

Graph the parabola given by the equation $x^2 - 2x - 8y - 7 = 0$.

Solution:

Transform this equation into the form $(x - h)^2 = 4p(y - k)$, since this equation is
degree two in x and degree one in y. We know this parabola opens either upward
or downward.

Complete the square on x: $\qquad\qquad$ $x^2 - 2x - 8y - 7 = 0$

   Isolate the x terms. $\qquad\qquad$ $x^2 - 2x = 8y + 7$

   Add 1 to both sides to complete the square. $\qquad$ $x^2 - 2x + 1 = 8y + 7 + 1$

   Write the left side as a perfect square. $\qquad$ $(x - 1)^2 = 8y + 8$

   Factor out the 8 on the right side. $\qquad$ $(x - 1)^2 = 8(y + 1)$

Compare with $(x - h)^2 = 4p(y - k)$ and $\qquad$ $(h, k) = (1, -1)$

identify (h, k) and p. $\qquad\qquad$ $4p = 8 \Rightarrow p = 2$

The vertex is at the point $(1, -1)$, and since $p = 2$ is positive, the parabola opens
upward. Since the parabola's vertex lies in quadrant IV and it opens upward, we
know there are two x-intercepts and one y-intercept. Use the general equation
$x^2 - 2x - 8y - 7 = 0$ to find the intercepts.

Find the y-intercept (set $x = 0$). $\qquad\qquad$ $-8y - 7 = 0$

   Solve for y. $\qquad\qquad\qquad$ $y = -\dfrac{7}{8}$

Find the x-intercepts (set $y = 0$). $\qquad\qquad$ $x^2 - 2x - 7 = 0$

   Solve for x. $\qquad$ $x = \dfrac{2 \pm \sqrt{4 + 28}}{2} = \dfrac{2 \pm \sqrt{32}}{2} = \dfrac{2 \pm 4\sqrt{2}}{2} = 1 \pm 2\sqrt{2}$

Label the following points and connect with a
smooth curve:

Vertex:	$(1, -1)$
y-intercept:	$\left(0, -\frac{7}{8}\right)$
x-intercepts:	$\left(1 - 2\sqrt{2}, 0\right)$ and $\left(1 + 2\sqrt{2}, 0\right)$

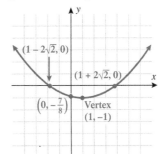

ANSWER

Vertex: $(-1, 1)$

x-intercepts: $x = -1 \pm 2\sqrt{2}$

y-intercept: $y = \dfrac{7}{8}$

▼

YOUR TURN   For the equation $x^2 + 2x + 8y - 7 = 0$, identify the vertex and
the intercepts, and graph.

EXAMPLE 6 Finding the Equation of a Parabola with Vertex (h, k)

Find the equation of a parabola whose vertex is located at the point $(2, -3)$ and whose focus is located at the point $(5, -3)$.

Solution:

Draw a Cartesian plane and label the vertex and focus. The vertex and focus share the same axis of symmetry $y = -3$ and indicate a parabola opening to the right.

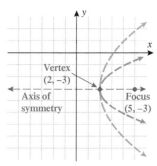

Write the standard (conic) equation of a parabola opening to the right.

$$(y - k)^2 = 4p(x - h) \quad p > 0$$

Substitute the vertex $(h, k) = (2, -3)$ into the standard equation.

$$[y - (-3)]^2 = 4p(x - 2)$$

Find p.

The general form of the vertex is (h, k), and the focus is $(h + p, k)$.

For this parabola, the vertex is $(2, -3)$, and the focus is $(5, -3)$.

Find p by taking the difference of the x-coordinates. $\quad p = 3$

Substitute $p = 3$ into $[y - (-3)]^2 = 4p(x - 2)$. $\qquad (y + 3)^2 = 4(3)(x - 2)$

Eliminate the parentheses. $\qquad y^2 + 6y + 9 = 12x - 24$

Simplify. $\qquad \boxed{y^2 + 6y - 12x + 33 = 0}$

▼ **ANSWER**

$y^2 + 6y + 8x - 7 = 0$

▼

YOUR TURN Find the equation of the parabola whose vertex is located at $(2, -3)$ and whose focus is located at $(0, -3)$.

B.2.3 Applications

B.2.3 SKILL

Solve applied problems that involve parabolas.

B.2.3 CONCEPTUAL

Understand that it is the focus that is the key to applications.

If we start with a parabola in the xy-plane and rotate it around its axis of symmetry, the result will be a three-dimensional paraboloid. Solar cookers illustrate the physical property that the rays of light coming into a parabola should be reflected to the focus. A flashlight reverses this process in that its light source at the focus illuminates a parabolic reflector to direct the beam outward.

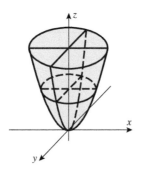

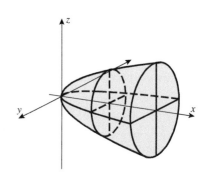

A satellite dish is in the shape of a paraboloid. Functioning as an antenna, the parabolic dish collects all of the incoming signals and reflects them to a single point, the focal point, which is where the receiver is located. In Examples 7 and 8, and in the Applications Exercises, the intention is not to find the three-dimensional equation of the paraboloid, but rather the equation of the plane parabola that's rotated to generate the paraboloid.

Satellite dish

EXAMPLE 7 **Finding the Location of the Receiver in a Satellite Dish**

A satellite dish is 24 feet in diameter at its opening and 4 feet deep in its center. Where should the receiver be placed?

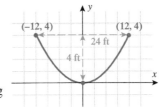

Solution:

Draw a parabola with a vertex at the origin representing the center cross section of the satellite dish.

Write the standard equation of a parabola opening upward with vertex at $(0, 0)$.

$$x^2 = 4py$$

The point $(12, 4)$ lies on the parabola, so substitute $(12, 4)$ into $x^2 = 4py$.

$$(12)^2 = 4p(4)$$

Simplify.

$$144 = 16p$$

Solve for p.

$$p = 9$$

Substitute $p = 9$ into the focus $(0, p)$.

$$\text{focus: } (0, 9)$$

The receiver should be placed 9 feet from the vertex of the dish.

Parabolic antennas work for sound in addition to light. Have you ever wondered how the sound of the quarterback calling audible plays is heard by the sideline crew? The crew holds a parabolic system with a microphone at the focus. All of the sound in the direction of the parabolic system is reflected toward the focus, where the microphone amplifies and records the sound.

EXAMPLE 8 **Finding the Equation of a Parabolic Sound Dish**

If the parabolic sound dish the sideline crew is holding has a 2-foot diameter at the opening and the microphone is located 6 inches from the vertex, find the equation that governs the center cross section of the parabolic sound dish.

Solution:

Write the standard equation of a parabola opening to the right with the vertex at the origin $(0, 0)$.

$$x = 4py^2$$

The focus is located 6 inches $\left(\frac{1}{2} \text{ foot}\right)$ from the vertex.

$$(p, 0) = \left(\frac{1}{2}, 0\right)$$

Solve for p.

$$p = \frac{1}{2}$$

Let $p = \frac{1}{2}$ in $x = 4py^2$.

$$x = 4\left(\frac{1}{2}\right)y^2$$

Simplify.

$$x = 2y^2$$

▶[SECTION B.2] SUMMARY

In this section, we discussed parabolas whose vertex is at the origin.

EQUATION	$y^2 = 4px$	$x^2 = 4py$
VERTEX	$(0, 0)$	$(0, 0)$
FOCUS	$(p, 0)$	$(0, p)$
DIRECTRIX	$x = -p$	$y = -p$
AXIS OF SYMMETRY	x-axis	y-axis
$p > 0$	opens to the right	opens upward
$p < 0$	opens to the left	opens downward
GRAPH		

For parabolas whose vertex is at the point (h, k):

EQUATION	$(y - k)^2 = 4p(x - h)$	$(x - h)^2 = 4p(y - k)$
VERTEX	(h, k)	(h, k)
FOCUS	$(p + h, k)$	$(h, p + k)$
DIRECTRIX	$x = -p + h$	$y = -p + k$
AXIS OF SYMMETRY	$y = k$	$x = h$
$p > 0$	opens to the right	opens upward
$p < 0$	opens to the left	opens downward

[SECTION B.2] EXERCISES

• **SKILLS**

In Exercises 1–4, match the parabola to the equation.

1. $y^2 = 4x$

2. $y^2 = -4x$

3. $x^2 = -4y$

4. $x^2 = 4y$

a.

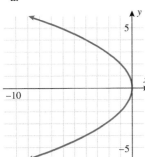

b.

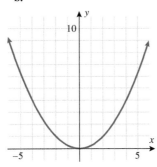

c.

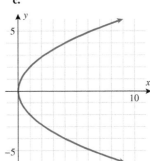

d.

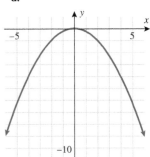

In Exercises 5–8, match the parabola to the equation.

5. $(y - 1)^2 = 4(x - 1)$

6. $(y + 1)^2 = -4(x - 1)$

7. $(x + 1)^2 = -4(y + 1)$

8. $(x - 1)^2 = 4(y - 1)$

a.

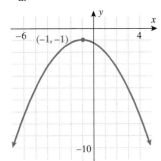

b.

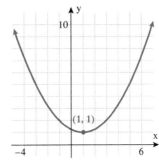

c.

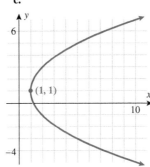

d.

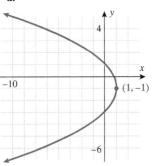

In Exercises 9–20, find an equation for the parabola described.

9. Vertex at $(0, 0)$; Focus at $(0, 3)$

10. Vertex at $(0, 0)$; Focus at $(2, 0)$

11. Vertex at $(0, 0)$; Focus at $(-5, 0)$

12. Vertex at $(0, 0)$; Focus at $(0, -4)$

13. Vertex at $(3, 5)$; Focus at $(3, 7)$

14. Vertex at $(3, 5)$; Focus at $(7, 5)$

15. Vertex at $(2, 4)$; Focus at $(0, 4)$

16. Vertex at $(2, 4)$; Focus at $(2, -1)$

17. Focus at $(2, 4)$; Directrix at $y = -2$

18. Focus at $(2, -2)$; Directrix at $y = 4$

19. Focus at $(3, -1)$; Directrix at $x = 1$

20. Focus at $(-1, 5)$; Directrix at $x = 5$

In Exercises 21–24, write an equation for each parabola.

21.

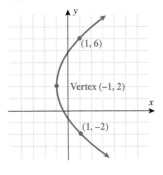

22.

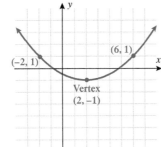

23.

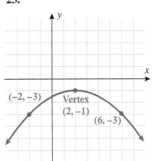

24.

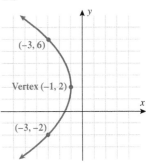

In Exercises 25–32, find the focus, vertex, directrix, and length of latus rectum and graph the parabola.

25. $x^2 = 8y$ **26.** $x^2 = -12y$ **27.** $y^2 = -2x$ **28.** $y^2 = 6x$

29. $x^2 = 16y$ **30.** $x^2 = -8y$ **31.** $y^2 = 4x$ **32.** $y^2 = -16x$

In Exercises 33–44, find the vertex and graph the parabola.

33. $(y - 2)^2 = 4(x + 3)$ **34.** $(y + 2)^2 = -4(x - 1)$ **35.** $(x - 3)^2 = -8(y + 1)$ **36.** $(x + 3)^2 = -8(y - 2)$

37. $(x + 5)^2 = -2y$ **38.** $y^2 = -16(x + 1)$ **39.** $y^2 - 4y - 2x + 4 = 0$ **40.** $x^2 - 6x + 2y + 9 = 0$

41. $y^2 + 2y - 8x - 23 = 0$ **42.** $x^2 - 6x - 4y + 10 = 0$ **43.** $x^2 - x + y - 1 = 0$ **44.** $y^2 + y - x + 1 = 0$

• APPLICATIONS

45. Satellite Dish. A satellite dish measures 8 feet across its opening and 2 feet deep at its center. The receiver should be placed at the focus of the parabolic dish. Where is the focus?

46. Satellite Dish. A satellite dish measures 30 feet across its opening and 5 feet deep at its center. The receiver should be placed at the focus of the parabolic dish. Where is the focus?

47. Eyeglass Lens. Eyeglass lenses can be thought of as very wide parabolic curves. If the focus occurs 2 centimeters from the center of the lens and the lens at its opening is 5 centimeters, find an equation that governs the shape of the center cross section of the lens.

48. Optical Lens. A parabolic lens focuses light onto a focal point 3 centimeters from the vertex of the lens. How wide is the lens 0.5 centimeter from the vertex?

Exercises 49 and 50 are examples of solar cookers. Parabolic shapes are often used to generate intense heat by collecting sun rays and focusing all of them at a focal point.

49. Solar Cooker. The parabolic cooker MS-ST10 is delivered as a kit, handily packed in a single carton, with complete assembly instructions and even the necessary tools.

Solar cooker, Ubuntu Village, Johannesburg, South Africa

Thanks to the reflector diameter of 1 meter, it develops an immense power: 1 liter of water boils in significantly less than half an hour. If the rays are focused 40 centimeters from the vertex, find the equation for the parabolic cooker.

50. Le Four Solaire at Font-Romeur "Mirrors of the Solar Furnace." There is a reflector in the Pyrenees Mountains that is eight stories high. It cost $2 million and took 10 years to build. Made of 9000 mirrors arranged in a parabolic formation, it can reach 6000°F just from the Sun hitting it! If the diameter of the parabolic mirror is 100 meters and the sunlight is focused 25 meters from the vertex, find the equation for the parabolic dish.

Solar furnace, Odeillo, France

51. Sailing under a Bridge. A bridge with a parabolic shape has an opening 80 feet wide at the base (where the bridge meets the water), and the height in the center of the bridge is 20 feet. A sailboat whose mast reaches 17 feet above the water is traveling under the bridge 10 feet from the center of the bridge. Will it clear the bridge without scraping its mast? Justify your answer.

52. Driving under a Bridge. A bridge with a parabolic shape reaches a height of 25 feet in the center of the road, and the width of the bridge opening at ground level is 20 feet combined (both lanes). If an RV is 10 feet tall and 8 feet wide, it won't make it under the bridge if it hugs the center line. Will it clear the bridge if it straddles the center line? Justify your answer.

53. Parabolic Telescope. The Arecibo radio telescope in Puerto Rico has an enormous reflecting surface, or radio mirror. The huge "dish" is 1000 feet in diameter and 167 feet deep and covers an area of about 20 acres. Using these dimensions, determine the focal length of the telescope. Find the equation for the dish portion of the telescope.

54. **Suspension Bridge.** If one parabolic segment of a suspension bridge is 300 feet and if the cables at the vertex are suspended 10 feet above the bridge, whereas the height of the cables 150 feet from the vertex reaches 60 feet, find the equation of the parabolic path of the suspension cables.

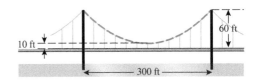

• CATCH THE MISTAKE

In Exercises 55 and 56, explain the mistake that is made.

55. Find an equation for a parabola whose vertex is at the origin and whose focus is at the point $(3, 0)$.

Solution:

Write the general equation for a parabola whose vertex is at the origin. $x^2 = 4py$

The focus of this parabola is $(p, 0) = (3, 0)$. $p = 3$

Substitute $p = 3$ into $x^2 = 4py$. $x^2 = 12y$

This is incorrect. What mistake was made?

56. Find an equation for a parabola whose vertex is at the point $(3, 2)$ and whose focus is located at $(5, 2)$.

Solution:

Write the equation associated with a parabola whose vertex is $(3, 2)$. $(x - h)^2 = 4p(y - k)$

Substitute $(3, 2)$ into $(x - h)^2 = 4p(y - k)$. $(x - 3)^2 = 4p(y - 2)$

The focus is located at $(5, 2)$; therefore, $p = 5$.

Substitute $p = 5$ into $(x - 3)^2 = 4p(y - 2)$. $(x - 3)^2 = 20(y - 2)$

This is incorrect. What mistake(s) was made?

• CONCEPTUAL

In Exercises 57–60, determine whether each statement is true or false.

57. The vertex lies on the graph of a parabola.

58. The focus lies on the graph of a parabola.

59. The directrix lies on the graph of a parabola.

60. The endpoints of the latus rectum lie on the graph of a parabola.

• CHALLENGE

61. Derive the standard equation of a parabola with its vertex at the origin, opening upward $x^2 = 4py$. [Calculate the distance d_1 from any point on the parabola (x, y) to the focus $(0, p)$. Calculate the distance d_2 from any point on the parabola (x, y) to the directrix $(-p, y)$. Set $d_1 = d_2$.]

62. Derive the standard equation of a parabola opening right, $y^2 = 4px$. [Calculate the distance d_1 from any point on the parabola (x, y) to the focus $(p, 0)$. Calculate the distance d_2 from any point on the parabola (x, y) to the directrix $(x, -p)$. Set $d_1 = d_2$.]

• TECHNOLOGY

63. With a graphing utility, plot the parabola $x^2 - x + y - 1 = 0$. Compare with the sketch you drew for Exercise 43.

64. With a graphing utility, plot the parabola $y^2 + y - x + 1 = 0$. Compare with the sketch you drew for Exercise 44.

65. In your mind, picture the parabola given by $(y + 3.5)^2 = 10(x - 2.5)$. Where is the vertex? Which way does this parabola open? Now plot the parabola with a graphing utility.

66. In your mind, picture the parabola given by $(x + 1.4)^2 = -5(y + 1.7)$. Where is the vertex? Which way does this parabola open? Now plot the parabola with a graphing utility.

67. In your mind, picture the parabola given by $(y - 1.5)^2 = -8(x - 1.8)$. Where is the vertex? Which way does this parabola open? Now plot the parabola with a graphing utility.

68. In your mind, picture the parabola given by $(x + 2.4)^2 = 6(y - 3.2)$. Where is the vertex? Which way does this parabola open? Now plot the parabola with a graphing utility.

69. Given is the parabola $y^2 - 4.5y - 4x - 8.9375 = 0$.
    a. Solve the equation for y and use a graphing utility to plot the parabola.
    b. Transform the equation into the form $(y - k)^2 = 4p(x - h)$. Find the vertex. Which way does the parabola open?
    c. Do (a) and (b) agree with each other?

70. Given is the parabola $2x^2 + 6.4x + y - 2.08 = 0$.
    a. Solve the equation for y and use a graphing utility to plot the parabola.
    b. Transform the equation into the form $(x - h)^2 = 4p(y - k)$. Find the vertex. Which way does the parabola open?
    c. Do (a) and (b) agree with each other?

B.3 THE ELLIPSE

SKILLS OBJECTIVES	CONCEPTUAL OBJECTIVES
▪ Find the equation of an ellipse centered at the origin. ▪ Find the equation of an ellipse centered at the point (h, k). ▪ Solve applied problems that involve ellipses.	▪ Understand that an ellipse is the set of all points in a plane, the sum of whose distances from two fixed points (foci) is constant. ▪ Employ completing the square to transform an equation into the standard form of an ellipse. ▪ Understand that the endpoints and the foci are important for application problems involving ellipses.

B.3.1 Ellipse with Its Center at the Origin

Definition of an Ellipse

B.3.1 SKILL

Find the equation of an ellipse centered at the origin.

B.3.1 CONCEPTUAL

Understand that an ellipse is the set of all points in a plane, the sum of whose distances from two fixed points (foci) is constant.

If we were to take a piece of string, tie loops at both ends, and tack the ends down so that the string had lots of slack, we would have the picture on the left. If we then took a pencil and pulled the string taut and traced our way around for one full rotation, the result would be an ellipse. See the second figure on the left.

> **DEFINITION** | **Ellipse**
>
> An **ellipse** is the set of all points in a plane, the sum of whose distances from two fixed points is constant. These two fixed points are called **foci** (plural of focus). A line segment through the foci called the **major axis** intersects the ellipse at the **vertices**. The midpoint of the line segment joining the vertices is called the **center**. The line segment that intersects the center and joins two points on the ellipse and is perpendicular to the major axis is called the **minor axis**.
>
>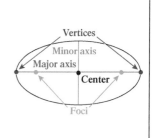

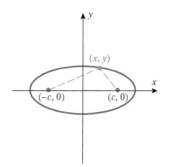

Let's start with an ellipse whose center is located at the origin. Using graph-shifting techniques, we can later extend the characteristics of an ellipse to one centered at a point other than the origin. Ellipses can vary from the shape of circles to something quite elongated, either horizontally or vertically, that resembles the shape of a racetrack. We say that the ellipse has either greater or lesser *eccentricity*; as we will see, there is a simple mathematical definition of *eccentricity*. It can be shown that the standard equation of an ellipse with its center at the origin is given by one of two forms, depending on whether the orientation of the major axis of the ellipse is horizontal or vertical. For $a > b > 0$, if the major axis is horizontal, then the equation is given by $\dfrac{x^2}{a^2} + \dfrac{y^2}{b^2} = 1$, and if the major axis is vertical, then the equation is given by $\dfrac{x^2}{b^2} + \dfrac{y^2}{a^2} = 1$.

Let's consider an ellipse with its center at the origin and the foci on the x-axis. Let the distance from the center to the focus be c. Therefore, the foci are located at the points $(-c, 0)$ and $(c, 0)$. The line segment containing the foci is called the major axis, and it lies along the x-axis. The sum of the two distances from the foci to any point (x, y) must be constant.

Derivation of the Equation of an Ellipse

WORDS	MATH
Calculate the distance from (x, y) to $(-c, 0)$ by applying the distance formula.	$\sqrt{[x - (-c)]^2 + y^2}$
Calculate the distance from (x, y) to $(c, 0)$ by applying the distance formula.	$\sqrt{(x - c)^2 + y^2}$
The sum of these two distances is equal to a constant ($2a$ for convenience).	$\sqrt{[x - (-c)]^2 + y^2} + \sqrt{(x - c)^2 + y^2} = 2a$
Isolate one radical.	$\sqrt{[x - (-c)]^2 + y^2} = 2a - \sqrt{(x - c)^2 + y^2}$
Square both sides of the equation.	$(x + c)^2 + y^2 = 4a^2 - 4a\sqrt{(x - c)^2 + y^2} + (x - c)^2 + y^2$
Square the binomials inside the parentheses.	$x^2 + 2cx + c^2 + y^2 = 4a^2 - 4a\sqrt{(x - c)^2 + y^2}$ $+ x^2 - 2cx + c^2 + y^2$
Simplify.	$4cx - 4a^2 = -4a\sqrt{(x - c)^2 + y^2}$
Divide both sides of the equation by -4.	$a^2 - cx = a\sqrt{(x - c)^2 + y^2}$
Square both sides of the equation.	$(a^2 - cx)^2 = a^2[(x - c)^2 + y^2]$
Square the binomials inside the parentheses.	$a^4 - 2a^2cx + c^2x^2 = a^2(x^2 - 2cx + c^2 + y^2)$
Distribute the a^2 term.	$a^4 - 2a^2cx + c^2x^2 = a^2x^2 - 2a^2cx + a^2c^2 + a^2y^2$
Group the x and y terms together, respectively, on one side and constants on the other side.	$c^2x^2 - a^2x^2 - a^2y^2 = a^2c^2 - a^4$
Factor out the common factors.	$(c^2 - a^2)x^2 - a^2y^2 = a^2(c^2 - a^2)$
Multiply both sides of the equation by -1.	$(a^2 - c^2)x^2 + a^2y^2 = a^2(a^2 - c^2)$
We can make the argument that $a > c$ in order for a point to be on the ellipse (and not on the x-axis). Thus, since a and c represent distances and therefore are positive, we know that $a^2 > c^2$, or $a^2 - c^2 > 0$. Hence, we can divide both sides of the equation by $a^2 - c^2$, since $a^2 - c^2 \neq 0$.	$x^2 + \dfrac{a^2y^2}{(a^2 - c^2)} = a^2$
Let $b^2 = a^2 - c^2$.	$x^2 + \dfrac{a^2y^2}{b^2} = a^2$
Divide both sides of the equation by a^2.	$\boxed{\dfrac{x^2}{a^2} + \dfrac{y^2}{b^2} = 1}$

The equation $\dfrac{x^2}{a^2} + \dfrac{y^2}{b^2} = 1$ represents an ellipse with its center at the origin with the foci along the x-axis, since $a > b$. The following box summarizes ellipses that have their center at the origin and foci along either the x-axis or y-axis:

EQUATION OF AN ELLIPSE WITH ITS CENTER AT THE ORIGIN

The **standard form of the equation of an ellipse** with its center at the origin is given by

ORIENTATION OF MAJOR AXIS	Horizontal (along the x-axis)	Vertical (along the y-axis)
EQUATION	$\dfrac{x^2}{a^2} + \dfrac{y^2}{b^2} = 1 \qquad a > b > 0$	$\dfrac{x^2}{b^2} + \dfrac{y^2}{a^2} = 1 \qquad a > b > 0$
FOCI	$(-c, 0)\quad(c, 0)$ where $c^2 = a^2 - b^2$	$(0, -c)\quad(0, c)$ where $c^2 = a^2 - b^2$
VERTICES	$(-a, 0)\quad(a, 0)$	$(0, -a)\quad(0, a)$
OTHER INTERCEPTS	$(0, b)\quad(0, -b)$	$(b, 0)\quad(-b, 0)$
GRAPH		

In both cases, the value of c, the distance along the major axis from the center to the focus, is given by $c^2 = a^2 - b^2$. The length of the major axis is $2a$, and the length of the minor axis is $2b$.

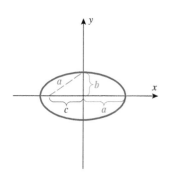

It is important to note that the vertices correspond to intercepts when an ellipse is centered at the origin. One of the first things we notice about an ellipse is its *eccentricity*. The **eccentricity**, denoted e, is given by $e = \dfrac{c}{a}$, where $0 < e < 1$. The circle is a limiting form of an ellipse, $c = 0$. In other words, if the eccentricity is close to 0, then the ellipse resembles a circle, whereas if the eccentricity is close to 1, then the ellipse is quite elongated, or eccentric.

Graphing an Ellipse with Its Center at the Origin

The equation of an ellipse in standard form can be used to graph an ellipse. Although an ellipse is defined in terms of the foci, the foci are not part of the graph. It is important to note that if the divisor of the term with x^2 is larger than the divisor of the term with y^2, then the ellipse is elongated horizontally.

EXAMPLE 1 **Graphing an Ellipse with a Horizontal Major Axis**

Graph the ellipse given by $\dfrac{x^2}{25} + \dfrac{y^2}{9} = 1$.

Solution:

Since $25 > 9$, the major axis is horizontal.	$a^2 = 25$	and	$b^2 = 9$
Solve for a and b.	$a = 5$	and	$b = 3$
Identify the vertices: $(-a, 0)$ and $(a, 0)$.	$(-5, 0)$	and	$(5, 0)$
Identify the endpoints (y-intercepts) on the minor axis: $(0, -b)$ and $(0, b)$.	$(0, -3)$	and	$(0, 3)$

Graph by labeling the points $(-5, 0)$, $(5, 0)$, $(0, -3)$, and $(0, 3)$ and connecting them with a smooth curve.

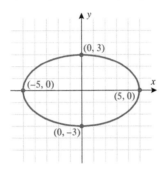

STUDY TIP

If the divisor of x^2 is larger than the divisor of y^2, then the major axis is horizontal along the x-axis, as in Example 1. If the divisor of y^2 is larger than the divisor of x^2, then the major axis is vertical along the y-axis, as you will see in Example 2.

If the divisor of x^2 is larger than the divisor of y^2, then the major axis is horizontal along the x-axis, as in Example 1. If the divisor of y^2 is larger than the divisor of x^2, then the major axis is vertical along the y-axis, as you will see in Example 2.

▶ **EXAMPLE 2** **Graphing an Ellipse with a Vertical Major Axis**

Graph the ellipse given by $16x^2 + y^2 = 16$.

Solution:

Write the equation in standard form by dividing by 16.	$\dfrac{x^2}{1} + \dfrac{y^2}{16} = 1$		
Since $16 > 1$, this ellipse is elongated vertically.	$a^2 = 16$	and	$b^2 = 1$
Solve for a and b.	$a = 4$	and	$b = 1$
Identify the vertices: $(0, -a)$ and $(0, a)$.	$(0, -4)$	and	$(0, 4)$
Identify the x-intercepts on the minor axis: $(-b, 0)$ and $(b, 0)$.	$(-1, 0)$	and	$(1, 0)$

Graph by labeling the points $(0, -4)$, $(0, 4)$, $(-1, 0)$, and $(1, 0)$ and connecting them with a smooth curve.

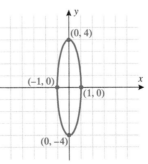

▼
ANSWER

a.

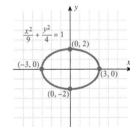

b.

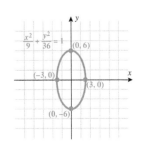

▼
YOUR TURN Graph the ellipses:

a. $\dfrac{x^2}{9} + \dfrac{y^2}{4} = 1$ **b.** $\dfrac{x^2}{9} + \dfrac{y^2}{36} = 1$

Finding the Equation of an Ellipse with Its Center at the Origin

What if we know the vertices and the foci of an ellipse and want to find the equation to which it corresponds? The axis on which the foci and vertices are located is the major axis. Therefore, we will have the standard equation of an ellipse, and a will be known (from the vertices). Since c is known from the foci, we can use the relation $c^2 = a^2 - b^2$ to determine the unknown b.

EXAMPLE 3 **Finding the Equation of an Ellipse Centered at the Origin**

Find the standard form of the equation of an ellipse with foci at $(-3, 0)$ and $(3, 0)$ and vertices $(-4, 0)$ and $(4, 0)$.

Solution:

The major axis lies along the x-axis, since it contains the foci and vertices.

Write the corresponding general equation of an ellipse.	$\dfrac{x^2}{a^2} + \dfrac{y^2}{b^2} = 1$

Identify a from the vertices:

Match vertices $(-4, 0) = (-a, 0)$ and $(4, 0) = (a, 0)$.	$a = 4$

Identify c from the foci:

Match foci $(-3, 0) = (-c, 0)$ and $(3, 0) = (c, 0)$.	$c = 3$
Substitute $a = 4$ and $c = 3$ into $b^2 = a^2 - c^2$.	$b^2 = 4^2 - 3^2$
Simplify.	$b^2 = 7$
Substitute $a^2 = 16$ and $b^2 = 7$ into $\dfrac{x^2}{a^2} + \dfrac{y^2}{b^2} = 1$.	$\dfrac{x^2}{16} + \dfrac{y^2}{7} = 1$

The equation of the ellipse is $\boxed{\dfrac{x^2}{16} + \dfrac{y^2}{7} = 1}$.

▼

ANSWER

$\dfrac{x^2}{11} + \dfrac{y^2}{36} = 1$

▼ **YOUR TURN** Find the standard form of the equation of an ellipse with vertices at $(0, -6)$ and $(0, 6)$ and foci $(0, -5)$ and $(0, 5)$.

B.3.2 Ellipse with Center (h, k)

B.3.2 SKILL

Find the equation of an ellipse centered at the point (h, k).

B.3.2 CONCEPTUAL

Employ completing the square to transform an equation into the standard form of an ellipse.

We can use graph-shifting techniques to graph ellipses that are centered at a point other than the origin. For example, to graph $\dfrac{(x - h)^2}{a^2} + \dfrac{(y - k)^2}{b^2} = 1$ (assuming h and k are positive constants), start with the graph of $\dfrac{x^2}{a^2} + \dfrac{y^2}{b^2} = 1$ and shift to the right h units and up k units. The center, the vertices, the foci, and the major and minor axes all shift. In other words, the two ellipses are identical in shape and size, except that the ellipse $\dfrac{(x - h)^2}{a^2} + \dfrac{(y - k)^2}{b^2} = 1$ is centered at the point (h, k).

The following table summarizes the characteristics of ellipses centered at a point other than the origin:

EQUATION OF AN ELLIPSE WITH ITS CENTER AT THE POINT (*h, k*)

The **standard form of equation of an ellipse** with its center at the point (h, k) is given by

ORIENTATION OF MAJOR AXIS	Horizontal (parallel to the *x*-axis)	Vertical (parallel to the *y*-axis)
EQUATION	$\dfrac{(x - h)^2}{a^2} + \dfrac{(y - k)^2}{b^2} = 1$	$\dfrac{(x - h)^2}{b^2} + \dfrac{(y - k)^2}{a^2} = 1$
FOCI	$(h - c, k) \quad (h + c, k)$	$(h, k - c) \quad (h, k + c)$
GRAPH		

In both cases, $a > b > 0$, $c^2 = a^2 - b^2$, the length of the major axis is $2a$, and the length of the minor axis is $2b$.

EXAMPLE 4 **Graphing an Ellipse with Center (*h, k*) Given the Equation in Standard Form**

Graph the ellipse given by $\dfrac{(x - 2)^2}{9} + \dfrac{(y + 1)^2}{16} = 1$.

Solution:

Write the equation in the form

$$\frac{(x - h)^2}{b^2} + \frac{(y - k)^2}{a^2} = 1. \qquad\qquad \frac{(x - 2)^2}{3^2} + \frac{[y - (-1)]^2}{4^2} = 1$$

Identify a, b, and the center (h, k). $a = 4$, $b = 3$, and $(h, k) = (2, -1)$

Draw a graph and label the center: $(2, -1)$.

Since $a = 4$, the vertices are up 4 and down four units from the center: $(2, -5)$ and $(2, 3)$.

Since $b = 3$, the endpoints of the minor axis are to the left and right three units: $(-1, -1)$ and $(5, -1)$.

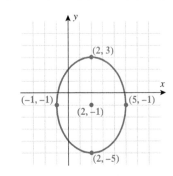

ANSWER

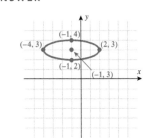

YOUR TURN Graph the ellipse given by $\dfrac{(x + 1)^2}{9} + \dfrac{(y - 3)^2}{1} = 1$.

All active members of the Lambda Chi fraternity are college students, but not all college students are members of the Lambda Chi fraternity. Similarly, all circles are ellipses, but not all ellipses are circles. When $a = b$, the standard equation of an ellipse simplifies to a standard equation of a circle. Recall that when we are given the equation of a circle in general form, we first complete the square in order to express the equation in standard form, which allows the center and radius to be identified. We use that same approach when the equation of an ellipse is given in a general form.

▶ **EXAMPLE 5** **Graphing an Ellipse with Center (h, k) Given an Equation in General Form**

Graph the ellipse given by $4x^2 + 24x + 25y^2 - 50y - 39 = 0$.

Solution:

Transform the general equation into standard form.

Group x terms together and y terms together and add 39 to both sides.
$$(4x^2 + 24x) + (25y^2 - 50y) = 39$$

Factor out the 4 common to the x terms and the 25 common to the y terms.
$$4(x^2 + 6x) + 25(y^2 - 2y) = 39$$

Complete the square on x and y.
$$4(x^2 + 6x + 9) + 25(y^2 - 2y + 1) = 39 + 4(9) + 25(1)$$

Simplify.
$$4(x + 3)^2 + 25(y - 1)^2 = 100$$

Divide by 100.
$$\frac{(x + 3)^2}{25} + \frac{(y - 1)^2}{4} = 1$$

Since $25 > 4$, this is an ellipse with a horizontal major axis.

Now that the equation of the ellipse is in standard form, compare to
$\frac{(x - h)^2}{a^2} + \frac{(y - k)^2}{b^2} = 1$ and
identify a, b, h, k.
$$a = 5, b = 2, \text{ and } (h, k) = (-3, 1)$$

Since $a = 5$, the vertices are five units to left and right of the center.
$$(-8, 1) \quad \text{and} \quad (2, 1)$$

Since $b = 2$, the endpoints of the minor axis are up and down two units from the center.
$$(-3, -1) \quad \text{and} \quad (-3, 3)$$

Graph.

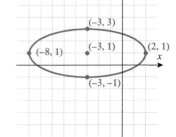

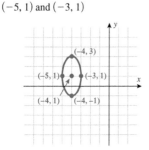

▼ ....

▼ YOUR TURN Write the equation $4x^2 + 32x + y^2 - 2y + 61 = 0$ in standard form. Identify the center, vertices, and endpoints of the minor axis, and graph.

B.3.3 Applications

There are many examples of ellipses all around us. On Earth we have racetracks, and in our solar system, the planets travel in elliptical orbits with the Sun as a focus. Satellites are in elliptical orbits around the Earth. Most communications satellites are in a *geosynchronous* (GEO) orbit—they orbit the Earth once each day. In order to stay over the same spot on Earth, a *geostationary* satellite has to be directly above the equator; it circles the Earth in exactly the time it takes the Earth to turn once on its axis, and its orbit has to follow the path of the equator as the Earth rotates.

If we start with an ellipse in the *xy*-plane and rotate it around its major axis, the result is a three-dimensional ellipsoid.

A football and a blimp are two examples of ellipsoids. The ellipsoidal shape allows for a more aerodynamic path.

PhotoDisc, Inc.

Peter Phipp/Age Fotostock America, Inc.

B.3.3 SKILL

Solve applied problems that involve ellipses.

B.3.3 CONCEPTUAL

Understand that the endpoints and the foci are important for application problems involving ellipses.

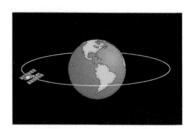

EXAMPLE 6   An Official NFL Football

A longitudinal section (that includes the two vertices and the center) of an official Wilson NFL football is an ellipse. The longitudinal section is approximately 11 inches long and 7 inches wide. Write an equation governing the elliptical longitudinal section.

Solution:

Locate the center of the ellipse at the origin and orient the football horizontally.

Write the general equation of a circle centered at the origin.	$\dfrac{x^2}{a^2} + \dfrac{y^2}{b^2} = 1$
The length of the major axis is 11 inches.	$2a = 11$
Solve for *a*.	$a = 5.5$
The length of the minor axis is 7 inches.	$2b = 7$
Solve for *b*.	$b = 3.5$
Substitute $a = 5.5$ and $b = 3.5$ into $\dfrac{x^2}{a^2} + \dfrac{y^2}{b^2} = 1$.	$\boxed{\dfrac{x^2}{5.5^2} + \dfrac{y^2}{3.5^2} = 1}$

[CONCEPT CHECK]

Would the foci of a football be closer to the center/vertices than the foci of a basketball?

▼

ANSWER Vertices

◉[SECTION B.3] SUMMARY

In this section, we first analyzed ellipses that are centered at the origin.

ORIENTATION OF MAJOR AXIS	Horizontal (along the x-axis)	Vertical (along the y-axis)
EQUATION	$\dfrac{x^2}{a^2} + \dfrac{y^2}{b^2} = 1 \quad a > b > 0$	$\dfrac{x^2}{b^2} + \dfrac{y^2}{a^2} = 1 \quad a > b > 0$
FOCI*	$(-c, 0) \quad (c, 0)$	$(0, -c) \quad (0, c)$
VERTICES	$(-a, 0) \quad (a, 0)$	$(0, -a) \quad (0, a)$
OTHER INTERCEPTS	$(0, -b) \quad (0, b)$	$(-b, 0) \quad (b, 0)$
GRAPH		

$*c^2 = a^2 - b^2$

For ellipses centered at the origin, we can graph an ellipse by finding all four intercepts.
For ellipses centered at the point (h, k), the major and minor axes and endpoints of the ellipse all shift accordingly.
It is important to note that when $a = b$, the ellipse is a circle.

[SECTION B.3] EXERCISES

• SKILLS

In Exercises 1–4, match the equation to the ellipse.

1. $\dfrac{x^2}{36} + \dfrac{y^2}{16} = 1$

2. $\dfrac{x^2}{16} + \dfrac{y^2}{36} = 1$

3. $\dfrac{x^2}{8} + \dfrac{y^2}{72} = 1$

4. $4x^2 + y^2 = 1$

a.

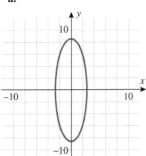

b.

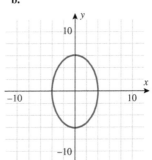

c.

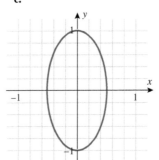

d.

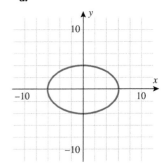

In Exercises 5–16, graph each ellipse. Label the center and vertices.

5. $\dfrac{x^2}{25} + \dfrac{y^2}{16} = 1$

6. $\dfrac{x^2}{49} + \dfrac{y^2}{9} = 1$

7. $\dfrac{x^2}{16} + \dfrac{y^2}{64} = 1$

8. $\dfrac{x^2}{25} + \dfrac{y^2}{144} = 1$

9. $\dfrac{x^2}{100} + y^2 = 1$

10. $9x^2 + 4y^2 = 36$

11. $\dfrac{4}{9}x^2 + 81y^2 = 1$

12. $\dfrac{4}{25}x^2 + \dfrac{100}{9}y^2 = 1$

13. $4x^2 + y^2 = 16$

14. $x^2 + y^2 = 81$

15. $8x^2 + 16y^2 = 32$

16. $10x^2 + 25y^2 = 50$

In Exercises 17–24, find the standard form of the equation of an ellipse with the given characteristics.

17. Foci: $(-4, 0)$ and $(4, 0)$ Vertices: $(-6, 0)$ and $(6, 0)$

18. Foci: $(-1, 0)$ and $(1, 0)$ Vertices: $(-3, 0)$ and $(3, 0)$

19. Foci: $(0, -3)$ and $(0, 3)$ Vertices: $(0, -4)$ and $(0, 4)$

20. Foci: $(0, -1)$ and $(0, 1)$ Vertices: $(0, -2)$ and $(0, 2)$

21. Major axis vertical with length of 8, minor axis length of 4, and centered at $(0, 0)$

22. Major axis horizontal with length of 10, minor axis length of 2, and centered at $(0, 0)$

23. Vertices $(0, -7)$ and $(0, 7)$ and endpoints of minor axis $(-3, 0)$ and $(3, 0)$

24. Vertices $(-9, 0)$ and $(9, 0)$ and endpoints of minor axis $(0, -4)$ and $(0, 4)$

In Exercises 25–28, match each equation with the ellipse.

25. $\dfrac{(x-3)^2}{4} + \dfrac{(y+2)^2}{25} = 1$

26. $\dfrac{(x+3)^2}{4} + \dfrac{(y-2)^2}{25} = 1$

27. $\dfrac{(x-3)^2}{25} + \dfrac{(y+2)^2}{4} = 1$

28. $\dfrac{(x+3)^2}{25} + \dfrac{(y-2)^2}{4} = 1$

a.

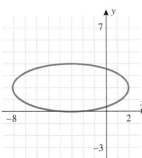

b.

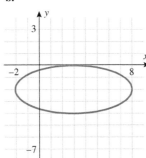

c.

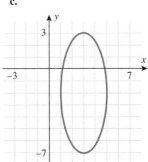

d.

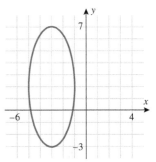

In Exercises 29–38, graph each ellipse. Label the center and vertices.

29. $\dfrac{(x-1)^2}{16} + \dfrac{(y-2)^2}{4} = 1$

30. $\dfrac{(x+1)^2}{36} + \dfrac{(y+2)^2}{9} = 1$

31. $10(x+3)^2 + (y-4)^2 = 80$

32. $3(x+3)^2 + 12(y-4)^2 = 36$

33. $x^2 + 4y^2 - 24y + 32 = 0$

34. $25x^2 + 2y^2 - 4y - 48 = 0$

35. $x^2 - 2x + 2y^2 - 4y - 5 = 0$

36. $9x^2 - 18x + 4y^2 - 27 = 0$

37. $5x^2 + 20x + y^2 + 6y - 21 = 0$

38. $9x^2 + 36x + y^2 + 2y + 36 = 0$

In Exercises 39–46, find the standard form of the equation of an ellipse with the given characteristics.

39. Foci: $(-2, 5)$ and $(6, 5)$ Vertices: $(-3, 5)$ and $(7, 5)$

40. Foci: $(2, -2)$ and $(4, -2)$ Vertices: $(0, -2)$ and $(6, -2)$

41. Foci: $(4, -7)$ and $(4, -1)$ Vertices: $(4, -8)$ and $(4, 0)$

42. Foci: $(2, -6)$ and $(2, -4)$ Vertices: $(2, -7)$ and $(2, -3)$

43. Major axis vertical with length of 8, minor axis length of 4, and centered at $(3, 2)$

44. Major axis horizontal with length of 10, minor axis length of 2, and centered at $(-4, 3)$

45. Vertices $(-1, -9)$ and $(-1, 1)$ and endpoints of minor axis $(-4, -4)$ and $(2, -4)$

46. Vertices $(-2, 3)$ and $(6, 3)$ and endpoints of minor axis $(2, 1)$ and $(2, 5)$

• **APPLICATIONS**

47. Carnival Ride. The Zipper, a favorite carnival ride, maintains an elliptical shape with a major axis of 150 feet and a minor axis of 30 feet. Assuming it is centered at the origin, find an equation for the ellipse.

Ingram Publishing/Photolibrary

48. Carnival Ride. A Ferris wheel traces an elliptical path with both a major and minor axis of 180 feet. Assuming it is centered at the origin, find an equation for the ellipse (circle).

Tina Buckman/Index Stock/Photolibrary

For Exercises 49 and 50, refer to the following information:

A high school wants to build a football field surrounded by an elliptical track. A regulation football field must be 120 yards long and 30 yards wide.

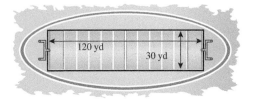

49. Sports Field. Suppose the elliptical track is centered at the origin and has a horizontal major axis of length 150 yards and a minor axis length of 40 yards.

  a. Write an equation for the ellipse.

  b. Find the width of the track at the end of the field. Will the track completely enclose the football field?

50. Sports Field. Suppose the elliptical track is centered at the origin and has a horizontal major axis of length 150 yards. How long should the minor axis be in order to enclose the field?

For Exercises 51 and 52, refer to orbits in our solar system:

The planets have elliptical orbits with the Sun as one of the foci. Pluto (orange), the planet furthest from the Sun, has a very elongated, or flattened, elliptical orbit, whereas the Earth (royal blue) has an almost circular orbit. Because of Pluto's flattened path, it is not always the planet furthest from the Sun.

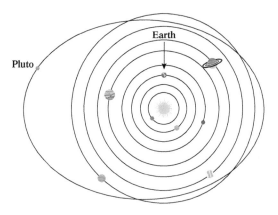

51. Planetary Orbits. The orbit of the dwarf planet Pluto has approximately the following characteristics (assume the Sun is the focus):

  ■ The length of the major axis $2a$ is approximately 11,827,000,000 kilometers.
  ■ The distance from the dwarf planet to the Sun is 4,447,000,000 kilometers.

Determine the equation for Pluto's elliptical orbit around the Sun.

52. Planetary Orbits. The Earth's orbit has approximately the following characteristics (assume the Sun is the focus):

  ■ The length of the major axis $2a$ is approximately 299,700,000 kilometers.
  ■ The distance from the Earth to the Sun is 147,100,000 kilometers.

Determine the equation for the Earth's elliptical orbit around the Sun.

For Exercises 53 and 54, refer to the following information:

Asteroids orbit the Sun in elliptical patterns and often cross paths with the Earth's orbit, making life a little tense now and again. A few asteroids have orbits that cross the Earth's orbit—called "Apollo asteroids" or "Earth-crossing asteroids." In recent years, asteroids have passed within 100,000 kilometers of the Earth!

53. **Asteroids.** Asteroid 433, or Eros, is the second largest near-Earth asteroid. The semimajor axis is 150 million kilometers and the eccentricity is 0.223, where eccentricity is defined as

$$e = \sqrt{1 - \frac{b^2}{a^2}},$$ where a is the semimajor axis or $2a$ is the major axis, and b is the semiminor axis or $2b$ is the minor axis. Find the equation of Eros's orbit. Round to the nearest million kilometers.

54. **Asteroids.** The asteroid Toutatis is the largest near-Earth asteroid. The semimajor axis is 350 million kilometers and the eccentricity is 0.634, where eccentricity is defined as

$$e = \sqrt{1 - \frac{b^2}{a^2}},$$ where a is the semimajor axis or $2a$ is the major axis and b is the semimajor axis or $2b$ is the minor axis. On September 29, 2004, it missed Earth by 961,000 miles. Find the equation of Toutatis's orbit.

55. **Halley's Comet.** The eccentricity of Halley's comet is approximately 0.967. If a comet had $e = 1$, what would its orbit appear to be from Earth?

56. **Halley's Comet.** The length of the semimajor axis is 17.8 AU (astronomical units) and the eccentricity is approximately 0.967. Find the equation of Halley's comet. (Assume 1 AU = 150,000,000 km.)

• **CATCH THE MISTAKE**

In Exercises 57 and 58, explain the mistake that is made.

57. Graph the ellipse given by $\dfrac{x^2}{6} + \dfrac{y^2}{4} = 1$.

Solution:

Write the standard form of the equation of an ellipse.	$\dfrac{x^2}{a^2} + \dfrac{y^2}{b^2} = 1$
Identify a and b.	$a = 6, b = 4$

Label the vertices and the endpoints of the minor axis, $(-6, 0)$, $(6, 0)$, $(0, -4)$, $(0, 4)$, and connect with an elliptical curve.

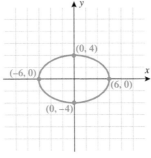

This is incorrect. What mistake was made?

58. Determine the foci of the ellipse $\dfrac{x^2}{16} + \dfrac{y^2}{9} = 1$.

Solution:

Write the general equation of a horizontal ellipse.	$\dfrac{x^2}{a^2} + \dfrac{y^2}{b^2} = 1$
Identify a and b.	$a = 4, b = 3$
Substitute $a = 4, b = 3$ into $c^2 = a^2 + b^2$.	$c^2 = 4^2 + 3^2$
Solve for c.	$c = 5$

Foci are located at $(-5, 0)$ and $(5, 0)$.

The points $(-5, 0)$ and $(5, 0)$ are located outside of the ellipse.

This is incorrect. What mistake was made?

• **CONCEPTUAL**

In Exercises 59–62, determine whether each statement is true or false.

59. If you know the vertices of an ellipse, you can determine the equation for the ellipse.

60. If you know the foci and the endpoints of the minor axis, you can determine the equation for the ellipse.

61. Ellipses centered at the origin have symmetry with respect to the x-axis, y-axis, and the origin.

62. All ellipses are circles, but not all circles are ellipses.

• CHALLENGE

63. The eccentricity of an ellipse is defined as $e = \dfrac{c}{a}$. Compare the eccentricity of the orbit of Pluto to that of the Earth (refer to Exercises 51 and 52).

64. The eccentricity of an ellipse is defined as $e = \dfrac{c}{a}$. Since $a > c > 0$, then $0 < e < 1$. Describe the shape of an ellipse when

    a. e is close to zero   **b.** e is close to one   **c.** $e = 0.5$

• TECHNOLOGY

65. Graph the following three ellipses: $x^2 + y^2 = 1$, $x^2 + 5y^2 = 1$, and $x^2 + 10y^2 = 1$. What can be said to happen to the ellipse $x^2 + cy^2 = 1$ as c increases?

66. Graph the following three ellipses: $x^2 + y^2 = 1$, $5x^2 + y^2 = 1$, and $10x^2 + y^2 = 1$. What can be said to happen to the ellipse $cx^2 + y^2 = 1$ as c increases?

67. Graph the following three ellipses: $x^2 + y^2 = 1$, $5x^2 + 5y^2 = 1$, and $10x^2 + 10y^2 = 1$. What can be said to happen to the ellipse $cx^2 + cy^2 = 1$ as c increases?

68. Graph the equation $\dfrac{x^2}{9} - \dfrac{y^2}{16} = 1$. Notice what a difference the sign makes. Is this an ellipse?

69. Graph the following three ellipses: $x^2 + y^2 = 1$, $0.5x^2 + y^2 = 1$, and $0.05x^2 + y^2 = 1$. What can be said to happen to ellipse $cx^2 + y^2 = 1$ as c decreases?

70. Graph the following three ellipses: $x^2 + y^2 = 1$, $x^2 + 0.5y^2 = 1$, and $x^2 + 0.05y^2 = 1$. What can be said to happen to ellipse $x^2 + cy^2 = 1$ as c decreases?

B.4 THE HYPERBOLA

SKILLS OBJECTIVES	CONCEPTUAL OBJECTIVES
▪ Find the equation of a hyperbola centered at the origin. ▪ Find the equation of a hyperbola centered at the point (h, k). ▪ Solve applied problems that involve hyperbolas.	▪ Understand that a hyperbola is the set of all points in a plane, the difference of whose distances from two fixed points (foci) is a positive constant. ▪ Employ completing the square to transform an equation into the standard form of a hyperbola. ▪ Understand that it is the difference in time of signals remaining constant that allows ships to navigate along a hyperbolic curve to shore.

B.4.1 Hyperbola with Its Center at the Origin

B.4.1 SKILL

Find the equation of a hyperbola centered at the origin.

B.4.1 CONCEPTUAL

Understand that a hyperbola is the set of all points in a plane, the difference of whose distances from two fixed points (foci) is a positive constant.

The definition of a hyperbola is similar to the definition of an ellipse. An ellipse is the set of all points, the *sum* of whose distances from two points (the foci) is constant. A *hyperbola* is the set of all points, the *difference* of whose distances from two points (the foci) is constant. What distinguishes their equations is a minus sign.

Ellipse centered at the origin: $\qquad \dfrac{x^2}{a^2} + \dfrac{y^2}{b^2} = 1$

Hyperbola centered at the origin: $\qquad \dfrac{x^2}{a^2} - \dfrac{y^2}{b^2} = 1$

DEFINITION Hyperbola

A **hyperbola** is the set of all points in a plane, the difference of whose distances from two fixed points is a positive constant. These two fixed points are called **foci**. The hyperbola has two separate curves called **branches**. The two points where the hyperbola intersects the line joining the foci are called **vertices**. The line segment joining the vertices is called the **transverse axis of the hyperbola**. The midpoint of the transverse axis is called the **center**.

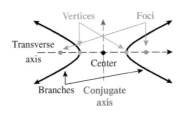

Let's consider a hyperbola with the center at the origin and the foci on the x-axis. Let the distance from the center to the focus be c. Therefore, the foci are located at the points $(-c, 0)$ and $(c, 0)$. The difference of the two distances from the foci to any point (x, y) must be constant. We then can follow a similar analysis as done with an ellipse.

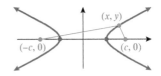

Derivation of the Equation of a Hyperbola

WORDS	MATH
The difference of these two distances is equal to a constant ($2a$ for convenience).	$\sqrt{[x - (-c)]^2 + y^2} - \sqrt{(x - c)^2 + y^2} = \pm 2a$
Following the same procedure that we did with an ellipse leads to:	$(c^2 - a^2)x^2 - a^2y^2 = a^2(c^2 - a^2)$
We can make the argument that $c > a$ in order for a point to be on the hyperbola (and not on the x-axis). Therefore, since a and c represent distances and therefore are positive, we know that $c^2 > a^2$, or $c^2 - a^2 > 0$. Hence, we can divide both sides of the equation by $c^2 - a^2$, since $c^2 - a^2 \neq 0$.	$x^2 - \dfrac{a^2y^2}{(c^2 - a^2)} = a^2$
Let $b^2 = c^2 - a^2$.	$x^2 - \dfrac{a^2y^2}{b^2} = a^2$
Divide both sides of the equation by a^2.	$\dfrac{x^2}{a^2} - \dfrac{y^2}{b^2} = 1$

The equation $\frac{x^2}{a^2} - \frac{y^2}{b^2} = 1$ represents a hyperbola with its center at the origin, with the foci along the x-axis. The following box summarizes hyperbolas that have their center at the origin and foci along either the x-axis or y-axis:

[CONCEPT CHECK]

TRUE OR FALSE In a hyperbola, the foci are always closer to the center than the vertices.

ANSWER False

EQUATION OF A HYPERBOLA WITH ITS CENTER AT THE ORIGIN

The **standard form of the equation of a hyperbola** with its center at the origin is given by

ORIENTATION OF TRANSVERSE AXIS	Horizontal (along the x-axis)	Vertical (along the y-axis)
EQUATION	$\frac{x^2}{a^2} - \frac{y^2}{b^2} = 1$	$\frac{y^2}{a^2} - \frac{x^2}{b^2} = 1$
FOCI	$(-c, 0)$ $(c, 0)$ where $c^2 = a^2 + b^2$	$(0, -c)$ $(0, c)$ where $c^2 = a^2 + b^2$
ASYMPTOTES	$y = \frac{b}{a}x$ $y = -\frac{b}{a}x$	$y = \frac{a}{b}x$ $y = -\frac{a}{b}x$
VERTICES	$(-a, 0)$ $(a, 0)$	$(0, -a)$ $(0, a)$
TRANSVERSE AXIS	Horizontal length $2a$	Vertical length $2a$
GRAPH	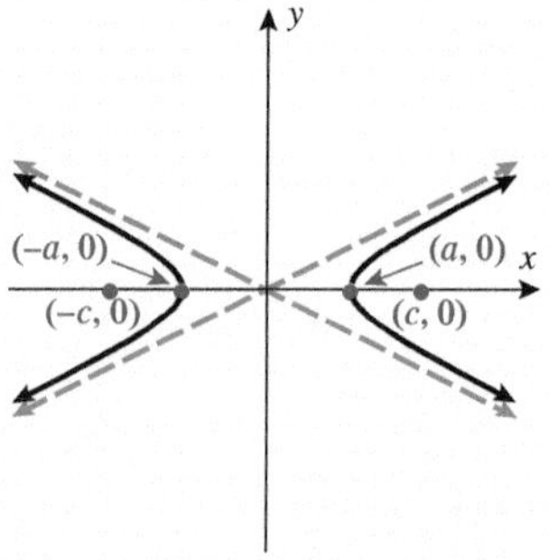	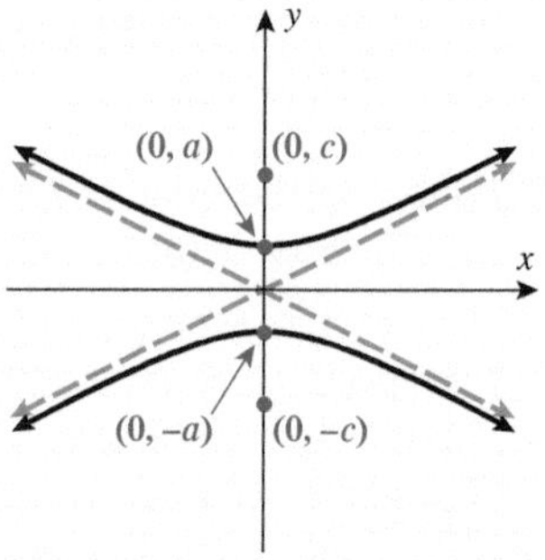

Note that for $\frac{x^2}{a^2} - \frac{y^2}{b^2} = 1$, if $x = 0$, then $-\frac{y^2}{b^2} = 1$, which yields an imaginary number for y. However, when $y = 0$, $\frac{x^2}{a^2} = 1$, and therefore $x = \pm a$. The vertices for this hyperbola are $(-a, 0)$ and $(a, 0)$.

EXAMPLE 1 **Finding the Foci and Vertices of a Hyperbola Given the Equation**

Find the foci and vertices of the hyperbola given by $\dfrac{x^2}{9} - \dfrac{y^2}{4} = 1$.

Solution:

Compare to the standard equation of a hyperbola,

$\dfrac{x^2}{a^2} - \dfrac{y^2}{b^2} = 1$.
$\qquad\qquad\qquad\qquad\qquad\qquad a^2 = 9,\ b^2 = 4$

Solve for a and b.
$\qquad\qquad\qquad\qquad\qquad\qquad a = 3,\ b = 2$

Substitute $a = 3$ into the vertices, $(-a, 0)$ and $(a, 0)$.
$\qquad (-3, 0)$ and $(3, 0)$

Substitute $a = 3,\ b = 2$ into $c^2 = a^2 + b^2$.
$\qquad c^2 = 3^2 + 2^2$

Solve for c.
$\qquad\qquad\qquad\qquad\qquad\qquad c^2 = 13$

$\qquad\qquad\qquad\qquad\qquad\qquad\qquad\qquad c = \sqrt{13}$

Substitute $c = \sqrt{13}$ into the foci, $(-c, 0)$ and $(c, 0)$. $\quad \left(-\sqrt{13}, 0\right)$ and $\left(\sqrt{13}, 0\right)$

The vertices are $\boxed{(-3, 0)}$ and $\boxed{(3, 0)}$, and the foci are $\boxed{\left(-\sqrt{13}, 0\right)}$ and $\boxed{\left(\sqrt{13}, 0\right)}$.

▼ **YOUR TURN** Find the vertices and foci of the hyperbola $\dfrac{y^2}{16} - \dfrac{x^2}{20} = 1$.

▼ **ANSWER**
Vertices: $(0, -4)$ and $(0, 4)$
Foci: $(0, -6)$ and $(0, 6)$

EXAMPLE 2 **Finding the Equation of a Hyperbola Given Foci and Vertices**

Find the equation of a hyperbola whose vertices are located at $(0, -4)$ and $(0, 4)$ and whose foci are located at $(0, -5)$ and $(0, 5)$.

Solution:

The center is located at the midpoint of the segment joining the vertices.
$\qquad \left(\dfrac{0 + 0}{2}, \dfrac{-4 + 4}{2}\right) = (0, 0)$

Since the foci and vertices are located on the y-axis, the standard equation is given by:
$\qquad \dfrac{y^2}{a^2} - \dfrac{x^2}{b^2} = 1$

The vertices $(0, \pm a)$ and the foci $(0, \pm c)$ can be used to identify a and c.
$\qquad a = 4,\ c = 5$

Substitute $a = 4,\ c = 5$ into $b^2 = c^2 - a^2$.
$\qquad b^2 = 5^2 - 4^2$

Solve for b.
$\qquad\qquad\qquad\qquad\qquad\qquad b^2 = 25 - 16 = 9$

$\qquad\qquad\qquad\qquad\qquad\qquad\qquad b = 3$

Substitute $a = 4$ and $b = 3$ into $\dfrac{y^2}{a^2} - \dfrac{x^2}{b^2} = 1$.
$\qquad \boxed{\dfrac{y^2}{16} - \dfrac{x^2}{9} = 1}$

▼ **YOUR TURN** Find the equation of a hyperbola whose vertices are located at $(-2, 0)$ and $(2, 0)$ and whose foci are located at $(-4, 0)$ and $(4, 0)$.

▼ **ANSWER**
$\dfrac{x^2}{4} - \dfrac{y^2}{12} = 1$

Graphing a Hyperbola with Its Center at the Origin

To graph a hyperbola, we use the vertices and asymptotes. The asymptotes are found by the equations $y = \pm\dfrac{b}{a}x$ or $y = \pm\dfrac{a}{b}x$, depending on whether the transverse axis is horizontal or vertical. An easy way to draw these graphing aids is to first draw the rectangular box that passes through the vertices and the points $(0, \pm b)$ or $(\pm b, 0)$. The **conjugate axis** is perpendicular to the transverse axis and has length $2b$. The asymptotes pass through the center of the hyperbola and the corners of the rectangular box.

$$\frac{x^2}{a^2} - \frac{y^2}{b^2} = 1 \qquad\qquad\qquad \frac{y^2}{a^2} - \frac{x^2}{b^2} = 1$$

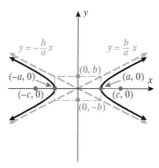
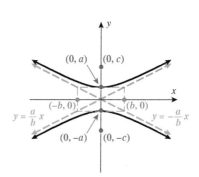

▶ **EXAMPLE 3** **Graphing a Hyperbola Centered at the Origin with a Horizontal Transverse Axis**

Graph the hyperbola given by $\dfrac{x^2}{4} - \dfrac{y^2}{9} = 1$.

Solution:

Compare $\dfrac{x^2}{2^2} - \dfrac{y^2}{3^2} = 1$ to the general equation $\dfrac{x^2}{a^2} - \dfrac{y^2}{b^2} = 1$.

Identify a and b. $\qquad\qquad\qquad\qquad\qquad\qquad a = 2 \quad$ and $\quad b = 3$

The transverse axis of this hyperbola lies on the x-axis.

Label the vertices $(-a, 0) = (-2, 0)$ and $(a, 0) = (2, 0)$ and the points $(0, -b) = (0, -3)$ and $(0, b) = (0, 3)$. Draw the rectangular box that passes through those points. Draw the asymptotes that pass through the center and the corners of the rectangle.

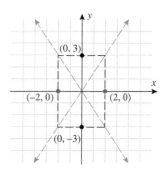

Draw the two **branches** of the hyperbola,
each passing through a vertex and guided
by the asymptotes.

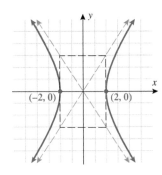

In Example 3, if we let $y = 0$, then $\dfrac{x^2}{4} = 1$ or $x = \pm 2$. Thus, the vertices are $(-2, 0)$
and $(2, 0)$, and the transverse axis lies along the x-axis. Note that if $x = 0$, $y = \pm 3i$.

▶ **EXAMPLE 4**   **Graphing a Hyperbola Centered at the Origin**
                   with a Vertical Transverse Axis

Graph the hyperbola given by $\dfrac{y^2}{16} - \dfrac{x^2}{4} = 1$.

Solution:

Compare $\dfrac{y^2}{4^2} - \dfrac{x^2}{2^2} = 1$ to the general equation $\dfrac{y^2}{a^2} - \dfrac{x^2}{b^2} = 1$.

Identify a and b. $a = 4$ and $b = 2$

The transverse axis of this hyperbola lies along
the y-axis.

Label the vertices $(0, -a) = \mathbf{(0, -4)}$ and
$(0, a) = \mathbf{(0, 4)}$, and the points $(-b, 0) = (-2, 0)$
and $(b, 0) = (2, 0)$. Draw the rectangular box
that passes through those points. Draw the
asymptotes that pass through the center and
the corners of the rectangle.

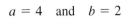

Draw the two **branches** of the hyperbola, each
passing through a vertex and guided by the
asymptotes.

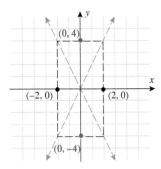

▼ **ANSWER**

a.

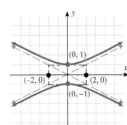

b.

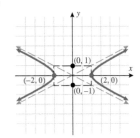

▼

YOUR TURN Graph the hyperbolas:

a. $\dfrac{y^2}{1} - \dfrac{x^2}{4} = 1$ **b.** $\dfrac{x^2}{4} - \dfrac{y^2}{1} = 1$

B.4.2 Hyperbola with Center (*h*, *k*)

B.4.2 **SKILL**

Find the equation of a hyperbola centered at the point (h, k).

B.4.2 **CONCEPTUAL**

Employ completing the square to transform an equation into the standard form of a hyperbola.

We can use graph-shifting techniques to graph hyperbolas that are centered at a point other than the origin—say, (h, k). For example, to graph $\dfrac{(x - h)^2}{a^2} - \dfrac{(y - k)^2}{b^2} = 1$, start with the graph of $\dfrac{x^2}{a^2} - \dfrac{y^2}{b^2} = 1$ and shift to the right h units and up k units. The center, the vertices, the foci, the transverse and conjugate axes, and the asymptotes all shift. The following table summarizes the characteristics of hyperbolas centered at a point other than the origin:

EQUATION OF A HYPERBOLA WITH ITS CENTER AT THE POINT (*h*, *k*)

The **standard form of the equation of a hyperbola** with its center at the point (h, k) is given by

ORIENTATION OF TRANSVERSE AXIS	Horizontal (parallel to the *x*-axis)	Vertical (parallel to the *y*-axis)
EQUATION	$\dfrac{(x - h)^2}{a^2} - \dfrac{(y - k)^2}{b^2} = 1$	$\dfrac{(y - k)^2}{a^2} - \dfrac{(x - h)^2}{b^2} = 1$
VERTICES	$(h - a, k)$ $(h + a, k)$	$(h, k - a)$ $(h, k + a)$
FOCI	$(h - c, k)$ $(h + c, k)$ where $c^2 = a^2 + b^2$	$(h, k - c)$ $(h, k + c)$ where $c^2 = a^2 + b^2$
GRAPH	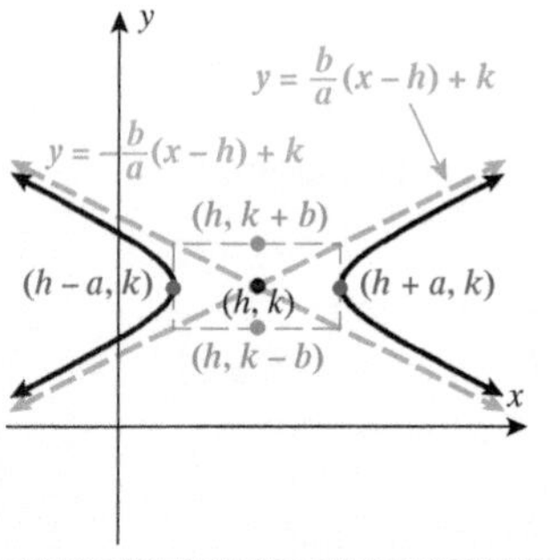	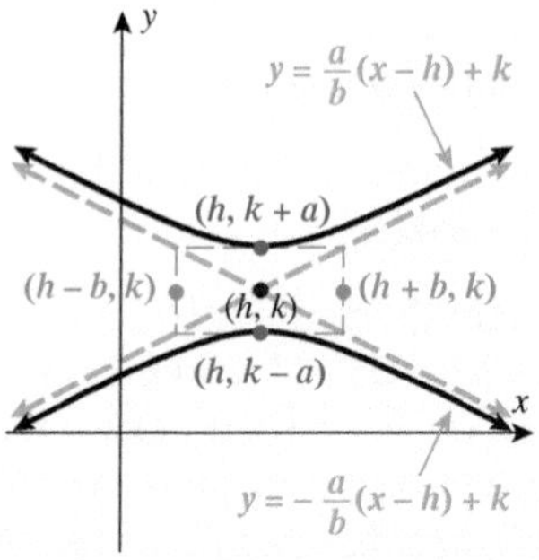

EXAMPLE 5   **Graphing a Hyperbola with Its Center Not at the Origin**

Graph the hyperbola $\dfrac{(y-2)^2}{16} - \dfrac{(x-1)^2}{9} = 1$.

Solution:

Compare $\dfrac{(y-2)^2}{4^2} - \dfrac{(x-1)^2}{3^2} = 1$ to the general equation $\dfrac{(y-k)^2}{a^2} - \dfrac{(x-h)^2}{b^2} = 1$.

Identify a, b, and (h, k). $a = 4$, $b = 3$, and $(h, k) = (1, 2)$

The transverse axis of this hyperbola lies along $x = 1$, which is parallel to the y-axis.

Label the vertices $(h, k - a) = (1, -2)$ and $(h, k + a) = (1, 6)$ and the points $(h - b, k) = (-2, 2)$ and $(h + b, k) = (4, 2)$. Draw the rectangular box that passes through those points. Draw the **asymptotes** that pass through the center $(h, k) = (1, 2)$ and the corners of the rectangle. Draw the two **branches** of the hyperbola, each passing through a vertex and guided by the asymptotes.

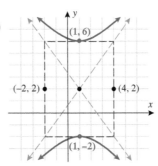

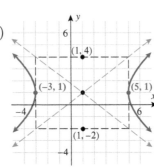

EXAMPLE 6   **Transforming an Equation of a Hyperbola to Standard Form**

Graph the hyperbola $9x^2 - 16y^2 - 18x + 32y - 151 = 0$.

Solution:

Complete the square on the x terms and y terms, respectively.

$$9(x^2 - 2x) - 16(y^2 - 2y) = 151$$
$$9(x^2 - 2x + 1) - 16(y^2 - 2y + 1) = 151 + 9 - 16$$
$$9(x - 1)^2 - 16(y - 1)^2 = 144$$
$$\frac{(x - 1)^2}{16} - \frac{(y - 1)^2}{9} = 1$$

Compare $\dfrac{(x-1)^2}{16} - \dfrac{(y-1)^2}{9} = 1$ to the general form $\dfrac{(x-h)^2}{a^2} - \dfrac{(y-k)^2}{b^2} = 1$.

Identify a, b, and (h, k). $a = 4$, $b = 3$, and $(h, k) = (1, 1)$

The transverse axis of this hyperbola lies along $y = 1$.

Label the vertices $(h - a, k) = (-3, 1)$ and $(h + a, k) = (5, 1)$ and the points $(h, k - b) = (1, -2)$ and $(h, k + b) = (1, 4)$. Draw the rectangular box that passes through these points. Draw the **asymptotes** that pass through the center $(1, 1)$ and the corners of the box. Draw the two **branches** of the hyperbola, each passing through a vertex and guided by the asymptotes.

[CONCEPT CHECK]

A hyperbola of the form

$Ax^2 - Bx - Cy^2 + Dy - E = 0$

where A, B, C, D, and E are all positive has a center that lies in which quadrant?

▼ ....

ANSWER Quadrant I

B.4.3 Applications

Nautical navigation is assisted by hyperbolas. For example, suppose that two radio stations on a coast are emitting simultaneous signals. If a boat is at sea, it will be slightly closer to one station than the other station, which results in a small time difference between the received signals from the two stations. If the boat follows the path associated with a constant time difference, that path will be hyperbolic.

The synchronized signals would intersect one another in associated hyperbolas. Each time difference corresponds to a different path. The radio stations are the foci of the hyperbolas. This principle forms the basis of a hyperbolic radio navigation system known as *loran* (**LO**ng-**RA**nge **N**avigation).

There are navigational charts that correspond to different time differences. A ship selects the hyperbolic path that will take it to the desired port, and the loran chart lists the corresponding time difference.

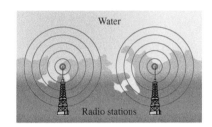

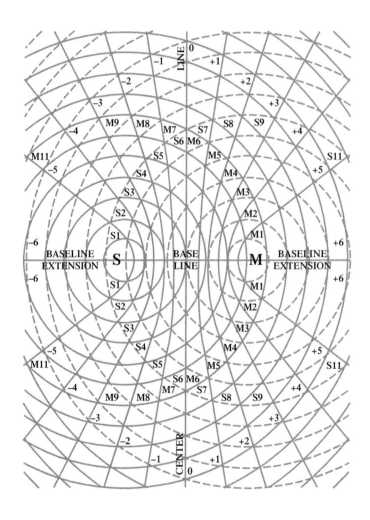

EXAMPLE 7   **Nautical Navigation Using Loran**

Two loran stations are located 200 miles apart along a coast. If a ship records a time difference of 0.00043 second and continues on the hyperbolic path corresponding to that difference, where does it reach shore?

Solution:

Draw the xy-plane and the two stations corresponding to the foci at $(-100, 0)$ and $(100, 0)$. Draw the ship somewhere in quadrant I.

The hyperbola corresponds to a path where the difference of the distances between the ship and the respective stations remains constant. The constant is $2a$, where $(a, 0)$ is a vertex. Find that difference by using $d = rt$. Assume that the speed of the radio signal is 186,000 miles per second.

Substitute $r = 186{,}000$ miles/second and $t = 0.00043$ second into $d = rt$.

$$d = (186{,}000 \text{ mi/sec})(0.00043 \text{ sec}) \approx 80 \text{ mi}$$

Set the constant equal to $2a$. $2a = 80$

Find a vertex $(a, 0)$. $(40, 0)$

> The ship reaches shore between the two stations, 60 miles from station B and 140 miles from station A.

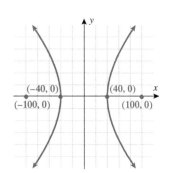

○[SECTION B.4]   SUMMARY

In this section, we discussed hyperbolas centered at the origin.

EQUATION	$\dfrac{x^2}{a^2} - \dfrac{y^2}{b^2} = 1$	$\dfrac{y^2}{a^2} - \dfrac{x^2}{b^2} = 1$
TRANSVERSE AXIS	Horizontal (x-axis), length $2a$	Vertical (y-axis), length $2a$
CONJUGATE AXIS	Vertical (y-axis), length $2b$	Horizontal (x-axis), length $2b$
VERTICES	$(-a, 0)$ and $(a, 0)$	$(0, -a)$ and $(0, a)$
FOCI	$(-c, 0)$ and $(c, 0)$ where $c^2 = a^2 + b^2$	$(0, -c)$ and $(0, c)$ where $c^2 = a^2 + b^2$
ASYMPTOTE	$y = \dfrac{b}{a}x$ and $y = -\dfrac{b}{a}x$	$y = \dfrac{a}{b}x$ and $y = -\dfrac{a}{b}x$
GRAPH		

For a hyperbola centered at (h, k), the vertices, foci, and asymptotes all shift accordingly.

[SECTION B.4] EXERCISES

● **SKILLS**

In Exercises 1–4, match each equation with the corresponding hyperbola.

1. $\dfrac{x^2}{36} - \dfrac{y^2}{16} = 1$ **2.** $\dfrac{y^2}{36} - \dfrac{x^2}{16} = 1$ **3.** $\dfrac{x^2}{8} - \dfrac{y^2}{72} = 1$ **4.** $4y^2 - x^2 = 1$

a.

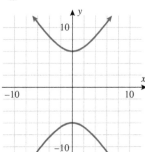

b.

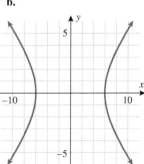

c.

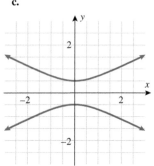

d.

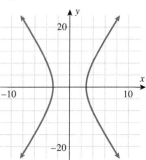

In Exercises 5–16, graph each hyperbola.

5. $\dfrac{x^2}{25} - \dfrac{y^2}{16} = 1$ **6.** $\dfrac{x^2}{49} - \dfrac{y^2}{9} = 1$ **7.** $\dfrac{y^2}{16} - \dfrac{x^2}{64} = 1$ **8.** $\dfrac{y^2}{144} - \dfrac{x^2}{25} = 1$ **9.** $\dfrac{x^2}{100} - y^2 = 1$ **10.** $9y^2 - 4x^2 = 36$

11. $\dfrac{4y^2}{9} - 81x^2 = 1$ **12.** $\dfrac{4}{25}x^2 - \dfrac{100}{9}y^2 = 1$ **13.** $4x^2 - y^2 = 16$ **14.** $y^2 - x^2 = 81$ **15.** $8y^2 - 16x^2 = 32$ **16.** $10x^2 - 25y^2 = 50$

In Exercises 17–24, find the standard form of an equation of the hyperbola with the given characteristics.

17. Vertices: $(-4, 0)$ and $(4, 0)$ Foci: $(-6, 0)$ and $(6, 0)$

18. Vertices: $(-1, 0)$ and $(1, 0)$ Foci: $(-3, 0)$ and $(3, 0)$

19. Vertices: $(0, -3)$ and $(0, 3)$ Foci: $(0, -4)$ and $(0, 4)$

20. Vertices: $(0, -1)$ and $(0, 1)$ Foci: $(0, -2)$ and $(0, 2)$

21. Center: $(0, 0)$; transverse: x-axis; asymptotes: $y = x$ and $y = -x$

22. Center: $(0, 0)$; transverse: y-axis; asymptotes: $y = x$ and $y = -x$

23. Center: $(0, 0)$; transverse axis: y-axis; asymptotes: $y = 2x$ and $y = -2x$

24. Center: $(0, 0)$; transverse axis: x-axis; asymptotes: $y = 2x$ and $y = -2x$

In Exercises 25–28, match each equation with the hyperbola.

25. $\dfrac{(x-3)^2}{4} - \dfrac{(y+2)^2}{25} = 1$ **26.** $\dfrac{(x+3)^2}{4} - \dfrac{(y-2)^2}{25} = 1$ **27.** $\dfrac{(y-3)^2}{25} - \dfrac{(x+2)^2}{4} = 1$ **28.** $\dfrac{(y+3)^2}{25} - \dfrac{(x-2)^2}{4} = 1$

a.

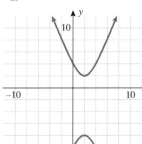

b.

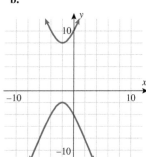

c.

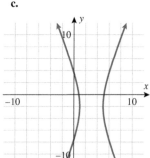

d.

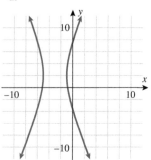

In Exercises 29–38, graph each hyperbola.

29. $\dfrac{(x-1)^2}{16} - \dfrac{(y-2)^2}{4} = 1$

30. $\dfrac{(y+1)^2}{36} - \dfrac{(x+2)^2}{9} = 1$

31. $10(y+3)^2 - (x-4)^2 = 80$

32. $3(x+3)^2 - 12(y-4)^2 = 36$

33. $x^2 - 4x - 4y^2 = 0$

34. $-9x^2 + y^2 + 2y - 8 = 0$

35. $-9x^2 - 18x + 4y^2 - 8y - 41 = 0$

36. $25x^2 - 50x - 4y^2 - 8y - 79 = 0$

37. $x^2 - 6x - 4y^2 - 16y - 8 = 0$

38. $-4x^2 - 16x + y^2 - 2y - 19 = 0$

In Exercises 39–42, find the standard form of the equation of a hyperbola with the given characteristics.

39. Vertices: $(-2, 5)$ and $(6, 5)$ Foci: $(-3, 5)$ and $(7, 5)$

40. Vertices: $(1, -2)$ and $(3, -2)$ Foci: $(0, -2)$ and $(4, -2)$

41. Vertices: $(4, -7)$ and $(4, -1)$ Foci: $(4, -8)$ and $(4, 0)$

42. Vertices: $(2, -6)$ and $(2, -4)$ Foci: $(2, -7)$ and $(2, -3)$

• APPLICATIONS

43. **Ship Navigation.** Two loran stations are located 150 miles apart along a coast. If a ship records a time difference of 0.0005 second and continues on the hyperbolic path corresponding to that difference, where will it reach shore?

44. **Ship Navigation.** Two loran stations are located 300 miles apart along a coast. If a ship records a time difference of 0.0007 second and continues on the hyperbolic path corresponding to that difference, where will it reach shore? Round to the nearest mile.

45. **Ship Navigation.** If the captain of the ship in Exercise 43 wants to reach shore between the stations and 30 miles from one of them, what time difference should he look for?

46. **Ship Navigation.** If the captain of the ship in Exercise 44 wants to reach shore between the stations and 50 miles from one of them, what time difference should he look for?

47. **Light.** If the light from a lamp casts a hyperbolic pattern on the wall due to its lampshade, calculate the equation of the hyperbola if the distance between the vertices is 2 feet and the foci are half a foot from the vertices.

48. **Special Ops.** A military special ops team is calibrating its recording devices used for passive ascertaining of enemy location. They place two recording stations, alpha and bravo, 3000 feet apart (alpha is due west of bravo). The team detonates small explosives 300 feet west of alpha and records the time it takes each station to register an explosion. The team also sets up a second set of explosives directly north of the alpha station. How many feet north of alpha should the team set off the explosives if it wants to record the same times as on the first explosion?

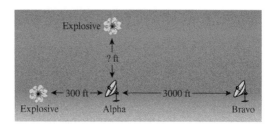

• CATCH THE MISTAKE

In Exercises 49 and 50, explain the mistake that is made.

49. Graph the hyperbola $\dfrac{y^2}{4} - \dfrac{x^2}{9} = 1$.

Solution:

Compare the equation to the standard form and solve for a and b. $a = 2, b = 3$

Label the vertices $(-a, 0)$ and $(a, 0)$. $(-2, 0)$ and $(2, 0)$

Label the points $(0, -b)$ and $(0, b)$. $(0, -3)$ and $(0, 3)$

Draw the rectangle connecting these four points, and align the asymptotes so that they pass through the center and the corner of the boxes. Then draw the hyperbola using the vertices and asymptotes.

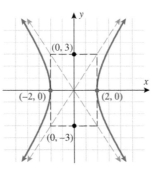

This is incorrect. What mistake was made?

50. Graph the hyperbola $\dfrac{x^2}{1} - \dfrac{y^2}{4} = 1$.

Solution:

Compare the equation to the general form and solve for a and b. $a = 2, b = 1$

Label the vertices $(-a, 0)$ and $(a, 0)$. $(-2, 0)$ and $(2, 0)$

Label the points $(0, -b)$ and $(0, b)$. $(0, -1)$ and $(0, 1)$

Draw the rectangle connecting these four points, and align the asymptotes so that they pass through the center and the corner of the boxes. Then draw the hyperbola using the vertices and asymptotes.

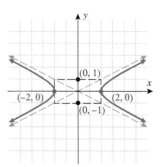

This is incorrect. What mistake was made?

In Exercises 51–54, determine whether each statement is true or false.

51. If you know the vertices of a hyperbola, you can determine the equation for the hyperbola.

52. If you know the foci and vertices, you can determine the equation for the hyperbola.

53. Hyperbolas centered at the origin have symmetry with respect to the x-axis, y-axis, and the origin.

54. The center and foci are part of the graph of a hyperbola.

55. Find the general equation of a hyperbola whose asymptotes are perpendicular.

56. Find the general equation of a hyperbola whose vertices are $(3, -2)$ and $(-1, -2)$ and whose asymptotes are the lines $y = 2x - 4$ and $y = -2x$.

57. Graph the following three hyperbolas: $x^2 - y^2 = 1$, $x^2 - 5y^2 = 1$, and $x^2 - 10y^2 = 1$. What can be said to happen to the hyperbola $x^2 - cy^2 = 1$ as c increases?

58. Graph the following three hyperbolas: $x^2 - y^2 = 1$, $5x^2 - y^2 = 1$, and $10x^2 - y^2 = 1$. What can be said to happen to the hyperbola $cx^2 - y^2 = 1$ as c increases?

59. Graph the following three hyperbolas: $x^2 - y^2 = 1$, $0.5x^2 - y^2 = 1$, and $0.05x^2 - y^2 = 1$. What can be said to happen to the hyperbola $cx^2 - y^2 = 1$ as c decreases?

60. Graph the following three hyperbolas: $x^2 - y^2 = 1$, $x^2 - 0.5y^2 = 1$, and $x^2 - 0.05y^2 = 1$. What can be said to happen to the hyperbola $x^2 - cy^2 = 1$ as c decreases?

B.5 ROTATION OF AXES

SKILLS OBJECTIVES

- Transform general second-degree equations into recognizable equations of conics by analyzing rotation of axes.
- Determine the angle of rotation that will transform a general second-degree equation into a familiar equation of a conic section.

CONCEPTUAL OBJECTIVES

- Understand how the equation of a conic section is altered by rotation of axes.
- Understand how the angle of rotation formula is derived.

B.5.1 Transforming Second-Degree Equations Using Rotation of Axes

B.5.1 SKILL

Transform general second-degree equations into recognizable equations of conics by analyzing rotations of axes.

B.5.1 CONCEPTUAL

Understand how the equation of a conic section is altered by rotation of axes.

In Sections B.1 through B.4, we learned the general equations of parabolas, ellipses, and hyperbolas that were centered at any point in the Cartesian plane and whose vertices and foci were aligned either along or parallel to either the x-axis or the y-axis. We learned, for example, that the general equation of an ellipse centered at the origin is given by

$$\frac{x^2}{a^2} + \frac{y^2}{b^2} = 1$$

where the major and minor axes are, respectively, either the x- or the y-axis, depending on whether a is greater than or less than b. Now let us look at an equation of a conic section whose graph is *not* aligned with the x- or y-axis: the equation $5x^2 - 8xy + 5y^2 - 9 = 0$.

This graph can be thought of as an ellipse that started with the major axis along the x-axis and the minor axis along the y-axis and then was rotated counterclockwise 45°. A new XY-coordinate system can be introduced such that this system has the same origin, but the XY-coordinate system is rotated by a certain amount from the standard xy-coordinate system. In this example, the major axis of the ellipse lies along the new X-axis, and the minor axis lies along the new Y-axis. We will see that we can write the equation of this ellipse as

$$\frac{X^2}{9} + \frac{Y^2}{1} = 1$$

We will now develop the *rotation of axes formulas*, which allow us to transform the generalized second-degree equation in xy, that is, $Ax^2 + Bxy + Cy^2 + Dx + Ey + F = 0$, into an equation in XY of a conic that is familiar to us.

Let the new XY-coordinate system be displaced from the xy-coordinate system by rotation through an angle θ. Let P represent some point a distance r from the origin.

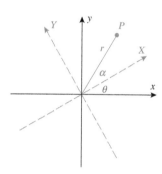

We can represent the point P as the point (x, y) or the point (X, Y).

We define the angle α as the angle r makes with the X-axis and $\alpha + \theta$ as the angle r makes with the x-axis.

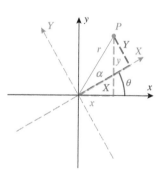

We can represent the point P in polar coordinates using the following relationships:

$$x = r\cos(\alpha + \theta)$$
$$y = r\sin(\alpha + \theta)$$
$$X = r\cos\alpha$$
$$Y = r\sin\alpha$$

WORDS	MATH
Start with the x-term and write the cosine identity for a sum.	$x = r\cos(\alpha + \theta)$ $\quad = r(\cos\alpha\cos\theta - \sin\alpha\sin\theta)$
Eliminate the parentheses and group r with the α-terms.	$x = (r\cos\alpha)\cos\theta - (r\sin\alpha)\sin\theta$
Substitute according to the relationships $X = r\cos\alpha$ and $Y = r\sin\alpha$.	$\boxed{x = X\cos\theta - Y\sin\theta}$
Start with the y-term and write the sine identity for a sum.	$y = r\sin(\alpha + \theta)$ $\quad = r(\sin\alpha\cos\theta + \cos\alpha\sin\theta)$
Eliminate the parentheses and group r with the α-terms.	$y = (r\sin\alpha)\cos\theta + (r\cos\alpha)\sin\theta$
Substitute according to the relationships $X = r\cos\alpha$ and $Y = r\sin\alpha$.	$\boxed{y = Y\cos\theta + X\sin\theta}$

By treating the highlighted equations for x and y as a system of linear equations in X and Y, we can then solve for X and Y in terms of x and y. The results are summarized in the following box:

ROTATION OF AXES FORMULAS

Suppose that the x- and y-axes in the rectangular coordinate plane are rotated through an acute angle θ to produce the X- and Y-axes. Then, the coordinates (x, y) and (X, Y) are related according to the following equations:

$$x = X\cos\theta - Y\sin\theta \qquad \qquad X = x\cos\theta + y\sin\theta$$
$$y = X\sin\theta + Y\cos\theta \qquad \text{or} \qquad Y = -x\sin\theta + y\cos\theta$$

EXAMPLE 1 Rotating the Axes

If the xy-coordinate axes are rotated $60°$, find the XY-coordinates of the point $(x, y) = (-3, 4)$.

Solution:

Start with the rotation formulas.

$$X = x\cos\theta + y\sin\theta$$
$$Y = -x\sin\theta + y\cos\theta$$

Let $x = -3$, $y = 4$, and $\theta = 60°$.

$$X = -3\cos 60° + 4\sin 60°$$
$$Y = -(-3)\sin 60° + 4\cos 60°$$

Simplify.

$$X = -3\underbrace{\cos 60°}_{\frac{1}{2}} + 4\underbrace{\sin 60°}_{\frac{\sqrt{3}}{2}}$$

$$Y = 3\underbrace{\sin 60°}_{\frac{\sqrt{3}}{2}} + 4\underbrace{\cos 60°}_{\frac{1}{2}}$$

$$X = -\frac{3}{2} + 2\sqrt{3}$$

$$Y = \frac{3\sqrt{3}}{2} + 2$$

The XY-coordinates are $\boxed{\left(-\frac{3}{2} + 2\sqrt{3}, \frac{3\sqrt{3}}{2} + 2\right)}$.

ANSWER
$\left(\frac{3\sqrt{3}}{2} - 2, -\frac{3}{2} - 2\sqrt{3}\right)$

YOUR TURN If the xy-coordinate axes are rotated $30°$, find the XY-coordinates of the point $(x, y) = (3, -4)$.

EXAMPLE 2 Rotating an Ellipse

Show that the graph of the equation $5x^2 - 8xy + 5y^2 - 9 = 0$ is an ellipse aligning with coordinate axes that are rotated by $45°$.

Solution:

Start with the rotation formulas.

$$x = X\cos\theta - Y\sin\theta$$
$$y = X\sin\theta + Y\cos\theta$$

Let $\theta = 45°$.

$$x = X\underbrace{\cos 45°}_{\frac{\sqrt{2}}{2}} - Y\underbrace{\sin 45°}_{\frac{\sqrt{2}}{2}}$$

$$y = X\underbrace{\sin 45°}_{\frac{\sqrt{2}}{2}} + Y\underbrace{\cos 45°}_{\frac{\sqrt{2}}{2}}$$

Simplify.

$$x = \frac{\sqrt{2}}{2}(X - Y)$$
$$y = \frac{\sqrt{2}}{2}(X + Y)$$

Substitute $x = \frac{\sqrt{2}}{2}(X - Y)$ and $y = \frac{\sqrt{2}}{2}(X + Y)$ into $5x^2 - 8xy + 5y^2 - 9 = 0$.

$$5\left[\frac{\sqrt{2}}{2}(X - Y)\right]^2 - 8\left[\frac{\sqrt{2}}{2}(X - Y)\right]\left[\frac{\sqrt{2}}{2}(X + Y)\right] + 5\left[\frac{\sqrt{2}}{2}(X + Y)\right]^2 - 9 = 0$$

Simplify. $$\frac{5}{2}(X^2 - 2XY + Y^2) - 4(X^2 - Y^2) + \frac{5}{2}(X^2 + 2XY + Y^2) - 9 = 0$$

$$\frac{5}{2}X^2 - 5XY + \frac{5}{2}Y^2 - 4X^2 + 4Y^2 + \frac{5}{2}X^2 + 5XY + \frac{5}{2}Y^2 = 9$$

Combine like terms. $$X^2 + 9Y^2 = 9$$

Divide by 9. $$\boxed{\frac{X^2}{9} + \frac{Y^2}{1} = 1}$$

This (as discussed earlier) is an ellipse whose major axis is along the X-axis.

The vertices are at the points $(X, Y) = (\pm 3, 0)$.

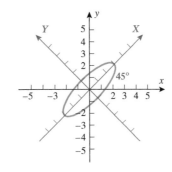

B.5.2 Determine the Angle of Rotation Necessary to Transform a General Second-Degree Equation into an Equation of a Conic

In Section B.1, we stated that the general second-degree equation

$$Ax^2 + Bxy + Cy^2 + Dx + Ey + F = 0$$

corresponds to a graph of a conic. Which type of conic it is depends on the value of the discriminant, $B^2 - 4AC$. In Sections B.2–B.4, we discussed graphs of parabolas, ellipses, and hyperbolas with vertices along either the axes or lines parallel (or perpendicular) to the axes. In all cases, the value of B was taken to be zero. When the value of B is nonzero, the result is a conic with vertices along the new XY-axes (or, respectively, parallel and perpendicular to them), which are the original xy-axes rotated through an angle θ. If given θ, we can determine the rotation equations as illustrated in Example 2, but how do we find the angle θ that represents the *angle of rotation*?

B.5.2 SKILL

Determine the angle of rotation that will transform a general second-degree equation into a familiar equation of a conic section.

B.5.2 CONCEPTUAL

Understand how the angle of rotation formula is derived.

To find the angle of rotation, let us start with a general second-degree polynomial equation:

$$Ax^2 + Bxy + Cy^2 + Dx + Ey + F = 0$$

We want to transform this equation into an equation in X and Y that does not contain an XY- term. Suppose we rotate our coordinates by an angle θ and then use the rotation equations

$$x = X\cos\theta - Y\sin\theta \qquad y = X\sin\theta + Y\cos\theta$$

in the general second-degree polynomial equation; then the result is

$$A(X\cos\theta - Y\sin\theta)^2 + B(X\cos\theta - Y\sin\theta)(X\sin\theta + Y\cos\theta)$$
$$+ C(X\sin\theta + Y\cos\theta)^2 + D(X\cos\theta - Y\sin\theta) + E(X\sin\theta + Y\cos\theta) + F = 0$$

If we expand these expressions and collect like terms, the result is an equation of the form

$$aX^2 + bXY + cY^2 + dX + eY + f = 0$$

where

$$a = A\cos^2\theta + B\sin\theta\cos\theta + C\sin^2\theta$$
$$b = B(\cos^2\theta - \sin^2\theta) + 2(C - A)\sin\theta\cos\theta$$
$$c = A\sin^2\theta - B\sin\theta\cos\theta + C\cos^2\theta$$
$$d = D\cos\theta + E\sin\theta$$
$$e = -D\sin\theta + E\cos\theta$$
$$f = F$$

WORDS	MATH
We do not want this new equation to have an XY-term, so we set $b = 0$.	$B(\cos^2\theta - \sin^2\theta) + 2(C - A)\sin\theta\cos\theta = 0$
We can use the double-angle formulas to simplify.	$B\underbrace{(\cos^2\theta - \sin^2\theta)}_{\cos(2\theta)} + (C - A)\underbrace{2\sin\theta\cos\theta}_{\sin(2\theta)} = 0$
Subtract the sine term.	$B\cos(2\theta) = (A - C)\sin(2\theta)$
Divide by $B\sin(2\theta)$.	$\dfrac{B\cos(2\theta)}{B\sin(2\theta)} = \dfrac{(A - C)\sin(2\theta)}{B\sin(2\theta)}$
Simplify.	$\boxed{\cot(2\theta) = \dfrac{A - C}{B}}$

ANGLE OF ROTATION FORMULA

To transform the equation of a conic

$$Ax^2 + Bxy + Cy^2 + Dx + Ey + F = 0$$

into an equation in X and Y without an XY-term, rotate the xy-axes by an acute angle θ that satisfies the equation

$$\cot(2\theta) = \frac{A - C}{B}$$

Notice that the trigonometric equation $\cot(2\theta) = \dfrac{A - C}{B}$ can be solved exactly for some values of θ (Example 3) and will have to be approximated with a calculator for other values of θ (Example 4).

EXAMPLE 3 **Determining the Angle of Rotation I: The Value of the Cotangent Function Is That of a Known (Special) Angle**

Determine the angle of rotation necessary to transform the following equation into an equation in X and Y with no XY-term.

$$3x^2 + 2\sqrt{3}xy + y^2 + 2x - 2\sqrt{3}y = 0$$

Solution:

Identify the A, B, and C parameters in the equation.

$$\underset{A}{3x^2} + \underset{B}{2\sqrt{3}}\,xy + \underset{C}{1y^2} + 2x - 2\sqrt{3}y = 0$$

Write the rotation formula.

$$\cot(2\theta) = \frac{A - C}{B}$$

Let $A = 3, B = 2\sqrt{3}$, and $C = 1$.

$$\cot(2\theta) = \frac{3 - 1}{2\sqrt{3}}$$

Simplify.

$$\cot(2\theta) = \frac{1}{\sqrt{3}}$$

Apply the reciprocal identity.

$$\tan(2\theta) = \sqrt{3}$$

From our knowledge of trigonometric exact values, we know that $2\theta = 60°$ or $\boxed{\theta = 30°}$.

EXAMPLE 4 **Determining the Angle of Rotation II: The Argument of the Cotangent Function Needs to Be Approximated with a Calculator**

Determine the angle of rotation necessary to transform the following equation into an equation in X and Y with no XY-term. Round to the nearest tenth of a degree.

$$4x^2 + 2xy - 6y^2 - 5x + y - 2 = 0$$

Solution:

Identify the A, B, and C parameters in the equation.

$$\underset{A}{4x^2} + \underset{B}{2xy} - \underset{C}{6y^2} - 5x + y - 2 = 0$$

Write the rotation formula.

$$\cot(2\theta) = \frac{A - C}{B}$$

Let $A = 4, B = 2$, and $C = -6$.

$$\cot(2\theta) = \frac{4 - (-6)}{2}$$

Simplify.

$$\cot(2\theta) = 5$$

Apply the reciprocal identity.

$$\tan(2\theta) = \frac{1}{5} = 0.2$$

Write the result as an inverse tangent function.

$$2\theta = \tan^{-1}(0.2)$$

With a calculator evaluate the right side of the equation.

$$2\theta \approx 11.31°$$

Solve for θ and round to the nearest tenth of a degree.

$$\boxed{\theta = 5.7°}$$

B.5.3 Graphing a Rotated Conic

Special attention must be given when evaluating the inverse tangent function on a calculator, as the result is always in quadrant I or IV. If 2θ turns out to be negative, then $180°$ must be added so that 2θ is in quadrant II (as opposed to quadrant IV). Then θ will be an acute angle lying in quadrant I.

Recall that we stated (without proof) in Section B.1 that we can identify a general equation of the form

$$Ax^2 + Bxy + Cy^2 + Dx + Ey + F = 0$$

as that of a particular conic depending on the discriminant.

Parabola	$B^2 - 4AC = 0$
Ellipse	$B^2 - 4AC < 0$
Hyperbola	$B^2 - 4AC > 0$

EXAMPLE 5   **Graphing a Rotated Conic**

For the equation $x^2 + 2xy + y^2 - \sqrt{2}x - 3\sqrt{2}y + 6 = 0$:

a. Determine which conic the equation represents.

b. Find the rotation angle required to eliminate the XY-term in the new coordinate system.

c. Transform the equation in x and y into an equation in X and Y.

d. Graph the resulting conic.

Solution (a):

Identify A, B, and C.
$$\underset{A}{1x^2} + \underset{B}{2xy} + \underset{C}{1y^2} - \sqrt{2}x - 3\sqrt{2}y + 6 = 0$$

$$A = 1, B = 2, C = 1$$

Compute the discriminant.
$$B^2 - 4AC = 2^2 - 4(1)(1) = 0$$

Since the discriminant equals zero, the equation represents a **parabola.**

Solution (b):

Write the rotation formula.
$$\cot(2\theta) = \frac{A - C}{B}$$

Let $A = 1, B = 2$, and $C = 1$.
$$\cot(2\theta) = \frac{1 - 1}{2}$$

Simplify.
$$\cot(2\theta) = 0$$

Write the cotangent function in terms of the sine and cosine functions.
$$\frac{\cos(2\theta)}{\sin(2\theta)} = 0$$

The numerator must equal zero.
$$\cos(2\theta) = 0$$

From our knowledge of trigonometric exact values, we know that $2\theta = 90°$ or $\boxed{\theta = 45°}$.

Solution (c):

Start with the equation
$x^2 + 2xy + y^2 - \sqrt{2}x - 3\sqrt{2}y + 6 = 0$,
and use the rotation formulas
with $\theta = 45°$.

$$x = X\cos 45° - Y\sin 45° = \frac{\sqrt{2}}{2}(X - Y)$$

$$y = X\sin 45° + Y\cos 45° = \frac{\sqrt{2}}{2}(X + Y)$$

Find x^2, xy, and y^2.

$$x^2 = \left[\frac{\sqrt{2}}{2}(X - Y)\right]^2 = \frac{1}{2}(X^2 - 2XY + Y^2)$$

$$xy = \left[\frac{\sqrt{2}}{2}(X - Y)\right]\left[\frac{\sqrt{2}}{2}(X + Y)\right] = \frac{1}{2}(X^2 - Y^2)$$

$$y^2 = \left[\frac{\sqrt{2}}{2}(X + Y)\right]^2 = \frac{1}{2}(X^2 + 2XY + Y^2)$$

Substitute the values
for x, y, x^2, xy,
and y^2 into the
original equation.

$$x^2 + 2xy + y^2 - \sqrt{2}x - 3\sqrt{2}y + 6 = 0$$

$$\frac{1}{2}(X^2 - 2XY + Y^2) + 2\left[\frac{1}{2}(X^2 - Y^2)\right]$$

$$+ \frac{1}{2}(X^2 + 2XY + Y^2) - \sqrt{2}\left[\frac{\sqrt{2}}{2}(X - Y)\right]$$

$$- 3\sqrt{2}\left[\frac{\sqrt{2}}{2}(X + Y)\right] + 6 = 0$$

Eliminate the parentheses
and combine like terms.

$$2X^2 - 4X - 2Y + 6 = 0$$

Divide by 2.

$$X^2 - 2X - Y + 3 = 0$$

Add Y.

$$Y = (X^2 - 2X) + 3$$

Complete the square on X.

$$Y = (X - 1)^2 + 2$$

Solution (d):

This is a parabola opening upward in the
XY-coordinate system shifted to the
right one unit and up two units.

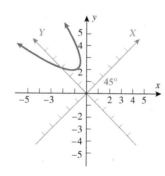

▶[SECTION B.5] **SUMMARY**

In this section, we found that the graph of the general second-degree equation

$$Ax^2 + Bxy + Cy^2 + Dx + Ey + F = 0$$

can represent conics in a system of rotated axes.

The following are the rotation formulas relating the xy-coordinate system to a rotated coordinate system with axes X and Y

$$x = X\cos\theta - Y\sin\theta$$
$$y = X\sin\theta + Y\cos\theta$$

where the rotation angle θ is found from the equation

$$\cot(2\theta) = \frac{A - C}{B}$$

[SECTION B.5] EXERCISES

- **SKILLS**

In Exercises 1–8, the coordinates of a point in the xy-coordinate system are given. Assuming that the XY-axes are found by rotating the xy-axes by an angle θ, find the corresponding coordinates for the point in the XY-system.

1. $(2, 4)$, $\theta = 45°$

2. $(5, 1)$, $\theta = 60°$

3. $(-3, 2)$, $\theta = 30°$

4. $(-4, 6)$, $\theta = 45°$

5. $(-1, -3)$, $\theta = 60°$

6. $(4, -4)$, $\theta = 45°$

7. $(0, 3)$, $\theta = 60°$

8. $(-2, 0)$, $\theta = 30°$

In Exercises 9–24, (a) identify the type of conic from the discriminant, (b) transform the equation in x and y into an equation in X and Y (without an XY-term) by rotating the x- and y-axes by an angle θ to arrive at the new X- and Y-axes, and (c) graph the resulting equation (showing both sets of axes).

9. $xy - 1 = 0$, $\theta = 45°$

10. $xy - 4 = 0$, $\theta = 45°$

11. $x^2 + 2xy + y^2 + \sqrt{2}x - \sqrt{2}y - 1 = 0$, $\theta = 45°$

12. $2x^2 - 4xy + 2y^2 - \sqrt{2}x + 1 = 0$, $\theta = 45°$

13. $y^2 - \sqrt{3}xy + 3 = 0$, $\theta = 30°$

14. $x^2 - \sqrt{3}xy - 3 = 0$, $\theta = 60°$

15. $7x^2 - 2\sqrt{3}xy + 5y^2 - 8 = 0$, $\theta = 60°$

16. $4x^2 + \sqrt{3}xy + 3y^2 - 45 = 0$, $\theta = 30°$

17. $3x^2 + 2\sqrt{3}xy + y^2 + 2x - 2\sqrt{3}y - 2 = 0$, $\theta = 30°$

18. $x^2 + 2\sqrt{3}xy + 3y^2 - 2\sqrt{3}x + 2y - 4 = 0$, $\theta = 60°$

19. $7x^2 + 4\sqrt{3}xy + 3y^2 - 9 = 0$, $\theta = \dfrac{\pi}{6}$

20. $37x^2 + 42\sqrt{3}xy + 79y^2 - 400 = 0$, $\theta = \dfrac{\pi}{3}$

21. $7x^2 - 10\sqrt{3}xy - 3y^2 + 24 = 0$, $\theta = \dfrac{\pi}{3}$

22. $9x^2 + 14\sqrt{3}xy - 5y^2 + 48 = 0$, $\theta = \dfrac{\pi}{6}$

23. $x^2 - 2xy + y^2 - \sqrt{2}x - \sqrt{2}y - 8 = 0$, $\theta = \dfrac{\pi}{4}$

24. $x^2 + 2xy + y^2 + 3\sqrt{2}x + \sqrt{2}y = 0$, $\theta = \dfrac{\pi}{4}$

In Exercises 25–38, determine the angle of rotation necessary to transform the equation in x and y into an equation in X and Y with no XY-term.

25. $x^2 + 4xy + y^2 - 4 = 0$

26. $3x^2 + 5xy + 3y^2 - 2 = 0$

27. $2x^2 + \sqrt{3}xy + 3y^2 - 1 = 0$

28. $4x^2 + \sqrt{3}xy + 3y^2 - 1 = 0$

29. $2x^2 + \sqrt{3}xy + y^2 - 5 = 0$

30. $2\sqrt{3}x^2 + xy + 3\sqrt{3}y^2 + 1 = 0$

31. $\sqrt{2}x^2 + xy + \sqrt{2}y^2 - 1 = 0$

32. $x^2 + 10xy + y^2 + 2 = 0$

33. $12\sqrt{3}x^2 + 4xy + 8\sqrt{3}y^2 - 1 = 0$

34. $4x^2 + 2xy + 2y^2 - 7 = 0$

35. $5x^2 + 6xy + 4y^2 - 1 = 0$

36. $x^2 + 2xy + 12y^2 + 3 = 0$

37. $3x^2 + 10xy + 5y^2 - 1 = 0$

38. $10x^2 + 3xy + 2y^2 + 3 = 0$

In Exercises 39–48, graph the second-degree equation. (*Hint:* Transform the equation into an equation that contains no xy-term.)

39. $21x^2 + 10\sqrt{3}xy + 31y^2 - 144 = 0$

40. $5x^2 + 6xy + 5y^2 - 8 = 0$

41. $8x^2 - 20xy + 8y^2 + 18 = 0$

42. $3y^2 - 26\sqrt{3}xy - 23x^2 - 144 = 0$

43. $3x^2 + 2\sqrt{3}xy + y^2 + 2x - 2\sqrt{3}y - 12 = 0$

44. $3x^2 - 2\sqrt{3}xy + y^2 - 2x - 2\sqrt{3}y - 4 = 0$

45. $37x^2 - 42\sqrt{3}xy + 79y^2 - 400 = 0$

46. $71x^2 - 58\sqrt{3}xy + 13y^2 + 400 = 0$

47. $x^2 + 2xy + y^2 + 5\sqrt{2}x + 3\sqrt{2}y = 0$

48. $7x^2 - 4\sqrt{3}xy + 3y^2 - 9 = 0$

- **CONCEPTUAL**

In Exercises 49–52, determine whether each statement is true or false.

49. The graph of the equation $x^2 + kxy + 9y^2 = 5$, where k is any positive constant less than 6, is an ellipse.

50. The graph of the equation $x^2 + kxy + 9y^2 = 5$, where k is any constant greater than 6, is a parabola.

51. The reciprocal function is a rotated hyperbola.

52. The equation $\sqrt{x} + \sqrt{y} = 3$ can be transformed into the equation $X^2 + Y^2 = 9$.

53. Determine the equation in X and Y that corresponds to

$\dfrac{x^2}{a^2} + \dfrac{y^2}{b^2} = 1$ when the axes are rotated through

   a. $90°$ **b.** $180°$

54. Determine the equation in X and Y that corresponds to

$\dfrac{x^2}{a^2} - \dfrac{y^2}{b^2} = 1$ when the axes are rotated through

   a. $90°$ **b.** $180°$

55. Identify the conic section with equation $y^2 + ax^2 = x$ for various values of a.

56. Identify the conic section with equation $x^2 - ay^2 = y$ for various values of a.

For Exercises 57–62, refer to the following:

To use a TI-83 or TI-83 Plus (function-driven software or graphing utility) to graph a general second-degree equation, you need to solve for y. Let us consider a general second-degree equation $Ax^2 + Bxy + Cy^2 + Dx + Ey + F = 0$.

   Group y^2 terms together, y terms together, and the remaining terms together.

$$Ax^2 + \underline{Bxy} + \underline{Cy^2} + Dx + \underline{Ey} + F = 0$$

$$Cy^2 + (Bxy + Ey) + (Ax^2 + Dx + F) = 0$$

Factor out the common y in the first set of parentheses.

$$Cy^2 + y(Bx + E) + (Ax^2 + Dx + F) = 0$$

Now this is a quadratic equation in y.

$$ay^2 + by + c = 0$$

Use the quadratic formula to solve for y.

$$Cy^2 + y(Bx + E) + (Ax^2 + Dx + F) = 0$$

$$a = C, \quad b = Bx + E, \quad c = Ax^2 + Dx + F$$

$$y = \frac{-b \pm \sqrt{b^2 - 4ac}}{2a}$$

$$y = \frac{-(Bx + E) \pm \sqrt{(Bx + E)^2 - 4(C)(Ax^2 + Dx + F)}}{2(C)}$$

$$y = \frac{-(Bx + E) \pm \sqrt{B^2x^2 + 2BEx + E^2 - 4ACx^2 - 4CDx - 4CF}}{2C}$$

$$y = \frac{-(Bx + E) \pm \sqrt{(B^2 - 4AC)x^2 + (2BE - 4CD)x + (E^2 - 4CF)}}{2C}$$

Case I: $B^2 - 4AC = 0 \rightarrow$ The second-degree equation $Ax^2 + Bxy + Cy^2 + Dx + Ey + F = 0$ is a parabola.

$$y = \frac{-(Bx + E) \pm \sqrt{(2BE - 4CD)x + (E^2 - 4CF)}}{2C}$$

Case II: $B^2 - 4AC < 0 \rightarrow$ The second-degree equation $Ax^2 + Bxy + Cy^2 + Dx + Ey + F = 0$ is an ellipse.

$$y = \frac{-(Bx + E) \pm \sqrt{(B^2 - 4AC)x^2 + (2BE - 4CD)x + (E^2 - 4CF)}}{2C}$$

Case III: $B^2 - 4AC > 0 \rightarrow$ The second-degree equation $Ax^2 + Bxy + Cy^2 + Dx + Ey + F = 0$ is a hyperbola.

$$y = \frac{-(Bx + E) \pm \sqrt{(B^2 - 4AC)x^2 + (2BE - 4CD)x + (E^2 - 4CF)}}{2C}$$

57. Use a graphing utility to explore the second-degree equation $3x^2 + 2\sqrt{3}xy + y^2 + Dx + Ey + F = 0$ for the following values of D, E, and F:

a. $D = 1, E = 3, F = 2$

b. $D = -1, E = -3, F = 2$

Show the angle of rotation to the nearest degree. Explain the differences.

58. Use a graphing utility to explore the second-degree equation $x^2 + 3xy + 3y^2 + Dx + Ey + F = 0$ for the following values of D, E, and F:

a. $D = 2, E = 6, F = -1$

b. $D = 6, E = 2, F = -1$

Show the angle of rotation to the nearest degree. Explain the differences.

59. Use a graphing utility to explore the second-degree equation $2x^2 + 3xy + y^2 + Dx + Ey + F = 0$ for the following values of D, E, and F:

a. $D = 2, E = 1, F = -2$

b. $D = 2, E = 1, F = 2$

Show the angle of rotation to the nearest degree. Explain the differences.

60. Use a graphing utility to explore the second-degree equation $2\sqrt{3}x^2 + xy + \sqrt{3}y^2 + Dx + Ey + F = 0$ for the following values of D, E, and F:

a. $D = 2, E = 1, F = -1$

b. $D = 2, E = 6, F = -1$

Show the angle of rotation to the nearest degree. Explain the differences.

61. Use a graphing utility to explore the second-degree equation $Ax^2 + Bxy + Cy^2 + 2x + y - 1 = 0$ for the following values of A, B, and C:

a. $A = 4, B = -4, C = 1$

b. $A = 4, B = 4, C = -1$

c. $A = 1, B = -4, C = 4$

Show the angle of rotation to the nearest degree. Explain the differences.

62. Use a graphing utility to explore the second-degree equation $Ax^2 + Bxy + Cy^2 + 3x + 5y - 2 = 0$ for the following values of A, B, and C:

a. $A = 1, B = -4, C = 4$

b. $A = 1, B = 4, D = -4$

Show the angle of rotation to the nearest degree. Explain the differences.

B.6 POLAR EQUATIONS OF CONICS

SKILLS OBJECTIVE

- Express equations of conics in polar form and graph.

CONCEPTUAL OBJECTIVE

- Define all conics in terms of a focus and a directrix.

B.6.1. Eccentricity

B.6.1 SKILL

Express equations of conics in polar form and graph.

B.6.1 CONCEPTUAL

Define all conics in terms of a focus and a directrix.

In Section B.1, we discussed parabolas, ellipses, and hyperbolas in terms of geometric definitions. Then in Sections B.2–B.4, we examined the rectangular equations of these conics. The equations for the conics when their centers are at the origin are simpler than when they are not (when conics are shifted). In Section B.5, we discussed polar coordinates and graphing of polar equations. In this section, we develop a more unified definition of the three conics in terms of a single focus and a directrix. You will see that if the *focus* is located at the origin, then equations of conics are simpler when written in polar coordinates.

Alternative Definition of Conics

Recall that when we work with rectangular coordinates we define a parabola (Sections B.1 and B.2) in terms of a fixed point (focus) and a line (directrix), whereas we define an ellipse and hyperbola (Sections B.1, B.3, and B.4) in terms of two fixed points (the foci). However, it is possible to define all three conics in terms of a single focus and a directrix.

The following alternative representation of conics depends on a parameter called *eccentricity*.

ALTERNATIVE DESCRIPTION OF CONICS

Let D be a fixed line (the **directrix**), F be a fixed point (a **focus**) not on D, and e be a fixed positive number (**eccentricity**). The set of all points P such that the ratio of the distance from P to F to the distance from P to D equals the constant e defines a conic section.

$$\frac{d(P, F)}{d(P, D)} = e$$

- If $e = 1$, the conic is a **parabola**.
- If $e < 1$, the conic is an **ellipse**.
- If $e > 1$, the conic is a **hyperbola**.

When $e = 1$, the result is a parabola, described by the same definition we used previously in Section B.1. When $e \neq 1$, the result is either an ellipse or a hyperbola. The major axis of an ellipse passes through the focus and is perpendicular to the directrix. The transverse axis of a hyperbola also passes through the focus and is perpendicular to the directrix. If we let c represent the distance from the focus to the center and a represent the distance from the vertex to the center, then eccentricity is given by

$$e = \frac{c}{a}$$

B.6.2 Equations of Conics in Polar Coordinates

In polar coordinates, if we locate the focus of a conic at the pole and the directrix is either perpendicular or parallel to the polar axis, then we have four possible scenarios:

- The directrix is *perpendicular* to the polar axis and p units to the *right* of the pole.
- The directrix is *perpendicular* to the polar axis and p units to the *left* of the pole.
- The directrix is *parallel* to the polar axis and p units *above* the pole.
- The directrix is *parallel* to the polar axis and p units *below* the pole.

Let us take the case in which the directrix is perpendicular to the polar axis and p units to the right of the pole.

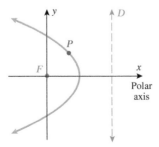

In polar coordinates (r, θ), we see that the distance from the focus to a point P is equal to r, that is, $d(P, F) = r$, and the distance from P to the closest point on the directrix is $d(P, D) = p - r \cos \theta$.

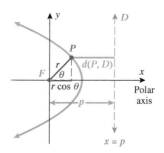

WORDS

Substitute $d(P, F) = r$ and $d(P, D) = p - r\cos\theta$

into the formula for eccentricity, $\dfrac{d(P, F)}{d(P, D)} = e.$

Multiply the result by $p - r\cos\theta$.

Eliminate the parentheses.

Add $er\cos\theta$.

Factor out the common r.

Divide by $1 + e\cos\theta$.

MATH

$$\frac{r}{p - r\cos\theta} = e$$

$$r = e(p - r\cos\theta)$$

$$r = ep - er\cos\theta$$

$$r + er\cos\theta = ep$$

$$r(1 + e\cos\theta) = ep$$

$$\boxed{r = \frac{ep}{1 + e\cos\theta}}$$

We need not derive the other three cases here, but note that if the directrix is perpendicular to the polar axis and p units to the *left* of the pole, the resulting polar equation is

$$r = \frac{ep}{1 - e\cos\theta}$$

If the directrix is parallel to the polar axis, the directrix is either above $(y = p)$ or below $(y = -p)$ the polar axis and we get the sine function instead of the cosine function, as summarized in the following box:

POLAR EQUATIONS OF CONICS

The following polar equations represent conics with one focus at the origin and with eccentricity e. It is assumed that the positive x-axis represents the polar axis.

EQUATION	DESCRIPTION
$r = \dfrac{ep}{1 + e\cos\theta}$	The directrix is *vertical* and p units to the *right* of the pole.
$r = \dfrac{ep}{1 - e\cos\theta}$	The directrix is *vertical* and p units to the *left* of the pole.
$r = \dfrac{ep}{1 + e\sin\theta}$	The directrix is *horizontal* and p units *above* the pole.
$r = \dfrac{ep}{1 - e\sin\theta}$	The directrix is *horizontal* and p units *below* the pole.

ECCENTRICITY	THE CONIC IS A ___	THE ___ IS PERPENDICULAR TO THE DIRECTRIX
$e = 1$	Parabola	Axis of symmetry
$e < 1$	Ellipse	Major axis
$e > 1$	Hyperbola	Transverse axis

EXAMPLE 1   **Finding the Polar Equation of a Conic**

Find a polar equation for a parabola that has its focus at the origin and whose directrix is the line $y = 3$.

Solution:

The directrix is horizontal and above the pole.

$$r = \frac{ep}{1 + e\sin\theta}$$

A parabola has eccentricity $e = 1$ and $p = 3$.

$$\boxed{r = \frac{3}{1 + \sin\theta}}$$

▼

YOUR TURN   Find a polar equation for a parabola that has its focus at the origin and whose directrix is the line $x = -3$.

▼ ANSWER

$$r = \frac{3}{1 - \cos\theta}$$

EXAMPLE 2   **Identifying a Conic from Its Equation**

Identify the type of conic represented by the equation $\dfrac{10}{3 + 2\cos\theta}$.

Solution:

To identify the type of conic, we need to rewrite the equation in the form.

$$r = \frac{ep}{1 \pm e\cos\theta}$$

Divide the numerator and denominator by 3.

$$= \frac{\dfrac{10}{3}}{\left(1 + \dfrac{2}{3}\cos\theta\right)}$$

Identify e in the denominator.

$$= \frac{\dfrac{10}{3}}{\left(1 + \underset{e}{\dfrac{2}{3}}\cos\theta\right)}$$

The numerator is equal to ep.

$$= \frac{\underset{p}{\dfrac{5}} \cdot \overset{e}{\dfrac{2}{3}}}{\left(1 + \underset{e}{\dfrac{2}{3}}\cos\theta\right)}$$

Since $e = \frac{2}{3} < 1$, the conic is an $\boxed{\text{ellipse}}$. The directrix is $x = 5$, so the major axis is along the x-axis (perpendicular to the directrix).

▼

YOUR TURN   Identify the type of conic represented by the equation

$$r = \frac{10}{2 - 10\sin\theta}$$

▼ ANSWER

hyperbola, $e = 5$, with transverse axis along the y-axis

B.6.3 Graphing a Conic Given in Polar Form

In Example 2, we found that the polar equation $r = \dfrac{10}{3 + 2\cos\theta}$ is an ellipse with major axis along the x-axis. We will graph this ellipse in Example 3.

EXAMPLE 3 **Graphing a Conic from Its Equation**

The graph of the polar equation $r = \dfrac{10}{3 + 2\cos\theta}$ is an ellipse.

a. Find the vertices. **b.** Find the center of the ellipse.

c. Find the lengths of the major and minor axes. **d.** Graph the ellipse.

Solution (a):

From Example 2 we see that $e = \frac{2}{3}$, which corresponds to an ellipse, and $x = 5$ is the directrix.

The major axis is perpendicular to the directrix. Therefore, the major axis lies along the polar axis. To find the vertices (which lie along the major axis), let $\theta = 0$ and $\theta = \pi$.

$\theta = 0:$

$$r = \frac{10}{3 + 2\cos 0} = \frac{10}{5} = 2$$

$\theta = \pi:$

$$r = \frac{10}{3 + 2\cos\pi} = \frac{10}{1} = 10$$

The vertices are the points $\boxed{V_1 = (2, 0)}$ and $\boxed{V_2 = (10, \pi)}$.

Solution (b):

The vertices in rectangular coordinates are $V_1 = (2, 0)$ and $V_2 = (-10, 0)$.

The midpoint (in rectangular coordinates) between the two vertices is the point $(-4, 0)$, which corresponds to the point $\boxed{(4, \pi)}$ in polar coordinates.

Solution (c):

The length of the major axis, $2a$, is the distance between the vertices. $\boxed{2a = 12}$

The length $a = 6$ corresponds to the distance from the center to a vertex.

Apply the formula $e = \dfrac{c}{a}$ with $a = 6$ and $e = \frac{2}{3}$ to find c.

$$c = ae = 6\left(\frac{2}{3}\right) = 4$$

Let $a = 6$ and $c = 4$ in $b^2 = a^2 - c^2$.

$$b^2 = 6^2 - 4^2 = 20$$

Solve for b.

$$b = 2\sqrt{5}$$

The length of the minor axis is $\boxed{2b = 4\sqrt{5}}$.

Solution (d):

Graph the ellipse.

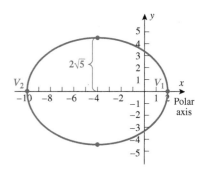

EXAMPLE 4   **Identifying and Graphing a Conic from Its Equation**

Identify and graph the conic defined by the equation $r = \dfrac{2}{2 + 3\sin\theta}$.

Solution:

Rewrite the equation in the form

$r = \dfrac{ep}{1 + e\sin\theta}.$
$$r = \frac{2}{2 + 3\sin\theta} = \frac{\overset{p}{\left(\dfrac{2}{3}\right)}\overset{e}{\left(\dfrac{3}{2}\right)}}{1 + \underset{e}{\left(\dfrac{3}{2}\right)}\sin\theta}$$

The conic is a *hyperbola* since $e = \frac{3}{2} > 1$.

The directrix is horizontal and $\frac{2}{3}$ unit above the pole (origin).

To find the vertices, let $\theta = \dfrac{\pi}{2}$ and $\theta = \dfrac{3\pi}{2}$.

$\theta = \dfrac{\pi}{2}$:
$$r = \frac{2}{2 + 3\sin\left(\dfrac{\pi}{2}\right)} = \frac{2}{5}$$

$\theta = \dfrac{3\pi}{2}$:
$$r = \frac{2}{2 + 3\sin\left(\dfrac{3\pi}{2}\right)} = \frac{2}{-1} = -2$$

The vertices in polar coordinates are $\left(\dfrac{2}{5}, \dfrac{\pi}{2}\right)$ and $\left(-2, \dfrac{3\pi}{2}\right)$.

The vertices in rectangular coordinates are $V_1 = \left(0, \dfrac{2}{5}\right)$ and $V_2 = (0, 2)$.

The center is the midpoint between the vertices: $\left(0, \dfrac{6}{5}\right)$.

The distance from the center to a focus is $c = \dfrac{6}{5}$.

Apply the formula $e = \dfrac{c}{a}$ with $c = \dfrac{6}{5}$ and $e = \dfrac{3}{2}$ to find a.
$$a = \frac{c}{e} = \frac{\dfrac{6}{5}}{\dfrac{3}{2}} = \frac{4}{5}$$

Let $a = \frac{4}{5}$ and $c = \frac{6}{5}$ in $b^2 = c^2 - a^2$.
$$b^2 = \left(\frac{6}{5}\right)^2 - \left(\frac{4}{5}\right)^2 = \frac{20}{25}$$

Solve for b.
$$b = \frac{2\sqrt{5}}{5}$$

The asymptotes are given by

$y = \pm\dfrac{a}{b}(x - h) + k$, where $a = \dfrac{4}{5}$,

$b = \dfrac{2\sqrt{5}}{5}$, and $(h, k) = \left(0, \dfrac{6}{5}\right)$.

$y = \pm\dfrac{2}{\sqrt{5}}x + \dfrac{6}{5}$

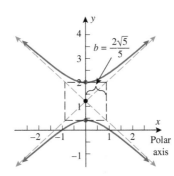

It is important to note that although we relate specific points (vertices, foci, etc.) to rectangular coordinates, another approach to finding a rough sketch is to simply point-plot the equation in polar coordinates.

EXAMPLE 5 **Graphing a Conic by Point-Plotting in Polar Coordinates**

Sketch a graph of the conic $r = \dfrac{4}{1 - \sin\theta}$.

Solution:

STEP 1 The conic is a parabola because the equation is in the form

$$r = \frac{(4)(1)}{1 - (1)\sin\theta}$$

Make a table with key values for θ and r.

θ	$r = \dfrac{4}{1 - \sin\theta}$	(r, θ)
0	$r = \dfrac{4}{1 - \sin 0} = \dfrac{4}{1} = 4$	$(4, 0)$
$\dfrac{\pi}{2}$	$r = \dfrac{4}{1 - \sin\left(\dfrac{\pi}{2}\right)} = \dfrac{4}{1 - 1} = \dfrac{4}{0}$	undefined
π	$r = \dfrac{4}{1 - \sin\pi} = \dfrac{4}{1} = 4$	$(4, \pi)$
$\dfrac{3\pi}{2}$	$r = \dfrac{4}{1 - \sin\left(\dfrac{3\pi}{2}\right)} = \dfrac{4}{1 - (-1)} = \dfrac{4}{2} = 2$	$\left(2, \dfrac{3\pi}{2}\right)$
2π	$r = \dfrac{4}{1 - \sin(2\pi)} = \dfrac{4}{1} = 4$	$(4, 2\pi)$

STEP 2 Plot the points on a polar graph and connect them with a smooth parabolic curve.

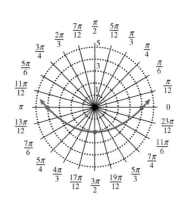

▶[SECTION B.6] SUMMARY

In this section, we found that we could graph polar equations of conics by identifying a single focus and the directrix. There are four possible equations in terms of eccentricity e:

EQUATION	DESCRIPTION
$r = \dfrac{ep}{1 + e\cos\theta}$	The directrix is *vertical* and p units to the *right* of the pole.
$r = \dfrac{ep}{1 - e\cos\theta}$	The directrix is *vertical* and p units to the *left* of the pole.
$r = \dfrac{ep}{1 + e\sin\theta}$	The directrix is *horizontal* and p units *above* the pole.
$r = \dfrac{ep}{1 - e\sin\theta}$	The directrix is *horizontal* and p units *below* the pole.

[SECTION B.6] EXERCISES

● **SKILLS**

In Exercises 1–14, find the polar equation that represents the conic described (assume that a focus is at the origin).

	Conic	Eccentricity	Directrix		Conic	Eccentricity	Directrix
1.	Ellipse	$e = \frac{1}{2}$	$y = -5$	**2.**	Ellipse	$e = \frac{1}{3}$	$y = 3$
3.	Hyperbola	$e = 2$	$y = 4$	**4.**	Hyperbola	$e = 3$	$y = -2$
5.	Parabola	$e = 1$	$x = 1$	**6.**	Parabola	$e = 1$	$x = -1$
7.	Ellipse	$e = \frac{3}{4}$	$x = 2$	**8.**	Ellipse	$e = \frac{2}{3}$	$x = -4$
9.	Hyperbola	$e = \frac{4}{3}$	$x = -3$	**10.**	Hyperbola	$e = \frac{3}{2}$	$x = 5$
11.	Parabola	$e = 1$	$y = -3$	**12.**	Parabola	$e = 1$	$y = 4$
13.	Ellipse	$e = \frac{3}{5}$	$y = 6$	**14.**	Hyperbola	$e = \frac{8}{5}$	$y = 5$

In Exercises 15–26, identify the conic (parabola, ellipse, or hyperbola) that each polar equation represents.

15. $r = \dfrac{4}{1 + \cos\theta}$

16. $r = \dfrac{3}{2 - 3\sin\theta}$

17. $r = \dfrac{2}{3 + 2\sin\theta}$

18. $r = \dfrac{3}{2 - 2\cos\theta}$

19. $r = \dfrac{2}{4 + 8\cos\theta}$

20. $r = \dfrac{1}{4 - \cos\theta}$

21. $r = \dfrac{7}{3 + \cos\theta}$

22. $r = \dfrac{4}{5 + 6\sin\theta}$

23. $r = \dfrac{40}{5 + 5\sin\theta}$

24. $r = \dfrac{5}{5 - 4\sin\theta}$

25. $r = \dfrac{1}{1 - 6\cos\theta}$

26. $r = \dfrac{5}{3 - 3\sin\theta}$

In Exercises 27–40, for the given polar equations: (a) identify the conic as either a parabola, ellipse, or hyperbola; (b) find the eccentricity and vertex (or vertices); and (c) graph.

27. $r = \dfrac{2}{1 + \sin\theta}$

28. $r = \dfrac{4}{1 - \cos\theta}$

29. $r = \dfrac{4}{1 - 2\sin\theta}$

30. $r = \dfrac{3}{3 + 8\cos\theta}$

31. $r = \dfrac{2}{2 + \sin\theta}$

32. $r = \dfrac{1}{3 - \sin\theta}$

33. $r = \dfrac{1}{2 - 2\sin\theta}$

34. $r = \dfrac{1}{1 - 2\sin\theta}$

35. $r = \dfrac{4}{3 + \cos\theta}$

36. $r = \dfrac{2}{5 + 4\sin\theta}$

37. $r = \dfrac{6}{2 + 3\sin\theta}$

38. $r = \dfrac{6}{1 + \cos\theta}$

39. $r = \dfrac{2}{5 + 5\cos\theta}$

40. $r = \dfrac{10}{6 - 3\cos\theta}$

• **APPLICATIONS**

For Exercises 41 and 42, refer to the following:

Planets travel in elliptical orbits around a single focus, the Sun. Pluto (orange), the dwarf planet furthest from the Sun, has a pronounced elliptical orbit, whereas Earth (royal blue) has an almost circular orbit. The polar equation of a planet's orbit can be expressed as

$$r = \frac{a(1 - e^2)}{(1 - e\cos\theta)}$$

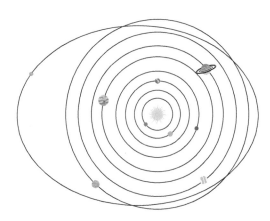

where e is the eccentricity and $2a$ is the length of the major axis. It can also be shown that the perihelion distance (minimum distance from the Sun to a planet) and the aphelion distance (maximum distance from the Sun to the planet) can be represented by $r = a(1 - e)$ and $r = a(1 + e)$, respectively.

41. Planetary Orbits. Pluto's orbit is summarized in the picture below. Find the eccentricity of Pluto's orbit. Find the polar equation that governs Pluto's orbit.

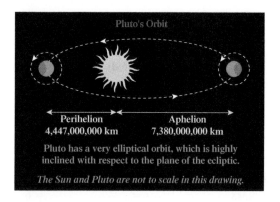

42. Planetary Orbits. Earth's orbit is summarized in the picture below. Find the eccentricity of Earth's orbit. Find the polar equation that governs Earth's orbit.

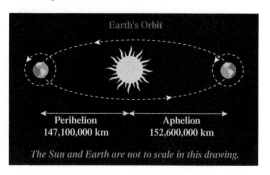

For Exercises 43 and 44, refer to the following:

Asteroids, meteors, and comets all orbit the Sun in elliptical patterns and often cross paths with Earth's orbit, making life a little tense now and again. Asteroids are large rocks (bodies under 1000 kilometers across), meteors range from sand particles to rocks, and comets are masses of debris. A few asteroids have orbits that cross the Earth's orbits—called Apollos or Earth-crossing asteroids. In recent years, asteroids have passed within 100,000 kilometers of Earth!

43. **Asteroids.** The asteroid 433 or Eros is the second largest near-Earth asteroid. The semimajor axis of its orbit is 150 million kilometers, and the eccentricity is 0.223. Find the polar equation of Eros's orbit.

44. **Asteroids.** The asteroid Toutatis is the largest near-Earth asteroid. The semimajor axis of its orbit is 350 million kilometers, and the eccentricity is 0.634. On September 29, 2004, it missed Earth by 961,000 miles. Find the polar equation of Toutatis's orbit.

● **CONCEPTUAL**

45. When $0 < e < 1$, the conic is an ellipse. Does the conic become more elongated or elliptical as e approaches 1 or as e approaches 0?

46. Show that $r = \dfrac{ep}{1 - e\sin\theta}$ is the polar equation of a conic with a horizontal directrix that is p units *below* the pole.

47. Convert from rectangular to polar coordinates to show that the equation of a hyperbola, $\dfrac{x^2}{a^2} - \dfrac{y^2}{b^2} = 1$, in polar form is

$$r^2 = -\dfrac{b^2}{1 - e^2\cos^2\theta}.$$

48. Convert from rectangular to polar coordinates to show that the equation of an ellipse, $\dfrac{x^2}{a^2} + \dfrac{y^2}{b^2} = 1$, in polar form is

$$r^2 = \dfrac{b^2}{1 - e^2\cos^2\theta}.$$

● **CHALLENGE**

49. Find the major diameter (length of the major axis) of the ellipse with polar equation $r = \dfrac{ep}{1 + e\cos\theta}$ in terms of e and p.

50. Find the minor diameter (length of the minor axis) of the ellipse with polar equation $r = \dfrac{ep}{1 + e\cos\theta}$ in terms of e and p.

51. Find the center of the ellipse with polar equation $r = \dfrac{ep}{1 + e\cos\theta}$ in terms of e and p.

52. Find the length of the latus rectum of the parabola with polar equation $r = \dfrac{p}{1 + \cos\theta}$. Assume that the focus is at the origin.

● **TECHNOLOGY**

53. Let us consider the polar equations $r = \dfrac{ep}{1 + e\cos\theta}$ and $r = \dfrac{ep}{1 - e\cos\theta}$ with eccentricity $e = 1$. With a graphing utility, explore the equations with $p = 1, 2,$ and 6. Describe the behavior of the graphs as $p \to \infty$ and also the difference between the two equations.

54. Let us consider the polar equations $r = \dfrac{ep}{1 + e\sin\theta}$ and $r = \dfrac{ep}{1 - e\sin\theta}$ with eccentricity $e = 1$. With a graphing utility, explore the equations with $p = 1, 2,$ and 6. Describe the behavior of the graphs as $p \to \infty$ and also the difference between the two equations.

55. Let us consider the polar equations $r = \dfrac{ep}{1 + e\cos\theta}$ and $r = \dfrac{ep}{1 - e\cos\theta}$ with $p = 1$. With a graphing utility, explore the equations with $e = 1.5, 3,$ and 6. Describe the behavior of the graphs as $e \to \infty$ and also the difference between the two equations.

56. Let us consider the polar equations $r = \dfrac{ep}{1 + e\sin\theta}$ and $r = \dfrac{ep}{1 - e\sin\theta}$ with $p = 1$. With a graphing utility, explore the equations with $e = 1.5, 3,$ and 6. Describe the behavior of the graphs as $e \to \infty$ and also the difference between the two equations.

57. Let us consider the polar equations $r = \dfrac{ep}{1 + e\cos\theta}$ and $r = \dfrac{ep}{1 - e\cos\theta}$ with $p = 1$. With a graphing utility, explore the equations with $e = 0.001, 0.5, 0.9,$ and 0.99. Describe the behavior of the graphs as $e \to 1$ and also the difference between the two equations. Be sure to set the window parameters properly.

58. Let us consider the polar equations $r = \dfrac{ep}{1 + e\sin\theta}$ and $r = \dfrac{ep}{1 - e\sin\theta}$ with $p = 1$. With a graphing utility, explore the equations with $e = 0.001, 0.5, 0.9,$ and 0.99. Describe the behavior of the graphs as $e \to 1$ and also the difference between the two equations. Be sure to set the window parameters properly.

59. Let us consider the polar equation $r = \dfrac{5}{5 + 2\sin\theta}$.

Explain why the graphing utility gives the following graphs with the specified window parameters:

a. $[-2, 2]$ by $[-2, 2]$ with θ step $= \dfrac{\pi}{2}$

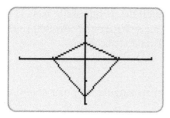

b. $[-2, 2]$ by $[-2, 2]$ with θ step $= \dfrac{\pi}{3}$

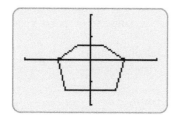

60. Let us consider the polar equation $r = \dfrac{2}{1 + \cos\theta}$.

Explain why a graphing utility gives the following graphs with the specified window parameters:

a. $[-2, 2]$ by $[-4, 4]$ with θ step $= \dfrac{\pi}{2}$

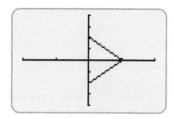

b. $[-2, 2]$ by $[-4, 4]$ with θ step $= \dfrac{\pi}{3}$

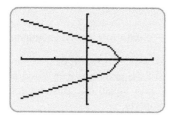

61. Let us consider the polar equation $r = \dfrac{6}{1 + 3\sin\theta}$.

Explain why a graphing utility gives the following graphs with the specified window parameters:

a. $[-8, 8]$ by $[-2, 4]$ with θ step $= \dfrac{\pi}{2}$

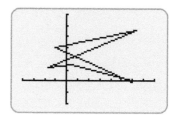

b. $[-4, 8]$ by $[-2, 6]$ with θ step $= 0.4\pi$

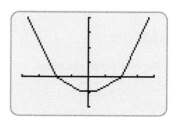

62. Let us consider the polar equation $r = \dfrac{2}{1 - \sin\theta}$.

Explain why a graphing utility gives the following graphs with the specified window parameters:

a. $[-4, 4]$ by $[-2, 4]$ with θ step $= \dfrac{\pi}{3}$

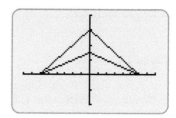

b. $[-4, 4]$ by $[-2, 6]$ with θ step $= 0.8\pi$

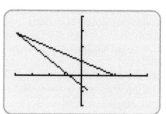

●[APPENDIX B REVIEW]

SECTION	CONCEPT	KEY IDEAS/FORMULAS
B.1	**Conic basics**	
	Classifying conic sections: Parabola, ellipse, and hyperbola	**Parabola:** Distance to a reference point (focus) and a reference line (directrix) remain constant (are equal). **Ellipse:** Sum of the distances between the point and two reference points (foci) is constant. **Hyperbola:** Difference of the distances between the point and two reference points (foci) is constant.
B.2	**The parabola**	
	Parabola with its vertex at the origin	
	Parabola with vertex (h, k)	(see table below)
B.3	**The ellipse**	
	Ellipse with its center at the origin	

Parabola with vertex (h, k):

EQUATION	$(y - k)^2 = 4p(x - h)$	$(x - h)^2 = 4p(y - k)$
VERTEX	(h, k)	(h, k)
FOCUS	$(p + h, k)$	$(h, p + k)$
DIRECTRIX	$x = -p + h$	$y = -p + k$
AXIS OF SYMMETRY	$y = k$	$x = h$
$p > 0$	opens to the right	opens upward
$p < 0$	opens to the left	opens downward

Parabola with vertex at the origin:

$x^2 = 4py$ Focus $(0, p)$ Directrix $y = -p$ Up: $p > 0$ Down: $p < 0$

$y^2 = 4px$ Focus $(p, 0)$ Directrix $x = -p$ Right: $p > 0$ Left: $p < 0$

Ellipse with its center at the origin:

$$\frac{x^2}{a^2} + \frac{y^2}{b^2} = 1$$
$$c^2 = a^2 - b^2$$

$$\frac{x^2}{b^2} + \frac{y^2}{a^2} = 1$$
$$c^2 = a^2 - b^2$$

SECTION	CONCEPT	KEY IDEAS/FORMULAS		
	Ellipse with center (h, k)	**ORIENTATION OF MAJOR AXIS**	Horizontal (parallel to the x-axis)	Vertical (parallel to the y-axis)
		EQUATION	$\dfrac{(x-h)^2}{a^2} + \dfrac{(y-k)^2}{b^2} = 1$	$\dfrac{(x-h)^2}{b^2} + \dfrac{(y-k)^2}{a^2} = 1$
		FOCI	$(h-c, k)$ $(h+c, k)$	$(h, k-c)$ $(h, k+c)$
		GRAPH		

B.4	**The hyperbola**			
	Hyperbola with its center at the origin	$$\frac{x^2}{a^2} - \frac{y^2}{b^2} = 1$$ $$c^2 = a^2 + b^2$$		$$\frac{y^2}{a^2} - \frac{x^2}{b^2} = 1$$ $$c^2 = a^2 + b^2$$

	Hyperbola with center (h, k)	**ORIENTATION OF TRANSVERSE AXIS**	Horizontal (parallel to the x-axis)	Vertical (parallel to the y-axis)
		EQUATION	$\dfrac{(x-h)^2}{a^2} - \dfrac{(y-k)^2}{b^2} = 1$	$\dfrac{(y-k)^2}{a^2} - \dfrac{(x-h)^2}{b^2} = 1$
		VERTICES	$(h-a, k)$ $(h+a, k)$	$(h, k-a)$ $(h, k+a)$
		FOCI	$(h-c, k)$ $(h+c, k)$ where $c^2 = a^2 + b^2$	$(h, k-c)$ $(h, k+c)$ where $c^2 = a^2 + b^2$
		GRAPH		

SECTION	CONCEPT	KEY IDEAS/FORMULAS
B.5	**Rotation of axes**	
	Transforming second-degree equations using rotation of axes	$x = X\cos\theta - Y\sin\theta$ $y = X\sin\theta + Y\cos\theta$
	Determining the angle of rotation necessary to transform a general second-degree equation into an equation of a conic	$\cot(2\theta) = \dfrac{A - C}{B}$
	Graphing a rotated conic	1. Determine which conic the equation represents. 2. Find the desired rotation angle. 3. Transform the equation from x, y to X, Y. 4. Graph the resulting conic.
B.6	**Polar equations of conics**	All three conics (parabolas, ellipses, and hyperbolas) are defined in terms of a single focus and a directrix.
	Eccentricity	$e = \dfrac{c}{a}$
	Equations of conics in polar coordinates	The directrix is *vertical* and p units to the *right* of the pole. $$r = \dfrac{ep}{1 + e\cos\theta}$$ The directrix is *vertical* and p units to the *left* of the pole. $$r = \dfrac{ep}{1 - e\cos\theta}$$ The directrix is *horizontal* and p units *above* the pole. $$r = \dfrac{ep}{1 + e\sin\theta}$$ The directrix is *horizontal* and p units *below* the pole. $$r = \dfrac{ep}{1 - e\sin\theta}$$
	Graphing a conic given in polar form	1. Determine which conic the equation represents. 2. Determine vertices, center, directrix, focus, etc. 3. Point-plot.

[APPENDIX B REVIEW EXERCISES]

B.1 Conic Basics

Determine whether each statement is true or false.

1. The focus is a point on the graph of the parabola.

2. The graph of $y^2 = 8x$ is a parabola that opens upward.

3. $\dfrac{x^2}{9} - \dfrac{y^2}{1} = 1$ is the graph of a hyperbola that has a horizontal transverse axis.

4. $\dfrac{(x + 1)^2}{9} + \dfrac{(y - 3)^2}{16} = 1$ is a graph of an ellipse whose center is $(1, 3)$.

B.2 The Parabola

Find an equation for the parabola described.

5. Vertex at $(0, 0)$; Focus at $(3, 0)$

6. Vertex at $(0, 0)$; Focus at $(0, 2)$

7. Vertex at $(0, 0)$; Directrix at $x = 5$

8. Vertex at $(0, 0)$; Directrix at $y = 4$

9. Vertex at $(2, 3)$; Focus at $(2, 5)$

10. Vertex at $(-1, -2)$; Focus at $(1, -2)$

11. Focus at $(1, 5)$; Directrix at $y = 7$

12. Focus at $(2, 2)$; Directrix at $x = 0$

Find the focus, vertex, directrix, and length of the latus rectum, and graph the parabola.

13. $x^2 = -12y$

14. $x^2 = 8y$

15. $y^2 = x$

16. $y^2 = -6x$

17. $(y + 2)^2 = 4(x - 2)$

18. $(y - 2)^2 = -4(x + 1)$

19. $(x + 3)^2 = -8(y - 1)$

20. $(x - 3)^2 = -8(y + 2)$

21. $x^2 + 5x + 2y + 25 = 0$

22. $y^2 + 2y - 16x + 1 = 0$

Applications

23. **Satellite Dish.** A satellite dish measures 10 feet across its opening and 2 feet deep at its center. The receiver should be placed at the focus of the parabolic dish. Where should the receiver be placed?

24. **Clearance Under a Bridge.** A bridge with a parabolic shape reaches a height of 40 feet in the center of the road, and the width of the bridge opening at ground level is 30 feet combined (both lanes). If an RV is 14 feet tall and 8 feet wide, will it make it through the tunnel?

B.3 The Ellipse

Graph each ellipse.

25. $\dfrac{x^2}{9} + \dfrac{y^2}{64} = 1$

26. $\dfrac{x^2}{81} + \dfrac{y^2}{49} = 1$

27. $25x^2 + y^2 = 25$

28. $4x^2 + 8y^2 = 64$

Find the standard form of an equation of the ellipse with the given characteristics.

29. Foci: $(-3, 0)$ and $(3, 0)$ Vertices: $(-5, 0)$ and $(5, 0)$

30. Foci: $(0, -2)$ and $(0, 2)$ Vertices: $(0, -3)$ and $(0, 3)$

31. Major axis vertical with length of 16, minor axis length of 6, and centered at $(0, 0)$

32. Major axis horizontal with length of 30, minor axis length of 20, and centered at $(0, 0)$

Graph each ellipse.

33. $\dfrac{(x - 7)^2}{100} + \dfrac{(y + 5)^2}{36} = 1$

34. $20(x + 3)^2 + (y - 4)^2 = 120$

35. $4x^2 - 16x + 12y^2 + 72y + 123 = 0$

36. $4x^2 - 8x + 9y^2 - 72y + 147 = 0$

Find the standard form of an equation of the ellipse with the given characteristics.

37. Foci: $(-1, 3)$ and $(7, 3)$ Vertices: $(-2, 3)$ and $(8, 3)$

38. Foci: $(1, -3)$ and $(1, -1)$ Vertices: $(1, -4)$ and $(1, 0)$

Applications

39. **Planetary Orbits.** Jupiter's orbit is summarized in the picture. Utilize the fact that the Sun is a focus to determine an equation for Jupiter's elliptical orbit around the Sun. Round to the nearest hundred thousand kilometers.

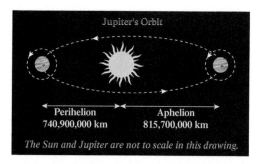

Jupiter's Orbit

Perihelion 740,900,000 km   Aphelion 815,700,000 km

The Sun and Jupiter are not to scale in this drawing.

40. **Planetary Orbits.** Mars's orbit is summarized in the picture that follows. Utilize the fact that the Sun is a focus to determine an equation for Mars's elliptical orbit around the Sun. Round to the nearest million kilometers.

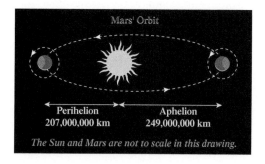

Mars' Orbit

Perihelion 207,000,000 km   Aphelion 249,000,000 km

The Sun and Mars are not to scale in this drawing.

B.4 The Hyperbola

Graph each hyperbola.

41. $\dfrac{x^2}{9} - \dfrac{y^2}{64} = 1$ **42.** $\dfrac{x^2}{81} - \dfrac{y^2}{49} = 1$

43. $x^2 - 25y^2 = 25$ **44.** $8y^2 - 4x^2 = 64$

Find the standard form of an equation of the hyperbola with the given characteristics.

45. Vertices: $(-3, 0)$ and $(3, 0)$ Foci: $(-5, 0)$ and $(5, 0)$

46. Vertices: $(0, -1)$ and $(0, 1)$ Foci: $(0, -3)$ and $(0, 3)$

47. Center: $(0, 0)$; Transverse: y-axis; Asymptotes: $y = 3x$ and $y = -3x$

48. Center: $(0, 0)$; Transverse axis: y-axis; Asymptotes: $y = \frac{1}{2}x$ and $y = -\frac{1}{2}x$

Graph each hyperbola.

49. $\dfrac{(y - 1)^2}{36} - \dfrac{(x - 2)^2}{9} = 1$

50. $3(x + 3)^2 - 12(y - 4)^2 = 72$

51. $8x^2 - 32x - 10y^2 - 60y - 138 = 0$

52. $2x^2 + 12x - 8y^2 + 16y + 6 = 0$

Find the standard form of an equation of the hyperbola with the given characteristics.

53. Vertices: $(0, 3)$ and $(8, 3)$ Foci: $(-1, 3)$ and $(9, 3)$

54. Vertices: $(4, -2)$ and $(4, 0)$ Foci: $(4, -3)$ and $(4, 1)$

Applications

55. Ship Navigation. Two loran stations are located 220 miles apart along a coast. If a ship records a time difference of 0.00048 second and continues on the hyperbolic path corresponding to that difference, where would it reach shore? Assume that the speed of radio signals is 186,000 miles per second.

56. Ship Navigation. Two loran stations are located 400 miles apart along a coast. If a ship records a time difference of 0.0008 second and continues on the hyperbolic path corresponding to that difference, where would it reach shore?

B.5 Rotation of Axes

The coordinates of a point in the xy-coordinate system are given. Assuming the X- and Y-axes are found by rotating the x- and y-axes by an angle θ, find the corresponding coordinates for the point in the XY-system.

57. $(-3, 2)$, $\theta = 60°$ **58.** $(4, -3)$, $\theta = 45°$

Transform the equation of the conic into an equation in X and Y (without an XY-term) by rotating the x- and y-axes through an angle θ. Then graph the resulting equation.

59. $2x^2 + 4\sqrt{3}xy - 2y^2 - 16 = 0$, $\theta = 30°$

60. $25x^2 + 14xy + 25y^2 - 288 = 0$, $\theta = \dfrac{\pi}{4}$

Determine the angle of rotation necessary to transform the equation in x and y into an equation in X and Y with no XY-term.

61. $4x^2 + 2\sqrt{3}xy + 6y^2 - 9 = 0$

62. $4x^2 + 5xy + 4y^2 - 11 = 0$

Graph the second-degree equation.

63. $x^2 + 2xy + y^2 + \sqrt{2}x - \sqrt{2}y + 8 = 0$

64. $76x^2 + 48\sqrt{3}xy + 28y^2 - 100 = 0$

B.6 Polar Equations of Conics

Find the polar equation that represents the conic described.

65. An ellipse with eccentricity $e = \frac{3}{7}$ and directrix $y = -7$

66. A parabola with directrix $x = 2$

Identify the conic (parabola, ellipse, or hyperbola) that each polar equation represents.

67. $r = \dfrac{6}{4 - 5\cos\theta}$ **68.** $r = \dfrac{2}{5 + 3\sin\theta}$

For the given polar equations, find the eccentricity and vertex (or vertices), and graph the curve.

69. $r = \dfrac{4}{2 + \cos\theta}$ **70.** $r = \dfrac{6}{1 - \sin\theta}$

Technology Exercises

Section B.2

71. In your mind, picture the parabola given by $(x - 0.6)^2 = -4(y + 1.2)$. Where is the vertex? Which way does this parabola open? Now plot the parabola with a graphing utility.

72. In your mind, picture the parabola given by $(y - 0.2)^2 = 3(x - 2.8)$. Where is the vertex? Which way does this parabola open? Now plot the parabola with a graphing utility.

73. Given is the parabola $y^2 + 2.8y + 3x - 6.85 = 0$.
   a. Solve the equation for y, and use a graphing utility to plot the parabola.
   b. Transform the equation into the form $(y - k)^2 = 4p(x - h)$. Find the vertex. Which way does the parabola open?
   c. Do (a) and (b) agree with each other?

74. Given is the parabola $x^2 - 10.2x - y + 24.8 = 0$.
   a. Solve the equation for y, and use a graphing utility to plot the parabola.
   b. Transform the equation into the form $(x - h)^2 = 4p(y - k)$. Find the vertex. Which way does the parabola open?
   c. Do (a) and (b) agree with each other?

Section B.3

75. Graph the following three ellipses: $4x^2 + y^2 = 1$, $4(2x)^2 + y^2 = 1$, and $4(3x)^2 + y^2 = 1$. What can be said to happen to ellipse $4(cx)^2 + y^2 = 1$ as c increases?

76. Graph the following three ellipses: $x^2 + 4y^2 = 1$, $x^2 + 4(2y)^2 = 1$, and $x^2 + 4(3y)^2 = 1$. What can be said to happen to ellipse $x^2 + 4(cy)^2 = 1$ as c increases?

Section B.4

77. Graph the following three hyperbolas: $4x^2 - y^2 = 1$, $4(2x)^2 - y^2 = 1$, and $4(3x)^2 - y^2 = 1$. What can be said to happen to hyperbola $4(cx)^2 - y^2 = 1$ as c increases?

78. Graph the following three hyperbolas: $x^2 - 4y^2 = 1$, $x^2 - 4(2y)^2 = 1$, and $x^2 - 4(3y)^2 = 1$. What can be said to happen to hyperbola $x^2 - 4(cy)^2 = 1$ as c increases?

Section B.5

79. With a graphing utility, explore the second-degree equation $Ax^2 + Bxy + Cy^2 + 10x - 8y - 5 = 0$ for the following values of A, B, and C:

    a. $A = 2, B = -3, C = 5$ **b.** $A = 2, B = 3, C = -5$

Show the angle of rotation to one decimal place. Explain the differences.

80. With a graphing utility, explore the second-degree equation $Ax^2 + Bxy + Cy^2 + 2x - y = 0$ for the following values of A, B, and C:

    a. $A = 1, B = -2, C = -1$ **b.** $A = 1, B = 2, C = 1$

Show the angle of rotation to the nearest degree. Explain the differences.

Section B.6

81. Let us consider the polar equation $r = \dfrac{8}{4 + 5\sin\theta}$. Explain why a graphing utility gives the following graph with the specified window parameters:

$$[-6, 6] \text{ by } [-3, 9] \text{ with } \theta \text{ step } = \frac{\pi}{4}$$

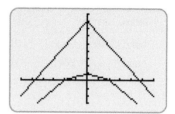

82. Let us consider the polar equation $r = \dfrac{9}{3 - 2\sin\theta}$. Explain why a graphing utility gives the following graph with the specified window parameters:

$$[-6, 6] \text{ by } [-3, 9] \text{ with } \theta \text{ step } = \frac{\pi}{2}$$

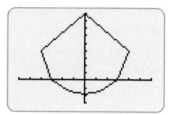

[APPENDIX B PRACTICE TEST]

Match the equation to the graph.

1. $x = 16y^2$
2. $y = 16x^2$
3. $x^2 + 16y^2 = 1$
4. $x^2 - 16y^2 = 1$
5. $16x^2 + y^2 = 1$
6. $16y^2 - x^2 = 1$

a.

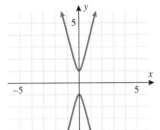

b.

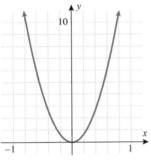

c.

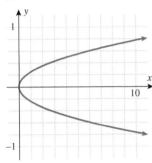

d.

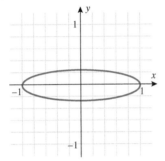

e.

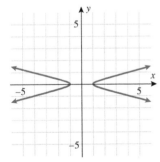

f.

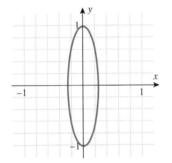

Find the equation of the conic with the given characteristics.

7. Parabola   vertex: $(0, 0)$ focus: $(-4, 0)$
8. Parabola   vertex: $(0, 0)$ directrix: $y = 2$
9. Parabola   vertex: $(-1, 5)$ focus: $(-1, 2)$
10. Parabola   vertex: $(2, -3)$ directrix: $x = 0$
11. Ellipse   center: $(0, 0)$ vertices: $(0, -4), (0, 4)$
            foci: $(0, -3), (0, 3)$
12. Ellipse   center: $(0, 0)$ vertices: $(-3, 0), (3, 0)$
            foci: $(-1, 0), (1, 0)$
13. Ellipse   vertices: $(2, -6), (2, 6)$ foci: $(2, -4), (2, 4)$
14. Ellipse   vertices: $(-7, -3), (-4, -3)$
            foci: $(-6, -3), (-5, -3)$

15. Hyperbola   vertices: $(-1, 0)$ and $(1, 0)$
              asymptotes: $y = -2x$ and $y = 2x$
16. Hyperbola   vertices: $(0, -1)$ and $(0, 1)$
              asymptotes: $y = -\frac{1}{3}x$ and $y = \frac{1}{3}x$
17. Hyperbola   foci: $(2, -6), (2, 6)$ vertices: $(2, -4), (2, 4)$
18. Hyperbola   foci: $(-7, -3), (-4, -3)$
              vertices: $(-6, -3), (-5, -3)$

Graph the following equations.

19. $9x^2 + 18x - 4y^2 + 16y - 43 = 0$
20. $4x^2 - 8x + y^2 + 10y + 28 = 0$
21. $y^2 + 4y - 16x + 20 = 0$
22. $x^2 - 4x + y + 1 = 0$

23. **Eyeglass Lens.** Eyeglass lenses can be thought of as very wide parabolic curves. If the focus occurs 1.5 centimeters from the center of the lens and the lens at its opening is 4 centimeters across, find an equation that governs the shape of the lens.

24. **Planetary Orbits.** The planet Uranus's orbit is described in the following picture, with the Sun as a focus of the elliptical orbit. Write an equation for the orbit.

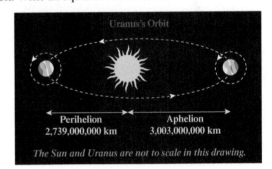

Uranus's Orbit

Perihelion
2,739,000,000 km

Aphelion
3,003,000,000 km

The Sun and Uranus are not to scale in this drawing.

25. Identify the conic represented by the equation
$$r = \frac{12}{3 + 2\sin\theta}.$$ State the eccentricity.

26. Use rotation of axes to transform the equation in x and y into an equation in X and Y that has no XY-term:
$6\sqrt{3}x^2 + 6xy + 4\sqrt{3}y^2 = 21\sqrt{3}$. State the rotation angle.

27. Given is the parabola $x^2 + 4.2x - y + 5.61 = 0$.

   a. Solve the equation for y and use a graphing utility to plot the parabola.

   b. Transform the equation into the form $(x - h)^2 = 4p(y - k)$. Find the vertex. Which way does the parabola open?

   c. Do (a) and (b) agree with each other?

28. With a graphing utility, explore the second-degree equation $Ax^2 + Bxy + Cy^2 + 10x - 8y - 5 = 0$ for the following values of A, B, and C:

   a. $A = 2, B = -\sqrt{3}, C = 1$ **b.** $A = 2, B = \sqrt{3}, C = -1$

Show the angle of rotation to one decimal place.
Explain the differences.

Answers to Odd-Numbered Exercises

Answers that require a proof, graph, or otherwise lengthy solution are not included. See *Instructor's Solutions Manual*.

CHAPTER 1

Section 1.1

1. $180°$ **3.** $-120°$ **5.** $300°$ **7.** $-288°$

9. a. $72°$ **b.** $162°$ **11. a.** $48°$ **b.** $138°$

13. a. $1°$ **b.** $91°$ **15.** $36°, 54°$

17. $60°, 120°$ **19.** $\gamma = 30°$

21. $\alpha = 120°, \beta = \gamma = 30°$

23. $103.1°$ **25.** $c = 5$ **27.** $b = 8$

29. $c = \sqrt{89}$ **31.** $b = 6\sqrt{2}$ **33.** $a = 3\sqrt{2}$

35. $10\sqrt{2}$ in.   **37.** 2 cm   **39.** 8 in.

41. 5 m, $5\sqrt{3}$ m, 10 m   **43.** $4\sqrt{3}$ yd, $8\sqrt{3}$ yd

45. 5 in., $5\sqrt{3}$ in.   **47.** $-120°$

49. $144°$ **51.** 1 hr

53. 85 ft   **55.** 241 ft   **57.** 9.8 ft

59. 17.3 ft   **61.** 48 ft $\times$ 28 ft

63. no, $10^2 + 15^2 \neq 20^2$ **65.** $10{,}200°$

67. In a $30°$-$60°$-$90°$ triangle, the length opposite the $60°$ angle has length $\sqrt{3} \times$ (shorter leg), not $2 \times$ (shorter leg).

69. false   **71.** true   **73.** true   **75.** true

77. $110°$ **79.** 3 **81.** 25

83. $45°$–$45°$–$90°$ triangles

85. 16.68 ft, 28.89 ft, 33.36 ft

Section 1.2

1. $B = 80°$ **3.** $F = 80°$ **5.** $B = 75°$

7. $B = 65°$ **9.** $A = 120° = D$

11. $A = 110°, G = 170°$

13. b   **15.** a   **17.** d

19. $f = 3$ **21.** $a = 15$ **23.** 11.55 km

25. $\frac{12}{25}$ in.   **27.** 38 ft   **29.** 200 ft

31. 120 m   **33.** 6 ft 5 in.   **35.** 2.8 ft

37. 0.87 in.   **39.** 7.8 cm or 78 mm

41. 4.0 cm or 40 mm

43. The ratio is set up incorrectly. It should be $\dfrac{A}{D} = \dfrac{B}{E}$.

45. true   **47.** false   **49.** false

51. false   **53.** $x = 2$

55. Triangles 1 and 2: $90°$ angle, vertical angles, same for triangles 3 and 4

59. $m(EF) = \sqrt{3}$ cm, $m(EG) = 2\sqrt{3}$ cm

Section 1.3

1. $\frac{4}{5}$ **3.** $\frac{5}{4}$ **5.** $\frac{4}{3}$ **7.** $\frac{\sqrt{5}}{5}$

9. $\sqrt{5}$ **11.** 2 **13.** $\frac{5\sqrt{34}}{34}$ **15.** $\frac{5}{3}$

17. $\frac{\sqrt{34}}{3}$ **19.** $30°$ **21.** $90° - x$ **23.** $60°$

25. $\cos(90° - x - y)$ **27.** $\sin(70° - A)$

29. $\tan(45° + x)$ **31.** 10 mi   **33.** $\frac{5}{13}$

35. $60°$ **37.** $\frac{3}{5}$ **39.** $\frac{\sqrt{33}}{7}$

41. $30°$ **43. a.** $\frac{2}{5}$ **b.** $\frac{5}{2}$

45. Opposite side has length 3, not 4.

47. $\sec x = \dfrac{1}{\cos x}$, not $\dfrac{1}{\sin x}$.

49. true   **51.** true

53. $\sin 30° = \frac{1}{2}, \cos 30° = \frac{\sqrt{3}}{2}$

55. $\tan 30° = \frac{\sqrt{3}}{3}, \tan 60° = \sqrt{3}$

57. $\sec 45° = \sqrt{2}, \csc 45° = \sqrt{2}$

59. $-1 \leq \sin\theta \leq 1$ and $-1 \leq \cos\theta \leq 1$

61. $\left|\sec\theta\right| \geq 1$ and $\left|\csc\theta\right| \geq 1$

63. $\cos\theta$ decreases from 1 to 0

65. a. 2.92398 **b.** 2.92380

67. a. 0.70274 **b.** 0.70281

Section 1.4

1. a   **3.** b   **5.** C

7. $\frac{\sqrt{3}}{3}$ **9.** $\sqrt{3}$ **11.** $\frac{2\sqrt{3}}{3}$

13. $\frac{2\sqrt{3}}{3}$ **15.** $\frac{\sqrt{3}}{3}$ **17.** $\sqrt{2}$

19. 0.6018 **21.** 0.1392 **23.** 1.3764

25. 1.0098 **27.** 1.0002 **29.** 0.7002

31. $68° 46' 4''$ **33.** $79° 40' 5''$ **35.** $47° 17' 5''$

37. $84° 42' 31''$ **39.** $33.33°$ **41.** $59.45°$

43. $27.754°$ **45.** $42.470°$ **47.** $15° 45'$

49. $22° 21'$ **51.** $30° 10' 30''$ **53.** $77° 32' 6''$

55. 0.1808 **57.** 0.4091 **59.** 2.1007

61. 1.09 angstroms   **63.** 2.405 **65.** 1.335

67. $4° 32'$ **69.** $\frac{23\sqrt{3}}{2}$ ft   **71.** $50\sqrt{2}$ ft

73. $40.3075°$ **75.** $\cos 60° = \frac{1}{2}$, not $\frac{\sqrt{3}}{2}$

77. false **79.** true

81. 0 **83.** 0 **85.** $\frac{2\sqrt{6}+2}{3}$

87. $\frac{-\sqrt{6}-\sqrt{2}}{4}$ **89.** 0.7575

91. **a.** 2.92398 **b.** 2.92380

93. 3.240

Section 1.5

1. three **3.** two **5.** 47° **7.** 55° **9.** 83°

11. 14 in. **13.** 18 ft **15.** 5.50 mi **17.** 12 km

19. 20.60 cm **21.** 50° **23.** 62°

25. 82.12 yd **27.** 10.6 km **29.** 19,293 km

31. $\beta = 58°, a \approx 6.4$ ft, $b \approx 10$ ft

33. $\beta = 146°, c \approx 3.6$ cm, $a \approx 2.5$ cm

35. $\beta = 30°, b \approx 3$ in., $a \approx 4$ in.

37. $\alpha = 18°, b \approx 9.2$ mm, $a \approx 3.0$ mm

39. $\beta = 35.8°, c \approx 137$ mi, $b \approx 80.1$ mi

41. $\alpha = 45°, a = 10.2$ km, $c \approx 14$ km

43. $\beta = 61.62°, \alpha \approx 28.38°, c \approx 1971$ ft, $a \approx 936.9$ ft

45. $\alpha \approx 56.0°, \beta \approx 34.0°, c \approx 51.3$ ft

47. $\alpha \approx 55.480°, \beta \approx 34.520°, b \approx 24,235$ km

49. 286 ft **51.** 88 ft **53.** 260 ft

55. 11.0°, too low **57.** 80 ft **59.** 170 m

61. 0.000016° **63.** 26 ft **65.** 4414 ft

67. N 34° W **69.** 8° **71.** 136.7°

73. 97 mi **75.** 3.5 ft **77.** 4.7 in.

79. Should be $\beta = \tan^{-1}(80)$, not $\tan(80)$.

81. true **83.** false **85.** false **87.** true

89. 4000 ft **91.** 3.4 ft **93.** 1.98 **95.** 40°

97. 0.8 **99.** $\theta, 0 \le \theta \le 90°$

Review Exercises

1. **a.** 62° **b.** 152° **3.** **a.** 55° **b.** 145°

5. $\gamma = 25°$ **7.** $\alpha = 140°, \gamma = 20°$

9. $8\sqrt{2}$ **11.** $\sqrt{65}$ **13.** $12\sqrt{2}$ yd

15. $3\sqrt{3}$ ft, 6 ft **17.** 150° **19.** $G = 75°$

21. $C = 75°$ **23.** $B = 75°$ **25.** $E = 4$

27. 147.6 km **29.** 32 m **31.** 48 in. or 4 ft

33. $\frac{2\sqrt{13}}{13}$ **35.** $\frac{\sqrt{13}}{2}$ **37.** $\frac{3}{2}$ **39.** 60°

41. 45° **43.** $\cos(x + 60°)$ **45.** $\sec(45° + x)$

47. b **49.** b **51.** c **53.** $\frac{\sqrt{3}}{3}$

55. $\sqrt{3}$ **57.** $\sqrt{2}$ **59.** $\frac{2\sqrt{3}}{3}$ **61.** 2

63. 1 **65.** 0.6691 **67.** 0.9548

69. 1.5399 **71.** 1.5477 **73.** 39.28°

75. 29.507° **77.** 42° 15′ **79.** 30° 10′ 30″

81. 0.6053 **83.** 0.4726 **85.** 1.50

87. 14 in. **89.** 14.5 mi **91.** 92.91 yd

93. $\beta = 60°, a \approx 11$ ft, $b \approx 18$ ft

95. $\beta = 41.5°, c \approx 287$ mi, $b \approx 190$ mi

97. $\alpha \approx 33.7°, \beta \approx 56.3°, c \approx 54.9$ ft

99. 75 ft **101.** N 58° E

103. legs: 41.32 ft, 71.57 ft, hypotenuse: 82.64 ft

105. **a.** 1.02041 **b.** 1.02085.

107. 2.612 **109.** -1

Practice Test

1. 20°, 60°, 100° **3.** 6000 ft

5. **a.** $\frac{3\sqrt{10}}{10}$ **b.** $\sqrt{10}$ **c.** $\frac{1}{3}$

7. 1.3629 **9.** 33.756°

11. 66° **13.** $\beta \approx 71.8°$

15. $\alpha \approx 27.0°$ **17.** 1.53

19. 44° 16′ 12″ **21.** $\frac{\sqrt{6}+\sqrt{2}}{4}$

23. 68° 29′ **25.** 620 yd

CHAPTER 2

Section 2.1

1. QI **3.** QII **5.** QIV **7.** y-axis **9.** x-axis

11. QIII **13.** QI **15.** QIII **17.** QII **19.** QII

21. 135° **23.** $-405°$ **25.** $-225°$

27. 330° **29.** 510° **31.** 840°

33. c **35.** e **37.** f **39.** 52° **41.** 268°

43. 330° **45.** 150° **47.** 154° **49.** 30° **51.** $-1200°$

53. 540° **55.** yes **57.** 1440°

59. horizontal: 50 ft, vertical: 87 ft

61. QII **63.** 2070°

65. Coterminal angles are not supplementary angles.

67. true **69.** false

71. $30° + n(360°), n = 0, 1, 2, \ldots$

73. $30° - 360°k, k = 0, 1, 2, 3, \ldots$

75. $d = \sqrt{(x_2 - x_1)^2 + (y_2 - y_1)^2}$

Section 2.2

1.

$\sin\theta$	$\cos\theta$	$\tan\theta$
$\frac{2\sqrt{5}}{5}$	$\frac{\sqrt{5}}{5}$	2
$\cot\theta$	$\sec\theta$	$\csc\theta$
$\frac{1}{2}$	$\sqrt{5}$	$\frac{\sqrt{5}}{2}$

3.

$\sin\theta$	$\cos\theta$	$\tan\theta$
$\frac{2\sqrt{5}}{5}$	$\frac{\sqrt{5}}{5}$	2
$\cot\theta$	$\sec\theta$	$\csc\theta$
$\frac{1}{2}$	$\sqrt{5}$	$\frac{\sqrt{5}}{2}$

5.

$\sin\theta$	$\cos\theta$	$\tan\theta$
$\frac{4\sqrt{41}}{41}$	$\frac{5\sqrt{41}}{41}$	$\frac{4}{5}$
$\cot\theta$	$\sec\theta$	$\csc\theta$
$\frac{5}{4}$	$\frac{\sqrt{41}}{5}$	$\frac{\sqrt{41}}{4}$

7.

$\sin\theta$	$\cos\theta$	$\tan\theta$
$\frac{2\sqrt{5}}{5}$	$-\frac{\sqrt{5}}{5}$	-2
$\cot\theta$	$\sec\theta$	$\csc\theta$
$-\frac{1}{2}$	$-\sqrt{5}$	$\frac{\sqrt{5}}{2}$

9.

$\sin\theta$	$\cos\theta$	$\tan\theta$
$-\frac{7\sqrt{65}}{65}$	$-\frac{4\sqrt{65}}{65}$	$\frac{7}{4}$
$\cot\theta$	$\sec\theta$	$\csc\theta$
$\frac{4}{7}$	$-\frac{\sqrt{65}}{4}$	$-\frac{\sqrt{65}}{7}$

11.

$\sin\theta$	$\cos\theta$	$\tan\theta$
$\frac{\sqrt{15}}{5}$	$-\frac{\sqrt{10}}{5}$	$-\frac{\sqrt{6}}{2}$
$\cot\theta$	$\sec\theta$	$\csc\theta$
$-\frac{\sqrt{6}}{3}$	$-\frac{\sqrt{10}}{2}$	$\frac{\sqrt{15}}{3}$

13.

$\sin\theta$	$\cos\theta$	$\tan\theta$
$-\frac{\sqrt{6}}{4}$	$-\frac{\sqrt{10}}{4}$	$\frac{\sqrt{15}}{5}$
$\cot\theta$	$\sec\theta$	$\csc\theta$
$\frac{\sqrt{15}}{3}$	$-\frac{2\sqrt{10}}{5}$	$-\frac{2\sqrt{6}}{3}$

15.

$\sin\theta$	$\cos\theta$	$\tan\theta$
$\frac{2\sqrt{29}}{29}$	$-\frac{5\sqrt{29}}{29}$	$-\frac{2}{5}$
$\cot\theta$	$\sec\theta$	$\csc\theta$
$-\frac{5}{2}$	$-\frac{\sqrt{29}}{5}$	$\frac{\sqrt{29}}{2}$

17.

$\sin\theta$	$\cos\theta$	$\tan\theta$
$\frac{2\sqrt{5}}{5}$	$\frac{\sqrt{5}}{5}$	2
$\cot\theta$	$\sec\theta$	$\csc\theta$
$\frac{1}{2}$	$\sqrt{5}$	$\frac{\sqrt{5}}{2}$

19.

$\sin\theta$	$\cos\theta$	$\tan\theta$
$\frac{\sqrt{5}}{5}$	$\frac{2\sqrt{5}}{5}$	$\frac{1}{2}$
$\cot\theta$	$\sec\theta$	$\csc\theta$
2	$\frac{\sqrt{5}}{2}$	$\sqrt{5}$

21.

$\sin\theta$	$\cos\theta$	$\tan\theta$
$-\frac{\sqrt{10}}{10}$	$\frac{3\sqrt{10}}{10}$	$-\frac{1}{3}$
$\cot\theta$	$\sec\theta$	$\csc\theta$
-3	$\frac{\sqrt{10}}{3}$	$-\sqrt{10}$

23.

$\sin\theta$	$\cos\theta$	$\tan\theta$
$\frac{2\sqrt{13}}{13}$	$-\frac{3\sqrt{13}}{13}$	$-\frac{2}{3}$
$\cot\theta$	$\sec\theta$	$\csc\theta$
$-\frac{3}{2}$	$-\frac{\sqrt{13}}{3}$	$\frac{\sqrt{13}}{2}$

25.

$\sin\theta$	$\cos\theta$	$\tan\theta$
1	0	undefined
$\cot\theta$	$\sec\theta$	$\csc\theta$
0	undefined	1

27.

$\sin\theta$	$\cos\theta$	$\tan\theta$
-1	0	undefined
$\cot\theta$	$\sec\theta$	$\csc\theta$
0	undefined	-1

29.

$\sin\theta$	$\cos\theta$	$\tan\theta$
1	0	undefined
$\cot\theta$	$\sec\theta$	$\csc\theta$
0	undefined	1

31.

$\sin\theta$	$\cos\theta$	$\tan\theta$
-1	0	undefined
$\cot\theta$	$\sec\theta$	$\csc\theta$
0	undefined	-1

33.

$\sin\theta$	$\cos\theta$	$\tan\theta$
-1	0	undefined

$\cot\theta$	$\sec\theta$	$\csc\theta$
0	undefined	-1

35.

$\sin\theta$	$\cos\theta$	$\tan\theta$
1	0	undefined

$\cot\theta$	$\sec\theta$	$\csc\theta$
0	undefined	1

37. $\theta = 810°$

$\sin\theta$	$\cos\theta$	$\tan\theta$
1	0	undefined

$\cot\theta$	$\sec\theta$	$\csc\theta$
0	undefined	1

39. $\theta = -810°$

$\sin\theta$	$\cos\theta$	$\tan\theta$
-1	0	undefined

$\cot\theta$	$\sec\theta$	$\csc\theta$
0	undefined	-1

41. $\frac{7}{24}$ **43.** 120 ft

45.

$\sin\theta$	$\cos\theta$	$\tan\theta$
$\frac{4\sqrt{17}}{17}$	$\frac{\sqrt{17}}{17}$	4

47. 20.1 ft   **49.** profit

51. 1.4 cm or 14 mm

53. Here, $r = \sqrt{1^2 + 2^2} = \sqrt{5}$, not 5.

55. false   **57.** true   **59.** $-\frac{3}{5}$

61. $\tan\theta = m$ **63.** $y = (\tan\theta)(x - a)$

65. $x = 90° + (180k)°$, k any integer

67. -1 **69.** undefined   **71.** 0

73. undefined   **75.** -1

Section 2.3

1. QIV   **3.** QII   **5.** QI   **7.** $-\frac{4}{5}$ **9.** $-\frac{60}{11}$

11. $\frac{4\sqrt{7}}{7}$ **13.** $-\frac{84}{85}$ **15.** 2 **17.** $\sqrt{3}$ **19.** $-\frac{\sqrt{11}}{5}$

21. 1 **23.** -1 **25.** 0 **27.** 1 **29.** 1

31. possible   **33.** not possible   **35.** possible

37. possible   **39.** $-\frac{1}{2}$ **41.** $-\frac{\sqrt{3}}{2}$

43. $\sqrt{3}$ **45.** $\frac{\sqrt{3}}{2}$ **47.** $\frac{\sqrt{3}}{3}$ **49.** 1 **51.** -2

53. 30° or 330°   **55.** 210° or 330°   **57.** 90° or 270°

59. 270°   **61.** -0.8387 **63.** -11.4301

65. 2.0627   **67.** -1.0711 **69.** -2.6051

71. 110°   **73.** 143°   **75.** 322°

77. 140°   **79.** 340°   **81.** 1.3

83. 12°   **85.** $\frac{\sqrt{3}}{2}$ **87.** 3 or 9

89. 15°; The lower leg is bent at the knee in a backward direction at an angle of 15°.

91. 75.5°

93. The reference angle is measured between the terminal side and the x-axis, not the y-axis.

95. true   **97.** false   **99.** $-\dfrac{a}{\sqrt{a^2 + b^2}}$

101. 30°   **103.** $-\frac{1}{2}$ or $\frac{1}{2}$ **105.** QI

107. QI   **109.** QI

Section 2.4

1. $\frac{8}{7}$ **3.** $-\frac{5}{3}$ **5.** $-\frac{1}{5}$ **7.** $\frac{2\sqrt{5}}{5}$ **9.** $-\frac{5\sqrt{7}}{7}$

11. $-\frac{\sqrt{3}}{3}$ **13.** $-\frac{4}{3}$ **15.** $\frac{4}{3}$ **17.** $\frac{\sqrt{11}}{5}$ **19.** $\frac{5}{64}$

21. -8 **23.** $-\frac{\sqrt{3}}{2}$ **25.** $-\frac{\sqrt{21}}{5}$ **27.** $-\sqrt{17}$

29. $-\sqrt{5}$ **31.** $-\frac{8\sqrt{161}}{161}$ **33.** $-\frac{15\sqrt{11}}{44}$ **35.** $-\frac{2\sqrt{13}}{13}$

37. $\sin\theta = \frac{4}{5}$, $\cos\theta = -\frac{3}{5}$ **39.** $\sin\theta = -\frac{\sqrt{26}}{26}$, $\cos\theta = \frac{5\sqrt{26}}{26}$

41. $\sin\theta = -\frac{2\sqrt{5}}{5}$, $\cos\theta = -\frac{\sqrt{5}}{5}$

43. $\sin\theta = -\frac{3\sqrt{34}}{34}$, $\cos\theta = -\frac{5\sqrt{34}}{34}$

45. $\dfrac{1}{\sin\theta}$ **47.** -1 **49.** $\dfrac{\cos^2\theta}{\sin\theta}$ **51.** $\dfrac{1}{\sin\theta}$

53. $1 + 2\sin\theta\cos\theta$ **55.** 10%   **57.** $\frac{5}{12}$

59. $r = 8\sin\theta - 8\sin^3\theta$

θ	r
30°	3
60°	$\sqrt{3}$
90°	0

61. $\tan\theta = \frac{4}{3} \approx \frac{1.33}{1}$; each dollar spent on costs produces \$1.33 in revenue.

63. Cosine is *negative* in QIII.

65. false   **67.** $\dfrac{b}{a}$, $a \neq 0$ **69.** 90°

71. $\cot\theta = \dfrac{\pm\sqrt{1 - \sin^2\theta}}{\sin\theta}$

73. $8|\cos\theta|$ **75.** $-\sqrt{x^2 - 1}$

77. **a.** 0.3746   **b.** 0.9272   **c.** 0.4040   **d.** 0.4040, yes

79. **a.** 0.5736   **b.** 0.1808   **c.** 0.5736   a and c

Review Exercises

1. QIV   **3.** QII

5. 120°   **7.** 450°

9. 280°   **11.** 120°

13.

$\sin\theta$	$\cos\theta$	$\tan\theta$
$-\frac{4}{5}$	$\frac{3}{5}$	$-\frac{4}{3}$
$\cot\theta$	$\sec\theta$	$\csc\theta$
$-\frac{3}{4}$	$\frac{5}{3}$	$-\frac{5}{4}$

15.

$\sin\theta$	$\cos\theta$	$\tan\theta$
$\frac{\sqrt{10}}{10}$	$-\frac{3\sqrt{10}}{10}$	$-\frac{1}{3}$
$\cot\theta$	$\sec\theta$	$\csc\theta$
-3	$-\frac{\sqrt{10}}{3}$	$\sqrt{10}$

17.

$\sin\theta$	$\cos\theta$	$\tan\theta$
$\frac{1}{2}$	$\frac{\sqrt{3}}{2}$	$\frac{\sqrt{3}}{3}$
$\cot\theta$	$\sec\theta$	$\csc\theta$
$\sqrt{3}$	$\frac{2\sqrt{3}}{3}$	2

19.

$\sin\theta$	$\cos\theta$	$\tan\theta$
$-\frac{\sqrt{2}}{2}$	$-\frac{\sqrt{2}}{2}$	1
$\cot\theta$	$\sec\theta$	$\csc\theta$
1	$-\sqrt{2}$	$-\sqrt{2}$

21.

$\sin\theta$	$\cos\theta$	$\tan\theta$
$-\frac{3}{5}$	$\frac{4}{5}$	$-\frac{3}{4}$
$\cot\theta$	$\sec\theta$	$\csc\theta$
$-\frac{4}{3}$	$\frac{5}{4}$	$-\frac{5}{3}$

23. $\theta = 270°$

$\sin\theta$	$\cos\theta$	$\tan\theta$
-1	0	undefined
$\cot\theta$	$\sec\theta$	$\csc\theta$
0	undefined	-1

25. $\theta = -1080°$

$\sin\theta$	$\cos\theta$	$\tan\theta$
0	1	0
$\cot\theta$	$\sec\theta$	$\csc\theta$
undefined	1	undefined

27. QIII **29.** QIV **31.** $\frac{4}{5}$ **33.** $-2\sqrt{2}$

35. 1 **37.** 0 **39.** impossible

41. possible **43.** $-\frac{1}{2}$ **45.** $-\frac{\sqrt{3}}{3}$

47. $-\frac{2\sqrt{3}}{3}$ **49.** -0.2419 **51.** 1.0355

53. -2.7904 **55.** -0.6494 **57.** $-\frac{11}{7}$

59. $-\frac{8}{15}$ **61.** $\frac{4}{27}$ **63.** $3\sqrt{3}$

65. $-\frac{7}{24}$ **67.** $-\frac{5}{12}$

69. $\sin\theta = -\frac{\sqrt{3}}{2}, \cos\theta = -\frac{1}{2}$

71. $\frac{\cos\theta}{\sin^2\theta}$ **73.** $\sin\theta$

75. $\frac{1}{\sin\theta} + \cos\theta$ **77.** -1 **79.** QIV

81. **a.** 0.2924 **b.** 0.4370 **c.** 0.2924 **a** and **c**

Practice Test

1.

	0°	**QI**	90°	**QII**	180°
$\sin\theta$	0	$+$	1	$+$	0
$\cos\theta$	1	$+$	0	$-$	-1
	QIII	270°	**QIV**	360°	
$\sin\theta$	$-$	-1	$-$	0	
$\cos\theta$	$-$	0	$+$	1	

3. 90°, 270° **5.** 220° **7.** $\frac{\sqrt{2}}{2}$ **9.** $\frac{\sqrt{2}}{2}$

11. 1 **13.** 1.1106 **15.** -0.3450

17. $\sin\theta = -\frac{\sqrt{35}}{6}, \tan\theta = -\sqrt{35}$

19. **a.** not possible **b.** possible

21. $\pm\frac{\sqrt{3}}{3}$ **23.** 250 yd **25.** $-\frac{2\sqrt{3}}{3}$

Cumulative Test

1. **a.** 53° **b.** 143° **3.** 210° **5.** 40 ft

7. $\cot 30°$ **9.** $\frac{\sqrt{3}}{2}$ **11.** $17°31'30''$

13. 54° **15.** QIII **17.** QIV

19. $-1170°$

$\sin\theta$	$\cos\theta$	$\tan\theta$
-1	0	undefined
$\cot\theta$	$\sec\theta$	$\csc\theta$
0	undefined	-1

21. $\frac{25}{24}$ **23.** $-\frac{\sqrt{3}}{2}$ **25.** $\frac{1}{\cos\theta} + \sin\theta$

CHAPTER 3

Section 3.1

1. 0.2 **3.** 0.18 **5.** 0.02 **7.** 0.125

9. 0.2 **11.** $\frac{\pi}{6}$ **13.** $\frac{\pi}{4}$ **15.** $\frac{7\pi}{4}$

17. $\frac{5\pi}{12}$ **19.** $\frac{17\pi}{18}$ **21.** $\frac{13\pi}{3}$ **23.** $-\frac{7\pi}{6}$

25. 30° **27.** 135° **29.** 67.5° **31.** 75°

33. 1620° **35.** 171° **37.** $-84°$ **39.** 229.18°

41. 48.70° **43.** $-160.37°$ **45.** 0.820

47. 1.95 **49.** 0.986 **51.** $\frac{\pi}{3}, 60°$ **53.** $\frac{\pi}{4}, 45°$

55. $\frac{5\pi}{12}, 75°$ **57.** $\frac{\pi}{3}, 60°$ **59.** $\frac{\sqrt{2}}{2}$

61. $-\frac{\sqrt{2}}{2}$ **63.** $-\frac{\sqrt{2}}{2}$ **65.** $-\frac{\sqrt{3}}{2}$ **67.** $-\frac{1}{2}$

69. $\frac{1}{2}$ **71.** $-\frac{\sqrt{3}}{3}$ **73.** $\frac{\sqrt{3}}{3}$ **75.** $\frac{\sqrt{3}}{3}$

77. 0 **79.** -1 **81.** $-\frac{\sqrt{2}}{2}$ **83.** $-\frac{\sqrt{3}}{2}$

85. $\frac{3\pi}{2}$ **87.** $\frac{\pi}{5}$ **89.** 5π **91.** $\frac{\pi}{16}$

93. 0.75 **95.** 157.1 rad **97.** $\frac{\pi}{4}$ or $45°$

99. $\frac{7\pi}{12}$ **101.** $\frac{1343\pi}{1800} \approx 2.3440$

103. The radius and the arc length must be converted to the same units prior to using the radian formula.

105. Note that the "45" in the expression tan 45 is in radians, not degrees.

107. false **109.** false **111.** $\frac{\pi}{2}$ rad **113.** 0.225 rad

115. $\frac{13\pi}{18}$ **117.** $-\sqrt{3}+5$ **119.** 42 rad $= \left(42 \cdot \frac{180}{\pi}\right)°$

Section 3.2

1. 12 mm **3.** $\frac{2\pi}{3}$ ft **5.** π m **7.** $\frac{11\pi}{5}$ μm

9. $\frac{200\pi}{3}$ km **11.** $\frac{32\pi}{5}$ cm **13.** 25 ft

15. 8 in. **17.** 3 yd **19.** 4 yd **21.** 12 mi.

23. $\frac{8}{11}$ km **25.** 1.3 mm **27.** 1.7 yd **29.** 150 mi

31. 2.2 μm **33.** 1300 km **35.** 22 ft

37. 12.8 ft^2 **39.** 2.85 km^2 **41.** 42.8 cm^2

43. 8.62 cm^2 **45.** 0.0236 ft^2 **47.** 1.35 mi^2

49. 5262 km **51.** 37 ft **53.** 628 ft

55. 50° **57.** 60 in **59.** 911 mi

61. 157 ft^2 **63.** 78 in.^{2} **65.** 75 rev/30 sec

67. 2.4 rev/sec **69.** $\frac{189\pi}{8} \cong 74$ sq mi

71. $\frac{(4.5 \times 10^{-9})}{2} \frac{\pi}{5} \cong 1.4 \times 10^{-9}$ m or 1.4 nm

73. The arc length formula for radian measure was used when the corresponding formula for degree measure should have been used.

75. false **77.** true **79.** $\theta_1\left(\frac{r_1}{r_2}\right)$

81. 11,800 ft^2 **83.** 133 ft^2

Section 3.3

1. $\frac{2}{5}$ m/sec **3.** 272 km/hr **5.** 7 nm/ms

7. $\frac{1}{64}$ in./min **9.** $\frac{3}{52}$ m/sec **11.** 9.8 m

13. 1.5 mi **15.** 15 mi **17.** 4,320,000 km

19. 4140 ft **21.** $\frac{5\pi}{2}$ rad/sec **23.** 20π rad/sec

25. $\frac{7\pi}{24}$ rad/sec **27.** $\frac{2\pi}{9}$ rad/sec **29.** $\frac{13\pi}{9}$ rad/sec

31. $\frac{10\pi}{7}$ rad/sec **33.** 6π in./sec

35. $\frac{\pi}{4}$ mm/sec **37.** $\frac{2\pi}{3}$ in./sec

39. $\frac{112\pi}{9}$ yd/sec **41.** 400π cm/sec

43. 26.2 cm **45.** 54.5 ft

47. 5650 m **49.** 236 cm **51.** 6.69 mi

53. 69.81 mph **55.** 1037.51 mph, 1040.35 mph

57. 15.71 ft/sec **59.** $\frac{200\pi}{3}$ rad/min

61. 12.05 mph **63.** 640 rev/min

65. 1.6 rotations/sec **67.** 17.59 m/sec

69. 0.01 m/sec

71. Angular velocity must be expressed in radians (not degrees) per second.

73. false **75.** $v_2 = \left(\frac{r_2}{r_1}\right)v_1$ **77.** 6.3 cm/sec

79. Linear speed for the red ball is twice that of the blue ball.

Section 3.4

1. $-\frac{\sqrt{3}}{2}$ **3.** $-\frac{\sqrt{3}}{2}$ **5.** $\frac{\sqrt{2}}{2}$ **7.** -1 **9.** $-\frac{2\sqrt{3}}{3}$

11. $-\sqrt{2}$ **13.** $\sqrt{3}$ **15.** $-\frac{\sqrt{3}}{2}$ **17.** $-\frac{\sqrt{3}}{2}$

19. $-\frac{\sqrt{2}}{2}$ **21.** $-\frac{\sqrt{3}}{2}$ **23.** $\frac{\sqrt{2}}{2}$ **25.** 1

27. $\frac{\sqrt{2}}{2}$ **29.** 0 **31.** $\frac{\pi}{6}, \frac{11\pi}{6}$ **33.** $\frac{4\pi}{3}, \frac{5\pi}{3}$

35. $\frac{\pi}{3}, \frac{5\pi}{3}$ **37.** $\frac{3\pi}{4}, \frac{5\pi}{4}$ **39.** $0, \pi, 2\pi, 3\pi, 4\pi$

41. $\pi, 3\pi$ **43.** $\frac{3\pi}{4}, \frac{7\pi}{4}$ **45.** $\frac{3\pi}{4}, \frac{5\pi}{4}$

47. $0, \pi, 2\pi$ **49.** $\frac{\pi}{2}, \frac{3\pi}{2}$ **51.** -0.4154

53. 1.3764 **55.** -0.7568 **57.** -0.7470

59. 22.9°F **61.** 99.1°F **63.** 2.6 ft

65. 135 lb **67.** 10,000 guests

69. 80°F **71.** -0.4 cm **73.** 14 in.

75. 10.7 μg/μL **77.** 35°C

79. should have used $\cos\left(\frac{5\pi}{6}\right) = -\frac{\sqrt{3}}{2}$ and $\sin\left(\frac{5\pi}{6}\right) = \frac{1}{2}$

81. true **83.** false **85.** odd **87.** $\frac{\pi}{4}, \frac{5\pi}{4}$

89. 25 **91.** $\frac{\pi}{4}, \frac{3\pi}{4}, \frac{5\pi}{4}, \frac{7\pi}{4}$

93. $\frac{5\pi}{6} < x < \frac{5\pi}{3}$

95. $\sin 423° \approx 0.891, \sin(-423°) \approx -0.891$

97. 0.5

Review Exercises

1. $\frac{3\pi}{4}$ **3.** $\frac{11\pi}{3}$ **5.** $\frac{6\pi}{5}$ **7.** $\frac{14\pi}{5}$ **9.** $-\frac{5\pi}{6}$

11. 60° **13.** 225° **15.** 100° **17.** 585°

19. $-50°$ **21.** $\frac{\pi}{4}$ **23.** $\frac{\pi}{6}$ **25.** 5.24 cm

27. 8.73 in. **29.** 0.5 **31.** $\frac{2\pi}{3}$

33. $\frac{1}{6}$ **35.** 1.6 in. **37.** 16.8 m **39.** 5 yd

41. 302 mi^2 **43.** 1885 m^2 **45.** $\frac{1}{3}$ ft/sec **47.** 5 mi/min

49. 360 mi **51.** 20 mi **53.** $\frac{2\pi}{3}$ rad/sec

55. $\frac{\pi}{16}$ rad/sec **57.** 10π m/sec **59.** 75π ft

61. 240π yd **63.** 754 in./min **65.** $-\frac{\sqrt{3}}{3}$

67. $-\frac{1}{2}$ **69.** 1 **71.** -1 **73.** -1

75. $\frac{1}{2}$ **77.** $-\frac{1}{2}$ **79.** $-\frac{1}{2}$ **81.** $\frac{7\pi}{6}, \frac{11\pi}{6}$

83. $0, \pi, 2\pi, 3\pi, 4\pi$ **85.** $100°46'48''$ **87.** -0.9659

Practice Test

1. 0.02 **3.** $\frac{13\pi}{9}$ rad **5.** $\frac{5\pi}{12}$ **7.** 1.0 mi

9. 164 ft^2 **11.** 52° **13.** 4.2 ft **15.** 12π in./sec

17. -1 **19.** 0 **21.** $\frac{23\pi}{36}$ **23.** $\frac{\pi}{6}, \frac{7\pi}{6}$

25. 60 units

Cumulative Test

1. 40°

3. $D = 105°, G = 75°$

5. $\frac{3}{5}$

7. $64°49'15''$

9. $\beta = 52.6$ degrees; $b \cong 173$ miles and $c \cong 217$ miles.

11. positive y-axis

13.

$\sin\theta$	$\cos\theta$	$\tan\theta$
0	-1	0
$\cot\theta$	$\sec\theta$	$\csc\theta$
undefined	-1	undefined

15. 1

17. $-\frac{3\sqrt{5}}{5}$

19. $\sin\theta = -\frac{6\sqrt{37}}{37}, \cos\theta = -\frac{\sqrt{37}}{37}$

21. $\frac{7\pi}{2}$ rad

23. 8.58 m

25. $\frac{\pi}{4}, \frac{5\pi}{4}$

CHAPTER 4

Section 4.1

1. c   **3.** a   **5.** h   **7.** b   **9.** e

	Amplitude	**Period**
11.	$\frac{3}{2}$	$\frac{2\pi}{3}$
13.	1	$\frac{2\pi}{5}$
15.	$\frac{2}{3}$	$\frac{4\pi}{3}$
17.	3	2
19.	5	6
21.	$\frac{4}{5}$	4π
23.	$\frac{1}{3}$	8π

25.   **27.**   **29.**

31.   **33.**   **35.**

37.   **39.**   **41.**

43.   **45.**   **47.**

49.   **51.**

53. $y = -\sin(2x)$ **55.** $y = \cos(\pi x)$

57. $y = -2\sin\left(\frac{\pi}{2}x\right)$ **59.** $y = \sin(8\pi x)$

61. 3.5 or 3500 widgets   **63.** 1 mg/L

65. amplitude: 4 cm, $m = 4$ g

67. $\frac{1}{4\pi}$ cycles/sec

69. amplitude: 0.005 cm, frequency: 256 Hz

71. amplitude: 0.008 cm, frequency: 375 Hz

73. 660 m/sec   **75.** 660 m/sec

77. The correct graph is reflected over the x-axis.

79. true   **81.** false   **83.** $(0, A)$

85. $x = \dfrac{n\pi}{B}$, n an integer

87.   **89.**

91. no   **93.** same

95. **a.** shifted to the left $\frac{\pi}{3}$

b. shifted to the right $\frac{\pi}{3}$

97. The graph of y_1 approaches the t-axis from above; the graph of y_2 oscillates, keeping its same form; the graph of y_3 oscillates, but dampens so that it approaches the t-axis from below and above.

Section 4.2

1. c   **3.** a   **5.** e   **7.** f

9.   **11.**   **13.**

15. **17.** **19.**

81. The phase shift is $-\dfrac{C}{B}$ so the phase shift is $\dfrac{\pi}{2}$ units, not π units.

21. **23.**

83. false   **85.** true

87. $x = \dfrac{(2n + 1)\pi}{\omega}$, n an integer

89. $\left[\dfrac{\phi}{\omega}, \dfrac{\phi}{\omega} + \dfrac{2\pi}{\omega}\right]$ **91.** $y = \sin\left(2x + \dfrac{\pi}{6}\right)$

93. close on $[0, 0.2]$ **95.**

	Amplitude	Period	Phase Shift
25.	$\frac{1}{3}$	2π	$\frac{\pi}{2}$
27.	2	2	$\frac{1}{\pi}$
29.	5	$\frac{2\pi}{3}$	$-\frac{2\pi}{3}$
31.	6	2	-2
33.	3	π	$\frac{3\pi}{8}$
35.	2	$\frac{2\pi}{3}$	$-\frac{\pi}{3}$

Section 4.3

1. b   **3.** h   **5.** c   **7.** d

	Period	Phase Shift
9.	π	π
11.	π	$\frac{\pi}{4}$
13.	$\frac{2\pi}{3}$	$\frac{\pi}{6}$
15.	2π	-4π
17.	1	$\frac{1}{2}$

37. **39.** **41.**

43. **45.**

19. **21.** **23.**

47. $y = 1 + \sin(\pi x)$ or $y = 1 + \cos\left[\pi\left(x - \frac{1}{2}\right)\right]$

49. $y = \sin\left[\pi\left(x - \frac{1}{2}\right)\right]$ or $y = \cos[\pi(x + 1)]$

25. **27.** **29.**

51. **53.** **55.**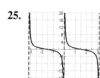

57. **59.** **61.**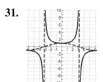

31. **33.** **35.**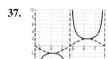

63. 4.56 mg/L

65. 0.098 million or \$98,000. Quarterly sales will vary by \$98,000 around \$387,000.

67. 0.387 million or \$387,000. This is the baseline around which the quarterly sales fluctuate.

37. **39.** **41.**

69. max: 220 amps, min: -220 amps, period: 0.1 sec, phase shift: 0.01 sec

71. $y = 59.3 + 20\sin\left(\frac{\pi}{6}x + \frac{4\pi}{3}\right)$

73. max height: 50 ft, traveled: 2400 ft

75. max: 600 deer, min: 400 deer. cycle: 8 yr

77. $y = 3\sin\left(\frac{\pi}{11}t\right) + 9$

79. $y = 25 + 25\sin\left[\frac{\pi}{2}(t - 1)\right]$ or $y = 25 - 25\cos\left(\frac{\pi t}{2}\right)$

43. **45.** **47.**

49.

51. domain: $x \neq n$, n an integer, range: $\mathbb{R}$

53. domain: $x \neq \dfrac{2n + 1}{10}\pi$, n an integer,

range: $(-\infty, -2] \cup [2, \infty)$

55. domain: $x \neq 2n\pi$, n an integer,

range: $(-\infty, 1] \cup [3, \infty)$

57.   **59.**

61. Forgot that the amplitude is 3, not 1. So, the guide function should have been $y = 3\sin(2x)$.

63. true **65.** true

67. all integer multiples of n

69. $-\pi, -\dfrac{\pi}{2}, 0, \dfrac{\pi}{2}, \pi$

71. $x \neq \dfrac{(2n - 1)\pi}{4}$, n an integer

73. 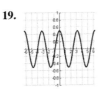 **75.** $\sqrt{2}$

$y_1 = \cos x$ (dotted bold),

$y_2 = \sin x$ (dashed bold),

$y_3 = \cos x + \sin x$ (solid bold)

Review Exercises

1. 2π **3.** $y = 4\cos x$ **5.** 5

7. 2π **9.** $y = \dfrac{1}{4}\sin x$ **11.** 2

	Amplitude	Period
13.	3	1
15.	$\dfrac{1}{5}$	$\dfrac{2\pi}{3}$

17.   **19.**

	Amplitude	Period	Phase Shift	Vertical Shift
21.	2	2π	$\dfrac{\pi}{2}$	2 (Up)
23.	4	$\dfrac{2\pi}{3}$	$-\dfrac{\pi}{4}$	−2 (Down)

	Amplitude	Period	Phase Shift	Vertical Shift
25.	1	2π	π	$\dfrac{1}{2}$ (Up)
27.	1	4π	-2π	0
29.	5	$\dfrac{\pi}{2}$	$\dfrac{\pi}{4}$	4 (Up)

31. 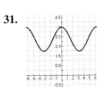 **33.** 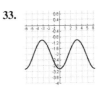 **35.**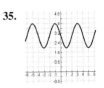

37. d **39.** a

41.   **43.**

45. b **47.** j **49.** g **51.** a **53.** i

55. domain: $x \neq n\pi$, n an integer, range: $\mathbb{R}$

57. domain: $x \neq \dfrac{2n + 1}{4}\pi$, n an integer,

range: $(-\infty, -3] \cup [3, \infty)$

59.   **61.** 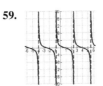 **63.**

65. **a.** shifted left $\dfrac{\pi}{6}$ units **b.** shifted right $\dfrac{\pi}{6}$ units

67. 5

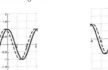

Practice Test

1. amplitude: 5, period: $\dfrac{2\pi}{3}$

3.   **5.**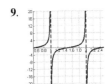

7. 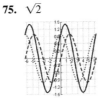 **9.**

11.   **13.**

15.

17. period: $\dfrac{2\pi}{\omega}$, phase shift: $-\dfrac{\phi}{\omega}$

19. $s = 31 + 14\cos\left(\frac{\pi}{6}t - \frac{\pi}{6}\right)$

21. a. 70,000 dollars **b.** 2020

23. $x = \dfrac{n\pi}{2}$, n an integer

25. a. true **b.** false **c.** true **d.** false **e.** true

Cumulative Test

1. leg: $8\sqrt{3} \approx 13.86$ in., hypotenuse: 16 in.

3. $\frac{5}{3}$ **5.** 63.5 ft

7.

$\sin\theta$	$\cos\theta$	$\tan\theta$
$\frac{5\sqrt{26}}{26}$	$\frac{\sqrt{26}}{26}$	5
$\cot\theta$	$\sec\theta$	$\csc\theta$
$\frac{1}{5}$	$\sqrt{26}$	$\frac{\sqrt{26}}{5}$

9. impossible **11.** $-\frac{2\sqrt{5}}{5}$ **13.** $\frac{3}{4}$

15. $\frac{\pi}{6}, 30°$ **17.** 2π cm **19.** $\frac{3\,\text{m}}{4\,\text{sec}}$

21. 136 lb

23. amplitude: 0.007, frequency: 425 Hz

25. domain: $x \neq n\pi$, n an integer, range: $\mathbb{R}$

CHAPTER 5

Section 5.1

1. 1 **3.** $\csc x$ **5.** 1

7. $\sin^2 x - \cos^2 x$ **9.** $\sec x$ **11.** $\sin^2 x$

13. $-\cos x$ **15.** 1 **17.** 0

19. $-2[\csc x + (\csc x)(\cot x)]$

47. contradiction **49.** identity

51. conditional **53.** contradiction

55. conditional **57.** identity

59. $|a|\cos\theta$ **61.** $y = -3\cos(2t)$, 1.57 sec

63. Simplified the two fractions in Step 2 incorrectly

65. Verifying the truth of the equation for a particular value of x (here, $x = \frac{\pi}{4}$) is not enough to prove that it is an identity.

67. false **69.** false **71.** QI, or QIV

73. no **75.** $a^2 + b^2$ **79.** $\cot x$ **81.** −

Section 5.2

1. $\dfrac{\sqrt{6} - \sqrt{2}}{4}$ **3.** $\dfrac{\sqrt{6} - \sqrt{2}}{4}$ **5.** $\dfrac{\sqrt{6} + \sqrt{2}}{4}$ **7.** $-2 + \sqrt{3}$

9. $\dfrac{\sqrt{2} + \sqrt{6}}{4}$ **11.** $2 + \sqrt{3}$ **13.** $2 + \sqrt{3}$ **15.** $\sqrt{2} - \sqrt{6}$

17. $\sqrt{2} - \sqrt{6}$ **19.** $\sqrt{2} - \sqrt{6}$ **21.** $\cos x$

23. $-\sin x$ **25.** $-2\cos(A - B)$

27. $\sin(3x)$ **29.** $-2\sin(A + B)$

31. $\tan 26°$ **33.** -1 **35.** $\dfrac{1 + 2\sqrt{30}}{12}$ **37.** $\dfrac{-6\sqrt{6} + 4}{25}$

39. $\dfrac{192 - 25\sqrt{15}}{-119}$ **41.** identity **43.** conditional

45. identity **47.** conditional **49.** identity

51. $y = \sin\left(x + \frac{\pi}{3}\right)$

53. $y = \cos\left(x - \frac{\pi}{4}\right)$

55. $y = -\sin(4x)$

61. $E = A\cos(kz)\cos(ct)$

63. a. $15\sin\theta$

   b. $15\sin\theta\cos 10° + 15\cos\theta\sin 10°$

65. $T(t) = 38 - 2.5\sin\left(\frac{\pi}{6}t\right)$

67. $\tan(A + B) \neq \tan A + \tan B$. Should have used
$$\tan(A + B) = \frac{\tan A + \tan B}{1 - \tan A \tan B}.$$

69. false **71.** true **73.** $\cos[(\alpha - \beta)x]$

77. $B = 2m\pi$, $A = 2n\pi$, n and m integers

81. a.   **b.**   **c.**

$y = \cos x$ as $h \to 0$

Section 5.3

1. $-\frac{4}{5}$ **3.** $\frac{120}{119}$ **5.** $\frac{120}{169}$ **7.** $-\frac{4}{3}$ **9.** $\frac{\sqrt{19}}{10}$

11. $\frac{119}{120}$ **13.** $\frac{\sqrt{3}}{3}$ **15.** $\frac{\sqrt{2}}{4}$ **17.** $\cos(4x)$

19. $-\frac{\sqrt{3}}{2}$ **21.** $\frac{\tan(8x)}{2}$

43.   **45.**

47. $C(t) = 2 + 10\cos(2t)$

49. $d = \sqrt{25 - 24\cos^2\theta}$

51. $T = 100.6 - 2\cos(2t)$

53. 22,565,385 lb

55. Should use $\sin x = -\frac{2\sqrt{2}}{3}$ since we are assuming that $\sin x < 0$

57. false **59.** false **61.** $\frac{1}{2}\sin(n\pi), 0$

63. $\frac{4\tan x(1 - \tan^2 x)}{(1 - \tan^2 x)^2 - 4\tan^2 x}$

65. cannot verify **67.** $x = 0, \frac{\pi}{3}, \pi, \frac{5\pi}{3}$

69. yes **71.** no

Section 5.4

1. $\frac{\sqrt{2 - \sqrt{3}}}{2}$ **3.** $-\frac{\sqrt{2 + \sqrt{3}}}{2}$ **5.** $\frac{\sqrt{2 - \sqrt{3}}}{2}$

7. $\frac{\sqrt{2 + \sqrt{3}}}{2}$ **9.** $\frac{\sqrt{2 - \sqrt{2}}}{2}$ **11.** $\sqrt{3 + 2\sqrt{2}}$

13. $-\frac{2}{\sqrt{2 + \sqrt{2}}}$ **15.** $\frac{-1}{\sqrt{3 + 2\sqrt{2}}}$ or $1 - \sqrt{2}$

17. $\frac{2\sqrt{13}}{13}$ **19.** $\frac{3\sqrt{13}}{13}$

21. $\frac{\sqrt{5} - 1}{2}$ or $\sqrt{\frac{3 - \sqrt{5}}{2}}$ **23.** $\sqrt{\frac{3 + 2\sqrt{2}}{6}}$

25. $-\frac{\sqrt{15}}{5}$ **27.** $\cos\left(\frac{5\pi}{12}\right)$ **29.** $\tan 75°$

45. **47.**

51. 0.57 **53.** 46.9 ft **55.** $\frac{1}{3}$

57. The wrong identity is used initially, should have used

$\sin\left(\frac{\pi}{2}\right) = \pm\sqrt{\frac{1 - \cos x}{2}}$. Also,

$\sin\left(\frac{x}{2}\right)$ is positive, not negative.

59. false **61.** false

65. $\sin 15° = \frac{\sqrt{6} - \sqrt{2}}{4} = \frac{\sqrt{2} - \sqrt{3}}{2}$

67. $\frac{\pm\sqrt{1 - \cos x}}{\sqrt{2} \pm \sqrt{1 + \cos x}}$

69. yes **71.** identity

Section 5.5

1. $\frac{1}{2}[\sin(3x) + \sin x]$ **3.** $\frac{5}{2}[\cos(2x) - \cos(10x)]$

5. $2[\cos x + \cos(3x)]$ **7.** $\frac{1}{2}[\cos x - \cos(4x)]$

9. $\frac{1}{2}[\cos(2x) + \cos\left(\frac{2x}{3}\right)]$ **11.** $\frac{1}{2}\left(\cos 75° + \frac{1}{2}\right)$

13. $2\cos(4x)\cos x$ **15.** $2\sin x\cos(2x)$

17. $\sin x \cos(7x)$ **19.** $-2\sin x \cos\left(\frac{3x}{2}\right)$

21. $2\cos\left(\frac{3x}{2}\right)\cos\left(\frac{5x}{6}\right)$ **23.** $2\sin(0.5x)\cos(0.1x)$

25. $-\tan x$ **27.** $\tan(2x)$

29. $\tan\left(\frac{A + B}{2}\right)$ **31.** $-\tan\left(\frac{A - B}{2}\right)$

33. $\tan\left(\frac{A + B}{2}\right)\cot\left(\frac{A - B}{2}\right)$

35. $\pm\sqrt{[1 - \cos(A + B)][1 + \cos(A - B)]}$

37. $\frac{\cos^2 A - \cos^2 B}{\cos A + \cos B}$ **39.** $P(t) = \sqrt{3}\cos\left(\frac{\pi}{6}t + \frac{4}{3}\pi\right)$

41. beat frequency: 102 Hz
average frequency: 443 Hz

43. $\sin\left[\frac{2\pi tc}{2}\left(\frac{1}{1.55} + \frac{1}{0.63}\right)10^6\right]$

$\cos\left[\frac{2\pi tc}{2}\left(\frac{1}{1.55} - \frac{1}{0.63}\right)10^6\right]$

45. $2\sin(1979\pi t)\cos(439\pi t)$

47. 5.98 ft^2

49. $\cos(A - B) + \frac{1}{2}[\sin(A + B) + \sin(A - B)]$

51. In the final step of the computation, $\cos A \cos B \neq \cos(AB)$ and $\sin A \sin B \neq \sin(AB)$. Should have used the product-to-sum identities.

53. false **55.** true

57. $\frac{1}{4}[\sin(A - B + C) + \sin(C - A + B)$
$- \sin(A + B + C) - \sin(A + B - C)]$

59. A or $B = \pi n$ for some integer n

61. $\frac{1}{2}(2\sin A \cos A) = \sin A \cos A$

63. $\sqrt{1 + \cos(A - B)}$ **65.** identity; $\sin(4x)$

Review Exercises

1. $\sec^2 x$ **3.** $\sec^2 x$ **5.** $\cos^2 x$ **13.** identity

15. conditional **17.** $\frac{\sqrt{2} - \sqrt{6}}{4}$ **19.** $\sqrt{3} - 2$

21. $\sin x$ **23.** $\tan x$ **25.** $\frac{117}{44}$ **27.** $-\frac{897}{1025}$

29. identity   **31.**

33. $\frac{7}{25}$ **35.** $\frac{671}{1800}$ **37.** $\frac{336}{625}$ **39.** $\frac{\sqrt{3}}{2}$

41. $\frac{3}{2}$ **49.** $\frac{-\sqrt{2-\sqrt{2}}}{2}$

51. $\dfrac{1}{\sqrt{3+2\sqrt{2}}}$ or $\sqrt{2}-1$ **53.** $\frac{7\sqrt{2}}{10}$

55. $-\frac{5}{4}$ **57.** $\sin\left(\frac{\pi}{12}\right)$

63.   **65.** $3[\sin(7x)+\sin(3x)]$

67. $-2\sin(4x)\sin x$ **69.** $2\sin\left(\frac{x}{3}\right)\cos x$ **71.** $\cot(3x)$

75. $+, y_2$

77. a.   **b.**   **c.**

$y = 3\cos(3x)$ as $h \to 0$

79. y_1 and y_3 **81.** y_1 and y_3 **83.** y_1 and y_3

Practice Test

3. conditional   **5.** $-\frac{\sqrt{2+\sqrt{3}}}{2}$ **7.** $-2-\sqrt{3}$

9. $\frac{23}{25}$ **11.** $-3\sqrt{7}$ **13.** $\frac{2\sqrt{10}-2}{9}$

15. $-\tan(2x)$ **17.** $\sin x + \sin 3$ **19.** $3|\cos x|$

21. $-\sqrt{\frac{6-\sqrt{33}}{12}}$ **23.** $-\frac{\sqrt{2}}{2}$

25. $\frac{1}{2}\left[2\cos\left(\frac{4\pi}{6}\right)\cos\left(\frac{2\pi}{6}\right)\right], -\frac{1}{4}$

Cumulative Test

1. 45 min   **3.** 0.8448   **5.** $45° + 360°k$, k any integer

7. 1   **9.** $-\sqrt{17}$ **11.** $\frac{\sqrt{3}}{2}$

13. 192 ft²   **15.** 37.7 cm

17.

19. $y = \cos(2x + \pi) + 2$ or $y = 2 - \cos(2x)$

21.   **23.** $-\frac{\sqrt{2}+\sqrt{6}}{4}$ **25.** $-\sqrt{2}-1$

CHAPTER 6

Section 6.1

1. $\frac{\pi}{4}$ **3.** $-\frac{\pi}{3}$ **5.** $\frac{3\pi}{4}$ **7.** $\frac{\pi}{6}$ **9.** $\frac{\pi}{6}$

11. $-\frac{\pi}{3}$ **13.** 60°   **15.** 45°   **17.** 120°   **19.** 30°

21. $-30°$ **23.** 135°   **25.** 57.10°   **27.** 7.02°

29. 62.18°   **31.** 48.10°   **33.** $-15.30°$

35. 166.70°   **37.** -0.63 **39.** 1.43

41. 0.92   **43.** 2.09   **45.** 1.28

47. 0.31   **49.** $\frac{5\pi}{12}$ **51.** undefined

53. $\frac{\pi}{6}$ **55.** $\frac{2\pi}{3}$ **57.** $\frac{\pi}{6}$ **59.** $\sqrt{3}$

61. $\frac{\pi}{3}$ **63.** undefined   **65.** 0

67. $-\frac{\pi}{4}$ **69.** $\frac{\sqrt{7}}{4}$ **71.** $\frac{12}{13}$ **73.** $\frac{3}{4}$ **75.** $\frac{5\sqrt{23}}{23}$

77. $\frac{4\sqrt{15}}{15}$ **79.** $\frac{11}{60}$ **81.** $\frac{24}{25}$ **83.** $\frac{56}{65}$ **85.** $\frac{24}{25}$

87. $\frac{120}{119}$ **89.** $\sqrt{1-u^2}$ **91.** $\frac{\sqrt{1-u^2}}{u}$

93. April and October   **95.** 3rd month

97. 0.026476 sec = 26 ms

99. $t \approx 173.4$, June 22   **101.** 11 yr

103. $\tan\theta = \dfrac{\tan\alpha + \tan\beta}{1 - \tan\alpha\tan\beta}, \dfrac{8x}{x^2 - 7}$

105. 0.70 m, 0.24 m

107. $\theta = \pi - \tan^{-1}\left(\dfrac{300}{200-x}\right) - \tan^{-1}\left(\dfrac{150}{x}\right)$

109. 53.7°   **111.** 68.7°

113. The identity $\sin^{-1}(\sin x) = x$ is valid only for x in the interval $\left[-\frac{\pi}{2}, \frac{\pi}{2}\right]$, not $[0, \pi]$.

115. In general, $\cot^{-1} x \neq \dfrac{1}{\tan^{-1} x}$.

117. false   **119.** false

121. $\frac{1}{2}$ is not in the domain.

123. a. $[0, \pi]$

 b. $f^{-1}(x) = \frac{\pi}{2} + \sin^{-1}\left(\frac{2-x}{4}\right), [-2, 6]$

125. $\dfrac{\sqrt{x^2-1}}{x}$

127. $\sin(\sin^{-1} x) = x$ only holds for $-1 \leq x \leq 1$

129. $\left[-\frac{\pi}{2}, 0\right) \cup \left(0, \frac{\pi}{2}\right]$

Section 6.2

1. $\frac{3\pi}{4}, \frac{5\pi}{4}$ **3.** $\frac{\pi}{3}, \frac{2\pi}{3}$ **5.** $\frac{\pi}{3}, \frac{5\pi}{3}$ **7.** $\frac{7\pi}{6}, \frac{11\pi}{6}, \frac{19\pi}{6}, \frac{23\pi}{6}$

9. $\theta = n\pi$, n an integer   **11.** $\frac{7\pi}{12}, \frac{11\pi}{12}, \frac{19\pi}{12}, \frac{23\pi}{12}$

13. $\theta = \dfrac{(7 + 12n)\pi}{3}, \dfrac{(11 + 12n)\pi}{3}$, n an integer

15. $\frac{\pi}{6}, \frac{2\pi}{3}, \frac{7\pi}{6}, \frac{5\pi}{3}, -\frac{\pi}{3}, -\frac{5\pi}{6}, -\frac{4\pi}{3}, -\frac{11\pi}{6}$

17. $-\frac{2\pi}{3}, -\frac{4\pi}{3}$ **19.** $\frac{\pi}{6}, \frac{\pi}{3}, \frac{7\pi}{6}, \frac{4\pi}{3}$

21. $\frac{\pi}{12}, \frac{7\pi}{12}, \frac{13\pi}{12}, \frac{19\pi}{12}$ **23.** $\frac{\pi}{3}, \frac{2\pi}{3}, \frac{4\pi}{3}, \frac{5\pi}{3}$

25. $\frac{\pi}{3}$ **27.** $\frac{\pi}{4}, \frac{3\pi}{4}, \frac{5\pi}{4}, \frac{7\pi}{4}$

29. $\frac{\pi}{2}, \frac{3\pi}{2}, \frac{\pi}{3}, \frac{5\pi}{3}$ **31.** $\frac{7\pi}{6}, \frac{11\pi}{6}, \frac{3\pi}{2}$

33. $\frac{\pi}{2}$ **35.** $\frac{\pi}{6}, \frac{5\pi}{6}, \frac{7\pi}{6}, \frac{11\pi}{6}$

37. 115.83°, 295.83°, 154.17°, 334.17°

39. 333.63°   **41.** 29.05°, 209.05°

43. 200.70°, 339.30°   **45.** 41.41°, 318.59°

47. 56.31°, 126.87°, 236.31°, 306.87°

49. 101.79°, 281.79°, 168.21°, 348.21°, 9.74°, 189.74°, 80.26°, 260.26°

51. 80.12°, 279.88°

53. 64.93°, 121.41°, 244.93°, 301.41°

55. 4th quarter of 2016, 2nd quarter of 2017, and 4th quarter of 2018

57. 9 pm   **59.** March

61. $x^2\left[\sin\theta + \dfrac{\sin(2\theta)}{2}\right]$ **63.** 2017

65. 24°   **67.** $x = \dfrac{\pi}{4} + \dfrac{n\pi}{2}$, n an integer   **69.** 35°

71. The given solution is correct, but the second solution in QIV (306.15°) was not given.

73. The value $\theta = \frac{3\pi}{2}$ does not satisfy the original equation.

75. false   **77.** false   **79.** $k + 1$ **81.** $\frac{\pi}{6}, \frac{5\pi}{6}, \frac{7\pi}{6}, \frac{11\pi}{6}$

83. $\left[0, \frac{\pi}{6}\right) \cup \left(\frac{5\pi}{6}, \frac{7\pi}{6}\right) \cup \left(\frac{11\pi}{6}, 2\pi\right)$

85. $0, \frac{\pi}{3}, \pi, \frac{5\pi}{3}$ **87.** $\frac{\pi}{6}, \frac{5\pi}{6}$ **89.** no solutions

91. infinitely many irrational solutions

Section 6.3

1. $\frac{\pi}{4}, \frac{5\pi}{4}$ **3.** π **5.** $\frac{\pi}{6}$ **7.** $\frac{\pi}{3}$

9. $\frac{\pi}{4}, \frac{3\pi}{4}, \frac{5\pi}{4}, \frac{7\pi}{4}$ **11.** $\frac{\pi}{2}, \frac{3\pi}{2}$ **13.** $0, \pi, \frac{\pi}{4}, \frac{7\pi}{4}$

15. $\frac{\pi}{6}, \frac{5\pi}{6}, \frac{3\pi}{2}, \frac{\pi}{2}, \frac{7\pi}{6}, \frac{11\pi}{6}$ **17.** $\frac{\pi}{6}, \frac{\pi}{3}, \frac{7\pi}{6}, \frac{4\pi}{3}$

19. $\frac{\pi}{6}, \frac{5\pi}{6}, \frac{7\pi}{6}, \frac{11\pi}{6}$ **21.** $\frac{3\pi}{2}$

23. $\frac{2\pi}{3}, \frac{4\pi}{3}$ **25.** $\frac{\pi}{3}, \frac{5\pi}{3}, \pi$ **27.** $\frac{3\pi}{4}, \frac{7\pi}{4}$

29. $0, \pi$ **31.** $\frac{\pi}{4}, \frac{5\pi}{4}$ **33.** $x = 0$

35. $x = \frac{\pi}{12}, \frac{5\pi}{12}, \frac{13\pi}{12}, \frac{17\pi}{12}$

37. 57.47°, 122.53°, 216.38°, 323.62°

39. 30°, 150°, 199.47°, 340.53°

41. 14.48°, 165.52°, 270°   **43.** 111.47°, 248.53°

45. 308.67°, 231.33°   **47.** 180°

49. $\frac{3}{4}$ sec   **51.** $\left(\frac{\pi}{3}, \frac{3}{2}\right), \left(\frac{5\pi}{3}, \frac{3}{2}\right), (\pi, -3)$

53. 2.55 min, 5.70 min, 8.84 min

55. (1.05, 2.73)   **57.** March and September

59. 1 am and 11 am

61. Cannot divide by $\cos x$ since it could be zero

63. true   **65.** true

67. 2 solutions

69. $\frac{\pi}{6}$ or 30°   **71.** 0

73. $\left[0, \frac{\pi}{12}\right] \cup \left[\frac{5\pi}{12}, \frac{7\pi}{12}\right] \cup \left[\frac{11\pi}{12}, \pi\right]$

75. 7.39°   **77.** 79.07°

Review Exercises

1. $\frac{\pi}{4}$ **3.** $\frac{\pi}{2}$ **5.** $-90°$ **7.** 60°

9. $-37.50°$ **11.** 22.50°   **13.** 1.75

15. -0.10 **17.** $-\frac{\pi}{4}$ **19.** $\sqrt{3}$ **21.** $\frac{\pi}{3}$

23. $\frac{60}{61}$ **25.** $\frac{7}{6}$ **27.** $\frac{6\sqrt{35}}{35}$ **29.** July

31. $\frac{2\pi}{3}, \frac{5\pi}{3}, \frac{5\pi}{6}, \frac{11\pi}{6}$ **33.** $-\frac{\pi}{2}, -\frac{3\pi}{2}$ **35.** $\frac{\pi}{3}, \frac{2\pi}{3}, \frac{4\pi}{3}, \frac{5\pi}{3}$

37. $\frac{3\pi}{8}, \frac{11\pi}{8}, \frac{7\pi}{8}, \frac{15\pi}{8}$ **39.** $0, \pi, \frac{3\pi}{4}, \frac{7\pi}{4}$

41. 80.46°, 260.46°, 170.46°, 350.46°

43. 90°, 270°, 138.59°, 221.41°

45. 17.62°, 162.38°   **47.** $\frac{\pi}{4}, \frac{5\pi}{4}$

49. $\pi, \frac{\pi}{3}$ **51.** $0, \pi, \frac{\pi}{6}, \frac{11\pi}{6}$ **53.** $\frac{3\pi}{2}$

55. π **57.** 90°, 270°, 135°, 315°

59. 0°   **61.** 90°, 270°, 60°, 300°

63. a. $-\frac{3}{5}$ b. -0.6 c. yes

65. 127.1°

Practice Exercises

1. $-1 \le x \le 1$ **3.** $0 \le x \le \pi$ **5.** $\frac{5\pi}{6}$

7. $\frac{\pi}{2}$ **9.** $-\frac{\pi}{6}$ **11.** $\frac{\sqrt{5}}{5}$

13. $\theta = \frac{4\pi}{3} + 2n\pi, \frac{5\pi}{3} + 2n\pi$, n an integer

15. 1.82, 4.46   **17.** $\pi, \frac{\pi}{3}, \frac{5\pi}{3}$

19. $\theta = \pi + 2n\pi$, n an integer

21. 14.48°, 165.52°, 90°, 270°

23. $\frac{3\pi}{8}, \frac{7\pi}{8}, \frac{11\pi}{8}, \frac{15\pi}{8}$

25. $\frac{\pi}{8}, \frac{3\pi}{8}, \frac{5\pi}{8}, \frac{7\pi}{8}, \frac{9\pi}{8}, \frac{11\pi}{8}, \frac{13\pi}{8}, \frac{15\pi}{8}$

27. no solution   **29.** 69.09°, 249.09°

Cumulative Test

1. $\frac{3}{4}$ **3.** QIII   **5.** $\frac{4}{3}$ **7.** $\frac{5}{6}$ rad

9. 225π in./min   **11.** $0, \pi, 2\pi, 3\pi, 4\pi$

13. 3   **15.**

19. -1 **21.** $\frac{7}{2}[\cos(7x) - \cos(3x)]$

23. $150°$ **25.** $0, \frac{2\pi}{3}, \frac{4\pi}{3}, 2\pi$

CHAPTER 7

Section 7.1

1. SSA   **3.** SSS   **5.** ASA

7. $\gamma = 75°$, $b \approx 12$ m, $c \approx 14$ m

9. $\beta = 62°$, $a \approx 163$ cm, $c \approx 215$ cm

11. $\beta = 116.1°$, $a \approx 80.2$ yd, $b \approx 256.6$ yd

13. $\gamma = 120°$, $a \approx 7$ m, $b \approx 7$ m

15. $\alpha = 97°$, $a \approx 118$ yd, $b \approx 52$ yd

17. $\gamma = 50°$, $a \approx 14$ cm, $b \approx 2.7$ cm

19. $\beta = 78°$, $a \approx 14$ m, $c \approx 58$ m

21. $\beta_1 \approx 20°$, $\gamma_1 \approx 144°$, $c_1 \approx 9$;
$\beta_2 \approx 160°$, $\gamma_2 \approx 4°$, $c_2 \approx 1$

23. $\alpha = 40°$, $\beta \approx 100°$, $b \approx 18$

25. no triangle

27. $\beta_1 \approx 77°$, $\alpha_1 \approx 63°$, $a_1 \approx 457$;
$\beta_2 \approx 103°$, $\alpha_2 \approx 37°$, $a_2 \approx 309$

29. $\alpha \approx 31°$, $\gamma \approx 43°$, $c \approx 2$

31. $\gamma_1 \approx 51°$, $\beta_1 \approx 102°$, $b_1 \approx 15$;
$\gamma_2 \approx 129°$, $\beta_2 \approx 24°$, $b_2 \approx 6.3$

33. $\alpha_1 \approx 61°$, $\gamma_1 \approx 67°$, $c_1 \approx 21$;
$\alpha_2 \approx 119°$, $\gamma_2 \approx 9°$, $c_2 \approx 3.6$

35. $\beta \approx 12°$, $\alpha \approx 148°$, $a \approx 15$

37. no triangle   **39.** no triangle   **41.** 1246 ft

43. 1.7 mi   **45.** 1.3 mi   **47.** 26 ft

49. 29 mi   **51.** 803 yd or 289 yd

53. 47 in.   **55.** 1.2 cm

57. The value of β is incorrect.

59. false   **61.** true

63. If $h < a < b$, then $\frac{1}{2}b < a < b$.

67. $1 + \sqrt{3}$

Section 7.2

1. C   **3.** S   **5.** S   **7.** C

9. $b \approx 5$, $\gamma \approx 33°$, $\alpha \approx 47°$

11. $a \approx 5$, $\gamma \approx 6°$, $\beta \approx 158°$

13. $a \approx 2$, $\beta \approx 80°$, $\gamma \approx 80°$

15. $b \approx 5$, $\alpha \approx 43°$, $\gamma \approx 114°$

17. $c \approx 4.9$, $\alpha \approx 138°$, $\beta \approx 30°$

19. $a \approx 151$, $\beta \approx 4.6°$, $\gamma \approx 5.4°$

21. $b \approx 7$, $\alpha \approx 30°$, $\gamma \approx 90°$

23. $\alpha \approx 93°$, $\beta \approx 39°$, $\gamma \approx 48°$

25. $\gamma \approx 77°$, $\beta \approx 51°$, $\alpha \approx 52°$

27. $\alpha \approx 96°$, $\beta \approx 25°$, $\gamma \approx 59°$

29. $\alpha \approx 131°$, $\beta \approx 35°$, $\gamma \approx 14°$

31. $\alpha \approx 75°$, $\beta \approx 57°$, $\gamma \approx 48°$

33. no triangle

35. $\gamma = 90°$, $\beta \approx 23°$, $\alpha \approx 67°$

37. $\gamma = 105°$, $b \approx 5$, $c \approx 9$

39. $\beta \approx 12°$, $\gamma \approx 137°$, $c \approx 16$

41. $\beta \approx 77°$, $\alpha \approx 66°$, $\gamma \approx 37°$

43. $\gamma \approx 2°$, $a \approx 13$, $\alpha \approx 168°$

45. 2710 mi   **47.** 63.7 ft   **49.** 16 ft

51. 26.0 cm   **53.** 97°   **55.** 56 ft

57. Beth: 36°, Tim: 26°

59. Should have used the smaller angle β in Step 2

61. false   **63.** true   **65.** $\alpha \approx 104°$ **69.** 97°

Section 7.3

1. 55.4   **3.** 0.5   **5.** 23.6   **7.** 54   **9.** 25

11. 6.4   **13.** 4408.4   **15.** 9.6   **17.** 97.4

19. 29   **21.** 0.39   **23.** 25.0   **25.** 26.7

27. 111.64   **29.** 111,632,076   **31.** no triangle

33. 2.3   **35.** 2.7   **37.** 312,297 nm^2

39. 10,591 ft^2 **41.** 16°   **43.** 312.4 mi^2

45. a. 41,842 ft^2 **b.** \$89,123

47. 47,128 ft^2 **49.** 23.38 ft^2

51. 31,913 ft^2 **53.** 60 in.^{2}

55. The semiperimeter is half the perimeter,
namely, $s = \dfrac{a + b + c}{2}$.

57. true   **59.** $\frac{x^2\sqrt{3}}{6}$ **63.** 21 cm^2

Section 7.4

1. $\sqrt{13}$ **3.** $5\sqrt{2}$ **5.** 25   **7.** $\sqrt{73}, 69.4°$

9. $\sqrt{26}, 348.7°$ **11.** $\sqrt{17}, 166.0°$

13. 8, 180°   **15.** $2\sqrt{3}, 60°$ **17.** $\frac{13}{15}, 23°$

19. $\langle -2, -2 \rangle$ **21.** $\langle -12, 9 \rangle$ **23.** $\langle 0, -14 \rangle$

25. $\langle -36, 48 \rangle$ **27.** $\langle 22, -41 \rangle$ **29.** $\langle 30, -61 \rangle$

31. $\langle 6.3, 3.0 \rangle$ **33.** $\langle -2.8, 15.8 \rangle$ **35.** $\langle 2.6, -3.1 \rangle$

37. $\langle 8.2, -3.8 \rangle$ **39.** $\langle -1, 1.7 \rangle$

41. $\langle -2.9, -0.78 \rangle$ **43.** $\langle -\frac{5}{13}, -\frac{12}{13} \rangle$

45. $\left\langle \frac{60}{61}, \frac{11}{61} \right\rangle$ **47.** $\left\langle \frac{24}{25}, -\frac{7}{25} \right\rangle$ **49.** $\left\langle -\frac{3}{5}, -\frac{4}{5} \right\rangle$

51. $\left\langle \frac{\sqrt{10}}{10}, \frac{3\sqrt{10}}{10} \right\rangle$ **53.** $\left\langle -\frac{2\sqrt{13}}{13}, \frac{3\sqrt{13}}{13} \right\rangle$

55. $7\vec{i} + 3\vec{j}$ **57.** $5\vec{i} - 3\vec{j}$ **59.** $-\vec{i} + 0\vec{j}$

61. $0.8\vec{i} - 3.6\vec{j}$ **63.** $2\vec{i} + 0\vec{j}$

65. $-5\vec{i} + 5\vec{j}$ **67.** $7\vec{i} + 0\vec{j}$

69. horizontal: 1905 ft/sec, vertical: 1100 ft/sec

71. 2801 lb

73. 11.7 mph, 31° west of due north

75. 303 mph, 52.41° east of due north

77. 492 mph, N 97° E

79. 228 mph or 270 mph   **81.** 250 lb

83. vertical: 51.4 ft/sec, horizontal: 61.3 ft/sec

85. 29.93 yd   **87.** 10.9°   **89.** 1156 lb

91. 1611 N; 19°

93. Magnitude cannot be negative.

95. false   **97.** true   **99.** vector

101. $\sqrt{a^2 + b^2}$ **103.** $5a, 307°$ **105.** $\sqrt{37}, 171°$

107. 127°   **109.** $\langle -13, 28 \rangle$ **111.** $\left\langle \frac{5}{13}, -\frac{12}{13} \right\rangle$

Section 7.5

1. 2   **3.** −3   **5.** 26   **7.** 46   **9.** 42   **11.** 11

13. −13a   **15.** −1.4   **17.** 98°

19. 109°   **21.** 3°   **23.** 51°   **25.** 30°

27. 105°   **29.** 180°   **31.** 163°   **33.** not orthogonal

35. orthogonal   **37.** not orthogonal

39. orthogonal   **41.** orthogonal

43. orthogonal   **45.** 400 ft-lb

47. 80,000 ft-lb   **49.** 1299 ft-lb

51. 148 ft-lb   **53.** 1607 lb

55. 694,593 ft-lb   **57.** 40°   **59.** 5362 ft-lb

61. The dot product of two vectors is a scalar, not a vector.

63. false   **65.** true   **67.** 17   **75.** $(a + b)^2$

77. $\frac{15\sqrt{2}}{2}$ **79.** −1083   **81.** 31.4°

Review Exercises

1. $\gamma = 150°$, $b \approx 8$, $c \approx 12$

3. $\gamma = 130°$, $a \approx 1$, $b \approx 9$

5. $\beta = 158°$, $a \approx 11$, $b \approx 22$

7. $\beta = 90°$, $a \approx \sqrt{2}$, $c \approx \sqrt{2}$

9. $\beta = 146°$, $b \approx 266$, $c \approx 178$

11. $\beta_1 \approx 26°$, $\gamma_1 \approx 134°$, $c_1 \approx 15$;
$\beta_2 \approx 154°$, $\gamma_2 \approx 6°$, $c_2 \approx 2$

13. $\gamma_1 \approx 29°$, $\alpha_1 \approx 127°$, $b_1 \approx 20$;
$\gamma_2 \approx 151°$, $\beta_2 \approx 5°$, $b_2 \approx 2$

15. no triangle

17. $\beta_1 \approx 15°$, $\gamma_1 \approx 155°$, $c_1 \approx 10$;
$\beta_2 \approx 165°$, $\gamma_2 \approx 5°$, $c_2 \approx 2$

19. $y \approx 21$ ft, $x \approx 31$ ft

21. $c \approx 46$, $\alpha \approx 42°$, $\beta \approx 88°$

23. $\gamma \approx 75°$, $\beta \approx 54°$, $\alpha \approx 51°$

25. $\gamma = 90°$, $\beta \approx 48°$, $\alpha \approx 42°$

27. $a \approx 4$, $\beta \approx 28°$, $\gamma \approx 138°$

29. $a \approx 11$, $\beta \approx 68°$, $\gamma \approx 22°$

31. $\gamma \approx 70°$, $\beta \approx 59°$, $\alpha \approx 51°$

33. $a \approx 26$, $\beta \approx 37°$, $\gamma \approx 43°$

35. $a \approx 28$, $\beta \approx 4°$, $\gamma \approx 166°$

37. no triangle

39. $\beta \approx 10°$, $\gamma \approx 155°$, $c \approx 10.3$

41. 6.2 mi   **43.** 141.8   **45.** 51.5

47. 89.8   **49.** 41.7   **51.** 5.2 in.

53. 13   **55.** 13   **57.** 26, 112.6°

59. 20, 323.1°   **61.** $\langle 2, 11 \rangle$ **63.** $\langle 38, -7 \rangle$

65. $\langle 2.6, 9.7 \rangle$ **67.** $\langle -3.1, 11.6 \rangle$ **69.** $\left\langle \frac{\sqrt{2}}{2}, -\frac{\sqrt{2}}{2} \right\rangle$

71. $5\vec{i} + \vec{j}$

73. 345 mph, 155° clockwise of north

75. −6   **77.** −9   **79.** 16   **81.** 59°

83. 49°   **85.** 166°   **87.** not orthogonal

89. orthogonal   **91.** not orthogonal

93. not orthogonal   **95.** 453 ft-lb

103. magnitude: 65, direction angle: −67.38014°

105. 40°

Practice Test

1. $\gamma = 110°$, $a \approx 7.8$, $c \approx 14.6$

3. $\gamma \approx 96.4°$, $\beta \approx 48.2°$, $\alpha \approx 35.4°$

5. $c \approx 8.8$, $\beta \approx 87°$, $\alpha \approx 50°$

7. no triangle   **9.** 57

11. 37,943 yd²   **13.** 13, 112.6°

15. $\left\langle -\frac{5\sqrt{29}}{29}, \frac{2\sqrt{29}}{29} \right\rangle$ **17. a.** $\langle -14, 5 \rangle$ **b.** −16

19. 59.5°   **21.** 153°

23. 26 mph, 54° clockwise of north

25. 1229 ft-lb

27. magnitude represents length

29. scalar, the same

Cumulative Test

1. a. 11° **b.** 101°   **3.** $-\frac{2\sqrt{5}}{5}$ **5.** $\frac{5\pi}{6}, \frac{11\pi}{6}$

7.

9. 1 **11.** conditional

15. $\sin\left(\frac{3\pi}{8}\right)$ **17.** $-\frac{\pi}{4}$ **19.** $\frac{\sqrt{17}}{4}$

21. 37.76°, 54.74°, 125.26°, 142.24°, 217.76°, 234.74°, 305.26°, 322.24°

23. $\alpha = 36.3°, b \approx 123.4$ m, $c \approx 78.1$ m

25. orthogonal

CHAPTER 8

Section 8.1

1. $4i$ **3.** $2i\sqrt{5}$ **5.** -4 **7.** $8i$

9. $3 - 10i$ **11.** $-10 - 12i$ **13.** $2 - 9i$

15. $10 - 14i$ **17.** $2 - 2i$ **19.** $-1 + 2i$

21. $-1 + 4i$ **23.** $96 - 60i$ **25.** $-48 + 27i$

27. $-102 + 30i$ **29.** $5 - i$ **31.** $13 + 41i$

33. $87 + 33i$ **35.** $-6 + 48i$ **37.** $\frac{56}{9} - \frac{11}{18}i$

39. $37 + 49i$ **41.** $\bar{z} = 4 - 7i, z\bar{z} = 65$

43. $\bar{z} = 2 + 3i, z\bar{z} = 13$ **45.** $\bar{z} = 6 - 4i, z\bar{z} = 52$

47. $\bar{z} = -2 + 6i, z\bar{z} = 40$ **49.** $-2i$

51. $\frac{3}{10} + \frac{1}{10}i$ **53.** $\frac{3}{13} - \frac{2}{13}i$ **55.** $\frac{14}{53} - \frac{4}{53}i$

57. $-i$ **59.** $-\frac{9}{34} + \frac{19}{34}i$ **61.** $\frac{18}{53} - \frac{43}{53}i$

63. $\frac{66}{85} + \frac{43}{85}i$ **65.** $-i$ **67.** 1

69. $21 - 20i$ **71.** $-5 + 12i$ **73.** $18 + 26i$

75. $-2 - 2i$ **77.** $8 - 2i$ ohms

79. Should have multiplied by the conjugate $4 + i$ (not $4 - i$).

81. true **83.** true **85.** $(x + i)^2(x - i)^2$

87. $41 - 38i$ **89.** $\frac{2}{125} + \frac{11}{125}i$

Section 8.2

1. – 8.

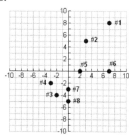

9. The head of the vector is the complex number. **11.** The head of the vector is the complex number.

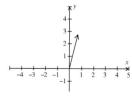

13. $\sqrt{2}(\cos 315° + i\sin 315°)$

15. $2(\cos 60° + i\sin 60°)$

17. $4\sqrt{2}(\cos 135° + i\sin 135°)$

19. $2\sqrt{3}(\cos 300° + i\sin 300°)$

21. $3(\cos 0° + i\sin 0°)$

23. $4\left[\cos\left(\frac{11\pi}{6}\right) + i\sin\left(\frac{11\pi}{6}\right)\right]$

25. $\left[\cos\left(\frac{2\pi}{3}\right) + i\sin\left(\frac{2\pi}{3}\right)\right]$

27. $\frac{1}{4}\left[\cos\left(\frac{11\pi}{6}\right) + i\sin\left(\frac{11\pi}{6}\right)\right]$

29. $\sqrt{58}[\cos(293.2°) + i\sin(293.2°)]$

31. $\sqrt{61}[\cos(140.2°) + i\sin(140.2°)]$

33. $13[\cos(112.6°) + i\sin(112.6°)]$

35. $10[\cos(323.1°) + i\sin(323.1°)]$

37. $2\sqrt{10}[\cos(2.82) + i\sin(2.82)]$

39. $5\sqrt{53}[\cos(6.00) + i\sin(6.00)]$

41. $1.83[\cos(1.15) + i\sin(1.15)]$

43. $12.9[\cos(3.40) + i\sin(3.40)]$

45. -5 **47.** $\sqrt{2} - \sqrt{2}i$ **49.** $-2 - 2\sqrt{3}i$

51. $-\frac{3}{2} + \frac{\sqrt{3}}{2}i$ **53.** $1 + i$ **55.** $\frac{5}{2} + \frac{5\sqrt{3}}{2}i$

57. $-\frac{3\sqrt{3}}{4} - \frac{3}{4}i$ **59.** $5 - 5i\sqrt{3}$

61. $2.1131 - 4.5315i$ **63.** $-0.5209 + 2.9544i$

65. $5.3623 - 4.4995i$ **67.** $-2.8978 + 0.7765i$

69. $-0.6180 - 1.9021i$ **71.** $5.54 + 2.30i$

73. Force A: 100, Force B: $60(\sqrt{3} + i)$, 212.6 lb

75. Force A: 80, Force B: $75(\sqrt{3} + i)$, 19.7°

77. Plane: $\vec{u} = 300(\cos 165° + i\sin 165°)$
Wind: $\vec{v} = 30(\cos 270° + i\sin 270°)$
294 mph, $294(\cos 171° + i\sin 171°)$

79. Boat: $\vec{u} = 15(\cos 140° + i\sin 140°)$
Current: $\vec{v} = 5(\cos 180° + i\sin 180°)$
19.10 mph, $19.10(\cos 150° + i\sin 150°)$

81. The point is in QIII, not QI. As such, you should add 180° to $\tan^{-1}\left(\frac{8}{3}\right)$.

83. true **85.** true **87.** 0° **89.** $|b|$

91. $a\sqrt{5}[\cos(296.6°) + i\sin(296.6°)]$

93. $(\sqrt{6} + \sqrt{2}) + (\sqrt{6} - \sqrt{2})i$

95. $6\sqrt{65}[\cos(2.09) + i\sin(2.09)]$

97.

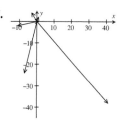

99. $\sqrt{2}(\cos 45° + i\sin 45°)$ **101.** $\sqrt{5}[\cos(26.57°) + i\sin(26.57°)]$

Section 8.3

1. $-6 + 6\sqrt{3}i$ **3.** $-4\sqrt{2} - 4\sqrt{2}i$ **5.** $0 + 8i$

7. $24\sqrt{3} + 24i$ **9.** $\frac{\sqrt{3}}{18} - \frac{1}{18}i$ **11.** $\frac{9\sqrt{2}}{2} + \frac{9\sqrt{2}}{2}i$

13. $0 + 12i$ **15.** $\frac{9}{2} - \frac{9\sqrt{3}}{2}i$ **17.** $-\frac{15\sqrt{3}}{2} + \frac{15}{2}i$

19. $-\frac{7\sqrt{2}}{16} + \frac{7\sqrt{2}}{16}i$ **21.** $\frac{3}{2} + \frac{3\sqrt{3}}{2}i$ **23.** $-\sqrt{2} + \sqrt{2}i$

25. $0 - 2i$ **27.** $-\frac{1}{2} + 0i$ **29.** $-\frac{3}{4} - \frac{3\sqrt{3}}{4}i$

31. $\frac{3}{2} + \frac{3\sqrt{3}}{2}i$ **33.** $-\frac{5}{2} - \frac{5\sqrt{3}}{2}i$ **35.** $-\frac{5}{2} + \frac{5\sqrt{3}}{2}i$

37. $-\frac{2\sqrt{3}}{3} - \frac{2}{3}i$ **39.** $\frac{3}{5} - \frac{3\sqrt{3}}{5}i$ **41.** $4 - 4i$

43. $-64 + 0i$ **45.** $-8 + 8\sqrt{3}i$ **47.** $1,048,576 + 0i$

49. $-1,048,576\sqrt{3} - 1,048,576i$

51. $2(\cos 150° + i\sin 150°)$,
$2(\cos 330° + i\sin 330°)$

53. $\sqrt{6}[\cos(157.5°) + i\sin(157.5°)]$,
$\sqrt{6}[\cos(337.5°) + i\sin(337.5°)]$

55. $2(\cos 20° + i\sin 20°)$,
$2(\cos 140° + i\sin 140°)$,
$2(\cos 260° + i\sin 260°)$

57. $\sqrt[3]{2}(\cos 110° + i\sin 110°)$,
$\sqrt[3]{2}(\cos 230° + i\sin 230°)$,
$\sqrt[3]{2}(\cos 350° + i\sin 350°)$

59. $2[\cos(78.75°) + i\sin(78.75°)]$,
$2[\cos(168.75°) + i\sin(168.75°)]$,
$2[\cos(258.75°) + i\sin(258.75°)]$,
$2[\cos(348.75°) + i\sin(348.75°)]$

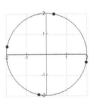

61. $2\sqrt{5}[\cos(82.5°) + i\sin(82.5°)]$,
$2\sqrt{5}[\cos(172.5°) + i\sin(172.5°)]$,
$2\sqrt{5}[\cos(262.5°) + i\sin(262.5°)]$,
$2\sqrt{5}[\cos(352.5°) + i\sin(352.5°)]$

63. $\pm 2, \pm 2i$

65. $-2, 1 - \sqrt{3}i, 1 + \sqrt{3}i$

67. $\sqrt{2} + \sqrt{2}i, -\sqrt{2} + \sqrt{2}i$,
$-\sqrt{2} - \sqrt{2}i, -\sqrt{2} - \sqrt{2}i$

69. $1, \frac{1}{2} + \frac{\sqrt{3}}{2}i, -\frac{1}{2} + \frac{\sqrt{3}}{2}i, -1$,
$-\frac{1}{2} - \frac{\sqrt{3}}{2}i, \frac{1}{2} - \frac{\sqrt{3}}{2}i$

71. $-\frac{\sqrt{2}}{2} + \frac{\sqrt{2}}{2}i, \frac{\sqrt{2}}{2} - \frac{\sqrt{2}}{2}i$ **73.** $2i, -\sqrt{3} - i, \sqrt{3} - i$

75. $(\cos 45° + i\sin 45°)$,
$(\cos 117° + i\sin 117°)$,
$(\cos 189° + i\sin 189°)$,
$(\cos 261° + i\sin 261°)$,
$(\cos 333° + i\sin 333°)$

77. $[\cos(50° + 60° k) + i\sin(50° + 60° k)]$,
$k = 0, 1, 2, 3, 4, 5$

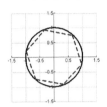

79. Reversed order of angles being subtracted

81. Should use DeMoivre's formula:
in general, $(a + b)^6 \neq a^6 + b^6$

83. true

85. $\frac{\sqrt{6n}}{2} + \frac{\sqrt{2n}}{2}i, -\frac{\sqrt{6n}}{2} - \frac{\sqrt{2n}}{2}i$

91. $\left[\cos\left(\frac{11\pi}{6(5)} + \frac{2k\pi}{5}\right) + i\sin\left(\frac{11\pi}{6(5)} + \frac{2k\pi}{5}\right)\right]$,
$k = 0, 1, 2, 3, 4$

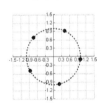

93. $\left[\cos\left(\frac{4\pi}{3(6)} + \frac{2k\pi}{6}\right) + i\sin\left(\frac{4\pi}{3(6)} + \frac{2k\pi}{6}\right)\right]$,
$k = 0, 1, 2, 3, 4, 5$

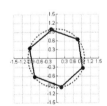

Section 8.4

1. – 10.

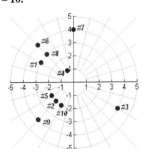

11. $\left(4, \frac{\pi}{3}\right)$ **13.** $\left(2, \frac{4\pi}{3}\right)$ **15.** $\left(4\sqrt{2}, \frac{3\pi}{4}\right)$

17. $(3, 0)$ **19.** $\left(2, \frac{7\pi}{6}\right)$ **21.** $(2, -2\sqrt{3})$

23. $\left(\frac{\sqrt{3}}{2}, -\frac{1}{2}\right)$ **25.** $(0, 0)$ **27.** $(0, -6)$

29. $(4, 4\sqrt{3})$ **31.** $(-1, -\sqrt{3})$ **33.** $\left(\frac{\sqrt{2}}{2}, -\frac{\sqrt{2}}{2}\right)$

35. $\left(\frac{5\sqrt{3}}{4}, -\frac{5}{4}\right)$ **37.** $(-3, 0)$ **39.** $\left(\frac{7\sqrt{2}}{2}, -\frac{7\sqrt{2}}{2}\right)$

41. d **43.** a

45. **47.** **49.**

51. **53.** **55.**

57. **59.** **61.**

63. $y = -2x + 1$, line

65. $(x - 1)^2 + y^2 = 9$, circle

67. $x = \frac{1}{2}(3 - y^2)$, parabola

69.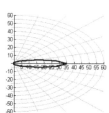

71. Graph of $r = \sqrt{\theta}$ (dotted) is more tightly wound than is the graph of $r = \theta$ (solid).

73. Graph of $r^2 = \frac{1}{4}\cos(2\theta)$ (dotted) is much closer to the origin than is the graph of $r^2 = 4\cos(2\theta)$ (solid).

75. **77. a. – c.**

79. The Point is in QIII, so needed to add π to the angle.

81. true **83.** $r = \dfrac{a}{\cos\theta}$ **85.** $(-a, \theta \pm 180°)$

87. $\left(1 - \sqrt{2}, \frac{\pi}{8}\right), \left(1 - \sqrt{2}, \frac{9\pi}{8}\right), \left(1 + \sqrt{2}, \frac{5\pi}{8}\right), \left(1 + \sqrt{2}, \frac{13\pi}{8}\right)$

89. x-axis symmetry: $r(-\theta) = r(\theta)$, for all θ
y-axis symmetry: $-r(-\theta) = r(\theta)$, for all θ
origin: $-r(\theta) = r(\theta)$, for all θ y-axis symmetry

91. beginning with $\theta = \frac{\pi}{2}$ and ending with $\frac{3\pi}{2}$

93. begins with $\theta = \frac{\pi}{2}$, then it crosses the origin (the first time) at $\theta = \cos^{-1}\left(-\frac{1}{3}\right)$, winds around, and eventually ends with $\theta = \frac{3\pi}{2}$

Section 8.5

1. **3.** **5.**

7. **9.** **11.**

13. **15.**  **17.**

19.

21. $y = \dfrac{1}{x^2}$ **23.** $y = x - 2$ **25.** $y = \sqrt{x^2 + 1}$

27. $x + y = 2$ **29.** $x + 4y = 8$

31. $y = \sin x$ **33.** $y = x^2 + 2$

35. $y = x$ **37.** 17.7 sec

39. does clear the fence (21.0938 ft)

41. horizontal: 13,261 ft, height: 5742 ft

43. 125 sec

45.

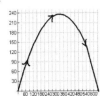

47.

t	x(t)	y(t)
0	A + B	0
$\frac{\pi}{2}$	0	A + B
π	−A − B	0
$\frac{3\pi}{2}$	0	−A − B
2π	A + B	0

49. 0.63 sec

51. $(-60, -60), (96, 21), (-94, 26)$

53. second racer

55. The original domain must be $t \geq 0$. Therefore, only the portion of the parabola where $y \geq 0$ is part of the plane curve.

57. false

59. $y = \sin\left(\dfrac{x}{n}\right)$

61. quarter circle in QI

63. $y^2 - x^2 = 1$

65.

$a = 2$ $a = 3$ $a = 4$
2π units of time

67.

$a = 2, b = 4$ $a = 4, b = 2$

$a = 1, b = 3$ $a = 3, b = 1$
2π units of time

Review Exercises

1. $-i$

3. $7i$

5. $6 + 3i$

7. $2 - 26i$

9. $-3i$

11. $-3 - 2i$

13. $-15 + 10i$

15. $2 \pm \sqrt{5}i$

17. $-\frac{3}{5} \pm \frac{4}{5}i$

19. -15

21. The head of the vector is the complex number.

23. – 24.

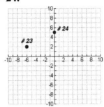

25. $2(\cos 315° + i\sin 315°)$

27. $8(\cos 270° + i\sin 270°)$

29. $61[\cos(169.6°) + i\sin(169.6°)]$

31. $17[\cos(28.1°) + i\sin(28.1°)]$

33. $3 - 3\sqrt{3}i$

35. $-1 + i$

37. $-3.7588 - 1.3681i$

39. $-12i$

41. $-\frac{21}{2} - \frac{21\sqrt{3}}{2}i$

43. $-\frac{\sqrt{3}}{2} + \frac{1}{2}i$

45. -6

47. -324

49. $16 - 16\sqrt{3}i$

51. $2(\cos 30° + i\sin 30°),$
$2(\cos 210° + i\sin 210°)$

53. $4(\cos 45° + i\sin 45°),$
$4(\cos 135° + i\sin 135°),$
$4(\cos 225° + i\sin 225°),$
$4(\cos 315° + i\sin 315°)$

55. $3 + 3\sqrt{3}i, -6, 3 - 3\sqrt{3}i$

57. $\frac{\sqrt{2}}{2} + \frac{\sqrt{2}}{2}i, -\frac{\sqrt{2}}{2} + \frac{\sqrt{2}}{2}i, -\frac{\sqrt{2}}{2} - \frac{\sqrt{2}}{2}i, \frac{\sqrt{2}}{2} - \frac{\sqrt{2}}{2}i$

59. $\left(2\sqrt{2}, \frac{3\pi}{4}\right)$

61. $\left(10, \frac{7\pi}{6}\right)$

63. $\left(2, \frac{3\pi}{2}\right)$

65. $\left(-\frac{3}{2}, \frac{3\sqrt{3}}{2}\right)$

67. $(1, \sqrt{3})$

69. $\left(-\frac{1}{2}, -\frac{\sqrt{3}}{2}\right)$

71.

73.

75.

77.

79. $x = 4 - y^2$

81. $y = 2x + 4$

83. $2868 - 6100i$

85. $12(\cos 246° + i\sin 246°)$

87. $\sqrt[4]{16}\left[\cos\left(\dfrac{120°}{4} + \dfrac{360° k}{4}\right) + i\sin\left(\dfrac{120°}{4} + \dfrac{360° k}{4}\right)\right]$, $k = 0, 1, 2, 3$

89. $\dfrac{\pi}{18}, \dfrac{5\pi}{18}, \dfrac{13\pi}{18}, \dfrac{17\pi}{18}, \dfrac{25\pi}{18}, \dfrac{29\pi}{18}$

91.

$a = 2, b = 3$ $a = 3, b = 2$
2π units of time

Practice Test

1. 0

3. $-\dfrac{4}{3} - \dfrac{2}{3}i$

5. $6(\cos 315° + i\sin 315°)$

7. modulus: 6, argument: $\dfrac{2\pi}{3}$

9. $8i$

11. $32,768\left(-1 + \sqrt{3}i\right)$

13. $(-a, \theta \pm 180°)$

15. $(5, 233°)$

17. **19.**

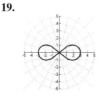

21. time: 5.3 sec, distance: 450 ft

23. Plane: $\vec{u} = 400(\cos 205° + i\sin 205°)$,
Wind: $\vec{v} = 24(\cos 270° + i\sin 270°)$,
$411(\cos 208° + i\sin 208°)$

25. $(96, 21)$

Cumulative Test

1. 12

3. $-\dfrac{11\sqrt{57}}{57}$

5. 75π km

7. amplitude: 0.009, frequency: 400 Hz

9.

11. $\tan\left(\dfrac{\pi}{3}\right)$

15. -1.56

17. 125.54°, 305.54°

19. no triangle

21. 3.1

23. $\dfrac{5}{13} + \dfrac{14}{13}i$

25.

APPENDIX A

Section A.1

1. $5(x + 5)$

3. $2(2t^2 - 1)$

5. $2x(x - 5)(x + 5)$

7. $3x(x^2 - 3x + 4)$

9. $x(x - 8)(x + 5)$

11. $2xy(2xy^2 + 3)$

13. $(x - 3)(x + 3)$

15. $(2x - 3)(2x + 3)$

17. $2(x - 7)(x + 7)$

19. $(15x - 13y)(15x + 13y)$

21. $(x + 4)^2$

23. $(x^2 - 2)^2$

25. $(2x + 3y)^2$

27. $(x - 3)^2$

29. $(x^2 + 1)^2$

31. $(p + q)^2$

33. $(t + 3)(t^2 - 3t + 9)$

35. $(y - 4)(y^2 + 4y + 16)$

37. $(2 - x)(4 + 2x + x^2)$

39. $(y + 5)(y^2 - 5y + 25)$

41. $(3 + x)(9 - 3x + x^2)$

43. $(x - 5)(x + 1)$

45. $(y - 3)(y + 1)$

47. $(2y + 1)(y - 3)$

49. $(3t + 1)(t + 2)$

51. $(-3t + 2)(2t + 1)$

53. $(x^2 + 2)(x - 3)$

55. $(a^3 - 8)(a + 2) = (a + 2)(a - 2)(a^2 + 2a + 4)$

57. $(3y - 5r)(x + 2s)$

59. $(4x - y)(5x + 2y)$

61. $(x - 2y)(x + 2y)$

63. $(3a + 7)(a - 2)$

65. prime

67. prime

69. $2(3x + 2)(x + 1)$

71. $(3x - y)(2x + 5y)$

73. $9(2s - t)(2s + t)$

75. $(ab - 5c)(ab + 5c)$

77. $(x - 2)(4x + 5)$

79. $x(3x + 1)(x - 2)$

81. $x(x - 3)(x + 3)$

83. $(x - 1)(y - 1)$

85. $(x^2 + 3)(x^2 + 2)$

87. $(x - 6)(x + 4)$

89. $x(x + 5)(x^2 - 5x + 25)$

91. $(x - 3)(x + 3)(x^2 + 9)$

93. $2(3x + 4)$

95. $(x + 2)(2x - 15)$

97. $-2(8t - 1)(t + 5)$

99. Last step. $(x - 1)(x^2 - 9) = (x - 1)(x - 3)(x + 3)$

101. false

103. true

105. $(a^n - b^n)(a^n + b^n)$

107. $8x^3 + 1$ and $(2x + 1)(4x^2 - 2x + 1)$

Section A.2

1. $(4, 2)$ **3.** $(-3, 0)$ **5.** $(0, -3)$

7. A: II   **B:** I   **C:** III
   D: IV   **E:** y-axis   **F:** x-axis

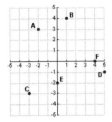

9.

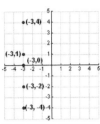

The line being described is $x = -3$.

11. $d = 4, M = (3, 3)$

13. $d = 4\sqrt{2}, M = (1, 2)$

15. $d = 3\sqrt{10}, M = \left(-\frac{17}{2}, \frac{7}{2}\right)$

17. $d = 5, M = \left(-5, \frac{1}{2}\right)$

19. $d = 4\sqrt{2}, M = (-4, -6)$

21. $d = 5, M = \left(\frac{3}{2}, \frac{11}{6}\right)$

23. $d = \frac{\sqrt{4049}}{60}, M = \left(-\frac{5}{24}, \frac{1}{15}\right)$

25. $d = 3.9, M = (0.3, 3.95)$

27. $d = 44.64, M = (1.05, -1.2)$

29. $d = 4\sqrt{2}, M = \left(\sqrt{3}, 3\sqrt{2}\right)$

31. $d = \sqrt{10 + 2\sqrt{2} + 4\sqrt{3}}, M = \left(\frac{1 - \sqrt{2}}{2}, \frac{-2 + \sqrt{3}}{2}\right)$

33. 21.84   **35.** right triangle

37. isosceles   **39.** 128.06 mi

41. 268 mi   **43.** \$330 million

45. Substituted the values incorrectly. It should have been
$\sqrt{(9 - 2)^2 + (10 - 7)^2}$.

47. Substituted the values incorrectly. It should have been
$\left(\frac{-3 + 7}{2}, \frac{4 + 9}{2}\right) = \left(2, \frac{13}{2}\right)$.

49. true   **51.** true

53. $d = \sqrt{2}\,|a - b|, M = \left(\dfrac{a + b}{2}, \dfrac{b + a}{2}\right)$

59. $d = 6.357, M = (0.7, 5.15)$

61. $d = 3.111, M = (2.2, 3.3)$

Section A.3

1. a. no   **b.** yes   **3. a.** yes   **b.** no

5. a. yes   **b.** no   **7. a.** yes   **b.** no

9.

(x, y)
$(-2, 0)$
$(0, 2)$
$(1, 3)$

11.

(x, y)
$(-1, 2)$
$(0, 0)$
$\left(\frac{1}{2}, -\frac{1}{4}\right)$
$(1, 0)$
$(2, 2)$

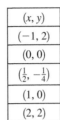

13.

(x, y)
$(1, 0)$
$(2, 1)$
$(5, 2)$
$(10, 3)$

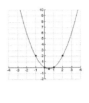

15.   **17.**

19.   **21.**

23. x-intercept: $(3, 0)$, y-intercept: $(0, -6)$

25. x-intercept: $(\pm 3, 0)$, y-intercept: $(0, -9)$

27. x-intercept: $(4, 0)$, no y-intercept

29. no x-intercept, y-intercept: $\left(0, \frac{1}{4}\right)$

31. x-intercept: $(\pm 2, 0)$, y-intercept: $(0, \pm 4)$

33. d   **35.** a   **37.** b   **39.** $(-1, -3)$

41. $(-7, 10)$ **43.** $(3, 2), (-3, 2), (-3, -2)$

45. x-axis   **47.** origin

49. x-axis   **51.** all   **53.** y-axis

55. y-axis   **57.** origin

59.   **61.**   **63.**

65.

67.

69.

71.

73.

75.

77.

break even: 2000 or 4000

profit: $2000 < x < 4000$, range: $y > 0$

79. The equation is not linear; you need more than two points to plot the graph.

81. To test for symmetry about the y-axis, one should replace x by $-x$, not y by $-y$.

83. false **85.** true **87.** origin

89. y-axis **91.** x-axis, y-axis, origin

93. x-axis, y-axis, origin

Section A.4

1. function **3.** not a function

5. function **7.** not a function

9. not a function **11.** function

13. not a function **15.** not a function

17. function **19.** not a function

21. function **23.** not a function

25. a. 5 **b.** 1 **c.** -3 **27. a.** 3 **b.** 2 **c.** 5

29. a. -5 **b.** -5 **c.** -5 **31. a.** 2 **b.** -8 **c.** -5

33. 1 **35.** $1, -3$ **37.** $[-4, 4]$

39. 6 **41.** -7 **43.** 6 **45.** -1 **47.** -33

49. $-\frac{7}{6}$ **51.** $\frac{2}{3}$ **53.** 4 **55.** $8 - x - a$

57. 2 **59.** 1 **61.** 2 **63.** 1

65. $(-\infty, \infty)$ **67.** $(-\infty, \infty)$

69. $(-\infty, 5) \cup (5, \infty)$

71. $(-\infty, -2) \cup (-2, 2) \cup (2, \infty)$

73. $(-\infty, \infty)$ **75.** $(-\infty, 7]$ **77.** $\left[-\frac{5}{2}, \infty\right)$

79. $(-\infty, -2] \cup [2, \infty)$ **81.** $(3, \infty)$ **83.** $(-\infty, \infty)$

85. $(-\infty, -4) \cup (-4, \infty)$ **87.** $\left(-\infty, \frac{3}{2}\right)$

89. $(-\infty, -2) \cup (3, \infty)$

91. $(-\infty, -4] \cup [4, \infty)$

93. $\left(-\infty, \frac{3}{2}\right)$ **95.** $(-\infty, \infty)$

97. $-2, 4$ **99.** $-1, 5, 6$

101. $y = 45x, \ x > 75$ **103.** 6 am: 64.8°F, noon: 90°F

105. \$634.50 (100 cards) \$641.66 (10 cards)

107. $V(x) = x(10 - 2x)^2, (0, 5)$

109. yes

111. (2000, 2000), (2005, 4000), (2010, 5000), (2015, 6000)

113. a. 0 **b.** 1000 **c.** 2000

115. Should apply the vertical line test to determine if the relationship describes a function. This is a function.

117. $f(x + 1) \neq f(x) + f(1)$, in general

119. $G(-1 + h) \neq G(-1) + G(h)$, in general

121. false **123.** false

125. 2 **127.** $C = -5, D = -2$

129. $(-\infty, -a) \cup (-a, a) \cup (a, \infty)$

131. noon: 90° F, too small to be considered temperatures in Florida

133. lowest: \$10, highest: \$642.38, yes

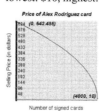

135. shifting two units right

Section A.5

1. neither **3.** even **5.** odd **7.** neither

9. odd **11.** even **13.** even **15.** neither

17. neither **19.** neither **21.** neither **23.** neither

25. a. $(-\infty, \infty)$ **b.** $[-1, \infty)$
 c. increasing: $(-1, \infty)$, decreasing: $(-3, -2)$,
   constant: $(-\infty, -3) \cup (-2, -1)$
 d. 0 **e.** -1 **f.** 2

27. a. $[-7, 2]$ **b.** $[-5, 4]$
 c. increasing: $(-4, 0)$, decreasing: $(-7, -4) \cup (0, 2)$,
   constant: nowhere
 d. 4 **e.** 1 **f.** -5

29. a. $(-\infty, \infty)$ **b.** $(-\infty, \infty)$
 c. increasing: $(-\infty, -3) \cup (4, \infty)$,
   decreasing: nowhere, constant: $(-3, 4)$
 d. 2 **e.** 2 **f.** 2

31. a. $(-\infty, \infty)$ **b.** $[-4, \infty)$
 c. increasing: $(0, \infty)$, decreasing: $(-\infty, 0)$,
   constant: nowhere
 d. -4 **e.** 0 **f.** 0

33. a. $(-\infty, 0) \cup (0, \infty)$ **b.** $(-\infty, 0) \cup (0, \infty)$
 c. increasing: $(-\infty, 0) \cup (0, \infty)$,
   decreasing: nowhere, constant: nowhere
 d. undefined **e.** 3 **f.** -3

35. a. $(-\infty, 0) \cup (0, \infty)$ **b.** $(-\infty, 5) \cup [7]$
 c. increasing: $(-\infty, 0)$, decreasing: $(5, \infty)$, constant: $(0, 5)$
 d. undefined **e.** 3 **f.** 7

37. $2x + h - 1$ **39.** $2x + h + 3$

41. $2x + h - 3$ **43.** $-6x - 3h + 5$

45. 13 **47.** 1 **49.** -2 **51.** -1

53. domain: $(-\infty, \infty)$, range: $(-\infty, 2]$,
 increasing: $(-\infty, 2)$, decreasing: nowhere,
 constant: $(2, \infty)$

55. domain: $(-\infty, \infty)$, range: $[0, \infty)$,
 increasing: $(0, \infty)$, decreasing: $(-1, 0)$,
 constant: $(-\infty, -1)$

57. domain: $(-\infty, \infty)$, range: $(-\infty, \infty)$,
 increasing: $(-\infty, \infty)$,
 decreasing: nowhere, constant: nowhere

59. domain: $(-\infty, \infty)$, range: $[1, \infty)$,
 increasing: $(1, \infty)$, decreasing: $(-\infty, 1)$,
 constant: nowhere

61. domain: $(-\infty, \infty)$, range: $[-1, 3]$,
 increasing: $(-1, 3)$, decreasing: nowhere,
 constant: $(-\infty, -1) \cup (3, \infty)$

63. domain: $(-\infty, \infty)$, range: $[1, 4]$,
 increasing: $(1, 2)$, decreasing: nowhere,
 constant: $(-\infty, 1) \cup (2, \infty)$

65. domain: $(-\infty, -2) \cup (-2, \infty)$,
 range: $(-\infty, \infty)$, increasing: $(-2, 1)$,
 decreasing: $(-\infty, -2) \cup (1, \infty)$,
 constant: nowhere

67. domain: $(-\infty, \infty)$, range: $[0, \infty)$,
 increasing: $(0, \infty)$, decreasing: nowhere,
 constant: $(-\infty, 0)$

69. domain: $(-\infty, \infty)$, range: $(-\infty, \infty)$,
 increasing: nowhere,
 decreasing: $(-\infty, 0) \cup (0, \infty)$,
 constant: nowhere

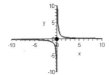

71. domain: $(-\infty, 1) \cup (1, \infty)$,
range: $(-\infty, -1) \cup (-1, \infty)$,
increasing: $(-1, 1)$,
decreasing: $(-\infty, -1) \cup (1, \infty)$,
constant: nowhere

73. domain: $(-\infty, \infty)$, range: $(-\infty, 2) \cup [4, \infty)$,
increasing: $(-\infty, -2) \cup (0, 2) \cup (2, \infty)$,
decreasing: $(-2, 0)$, constant: nowhere

75. domain: $(-\infty, 1) \cup (1, \infty)$,
range: $(-\infty, 1) \cup (1, \infty)$,
increasing: $(-\infty, 1) \cup (-\infty, 1)$,
decreasing: nowhere, constant: nowhere

77. $C(x) = \begin{cases} 10x, & 0 \le x \le 50 \\ 9x, & 50 < x \le 100 \\ 8x, & x > 100 \end{cases}$

79. $C(x) = \begin{cases} 250x, & 0 \le x \le 10 \\ 175x + 750, & x > 10 \end{cases}$

81. $C(x) = \begin{cases} 1000 + 35x, & 0 \le x \le 100 \\ 2000 + 25x, & x > 100 \end{cases}$

83. $R(x) = \begin{cases} 50{,}000 + 3x, & 0 \le x \le 100{,}000 \\ -50{,}000 + 4x, & x > 100{,}000 \end{cases}$

85. $P(x) = 65x - 800$

87. $f(x) = 0.98 + 0.22[[x]], x \ge 0$

89. $f(t) = 3(-1)^{[[t]]}, t \ge 0$

91. a. 20 millions of tons per yr   **b.** 110 millions of tons per yr

93. 0 ft/sec

95. Should exclude the origin since $x = 0$ is not in the domain. The range should be $(0, \infty)$.

97. The portion of $C(x)$ for $x > 30$ should be:

$15 + \underbrace{x - 30}_{\substack{\text{Number miles} \\ \text{beyond first 30}}}$

99. true   **101.** false   **103.** $a = 2b$

105. odd   **107.** odd

109. domain: $\mathbb{R}$ range: the set of integers

Section A.6

1. l   **3.** a   **5.** b   **7.** i   **9.** c   **11.** g

13. $y = |x| + 3$ **15.** $y = |x|$

17. $y = 3|x|$ **19.** $y = x^3 - 4$

21. $y = (x + 1)^3 + 3$ **23.** $y = -x^3$

25.   **27.**

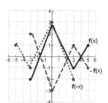

29.   **31.**

33.   **35.**

37.   **39.**

41.   **43.**

45.   **47.**

49.

51.

53.

55.

57.

59.

61.

63.

65.

67.

69.

71.

73.

75. $f(x) = (x - 3)^2 + 2$

77. $f(x) = -(x + 1)^2 + 1$ **79.** $f(x) = 2(x - 2)^2 - 5$

81. $S(x) = 10x, \overline{S}(x) = 10x + 50$

83. $T(x) = 0.33(x - 6500)$

85. (b) is wrong – shift right 3 units.

87. (b) should be deleted since $|3 - x| = |x - 3|$.

89. true   **91.** true   **93.** $(a + 3, b + 2)$

95. a.

b.

Graph that is below x-axis is reflected above it.

97. a.

b.

$a > 1$: horizontal compression
$0 < a < 1$: horizontal expansion

99.

stretched by factor of 2, graph below
x-axis reflected above it, vertical
shift up one unit

Section A.7

1.

	$f(x) + g(x)$	$f(x) - g(x)$
	$x + 2$	$3x$
domain	$(-\infty, \infty)$	$(-\infty, \infty)$
	$f(x) \cdot g(x)$	$f(x)/g(x)$
	$-2x^2 + x + 1$	$\dfrac{2x + 1}{1 - x}$
domain	$(-\infty, \infty)$	$(-\infty, 1) \cup (1, \infty)$

3.

	$f(x) + g(x)$	$f(x) - g(x)$
	$3x^2 - x - 4$	$x^2 - x + 4$
domain	$(-\infty, \infty)$	$(-\infty, \infty)$
	$f(x) \cdot g(x)$	$f(x)/g(x)$
	$2x^4 - x^3 - 8x^2 + 4x$	$\dfrac{2x^2 - x}{x^2 - 4}$
domain	$(-\infty, \infty)$	$(-\infty, -2) \cup (-2, 2) \cup (2, \infty)$

5.

	$f(x) + g(x)$	$f(x) - g(x)$
	$\dfrac{1 + x^2}{x}$	$\dfrac{1 - x^2}{x}$
domain	$(-\infty, 0) \cup (0, \infty)$	$(-\infty, 0) \cup (0, \infty)$
	$f(x) \cdot g(x)$	$f(x)/g(x)$
	1	$\dfrac{1}{x^2}$
domain	$(-\infty, 0) \cup (0, \infty)$	$(-\infty, 0) \cup (0, \infty)$

7.

	$f(x) + g(x)$	$f(x) - g(x)$
	$3\sqrt{x}$	$-\sqrt{x}$
domain	$[0, \infty)$	$[0, \infty)$
	$f(x) \cdot g(x)$	$f(x)/g(x)$
	$2x$	$\dfrac{1}{2}$
domain	$[0, \infty)$	$(0, \infty)$

9.

	$f(x) + g(x)$	$f(x) - g(x)$
	$\sqrt{4 - x} + \sqrt{x + 3}$	$\sqrt{4 - x} - \sqrt{x + 3}$
domain	$[-3, 4]$	$[-3, 4]$
	$f(x) \cdot g(x)$	$f(x)/g(x)$
	$\sqrt{4 - x} \cdot \sqrt{x + 3}$	$\dfrac{\sqrt{4 - x}\,\sqrt{x + 3}}{x + 3}$
domain	$[-3, 4]$	$(-3, 4]$

11. $(f \circ g)(x) = 2x^2 - 5$

$(g \circ f)(x) = 4x^2 + 4x - 2$

$dom(f \circ g) = (-\infty, \infty) = dom(g \circ f)$

13. $(f \circ g)(x) = \dfrac{1}{x + 1}, (g \circ f)(x) = \dfrac{2x - 1}{x - 1}$

$dom(f \circ g) = (-\infty, -1) \cup (-1, \infty)$

$dom(g \circ f) = (-\infty, 1) \cup (1, \infty)$

15. $(f \circ g)(x) = \dfrac{1}{|x - 1|}, (g \circ f)(x) = \dfrac{1}{|x| - 1}$

$dom(f \circ g) = (-\infty, 1) \cup (1, \infty)$

$dom(g \circ f) = (-\infty, -1) \cup (-1, 1) \cup (1, \infty)$

17. $(f \circ g)(x) = \sqrt{x + 4}, (g \circ f)(x) = \sqrt{x - 1} + 5$

$dom(f \circ g) = [-4, \infty), dom(g \circ f) = [1, \infty)$

19. $(f \circ g)(x) = x, (g \circ f)(x) = x$

$dom(f \circ g) = (-\infty, \infty) = dom(g \circ f)$

21. 15   **23.** 13   **25.** $26\sqrt{3}$

27. $\dfrac{110}{3}$ **29.** 11   **31.** $3\sqrt{2}$

33. undefined   **35.** undefined   **37.** 13

39. $f(g(1)) = \frac{1}{3}, g(f(2)) = 2$

41. both undefined

43. $f(g(1)) = \frac{1}{3}, g(f(2)) = 4$

45. $f(g(1)) = \sqrt{5}, g(f(2)) = 6$

47. both undefined

49. $f(g(1)) = \sqrt[3]{3}, g(f(2)) = 4$

61. $f(x) = 2x^2 + 5x$ $g(x) = 3x - 1$

63. $f(x) = \dfrac{2}{|x|}$ $g(x) = x - 3$

65. $f(x) = \dfrac{3}{\sqrt{x} - 2}$ $g(x) = x + 1$

67. $F(C(K)) = \frac{9}{5}(K - 273.15) + 32$

69. **a.** $(A \circ l)(x) = \left(\dfrac{x}{4}\right)^2$ **b.** 625 ft^2 **c.** 2500 ft^2

71. **a.** $C(x(p)) = 62{,}000 - 20p$

    b. $R(x(p)) = 600{,}000 - 200p$

    c. $P(x(p)) = 538{,}000 - 180p$

73. $22{,}500\pi t$ ft^2 **75.** $\sqrt{h^2 + 4}$

77. Must exclude -2 from the domain

79. $(f \circ g)(x) = f(g(x))$, not $f(x) \cdot g(x)$

81. Didn't distribute " $-$ " to all parts of $g(x)$.

83. false   **85.** true

87. $\dfrac{1}{x}; x \neq 0, a$ **89.** $x; [-a, \infty)$

91. $[-7, 9]$

93. $(-\infty, 3) \cup (-3, -1] \cup [4, 6) \cup (6, \infty)$

Section A.8

1. function, no   **3.** function, yes

5. function, yes   **7.** not a function

9. function, no   **11.** function, no

13. function, yes

15. function, no

17. not one-to-one

19. one-to-one

21. not one-to-one

23. one-to-one

25.

27.

29.

31.

33.

35.

37.

39.

41.

43. $f^{-1}(x) = x + 1$
$dom(f) = rng(f^{-1}) = (-\infty, \infty)$
$rng(f) = dom(f^{-1}) = (-\infty, \infty)$

45. $f^{-1}(x) = -\frac{1}{3}x + \frac{2}{3}$
$dom(f) = rng(f^{-1}) = (-\infty, \infty)$
$rng(f) = dom(f^{-1}) = (-\infty, \infty)$

47. $f^{-1}(x) = \sqrt[3]{x - 1}$
$dom(f) = rng(f^{-1}) = (-\infty, \infty)$
$rng(f) = dom(f^{-1}) = (-\infty, \infty)$

49. $f^{-1}(x) = x^2 + 3$
$dom(f) = rng(f^{-1}) = [3, \infty)$
$rng(f) = dom(f^{-1}) = [0, \infty)$

51. $f^{-1}(x) = \sqrt{x + 1}$
$dom(f) = rng(f^{-1}) = [0, \infty)$
$rng(f) = dom(f^{-1}) = [-1, \infty)$

53. $f^{-1}(x) = -2 + \sqrt{x + 3}$
$dom(f) = rng(f^{-1}) = [-2, \infty)$
$rng(f) = dom(f^{-1}) = [-3, \infty)$

55. $f^{-1}(x) = \frac{2}{x}$
$dom(f) = rng(f^{-1}) = (-\infty, 0) \cup (0, \infty)$
$rng(f) = dom(f^{-1}) = (-\infty, 0) \cup (0, \infty)$

57. $f^{-1}(x) = 3 - \frac{2}{x}$
$dom(f) = rng(f^{-1}) = (-\infty, 3) \cup (3, \infty)$
$rng(f) = dom(f^{-1}) = (-\infty, 0) \cup (0, \infty)$

59. $f^{-1}(x) = \frac{5x - 1}{x + 7}$
$dom(f) = rng(f^{-1}) = (-\infty, 5) \cup (5, \infty)$
$rng(f) = dom(f^{-1}) = (-\infty, -7) \cup (-7, \infty)$

61. not one-to-one

63. one-to-one

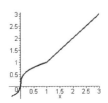

65. $f^{-1}(x) = \frac{5}{9}(x - 32)$

67. $C(x) = \begin{cases} 250x, & 0 \le x \le 10 \\ 2500 + 175(x - 10), & x > 10 \end{cases}$

$C^{-1}(x) = \begin{cases} \dfrac{x}{250}, & 0 \le x \le 2500 \\ \dfrac{x - 750}{175}, & x > 2500 \end{cases}$

69. $E(x) = 5.25x, E^{-1}(x) = \dfrac{x}{5.25}, x \ge 0$

71. Not a function since the graph does not pass the vertical line test

73. Must restrict the domain to a portion on which f is one-to-one, say $x \ge 0$

75. false

77. false

79. $(b, 0)$

81. $f^{-1}(x) = \sqrt{1 - x^2}, 0 \le x \le 1$, range $[0, 1]$

83. $m \ne 0$

85. not one-to-one

87. not one-to-one

89. no

91. yes

Review Exercises

1. $2xy^2(7x - 5y)$

3. $(x + 5)(2x - 1)$

5. $(4x + 5)(4x - 5)$

7. $(x + 5)(x^2 - 5x + 25)$

9. $2x(x + 5)(x - 3)$

11. $(x^2 - 2)(x + 1)$

13. – 16.

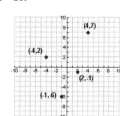

(−4, 2) Quadrant II

(4, 7) Quadrant I

(−1, −6) Quadrant III

(2, −1) Quadrant IV

17. $3\sqrt{5}$ **19.** $\sqrt{205}$ **21.** $\left(\frac{5}{2}, 6\right)$

23. (3.85, 5.3)   **25.** 52.20

27. x-intercepts: $(\pm 2, 0)$, y-intercepts: $(0, \pm 1)$

29. x-intercepts: $(\pm 3, 0)$, no y-intercepts

31. y-axis   **33.** origin

35.   **37.**

39.   **41.**

43. yes   **45.** yes   **47.** no   **49.** yes   **51.** no

53. a. 2   **b.** 4   **c.** $x = -3, 4$

55. a. 0   **b.** −2   **c.** $x \approx -5, 2$

57. 5   **59.** −665   **61.** −2   **63.** 4

65. $(-\infty, \infty)$ **67.** $(-\infty, -4) \cup (-4, \infty)$

69. $[4, \infty)$ **71.** $D = 18$ **73.** neither

75. odd   **77.** neither   **79.** odd

81. domain: $[-4, 7]$, range: $[-2, 4]$,
increasing: $(3, 7)$, decreasing: $(0, 3)$,
constant: $(-4, 0)$

83. −2

85. domain: $(-\infty, \infty)$, range: $(0, \infty)$

87. domain: $(-\infty, \infty)$, range: $[-1, \infty)$

89. $C(x) = \begin{cases} 25, & x \le 2, \\ 25 + 10.50(x - 2), & x > 2 \end{cases}$

91.   **93.**

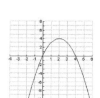

95.   **97.**

     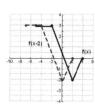

99.

101. $y = \sqrt{x + 3}, [-3, \infty)$

103. $y = \sqrt{x - 2} + 3, [2, \infty)$

105. $y = 5\sqrt{x} - 6, [0, \infty)$

107. $y = (x + 2)^2 - 12$

109. $g(x) + h(x) = -2x - 7$
$g(x) - h(x) = -4x - 1$
$g(x) \cdot h(x) = -3x^2 + 5x + 12$
$\dfrac{g(x)}{h(x)} = \dfrac{-3x - 4}{x - 3}$

$\left.\begin{array}{l} dom(g + h) \\ dom(g - h) \\ dom(gh) \end{array}\right\} = (-\infty, \infty)$

$dom\left(\dfrac{g}{h}\right) = (-\infty, 3) \cup (3, \infty)$

111. $g(x) + h(x) = \dfrac{1}{x^2} + \sqrt{x}$

$g(x) - h(x) = \dfrac{1}{x^2} - \sqrt{x}$

$g(x) \cdot h(x) = \dfrac{1}{x^{3/2}}, \dfrac{g(x)}{h(x)} = \dfrac{1}{x^{5/2}}$

$\left.\begin{array}{l} dom(g + h) \\ dom(g - h) \\ dom(gh) \\ dom\left(\dfrac{g}{h}\right) \end{array}\right\} = (0, \infty)$

113. $g(x) + h(x) = \sqrt{x-4} + \sqrt{2x+1}$

$g(x) - h(x) = \sqrt{x-4} - \sqrt{2x+1}$

$g(x) \cdot h(x) = \sqrt{x-4} \cdot \sqrt{2x+1}$

$\dfrac{g(x)}{h(x)} = \dfrac{\sqrt{x-4}}{\sqrt{2x+1}}$

$\left. \begin{array}{r} dom\,(f+g) \\ dom\,(f-g) \\ dom\,(fg) \\ dom\left(\dfrac{g}{h}\right) \end{array} \right\} = [4, \infty)$

115. $(f \circ g)(x) = 6x - 1, (g \circ f)(x) = 6x - 7$

$dom(f \circ g) = (-\infty, \infty) = dom(g \circ f)$

117. $(f \circ g)(x) = \dfrac{8 - 2x}{13 - 3x}$

$(g \circ f)(x) = \dfrac{x + 3}{4x + 10}$

$dom(f \circ g) = (-\infty, 4) \cup \left(4, \frac{13}{3}\right) \cup \left(\frac{13}{3}, \infty\right)$

$dom(g \circ f) = (-\infty, -3) \cup \left(-3, -\frac{5}{2}\right) \cup \left(-\frac{5}{2}, \infty\right)$

119. $(f \circ g)(x) = \sqrt{(x-3)(x+3)}$

$(g \circ f)(x) = x - 9$

$dom(f \circ g) = (-\infty, -3] \cup [3, \infty)$

$dom(g \circ f) = [5, \infty)$

121. $f(g(3)) = 857, g(f(-1)) = 51$

123. $f(g(3)) = \frac{17}{31}, g(f(-1)) = 1$

125. $f(g(3)) = 12, g(f(-1)) = 2$

127. $f(x) = 3x^2 + 4x + 7, g(x) = x - 2$

129. $f(x) = \dfrac{1}{\sqrt{x}}, g(x) = x^2 + 7$

131. $625\pi(t + 2) \text{ in.}^2$ **133.** yes   **135.** no

137. yes   **139.** yes   **141.** yes

143.   **145.**

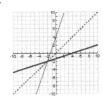

147. $f^{-1}(x) = \dfrac{x-1}{2}$

$dom(f) = rng(f^{-1}) = (-\infty, \infty)$

$rng(f) = dom(f^{-1}) = (-\infty, \infty)$

149. $f^{-1}(x) = x^2 - 4$

$dom(f) = rng(f^{-1}) = [-4, \infty)$

$rng(f) = dom(f^{-1}) = [0, \infty)$

151. $f^{-1}(x) = \dfrac{6 - 3x}{x - 1}$

$dom(f) = rng(f^{-1}) = (-\infty, -3) \cup (-3, \infty)$

$rng(f) = dom(f^{-1}) = (-\infty, 1) \cup (1, \infty)$

153. $s^{-1}(x) = \dfrac{x - 22,000}{0.08}$, sales required to earn a desired income

Practice Test

1. $(x - 4)(x + 4)$ **3.** $(2x + 3y)^2$

5. $(2x + 1)(x - 1)$ **7.** $t(t + 1)(2t - 3)$

9. $(x - 4y)(x + 3y)$ **11.** $3(3 + x)(9 - 3x + x^2)$

13. $\sqrt{82}$ **15.** $d = \sqrt{29}, M = \left(\frac{1}{2}, 5\right)$

17. $y = 1, 9$ **19.** x-axis

21.

23. b   **25.** c   **27.** $\dfrac{\sqrt{x-2}}{x^2 + 11}, [2, \infty)$

29. $x + 9, (2, \infty)$ **31.** 4   **33.** neither

35. domain: $[3, \infty)$ range: $(-\infty, 2]$

37. domain: $(-\infty, -1) \cup (-1, \infty)$ range: $[1, \infty)$

39. a. -2 **b.** 4   **c.** -3 **d.** $x = -3, 2$

41. $6x + 3h - 4$

43. $f^{-1}(x) = x^2 + 5$

$dom(f) = rng(f^{-1}) = [5, \infty)$

$rng(f) = dom(f^{-1}) = [0, \infty)$

45. $f^{-1}(x) = \dfrac{5x - 1}{x + 2}$

$dom(f) = rng(f^{-1}) = (-\infty, 5) \cup (5, \infty)$

$rng(f) = dom(f^{-1}) = (-\infty, -2) \cup (-2, \infty)$

47. $[0, \infty)$ **49.** $S(x) = 0.42x$ **51.** both

53. yes

APPENDIX B

Section B.1

1. hyperbola   3. circle   5. hyperbola

7. ellipse   9. parabola   11. circle

Section B.2

1. c   3. d   5. c   7. a

9. $x^2 = 12y$ 11. $y^2 = -20x$

13. $(x-3)^2 = 8(y-5)$

15. $(y-4)^2 = -8(x-2)$

17. $(x-2)^2 = 4(3)(y-1) = 12(y-1)$

19. $(y+1)^2 = 4(1)(x-2) = 4(x-2)$

21. $(y-2)^2 = 8(x+1)$

23. $(x-2)^2 = -8(y+1)$

25. vertex: $(0, 0)$
    focus: $(0, 2)$
    directrix: $y = -2$
    length of latus rectum: 8

27. vertex: $(0, 0)$
    focus: $\left(-\frac{1}{2}, 0\right)$
    directrix: $x = \frac{1}{2}$
    length of latus rectum: 2

29. vertex: $(0, 0)$
    focus: $(0, 4)$
    directrix: $y = -4$
    length of latus rectum: 16

31. vertex: $(0, 0)$
    focus: $(1, 0)$
    directrix: $x = -1$
    length of latus rectum: 4

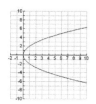

33. vertex: $(-3, 2)$

35. vertex: $(3, -1)$

37. vertex: $(-5, 0)$

39. vertex: $(0, 2)$

41. vertex: $(-3, -1)$

43. vertex: $\left(\frac{1}{2}, \frac{5}{4}\right)$

45. $(0, 2)$ receiver placed 2 ft from vertex

47. opens up: $y = \frac{1}{8}x^2$, for any x in $[-2.5, 2.5]$
    opens right: $x = \frac{1}{8}y^2$, for any y in $[-2.5, 2.5]$

49. $x^2 = 4(40)y = 160y$

51. yes, opening height 18.75 ft, mast 17 ft

53. 374.25 ft, $x^2 = 1497y$

55. If the vertex is at the origin and the focus is at $(3, 0)$, then the parabola must open to the right. So, the general equation is $y^2 = 4px$, for some $p > 0$.

57. true   59. false

61. Equate d_1 and d_2 and simplify:
$$\sqrt{(x-0)^2 + (y-p)^2} = |y + p|$$
$$x^2 = 4py$$

63.

65. vertex: $(2.5, -3.5)$
    opens righ

67. vertex: $(1.8, 1.5)$
    opens left

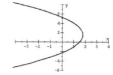

Section B.3

1. d **3.** a

5. center: $(0, 0)$
vertices: $(\pm 5, 0), (0, \pm 4)$

7. center: $(0, 0)$
vertices: $(\pm 4, 0), (0, \pm 8)$

9. center: $(0, 0)$
vertices: $(\pm 10, 0), (0, \pm 1)$

11. center: $(0, 0)$
vertices: $\left(\pm \frac{3}{2}, 0\right), \left(0, \pm \frac{1}{9}\right)$

13. center: $(0, 0)$
vertices: $(\pm 2, 0), (0, \pm 4)$

15. center: $(0, 0)$
vertices: $(\pm 2, 0), (0, \pm \sqrt{2})$

17. $\dfrac{x^2}{36} + \dfrac{y^2}{20} = 1$ **19.** $\dfrac{x^2}{7} + \dfrac{y^2}{16} = 1$

21. $\dfrac{x^2}{4} + \dfrac{y^2}{16} = 1$ **23.** $\dfrac{x^2}{9} + \dfrac{y^2}{49} = 1$

25. c **27.** b

29. center: $(1, 2)$
vertices: $(-3, 2), (5, 2), (1, 0), (1, 4)$

31. center: $(-3, 4)$
vertices: $\left(-2\sqrt{2} - 3, 4\right), \left(2\sqrt{2} - 3, 4\right),$
$\left(-3, 4 + 4\sqrt{5}\right), \left(-3, 4 - 4\sqrt{5}\right)$

33. center: $(0, 3)$
vertices: $(-2, 3), (2, 3), (0, 3), (0, 4)$

35. center: $(1, 1)$
vertices: $\left(1 \pm 2\sqrt{2}, 1\right), (1, 3), (1, -1)$

37. center: $(-2, -3)$
vertices: $\left(-2 \pm \sqrt{10}, -3\right), \left(-2, -3 \pm 5\sqrt{2}\right)$

39. $\dfrac{(x-2)^2}{25} + \dfrac{(y-5)^2}{9} = 1$

41. $\dfrac{(x-4)^2}{7} + \dfrac{(y+4)^2}{16} = 1$

43. $\dfrac{(x-3)^2}{4} + \dfrac{(y-2)^2}{16} = 1$

45. $\dfrac{(x+1)^2}{9} + \dfrac{(y+4)^2}{25} = 1$

47. $\dfrac{x^2}{225} + \dfrac{y^2}{5625} = 1$

49. a. $\dfrac{x^2}{5625} + \dfrac{y^2}{400} = 1$

b. width at end of field is 24 yards; no such width of football field is 30 yards wide.

51. $\dfrac{x^2}{5{,}914{,}000{,}000^2} + \dfrac{y^2}{5{,}729{,}000{,}000^2} = 1$

53. $\dfrac{x^2}{150{,}000{,}000^2} + \dfrac{y^2}{146{,}000{,}000^2} = 1$

55. straight line

57. It should be $a^2 = 6$, $b^2 = 4$, so that $a = \pm\sqrt{6}$, $b = \pm 2$.

59. false   **61.** true

63. Pluto: $e \cong 0.25$ Earth: $e \cong 0.02$

65. As c increases, the ellipse becomes more elongated.

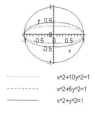

$$x\text{^}2 + 10y\text{^}2 = 1$$
$$x\text{^}2 + 5y\text{^}2 = 1$$
$$x\text{^}2 + y\text{^}2 = 1$$

67. As c increases, the circle gets smaller.

$$x\text{^}2 + y\text{^}2 = 1$$
$$5x\text{^}2 + 5y\text{^}2 = 1$$
$$10x\text{^}2 + 10y\text{^}2 = 1$$

69. As c decreases, the major axis becomes longer.

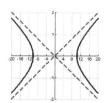

Section B.4

1. b   **3.** d

5.

7.

9.

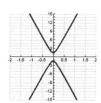

11.

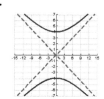

13.

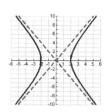

15.

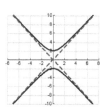

17. $\dfrac{x^2}{16} - \dfrac{y^2}{20} = 1$ **19.** $\dfrac{y^2}{9} - \dfrac{x^2}{7} = 1$

21. $x^2 - y^2 = a^2$ **23.** $\dfrac{y^2}{4} - x^2 = b^2$

25. c   **27.** b

29.

31.

33.

35.

37.

39. $\dfrac{(x-2)^2}{16} + \dfrac{(y-5)^2}{9} = 1$ **41.** $\dfrac{(y+4)^2}{9} - \dfrac{(x-4)^2}{7} = 1$

43. Ship will come ashore between the two stations, 28.5 miles from one and 121.5 miles from the other.

45. 0.000484 sec   **47.** $y^2 - \frac{4}{5}x^2 = 1$

49. The transverse axis should be vertical. The points are $(3, 0)$, $(-3, 0)$ and the vertices are $(0, 2)$, $(0, -2)$.

51. false   **53.** true

55. $\dfrac{x^2}{a^2} - \dfrac{y^2}{a^2} = 1$, which is equivalent to $x^2 - y^2 = a^2$.

57. As c increases, the graphs become more squeezed down toward the x-axis.

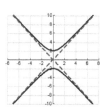

- Innermost curve
- Outermost Curve
- Middle Curve

59. As c decreases, the vertices are located at $\left(\pm\frac{1}{c}, 0\right)$ and are moving away from the origin.

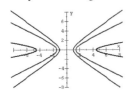

Section B.5

1. $(3\sqrt{2}, \sqrt{2})$

3. $\left(-\frac{3\sqrt{3}}{2} + 1, \frac{3}{2} + \sqrt{3}\right)$

5. $\left(-\frac{1 + 3\sqrt{3}}{2}, \frac{\sqrt{3} - 3}{2}\right)$

7. $\left(\frac{3\sqrt{3}}{2}, \frac{3}{2}\right)$

9. a. hyperbola

   b. $\dfrac{X^2}{2} - \dfrac{Y^2}{2} = 1$

   c.

11. a. parabola

   b. $2X^2 - 2Y^2 - 1 = 0$

   c.

13. a. hyperbola

   b. $\dfrac{X^2}{6} - \dfrac{Y^2}{2} = 1$

   c.

15. a. ellipse

   b. $\dfrac{X^2}{2} + \dfrac{Y^2}{1} = 1$

   c.

17. a. parabola

   b. $2X^2 - 2Y - 1 = 0$

   c.

19. a. ellipse

   b. $\dfrac{X^2}{1} + \dfrac{Y^2}{9} = 1$

   c.

21. a. hyperbola

   b. $\dfrac{X^2}{3} - \dfrac{Y^2}{2} = 1$

   c.

23. a. parabola

   b. $Y^2 - X - 4 = 0$

   c.

25. $45°$ **27.** $60°$ **29.** $30°$

31. $45°$ **33.** $15°$ **35.** $40.3°$ **37.** $50.7°$

39.   **41.**

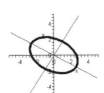

43.   **45.**

47.

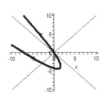

49. true   **51.** true

53. a. $\dfrac{x^2}{b^2} + \dfrac{y^2}{a^2} = 1$

   b. the original equation

55. $a < 0$: hyperbola; $a = 0$: parabola;
$a > 0$: ellipse; $a = 1$: circle

57. a.   **b.**

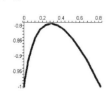

rotation

59. a.   **b.**

vertices are separating

61. a. $-26.57°$ **b.** $19.33°$

c. $-26.57°$

Section B.6

1. $r = \dfrac{5}{2 - \sin\theta}$ **3.** $r = \dfrac{8}{1 + 2\sin\theta}$

5. $r = \dfrac{1}{1 + \cos\theta}$ **7.** $r = \dfrac{6}{4 + 3\cos\theta}$

9. $r = \dfrac{12}{3 - 4\cos\theta}$ **11.** $r = \dfrac{3}{1 - \sin\theta}$

13. $r = \dfrac{18}{5 + 3\sin\theta}$ **15.** parabola

17. ellipse   **19.** hyperbola

21. ellipse   **23.** parabola   **25.** hyperbola

27. a. parabola
  b. $e = 1, (0, 1)$
  c.

29. a. hyperbola
  b. $e = 2, (0, -4), \left(0, -\frac{4}{3}\right)$
  c.

31. a. ellipse
  b. $e = \frac{1}{2}, \left(0, \frac{1}{2}\right), (0, -2)$
  c.

33. a. parabola
  b. $e = 1, \left(0, -\frac{1}{4}\right)$
  c.

35. a. ellipse
  b. $e = \frac{1}{3}, (1, 0), (-2, 0)$
  c.

37. a. hyperbola
  b. $e = \frac{3}{2}, \left(0, \frac{6}{5}\right), (0, 6)$
  c.

39. a. parabola
  b. $e = 1, \left(0, \frac{1}{5}\right)$
  c.

41. $0.248, r = \dfrac{5{,}913{,}500{,}000(1 - 0.248^2)}{1 - 0.248\cos\theta}$

43. $r = \dfrac{150{,}000{,}000(1 - 0.223^2)}{1 - 0.223\cos\theta}$

45. $e \to 1$: elongated (more elliptical),
$e \to 0$: elliptical (circular)

49. $\dfrac{2ep}{1 - e^2}$

51. $\left(-\dfrac{\dfrac{ep}{1 + e} - \dfrac{ep}{1 - e}}{2}, \pi \right)$

53. $r = \dfrac{p}{1 + \cos\theta}$ $\qquad r = \dfrac{p}{1 - \cos\theta}$

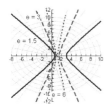 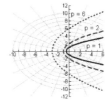

As $p \to \infty$, the graphs get wider and
open in opposite directions.

55. $r = \dfrac{e}{1 + e\cos\theta}$ $\qquad r = \dfrac{e}{1 - e\cos\theta}$

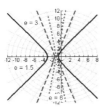 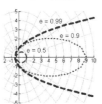

As $e \to \infty$, the graphs get wider, and the latter
family of curves is shifted to the left.

57. $r = \dfrac{e}{1 + e\cos\theta}$ $\qquad r = \dfrac{e}{1 - e\cos\theta}$

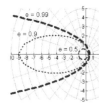

As $e \to 1$, graphs get larger and centers
move accordingly; centers for the latter
family of curves move to the right, while
centers for the first family move left.

Review Exercises

1. false $\qquad\qquad$ **3.** true

5. $y^2 = 12x$

7. $y^2 = -20x$

9. $(x - 2)^2 = 8(y - 3)$

11. $(x - 1)^2 = -4(y - 6)$

13. F: $(0, -3)$, V: $(0, 0)$, D: $y = 3$, LR: 12

15. F: $\left(\tfrac{1}{4}, 0\right)$, V: $(0, 0)$, D: $x = -\tfrac{1}{4}$, LR: 1

17. F: $(3, -2)$, V: $(0, 0)$, D: $x = 1$, LR: 4

19. F: $(-3, -1)$, V: $(-3, 1)$, D: $y = 3$, LR: 8

21. F: $\left(-\tfrac{5}{2}, -\tfrac{79}{8}\right)$, V: $\left(-\tfrac{5}{2}, -\tfrac{75}{8}\right)$, D: $y = -\tfrac{71}{8}$, LR: 2

23. 3.125 ft from center

25. $\qquad\qquad\qquad\qquad$ **27.**

29. $\dfrac{x^2}{25} + \dfrac{y^2}{16} = 1$ $\qquad$ **31.** $\dfrac{x^2}{9} + \dfrac{y^2}{64} = 1$

33. $\qquad\qquad\qquad\qquad$ **35.**

37. $\dfrac{(x - 3)^2}{25} + \dfrac{(y - 3)^2}{9} = 1$

39. $\dfrac{[x - (3.74 \times 10^7)]^2}{6.058 \times 10^{17}} + \dfrac{(y - 0)^2}{6.044 \times 10^{17}} = 1$

41. $\qquad\qquad\qquad\qquad$ **43.**

45. $\dfrac{x^2}{9} - \dfrac{y^2}{16} = 1$ $\qquad$ **47.** $\dfrac{y^2}{9} - x^2 = 1$

49.

51.

53. $\dfrac{(x-4)^2}{16} - \dfrac{(y-3)^2}{9} = 1$

55. between the stations: 65.36 mi from one, 154.64 mi from the other

57. $\left(-\dfrac{3}{2} + \sqrt{3}, \dfrac{3\sqrt{3}}{2} + 1\right)$

59. $\dfrac{x^2}{4} - \dfrac{y^2}{4} = 1$

61. $60°$

63.

65. $r = \dfrac{21}{7 - 3\sin\theta}$

67. hyperbola

69. $e = \dfrac{1}{2}, \left(\dfrac{4}{3}, 0\right), (-4, 0)$

71. $(0.6, -1.2)$, down

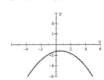

73. a. $y = -1.4 \pm \sqrt{-3x + 8.81}$

b. $[y - (-1.4)^2] = 4\left(-\dfrac{3}{4}\right)(x - 2.937), (2.937, -1.4)$, left

c. yes

75. increases minor axis (x-axis)

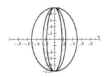

77. move toward origin
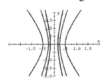

79. a. 1.2 rad

b. 0.2 rad

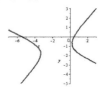

Practice Test

1. c

3. d

5. f

7. $y^2 = -16x$

9. $(x + 1)^2 = -12(y - 5)$

11. $\dfrac{x^2}{7} + \dfrac{y^2}{16} = 1$

13. $\dfrac{(x-2)^2}{20} + \dfrac{y^2}{36} = 1$

15. $x^2 - \dfrac{y^2}{4} = 1$

17. $\dfrac{y^2}{16} - \dfrac{(x-2)^2}{20} = 1$

19.

21.

23. $x^2 = 6y$

25. ellipse, $e = \dfrac{2}{3}$

27. a. $y = x^2 + 4.2x + 5.61$

b. $[x - (-1.2)]^2 = 4\left(\dfrac{1}{4}\right)(y - 1.2), (-1.2, 1.2)$

c. yes

Applications Index

687

Subject Index

RIGHT TRIANGLE TRIGONOMETRY

$$\sin\theta = \frac{\text{opp}}{\text{hyp}} \qquad \csc\theta = \frac{\text{hyp}}{\text{opp}}$$

$$\cos\theta = \frac{\text{adj}}{\text{hyp}} \qquad \sec\theta = \frac{\text{hyp}}{\text{adj}}$$

$$\tan\theta = \frac{\text{opp}}{\text{adj}} \qquad \cot\theta = \frac{\text{adj}}{\text{opp}}$$

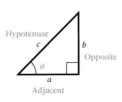

EXACT VALUES OF TRIGONOMETRIC FUNCTIONS

x degrees	x radians	sin x	cos x	tan x
0°	0	0	1	0
30°	$\frac{\pi}{6}$	$\frac{1}{2}$	$\frac{\sqrt{3}}{2}$	$\frac{\sqrt{3}}{3}$
45°	$\frac{\pi}{4}$	$\frac{\sqrt{2}}{2}$	$\frac{\sqrt{2}}{2}$	1
60°	$\frac{\pi}{3}$	$\frac{\sqrt{3}}{2}$	$\frac{1}{2}$	$\sqrt{3}$
90°	$\frac{\pi}{2}$	1	0	—

SPECIAL RIGHT TRIANGLES

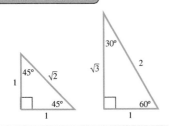

TRIGONOMETRIC FUNCTIONS IN THE CARTESIAN PLANE

$$\sin\theta = \frac{y}{r} \qquad \csc\theta = \frac{r}{y}$$

$$\cos\theta = \frac{x}{r} \qquad \sec\theta = \frac{r}{x}$$

$$\tan\theta = \frac{y}{x} \qquad \cot\theta = \frac{x}{y}$$

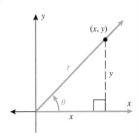

ANGLE MEASUREMENT

π radians $= 180°$

$s = r\theta \quad A = \frac{1}{2}r^2\theta \quad (\theta \text{ in radians})$

To convert from degrees to radians, multiply by $\frac{\pi}{180°}$.

To convert from radians to degrees, multiply by $\frac{180°}{\pi}$.

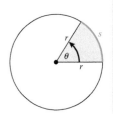

OBLIQUE TRIANGLE

Law of Sines

In any triangle,
$$\frac{\sin\alpha}{a} = \frac{\sin\beta}{b} = \frac{\sin\gamma}{c}.$$

Law of Cosines

$$a^2 = b^2 + c^2 - 2bc\cos\alpha$$
$$b^2 = a^2 + c^2 - 2ac\cos\beta$$
$$c^2 = a^2 + b^2 - 2ab\cos\gamma$$

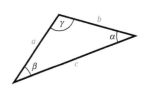

CIRCULAR FUNCTIONS $(\cos\theta, \sin\theta)$

Reciprocal Identities

$$\csc x = \frac{1}{\sin x} \qquad \sec x = \frac{1}{\cos x} \qquad \cot x = \frac{1}{\tan x}$$

Quotient Identities

$$\tan x = \frac{\sin x}{\cos x} \qquad \cot x = \frac{\cos x}{\sin x}$$

Sum Identities

$$\sin(x + y) = \sin x \cos y + \cos x \sin y$$

$$\cos(x + y) = \cos x \cos y - \sin x \sin y$$

$$\tan(x + y) = \frac{\tan x + \tan y}{1 - \tan x \tan y}$$

Difference Identities

$$\sin(x - y) = \sin x \cos y - \cos x \sin y$$

$$\cos(x - y) = \cos x \cos y + \sin x \sin y$$

$$\tan(x - y) = \frac{\tan x - \tan y}{1 + \tan x \tan y}$$

Double-Angle Identities

$$\sin(2x) = 2 \sin x \cos x$$

$$\cos(2x) = \begin{cases} \cos^2 x - \sin^2 x \\ 1 - 2\sin^2 x \\ 2\cos^2 x - 1 \end{cases}$$

$$\tan(2x) = \frac{2 \tan x}{1 - \tan^2 x} = \frac{2 \cot x}{\cot^2 x - 1} = \frac{2}{\cot x - \tan x}$$

Half-Angle Identities

$$\sin\left(\frac{x}{2}\right) = \pm\sqrt{\frac{1 - \cos x}{2}}$$

$$\cos\left(\frac{x}{2}\right) = \pm\sqrt{\frac{1 + \cos x}{2}}$$

Sign $(+/-)$ is determined by quadrant in which $x/2$ lies

$$\tan\left(\frac{x}{2}\right) = \frac{1 - \cos x}{\sin x} = \frac{\sin x}{1 + \cos x} = \pm\sqrt{\frac{1 - \cos x}{1 + \cos x}}$$

Identities for Reducing Powers

$$\sin^2 x = \frac{1 - \cos(2x)}{2} \qquad \cos^2 x = \frac{1 + \cos(2x)}{2}$$

$$\tan^2 x = \frac{1 - \cos(2x)}{1 + \cos(2x)}$$

Identities for Negatives

$$\sin(-x) = -\sin x \qquad\qquad \cos(-x) = \cos x$$

$$\tan(-x) = -\tan x$$

Pythagorean Identities

$$\sin^2 x + \cos^2 x = 1 \qquad\qquad \tan^2 x + 1 = \sec^2 x$$

$$1 + \cot^2 x = \csc^2 x$$

Cofunction Identities

$$\left(\text{Replace } \frac{\pi}{2} \text{ with } 90° \text{ if } x \text{ is in degree measure.} \right)$$

$$\sin\left(\frac{\pi}{2} - x\right) = \cos x \qquad \cos\left(\frac{\pi}{2} - x\right) = \sin x$$

$$\tan\left(\frac{\pi}{2} - x\right) = \cot x \qquad \cot\left(\frac{\pi}{2} - x\right) = \tan x$$

$$\sec\left(\frac{\pi}{2} - x\right) = \csc x \qquad \csc\left(\frac{\pi}{2} - x\right) = \sec x$$

Product-to-Sum Identities

$$\sin x \cos y = \tfrac{1}{2}[\sin(x + y) + \sin(x - y)]$$

$$\cos x \sin y = \tfrac{1}{2}[\sin(x + y) - \sin(x - y)]$$

$$\sin x \sin y = \tfrac{1}{2}[\cos(x - y) - \cos(x + y)]$$

$$\cos x \cos y = \tfrac{1}{2}[\cos(x + y) + \cos(x - y)]$$

Sum-to-Product Identities

$$\sin x + \sin y = 2 \sin\left(\frac{x + y}{2}\right) \cos\left(\frac{x - y}{2}\right)$$

$$\sin x - \sin y = 2 \cos\left(\frac{x + y}{2}\right) \sin\left(\frac{x - y}{2}\right)$$

$$\cos x + \cos y = 2 \cos\left(\frac{x + y}{2}\right) \cos\left(\frac{x - y}{2}\right)$$

$$\cos x - \cos y = -2 \sin\left(\frac{x + y}{2}\right) \sin\left(\frac{x - y}{2}\right)$$

GRAPHS OF THE TRIGONOMETRIC FUNCTIONS

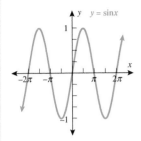

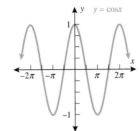

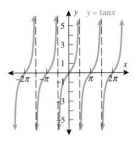

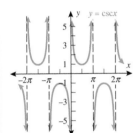

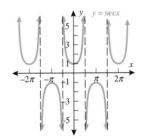

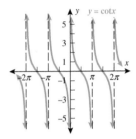

AMPLITUDE, PERIOD, AND PHASE SHIFT

$$y = A\sin(Bx + C) \qquad\qquad y = A\cos(Bx + C)$$

$$\text{Amplitude} = |A| \qquad \text{Period} = \frac{2\pi}{B}$$

$$\text{Phase shift} = \frac{C}{B}\begin{cases} \text{left} & \text{if } \dfrac{C}{B} > 0 \\[2ex] \text{right} & \text{if } \dfrac{C}{B} < 0 \end{cases}$$

$$y = A\tan(Bx + C) \qquad\qquad y = A\cot(Bx + C)$$

$$\text{Period} = \frac{\pi}{B}$$

$$\text{Phase shift} = \frac{C}{B}\begin{cases} \text{left} & \text{if } \dfrac{C}{B} > 0 \\[2ex] \text{right} & \text{if } \dfrac{C}{B} < 0 \end{cases}$$

POLAR COORDINATES

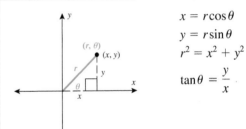

$$x = r\cos\theta$$
$$y = r\sin\theta$$
$$r^2 = x^2 + y^2$$

$$\tan\theta = \frac{y}{x}.$$

Complex Numbers

$x + iy$	$=$	$r(\cos\theta + i\sin\theta)$
Rectangular form		Trigonometric (polar) form

POWERS AND ROOTS OF COMPLEX NUMBERS

$$\begin{aligned} z^n &= (x + iy)^n \\ &= [r(\cos\theta + i\sin\theta)]^n \\ &= r^n[\cos(n\theta) + i\sin(n\theta)] \\ & n = 1, 2, \ldots \end{aligned}$$

$$\begin{aligned} \sqrt[n]{z} &= (x + iy)^{1/n} \\ &= [r(\cos\theta + i\sin\theta)]^{1/n} \\ &= r^{1/n}\left[\cos\left(\frac{\theta + 2k\pi}{n}\right) + i\sin\left(\frac{\theta + 2k\pi}{n}\right)\right] \\ & k = 0, 1, 2, \ldots, n - 1 \end{aligned}$$

HERON'S FORMULA FOR AREA

If the semiperimeter s of a triangle is

$$s = \frac{a + b + c}{2}$$

then the area of that triangle is

$$A = \sqrt{s(s - a)(s - b)(s - c)}$$

$y = \sin^{-1}x$	$x = \sin y$	$-\dfrac{\pi}{2} \leq y \leq \dfrac{\pi}{2}$	$-1 \leq x \leq 1$
$y = \cos^{-1}x$	$x = \cos y$	$0 \leq y \leq \pi$	$-1 \leq x \leq 1$
$y = \tan^{-1}x$	$x = \tan y$	$-\dfrac{\pi}{2} < y < \dfrac{\pi}{2}$	x is any real number
$y = \cot^{-1}x$	$x = \cot y$	$0 < y < \pi$	x is any real number
$y = \sec^{-1}x$	$x = \sec y$	$0 \leq y \leq \pi, y \neq \dfrac{\pi}{2}$	$x \leq -1$ or $x \geq 1$
$y = \csc^{-1}x$	$x = \csc y$	$-\dfrac{\pi}{2} \leq y \leq \dfrac{\pi}{2}, y \neq 0$	$x \leq -1$ or $x \geq 1$

GRAPHS OF THE INVERSE TRIGONOMETRIC FUNCTIONS

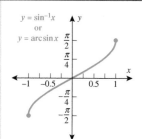

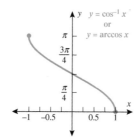

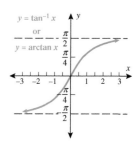

VECTORS

A Vector $\mathbf{v} = \overrightarrow{AB}$

Vector Addition

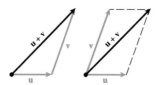

Scalar Multiplication

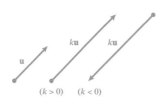

$(k > 0)$ $(k < 0)$

For vectors $\mathbf{u} = \langle a, b \rangle$ and $\mathbf{v} = \langle c, d \rangle$, and real number k,

$$\mathbf{u} = a\mathbf{i} + b\mathbf{j}$$

$$|\mathbf{u}| = \sqrt{a^2 + b^2}$$

$$\mathbf{u} + \mathbf{v} = \langle a + c, b + d \rangle$$

$$k\mathbf{u} = \langle ka, kb \rangle$$

$$\mathbf{u} \cdot \mathbf{v} = ac + bd$$

$$\cos\theta = \frac{\mathbf{u} \cdot \mathbf{v}}{|\mathbf{u}||\mathbf{v}|}$$

$$\text{Comp}_\mathbf{v}\, \mathbf{u} = |\mathbf{u}|\cos\theta = \frac{\mathbf{u} \cdot \mathbf{v}}{|\mathbf{u}|}$$

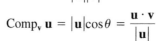

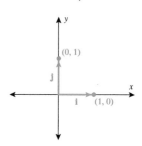

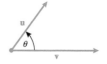